D0788247

PRINCIPLES OF ELECTRONIC DEVICES AND CIRCUITS

David E. LaLond
and
John A. Ross

Delmar Publishers Inc.™

I(T)P™

NOTICE TO THE READER

Publisher does not warrant or guarantee any of the products described herein or perform any independent analysis in connection with any of the product information contained herein. Publisher does not assume, and expressly disclaims, any obligation to obtain and include information other than that provided to it by the manufacturer.

The reader is expressly warned to consider and adopt all safety precautions that might be indicated by the activities described herein and to avoid all potential hazards. By following the instructions contained herein, the reader willingly assumes all risks in connection with such instructions.

The publisher makes no representations or warranties of any kind, including but not limited to, the warranties of fitness for particular purpose or merchantability, nor are any such representations implied with respect to the material set forth herein, and the publisher takes no responsibility with respect to such material. The publisher shall not be liable for any special, consequential or exemplary damages resulting, in whole or in part, from the readers' use of, or reliance upon, this manual.

Delmar Staff:
Administrative Editor: Wendy Welch
Associate Editor: Christine E. Worden
Developmental Editor: Catherine Sustana
Project Editor: Judith Boyd Nelson, Barbara A. Riedell
Production Supervisor: Teresa Luterbach
Production Coordinator: Karen Smith
Art and Design Coordinator: Lisa L. Bower

Cover photo by Uniphoto Inc.

For information, address

Delmar Publishers Inc.
3 Columbia Circle, Box 15015,
Albany, NY 12203-5015

Copyright © 1994 by Delmar Publishers Inc.
The trademark ITP is used under license.

Printed in the United States of America
Published simultaneously in Canada
by Nelson Canada,
a division of the Thomson Corporation

1 2 3 4 5 6 7 8 9 10   XXX   00 99 98 97 96 95 94

Library of Congress Cataloging-in-Publication Data

LaLond, David, 1950–
    Principles of electronic devices and circuits/David E. LaLond and John A. Ross.
        p.   cm.
    Includes index.
    ISBN 0-8273-4663-8 (textbook)
    1. Electronics—Programmed instruction.   2. Transistor circuits—Programmed instruction.   3. Semiconductors—Programmed instruction.
    I. Ross, John A. (John Allan), 1955–   ,   II. Title.
    TK7816.L35   1994
    621.3815′07′7—dc20   92–36004
                                                          CIP

TABLE OF CONTENTS

Preface/vii
Acknowledgments/xiv

Chapter 1 **CIRCUIT ANALYSIS**/2
1.1 Ohm's Law/3
1.2 Voltage Dividers/6
1.3 Current Dividers and Kirchhoff's Current Law/15
1.4 Thevenin's Theorem/22
1.5 Kirchhoff's Voltage Law/35
1.6 Superposition Theorem/40

Chapter 2 **SEMICONDUCTOR THEORY**/58
2.1 Back to the Basics/60
2.2 Semiconductor Materials/64
2.3 Making the Semiconductor a Better Conductor/66
2.4 Combining the Semiconductor Materials/70
2.5 Conduction Within the Diode/74
2.6 Diode Characteristics/77
2.7 Less-Than-Perfect Diode Operation/80

Chapter 3 **DIODE CIRCUIT APPLICATIONS**/86
3.1 Introducing the Power Supply/88
3.2 Diodes as Rectifiers/91
3.3 Power Supply Input Filters/102
3.4 Regulating the Power Supply/107
3.5 Voltage Multipliers/111
3.6 Diodes Used as Signal Clippers/115
3.7 Breakdown Voltage Limiting/120
3.8 Signal Clampers/123
3.9 Unbalanced Loads and the DC Return/129

Chapter 4 **BIPOLAR JUNCTION TRANSISTORS**/140
4.1 What's Inside the Bipolar Junction Transistor/142
4.2 The Transistor as Two Back-To-Back Diodes/143
4.3 Alpha and Beta/146
4.4 More About Transistor Characteristics/149
4.5 Using Base Bias/156
4.6 Minimizing the Effects of Beta/164
4.7 Using Voltage Divider Bias to Minimize Effects of Beta/179
4.8   The Biasing of the PNP Transistor/189
4.9 Collector Curves/195

Chapter 5 **TRANSISTORS WORKING AS SMALL-SIGNAL AMPLIFIERS**/212
5.1 Some Fundamental Concepts/214
5.2 Equivalent Circuits/218
5.3 Exploring the Common-Emitter Configuration/226
5.4 Analyzing the Input Impedance of a Common-Emitter Circuit That Uses Voltage Divider Bias/238

5.5   Analyzing a Common-Emitter Circuit That Uses Collector Feedback Bias/244
5.6   Exploring the Common-Collector Configuration/248
5.7   Exploring the Common-Base Configuration/254
5.8   Stage Gain Versus Transistor Gain/259
5.9   Multistage Amplifiers/260

Chapter 6   POWER AMPLIFIERS/274
6.1   Using DC and AC Load Lines to Understand Large-Signal Amplifiers/276
6.2   Classifying Amplifiers/281
6.3   Class A Amplifiers/282
6.4   Class B Amplifiers/295
6.5   Class AB Amplifiers/302
6.6   Class C Amplifiers/312
6.7   Coupling Amplifier Stages Together/324

Chapter 7   JUNCTION FIELD EFFECT TRANSISTORS/336
7.1   Basic JFET Operation/338
7.2   Biasing the JFET/346
7.3   Current Source Bias/359
7.4   Transconductance/361
7.5   Using JFETs as Amplifiers/364
7.6   Other Applications for JFETs/374

Chapter 8   MOSFETS AND MOSFET CIRCUITS/384
8.1   Advantages of the MOSFET/386
8.2   Types of MOSFETs/387
8.3   The Enhancement MOSFET, or E MOSFET/391
8.4   D MOSFET Biasing and Circuit Analysis/395
8.5   E MOSFET Biasing and Circuit Analysis/400
8.6   MOSFET Applications/406

Chapter 9   TRANSISTOR SWITCHING/416
9.1   Transistor Switching Operations/418
9.2   Improving the Switching Speed of a Transistor/426
9.3   Transistor Switching Applications/432
9.4   FET Switching/439
9.5   UJT Operation/441
9.6   Generating Signals/449
9.7   More Switching Applications/458

Chapter 10   MORE SMALL-SIGNAL AMPLIFICATION/468
10.1   Negative Feedback/470
10.2   Small-Signal Transistor Parameters/470
10.3   Finding the Bandwidth of BJT and JFET Amplifiers/473
10.4   Differential Amplifiers/488

Chapter 11 OPERATIONAL AMPLIFIER FUNDAMENTALS/504

11.1   Beginning Op-Amp Information/507
11.2   Electrical Parameters and Impedances of Op-Amps/512
11.3   Frequency-Dependent Characteristics of Amps/524

Chapter 12 BASIC OPERATIONAL AMPLIFIER CIRCUITS/536

12.1   Four Methods of Feedback/538
12.2   Noninverting Voltage Amplifier/542
12.3   Op-Amp Circuits Using Parallel-Parallel Negative Feedback/558
12.4   The Inverting Current Amplifier/567
12.5   A Voltage-to-Current Transducer/573
12.6   Op-Amp Applications/577

Chapter 13 WAVE GENERATION AND WAVESHAPING CIRCUITS/594

13.1   General Oscillator Theory/596
13.2   Frequency Selection Circuits/598
13.3   Wien-Bridge Oscillator/613
13.4   Twin-T Oscillator/616
13.5   LC Oscillators/618
13.6   Relaxation Oscillators/626
13.7   Waveshaping Circuits/631

Chapter 14 ACTIVE DIODE CIRCUITS AND COMPARATORS/650

14.1   Active Diode Circuits/652
14.2   Comparators and Schmitt Triggers/664
14.3   Window Comparators/689

Chapter 15 ACTIVE FILTERS/706

15.1   Defining Filter Performance/708
15.2   Classifying Filters/715
15.3   Active Filters/728
15.4   Active Filters Using Miller-Effect Feedback/741

Chapter 16 ANALOG AND DIGITAL SIGNALS/760

16.1   Transducers/762
16.2   Differential and Instrumentation Amplifiers/773
16.3   Converting the Differential Output Signal to Data/784
16.4   Applications/792

Chapter 17 MULTIVIBRATORS, OSCILLATORS, AND TIMERS/800

17.1   Astable Multivibrators/802
17.2   Bistable Multivibrators/807
17.3   Monostable, or One-Shot, Multivibrators/811
17.4   The 555 Timer/820
17.5   Multivibrator and Timer Applications/826

Chapter 18   **THYRISTORS AND THYRISTOR DEVICES**/838
     18.1   The Shockley Diode/840
     18.2   Silicon (SCRs) Controlled Rectifiers/843
     18.3   Triacs/857
     18.4   Other Thyristors/864
     18.5   Unijunction Transistors/867
     18.6   Thyristor Applications/877

Chapter 19   **REGULATORS**/886
     19.1   Voltage Regulation/888
     19.2   Linear Regulators/894
     19.3   Switching Regulators/900
     19.4   Foldback Current Limiting/906
     19.5   Regulating Current/909
     19.6   Integrated Circuit Voltage Regulators/913
     19.7   Voltage Regulator Applications/921

Chapter 20   **COMMUNICATIONS CONCEPTS**/928
     20.1   Modulation/930
     20.2   Oscillators, Multipliers, and Modulators/938
     20.3   Detectors/946
     20.4   Phase-Locked Loops/950
     20.5   High-Frequency Diodes/955

GLOSSARY/967

APPENDIX A   **MATH DERIVATIONS**/975

APPENDIX B   **DATA SHEETS**/986

APPENDIX C   **ANSWERS TO SELECTED PROBLEMS**/1012

APPENDIX D   **LIST OF ABBREVIATIONS**/1033

INDEX/1034

PREFACE

THE BOOK'S AIM

There are many ways to teach a course in solid state electronics. At one extreme is the highly mathematical approach that relies on many formulas. At the other extreme is the show-and-tell approach, which may result in you, the student, not really obtaining enough information to truly understand how a particular device or circuit operates. *Principles of Electronic Devices and Circuits* is aimed at the middle ground between the rigorous mathematical approach and the show-and-tell approach.

First, *Principles of Electronic Devices and Circuits* treats the subject as a continuation of the material you learned in your beginning DC/AC class. In particular, DC/AC theorems and laws that you learned, and committed to long-term memory, can be used to analyze new circuits and new devices. Thus, you need not memorize new circuits. Instead, each new circuit can be fully understood by applying the circuit theorems and laws that you already know.

The second way this book reaches the middle ground is by simplifying and providing the explanations for many of the circuit equations that are necessary in electronics. This textbook shows you that many of the circuit equations for transistor circuits are nothing more than Ohm's law, voltage divider applications, or variations of Kirchhoff's voltage law—concepts you have already learned.

Many graphs and diagrams are provided to support the approach that most new circuit concepts can be simplified to a basic equation learned in DC circuit analysis. This is done in the belief that most people are able to understand a new concept if they can see that concept illustrated in graphical form.

THE BOOK'S APPROACH

Several important strategies are used in this book's presentation. Some of these are:

- Building your critical-thinking skills in analyzing circuits rather than just memorizing formulas
- Making liberal use of pictorial, graphical, and diagramatic information to facilitate understanding
- Beginning each chapter with a list of objectives and offering checkpoints within the chapter so you know when a particular objective has been completed
- Ending each chapter with a summary of the key topics covered to help reinforce the material
- Offering clear and frequent examples to demonstrate new concepts
- Reinforcing concepts with frequent practice problems to help students internalize the material
- Dividing the chapter into sections, with each section ending with additional questions and problems, to make the material easier to understand

- Providing troubleshooting sections containing problems, as well as hints, to practice the skills learned
- Providing end-of-chapter problem sets to review the material covered in the chapter.

THE BOOK'S CONTENT

Principles of Electronic Devices and Circuits begins with a review of basic theories learned in a beginning electronics course. These theorems are the tools the electronics technician uses to predict circuit behavior. They are the foundation skills used extensively throughout this text.

Chapters 2 and 3 introduce concepts related to solid state devices, diode operation, and circuitry. An introduction to power supplies is found in Chapter 3.

Chapter 4 begins an explanation of transistor behavior, then introduces the beta box model, which is simply a visual representation of equations derived from Kirchhoff's voltage law. The beta box model is used to analyze the DC operation of transistor circuits. Various ways in which transistors are biased (activated) are discussed and examined.

Chapter 5 covers the three basic AC transistor circuits and shows how the beta box model can be used to predict both DC and AC behavior of a transistor circuit. The gain loop model is introduced and used to predict the AC output voltage of a transistor circuit.

Chapter 6 studies how transistors are used as power amplifiers. This chapter introduces and explains class A, class B, and class C operation.

Chapters 7 and 8 discuss the different types of transistors, field effect transistors (FETs), and metal-oxide semiconductor field effect transistors (MOSFETs). The different types of FETs and MOSFETs are discussed, along with universal transconductance curves for FET and MOSFET circuits used in these chapters.

Chapter 9 is devoted to the topic of how transistors can be used as electronic switches. Switching concepts and terms are explained. Transistor circuits that produce nonsinusoidal waves are discussed and analyzed in this chapter.

Chapter 10 shows you how to determine the bandwidth of a transistor amplifier, discusses h parameters, and looks at the impact of negative feedback on circuit operation. The chapter concludes with a discussion of differential amplifiers, which lays the groundwork for further study of operational amplifiers.

Chapters 11 and 12 provide an in-depth discussion of operational amplifiers, showing how to identify open-loop and closed-loop operation, how to determine output voltage, calculate signal-to-noise ratio, and how to use negative feedback to identify an op-amp circuit.

Chapter 13 discusses wave generation and waveshaping circuits, explaining different oscillators such as Wien-Bridge, twin-T, and LC oscillators.

Chapter 14 persents active diode circuits and comparators.

Chapter 15 provides information on active filters and their performance, covering low-pass, high-pass, band-pass, band-reject, Butterworth, Chebyshev, Bessel, and other filters. Analog and digital signals are covered in Chapter 16, and Chapter 17 explains bistable, astable, and multistable vibrators and their applications. The chapter also includes coverage of oscillators and timers.

Thyristors are covered in Chapter 18, and Chapter 19 provides information on power supply circuits. Finally, Chapter 20 introduces special devices such as detectors and phase-locked loops.

USING THE BETA BOX

Figure 1 is a schematic of a base-biased transistor circuit, a circuit that is studied in this textbook. The part of the circuit that is highlighted is the part that "controls" what happens in the circuit. The beta box model is used to simplify this part of the circuit in such a way that the highlighted loop seems to be a simple series circuit, which you learned in your beginning DC/AC course. Thus, you can use the beta box model to convert transistor circuits into "equivalent" series circuits that can be solved using Ohm's law.

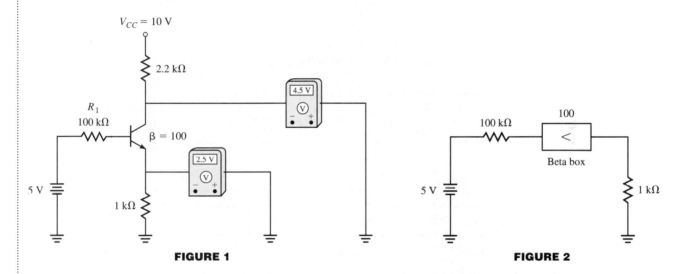

FIGURE 1   **FIGURE 2**

First, visualize the beta box as separating the base and emitter circuits (Figure 2). The < (less than) sign indicates on which side of the box current is smaller. The ratio (value of beta) is placed above the box. The adjustment for I_E is shown in Figure 3. If the current suddenly becomes beta times bigger, it is equivalent to the resistor's becoming beta times smaller. Once the beta box has been equalized for a current, you can replace the model with a simple series circuit. This is shown in Figure 4. You can now solve for I_E by applying Ohm's law. It turns out that this current is a very close approximation of the current flowing through the transistor.

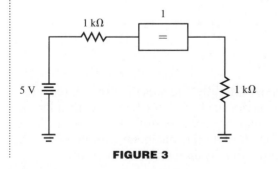

FIGURE 3

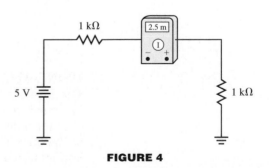

FIGURE 4

The flowchart in Figure 5 shows the steps for using the beta box model.

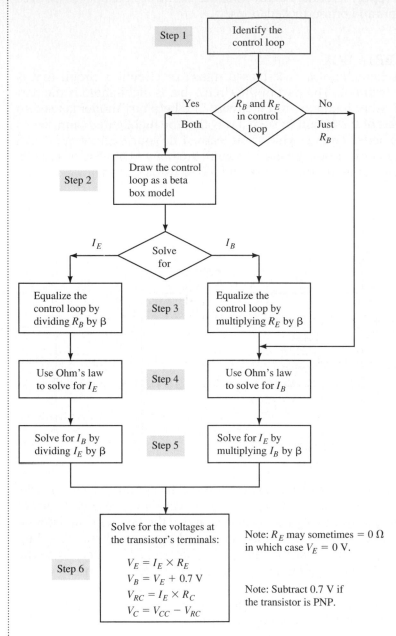

FIGURE 5

USING THE BOOK

The format of this textbook is "reader-friendly" because it uses an easy-to-understand, conversational writing style. Each chapter begins with a list of chapter objectives. As you read the chapter, checkpoints marked by an icon in the margin indicate when a particular objective has been completed. Each application of the beta box is also indicated by an icon in the margin.

Major sections within each chapter finish with a set of review questions. Examples and practice problems are provided throughout each chapter for review. Numerous diagrams and circuit schematics are used to help you visualize key points within the circuits being explained.

Each chapter ends with a section on troubleshooting and a set of problems to help reinforce the material presented.

ACCURACY AND DEVELOPMENT

Great care has been taken to ensure that this textbook meets the needs of both instructors and students who have chosen electronics as a career. Through an exhaustive review process, we have tailored the book's organization and teaching tools to our demanding audience. Extensive reviewing and accuracy testing during development have made this book an efficient, effective teaching and learning tool.

THE LEARNING PACKAGE

Principles of Electronic Devices has a well-coordinated lab manual that provides hands-on learning experiences to reinforce the concepts covered in this textbook. With more than 75 experiments, the lab manual helps the student consolidate knowledge gained from each experiment. It challenges the student to expand on that knowledge, using it in a new or different application. Flashcards also are available to help the student internalize new concepts and practice new applications.

This package includes an instructor's guide for the text and lab manual, full-color transparencies of important figures from the book, a computerized test bank, and a printed test bank. A complete package for the instructor is available from Delmar Publishers Inc.

SCHEDULING INSTRUCTIONS

Ideally, your program can use all the material provided in the textbook and the correlated materials to maximize learning and practice opportunities. The following suggestions are provided as possible scenarios for those whose time is limited:

- For students with strong DC and AC skills, quickly review Chapter 1 to familiarize yourself with the format of the text while reviewing material with which the students are familiar. Provide just an overview of common base amplifiers covered in Chapter 5.
- For students with a separate radio or electronic communications class, give light treatment to class C amplifiers covered in Chapter 6 and the material in Chapter 20.
- For students with a separate pulse and waveshaping course, cover only those sections of Chapter 9 that are not covered in your pulse and waveshaping course. In Chapter 10, minimize the amount of time spent in reducing circuit bandwidth and delete the section on h parameters.
- Chapter 18 may be covered lightly for students who have separate courses in industrial controls.

Determine the values of the emitter, collector, and base currents and the emitter, base, and collector voltages for the voltage divider circuit shown in Figure 4.53. Determine if the voltage divider is stiff or firm.

REVIEW SECTION 4.7
1. Of base bias, collector feedback bias, and voltage divider bias, which method provides the most circuit stability?
2. R_t and R_b are used as a _____.
3. When we thevenize to find the thevenized resistance, R_t and R_b are in (series, parallel).
4. If beta varies from 50 to 300, what value of beta should be used to determine whether a voltage divider is stiff or firm?
5. Explain how to calculate the values of the base, emitter, and collector voltages and currents for a stiff voltage divider circuit.
6. Compare and contrast voltage divider bias and the other bias circuits that we have studied.

Progress checks in the margin let you know when you have completed a chapter objective.

✓ **Progress Check**
You have now completed objective 12.

4.8 THE BIASING OF THE PNP TRANSISTOR

The voltages used to bias PNP transistors have the opposite polarity from those used to bias NPN transistors. Even though current flows in the opposite direction, we can still use the same skills and procedures that we used to solve NPN transistor circuits.

Figure 4.54 represents the beta box model for a PNP transistor. Notice that the collector and base currents flow into the node inside the beta box. These two currents combine to form the emitter current, which flows out the emitter. Also, note the polarity of the diode representing the base-emitter junction. The cathode, or negative side, of the diode attaches to the node within the beta box. To forward bias the transistor junction, the base must be negative with respect to the emitter.

In Figure 4.55, we see the base bias and emitter bias configurations for PNP transistors. In the base bias circuit, the collector supply voltage is now negative with respect to ground. To see how we can solve this circuit, consider the following example.

The Beta Box icon appears in the margin to indicate an application of the Beta Box Model introduced in Chapter 4.

FIGURE 4.54 A beta box bias model for a PNP transistor. The arrows point in the direction of electron flow.

---

d in base-bias circuits. This difference is important. If the base resistor h emitter-bias circuit has a small value, it will have a minimal effect on the DC operation of the circuit. A control loop model will show the small effect that the base resistor value has on circuit calculations.

In Figure 4.31, we see the control loop of a beta box model for an emitter-bias circuit. Because there are resistors on each side of the beta box, we must equalize the model. We have already solved for the emitter current in base-bias circuits that have an emitter resistor. We can likewise solve for the emitter current in an emitter-bias circuit. In other words, we are learning to use a consistent procedure for solving transistor circuits. When we equalize the model, we will reduce the base resistance by dividing it by beta.

FIGURE 4.31 Beta box model of the control loop for the emitter-bias circuit shown in Figure 4.30

EXAMPLE 4.9

Using the circuit shown in Figure 4.30, solve for the emitter, collector, and base currents and the emitter, collector, and base voltages.

Figure 4.32a shows the control loop taken from the circuit in Figure 4.30 and Figure 4.32b illustrates the equalized model. Notice how the effective base resistance is reduced. As shown in the equivalent series circuit in Figure 4.32c, the effective base resistance is insignificant when compared with the value of the emitter resistance. In this case, the emitter resistance is

Step-by-step schematics demonstrate the Beta Box flowchart (see Figure 5 of the Preface).

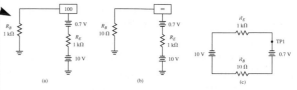

FIGURE 4.32 (a) Control loop model for the circuit shown in Figure 4.30. (b) Equalized model of the circuit. (c) Equivalent series circuit.

REVIEW SECTION 4.9

Given the characteristic curves and circuit shown in the accompanying figure:
1. Determine the load line (reproduce the curves on a separate sheet of paper and draw in the load line).
2. Determine V_{CEQ} and I_{CQ}.
3. What is the value of I_E?

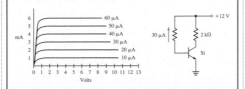

TROUBLESHOOTING

In the next chapter, we will investigate how transistors amplify AC signals. Often, the failure of a transistor to amplify properly occurs because some part of the DC circuit operation has broken down. In this section, we will discuss some of the failures that can happen in the DC circuit operation. Common practice dictates that we begin any troubleshooting process by measuring DC values and checking those measurements against the values that we predict. From there, we can move our problem solving to the AC operation of the circuit.

DON'T OVERLOOK THE POWER SUPPLY
Check the power supply first. Make sure that the voltage level coming out of the power supply is correct. Then check for proper distribution of the voltages around the circuit. If the voltages are missing, turn off the power, and recheck the power supply. Blown fuses may indicate a short circuit in the rectifier circuits. If the rectifiers check out, use an ohmmeter and test from the collector voltage point to ground for a short circuit. A short between the collector voltage supply and ground can cause fuses to blow.

USE YOUR EYES AND NOSE
A technician should complete the repair properly and quickly. A careful visual inspection of the circuitry may show which part has failed and may save the time needed for calculations or tests. Look for broken leads, bad solder

Each chapter ends with a troubleshooting section that discusses common problems and provides tips for solving them.

The WHAT'S WRONG WITH THESE CIRCUITS? section gives you a chance to apply the concepts from the troubleshooting section and to test your own troubleshooting skills.

WHAT'S WRONG WITH THESE CIRCUITS?

Answer questions 1–5 using the following circuit.
1. If R_{B1} was reduced in value, would the operating point on the DC load line move toward saturation or cutoff?

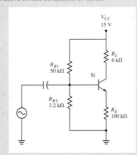

2. If R_E opened, V_{CE} would equal approximately _____.
3. If R_{B1} opened, V_C would equal approximately _____.
4. What resistor(s) compensate(s) for beta?
5. What resistor(s) compensate(s) for thermal runaway?

Determine the cause of the failure of the following base-biased circuit with an emitter resistor for each of the readings in 6–12. The normal readings are M1 = 0.54 V, M2 = 1.23 V, and M3 = 6.64 V.

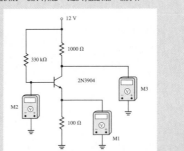

ACKNOWLEDGMENTS

Many individuals have helped in shaping this textbook. We wish to thank the following reviewers for their time and valuable insights: Richard G. Anthony, O. Dean Bingham, Richard J. Blakely, James E. Boyer, Myron L. Brignoli, Etzel C. Brower, Phillip J. Chiarelli, Orville R. Detraz, Barbara J. Dettman, James Feasel, Ralph Frese, Arnie Garcia, Geoffrey Geisz, Thomas Gibson, Timothy Goulden, Kevin Gray, Leigh A. Hargis, Charles A. Heskett, Ronald Hessman, Richard A. Honeycutt, Gerald E. Jensen, Wayne Keesling, Joseph E. McLaughlin, Daniel Morrow, Maurice J. Nadeau, Robert Reaves, Paul Rosenberg, Lee Rosenthal, Michael Sanderson, Gerald Schickman, James J. Schreiber, Peter J. Smith, Roger Thielking, Richard D. Thomson, Innocent Usoh, John Waleski. For their assistance with the technical accuracy check, we thank Mark Kirkegaard and Steven J. Yelton. In addition, we thank Mark Kirkegaard and Noel Henry for providing answers for the problems in the text.

We particularly thank Larry Parker for his invaluable help in completing this textbook. We are grateful to the following individuals who contributed their time and expertise: Steven Baltazar, Clifford Cooper, Frank Dungan, David Fridenmaker, Charles A. Heskett, Ronald Scott Rohleder, Paul Rosenberg, and Richard D. Thomson.

We also thank the staff at Delmar Publishers, whose patience and expertise helped make this book possible. Thanks in particular to Wendy Welch, who helped get the project started and supported the concept of the beta box model, Christine Worden and Catherine Sustana, whose development of this text made it a reality, and Judith Boyd Nelson, Karen Smith, and Teresa Luterbach, whose diligent efforts guided it successfully through production. We owe much to each of them. Thank you for seeing this through to completion.

DEDICATIONS

To Jack and Kyle LaLond
Live long and prosper.
DEL

To John C. and Lorraine Ross
I couldn't have done this without you.
JAR

1

CIRCUIT ANALYSIS

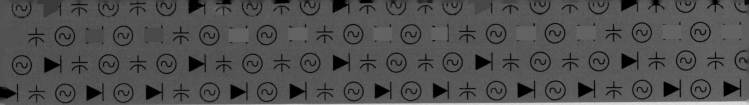

✓ **As you read this chapter, concentrate on learning how to:**

1. Use Ohm's law for simplification of circuit analysis
2. Use the voltage divider concept to predict voltages across components in series circuits
3. Determine the loading effects on voltage dividers and how to analyze and prevent them
4. Use the current divider concept and Kirchhoff's current law to predict currents through components in electronic circuits
5. Use Thevenin's theorem to develop Thevenin equivalent circuits
6. Use Kirchhoff's voltage law for checking answers
7. Use Kirchhoff's current law to develop simple equations
8. Use the superposition theorem to analyze electronic circuits having more than one voltage source
9. Recognize limitations of test equipment and how to overcome them when troubleshooting electronic circuits

INTRODUCTION

This chapter is devoted to reviewing the analysis of electronic circuits, using techniques and theorems studied in courses on basic DC-AC circuits.

The techniques and theorems reviewed here are the fundamental tools you will use to analyze new circuits and how they work. You will learn ways to use voltage and current dividers, Thevenin's theorem, the superposition theorem, and Kirchhoff's voltage and current laws to develop equations for most of the circuits studied in this text. You will also learn how many of those equations can be reduced to basic Ohm's law equations to simplify calculations.

1.1 OHM'S LAW

Ohm's law states the relationship between current (I) in amperes (A), voltage (V) in volts (V), and resistance (R) in ohms (Ω). In a DC circuit, the relationship is interdependent. If two of the quantities are known, the third can be found by one of three simple equations:

$$I = \frac{V}{R}$$
$$V = IR$$
$$R = \frac{V}{I}$$

These equations show that voltage is directly proportional to current and resistance and that current is inversely proportional to resistance. These are important concepts to remember.

These principles are consistent and apply to all DC circuits. They also apply to AC circuits, but a few other factors come into play for AC circuits. The two main factors that affect AC circuits that do not affect DC circuits are inductance and capacitance. As we go through this text, these factors will be covered in depth. For now, we will simply say that there are different types of impedance that occur in AC circuits but not in DC circuits.

EXAMPLE 1.1

Figure 1.1 shows a simple DC series circuit consisting of a battery and a resistor. Note that the current in this circuit flows from the negative (−) battery terminal, through the external circuit, back into the positive (+) battery terminal. This convention is referred to as **electron current flow** and will be used throughout this text.

Also note that current can flow through only one path in this simple series circuit. This single path is one of the rules of series circuits: The current has the same value in all parts of a series circuit.

According to this rule all of the current must flow through the resistor. The following solutions serve as a review of the Ohm's law relationships.

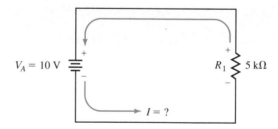

FIGURE 1.1 A simple DC series circuit

Solutions
How do we find how much current flows in the circuit and consequently through the resistor? We use Ohm's law. The applied voltage (V_A) is 10 V, and the resistance in resistor R_1 is 5 kΩ. Therefore:

$$I = \frac{V_A}{R_1} = \frac{10\,V}{5\,k\Omega} = 2\,mA$$

Figure 1.2 shows a series circuit with two resistors and a battery. The current flowing through the circuit is 0.002 A or 2 mA (2 milliamperes). To find how much voltage is applied, we use the relationship V = IR. We first find the total resistance (R_T) by adding the values of the two resistances:

$$R_T = R_1 + R_2 = 20 \text{ k}\Omega + 100 \text{ k}\Omega = 120 \text{ k}\Omega$$

Then we solve for V_A:

$$V_A = IR_T = 2 \text{ mA} \times 120 \text{ k}\Omega = 240 \text{ V}$$

We can find the resistance value in the circuit in Figure 1.3 by using the relationship R = V/I:

$$R = \frac{V_A}{I} = \frac{100 \text{ V}}{1.2 \text{ mA}} = 83.3 \text{ k}\Omega$$

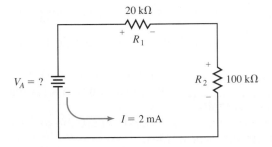

FIGURE 1.2   A series circuit with unknown voltage

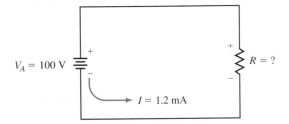

FIGURE 1.3   A series circuit with unknown resistance

PRACTICE PROBLEM 1.1

Find I flowing through R in Figure 1.4a.
Find R in Figure 1.4b.
Find V_A in Figure 1.4c.

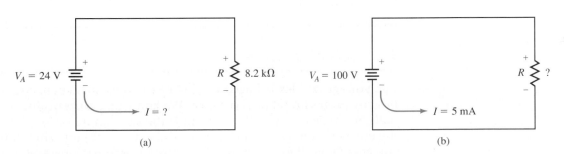

(a)   (b)

FIGURE 1.4   Circuits for practice problem 1.1   (continued on next page)

FIGURE 1.4   (continued)

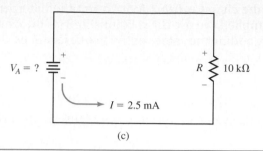

(c)

Progress Check

You have now completed objective 1.

REVIEW SECTION 1.1

1. _____ states the relationship between voltage, current, and resistance.
2. Write the three equations for the relationships referred to in question 1.
3. Voltage is _____ proportional to resistance.
4. Current is _____ proportional to resistance.
5. Define electron current flow.
6. The current in a series circuit is divided among the resistive elements in the circuit. True or false?
7. In a series circuit, the most current flows through the least resistance. True or false?
8. In a series circuit R = 100 Ω and V = 20 V. I in the circuit must equal:
   (a) 0.02 A
   (b) 0.2 A
   (c) 2.0 A
   (d) 5.0 A
9. In a series circuit V = 24 V and I = 2.4 mA. R in the circuit must equal:
   (a) 0.1 kΩ
   (b) 1 kΩ
   (c) 10 kΩ
   (d) 100 kΩ
10. In a series circuit I = 25 mA and R = 3.3 kΩ. V in the circuit must equal:
    (a) 825 V
    (b) 82.5 V
    (c) 8.25 V
    (d) 825 mV

1.2   VOLTAGE DIVIDERS

The **voltage divider** is based on the idea that in any series circuit voltage is directly proportional to resistance. We know that current flow has only one path in a series circuit. The current is the same in all parts of the circuit, so the current is the same in all the resistors in that circuit. Therefore, when we consider Ohm's law, we make the following assumptions: If R_1 causes a voltage V_1, and R_2 is twice the value of R_1, then V_2, the voltage drop across R_2, must be twice the value of V_1. Ignoring any wire resistance in the series circuit, the total voltage drop V_T must equal V_A, the applied voltage. Those assumptions constitute the voltage divider concept. The following examples illustrate how the voltage divider concept is used in DC series circuits.

EXAMPLE 1.2

Figure 1.5 shows a series circuit with three resistors, constituting a voltage divider. The total resistance (R_T) is equal to the sum of all resistances:

$$R_T = R_1 + R_2 + R_3 = 2\,k\Omega + 3\,k\Omega + 5\,k\Omega = 10\,k\Omega$$

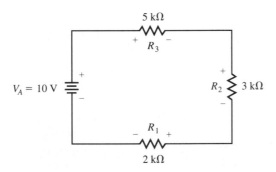

FIGURE 1.5 A voltage divider circuit

Using Ohm's law to solve for the current in the circuit, we have:

$$I = \frac{V_A}{R_T} = \frac{10\,V}{10\,k\Omega} = 1\,mA$$

As previously stated, the current has the same value in all parts of the circuit. Therefore, voltage drops for each resistor can be calculated by using Ohm's law.

$$V_1 = 1\,mA \times 2\,k\Omega = 2\,V$$
$$V_2 = 1\,mA \times 3\,k\Omega = 3\,V$$
$$V_3 = 1\,mA \times 5\,k\Omega = 5\,V$$

We stated that V_T must equal V_A. To prove this we add all voltage drops in the circuit:

$$V_T = V_1 + V_2 + V_3 = 2\,V + 3\,V + 5\,V = 10\,V$$

the same voltage as V_A.

This example confirms that Ohm's law works in a series circuit. But suppose you are not interested in circuit current and the voltage across each resistor. Instead, you want to know the voltage across only a single resistor in the circuit. As you will see later, this situation arises frequently in transistor circuits that are activated by a voltage divider circuit. The solution that follows uses the voltage divider concept instead of Ohm's law to find V_1 in the circuit in Figure 1.5.

Solution
In a series circuit, voltages are directly proportional to resistances. We can use a ratio to compare the percent of R_T that R_1 represents:

$$\frac{R_1}{R_T} = \frac{2\,k\Omega}{10\,k\Omega} = \frac{1}{5}, \text{ or } 20\%$$

Now we can compare the value of V_1 with V_T:

$$V_1 = 0.20 \times V_T = 0.20 \times 10\ V = 2\ V$$

For those who prefer to work from an equation, we can use the **general voltage divider equation.** The voltage drop (V_x) across any resistance or combination of resistances in a series circuit is equal to the ratio of that resistance value to the total resistance (R_x/R_T) times the applied voltage (V_A).

In equation form:

$$V_x = \frac{R_x}{R_T} \times V_A$$

In this example we have:

$$V_1 = \frac{R_1}{R_T} \times V_T = \frac{2\ k\Omega}{10\ k\Omega} \times 10\ V = 2\ V$$

These calculations show the true direct relationship between the ratios of resistances and voltages:

$$\frac{R_1}{R_T} = \frac{V_1}{V_T}, \quad \text{or} \quad R_1 : R_2 = V_1 : V_2$$

$$\frac{2\ k\Omega}{10\ k\Omega} = \frac{2\ V}{10\ V} \quad \text{or} \quad 2\ k\Omega : 10\ k\Omega = 2\ V : 10\ V$$

In Figure 1.6, use the voltage divider concept to determine the voltage drop across R_2 and R_4.

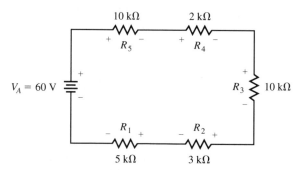

FIGURE 1.6   Circuit for practice problem 1.2

Show the ratio relationship between the values of R_1 and R_T and between V_1 and V_A.

EXAMPLE 1.3

Figure 1.7 shows a parallel configuration. When we connect two or more components across a single power source we form **parallel branches,** in which the applied voltage appears across each branch. In our circuit the branch on the right contains two resistors. Branches that have resistors in series are often referred to as **series strings.** Hence, in this circuit we treat the two-resistor branch as a series string in parallel with the single-resistor branch.

Resistors R_2 and R_3 must share the voltage across the branch. We will now find the voltage drops across R_2 and R_3 by the voltage divider concept.

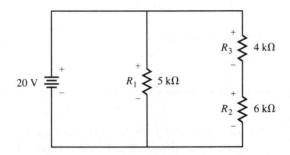

FIGURE 1.7   Applying the voltage divider concept to a series string

Solution

The total resistance of the two-resistor branch is
$$R_{b2} = R_2 + R_3 = 6\,k\Omega + 4\,k\Omega = 10\,k\Omega$$
The ratio of R_2 to R_T is

$$\frac{R_2}{R_{b2}} = \frac{6\,k\Omega}{10\,k\Omega} = 0.6 = 60\%$$
$$V_2 = 0.6 \times 20 \ \ V = 12\,V$$

Now we can find V_3 by simple subtraction:
$$V_3 = V_A - V_2 = 20\,V - 12\,V = 8\,V$$

If we want to solve this problem by Ohm's law we must first find the branch current (I_{b2}) of the two-resistor branch:

$$I_{b2} = \frac{V_A}{R_{b2}} = \frac{20\,V}{10\,k\Omega} = 2\,mA$$

We know that the same current flows through both resistors in the series string, so we can now find V_2 and V_3.
$$V_2 = I_2 \times R_2 = 2\,mA \times 6\,k\Omega = 12\,V$$
$$V_3 = I_3 \times R_3 = 2\,mA \times 4\,k\Omega = 8\,V$$

This example illustrates that it may be easier to analyze a circuit by using the voltage divider concept than Ohm's law.

In Figure 1.8 find V_2 and V_3 by using the voltage divider theorem and Ohm's law.

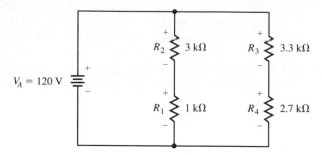

FIGURE 1.8   Circuit for practice problem 1.3

EXAMPLE 1.4

Figure 1.9 shows a voltage divider circuit connected to an unknown circuit or device (a black box). In a practical circuit, the black box would draw some current, but for our purposes assume that the box draws no current. At this time we do not care what is inside the black box.

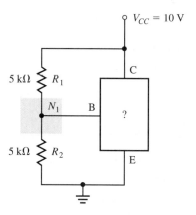

FIGURE 1.9   Voltage divider circuit with an unknown load

In a circuit that has two or more branches, any point that is common to at least two different branches is called a **node.** In this circuit we want to know what the voltage is at the junction of R_1 and R_2, or node 1 (N_1). The black box seems to require two different voltage levels to function: one level directly from the power source (V_{CC}) to point C, the other from the voltage divider node N_1 to point B. (Designations V_{CC}, C, B, and E follow standard labeling practices for transistor circuits and will be used later in the text.)

It is possible to obtain the two separate voltages by using two separate power supplies or batteries, but cost constraints usually prohibit such practices. Therefore, practical circuits are designed so that the power supply for the circuit is tapped (tied into) and set to the desired level with a voltage divider, as is done with this circuit.

Solution

The total resistance of the voltage divider is 10 kΩ, evenly divided into two 5-kΩ resistors. The center between the two resistors provides the tap-off (tie-in) point for N_1 voltage. Since $R_1 = R_2$, the voltage is equally divided between the two resistors. Therefore, 50% of the voltage is dropped across R_1, and

$$0.5 \times 10\,\text{V} = 5\,\text{V}$$

is present at node N_1.

PRACTICE PROBLEM 1.4

In Figure 1.10 find the value of the node voltage at N_1 for the case where $V_{CC} = 10$ V and where $V_{CC} = 15$ V.

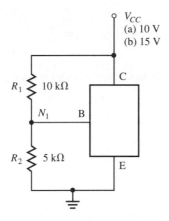

FIGURE 1.10 Circuit for practice problem 1.4

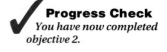
Progress Check
You have now completed objective 2.

LOADED VOLTAGE DIVIDERS

The unknown circuits in Figures 1.9 and 1.10 draw energy from node N_1 in the form of current. If the current draw becomes excessive, it can upset the ratio established by the voltage divider and thus can change the value of the voltage at the voltage divider node.

EXAMPLE 1.5

Figure 1.11a shows a voltage divider feeding energy to a 5-kΩ load resistor (R_L). If we were to assume that the voltage at N_1 was still 5 V as in example 1.4, we would be making a serious error. In this case, the resistance of the load is so small that it disturbs the balance of the voltage divider and alters the voltage appearing at node N_1. This phenomenon occurs because of the parallel resistances of R_2 and R_L. The equivalent resistance (R_{eq}) of parallel resistances can be determined in several ways. The designator for parallel resistors is two parallel vertical lines ($\|$). In this case $R_{eq} = R_2 \| R_L$. However, this representation for parallel resistance does not tell you how to find R_{eq}. The following methods can be used: (a) **product over the sum;** (b) **reciprocal of the sum of the reciprocals;** and (c) in the case of equal resistances, division of one resistor value by the total number of resistors in the parallel branches.

We will use each of these methods to find R_{eq} in Figure 1.11a.

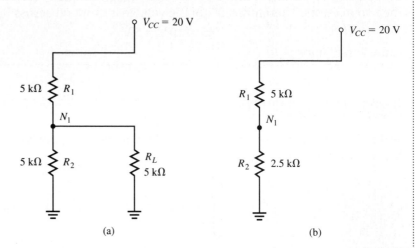

FIGURE 1.11 (a) Voltage divider with load. (b)Equivalent circuit for loaded voltage divider.

Solution

(a)

$$R_{eq} = \frac{(R_2 \times R_L)}{(R_2 + R_L)}$$

$$R_{eq} = \frac{(5 \text{ k}\Omega \times 5 \text{ k}\Omega)}{5 \text{ k}\Omega + 5 \text{ k}\Omega}$$

$$R_{eq} = \frac{25 \text{ M}\Omega}{10 \text{ k}\Omega}$$

$$R_{eq} = 2.5 \text{ k}\Omega$$

(b)

$$R_{eq} = \frac{1}{1/R_2 + 1/R_L}$$

$$R_{eq} = \frac{1}{1/5 \text{ k}\Omega + 1/5 \text{ k}\Omega}$$

$$R_{eq} = \frac{1}{200 \text{ }\mu\text{S} + 200 \text{ }\mu\text{S}}$$

$$R_{eq} = \frac{1}{400 \text{ }\mu\text{S}}$$

$$R_{eq} = 2.5 \text{ k}\Omega$$

(c)

$$R_{eq} = \frac{R_2}{2} = \frac{5 \text{ k}\Omega}{2}$$

$$R_{eq} = 2.5 \text{ k}\Omega$$

The resulting equivalent resistance is shown in Figure 1.11b. In terms of the power supply, we now have a total of 7.5 kΩ to ground, instead of the 10 kΩ we had in Figure 1.9.

Using the voltage divider ratio method to determine the voltage seen at N_1 under these conditions we have:

$$\frac{R_{eq}}{R_T} = \frac{2.5\ k\Omega}{7.5\ k\Omega} = 0.3333 = 33\%$$

The voltage at N_1 in this case is:

$$V_{N_1} = 0.3333 \times 10\ V = 3.33\ V$$

Compare this with the circuit if the load were removed:

$$\frac{5\ k\Omega}{10\ k\Omega} = 0.50 = 50\%$$
$$V_{N_1} = 0.50 \times 10\ V = 5\ V$$

This example illustrates a typical result: When a load is attached to a voltage divider the voltage at the node where the load is connected will frequently drop.

PRACTICE PROBLEM 1.5

In Figure 1.12a find the value of voltage at node N_1 with the load attached and with the load removed.

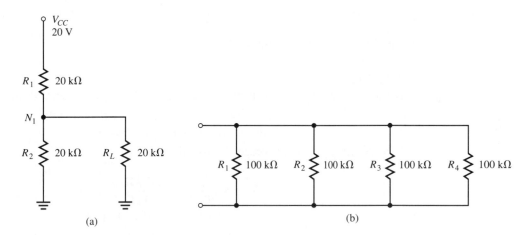

FIGURE 1.12 Circuits for practice problem 1.5

In Figure 1.12b find R_{eq} of the parallel branches using two of the three methods discussed in example 1.5.

PREVENTING LOADING EFFECTS

It is sometimes essential for the voltage provided by a voltage divider to remain fixed; that is, the voltage must not drop significantly when a load is attached. To prevent such loading effects, we use loads of sufficiently large values that the divider circuit voltage is altered by the least amount possible. This is accomplished by using what are called **firm** or **stiff voltage dividers.**

In a firm divider the load is at least 10 times the resistive value of the resistance across which it is connected. The load in a stiff divider is at least 100 times the resistive value of the resistance across which it is connected. The concepts of firm and stiff voltage dividers are important (1) for selecting the proper test equipment for measuring voltage in electronic circuits and (2) for studying transistor circuits.

EXAMPLE 1.6

Figure 1.13 shows a firm voltage divider. Unloaded, the voltage divider provides 5 V at node N_1. Loaded, as shown, the voltage will drop a certain amount. We want to know how much this drop is, expressed as a percent.

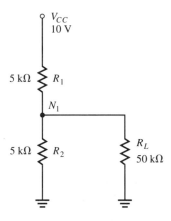

FIGURE 1.13   A firm voltage divider

Solution
Total resistance is the sum of R_1 and the parallel resistance of R_2 and R_L:

$$R_T = R_1 + (R_2 \,\|\, R_L) = 5\ k\Omega + (5\ k\Omega \,\|\, 50\ k\Omega)$$
$$R_T = 5\ k\Omega + 4.55\ k\Omega = 9.55\ k\Omega$$

The voltage at node N_1 under load is then:

$$V_{N_1} = \frac{R_{eq}}{R_T} = \frac{4.55\ k\Omega}{9.55\ k\Omega}$$
$$V_{N_1} = 0.476 \times 10\ V = 4.76\ V$$

The voltage drop at N_1 under load is:

$$5\ V - 4.76\ V = 0.24\ V$$

The percent change is:

$$\frac{0.24\ V}{5\ V} \times 100 = 4.8\%$$

PRACTICE PROBLEM 1.6

In Figure 1.14 find $R_2 \,\|\, R_L$; R_T of the loaded circuit; the voltage at node N_1 under load; and the percent change in voltage at note N_1, loaded compared with unloaded.

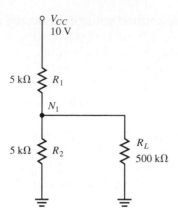

FIGURE 1.14 Stiff voltage divider for practice problem 1.6

REVIEW SECTION 1.2

1. In a voltage divider, the ratio of voltages is _____ to the ratio of resistances.
2. Write the general voltage divider equation.
3. In a circuit that has two or more branches, any point that is common to at least two different branches is called a _____.
4. In an unloaded voltage divider the voltage across any resistance equals the source voltage. True or false?
5. In a loaded voltage divider, the voltage reading at a node will be _____ that in an unloaded circuit.
   (a) Larger than
   (b) Smaller than
   (c) The same as
6. What steps can you take to prevent loading effects?
7. Define firm voltage divider.
8. Define stiff voltage divider.
9. Any parallel resistive circuit can be used as a voltage divider. True or false?
10. Show the symbol used to designate parallel resistances.

1.3 CURRENT DIVIDERS AND KIRCHHOFF'S CURRENT LAW

A **current divider** can be an important circuit analysis tool. If you learn to use it proficiently, it will increase your ability to analyze the transistor circuits studied later in this text. But what is a current divider, and how does it relate to what we have already studied?

We have learned that a series resistance string is a voltage divider and that current is the same in all parts of the series circuit. We found that the voltage across parallel branches is the same in each branch. From Ohm's law we know that total current is equal to the applied voltage divided by the total resistance, or

$$I_T = \frac{V_A}{R_T}$$

This relationship also tells us that current is inversely proportional to resistance; that is, with applied voltage held constant, the larger the value of resistance, the lower the value of current flowing through that resistance. With that knowledge we can conclude the following: (1) total current in a parallel circuit must be divided among the branches of that circuit according to the branch resistances; and (2) currents will follow the path of least resistance. In other words, the least current flows through the branch with the highest resistance. That principle is one basis for the current divider concept. Another is **Kirchhoff's current law,** which states that the sum of currents entering a node is equal to the sum of currents leaving that node. The current divider concept enables us to predict how much current will flow through each branch in a parallel circuit.

EXAMPLE 1.7

Figure 1.15 shows a two-branch parallel circuit. Each branch has one resistor. Total current is indicated by I_T, and branch currents by I_1 and I_2. At node N_1, the current splits, with I_1 flowing through resistor R_1, and I_2 flowing through resistor R_2. These two branch currents are recombined at node N_2. Two ways to analyze the currents in this circuit follow.

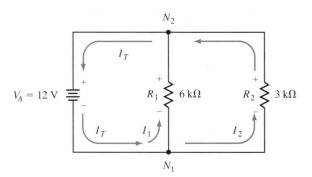

FIGURE 1.15 Current divider.
Arrows indicate direction of electron flow.

Solution 1
Since the voltages across all branches in a parallel circuit are the same we can determine the current flow through each branch:

$$I_1 = \frac{V_A}{R_1} = \frac{12\,V}{6\,k\Omega} = 2\,mA$$

$$I_2 = \frac{V_A}{R_2} = \frac{12\,V}{3\,k\Omega} = 4\,mA$$

The total current in the circuit is:

$$I_T = I_1 + I_2 = 2\ mA + 4\ mA = 6\ mA$$

These results are based on Kirchhoff's current law. The current entering N_1 equals the current leaving N_1, and the combined currents entering node N_2 equal the current leaving node N_2.

Solution 2

Determine the total resistance in the circuit, $R_T = R_1 \parallel R_2$. Using the product over the sum method:

$$R_T = \frac{R_1 \times R_2}{R_1 + R_2} = \frac{6\,k\Omega \times 3\,k\Omega}{6\,k\Omega + 3\,k\Omega}$$

$$R_T = \frac{18\,M\Omega}{9\,k\Omega} = 2\,k\Omega$$

Using the reciprocal of the sum of the reciprocals:

$$R_T = \frac{1}{1/R_1 + 1/R_2} = \frac{1}{1/6\,k\Omega + 1/3\,k\Omega}$$

$$R_T = \frac{1}{167\,\mu S + 333\,\mu S} = \frac{1}{500\,\mu S} = 2\,k\Omega$$

Use Ohm's law to determine I_T:

$$I_T = \frac{V_A}{R_T} = \frac{12\,V}{2\,k\Omega} = 6\,mA$$

How do we find the branch currents using this information? One way is by using a ratio. Add the branch resistances:

$$R_{sum} = R_1 + R_2 = 6\,k\Omega + 3\,k\Omega = 9\,k\Omega$$

Use the result to set up ratios:

$$\frac{R_1}{R_{sum}} = \frac{6\,k\Omega}{9\,k\Omega} = 0.667,\ \text{or}\ \frac{2}{3}$$

$$\frac{R_2}{R_{sum}} = \frac{3\,k\Omega}{9\,k\Omega} = 0.333,\ \text{or}\ \frac{1}{3}$$

Remember that in the voltage divider, voltage is directly proportional to resistance, and, in the current divider, current is inversely proportional to resistance. This means that the largest current flows through the least resistance. You can now calculate I_1 and I_2.

$$I_1 = \frac{1}{3} \times I_T = \frac{1}{3} \times 6\,mA = 2\,mA$$

$$I_2 = \frac{2}{3} \times I_T = \frac{2}{3} \times 6\,mA = 4\,mA$$

Another way to determine branch currents is by using the inverse current divider equation (invert R_x and R_T).

$$I_1 = \frac{R_T}{R_1} \times I_T = \frac{2\,k\Omega}{6\,k\Omega} \times 6\,mA = 2\,mA$$

$$I_2 = \frac{R_T}{R_2} \times I_T = \frac{2\,k\Omega}{3\,k\Omega} \times 6\,mA = 4\,mA$$

Of course, when there are only two branches to deal with, you can solve for the current flow in one branch, then subtract that current from total current to solve for current flow in the other branch.

For the circuit in Figure 1.16 use all methods discussed in example 1.7 to find I_T, I_1, and I_2.

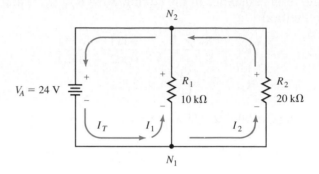

FIGURE 1.16 Circuit for practice problem 1.7

Given a choice, most people would solve for the currents in example 1.7 by using solution 1. It is easier. So why bother with solution 2? Because there are times when you may not have enough information to use solution 1.

EXAMPLE 1.8

A **constant current source** is an energy source that provides a constant value of current to a load. Because transistors can be used for constant current sources, this is an important concept in circuit analysis.

Figure 1.17 shows a parallel circuit driven by a constant current source. At this point, we do not know the voltage across each branch. Therefore, we cannot apply the method used in solution 1 in example 1.7 to determine the branch currents. We obtain a solution by using the current divider equation and Kirchhoff's current law.

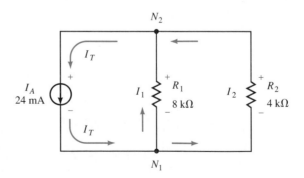

FIGURE 1.17 Constant current source–driven current divider.
The arrow in the constant current source indicates the direction of current flow.

Solution

Applied current $I_A = I_T = 24$ mA. I_T splits at node N_1 into I_1 and I_2 and recombines at node N_2. We start by finding R_T:

$$R_T = R_1 \parallel R_2 = \frac{1}{1/R_1 + 1/R_2}$$

$$R_T = \frac{1}{(1/8\text{ k}\Omega + 1/4\text{ k}\Omega)} = \frac{1}{(125\ \mu\text{S} + 250\ \mu\text{S})}$$

$$R_T = \frac{1}{375\ \mu\text{S}} = 2.67\text{ k}\Omega$$

Next, use the inverse current divider equation:

$$I_1 = \frac{R_T}{R_1} \times I_T = \frac{2.67\text{ k}\Omega}{8\text{ k}\Omega} \times 24\text{ mA} = 8\text{ mA}$$

$$I_2 = \frac{R_T}{R_2} \times I_T = \frac{2.67\text{ k}\Omega}{4\text{ k}\Omega} \times 24\text{ mA} = 16\text{ mA}$$

From Kirchhoff's current law, we know that current entering node N_1 must also leave node N_1. Therefore, with 24 mA entering node N_1 and 8 mA leaving through R_1 (or 16 mA leaving through R_2), we can use simple subtraction to find the remaining current:

$$I_1 = I_T - I_2 = 24\text{ mA} - 16\text{ mA} = 8\text{ mA}$$
$$I_2 = I_T - I_1 = 24\text{ mA} - 8\text{ mA} = 16\text{ mA}$$

This example illustrates why it is important to understand the current divider and Kirchhoff's current law. As stated previously, the transistor can be a constant current source. Transistors are studied in depth later in the text.

PRACTICE PROBLEM 1.8

For the circuit in Figure 1.18 use the current divider and Kirchhoff's current law to find I_1 and I_2.

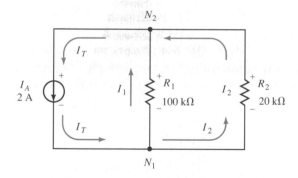

FIGURE 1.18 Circuit for practice problem 1.8

EXAMPLE 1.9

Figure 1.19a shows a three-branch parallel circuit, a new situation. We see that I_T enters node N_1 and splits; I_1 goes through R_1, and $I_2 + I_3$ flow toward node N_2. At node N_2, the current splits again; I_2 goes through R_2, and I_3 goes

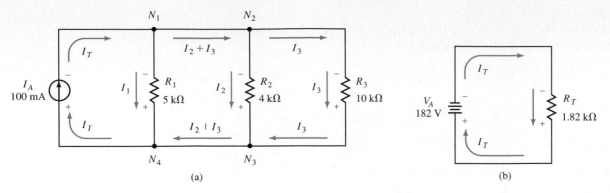

FIGURE 1.19 (a) Multiple branch constant current source divider. (b) Equivalent circuit.

through R_3. At node N_3, I_2 and I_3 recombine and enter node N_4, where all three currents are recombined. Using what you have already learned, we can solve for branch currents.

Solution

We can use Ohm's law to find V_A. First, find R_T by using the reciprocal of the sum of the reciprocals:

$$R_T = \frac{1}{1/R_1 + 1/R_2 + 1/R_3}$$

$$R_T = \frac{1}{1/5\,k\Omega + 1/4\,k\Omega + 1/10\,k\Omega}$$

$$R_T = \frac{1}{200\,\mu S + 250\,\mu S + 100\,\mu S}$$

$$R_T = \frac{1}{550\,\mu S} = 1818\,\Omega$$

We round the value (go to the next full number) from $1818\,\Omega$ to $1.82\,k\Omega$. Next, find V_A:

$$V_A = I_T \times R_T = 100\,mA \times 1.82\,k\Omega = 182\,V$$

Figure 1.19b shows the resultant equivalent circuit. We know that V_A is the same across all branches in a parallel circuit, so now we can solve for the branch currents:

$$I_1 = \frac{V_1}{R_1} = \frac{182\,V}{5\,k\Omega} = 36.4\,mA$$

$$I_2 = \frac{V_2}{R_2} = \frac{182\,V}{4\,k\Omega} = 45.5\,mA$$

$$I_3 = \frac{V_3}{R_3} = \frac{182\,V}{10\,k\Omega} = 18.2\,mA$$

Now verify the current:

$$I_A = I_T = I_1 + I_2 + I_3$$
$$I_A = 36.4\,mA + 45.5\,mA + 18.2\,mA = 100.1\,mA$$

By rounding R_T in calculating V_A, we find that I_T is slightly off, but well within acceptable tolerances for such calculations.

For the circuit in Figure 1.20 find R_T using the reciprocal of the sum of the reciprocals. Use Ohm's law to find V_A, I_1, I_2, I_3, and I_4. Finally, determine the tolerance in I_T based on the calculations for the branch currents.

Determine the current entering node N_1, node N_2, and node N_4.

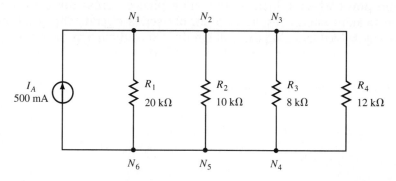

Progress Check
You have now completed objective 5.

FIGURE 1.20 Circuit for practice problem 1.9

REVIEW SECTION 1.3
1. The current divider concept is used in circuits where the resistances are in _____.
2. Write the general current divider equation.
3. State Kirchhoff's current law.
4. What predictions can be made for electronic circuits by using the current divider concept?
5. Define constant current source.
6. Draw the symbol for a constant current source, and indicate the direction of electron current flow.
7. In a circuit that has just two nodes, the current flowing into node 1 must equal the current flowing into node 2. True or false?
8. A parallel circuit has two equal-resistance branches. The total circuit current is 500 mA. The current flowing through each branch must equal:
   (a) 500 mA
   (b) 50 mA
   (c) 150 mA
   (d) 250 mA
9. A circuit has R_1 in series with $R_2 \parallel R_3$. $V_A = 50$ V; $R_1 = 2.2$ kΩ; $R_2 = 3.3$ kΩ; and $R_3 = 4.7$ kΩ. I_T must equal:
   (a) 1.21 mA
   (b) 12.1 mA
   (c) 121 mA
   (d) 1.21 A
10. For the circuit in question 9, use the inverse current divider equation to solve for I_3.

Thevenin's theorem is based on the idea that any circuit of any complexity, such as that shown in Figure 1.21a, can be reduced to an equivalent circuit consisting of a single resistance, identified as R_{TH}, connected to a single voltage source, identified as V_{TH}. This combination is a **Thevenin equivalent circuit,** as shown in Figure 1.21b.

Thevenin's theorem is important because it simplifies complex circuits to the point where a load of any value placed across the open circuit of the equivalent circuit results in a simple series circuit. The resulting circuit is easily solved by using the voltage divider and current divider equations and Ohm's law. In other words, Thevenin's theorem provides an easy way to determine what happens if the value of the load is changed. The following example illustrates this concept.

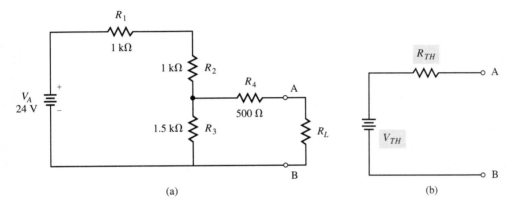

(a) (b)

FIGURE 1.21 (a) Complex circuit for Thevenin's theorem analysis. (b) A Thevenin equivalent circuit.

EXAMPLE 1.10

Figure 1.22 shows a Thevenin equivalent circuit. First we will apply a 10-kΩ load and solve for V_L and I_L. Then we will apply a 1-kΩ load and repeat the calculations.

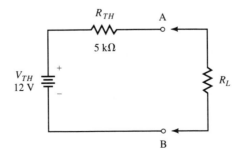

FIGURE 1.22 A Thevenin equivalent circuit for example 1.10

Solutions

Solve for V_L using the voltage divider equation:

$$V_L = \frac{R_L}{R_{TH} + R_L} \times V_{TH}$$

$$V_L = \frac{10\ k\Omega}{5\ k\Omega + 10\ k\Omega} \times 12\ V$$

$$V_L = 0.0667 \times 12\ V = 8\ V$$

Solve for I_L using Ohm's law:

$$I_L = \frac{V_L}{R_L} = \frac{8\ V}{10\ k\Omega} = .80\ mA$$

Now recalculate with $R_L = 1\ k\Omega$:

$$V_L = \frac{R_L}{R_{TH} + R_L} \times V_{TH}$$

$$V_L = \frac{1\ k\Omega}{5\ k\Omega + 1\ k\Omega} \times 12\ V$$

$$V_L = 0.1667 \times 12\ V = 2\ V$$

$$I_L = \frac{V_L}{R_L} = \frac{2\ V}{1\ k\Omega} = 2\ mA$$

This example illustrates how easily the effects of a changed load can be calculated when using a Thevenin equivalent circuit.

PRACTICE PROBLEM 1.10

In Figure 1.22, determine V_L and I_L with $R_L = 2.2\ k\Omega$ and $R_L = 3.3\ k\Omega$

EXAMPLE 1.11

Figure 1.23 shows a series-parallel circuit driven by a 12-V source. Finding the voltage and current for R_L without using Thevenin's theorem is somewhat involved. In this example, we will solve for V_L and I_L without Thevenin's equivalent circuit.

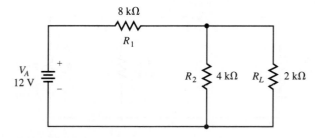

FIGURE 1.23 A series-parallel circuit

Solutions

Solving for R_T:

$$R_T = R_1 + \frac{R_2 \times R_L}{R_2 + R_L}$$

$$R_T = 8\,k\Omega + \frac{4\,k\Omega \times 2\,k\Omega}{4\,k\Omega + 2\,k\Omega}$$

$$R_T = 8\,k\Omega + 1.33\,k\Omega$$

$$R_T = 9.33\,k\Omega$$

Solving for I_T:

$$I_T = \frac{V_A}{R_T} = \frac{12\,V}{9.333\,k\Omega} = 1.29\,mA$$

Solving for I_L by using a ratio:

$$R_{sum} = R_2 + R_L = 6\,k\Omega$$

$$\frac{R_L}{R_{sum}} = \frac{2\,k\Omega}{6\,k\Omega} = \frac{1}{3}$$

From previous examples we know that 2/3 of I_T must flow through R_L:

$$I_L = \frac{2}{3} \times I_T = \frac{2}{3} \times 1.29\,mA = 860\,\mu A$$

Solving for V_L:

$$V_L = I_L \times R_L = 860\,\mu A \times 2\,k\Omega = 1.72\,V$$

EXAMPLE 1.12

In this example we will demonstrate how to **thevenize** (solve for V_{TH} and R_{TH}) the circuit shown in Figure 1.23. Then we will use the equivalent circuit to solve for V_L and I_L.

Step 1 in applying Thevenin's theorem is to remove (open) the component designated as the load and mark the open terminals, points A and B in Figure 1.24. The result is that this circuit has been reduced to a simple series circuit. This may not always be the case, as you will learn in further analytical studies. However, at this point, we want to know the voltage between points A and B, now an open circuit. The resultant **open circuit voltage** is V_{TH}.

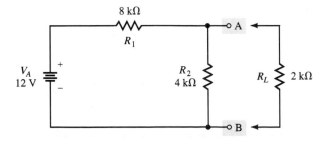

FIGURE 1.24 Step 1. Thevenizing the series-parallel circuit in Figure 1.23

Solution (V_{TH})

Applying the voltage divider equation to the modified circuit:

$$V_{TH} = \frac{R_2}{R_1 + R_2} \times V_A$$

$$V_{TH} = \frac{4\ k\Omega}{8\ k\Omega + 4\ k\Omega} \times 12\ V$$

$$V_{TH} = 0.333 \times 12\ V = 4\ V$$

Step 2 is to determine the equivalent resistance between the open terminals. This is easily done by replacing the voltage source with a short circuit, as shown in Figure 1.25a. The resistance looking back into the circuit, with R_L removed and V_A reduced to zero, is R_{TH}.

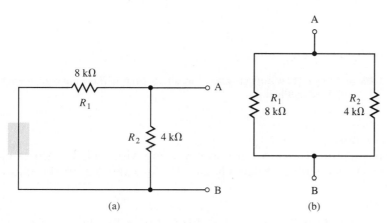

FIGURE 1.25 (a) Step 2. Circuit for finding RTH. Source replaced with a piece of wire. (b) Thevenin terminal circuit.

Solution (R_{TH})

The circuit in Figure 1.25a may be redrawn as in Figure 1.25b, showing that the resistors are in parallel. You then can solve for R_T, which is R_{TH} as seen by the terminals, using either the product over the sum or the reciprocal of the sum of the reciprocals.

Solve by the product over the sum:

$$R_T = R_{TH} = \frac{R_1 \times R_2}{R_1 + R_2}$$

$$R_T = \frac{8\ k\Omega \times 4\ k\Omega}{8\ k\Omega + 4\ k\Omega} = 2.67\ k\Omega$$

Solve by the reciprocal of the sum of the reciprocals:

$$R_T = R_{TH} = \frac{1}{1/8\ k\Omega + 1/4\ k\Omega}$$

$$R_T = \frac{1}{125\ \mu S + 250\ \mu S}$$

$$R_T = \frac{1}{375\ \mu S} = 2.67\ k\Omega$$

Step 3 is drawing the Thevenin equivalent circuit, as shown in Figure 1.26a, in which V_A has become V_{TH} and R_T has become R_{TH}.

Step 4 is to reinsert the load resistance R_L, as shown in Figure 1.26b. The key point here is that this equivalent circuit is not a "real" circuit, but it is **electrically identical** to the real (original) circuit shown in Figure 1.23.

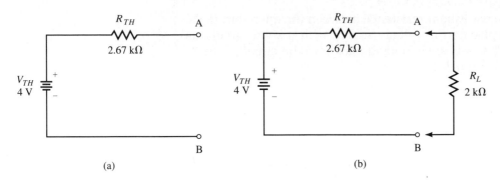

(a) (b)

FIGURE 1.26 (a) Step 3. Thevenin equivalent circuit. (b) Step 4. Thevenin equivalent circuit with load applied.

Solution (V_L and I_L)

As shown in Figure 1.26b, we are now working with a simple series circuit, which makes calculations relatively simple (for example, in the equivalent circuit, $I_L = I_T$).

Solve for R_T:

$$R_T = R_{TH} + R_L = 2.67 \text{ k}\Omega + 2 \text{ k}\Omega = 4.67 \text{ k}\Omega$$

Solve for I_L:

$$I_T = I_L = \frac{V_{TH}}{R_T} = \frac{4 \text{ V}}{4.67 \text{ k}\Omega} = 857 \text{ }\mu\text{A}$$

Solve for V_L:

$$V_L = I_L \times R_L = 857 \text{ }\mu\text{A} \times 2 \text{ k}\Omega = 1.71 \text{ V}$$

Another way to solve for these values:

$$V_L = \frac{R_L}{R_L + R_{TH}} \times V_{TH}$$

$$V_L = \frac{2 \text{ k}\Omega}{2 \text{ k}\Omega + 2.67 \text{ k}\Omega} \times 4 \text{ V}$$

$$V_L = 0.4283 \times 4 \text{ V} = 1.71 \text{ V}$$

$$I_L = \frac{V_L}{R_L} = \frac{1.71 \text{ V}}{2 \text{ k}\Omega} = 855 \text{ }\mu\text{A}$$

The difference between the results of the two methods is well within acceptable tolerances. Also, compare the V_L and I_L results here with the results found in example 1.10. We can deduce from this example that any circuit, no matter how complex, can be reduced to a two-resistor series circuit where $V_A = V_{TH}$; $R_1 = R_{TH}$; and $R_2 = R_L$.

Thevenize the circuit in Figure 1.27 where $R_4 = R_L$. Use the voltage divider equation to solve for V_{TH}; Use the product over the sum and the reciprocal of the sum of the reciprocals to solve for R_{TH}; Draw the terminal diagram and Thevenin equivalent circuit.

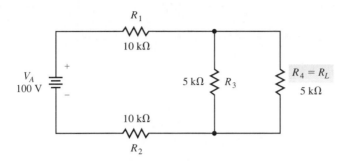

FIGURE 1.27 Circuit for practice problem 1.11

Draw the Thevenin equivalent circuit with R_L inserted. Use Ohm's law to solve for I_L and V_L. Use the voltage divider equation to solve for V_L, and use Ohm's law to solve for I_L.

REDRAWING COMPLEX CIRCUITS

Circuit diagrams are often drawn with the voltage source on the left and the circuit components on the right, the procedure we have used to this point. However, in your career as an electronic technician you will seldom encounter such simple schematic diagrams. Therefore it is important for you to learn to look at a circuit diagram from different perspectives and how to redraw them in a more simplified version. Once redrawn, the circuit can be thevenized into a simple two-resistance series circuit. The next example illustrates a procedure for simplifying more complex circuits.

EXAMPLE 1.13

Figure 1.28a shows a slightly more complex circuit. For evaluation purposes we will let $R_4 = R_L$. When R_L is removed (Figure 1.28b), V_A is looking at only one complete path for current flow; R_3 does not exist as far as V_A is concerned; electrically it is not there. Although it is true that one end of R_3 is connected to the circuit at node N_1, the other end is connected to terminal A, which is an electrically open circuit. We can therefore conclude that the values of voltages at node N_1 and at terminal A are the same. This is so because, with no current flow through R_3, there can be no voltage drop across R_3. This is easily proven by Ohm's law (remember from basic mathematics that any value multiplied by 0 is 0):

$$V_3 = I_3 \times R_3 = 0 \times 1.8\,k\Omega = 0$$

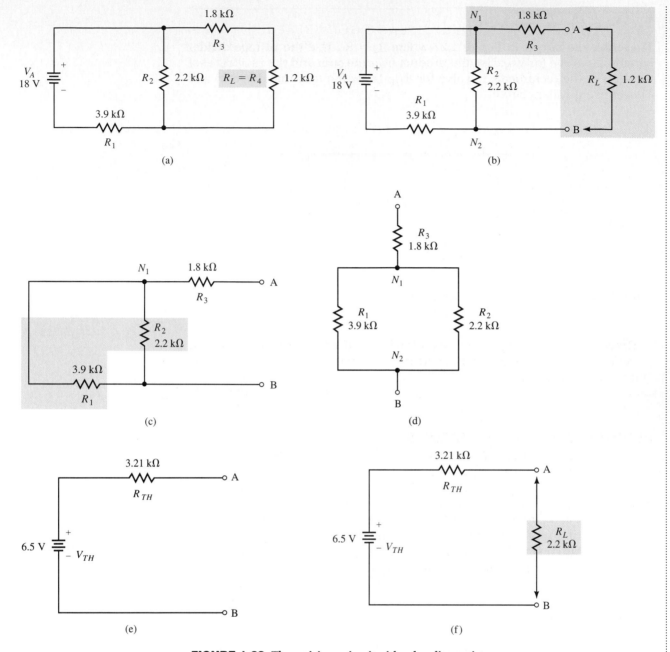

FIGURE 1.28 Thevenizing a circuit with a dangling resistor

Since there is no voltage drop across R_3, the potential difference between node N_1 and terminal A is zero. This phenomenon makes node N_1 and terminal A **equipotential points** (two points having the same voltage level). *It is a grave error, however, to conclude that terminal A must be at 0 V.*

For example, in this circuit, node N_1 is tied directly to the positive terminal of the 18-V source. Following the standard practice for electrical circuits, the negative terminal of V_A is the ground reference point. Therefore, measuring the voltage between node N_1, or from terminal A, to ground will show the same 18 V, because you are, in effect, measuring directly across the source.

Solution (V_{TH})

Looking into the circuit from V_A, with R_L removed and R_3 electrically out of the circuit, we have a simple series circuit consisting of V_A, R_1, and R_2, as shown in Figure 1.28c. We can use the voltage divider equation to find V_2, which equals V_{TH}:

$$V_{TH} = V_2 = \frac{R_2}{R_1 + R_2} \times V_A$$

$$V_{TH} = \frac{2.2 \text{ k}\Omega}{3.9 \text{ k}\Omega + 2.2 \text{ k}\Omega} \times 18 \text{ V} = 6.5 \text{ V}$$

Solution (R_{TH})

Looking back into the circuit from terminals A and B, with the source short circuited as shown in Figure 1.28c, R_3 re-enters our calculations. Figure 1.28d shows how R_3 is in series with parallel resistors R_1 and R_2. Solving for the equivalent total resistance R_T produces R_{TH} for the Thevenin equivalent circuit:

$$R_{TH} = R_T = R_3 + \frac{R_1 \times R_2}{R_1 + R_2}$$

$$R_{TH} = 1.8 \text{ k}\Omega + \frac{3.9 \text{ k}\Omega \times 2.2 \text{ k}\Omega}{3.9 \text{ k}\Omega + 2.2 \text{ k}\Omega}$$

$$R_{TH} = 3.21 \text{ k}\Omega$$

Figure 1.28e shows the Thevenin equivalent circuit using the calculated V_{TH} and R_{TH}. With R_L inserted, as in Figure 1.28f, we can now determine V_L and I_L.

Using Ohm's law:

$$I_L = I_T = \frac{V_{TH}}{R_{TH} + R_L} = \frac{6.5 \text{ V}}{3.21 \text{ k}\Omega + 2.2 \text{ k}\Omega}$$

$$I_L = 1.2 \text{ mA}$$

$$V_L = I_L \times R_L = 1.2 \text{ mA} \times 2.2 \text{ k}\Omega = 2.64 \text{ V}$$

Using the voltage divider equation:

$$V_L = \frac{R_L}{R_{TH} + R_L} \times V_{TH} = \frac{2.2 \text{ k}\Omega}{3.21 \text{ k}\Omega + 2.2 \text{ k}\Omega} \times 6.5 \text{ V}$$

$$V_L = 2.64 \text{ V}$$

$$I_L = \frac{V_L}{R_L} = \frac{2.64 \text{ V}}{2.2 \text{ k}\Omega} = 1.2 \text{ mA}$$

Again, we see that calculations using the two methods are very close, well within acceptable tolerance.

PRACTICE PROBLEM 1.12

Thevenize the circuit in Figure 1.29 where $R_7 = R_L$. Use the voltage divider equation to solve for V_{TH}. Use the product over sum and the reciprocal of the sum of the reciprocals to solve for R_{TH}. Draw the terminal diagram and Thevenin equivalent circuit.

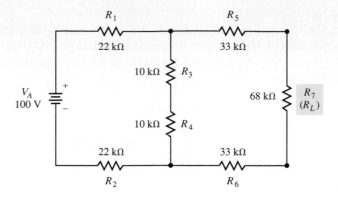

FIGURE 1.29 Circuit for practice problem 1.12

Draw the Thevenin equivalent circuit with the load inserted. Use Ohm's law to solve for I_L. Use Ohm's law to solve for V_L. Use the voltage divider equation to solve for V_L, and use Ohm's law to solve for I_L.

EXAMPLE 1.14

When you thevenize a circuit, it is not necessary to make the circuit load a resistor. Figure 1.30a shows a capacitor connected to a resistive network driven by a 15-V power source. The capacitor will eventually change to 5 V in 5 RC time constants because it is across a 5-V drop. (1 **RC time constant** is the product of the Thevenin resistance and the capacitance: $R_{TH} \times C$.) In this circuit, we must find the resistance of the circuit as seen by the capacitor; that is, by looking back into the circuit. To do so, we apply Thevenin's theorem.

Solution

Figure 1.30b shows the capacitor load removed from the circuit, the voltage source shorted, and the terminals marked. From this diagram, and looking back into the circuit from terminals A and B, we can draw the terminal diagram as shown in Figure 1.30c.

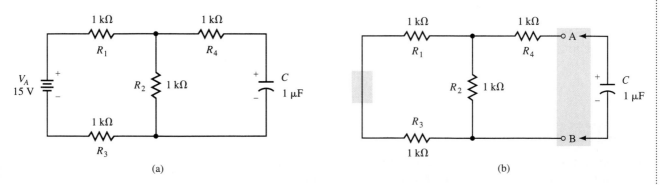

(a)   (b)

FIGURE 1.30 (a) A circuit with a capacitive load. (b) First step in obtaining R_{TH}.

(continued on next page)

FIGURE 1.30 (c) Terminal diagram of thevenized circuit. (continued)

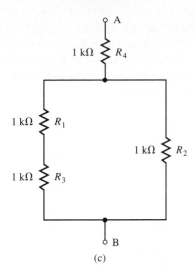

(c)

The terminal diagram shows a series-parallel combination. First we solve for the parallel branch:

$$R_{eq} = \frac{(R_1 + R_3) \times R_2}{R_1 + R_3 + R_2}$$

$$R_{eq} = \frac{2\,k\Omega \times 1\,k\Omega}{2\,k\Omega + 1\,k\Omega}$$

$$R_{eq} = \frac{2\,M\Omega}{3\,k\Omega} = 667\,\Omega$$

Solving for R_{TH}:

$$R_{TH} = R_4 + R_{eq} = 1\,k\Omega + 667\,\Omega = 1.667\,k\Omega$$

We can now calculate how long it will take for C to charge to V_{TH}, using 5 RC time constants:

$$C_{charge} = R_{TH} \times C \times 5 = 1.667\,k\Omega \times 1\,\mu F \times 5 = 8.33\,ms$$

PRACTICE PROBLEM 1.13

For the circuit in Figure 1.31, assume the capacitor is the load. Thevenize the circuit. Determine the time required for the capacitor to become fully charged.

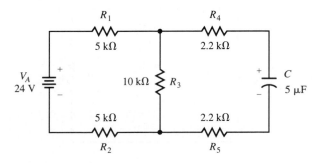

FIGURE 1.31 Circuit for practice problem 1.13

EXAMPLE 1.15

This example illustrates how thevenizing can reduce a complex circuit such as the **unbalanced bridge circuit** (the ratio of R_1 to R_2 is not equal to the ratio of R_3 to R_4) shown in Figure 1.32a. Bridge circuits are used in an instrument such as an ohmmeter. The instrument uses a **galvonometer** (a sensitive device used to detect the presence of electrical currents and to measure the

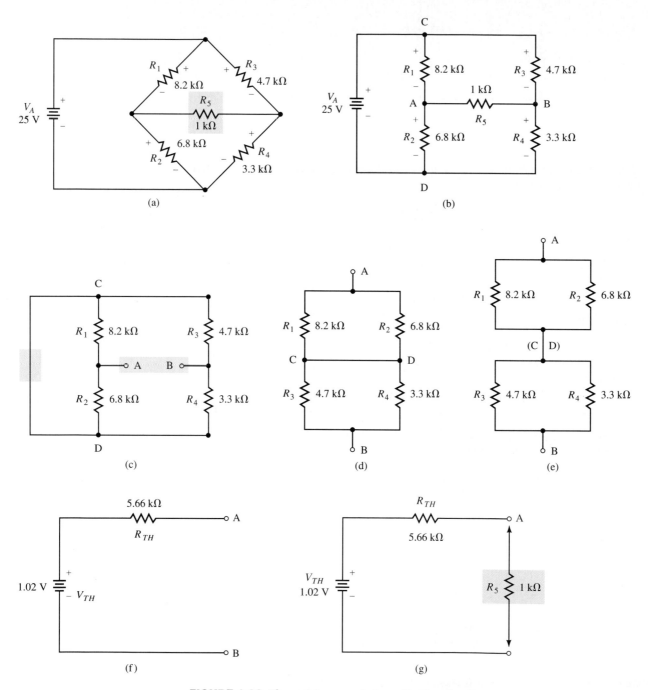

FIGURE 1.32 Thevenizing an unbalanced bridge circuit

magnitude and direction of those currents) in place of R_5 in this circuit. Note that we did not specify the direction of current flow in R_5. At this point we do not know that direction.

Solution

The first step is to simplify the diagram and label connection points, as shown in Figure 1.32b. Next, open R_5 and short V_A, as in Figure 1.32c. We now have two parallel circuits, each with a two-resistor series string, shown in Figures 1.32d, e. We show two versions of the terminal diagram so that you can learn that there are several ways to redraw a circuit, thus making it easier to evaluate the circuit.

Solution (R_{TH})

To solve for R_{TH}, first find the equivalent resistance of R_1 and R_2 (R_{eq1}) and R_3 and R_4 (R_{eq2}), then add the results. We will use the reciprocal of the sum of the reciprocals method:

$$R_{eq1} = \frac{1}{1/R_1 + 1/R_2}$$

$$R_{eq1} = \frac{1}{1/8.2\ k\Omega + 1/6.8\ k\Omega}$$

$$R_{eq1} = \frac{1}{122\ \mu S + 147\ \mu S}$$

$$R_{eq1} = \frac{1}{269\ \mu S} = 3.72\ k\Omega$$

$$R_{eq2} = \frac{1}{1/(R_3 + 1/R_4)}$$

$$R_{eq2} = \frac{1}{1/4.7\ k\Omega + 1/3.3\ k\Omega}$$

$$R_{eq2} = \frac{1}{213\ \mu S + 303\ \mu S}$$

$$R_{eq2} = \frac{1}{516\ \mu S} = 1.94\ k\Omega$$

Adding these results produces R_{TH}:

$$R_{TH} = R_{eq1} + R_{eq2} = 3.72\ k\Omega + 1.94\ k\Omega = 5.66\ k\Omega$$

Solution (V_{TH})

The next step is to find V_{TH}. First find the voltage at terminal A, then at terminal B. The difference between these two voltages is V_{TH}.

Find the voltage at terminal A (across R_2) by applying the voltage divider equation:

$$V_2 = \frac{R_2}{R_{eq1}} \times V_A = \frac{6.8\ k\Omega}{15\ k\Omega} \times 25\ V$$

$$V_2 = 11.33\ V \text{ at terminal A}$$

Use the same procedure to find the voltage at terminal B (across R_4):

$$V_4 = \frac{R_4}{R_{eq2}} \times V_A = \frac{3.3\ k\Omega}{8\ k\Omega} \times 25\ V$$

$$V_4 = 10.31\ V \text{ at terminal B}$$

With different voltages at terminals A and B, we are assured that current will flow through R_5 when it is inserted in the circuit.

The next step is to find V_{TH}:

$$V_{TH} = V_2 - V_4 = 11.33 \text{ V} - 10.31 \text{ V} = 1.02 \text{ V}$$

We can now draw the Thevenin equivalent circuit, as shown in Figure 1.32f. When we reinsert R_5 as R_L, as shown in Figure 1.32g, we have a simple two-resistor series circuit. This circuit is electrically identical to the original bridge circuit as far as R_5 is concerned.

Solution (V_L and I_L)

With R_L reinserted, we use the voltage divider equation and Ohm's law to find V_5 and I_5.

$$V_L = \frac{R_L}{(R_L + R_{TH})} \times V_{TH}$$

$$V_L = \frac{1 \text{ k}\Omega}{1 \text{ k}\Omega + 5.66 \text{ k}\Omega} \times 1.02 \text{ V}$$

$$V_L = 153.2 \text{ mV}$$

$$I_L = \frac{V_L}{R_L} = \frac{153.2 \text{ mV}}{1 \text{ k}\Omega} = 153.2 \text{ }\mu\text{A}$$

PRACTICE PROBLEM 1.14

Thevenize the circuit in Figure 1.33, where $R_5 = R_L$. Simplify the circuit, and label the connecting points. Solve for V_{TH}. Solve for R_{TH}. Draw the Thevenin equivalent circuit.

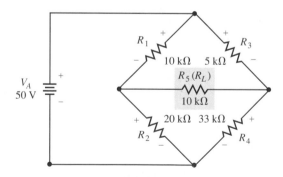

FIGURE 1.33 Circuit for practice problem 1.14

Progress Check
You have now completed objective 5.

Insert R_L and use the voltage divider equation and Ohm's law to solve for V_L and I_L.

REVIEW SECTION 1.4

1. State Thevenin's theorem.
2. A Thevenin equivalent circuit consists of _____ and _____.
3. Thevenin's theorem can be used to reduce any circuit to a simple two-resistor, one-power-source circuit. True or false?
4. State the steps that must be taken to thevenize a circuit.
5. Does decreasing the load resistance in a Thevenin equivalent circuit increase or decrease the voltage drop across the load?
6. To the load, a thevenized circuit is electrically different from the original circuit. True or false?
7. The load in a thevenized circuit must always be a resistor. True or false?
8. Define equipotential points in a circuit.
9. What is the voltage drop across a resistor having +5 V from either end to ground?
10. A resistor such as the one described in question 9 is:
    (a) open
    (b) shorted
    (c) superimposed
    (d) dangling
11. Current will not flow through the bridging resistor in an unbalanced bridge circuit. True or false?

1.5 KIRCHHOFF'S VOLTAGE LAW

Kirchhoff's voltage law states that the sum of the voltage drops around any closed-loop circuit must equal zero, and that sum must equal, but be of opposite polarity to, the source voltage. In this section, we will apply Kirchhoff's voltage law to closed-loop circuits. We will also illustrate how the sum of voltages must include the voltage of the source and that polarity must be taken into account. In other words, equal negative and positive voltages total zero. Using Kirchhoff's voltage law is an excellent way to check answers derived from closed-loop calculations. *All loops must be considered, not just the ones with voltage sources.*

EXAMPLE 1.16

Figure 1.34a shows a three-loop, series-parallel combination circuit. Loop 1 consists of V_A, R_1, and R_2. Loop 2 consists of R_3, R_4, and R_2. Loop 3 consists of V_A, R_3, R_4, and R_1. Each loop is evaluated as a series string. We will first evaluate the circuit using Ohm's law. Then we will check the answers using Kirchhoff's voltage law.

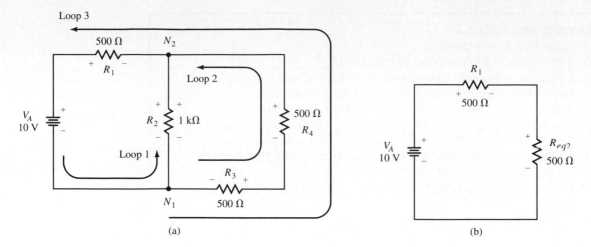

FIGURE 1.34 (a) Three-loop, series-parallel combination circuit. (b) Revised circuit with series-string equivalent.

Solution

We first find the equivalent resistance in loop 2 (R_{eq2}), which is $R_3 + R_4$ in parallel with R_2:

$$R_3 + R_4 = 500\ \Omega + 500\ \Omega = 1\ k\Omega$$

We now have two 1-kΩ resistors in parallel, so we can divide:

$$R_{eq2} = \frac{1\ k\Omega}{2} = 500\ \Omega$$

Figure 1.34b shows the revised circuit with a series string consisting of two 500-Ω resistors connected across a 10-V V_A. From this circuit, we use Ohm's law to calculate I_T:

$$I_T = \frac{V_A}{R_1 + R_{eq2}} = \frac{10\ V}{500\ \Omega + 500\ \Omega} = 10\ mA$$

Since we have equal resistances in the revised circuit, V_A is equally divided between the two:

$$V_1 = V_{eq2} = \frac{10\ V}{2} = 5\ V$$

The 10-mA current flow in the original circuit splits at node N_1, with 5 mA flowing through R_2 and 5 mA flowing through the series string, R_3 and R_4. We can now find V_2, V_3, and V_4:

$$V_2 = I_2 \times R_2 = 5\ mA \times 1\ k\Omega = 5\ V$$

This same 5 V appears in the parallel branch in loop 2, where $R_3 = R_4$, so we can again divide:

$$V_3 = V_4 = \frac{5\ V}{2} = 2.5\ V$$

I_1 and I_2 recombine at node N_2 and flow through R_1, where the voltage reconfirms V_1:

$$V_1 = I_T \times R_1 = 10 \text{ mA} \times 500 \ \Omega = 5 \text{ V}$$

USING KIRCHHOFF'S VOLTAGE LAW TO CHECK ANSWERS

We will now analyze the loops in Figure 1.35a, which is the same circuit as Figure 1.34a but with the voltage drops added. Evaluation requires that particular attention be paid to the polarities across the resistors. Also, remember to include every loop, even those without a voltage source, such as loop 2 in Figures 1.34a and 1.35a. In analyzing each loop, we will start and end at node N_1, moving counterclockwise.

Loop 1: $- 5 \text{ V} - 5 \text{ V} + 10 \text{ V} = 0$
Loop 2: $- 2.5 \text{ V} - 2.5 \text{ V} + 5 \text{ V} = 0$
Loop 3: $- 2.5 \text{ V} - 2.5 \text{ V} - 5 \text{ V} + 10 \text{ V} = 0$

The result verifies the calculations previously made. However, it is not necessary to go in a counterclockwise direction to use Kirchhoff's voltage law. In analyzing each loop in Figure 1.35b, we will start and end at node N_2, moving in a clockwise direction.

Loop 1: $+ 5 \text{ V} - 10 \text{ V} + 5 \text{ V} = 0$
Loop 2: $+ 2.5 \text{ V} + 2.5 \text{ V} - 5 \text{ V} = 0$
Loop 3: $+ 2.5 \text{ V} + 2.5 \text{ V} - 10 \text{ V} + 5 \text{ V} = 0$

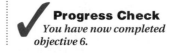

Progress Check
You have now completed objective 6.

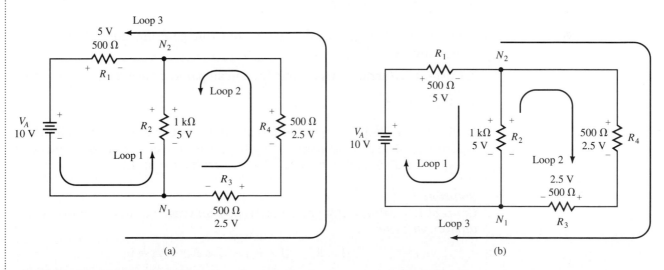

FIGURE 1.35 (a) Using Kirchhoff's voltage law to analyze loops. (b) Repeat of 1.35a but moving clockwise.

Use Kirchhoff's voltage law to analyze the circuit in Figure 1.36. Find the current flow through each resistor and the IR drop for each resistor.

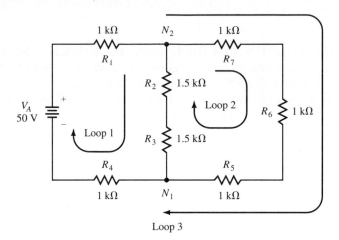

FIGURE 1.36 Circuit for practice problem 1.15

EXAMPLE 1.17

This example will demonstrate how useful Kirchhoff's voltage law can be, especially if you do not know, or cannot remember, the equation that applies when analyzing a particular circuit.

Figure 1.37 shows a simple series circuit from which we will illustrate another idea used in this text. In most cases, you will be able to develop equations on your own by using algebra and basic laws, such as Ohm's law and Kirchhoff's voltage law. For this example, assume you are familiar with just these three basic concepts:

1. Kirchhoff's voltage law: the sum of the voltage drops around any closed loop must equal zero.

2. Current in a series circuit is the same in all parts of the circuit.

3. Ohm's law: V = IR.

Could you develop an equation to calculate the current I_T in the circuit in Figure 1.37? The answer is yes. Using the basic concepts that you know, you can develop an equation for solving total circuit current (I_T).

Solution
Concept 1: Starting at test point 1 (TP1) and moving in a counterclockwise direction, we have:

$$- (I \times R_1) - (I \times R_2) - (I \times R_3) + V_A = 0$$

where the I $\times$ R values are the voltage drops across the resistors.
Rewriting the equation and inserting values, we have:

$$15\,V = (I \times 8\,k\Omega) + (I \times 3\,k\Omega) + (I \times 5\,k\Omega)$$

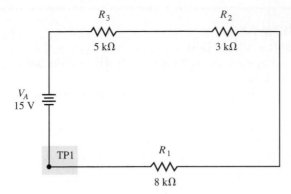

FIGURE 1.37 Circuit for developing an equation

Concept 2: Factor out I and rewrite:

$$15\,V = I(8\,k\Omega + 3\,k\Omega + 5\,k\Omega)$$

Divide both sides by the resistance and rewrite:

$$I = \frac{15\,V}{8\,k\Omega + 3\,k\Omega + 5\,k\Omega} = \frac{15\,V}{16\,k\Omega} = 938\,\mu A$$

Concept 3: The result of the above calculation is the I = V/R form of Ohm's law.

This example illustrates how you can analyze a circuit without knowing too many equations. From that analysis, you can develop an equation that can easily be solved by Ohm's law. We have just analyzed a circuit using a few basic facts and have developed our own equation. These accomplishments are not trivial. Our goal throughout this text is to find ways in which we can apply simple concepts and laws to analyze circuits by using loops to develop an equation that is some form of Ohm's law. This skill will free you from memorizing many different equations. It will also help you to develop circuit analysis skills.

PRACTICE PROBLEM 1.16

Use the three basic concepts discussed in example 1.17 to develop an equation for the circuit in Figure 1.38. Solve for I_T in the circuit, current flow through each resistor, and IR drops across each resistor.

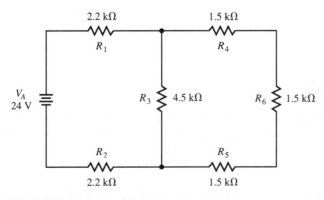

FIGURE 1.38 Circuit for practice problem 1.16

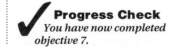

Progress Check
You have now completed objective 7.

1.6 SUPERPOSITION THEOREM

The **superposition theorem** is very useful for analyzing circuits that have more than one voltage source. When we begin working with transistor amplifiers later in this text, we will see many circuits that have both DC and AC voltage sources. The superposition theorem works in circuits with both types of sources. Therefore, the concept of superpositioning will be important in our study of transistors.

Simply stated, the superposition theorem says that in a circuit having more than one source, determine the effect of one source at a time, then superimpose (algebraically add) the results of all the sources. This is how it works. Find the total current in any branch by first determining the currents produced in that branch by each source acting separately; that is, reduce all other sources to zero (replace by their internal resistance, ideally 0 Ω, or short circuit). The total current in the branch is the sum of the individual currents, added algebraically. We will work through some examples to illustrate the superposition theorem in circuits containing both DC and AC sources.

EXAMPLE 1.18

Figure 1.39a shows a circuit having two DC voltage sources. We will follow a step-by-step superpositioning procedure to find the currents flowing in the circuit.

Solution
Step 1: Reducing V_{A2} to zero (short circuiting) produces a series-parallel combination as shown in Figure 1.39b.

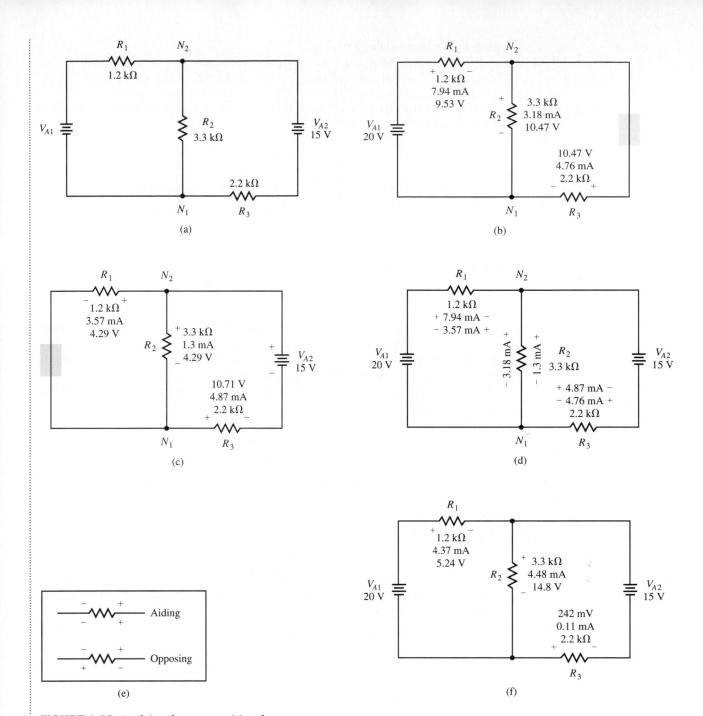

FIGURE 1.39 Applying the superposition theorem

Step 2: Find the total resistance in the circuit.

$$R_T = R_1 + \frac{R_2 \times R_3}{R_2 + R_3}$$

$$R_T = 1.2 \text{ k}\Omega + \frac{3.3 \text{ k}\Omega \times 2.2 \text{ k}\Omega}{3.3 \text{ k}\Omega + 2.2 \text{ k}\Omega}$$

$$R_T = 1.2 \text{ k}\Omega + 1.32 \text{ k}\Omega = 2.52 \text{ k}\Omega$$

Step 3: Use Ohm's law to find I_{T1}.

$$I_{T1} = \frac{V_{A1}}{R_T} = \frac{20\,V}{2.52\,k\Omega} = 7.94\,mA$$

Step 4: Use the current divider equation to determine the current flow through each resistor.

$$I_3 = \frac{R_2}{R_3 + R_2} \times I_{T1}$$

$$I_3 = \frac{3.3\,k\Omega}{2.2\,k\Omega + 3.3\,k\Omega} \times 7.94\,mA = 4.76\,mA$$

$$I_2 = I_{T1} - I_3 = 7.94\,mA - 4.76\,mA = 3.18\,mA$$

$$I_1 = I_{T1} = 7.94\,mA$$

Step 5: Use Ohm's law and Kirchhoff's voltage law to find the voltage drops across each resistor.

$$V_1 = I_1 \times R_1 = 7.94\,mA \times 1.2\,k\Omega = 9.53\,V$$

$$V_2 = V_3 = V_{A1} - V_1 = 20\,V - 9.53\,V = 10.47\,V$$

Step 6: Mark the direction of current flow, and enter the current and voltage values found. (NOTE: It is important to mark the direction of current flow produced by each source because it tells whether to add or subtract currents after completion of analysis of all sources in the circuit.)

Step 7: Restore V_{A2} and short circuit V_{A1}, thus producing the series-parallel circuit shown in Figure 1.39c.

Step 8: Find R_T.

$$R_T = R_3 + \frac{R_2 \times R_1}{R_2 + R_1}$$

$$R_T = 2.2\,k\Omega + \frac{3.3\,k\Omega \times 1.2\,k\Omega}{3.3\,k\Omega + 1.2\,k\Omega}$$

$$R_T = 2.2\,k\Omega + 880\,\Omega = 3.08\,k\Omega$$

Step 9: Use Ohm's law to find I_{T2}.

$$I_{T2} = \frac{V_{A2}}{R_T} = \frac{15\,V}{3.08\,k\Omega} = 4.87\,mA$$

Step 10: Use the current divider equation to determine the current flow through each resistor.

$$I_3 = I_{T2} = 4.87\,mA$$

$$I_2 = \frac{R_1}{R_2 + R_1} \times I_{T2}$$

$$I_2 = \frac{1.2\,k\Omega}{3.3\,k\Omega + 1.2\,k\Omega} \times 4.87\,mA = 1.3\,mA$$

$$I_1 = I_{T2} - I_1 = 4.87\,mA - 1.3\,mA = 3.57\,mA$$

Step 11: Use Ohm's law and Kirchhoff's voltage law to find the voltage drops across each resistor.

$$V_3 = I_{T2} \times R_3 = 4.87\,mA \times 2.2\,k\Omega = 10.71\,V$$

$$V_1 = V_2 = V_{A2} - V_3 = 15\,V - 10.71\,V = 4.29\,V$$

Step 12: Mark the direction of current flow, and enter the current and voltage values derived from the analysis of each source as shown in Figure 1.39d.

Step 13: Algebraically total the result. Figure 1.39e shows aiding and oppos-ing currents. You *add aiding currents* and *subtract opposing currents*.

$$I_1 = 7.94 \text{ mA} - 3.57 \text{ mA} = 4.37 \text{ mA}$$
$$I_2 = 3.18 \text{ mA} + 1.3 \text{ mA} = 4.48 \text{ mA}$$
$$I_3 = 4.87 \text{ mA} - 4.76 \text{ mA} = 0.11 \text{ mA}$$
$$V_1 = 9.53 \text{ V} - 4.29 \text{ V} = 5.24 \text{ V}$$
$$V_2 = 10.47 \text{ V} + 4.29 \text{ V} = 14.8 \text{ V}$$
$$V_3 = 10.71 \text{ V} - 10.47 \text{ V} = 240 \text{ mV}$$

Step 14: Redesignate the actual currents and voltage drops in the circuit as shown in Figure 1.39f.

PRACTICE PROBLEM 1.17

Follow the step-by-step superposition procedure outlined in example 1.18 to analyze the circuit in Figure 1.40.

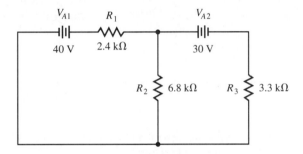

FIGURE 1.40 Circuit for practice problem 1.17

EXAMPLE 1.19

In this example we will illustrate the superposition theorem for a circuit con-taining a 10-VDC source and a 5-v_p AC source as shown in Figure 1.41a. Re-member from your AC theory that **peak voltage** (v_p) is the maximum instan-taneous voltage of a sine wave. The difference between the maximum instantaneous positive voltage and the minimum instantaneous negative voltage of a sine wave is called the **peak-to-peak voltage** (V_{p-p}). For a more complete review of AC theory, refer to Chapter 3. You will commonly en-counter this situation when analyzing transistor amplifiers. (Note the labels on the sources: capital V for the DC source and lower case v for the AC source. This is the standard labeling practice for DC and AC voltages.)

If you follow the step-by-step procedure outlined here, you should have no difficulty in analyzing the transistor amplifiers studied later in this text.

Solution
Step 1: Determine how the DC source affects the circuit by reducing the AC source to zero. This results in the DC series-parallel circuit shown in Figure 1.41b.

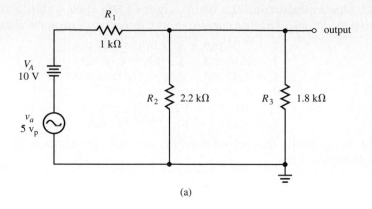

(a)

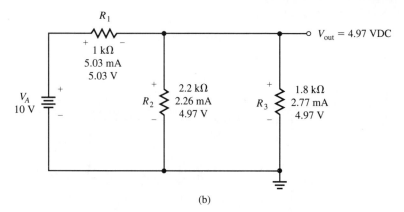

(b)

FIGURE 1.41 Applying the superposition theorem to a circuit having both a DC and an AC power source (continued on next page)

Step 2: Find the total resistance in the circuit.

$$R_T = R_1 + \frac{R_2 \times R_3}{R_2 + R_3}$$

$$R_T = 1\ k\Omega + \frac{2.2\ k\Omega \times 1.8\ k\Omega}{2.2\ k\Omega + 1.8\ k\Omega}$$

$$R_T = 1\ k\Omega + 0.99\ k\Omega = 1.99\ k\Omega$$

Step 3: Use Ohm's law to find I_T.

$$I_T = \frac{V_A}{R_T} = \frac{10\ V}{1.99\ k\Omega} = 5.03\ mA$$

Step 4: Use the current divider equation to determine the current flow through each resistor.

$$I_3 = \frac{R_2}{R_3 + R_2} \times I_{T1}$$

$$I_3 = \frac{2.2\ k\Omega}{2.2\ k\Omega + 1.8\ k\Omega} \times 5.03\ mA = 2.77\ mA$$

$$I_2 = I_T - I_3 = 5.03\ mA - 2.77\ mA = 2.26\ mA$$

$$I_1 = I_T = 5.03\ mA$$

FIGURE 1.41 (continued)

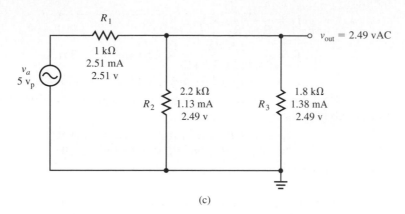

(c)

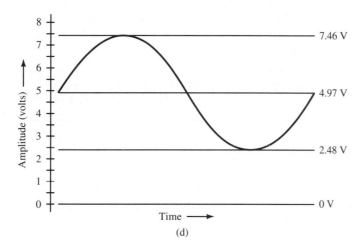

(d)

Step 5: Use Ohm's law and Kirchhoff's voltage law to find the voltage drops across each resistor.

$$V_1 = I_1 \times R_1 = 5.03 \text{ mA} \times 1 \text{ k}\Omega = 5.03 \text{ V}$$
$$V_2 = V_3 = V_A - V_1 = 10 \text{ V} - 5.03 \text{ V} = 4.97 \text{ V}$$

The output voltage is taken across R_3 and is 4.97 VDC.

Step 6: Mark the direction of current flow, and enter the current and voltage values found.

Step 7: Restore v_a and short circuit V_A, thus producing the AC series-parallel circuit shown in Figure 1.41c. The AC source sees the same circuit the DC source saw, thereby shortening the steps required to analyze the AC circuit. All AC values in the following calculations are peak values.

Step 8: Use Ohm's law to find i_t (lowercase letters indicate AC values). R_T remains the same, so:

$$i_t = \frac{5 \text{ v}}{1.99 \text{ k}\Omega} = 2.51 \text{ mA}$$

Step 9: Use the current divider equation to determine the current flow through each resistor.

$$i_1 = i_t = 2.51 \text{ mA}$$
$$i_2 = \frac{R_3}{R_2 + R_3} \times i_t$$
$$i_2 = \frac{1.8 \text{ k}\Omega}{2.2 \text{ k}\Omega + 1.8 \text{ k}\Omega} \times 2.51 \text{ mA} = 1.13 \text{ mA}$$
$$i_3 = i_t - i_1 = 2.51 \text{ mA} - 1.13 \text{ mA} = 1.38 \text{ mA}$$

Step 10: Use Ohm's law and Kirchhoff's voltage law to find the IR drops across each resistor.

$$v_1 = i_t \times R_1 = 2.51 \text{ mA} \times 1 \text{ k}\Omega = 2.51 \text{ V}$$
$$v_2 = v_3 = v_a - v_1 = 5 \text{ v} - 2.51 \text{ v} = 2.49 \text{ v}$$

The output voltage is taken across R_3 and is 2.49 v.

Step 11: Combine the results obtained from the calculations of the DC source circuit and the AC source circuit. With both sources back in the circuit as shown in Figure 1.41a, the output signal representation is shown in Figure 1.41d. We see that the sine wave is shifted from ground reference to +4.97 VDC; the DC voltage level is still 4.97 V, but the 2.49 v voltage is "riding" (is superimposed) on the DC voltage level.

$$\text{VDC level} = 4.97 \text{ V}$$
$$\text{v high level} = 4.97 \text{ V} + 2.49 \text{ v}_p = 7.46 \text{ V}$$
$$\text{v low level} = 4.97 \text{ V} - 2.49 \text{ v}_p = 2.48 \text{ V}$$

✓ Progress Check
You have now completed objective 8.

PRACTICE PROBLEM 1.18

Use the step-by-step superposition procedure to analyze the DC-AC source circuit shown in Figure 1.42. Draw the output signal representation, showing combined voltages.

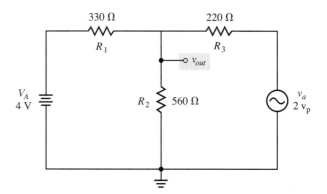

FIGURE 1.42 Circuit for practice problem 1.18

REVIEW SECTION 1.6
1. The superposition theorem is very useful in analyzing circuits that have _____ voltage source(s).
2. State the superposition theorem.
3. The superposition theorem cannot be used for analyzing circuits with both DC and AC sources. True or false?

4. Briefly state the procedure for applying the superposition theorem to circuits having multiple power sources.
5. When algebraically adding currents, you _____ aiding currents and _____ opposing currents.
6. What is the result of superimposed voltages in circuits having both DC and AC power sources?
7. Explain the meaning of the phrase "an AC voltage riding on a DC voltage."
8. A circuit has $v_a = 1.75\,v_p$ and $V_A = 3.25$ VDC. The result produces an output voltage range of:
   (a) $+1.5$ V to -0.25 V
   (b) $+3.5$ V to -3.5 V
   (c) $+5.25$ V to $+1.75$ V
   (d) $+1.75$ V to -1.75 V
9. Draw the output waveform for the circuit described in question 8.
10. Draw the output waveform for a circuit where VDC = 5 V and $v = 5\,v_{p\text{-}p}$.

TROUBLESHOOTING

The skills you learned in this chapter enable you to analyze circuits from their diagrams. But analysis is simply a way to evaluate how a circuit *should* work. As an electronic technician, your working life will involve far more than simple circuit analysis. It will involve *checking the operation of the actual circuit* against the calculated operation.

The operational check of real circuits is done with test instruments, primarily analog volt-ohm-milliammeters (VOMs) or digital multimeters (DMMs) and oscilloscopes. See Figure 1.43. VOMs and DMMs can both be used as an ohmmeter, a voltmeter, a milliammeter, or an ammeter (current meter for measuring higher current values than milliamperes). In general, only two test instruments, the multimeter and the oscilloscope, will provide most of the information you will need to evaluate a real circuit. Your goal should be to gain a thorough understanding of the proper use of these instruments. Test instruments have limitations, and improper use of an instrument can damage circuits that were functioning properly or the test instruments themselves.

BASIC CONCEPTS

An **ohmmeter** works on the principle of applying a DC voltage (usually 3 V) *across a resistive component* and calculating the resistance. It is used simply to check the resistance across a component or of a circuit. *Never use an ohmmeter in a circuit to which power is applied or you may destroy the ohmmeter.*

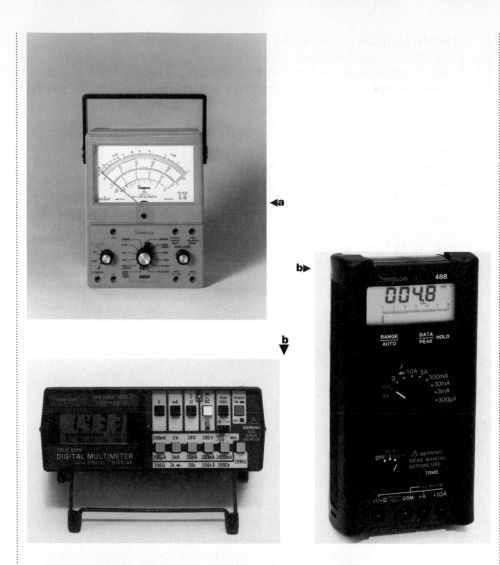

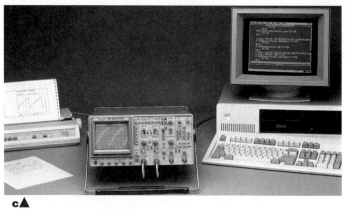

FIGURE 1.43 (a) A volt-ohm-milliammeter. (b) Digital multimeters. (c) A digitizing oscilloscope. *(a and b courtesy of Simpson Electric Company; c courtesy of Tektronix, Inc.)*

Before an analog ohmmeter is used, it must be zeroed by connecting together the two probes (positive and negative) and manipulating the **zero adjust** knob until the meter dial is set at zero ohms. Measurements are made across resistive components (resistors, coils, and so on), but not across capacitors, which may be holding an electrical charge that can damage the meter. In series circuits, resistances can be measured with resistors in the circuit. In parallel circuits, to get a correct meter reading, the resistor under measurement must be disconnected at one end.

A **voltmeter** is used to measure voltage drops *across components* in a circuit to which power is applied. To obtain an accurate reading, the voltmeter must have an input impedance (resistance) of at least 10 times, and preferably 100 times, the value of the resistive element across which the measurement is being taken. The meter across a component sets up a parallel resistance, and if the input impedance is too small, erroneous readings will result. If measuring an open circuit, the voltmeter establishes a path for current flow, resulting in an erroneous reading.

A **milliammeter** is used to measure small currents in a circuit. *The circuit must be opened, and the meter must be inserted in series with the component to be tested.* The meter must *never* be placed *across* a component under test.

An **ammeter,** used for measuring higher currents, is treated the same way as the milliammeter; that is, it must be inserted in series with, never across, the component or circuit under test.

An **oscilloscope** is used to measure and display on a cathode ray tube (CRT) the response of various elements operating in a circuit. The display can show, for example, DC voltage levels, frequency responses, and AC signal waveforms. The oscilloscope can be used across virtually any part, or all, of a powered circuit. Many oscilloscopes have dual displays so that two parts of an operating circuit, such as the input and the output, can be compared.

EXAMPLE 1.20

Figure 1.44 shows a simple two-resistor series circuit. Observation tells us that each resistor should have an IR drop of 15 V. Yet, when we place a voltmeter across R_2, we measure only 10 V. This does not appear to be correct. Is the circuit defective? Has overheating caused R_2 to increase in value? Maybe. The decision you make at this point is critical.

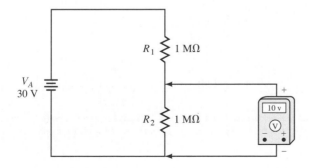

FIGURE 1.44 Circuit with test instrument loading

Solution

Turn off the power, and check the resistance with an ohmmeter. If that is correct, check R_1. If that is also correct, turn on the power, and check the power supply with a voltmeter. That seems to check out, too. So what is wrong?

Check the input impedance of the voltmeter! You find that the input impedance is 1 MΩ, satisfactory for many circuits, but not for this circuit. The 1-MΩ input impedance is across a 1-MΩ component, thus creating a parallel resistance circuit. Ohm's law tells you that two 1-MΩ resistors in parallel becomes 500 kΩ:

$$R_{eq} = \frac{1\ M\Omega}{2} = 500\ k\Omega$$
$$R_T = R_1 + R_{eq} = 1\ M\Omega + 0.5\ M\Omega = 1.5\ M\Omega$$

Now we have a circuit with 30 V applied to 1.5 MΩ, so:

$$I_T = \frac{V_A}{R_T} = \frac{30\ V}{1.5\ M\Omega} = 20\ \mu A$$

Now we can find $I_{R_{eq}}$:

$$I_{R_{eq}} = I_T \times R_{eq} = 20\ \mu A \times 500\ k\Omega = 10\ V$$

Therefore, we can conclude that the meter in use is not the right meter for the job at hand. When possible, always use a meter with an *input impedance of at least 10 times* the value of the resistive component under test.

EXAMPLE 1.21

Figure 1.45 shows an oscilloscope measuring the output across R_2. Ideally, the signal would be a 50-v_p symmetrical sine wave, but the sine wave is lower than expected. A check of the input impedance of the oscilloscope shows 100 MΩ, so that is acceptable. However, the upper frequency limit of the oscilloscope in use is only 20 MHz. Since the circuit input signal frequency is 50 MHz, obtaining a proper output signal requires an oscilloscope with at least a 50-MHz frequency response. Many oscilloscopes have frequency responses of 60 MHz and higher.

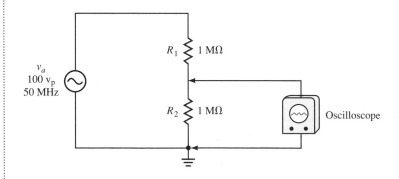

FIGURE 1.45 Circuit loaded by a test instrument with frequency response deficiency

Figure 1.46 shows a series-parallel circuit. On the circuit diagram: show how you would connect a milliammeter to measure I_3; show how you would connect a voltmeter to measure V_4; and explain what you would do to measure the resistance of R_5.

Replace the DC source with an AC source of 100 v_p. Show how you would connect an oscilloscope to measure the output signal at each of the test points.

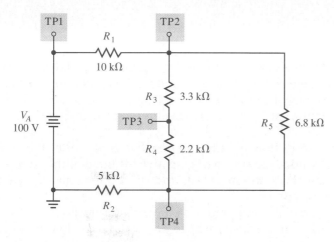

FIGURE 1.46 Circuit for practice problem 1.19

Progress Check
You have now completed objective 9.

TROUBLESHOOTING QUESTIONS

1. The ultimate goal for any electronic technician must be _____ _____.

2. Troubleshooting involves the _____ _____ against the calculated operation.

3. The operational check of real circuits is done with _____ _____.

4. Resistance is measured with a(an) _____ connected _____ a component with power to the circuit turned _____.

5. Voltage is measured with a(an) _____ connected _____ a component with power to the circuit turned _____.

6. Current in a circuit is measured with a(an) _____ connected _____ a component with power to the circuit turned _____.

7. Test instruments have no limitations. True or false?

8. A(an) _____ is used to measure and display the response of various elements operating in an electronic circuit.

9. What will be the result of measuring a circuit load of 500 kΩ with a voltmeter having an input impedance of 1 MΩ?
10. What will be the result of measuring the frequency response of a circuit where the source frequency is 100 MHz and the upper frequency limit of the oscilloscope is 60 MHz?

SUMMARY

Ohm's law states the relationship between current, voltage, and resistance. It states that voltage is directly proportional to current and resistance and that current is inversely proportional to resistance. Circuit analysis, by whatever concept, theorem, or law, usually can be reduced to Ohm's law for ease of calculation.

Electron current flow is from the negative terminal of the power source, through the external circuit, to the positive terminal of the source. In a series circuit, current has only one path, and the current has the same value in every part of the circuit.

In a voltage divider, the ratio of voltages is proportional to the ratio of resistances. This ratio provides a way to predict voltages in a series circuit without having to calculate currents. A loaded voltage divider can lower the output voltage of the circuit. Firm or stiff dividers can help to prevent this lower output voltage.

In a current divider, total current is divided among the branches of a parallel circuit, the most current flowing through the least resistance. Kirchhoff's current law says that the sum of currents flowing into a node (connecting point) is equal to the current flowing out of that node. The current divider and Kirchhoff's current law provide a way to predict how much current will flow through each branch in a parallel circuit, particularly when the circuit is driven by a constant current source (an energy source that provides a constant value of current to a load).

Thevenin's theorem can be used to reduce any circuit, of any complexity, to a single-resistance (R_{TH}) one-source (V_{TH}) series circuit. The V_{TH}-R_{TH} circuit is the Thevenin equivalent circuit. It is not necessary that the load be a resistor.

Kirchhoff's voltage law states that the sum of voltages around a closed loop must equal zero and that the sum of voltage drops must be equal to, but of opposite polarity to, the source voltage. All loops must be considered, even those without a power source. This law can be used to check answers derived from closed-loop calculations and for help in developing simple equations.

The superposition theorem can be used to analyze circuits that have more than one power source. In a circuit having more than one source, determine the effect of one source at a time, then superimpose (algebraically add) the results of all sources.

Troubleshooting and repairing electronic equipment must be the ultimate goal for any electronic technician. You must know what test instruments to use, what their limitations are, and how to overcome those limitations.

1. Write the three equations for Ohm's law.
2. State the proportional relationship between voltage, resistance, and current.
3. Define electron current flow as it relates to an electronic circuit.
4. State the rule for current in a series circuit.
5. Find the voltages across all resistors in a series circuit where $V_A = 100$ V, $R_1 = 3.3$ kΩ, $R_2 = 4.7$ kΩ, $R_3 = 5.6$ kΩ, and $R_4 = 1$ kΩ.
6. Find I_T in the circuit described in problem 5.
7. Find V_A in a series circuit where $I_T = 2$ A; $R_1 = R_2 = 2.2$ kΩ; $R_3 = 8.2$ kΩ; and $R_4 = 33$ kΩ.
8. Find R_T in a series circuit where $V_A = 24$ V and $I_T = 25$ mA.
9. In Figure 1.6, assume $V_A = 240$ V. Find I_T.
10. Using I_T found in problem 9, find the ratio relationships between R_3 and R_T and between V_3 and V_A in Figure 1.6.
11. In Figure 1.8, assume $V_A = 48$ V. Find (a) R_T, (b) I_T, and (c) voltage drops across each resistor.
12. In Figure 1.47, assume R_L is open. Find the voltage (unloaded) at node N_1.
13. Find the loaded voltage at node N_1 in Figure 1.47 if $R_L = 22$ kΩ.
14. What value must R_L in Figure 1.47 be to assure that the loaded voltage changes by no more than 1% from the unloaded voltage?
15. Define (a) *firm* voltage divider and (b) *stiff* voltage divider.
16. In Figure 1.48 find I_1.
17. In Figure 1.48 find (a) V_2 and (b) V_3.
18. With R_1 open in Figure 1.48, find I_3.
19. Open the circuit, and place a milliammeter between TP1 and TP2 in Figure 1.48. What current value will you read?

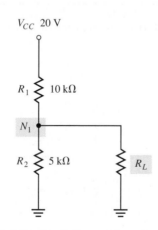

FIGURE 1.47 Circuit for problems 12, 13, and 14

20. In Figure 1.49 assume $R_2 = R_L$. Thevenize and draw the equivalent circuit as seen by R_L.
21. In Figure 1.49 assume $R_1 = R_L$. Thevenize and draw the equivalent circuit as seen by R_L.
22. In Figure 1.49 assume $R_3 = R_L$. Thevenize and draw the equivalent circuit as seen by R_L.
23. Open the circuit and place a 1-kΩ resistor between TP1 and TP2 in Figure 1.49. Assume $R_2 = R_L$. Thevenize and draw the equivalent circuit as seen by R_L.

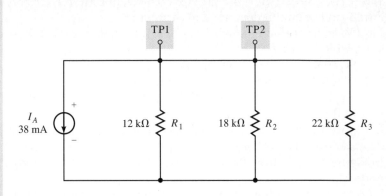

FIGURE1.48 Circuit for problems 16, 17, 18, and 19

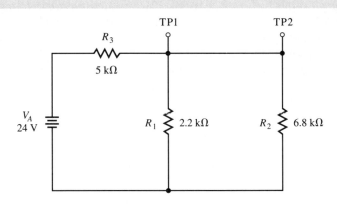

FIGURE 1.49 Circuit for problems 20, 21, 22, and 23

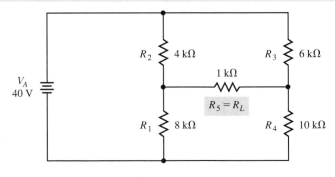

FIGURE 1.50 Circuit for problems 24 and 25

24. Thevenize the circuit in Figure 1.50 where $R_5 = R_L$. Determine I_L and the direction of current flow through R_L.
25. How many closed loops are there in Figure 1.50?
26. Write a loop equation for loop L_2 in Figure 1.51.
27. Write a loop equation for loop L_4 in Figure 1.51.
28. In Figure 1.52 find I_1, I_2, and V_3.
29. Assume the polarity of V_{A2} is reversed in Figure 1.52. Using the superposition theorem, find I_2.
30. Find VDC output in Figure 1.53.
31. Find vAC peak output in Figure 1.53.
32. In Figure 1.53 (a) use the superposition theorem to determine the output signal, and (b) draw the output waveform, labeling the values for VDC and vAC (peaks) with reference to ground.
33. In Figure 1.54, the digital voltmeter (DVM) connected across R_1 has an input impedance of 1 MΩ. (a) Determine what voltage reading the DVM will indicate and (b) what the total current draw of the circuit will be.

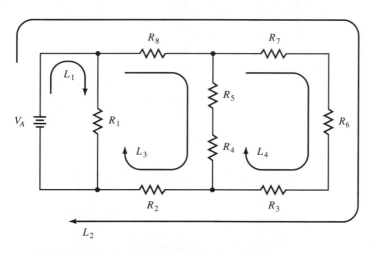

FIGURE 1.51 Circuit for problems 26 and 27

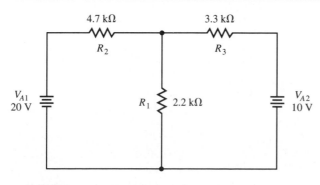

FIGURE 1.52 Circuit for problems 28 and 29

34. What input impedance must the DVM have if you want to have a reading within 1% of the actual voltage value?
35. With the DVM in Figure 1.54 disconnected, find I_T for the circuit.

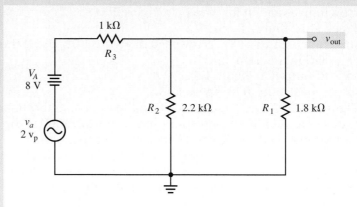

FIGURE 1.53 Circuit for problems 30, 31, and 32

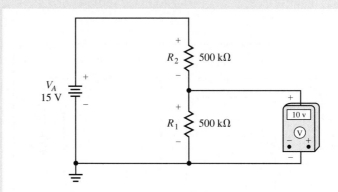

FIGURE 1.54 Circuit for problems 33, 34, and 35

2

SEMICONDUCTOR THEORY

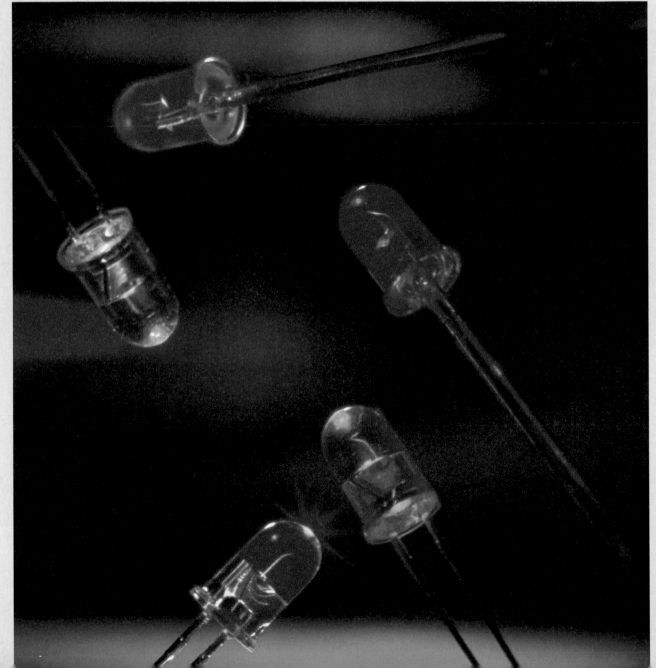

✔ **As you read this chapter, concentrate on learning how to:**

1. Describe basic atomic theory
2. Define the flow of current
3. Identify and describe N-type and P-type semiconductor materials
4. Describe the process of making a PN junction diode
5. Identify the cathode and anode of a diode on a schematic diagram and on an actual device
6. Define forward bias and reverse bias
7. Define peak inverse voltage
8. Describe the operating characteristics of a diode
9. Describe how a zener diode functions

INTRODUCTION

The devices that we will study throughout this text are often called **solid state devices.** In a solid state device, current flows through a solid substance such as silicon or germanium. Properties of the particular substance determine the amount of current flow, the type of current, the characteristics of the device, and the intended applications of the device.

Circuits covered in this and the following chapter use solid state devices called **diodes.** A diode is a device that allows current to flow in only one direction. In a very real sense, the diode acts as a simple switch. That is, the diode works as an open switch for one direction and as a closed switch for the other direction of current flow. In comparison, a resistor will allow current to flow in both directions.

Even with the limitations of construction, diodes have many diverse applications. In some circuits, diodes convert AC voltage to pulsating DC voltage. Other circuits employ diodes as voltage regulators. In another application, diode circuits increase the maximum amount of output voltage without the use of a transformer. Also, diodes may function as waveshaping devices called clippers or as positive or negative voltage clampers.

Before studying diodes, let's spend some time looking at some of the fundamental properties of matter and how they relate to electricity and electronics. A firm knowledge of these fundamentals will strengthen any understanding of how solid state devices work.

Matter is anything that takes up space and has mass. It may take the state of a gas, a liquid, or a solid. We breathe gaseous matter called air, write on solid matter called paper, and drink liquid matter called water. If we could see matter broken down into its smallest particles, we would first see **molecules.** A molecule is the smallest particle of a substance that retains all the properties of that substance. Millions of molecules of water make up one raindrop. Each molecule is made up of atoms. For instance, three atoms—one oxygen and two hydrogen—are in each molecule of the raindrop. An **atom** is the smallest particle of an **element** (a fundamental substance that consists of atoms of only one kind). Oxygen and hydrogen are elements; water is not an element.

ELECTRONS CARRY NEGATIVE CHARGES

Each atom has a nucleus of protons and neutrons that is surrounded by electrons. A popular analogy compares the structure of the atom to our sun and the orbiting planets. As we can see in Figure 2.1, the electrons orbit the nucleus. Figure 2.1 shows an oxygen atom. Eight electrons move around a nucleus of eight protons and eight neutrons. A negative sign marks each electron, and a positive sign marks each proton. Each **electron** has a negative electrical charge, while each **proton** has an equal positive charge. **Neutrons** do not have an electrical charge and remain neutral.

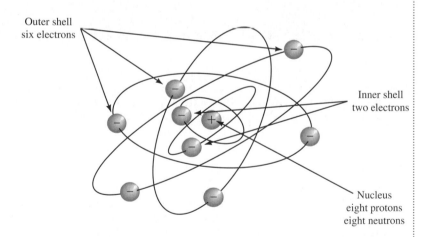

Outer shell
six electrons

Inner shell
two electrons

Nucleus
eight protons
eight neutrons

FIGURE 2.1　Structure of an oxygen atom

In an atom, the positively charged protons combine in the nucleus, while the negatively charged electrons remain in orbit. A neutral atom contains the same numbers of electrons and protons. For matter to remain electrically neutral, the sum of the positive charges must equal the sum of the negative charges. According to the atomic theory of electron shells, electrons exist in shells around the nucleus. The first shell will hold no more than two electrons, while the second shell may hold no more than eight electrons. When the shells fill to the limit, a new shell begins to form. Figure 2.2 shows the maximum number of electrons that can exist in each of the first four shells. Although an atom may have as many as seven shells, we will be discussing only materials that have two to four shells.

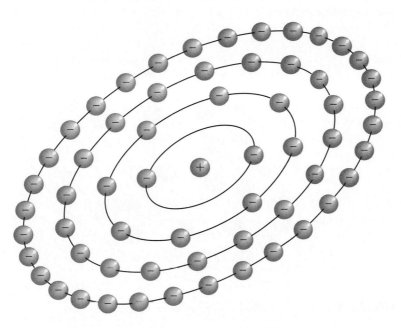

FIGURE 2.2 Maximum number of electrons that can exist in a shell

All forms of energy, including heat and electrical energy, encourage the movement of electrons. Absorbing sufficient energy, such as heat or light, can cause electrons to break away from their outer orbit. These electrons become free electrons and can wander until attracted by a positively charged atom. When a neutral atom loses or gains one or more electrons, it becomes a charged atom. We can define an atom that loses electrons as a **positive ion.** Conversely, a **negative ion** is an atom that gains electrons.

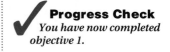

Progress Check
You have now completed objective 1.

HOLES CARRY POSITIVE CHARGES

As an electron moves from one atom to another, it leaves a vacancy in the outer orbital shell of the atom that it left. This vacancy is called a **hole.** Every negatively charged electron produces a hole that behaves as an equal positive charge when it leaves its orbital shell. With the positive charge of the hole acting as an attractive force, an electron from another atom may "jump" to fill the hole.

All this activity is based on a fundamental **law of electrical conduction:** Like charges repel, unlike charges attract.

As an example, a copper atom loses the one electron from its outermost shell, or **valence shell,** and becomes a positive ion. Figure 2.3 shows how a normal copper atom would appear. If the atom lost one electron (giving it 28 instead of 29) it would have a positive charge. Any negatively charged free electron traveling through the material would become attracted to the positive ion. Once the empty electron space of the positive ion has been filled with the free electron, it becomes a neutral atom. Figure 2.4 represents a copper wire. In this figure, electrons continually move from one atom to another. Copper atoms share the valence electrons but do not tightly hold the free electrons. Since the electrons can move freely through the material, copper is a good conductor of electricity. Copper has many free electrons and few valence electrons.

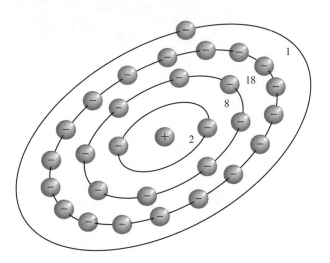

FIGURE 2.3   Representation of a copper atom, with 29 electrons

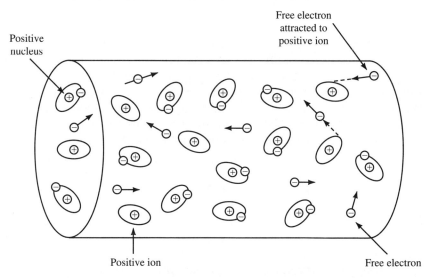

FIGURE 2.4   Atoms in a copper wire. Because of thermal energy, atoms in any material are active. Copper has only one electron in its valence shell, and that electron may move into the conduction band easily. At any given point in time, a few atoms will be short one electron. Each of those atoms will have a net positive charge. That charged atom is called a positive ion. There will also be free electrons in the conduction band. The free electrons travel in random directions.

Electricity is the flow of free electrons in a given direction. In Figure 2.4, the free electrons can move in all directions. The net effect is no flow of free electrons in a given direction, hence, no electricity. However, attaching a battery to each end of the copper wire provides the energy needed for electron movement. In accordance with the rule of repulsion and attraction, the free electrons move away from the negative terminal and toward the positive terminal of the battery.

Figure 2.5 shows the effect of attaching a carbon-zinc battery to the copper wire, or conductor. A carbon-zinc, or flashlight, battery consists of a center carbon post, a zinc casing, and chemical paste. As these parts react to one another, the zinc has an excess of electrons, and the carbon has a deficiency of electrons. Connecting the copper wire to the very negative zinc and the very positive carbon causes a movement of electrons. The excess electrons are repulsed by the negative zinc and attracted to the positive carbon. In addition, the copper wire adds its own free electrons and positive and negative ions to the equation. Many free electrons drifting in one direction through a conductor constitute electrical current.

Progress Check
You have now completed objective 2.

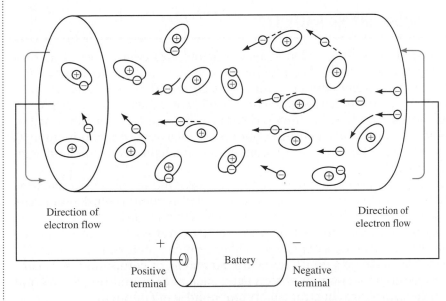

Direction of electron flow

Direction of electron flow

+ Positive terminal

Battery

− Negative terminal

FIGURE 2.5 Electron flow in a copper wire. An attached battery provides a positive attraction on one end of a copper wire and a negative repulsion at the other end. The free electrons in the wire are both repelled by the negative battery terminal and attracted by the positive battery terminal. The result is a negative-to-positive flow of electrons through the copper wire.

Any amount of current flow depends on the conductor material and size. These factors have a direct relationship to the resistance that the material has against current flow. Some materials, such as copper, silver, and gold, have more free electrons and are good conductors of electrical current. Materials with fewer free electrons, such as wood or glass, work well as insulators and resist the flow of electrical current. An amount of current also relies on the amount of force provided by the energy source.

2.2   SEMICONDUCTOR MATERIALS

One group of materials, called semiconductors, has conductive properties that range between those of conductors and insulators. A **semiconductor** is a substance that neither conducts nor resists current flow very well. Literally, *semiconductor* means "half-conductor." While a conductor presents low resistance to electron flow in either direction, a semiconductor exhibits higher resistance to the flow of electrons. Conductivity of a pure semiconductor material can be varied by changes of temperature and light and by the introduction of impurities.

Several types of semiconductor materials exist. One set of semiconductors, which includes silicon and germanium, is labeled **elemental semiconductors** because materials in this set are elements (contain a single kind of atom). The majority of electronic devices such as rectifiers, transistors, and integrated circuits are constructed from silicon. Other semiconductors, including gallium arsenide, make up a group called **intermetallic,** or **compound, semiconductors.** Devices that emit or absorb light, such as light-emitting diodes and infrared detectors, are constructed of compound semiconductors.

Adding impurities to the pure semiconductor material dramatically affects the electronic and optical properties of the semiconductor. The addition of an impurity to a semiconductor can cause conductivity to vary over wide ranges. Adding an impurity concentration of one part per million to a sample of silicon can change the semiconductor from a poor conductor to a good current conductor. In addition, the added impurity can change the process of conduction so that the material conducts either positive or negative charge carriers.

SEMICONDUCTOR CRYSTALS

Semiconductor materials have atoms that form into single crystals. A **crystal** is a chemical combination that can grow to indefinite proportions. Figure 2.6a illustrates one of 14 possible crystal arrangements. Notice that the arrangement could repeat itself indefinitely, limited only by the amount of material available. At any point, the crystal appears to have the same shape.

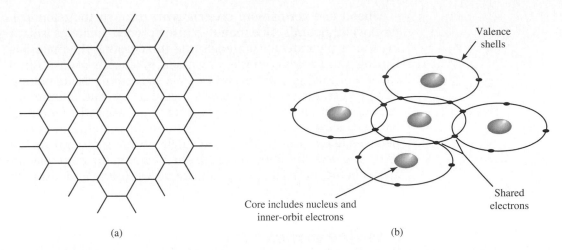

FIGURE 2.6   (a) One of the 14 possible crystalline lattice structures. (b) Covalent bonding in a silicon crystal.

The organized arrangement of atoms is referred to as a **lattice.** Semiconductor materials such as silicon, germanium, and carbon form macromolecular crystals. For these materials, the lattice has a diamondlike appearance. Each individual atom links with other atoms by sharing electrons to make up the entire lattice. Forces attracting the atoms and holding them apart determine the distances between the atoms.

COVALENT BONDING

Atoms in the lattice link with each other by sharing electrons in the valence shells. Each pair of shared valence electrons creates a **covalent bond.** Figure 2.6b shows covalent bonding in a silicon crystal. The central atom in the crystal has eight electrons in the outer shell. Four of those electrons are owned by the atom and the other four are shared with the adjoining atoms. If other atoms were available, they would be attached to the exterior of the crystal. Then they would share their electrons with their adjoining atoms. Every atom in the crystal has eight electrons in the outer shell. Those eight electrons are four owned electrons and four shared electrons. Crystals formed by covalent bonding can also be referred to as **valence crystals.**

If an atom has eight electrons in its valence band, it is called **chemically stable.** Although explaining chemical stability is beyond the scope of this book, the technician should understand that a chemically stable atom does not make any further combinations with other atoms. In Figure 2.6b the central atom will not share its electrons with more than four other atoms because it is chemically stable. However, the exterior atoms are not chemically stable and will accept other atoms. If another atom came near the lattice structure, it would be joined to the lattice atoms. Also notice that the lattice is electrically balanced because it has the same number of electrons and protons.

GROWTH OF SEMICONDUCTOR CRYSTALS

Solid state technology depends on new design concepts and the quality of semiconductor materials. Those materials must be available in large single crystals that have extremely close tolerances of purity. Several methods for producing or growing semiconductor crystals exist.

Silicon and germanium crystals grow through the chemical decomposition of compounds. One manufacturing process includes isolating and melting the semiconductor materials and then casting the materials into ingots. During the process, checks are made to ensure purity. Another process, called epitaxial growth, grows crystals well below their melting point. With this process, the semiconductor material mixes with another element. Melting occurs at the melting temperature of the second element. If the melting happens slowly enough, the semiconductor material does not liquefy, but behaves like a plastic. Then, the single crystal grows on the parent crystal. With a lowered melting point, the semiconductor materials are purer. The atoms in the material are then able to move around and form crystals.

REVIEW SECTION 2.2

1. A _____ is a substance that neither conducts nor resists current flow very well.
2. List two semiconductor materials.
3. List three factors that affect the conductivity of a semiconductor.
4. Explain how a semiconductor changes from a poor conductor to a good conductor.
5. What is the maximum number of electrons in the valence shell of atoms having more than one shell?
6. Define covalent bonding.
7. Pure silicon is a(n) _____ semiconductor.
8. Because of covalent bonding a semiconductor appears to have _____ valence electrons.

2.3 MAKING THE SEMICONDUCTOR A BETTER CONDUCTOR

INTRINSIC SEMICONDUCTOR CRYSTALS

An **intrinsic semiconductor** crystal does not contain atoms from other materials and has no defects in the lattice arrangement. At absolute zero (−273°C), no charge carriers exist in the intrinsic crystal. Higher temperatures cause minimal amounts of electron-and-hole pair generation.

These electron-hole pairs are the only charge carriers in intrinsic material. However, the slight movement of the electron-hole pairs does not create current flow. Since the electrons and holes are created in pairs, an equal number of electrons and holes exists. When the electrons and holes recombine, the pairs disappear. All this activity within the intrinsic crystal is directly proportional to temperature. A higher temperature results in a higher production of carriers but also in a higher rate of recombination.

EXTRINSIC SEMICONDUCTOR CRYSTALS

As you may suspect, the equal number of holes and electrons produced by thermal excitation prevents any appreciable amount of current from flowing in an intrinsic semiconductor. To increase the conductive abilities of semiconductors, manufacturers add impurities to the intrinsic material. We call this process **doping.** If an impurity has been introduced, then the crystal is said to be an **extrinsic semiconductor.** Depending on the type of introduced impurity, the extrinsic semiconductor will have either more holes or more free electrons.

N-TYPE SEMICONDUCTOR MATERIALS

Adding an impurity with more free electrons than the intrinsic material changes the electron-hole balance within the intrinsic material. Introducing an impurity places the energy level of the impurity next to the conductive property of the pure material. At absolute zero, electrons fill the energy level of the impurity. Very little energy is required to move the excess electrons to the semiconductor material. At $-223°C$ to $-173°C$, almost all the electrons from the impurity have moved to the semiconductor material. A material doped with impurities can have a concentration of electrons at a temperature too low for intrinsic electron-hole pairing.

Materials such as arsenic, antimony, and phosphorus can donate free electrons. We refer to those materials as **donors.** If more electrons exist in the outer orbits of the impurity atoms than exist in the outer orbits of the semiconductor atoms, the excess electrons become free electrons. A **free electron** is an electron not needed for chemical stability. Only a little energy will cause that electron to leave its orbit and be free of the atom.

Figure 2.7 shows how doping with a donor impurity affects a semiconductor. In this figure, we see an atom with five electrons in its outer shell. Using the Greek prefix *penta*, meaning "five," we refer to those atoms as **pentavalent atoms.** As the impure pentavalent atom shares four of its electrons with nearby silicon atoms, covalent bonding occurs. Because the remaining electron does not fit the old valence shell, it becomes freer than the bonded electrons. A new shell is formed with one valence electron and many free electrons. Applying a voltage to the lattice structure easily dislodges the excess electron and produces current flow.

The *majority* charge carrier is the electron. More free electrons—with negative charges—exist in the material. Because of this negative charge, semiconductor material doped with pentavalent atoms is defined as an **N-type semiconductor.** Obviously, the amount of applied voltage influences the amount of produced current flow. Nevertheless, the number of loose charge carriers within the crystal also determines the amount of current flow. Increasing the doping level aids the production of current flow.

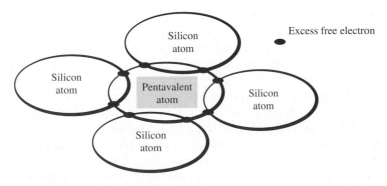

FIGURE 2.7 Pentavalent doping in a silicon crystal

P-TYPE SEMICONDUCTOR MATERIALS

Some impurities have fewer than four electrons in their valence shell. Introducing this type of impurity places an empty energy level next to the semiconductor material valence shell. Thermal energy causes the movement of electrons from the valence shell of the semiconductor atoms to the impurity atoms. The impurity atoms take electrons away from the semiconductor atoms. Since impurities such as aluminum, indium, boron, and gallium accept electrons, we call them **acceptor impurities.** Losing electrons, the semiconductor materials have more holes than electrons. Since the holes behave as positive charges, we call the semiconductor a **P-type semiconductor.**

Figure 2.8 shows an impurity atom with only three atoms in the valence shell. Just as we used the label *pentavalent* in the previous example, we use the name *trivalent* for this type of impurity atom. The prefix *tri-* means "three." Placed within the silicon atom lattice, a trivalent atom cannot provide enough free electrons for completion of covalent bonding. Only seven valence electrons exist in the valence shell. Four electrons are owned by the silicon atom and three are shared. But because the acceptor atom has only three electrons in its valence band, it cannot share an electron with the silicon atom. Each incomplete bond lacks an electron. Removing a valence electron from a covalent bond leaves a vacancy, or a hole.

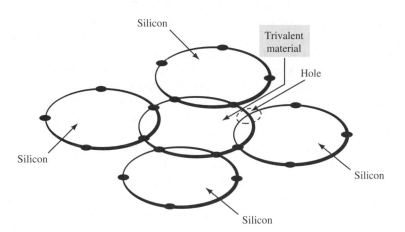

FIGURE 2.8 Trivalent doping

With the added impurity of Figure 2.8, the majority charge carriers are holes. More free holes—with positive charges—exist in the material. Applying a voltage across the doped crystal pushes electrons away from the silicon atom and toward the holes of the acceptor material. As the valence electrons jump from hole to hole, another hole exists from the broken covalent bond. New holes appear. Hole movement continues until the hole reaches the negative side of the applied voltage source. At the negative voltage terminal, the hole recombines with an electron and vanishes.

At the positive voltage terminal, another electron is pulled from the semiconductor crystal. This creates another new hole. The applied voltage causes the hole—with its positive charge—to move toward the negative voltage terminal. Figure 2.9 shows the movement of the holes and electrons.

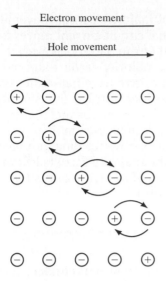

Electron movement

Hole movement

FIGURE 2.9   Electron flow and hole flow are equal but opposite.

RECOMBINATION

Excess current carriers produced by the impurity atom provide greater current conduction. Those excess current carriers are called **majority carriers.** In N-type semiconductors, the majority carriers are free electrons, and minority carriers are the holes. In P-type semiconductors, the majority carriers are holes, and minority carriers are the electrons.

Figure 2.10a shows a voltage applied to a block of N-type semiconductor material. As shown by the arrows, current I leaves the negative terminal of the battery. Because of the movement of the majority carriers, the current passes through the semiconductor material and returns to the positive terminal of the voltage source. A large arrow shows the movement of the majority carrier electrons within the block. Simultaneously, the holes, or, in this case, **minority carriers,** move in the opposite direction. When the holes reach the negative terminal, they meet electrons entering the same terminal. Upon combining with each other **(recombination),** both holes and electrons stop carrying current.

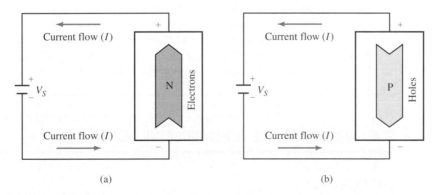

(a)   (b)

FIGURE 2.10   (a) Voltage applied to a block of N-type semiconductor material. (b) Direction of majority carrier movement in P-type material.

Figure 2.10b shows the direction of majority carrier movement in P-type material. As in Figure 2.10a, the external circuit current moves from the negative terminal to the positive terminal of the battery. However, since we are viewing P-type material, holes are the majority carriers and move in the direction of the large arrow. While the positive terminal of the semiconductor repulses positive holes, the negative terminal attracts the majority carriers.

Conduction by holes outside the semiconductor material is not possible. When the holes reach the negative terminal, recombination occurs. Electrons leaving the semiconductor at the positive terminal produce more holes at that location. These holes replace the majority carriers lost to recombination. In Figure 2.10b, hole conduction, along with a small amount of free electron flow, causes the flow of semiconductor current.

✓ **Progress Check**
You have now completed objective 3.

REVIEW SECTION 2.3

1. If more impurities are added to silicon, will its conductivity increase, decrease, or remain constant?
2. The process of adding impurities to silicon is called _____.
3. After an intrinsic material is doped it is said to be _____.
4. N-type semiconductors have more free _____ than holes.
5. Pentavalent materials have _____ valence electrons.
6. Semiconductor materials doped with pentavalent atoms are called _____ semiconductors.
7. As the doping level increases, the ability to produce current flow will _____ (increase, decrease, remain constant) in an N-type semiconductor.
8. List three types of acceptor impurities.
9. Materials with three valence electrons are referred to as _____ (trivalent, pentavalent).
10. Removing a valence electron from a covalent bond leaves a _____.
11. The carriers in a P-type material are _____ (holes, electrons).
12. Hole flow is from _____ to _____ (negative to positive, positive to negative).
13. The majority carriers in an N-type semiconductor are _____ .
14. The majority carriers in a P-type semiconductor are _____.
15. Minority carriers are _____ in N-type and _____ in P-type materials.

2.4   COMBINING THE SEMICONDUCTOR MATERIALS

DIFFUSION

Doping creates excess charge carriers that vary their positions and concentrations within the semiconductor material. The charge carriers move from regions of high carrier concentration to regions of low carrier concentration. This movement is called **diffusion.**

When someone opens a bottle of perfume in one corner of a closed room, the scent spreads throughout the room. Without the effect of air motion, the scent spreads by diffusion. Individual molecules have random motion. Each molecule moves in an arbitrary motion until it collides with another molecule. Then the molecule moves in a new direction. With the continued movement of molecules, the net area covered by the molecules also increases. This process continues until the molecules are distributed evenly around the room. For a given time period, any particular area gains as many molecules as it loses.

In a semiconductor, carriers diffuse because of random thermal motion and the action of impurities. A pulse of electrons will spread out in a given amount of time. Initially, the excess electrons concentrate. As time passes, the excess electrons continue to diffuse to regions of low concentration until the electrons are evenly distributed across the region (total diffusion).

If we add an electrical field to the charge carriers, drift and diffusion influence current densities. An electrical potential prompts charge carriers to move in a certain direction. Considering their charges, holes drift with the potential; electrons drift away from the potential. If the potential is in the direction of decreasing hole concentration, the current adds to hole flow and subtracts from electron flow. The total current relies on the flow and concentrations of charge carriers, spacing between carriers, the size of the electrical potential, and the direction of the electrical potential.

THE PN JUNCTION

While growing a single crystal, donor and acceptor impurities are added, forming a **PN junction.** Figure 2.11 illustrates how the PN junction is formed. In Figure 2.11a the P-type material is also electrically neutral but chemically unstable. Remember that P-type material has an excess of holes. The N-type material is also electrically neutral but chemically unstable. The N-type material has an excess of electrons. Figure 2.11b illustrates that at the junction of the N- and P-type materials, electrons and holes recombine. At that junction the material becomes chemically stable, as there are now eight electrons in the valence orbits. However, this chemical stability has caused an electrical imbalance. Now, the acceptor (P-type) material has gained electrons while the number of protons has remained the same, resulting in a net negative charge. The donor (N-type) material has lost electrons while the number of protons has remained the same, resulting in a net positive charge. Since unlike charges attract, a surface charge is created at the junction of the P- and N-type materials. This surface charge creates a barrier that stops any more recombining from occurring in the crystal. The barrier is called the *depletion region* because there are no current carriers. The surface charge is called the *barrier potential.* The final result is that P-type material is on one side of the crystal, N-type material is on the other side, and the depletion region separates the two materials. This basic joining of the two opposite semiconductor materials and the formation of a PN junction give us a simple **PN junction diode.**

Construction of the junction diode may vary; but, in every instance, the two semiconductor materials bind closely together to form a single piece. Majority carriers near the PN junction become attracted to each other, com-

Progress Check
You have now completed objective 4.

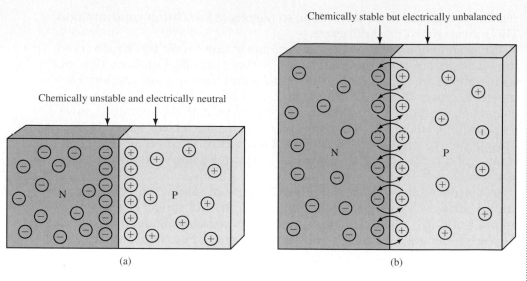

Chemically unstable and electrically neutral

Chemically stable but electrically unbalanced

(a)

(b)

FIGURE 2.11 The joining of two semiconductor materials

bine, and cancel out. Because of recombination, the N-type material loses electrons and contains positive ions, while the P-type material loses holes and contains negative ions. Remember, this action occurs only close to the junction.

THE DEPLETION REGION

The area around the junction is referred to as the **depletion region.** This region, shown in Figure 2.12, has no carriers. Before the joining of the N- and P-type materials, the N-type material has a large concentration of electrons, while the P-type material has a large concentration of holes. With diffusion, the holes move into the N-material, while the electrons move into the P-material. The conduction of **diffusion current** occurs.

Electrons diffusing from the N side leave positive ions; holes diffusing from the P side leave negative ions behind. A positive charge builds along the N side of the junction, and a negative charge builds along the P side. An electrical field exists from the positive charge toward the negative charge. Effec-

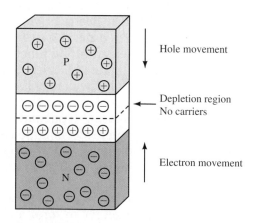

Hole movement

Depletion region
No carriers

Electron movement

FIGURE 2.12 The depletion region

tively, the field opposes the movement of charge carriers. The diffusion current cannot increase indefinitely, nor can it flow across the junction because an opposing electrical field, or barrier potential, builds at the junction.

THE BARRIER POTENTIAL

Since the N-type material retains a positive charge and the P-type material has a negative charge near the junction, a difference of potential is created. We call this difference of potential a **barrier potential.** For silicon semiconductor materials, the difference of potential equals approximately 0.7 V. Germanium semiconductor materials have a difference of potential of approximately 0.3 V.

Look at Figure 2.13. Not surprisingly, the positive charge of the N-type material repels the holes of the P-type material away from the junction. Also, the negative charge of the P-type material repels the electrons of the N-type material. In effect, the region near the junction forms a barrier that prevents the additional recombination of majority carriers.

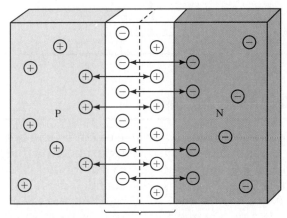

Barrier potential at PN junction—0.7 V for Si, 0.3 V for Ge

FIGURE 2.13 Formation of the barrier potential prevents the movement of charge carriers across the PN junction.

Figure 2.14a shows the schematic diagram for a diode. The straight, vertical line in the figure represents the cathode, or the N-type semiconductor material. A triangle represents the anode, or the P-type material. Electron flow in a diode is always from cathode to anode, that is, against the arrow. Figure 2.14b shows a typical diode. At one end of the diode body, a ring marks the cathode lead.

Progress Check
You have now completed objective 5.

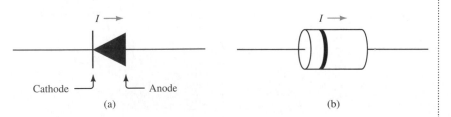

(a)

(b)

FIGURE 2.14 (a) Schematic diagram of a diode. (b) Pictorial representation of a typical diode.

2.5 CONDUCTION WITHIN THE DIODE

To produce current within the diode, the free electrons and holes must have enough energy to overcome the barrier potential and cross the junction. An applied voltage source provides the energy needed for the crossing. In Figure 2.15, a battery voltage is applied across a PN junction diode. The positive terminal of the battery connects to the P-type material, and the negative terminal connects to the N-type material. Connecting the battery in this way causes the holes in the P-type material and the electrons in the N-type material to repel toward the junction.

Near the junction, the holes and electrons combine. At the other end of the N-type material, electrons arrive and replace electrons lost to recombination. At the other end of the P-type material, electrons flow toward the

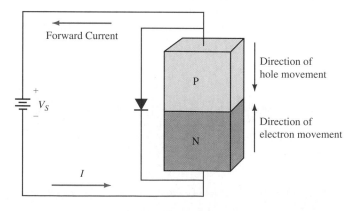

FIGURE 2.15 Forward bias voltage attached to a P-N junction diode

positive terminal of the battery. As the electrons move out of the P-type material, holes form and replace those canceled during recombination. All this movement leads to conduction by majority carriers within the diode.

FORWARD BIASING

Figure 2.15 illustrates how a diode is forward biased. **Biasing** means to apply a voltage in order to give a particular direction to the circuit's current. When a battery is attached to a diode as in Figure 2.15, current will flow in the direction indicated. Notice that the negative battery terminal is attached to the N-type material and the positive battery terminal is attached to the P-type material. If the voltage is great enough, the barrier region will be reduced, allowing the current to flow, Figure 2.16a and b.

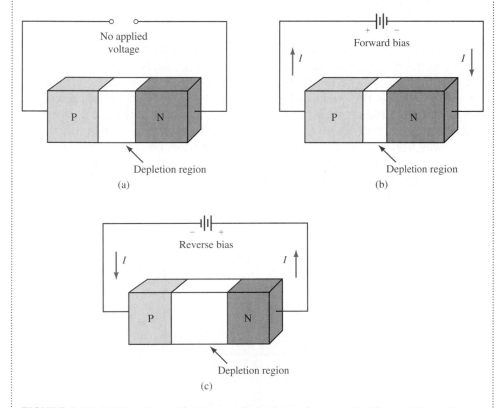

FIGURE 2.16 P-N junction with (a) no applied voltage, (b) an applied forward bias voltage, and (c) an applied reverse bias voltage. Note that the P and N regions remain the same, and the depletion region narrows and widens.

This is how **forward biasing** allows the current to flow. When a negative potential is applied to N-type material, the free electrons in that material are repelled and pushed toward the barrier region. Some of those electrons join the positive ions in the barrier region, causing those ions to become neutral atoms. A positive potential is also applied to the P-type material. The positive potential will repel and push the holes in the P-type material toward the barrier region. Some of these holes will join the negative ions, causing them to become neutral atoms. With electron flow in the N region and hole flow in the P region, the current in the diode is called **forward current.** Once most of these negative and positive ions have been neutralized, the barrier region will have effectively disappeared. The battery continues to push electrons and holes toward the PN junction, where they recombine. Each time an electron and a hole combine a new hole is formed in the P-type material and an electron is introduced in the N-type material. The battery provides the electrons to the N-type material, and when the electron-hole pair is created in the P-type material, the battery draws the electron toward its positive terminal. The result is current flowing from the negative terminal of the battery, through the diode, and back to the positive terminal of the battery.

REVERSE BIASING

Figure 2.17 illustrates **reverse biasing.** Notice that the battery is connected opposite to the way it is in Figure 2.15. That is, the positive battery terminal is connected to the N-type material, and the negative battery terminal is connected to the P-type material. The result of reverse bias is to change (reverse) the current through the diode. However, very little current can get through the diode. Here is the reason why. The positive battery potential applied to the N-type material attracts the free electrons away from the barrier region, and the negative battery potential applied to the P-type material attracts the holes away from the barrier region. As there are now fewer free electrons and holes, more negative and positive ions are formed, causing the barrier region to effectively become thicker. The thicker the barrier region the less current can pass through the diode, Figure 2.16c.

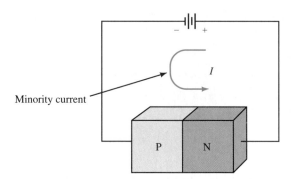

Minority current

I

P   N

FIGURE 2.17 PN diode biased in the reverse direction. The reverse current flows in the opposite direction of normal current.

Some current does pass. That current is called **minority current.** Thermal energy will cause some holes to appear in the N-type material and some free electrons to appear in the P-type material. These holes and electrons are called **minority current carriers.** The battery pushes the minority current carriers toward the barrier region. There they recombine, and a small current results. This small current produced by minority carrier action is called **reverse current.**

Progress Check
You have now completed objective 6.

1. Forward bias causes the depletion region to get _____ (narrower, wider).
2. For a diode to be reverse biased the cathode must be _____ (positive, negative) with respect to the anode.
3. Explain why current does not flow in a reverse-biased diode.
4. Is this diode forward- or reverse-biased?

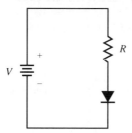

5. As a diode is forward biased its resistance will _____ (increase, decrease, remain constant).

2.6 DIODE CHARACTERISTICS

Even though the diode is the most basic of all the semiconductor devices, its characteristics make it available for a wide range of uses. Some applications use the resistance of the diode. Other applications may use the variable capacitance, the breakdown voltage characteristic, or the switching characteristic. Figure 2.18 lists a few of the specifications for a common 1N1206A diode:

Maximum working peak reverse voltage	600 V
Average rectified current at 150°C	12 A
Forward voltage at 12 A	1.1 V
Operating temperature range	65°C to 200°C
Maximum forward surge current	240 A

FIGURE 2.18 Specifications of a diode

The **peak reverse voltage rating** indicates the maximum range allowed for the diode to work under reverse voltage conditions. Exceeding the peak reverse voltage rating would damage the diode. Another rating shows the temperature rating for diode operation. Since the characteristics of a diode are affected by temperature, the specification sheet shows the minimum and maximum operating conditions.

CHARACTERISTIC CURVES OF A DIODE
The characteristic curve, Figure 2.19, graphically depicts the voltage-current relationships of the diode. In the figure, distance along the horizontal axis gives changes of voltage applied to the diode terminals. Distances along the

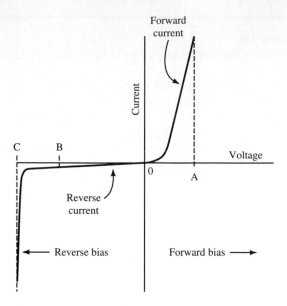

FIGURE 2.19 Characteristic curve of an ideal diode

vertical axis are changes of current through the diode. To the right of point 0, we see forward bias voltages. Reverse bias voltages lie to the left of point 0. At point 0, the bias is zero. As we can see, the bias goes down or up as we move either to the left or the right of point 0. Increases in distance above the horizontal line indicate an increase in forward current, while increases in distance below the line indicate reverse current increases.

Notice the steep upward slope of the curve in the region from point 0 to point A. Forward current increases rapidly with increases in forward bias. An opposite effect occurs in the region from point 0 to point B. Here, the reverse bias grows to a high value, but the reverse current increases very little. At point B, we see a knee developing in the curve. Diode resistance is suddenly overcome. From point B to point C, a small increase of reverse bias causes a rapid increase of reverse current.

Figure 2.20 shows four characteristic curves for approximations of diode operation. The curve shown in Figure 2.20a shows the ideal diode. Once the terminal voltage reaches zero, the ideal diode conducts instantly with no opposition to current flow. The vertical line in the plot indicates an electrical short. If we reverse the diode connections, no current flows. The horizontal line in the plot indicates an open circuit. Another curve, shown in Figure 2.20b, shows the effects of the barrier potential in a diode. As the energy in the diode allows the electrons to jump the barrier potential, the diode begins to conduct. Adding more detail, Figure 2.20c represents diode action when reaching reverse breakdown (see below). Finally, the characteristic curve shown in Figure 2.20d depicts the small amount of resistance seen when an ideal diode becomes forward-biased.

To compare the performance of a typical diode with that of the ideal diode, look at Figure 2.21, which shows the current-voltage curve for the typical diode. The area of the curve near the diode knee voltage has a large amount of curve. At voltages that exceed the diode knee voltage, the curvature lessens. Because of the variance, we cannot directly apply Ohm's law when trying to calculate the diode resistance.

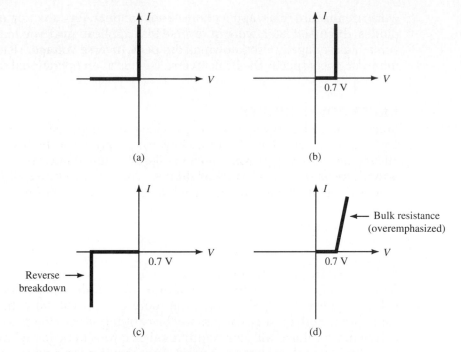

(a) (b)

(c) (d)

FIGURE 2.20 (a) Curve of a silicon diode that conducts instantly with no opposition to current flow (ideal diode). (b) Curve showing the effects of the barrier potential within the diode. (c) Curve showing diode action when reaching reverse breakdown. (d) Curve showing small amount of resistance present when a diode is forward biased.

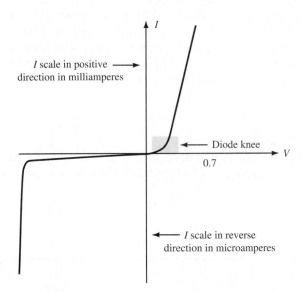

FIGURE 2.21 The current-voltage curve of a typical diode

BREAKDOWN VOLTAGE OR PEAK INVERSE VOLTAGE

The voltage at point B in Figure 2.19 is called the **zener,** or **breakdown, voltage.** The zener voltage is that point where a small increase in reverse bias causes a large increase in reverse current. Semiconductor diodes generally work in forward bias applications and conduct forward current. (Zener diodes, discussed later, work in reverse bias applications.) The maximum reverse bias voltage is also known as the **peak inverse voltage.** Usually, diode manufacturers specify the peak inverse voltage as an operational safety limit.

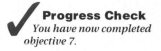

Progress Check
You have now completed objective 7.

BREAKDOWN DIODES

Most diodes, when reverse biased, provide a very high resistance and, hence, very small current flow. There is, however, a group of diodes designed to allow large current flow when reverse-biased. These diodes are called **breakdown diodes** or **reverse current diodes.** The manufacturers of breakdown diodes dope the diodes so that they have a large amount of minority charge carriers. Later in this chapter, the best known of the breakdown diodes, the zener diode, will be discussed.

AVALANCHE BREAKDOWN

All diodes have limits as to the amount of current they can carry. The more voltage applied across the diode, the more current will flow through it. At some point all the charge carriers will be carrying all the charge they can. Applying more voltage will cause additional electrons to be forced from the valence band into the conduction band. That requires the dissipation of a lot of energy. The temperature of the diode rises rapidly, causing more electrons to move from the valence band to the conduction band, which causes the temperature to rise even more. Forcing electrons from the crystal's valence band damages the diode and will quickly destroy it. This phenomenon, known as **avalanche breakdown,** should be avoided in most applications.

There is a special type of diode that is designed to operate in the avalanche breakdown region. Such diodes are called **controlled avalanche rectifiers.** The damage that can be caused by avalanche breakdown is controlled by rapidly dissipating the heat generated during breakdown. These diodes are used in high-voltage applications.

REVIEW SECTION 2.6
1. Draw a diode characteristic curve. Indicate forward and reverse bias.
2. Compare the performance of a typical diode with that of an ideal diode.
3. What type of diode is designed to operate in reverse bias applications?
4. What is avalanche breakdown?

2.7 LESS-THAN-PERFECT DIODE OPERATION

DIODE RESISTANCE

Even when conducting, diodes offer some resistance to current flow. We can consider several types of resistances when studying a forward-biased diode. Although its resistance is considered minimal, the material used to construct the diode does offer bulk resistance. This resistance may vary from 0.1 to 2 Ω. Doping lowers the bulk resistance of the material, but some resistance still remains.

In Figure 2.21 the effect of bulk resistance may be seen in the curve above the diode knee. If the bulk resistance were zero, the curve above the diode knee would be vertical. Since it is not zero, the curve has a small horizontal component. The greater the bulk resistance of a diode, the more horizontal the curve will be.

VOLTAGE DROPS IN DIODES

We assume that the voltage applied to a diode appears entirely across the junction, and we ignore any voltage drop in the neutral regions or at the external contacts. For most devices, the high amount of doping negates any resistance for each neutral region and decreases any voltage drop. Also, the typical area of a diode is large compared with its length. Yet, some diodes exhibit a significant amount of resistance that causes changes in the expected current-voltage characteristics. As we know, *any difference of potential across a resistance is a voltage drop.*

There are several voltage drops in a forward-biased diode. First and greatest is the voltage required to overcome the barrier region. For silicon diodes that voltage is approximately 0.7 V. Also, a very small voltage drop occurs at the connection of the leads and the semiconductor material. The third type of voltage drop occurs over the N-type and P-type materials. Although many free electrons and holes are available, a small voltage drop occurs while moving them toward the barrier region. All the voltage drops discussed are within the diode, and all but the 0.7 V are very small. As a result this text will assume an approximate voltage drop of 0.7 V as the voltage dropped across a diode.

The manufacturers of diodes use a variety of design techniques to ensure that the diode does not have an excessive voltage drop. Some of these design techniques are doping material and quantity, lead connecting techniques, and junction geometry.

CAPACITANCE IN DIODES

A **capacitor** exists whenever two conductors are separated by a dielectric. The P-type material and the N-type material are conductors, and the barrier region is a dielectric since it prevents current flow. Together they form a **junction capacitor.** When the bias on the diode is changed, the barrier region will change in thickness. The effect is to change the thickness of the dielectric. The amount of **capacitance** (ability to store electrical charge) of a capacitor is inversely proportional to the thickness of the dielectric. Thus, the amount of a diode's capacitance is changed by varying the bias voltage applied. The modern digital radio uses a special diode called a **varactor** in its tuning circuits. Pushing a selector button changes the bias voltage on this special diode, allowing you to select the station of your choice.

Progress Check
You have now completed objective 8.

HOW DOES A ZENER DIODE WORK?

A special kind of diode, called the **zener diode,** operates in the breakdown region. That is, the zener diode is normally used in reverse-biased mode. The voltage required to begin conduction in the reverse direction is called the

zener knee. Building momentum, newly freed charge carriers continue to collide with other charge carriers. Soon, an appreciable amount of current flows in a direction opposite the direction of current flow in a normal diode. Figure 2.22 shows the schematic symbol for the zener diode. The triangle points in the direction of the electron flow during breakdown. The colored band at one end of an actual zener diode designates the cathode end. *When operating zener diodes, be sure the cathode is positive with respect to the anode.*

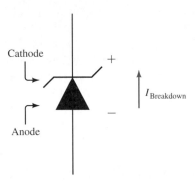

FIGURE 2.22 Schematic diagram of a zener diode

A zener diode is useful because it has the ability to maintain the breakdown, or knee, voltage over a wide range of currents. This ability allows the zener diode to function as a constant voltage source for voltage regulation applications. In later chapters, you will see that most DC power supplies do not provide a constant voltage but rather the voltage fluctuates. The zener diode can reduce these fluctuations. That process is called **voltage regulation.**

An ideal zener diode would maintain a precise voltage regardless of current and temperature changes. However, temperature and current changes do affect the breakdown voltage of a zener diode. Manufacturers of zener diodes establish the location of the zener knee. Figure 2.23 shows the specifications for a 1N758A zener diode.

Power dissipation (P_d)	400 mW
Breakdown voltage (B_V)	10 V ± 5%
Maximum zener current (I_{zm})	35 mA
Test current (I_{zt})	20 mA
Zener impedance (Z_{zt})	17 Ω
Knee impedance (Z_{zk})	Not specified
Temperature coefficient of breakdown voltage	+0.060%/degree C
Leakage current (I_R)	0.1 μA

FIGURE 2.23 Specifications for a 1N758A zener diode

Power dissipation ratings are specified for different temperature ranges. Generally, manufacturers will use 25°C and 50°C as typical ranges. When considering power dissipation, manufacturers specify a number of breakdown voltage values for particular power ratings, such as 400 mW, 1 W, 10 W, and 50 W.

Nominal breakdown voltages fit within minimum and maximum voltage limits and have normal tolerances of ±20%, ±10%, and ±5%. The maximum current rating relates to the power rating. If we have a 10-W, 10-V zener diode, the formula

$$I_{zm} = \frac{P_d}{B_v}$$

gives us the maximum current rating of 1 A.

Looking at the zener test current, we see the current rating for measuring and specifying breakdown voltage, zener impedance, and temperature coefficient. Most test currents operate a zener diode at one-quarter of its rated power.

Zener impedance measures the slope or rate of change of the breakdown voltage characteristic for the test current. With the zener knee impedance, we not only measure the slope of the breakdown voltage characteristic, but also define the sharpness of the knee. Figure 2.24 shows the differences between a sharp knee and a soft knee. Voltage regulation for low-level applications requires a sharp knee.

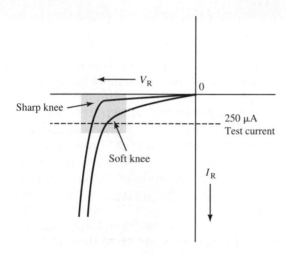

FIGURE 2.24 Comparing a zener soft knee with a sharp knee

Temperature also affects the breakdown voltage. The temperature coefficient of breakdown voltage measures the change in breakdown voltage caused by temperature change. Manufacturers determine the coefficient by recording the breakdown voltage at test current at 25°C and 125°C.

Finally, leakage current within the zener diode comes into play. Although zener diodes operate in the breakdown region, some leakage current always exists. Some zener diode applications, such as circuit protection, require that the leakage current stay at a low level. In protection applications, the diode is biased in the blocking region of its characteristic; it will conduct only under transient overvoltage conditions. A higher-than-normal leakage current may either interfere with circuit operation or drain power from the circuit.

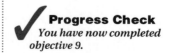

Progress Check
You have now completed objective 9.

SUMMARY

In this chapter we first looked at the basic structure of matter. Protons, neutrons, and electrons make up atoms. Electrons have negative electrical charges, protons have equal positive charges, and neutrons have no electrical charge. Electrons that escape from their orbits are called free electrons.

Because of the fundamental law of repulsion and attraction, electrons move away from a negative voltage source and toward a positive voltage source. The movement of many free electrons in one direction is called current. Current flow depends on the type and size of the conductor.

A semiconductor neither conducts nor resists current flow very well. Some semiconductor materials, such as silicon and germanium, form into crystals. When the semiconductor atoms link from crystal to crystal, they also share electrons. Energy can free the electrons and allow them to flow from atom to atom.

When a negatively charged electron moves from one atom to another, it leaves a positively charged vacancy called a hole. The positive charge of the hole attracts other negatively charged electrons. Applying a voltage can cause the movement of electrons from hole to hole.

Intrinsic crystals are a pure form of a semiconductor material. An intrinsic semiconductor does not conduct current well. Doping is the process of adding impurities to the intrinsic semiconductor. A doped semiconductor is said to be extrinsic and will conduct electricity better than an intrinsic semiconductor.

An extrinsic semiconductor with an excess of holes is called a P-type semiconductor. An extrinsic semiconductor with an excess of electrons is called an N-type semiconductor. In each case, the excess current carriers, called majority carriers, provide greater current conduction.

Diffusion is the process of spreading excess charge carriers throughout a semiconductor material. When a semiconductor crystal has both P-type and N-type materials, a junction is formed between the materials. The electrons of the N-type material diffuse into the P-type material, and the holes of the P-type material diffuse in the N-type material. This diffusion causes positive ions to be formed in the N-type material and negative ions in the P-type material. An electrostatic field is then formed at the junction. The electrostatic field is called the barrier potential. The barrier potential bars any more charge carriers from crossing the junction. The area around the junction that has no charge carriers is called the depletion region.

A P-N junction diode requires sufficient energy to allow its charge carriers to cross the barrier and the junction. Applying a voltage to the junction of the diode is called biasing. Forward biasing connects the positive voltage terminal to the P-type material and the negative voltage terminal to the N-type material. Reverse biasing has the opposite voltage connections.

Characteristic curves graphically represent voltage-current relationships of diodes. Diodes exhibit forward and reverse current, resistance, capacitance, and voltage drops. Special types of diodes, called zeners, operate in the reverse voltage region and work well for some circuit applications such as power supply regulation.

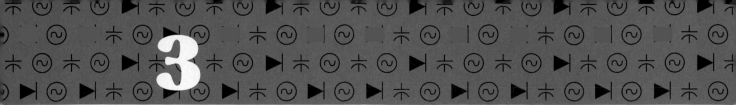

3

DIODE CIRCUIT APPLICATIONS

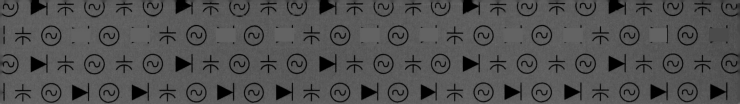

✔ **As you read this chapter, concentrate on learning how to:**

1. Determine the output voltage of a half-wave rectifier
2. Determine the output voltage of a full-wave rectifier
3. Determine the output voltage of a bridge rectifier
4. Describe the effect of filtering on the ripple voltage
5. Determine the output voltage from a zener diode regulator circuit
6. Identify voltage multiplier circuits and determine the output voltage
7. Identify signal clipper circuits and draw the output waveform
8. Identify signal clamper circuits and draw the output waveform
9. Identify and resolve an unbalanced load condition
10. Troubleshoot electronic circuits safely
11. Use basic meters to troubleshoot power supplies

INTRODUCTION

In Chapter 2, we studied both atomic and semiconductor theory. We discovered how a simple law of physics—attraction and repulsion—affects the building of semiconductor devices. Now, as we move into circuit applications for the semiconductor diode, we will see how the same law works on a larger scale.

In this chapter, we will study diode applications, such as power supplies. We will examine how the complete power supply works. Our discussion of power supplies will cover terms such as *rectification, filtering,* and *regulation.*

As we study diodes and power supplies, we will also learn about applications of other kinds of diodes. Zener diodes regulate voltages in many circuits. Other types of diodes are used in circuits that multiply voltages for specific applications. Still other types of diodes modify waveforms. Concluding the chapter, we will discuss troubleshooting power supplies.

Many pieces of electronic equipment utilize the AC power line. However, most electronic equipment requires DC operating voltages. In addition, the equipment may also require DC operating voltages that range higher and lower than the AC line voltage. To meet the requirements of today's electronic equipment, power supplies feature a number of different components. Figure 3.1 shows a basic block diagram of an AC-operated power supply. A transformer can step up or step down the AC line voltage to the needed value. Then, one or more diodes convert, or **rectify,** the AC voltage to a pulsating DC voltage source. Then, the filter smooths the pulsating voltage into a constant DC voltage source. Last, the regulator, a circuit made of several different components, provides a constant voltage to the load resistance.

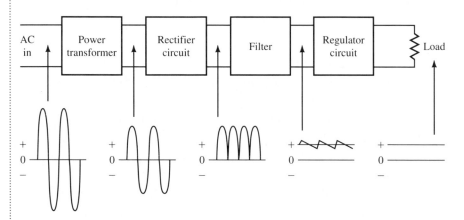

FIGURE 3.1   Basic block diagram of an AC-operated power supply

AC THEORY REVIEW

Since the DC power supply converts AC into DC power, let's review what we know about AC. Figure 3.2 illustrates the sinusoidal waveform of AC line voltage and current. One cycle is completed when the voltage or current has gone 360 degrees. The top half-cycle is called the **positive alternation,** and the bottom half-cycle is called the **negative alternation.** The wave is symmetrical. That is, the area and shape of the bottom alternation reflect the area and shape of the top alternation. The time that is required for a sine wave to complete one cycle is called its **period.** The reciprocal of period is frequency.

$$\text{\textit{frequency}}\,(\text{hertz}) = \frac{1}{\text{\textit{period}}\,(\text{seconds})}$$

Let's review the ways we measure AC waveforms. In Figure 3.2 look at the top of the positive alternation. That 90-degree point has an instantaneous voltage called $+V_{peak}$ **(peak positive voltage).** The negative alternation also has a peak voltage, called $-V_{peak}$. This **peak negative voltage** occurs at the 270-degree point in Figure 3.2. To find the average of one alternation of a sine wave we take the highest point, which is the peak, and multiply it by 0.637. Since the waveform in Figure 3.2 can represent either voltage or current, we can solve for average voltage (V_{avg}) or average current (I_{avg}) by multiplying the V_{peak} or I_{peak} by 0.637.

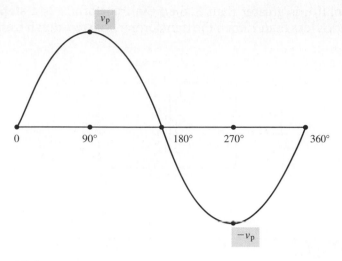

FIGURE 3.2   AC sine wave

$$V_{peak} \times 0.637 = V_{avg}$$
$$I_{peak} \times 0.637 = I_{avg}$$

The last measurement to review is **root-mean-square (rms).** The term most often associated with rms is VAC or V_{AC}. In Figure 3.3 either a 120-V battery or a 120-V_{rms} generator may be connected to the 10-Ω load resistor. Whether the generator is connected to the load or to the battery, the resistor will dissipate the same amount of heat (12 W). An rms measurement then has the same power rating as DC. V_{rms} is calculated by multiplying V_{peak} by 0.707.

$$170\,V_{peak} \times 0.707 = 120\,V_{rms}$$

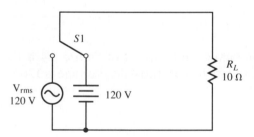

FIGURE 3.3   Equivalent power-generating ability of AC and DC sources

TRANSFORMER REVIEW

Many designs of power supplies use transformers to step up or step down voltage. In Figure 3.4 line power is connected to transformer T_1, which is connected to load resistor R_L. The line power is connected to the primary side of the transformer, and the load resistor is connected to the secondary side. Whether a transformer will step up or step down voltage is a function of the turns ratio. The **turns ratio** is the number of secondary coil turns (N_S) to the number of primary coil turns (N_P). The Greek letter η (eta) is used for turns ratio. If η is greater than 1, then the transformer is a **step-up trans-former.** If η is less than 1, then the transformer is a **step-down transformer.**

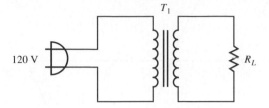

FIGURE 3.4   Simplified transformer circuit

EXAMPLE 3.1

Use Figure 3.4. The primary of T_1 has 200 turns. The secondary of T_1 has 20 turns. What is η?

Solution
Divide the secondary by the primary.

$$\eta = \frac{N_S}{N_P} = \frac{20 \text{ turns}}{200 \text{ turns}} = 0.1$$

By using the turns ratio, we can determine the secondary voltage. The primary voltage is multiplied by the turns ratio η.

In a transformer, if voltage is stepped down, then current is stepped up. That is, current is inverse to voltage. Secondary current may be solved by using the turns ratio. Divide primary current by η to find secondary current.

EXAMPLE 3.2

What are the secondary voltage and current of Figure 3.4 if the primary of T_1 has 200 turns, the secondary has 20 turns, line voltage is 120 VAC, and R_L is 10 Ω.

Solution
First solve for η.

$$\eta = \frac{N_S}{N_P} = \frac{20 \text{ turns}}{200 \text{ turns}} = 0.1$$

Then solve for the secondary voltage (V_S).

$$V_S = \eta \times V_P = 0.1 \times 120 \, V_P = 12 \text{ V}$$

Then solve for the secondary current (I_S). There are several ways. Depending on the information one way will be easier than the others. In this case Ohm's law would be easiest.

$$I = \frac{V_S}{R_L} = \frac{12 \text{ V}}{10 W} = 1.2 \text{ A}$$

1. Frequency is the _____ of period.
2. The maximum instantaneous voltage of any one alternation is called _____ .
3. How is V_{avg} for any one alternation determined?
4. RMS means _____ .
5. How is V_{ac} determined?
6. Solve for the turns ratio of a transformer whose primary has 1000 turns and whose secondary has 200 turns.
7. If line voltage is 120 VAC and the transformer has a turns ratio of 0.5, what is the secondary voltage?
8. Most electronic equipment requires _____ to operate.
9. Which block of Figure 3.1 is used to step up or step down AC voltages?
10. The rectifier circuit converts _____ to pulsating DC voltage.
11. What are the functions of the filter and regulator circuit in Figure 3.1?

3.2 DIODES AS RECTIFIERS

As we saw in Chapter 2, the most obvious quality of a P-N junction diode is its unilateral current characteristic: A diode conducts current in only one direction. If the diode is forward-biased, the electron current flows from the negative N-type material to the positive P-type material. Little or no current will flow if the diode is reverse-biased and the P-type material is negative with respect to the N-type material. We can think of a forward-biased diode as a short circuit and a reverse-biased diode as an open circuit. All these characteristics make the diode useful as a rectifier. Two basic types of rectifier circuits exist, the half-wave rectifier and the full-wave rectifier.

HALF-WAVE RECTIFIER CIRCUITS

Figure 3.5 shows a **half-wave rectifier circuit.** A transformer labeled T_1 supplies the AC voltage to the diode. Diode D_1 connects in series with the T_1 secondary and the load, designated as R_L. Common practice allows us to represent various circuits connected to a power supply as R_L.

We already know that a diode allows current to flow in only one direction. Conduction occurs when the anode is positive with respect to the cathode. Let us consider the action that takes place in the circuit shown in Figure 3.5.

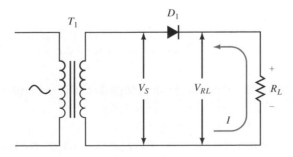

FIGURE 3.5 A half-wave rectifier circuit

When the AC voltage causes the top of the T_1 secondary to swing positive with respect to the bottom, D_1 conducts. Current flows from the bottom of T_1, then through the load, through the diode, producing a voltage drop across R_L that follows the input voltage. However, when the AC voltage changes polarity, the top of the T_1 secondary goes negative with respect to the bottom. Now, the diode is reverse-biased. Because the diode is reverse-biased, it does not conduct. No current flows through the load.

To further illustrate this action, Figure 3.6 shows how an oscilloscope waveform across the load would appear. We call this type of waveform a **half-wave signal.** D_1 conducts on each positive alternation. The current through the load consists of positive half-cycle pulses. Another way to look at this is to say that current flows through the load for only 180 degrees of the input voltage. Because the direction of current remains constant while the amount of current changes as the positive input voltage changes, the circuit provides pulsating DC voltage. In some applications, pulsating DC voltage answers the circuit requirements. Most electronic circuits, though, need a smoother DC current. Passing the pulsating DC output of the rectifier through a filter circuit provides the smoother output.

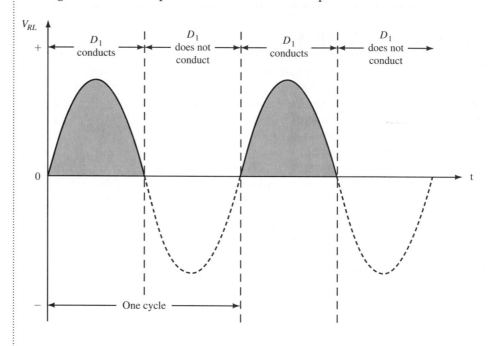

EXAMPLE 3.3

Determine V_{avg} and I_{avg} for the half-wave rectifier in Figure 3.5. The input voltage is line voltage, 120 VAC. T_1 has a turns ratio of 0.2. R_L is 100 Ω.

Solution
First we need to determine the voltage developed on the secondary of T_1.

$$V_s = V_p \times \text{turns ratio}$$
$$V_s = 120 \text{ VAC} \times 0.2 = 24 \text{ VAC}$$

Second, we need to determine the secondary peak voltage. We know that

$$\text{VAC} = V_{peak} \times 0.707$$

Since we know the VAC of the secondary, we can rewrite the formula to solve for V_{peak}.

$$V_{peak} = \frac{VAC}{0.707} = \frac{24\ VAC}{0.707} = 34\ V_{peak}$$

Now we can solve for V_{avg}. We have learned that V_{avg} for one alternation is $V_{avg} = V_{peak} \times 0.637$. But with power supplies, both alternations must be considered. The second alternation provides no voltage. Thus, the average voltage for two alternations is one-half of 0.637, which is 0.318. Hence,

$$V_{avg} = V_{peak} \times 0.318 = 34\ V \times 0.318 = 10.8\ V$$

Average current is found by using Ohm's law.

$$I_{avg} = \frac{V_{avg}}{Load\ Resistance} = \frac{10.8\ V}{100\ \Omega} = 0.108\ A$$

Progress Check
You have now completed objective 1.

PRACTICE PROBLEM 3.1

Determine V_{avg} and I_{avg} for the half-wave rectifier in Figure 3.5. The input voltage is line voltage, 440 VAC. T_1 has a turns ratio of 0.25. R_L is 150 Ω.

FULL-WAVE RECTIFIER CIRCUITS

Half-wave rectifier circuits have one major disadvantage: Only one current pulse develops in the load during each cycle of the AC power supply. Even though additional inductive and capacitive filter circuitry helps to smooth the pulsating DC, the smoothing effect does not accomplish everything needed for an electronic circuit. Another rectifier circuit, the **full-wave rectifier** circuit, develops two current pulses in the load during each cycle of the AC power supply. There are two major types of full-wave rectifier circuits: the center-tapped full-wave rectifier and the bridge rectifier.

CENTER-TAPPED FULL-WAVE RECTIFIER CIRCUITS

In this section, we will look at the **center-tapped full-wave rectifier.** Figure 3.7 shows the primary side of a transformer connected to the AC line. A fuse

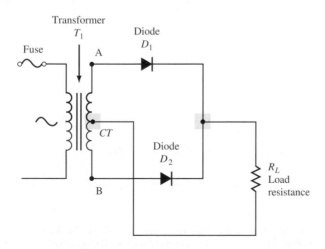

FIGURE 3.7   **A center-tapped full-wave rectifier**

protects the primary side of the transformer from possible overload damage. While one side of the load connects to the cathodes of rectifier diodes D_1 and D_2, the other side fastens directly to the center of the transformer secondary.

In Figure 3.8, we see that the top of the transformer secondary (point A) has a positive polarity with respect to the bottom (point B). Since the circuit features a center-tap (CT) on the transformer, we must also consider the polarity differences at point CT. The top of the secondary has a positive polarity, and the bottom of the secondary has a negative polarity with respect to the center-tap. At this time, the anode of D_1 goes positive with respect to the D_1 cathode. D_1 conducts current through the load resistance as shown by the I_{D1} arrow. Since the D_2 anode has a negative polarity with respect to its cathode, the diode will not conduct.

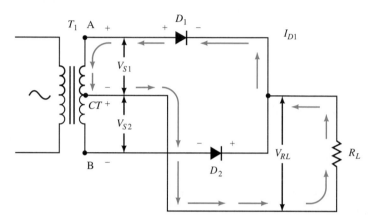

FIGURE 3.8   Circuit action in a full-wave center-tapped rectifier during the first alternation. Note the positive polarity at the top of the transformer secondary.

Figure 3.9 illustrates circuit action when the voltage across the transformer secondary reverses polarity. From the perspective of the center-tap, the top of the secondary is negative while the bottom is positive. Diode D_2 becomes forward-biased and conducts current and diode D_1 becomes reverse-biased. Consequently, D_1 does not conduct current.

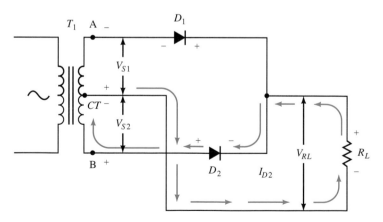

FIGURE 3.9   Circuit action in a full-wave center-tapped rectifier during the second alternation. Note the negative polarity at the top of the transformer secondary.

In both cases, current flows through the load resistance in the same direction, from the bottom of R_L to the top. Since the load current does not change direction, pulsating DC voltage develops. The oscilloscope waveform of the diode input voltages in Figure 3.10 shows how the diodes conduct on opposite AC alternations. Current flows through the load during the positive and negative alternations of the input.

V_{S1} represents the input waveform of D_1, while V_{S2} represents the input waveform of diode D_2. V_{S1} goes positive when V_{S2} goes negative during one half-cycle. Opposite conditions occur during the next half-cycle. Moving to Figure 3.11, we see how the waveform of the load resistance voltage would

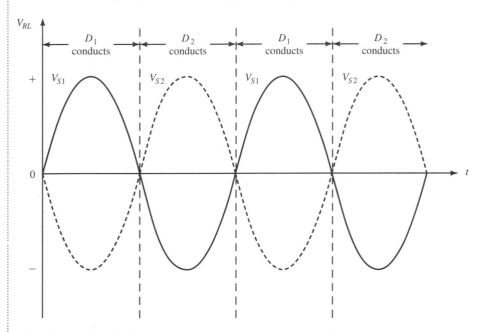

FIGURE 3.10 Oscilloscope waveform of center-tap full-wave rectifier diode input voltages

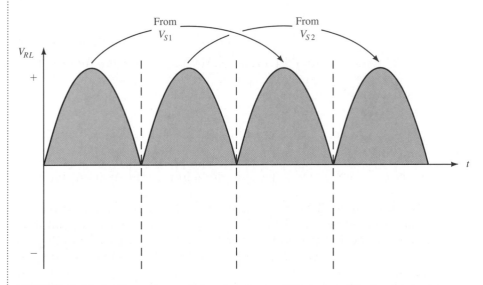

FIGURE 3.11 Oscilloscope waveform of center-tap full-wave rectifier load resistance voltage

appear. Compare this waveshape with the one seen at the output of the half-wave rectifier circuit (Figure 3.6). A pulsating DC voltage is delivered to the load for the full 360 degrees of the input cycle. Since two current pulses occur during each cycle, filtering the pulsating DC becomes much easier.

EXAMPLE 3.4

Determine the average DC voltage from a center-tapped full-wave rectifier. For this example, let us say that the transformer in Figure 3.7 has a 10-to-1 turns ratio. The load resistor (R_L) has a 560-Ω value. Assume the ideal diode approximation.

Solution

Since the primary connects to the AC line, 120 V_{rms} is applied to the circuit. The peak voltage of the input would equal approximately 170 V_p:

$$120\,V_{rms} \times 1.414 = 169.68\,V_p$$

The step-down transformer reduces this peak voltage to 17 V.

$$\frac{170\,V}{10} = 17\,V_p$$

This 17 V drops across the entire secondary. If we consider the diode as a short when it conducts, the load resistor will see 8.5 V_p because of the center-tap. To find the average DC load voltage, multiply the peak load voltage by 0.637.

$$8.5\,V_p \times 0.637 = 5.41\,V$$

Determining the peak current, we get:

$$\frac{8.5\,V_P}{560\ \Omega} = 15.18\,mA_P$$

In Figure 3.7 the conversion factor used with a full-wave out is 0.637. Appendix A gives an explanation of this conversion factor. Completing the calculation, we find that:

$$15.18\,mA_p \times 0.637 = 9.66\,mADC$$

With the result we obtained in the example, we can decide whether the load resistor can withstand the delivered power. In addition, we can verify that the diodes will work without damage.

EXAMPLE 3.5

Determine the peak inverse voltage of a diode in a center-tapped full-wave rectifier.

Let us consider Figure 3.8. In this figure, the top diode is forward-biased, while the bottom diode is switched off because of reverse bias. We can treat the top diode as a short circuit while it conducts. We can use Kirchhoff's voltage law to analyze the T_1 secondary's effect on D_2. Draw a closed loop starting at point A. Go through D_1. Remember the voltage drop of D_1 is zero. Now continue the loop through D_2 to point B. Complete the loop by continuing through the secondary back to point A. If the secondary is sourcing 17 V and

D_1 is dropping 0 V, then D_2 must be dropping 17 V. Consequently, in a center-tapped full-wave circuit, each diode must be able to handle the entire peak secondary voltage when reverse-biased.

Solution
The peak inverse voltage (PIV) equals the entire voltage across the secondary windings. Because of this factor, we must choose diodes that will not go into reverse breakdown below 17 V.

PRACTICE PROBLEM 3.2

In Figure 3.8, if the turns ratio is 1:5 (step up) and $R_L = 1$ kΩ, determine the following:

Peak load voltage = _____ V
Peak load current = _____ A
Peak inverse voltage = _____ V

Progress Check
You have now completed objective 2.

FULL-WAVE BRIDGE RECTIFIER CIRCUITS

Figure 3.12 shows the schematic of another type of full-wave rectifier called the **bridge rectifier.** Instead of using a transformer with a center-tapped secondary, the circuit utilizes four rectifier diodes for full-wave rectification. In the figure the top of the transformer secondary (point A) has a positive polarity with respect to the bottom (point B). With point A positive, the cathode of diode D_1 and the anode of diode D_4 are positive. At the same time, because of the negative polarity of the secondary bottom, the cathode of diode D_3 and the anode of diode D_2 go to a negative value. A diode is forward-biased when its cathode is negative with respect to its anode. Both D_4 and D_3 are forward-biased and will conduct. However, D_1 and D_2 are reverse-biased and will not conduct. The arrows in Figure 3.12 represent the current path. The current path starts at point B and goes to D_2 and D_3. D_2 is reverse-biased, and D_3 is forward-biased. The current will go through D_3 to the junction of the load resistor and D_1. D_1 is reverse-biased. All the current goes through the load resistor to the junction of D_2 and D_4. D_2 is reverse-biased, and D_4 is forward-bi-

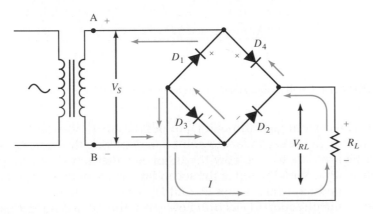

FIGURE 3.12 Full-wave bridge rectifier circuit when point A is positive to point B

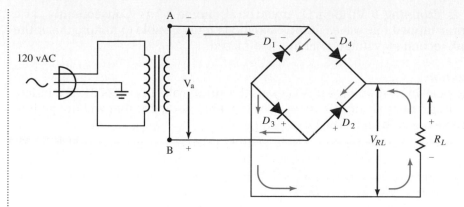

FIGURE 3.13 Full-wave bridge rectifier when point A is negative to point B

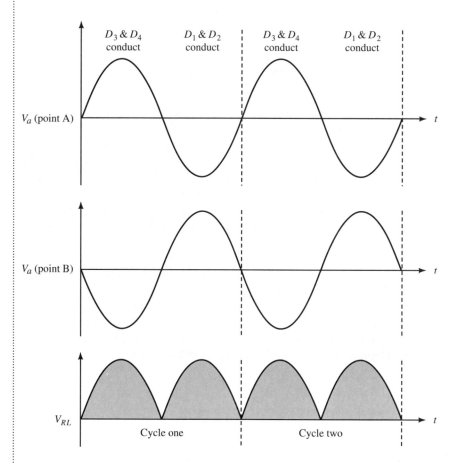

FIGURE 3.14 Oscilloscope waveform of the bridge output

ased. The current goes through D_4 to point A on T_1's secondary. The current flows through the secondary to point B, completing the current path.

In Figure 3.13 we see how the other AC alternation affects polarity and electron flow. With the top of the secondary at a negative potential, diodes D_1 and D_2 conduct, while diodes D_3 and D_4 remain nonconductive. The arrows indicate the direction of electron flow. Even though a different pair of diodes conducts, the direction of electron flow remains the same through the load.

One might think of the diodes in the bridge as "traffic police" directing the flow of traffic. With the load as a one-way street, the traffic always steers in the same direction.

Since a bridge rectifier does not require a tapped transformer, it rectifies the full secondary voltage on each alternation. Figure 3.14 shows the bridge output waveform. Each cycle of the AC supply produces two current pulses to provide full-wave rectification. The pulsating DC output is about equal to the peak voltage across the entire secondary. Compared to a center-tapped full-wave rectifier, a bridge rectifier will produce twice as much DC output voltage. This is because a bridge rectifier uses the entire secondary voltage. Since the diodes lie in series, the peak inverse voltage equals the peak value of the supply voltage. In addition, the higher frequency of pulses makes filtering the pulsating DC voltage output much easier.

EXAMPLE 3.6

Determine the average DC load voltage and the DC load current.

Let the voltage across the secondary winding of the transformer shown in Figure 3.15a equal 20 V_p. The load resistor is 1.5 kΩ. As before, treat the conducting diodes as shorts.

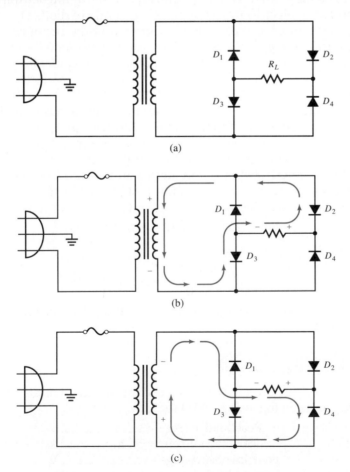

FIGURE 3.15 (a) A full-wave bridge rectifier circuit. (b) Example control loop. Bridge rectifier with D_3 and D_2 conducting. (c) Example analysis. Bridge rectifier with D_1 and D_4 conducting.

Solution

Convert the peak input to an equivalent DC value. Use the full-wave conversion factor of 0.637.

$$20 \, V_p \times 0.637 = 12.7 \, VDC$$

All the secondary voltage reaches the load because we do not have a center-tap. Refer to Figure 3.15b or 3.15c to analyze the complete loop. As you should see, the voltage across the load resistor must equal the secondary voltage to satisfy Kirchhoff's voltage law. Now, apply Ohm's law to obtain the current through the load.

$$\frac{12.7 \, VDC}{1.5 \, k\Omega} = 8.48 \, mA$$

EXAMPLE 3.7

Determine the peak inverse voltage across the diodes as used in the circuit shown in Figure 3.15a, b, and c and in example 3.6. Determine the maximum average DC current through the diodes.

In Figure 3.15b, the polarity is positive at this time; D_3 is forward-biased and D_1 is reverse-biased. Treating diode D_3 as a short, the bottom terminal of the secondary is electrically the same as the anode of diode D_1. The cathode of D_1 connects directly across the top terminal of the secondary. Therefore, the entire peak voltage of the secondary is applied to diode D_1 when the diode is reverse-biased. This also occurs for the other three diodes in the bridge.

Solution

To analyze what happens during this situation, let us move to Figure 3.15c. In this figure, diodes D_1 and D_4 conduct. The two diodes are in series with the load and with each other. During conduction time, they are in series because they share a common current. When the polarity of the secondary reverses, both D_1 and D_4 will stop conducting and D_2 and D_3 will start conducting. However, current flowed for only half the cycle through D_1 and D_4. Therefore, we can use the conversion factor for half-wave voltages:

$$20 \, V_p \times 0.318 = 6.36 \, VDC$$

and

$$\frac{6.36 \, VDC}{1.5 \, k\Omega} = 4.24 \, mADC$$

We must select diodes that do not have an average DC current rating greater than 4.24 mADC.

PRACTICE PROBLEM 3.3

In Figure 3.15, line power is 120 V, and the transformer has an 8:1 turns ratio. Calculate the following, using 1.5 kΩ for R_L:

Peak load voltage = _____ V

Peak load current = _____ A

Peak inverse voltage = _____ V

Average load voltage = _____ V

Average load current = _____ A

EXAMPLE 3.8

Determine the effects of diode drops on rectifier operation.

As we saw in example 3.7, one pair of diodes conducts for each voltage alternation. When conducting, the two diodes in each pair are in series with the load and with each other. Because of diode voltage drops, 1.4 V is lost. To calculate the effect on the circuit, we subtract the voltage drop value from the peak voltage delivered to the load. We will use the values from the previous example.

Solution

$$(20 \text{ V} - 1.4 \text{ V}) \times 0.637 = 11.8 \text{ VDC}$$

with a current of

$$\frac{11.8 \text{ VDC}}{1.5 \text{ k}\Omega} = 7.87 \text{ mADC}$$

Progress Check
You have now completed objective 3.

REVIEW SECTION 3.2

1. For a diode to conduct, the _____ must be positive with respect to the _____ .
2. A conducting diode acts like a(n) _____ , while a nonconducting diode acts like a(n) _____ .
3. A half-wave rectifier circuit uses _____ (1, 2, 4) diode(s).
4. In a half-wave rectifier, if the diode is not conducting (is off) the current through the load is _____ (minimum, maximum).
5. Current flows through the load for _____ (0 degrees, 180 degrees, 360 degrees) of input signal in a half-wave rectifier.
6. In Figure 3.5, the voltage at V_s is _____ (AC, pulsating, DC), while the voltage at V_{RL} is _____ (AC, pulsating, DC).
7. Explain how the circuit in Figure 3.5 operates.
8. A fuse is connected to the _____ (primary, secondary) side of the transformer to protect against excess current.
9. In Figure 3.7 if point A of T_1 is positive with respect to CT, point B will be _____ (positive, negative).
10. In Figure 3.7, if point A of T_1 is negative with respect to CT, D_1 will conduct and D_2 will not conduct. True or false?
11. In a full-wave center-tapped rectifier output voltage is for _____ (0, 180, 360) degrees of input signal.
12. Draw the waveforms as seen across R_L in Figures 3.5 and 3.7.
13. Explain how the circuit in Figure 3.7 converts AC to pulsating DC.
14. In Figure 3.12, if point A is positive with respect to point B, diodes D _____ and D _____ are conducting while diodes D _____ and D _____ are not conducting.
15. In Figure 3.12, D_3 and D_2 can both be conducting at the same time. True or false?
16. A full-wave bridge uses _____ (1, 2, 4) diode(s).

17. A full-wave bridge produces a voltage _____ the voltage output of a full-wave center-tap. (Assume the same number of turns in the secondary.)
    a. Equal to
    b. $1/2$ the value of
    c. 2 times
    d. 4 times
18. In a full-wave bridge the peak inverse voltage equals the _____ (p, p-p) value of secondary.
19. Compare and contrast the characteristics of a half-wave rectifier, a full-wave center-tap, and a full-wave bridge.

3.3 POWER SUPPLY INPUT FILTERS

So far, we have studied rectifier circuits most often found in power supplies. All three types—half-wave, center-tapped full-wave, and full-wave bridge—convert the AC line voltage into a usable form of DC voltage. However, most power supplies will consist of more than the rectifier circuit. In this section, we will study filter systems used to smooth out the pulsating DC voltage output of the rectifiers.

FILTERING A HALF-WAVE RECTIFIER CIRCUIT WITH A CAPACITOR

Figure 3.16 shows a half-wave rectifier circuit with a capacitor (C_1) connected across the load. As the first rectifier current pulse rises, electrons flow onto the lower plate of the capacitor. From there, the electrons flow off of the upper plate, through diode D_1 and the secondary of the transformer. Because of this action, the capacitor charges to the peak of the secondary voltage, represented by V_s.

As V_s decreases, D_1 is reverse-biased. The capacitor has charged to V_s (peak) minus the voltage drop across the diode, but V_s is now less than the secondary peak voltage, which reverse biases D_1. As D_1 is reverse-biased, the capacitor can discharge only through the load resistor. Therefore, the capacitor discharge current flows from the lower plate of the capacitor, through the load, to the upper plate of the capacitor. The capacitor has a value that will

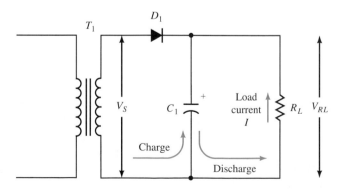

FIGURE 3.16 Half-wave rectifier circuit with a capacitor connected across the load

not allow it to discharge as rapidly as the secondary voltage decreases. Because of this, the load current and the capacitor voltage decrease lag the decrease of secondary voltage. Once the secondary voltage becomes less than the voltage across the capacitor, diode D_1 reverse biases. D_1 conducts only long enough to charge the capacitor. When the next voltage pulse arrives, the partially discharged capacitor becomes recharged by the peak of the secondary voltage.

With all this action, the load voltage pulses no longer resemble the pulses seen in Figure 3.6. Figure 3.17 shows the changes that occur. A dashed-line curve represents the secondary voltage, V_s, while a solid-line curve represents how the voltage across C_1 and the load varies. The load current has the same waveform as the solid-line voltage curve. The varying voltage is called **ripple voltage**, a part of the output voltage. As the secondary voltage V_s alternates, it causes the diode anode to go positive and the capacitor to charge until reaching the peak value. The shaded areas in the figure show the time during which the diode conducts and the capacitor charges. The unshaded areas show the capacitor discharging through the load until the next positive voltage pulse exceeds the voltage remaining on the capacitor plates. Then the process of conducting and charging starts again.

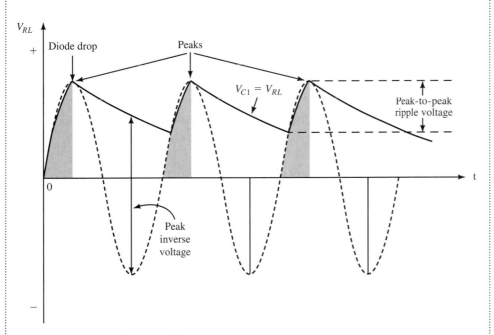

FIGURE 3.17 Load voltage pulses on a filtered circuit. Compare with the waveform of an unfiltered circuit shown in Figure 3.6.

Adding a filter circuit to the rectifier affects PIV requirements for the diode. In Figure 3.18 C_1 has been charged to V_{peak} by the positive alternation. When the secondary voltage goes to negative V_{peak}, at that instant the secondary and C_1 act like batteries trying to push current through the same circuit. When two sources are pushing current in the same direction, it is called **series aiding**. If two sources are pushing current in opposite directions, it is called **series opposing**. Figure 3.19 is the equivalent DC circuit of the highlighted area in Figure 3.18 when the V_s is at its maximum negative peak. Notice they are trying to push current in the same direction, which means they are series aiding. However, D_1 is reverse-biased and will not allow current flow. If we did a Kirchhoff's loop to include C_1, the secondary winding, and D_1, notice that the PIV of D_1 would have to equal or exceed the sum of the voltages of C_1 and the secondary winding. The maximum reverse bias potential occurs when the capacitor is fully charged and the secondary is at negative V_{peak}. Therefore, the PIV rating of the rectifier diode must be at least two times the secondary peak voltage.

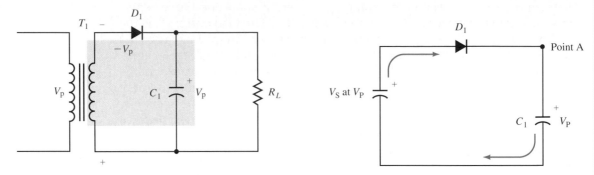

FIGURE 3.18 Diode PIV condition   **FIGURE 3.19** Equivalent PIV circuit

FILTERING A FULL-WAVE RECTIFIER CIRCUIT WITH A CAPACITOR

Placing a capacitor across the load terminals of a full-wave rectifier circuit gives similar results. Again, the capacitor charges during each rising positive pulse and then discharges through the load until the next voltage pulse arrives. As Figure 3.20 shows, the lower plate of the capacitor connects directly to the center-tap of the transformer secondary. For each consecutive AC voltage alternation, the transformer secondary top and then the transformer secondary bottom go positive with respect to the center-tap. With two alternations in each voltage cycle, two positive pulses charge the capacitor during each voltage cycle.

Figure 3.21 demonstrates circuit action during the applied voltage cycles. A dashed-line curve represents the series of charging voltage pulses applied across the capacitor and the load. Shaded areas represent the time taken for the capacitor to charge. As with the waveforms seen in the half-wave rectifier circuit, the solid-line curve shows that the capacitor charges to the peak volt-

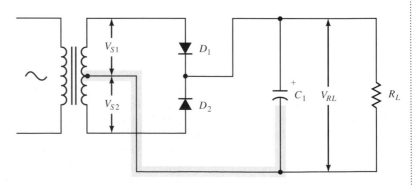

FIGURE 3.20 Using a capacitor to filter the output of a full-wave rectifier circuit

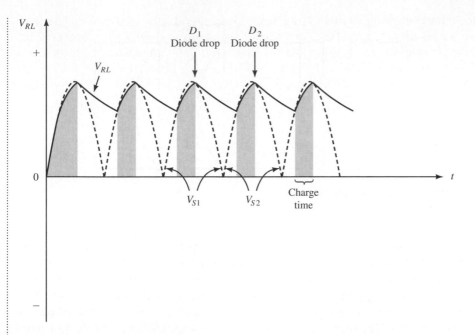

FIGURE 3.21 Full-wave filtering

age of each pulse and then discharges through the load until the next pulse arrives. In effect, the solid-line curve represents the voltage across the capacitor, the load, and the current in the load. Because charging the capacitor requires time, the solid charging line lags behind the dashed supply line.

Going back to the half-wave circuit for a moment, a major difference becomes apparent. Given the same supply frequency, the charging pulses in the full-wave rectifier circuit occur twice as often. As a result, the capacitor in the full-wave rectifier circuit has a shorter time interval for discharging. Consequently, the amplitude of the ripple voltage is significantly less in the full-wave rectifier than in the half-wave rectifier.

ADDING A RESISTOR IN SERIES WITH FILTER CAPACITOR

Some applications require the presence of a very low ripple voltage and a small load current. Placing a resistor in series between the rectifier and the filter capacitor, which is in parallel with the load resistor, will provide the additional filtering and lower the maximum possible load current. In Figure 3.22 rectifier D_1 provides a pulsating DC voltage. The first filter capacitor, C_1, charges to near V_{peak} of the secondary. As you have learned, there is a ripple voltage present. C_1's voltage is applied to series RC filter R_1 and C_2. A resistor is unaffected by frequency, while a capacitor blocks DC and passes AC. C_2 acts like a near short to the ripple voltage and as an open to the DC voltage supplied by C_1. At ripple voltage frequencies the resistance of R_1 is much greater than the reactance of C_2. The result is that most of the ripple voltage is dropped by R_1 and very little is present at the load. At the DC frequency of zero hertz the reactance of C_2 is much greater than the resistance of R_1. Very little of the DC voltage is dropped across R_1; while most of the voltage is developed across C_2. C_2, whose voltage is only a little less than C_1, then discharges through the load resistor, R_L.

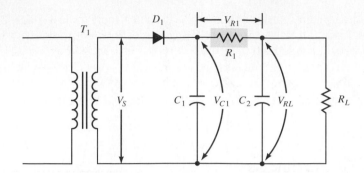

FIGURE 3.22 Adding a filter resistor

From the perspective of the load, the direct voltage drop V_{C1} across the resistor is lost. Because of this lost voltage, it is desirable to keep the voltage drop as low as possible with the direct voltage V_{C2} going across the filter to the load. Utilizing Ohm's law, a given current I_{R1} exists in direct proportion to the given resistance of resistor R_1. Filter resistors have the advantages of being inexpensive, small, and lightweight for applications that do not need large current from the power supply.

DEFINING TYPES OF FILTERS

A **filter** is a device that receives energy in short pulses of rectified current, stores the energy, and then provides that energy to the load as a steady current. Filters are defined in two ways: by their schematic design and by their function. In a schematic design, filters are named for the way they look. For example, look at Figure 3.23a. Figure 3.23a is called a π **(Greek letter pi) filter.** Notice that the two capacitors look like the legs and the resistor looks like the cross member of the π. Figure 3.23b is called a **double L** or **inverted double L filter.** R_1 and C_1 form one L of the filter and R_2 and C_2 form the second L. Many filter designs have become obsolete with the invention of three-terminal regulators and switching power supplies, which we will discuss later in this book. The most common filter used today is the **capacitor input filter.** A single capacitor is connected directly to the output of the rectifier and is in parallel with the load. This is the basic type of filter that we have been discussing.

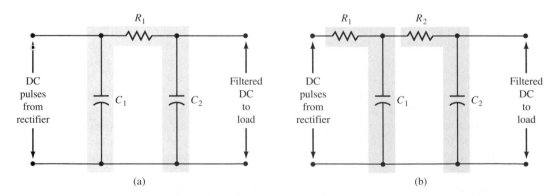

FIGURE 3.23 (a) π filter. (b) Double L filter.

The second way to define filters is by their function: low-pass, band-pass, and high-pass filters. A **low-pass filter** passes low frequencies, a **band-pass filter** passes a range of frequencies, and a **high-pass filter** passes only high frequencies. Since the purpose of a DC power supply is to produce power at zero hertz, the filters used with rectifiers are low-pass filters. A low-pass filter is identified by the filter capacitor being in parallel with the load.

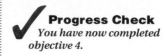

Progress Check
You have now completed objective 4.

REVIEW SECTION 3.3

1. What section of a power supply converts pulsating DC to a more constant DC?
2. A filter capacitor charges faster than it discharges. True or false?
3. When a filter is added, the varying voltage across R_L is called _____ voltage.
4. Which is closer to a purely DC voltage, Figure 3.5 or Figure 3.11? Explain.
5. The PIV on the diode in Figure 3.16 is _____ times the value of the peak secondary voltage.
6. What is easier to filter, a half-wave or a full wave? Explain.
7. The capacitor input filter is connected in series with the rectifier diodes and in parallel with the load. True or false?
8. If the output DC voltage of an unfiltered power supply is V_{avg}, what is the output DC voltage of a filtered power supply?
9. Increasing the capacitance of the filter capacitor will increase the amplitude of the ripple voltage. True or false?
10. Explain why the rectifier diodes conduct for a shorter time when a capacitor input filter is used than when no filter is used.
11. In Figure 3.22 why does adding a resistor to the capacitor filter reduce the ripple voltage amplitude?
12. A filter stores energy to be used a short time later. True or false?
13. The circuit in Figure 3.23a is called a _____ filter.
14. The filters in Figure 3.23a, b block _____ (AC, DC) and pass _____ (AC, DC).
15. What is the most common filter used today?
16. Referring to Figure 3.18b, draw input and output waveforms.

3.4 REGULATING THE POWER SUPPLY

WHAT IS REGULATION?

Many electronic applications require a power supply output voltage that remains nearly constant even if the load current varies during normal operation. We call this important power supply characteristic **regulation.** In practice, the design of a power supply provides a certain maximum output known as the full-load current. Under no-load conditions, the voltage at the output goes to its highest value. At the rated full-load current, the voltage reduces to some definite value. Because most circuits work best at some particular value, good regulation is desirable.

USING A ZENER DIODE AS A REGULATOR

As we learned in Chapter 2, several characteristics make the zener diode desirable for power supply regulation. **Zener impedance** is the basic measure of the quality of a zener diode as a voltage regulator. An ideal regulator would not change its voltage with changes in current and would have a zener impedance of zero. With the zener impedance rating, we can determine the worst-case change in voltage and make allowances for operating the zener diode either above or below the test current rating.

Figure 3.24 shows a simple zener diode regulator circuit. To achieve reverse breakdown, the voltage applied across the zener diode should be larger than the zener knee of the diode. In the figure, the voltage is positive with respect to ground and places a positive voltage on the zener diode cathode. When the zener diode reaches reverse conduction, we see a constant voltage across the load that equals the zener breakdown voltage.

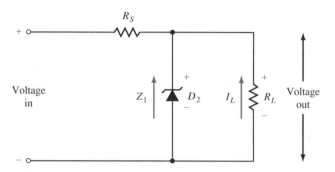

FIGURE 3.24 Simple zener diode regulator circuit

EXAMPLE 3.9

Determine the value of the series resistor needed to protect the zener when it is not connected to the load.

Figure 3.25 shows a zener regulator circuit with no load. In this case, the DC input to the zener may vary between 7 and 10 VDC. The zener knee, or the point at which the diode will break down, is 5.2 V.

Solution

Since the maximum voltage to be dropped across the source resistor will occur when the DC input voltage is maximum, calculate source resistor value

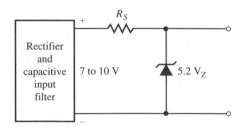

FIGURE 3.25 A zener regulator circuit with no load

using maximum DC input voltage. The maximum DC input voltage will equal 10 V. If the zener diode drops 5.2 V, we can calculate the voltage drop across the source resistor as

$$10 \text{ V} - 5.2 \text{ V} = 4.8 \text{ V}$$

At this point, we know the amount of maximum voltage that can appear across the source resistor. To choose the proper value of the resistor, we must also know the maximum current that the zener can handle before breakdown occurs. Suppose we find that the zener diode we have chosen can handle up to 100 mA. We can calculate

$$R_s\left(\text{min}\right) = \frac{4.8 \text{ V}}{100 \text{ mA}} = 48 \text{ }\Omega$$

If we choose the next largest standard value for resistors, 56 Ω, we guarantee that the current flowing through the zener diode will equal less than 100 mA.

EXAMPLE 3.10

With this example, we will look at the effect of placing a load across the zener diode. Determine the minimum load usable in a zener regulator.

We know that the zener breakdown voltage must be exceeded before a zener diode can conduct. Once the breakdown voltage has been exceeded, the zener diode requires a minimum amount of current to stay in the breakdown region, or the zener knee current. Let's continue to use the zener diode from example 3.9. The diode needs at least 20 mA to stay in a reverse breakdown state. Determining the minimum load resistance begins with finding how much current is delivered to the zener diode when the DC input voltage drops to the minimum value. We will assume that the zener diode has no impedance. From example 3.9, we have

$$7 \text{ V(min)} - 5.2 \text{ V}_z \text{ (or zener breakdown voltage)} = 1.8 \text{ V}$$

This voltage drops across the source resistor so that

$$\frac{1.8 \text{ V}}{56 \text{ }\Omega} = 32.15 \text{ mA}$$

The value 32.15 mA is the lowest possible available current that can flow into the node connecting the zener diode cathode and the load resistor.

Solution
In Figure 3.25, 20 mA of the 32.15 mA must be reserved for the zener. We can divert the remainder to the load resistor:

$$32.15 \text{ mA} - 20 \text{ mA} = 12.15 \text{ mA(max)}$$

If the zener diode is in breakdown, the load voltage would equal 5.2 V. Then:

$$R_L\left(\text{min}\right) = \frac{V_z}{I_1\left(\text{max}\right)} = \frac{5.2 \text{ V}}{12.15 \text{ mA}} = 428 \text{ }\Omega$$

If the load resistance drops below 428 Ω, regulation cannot be guaranteed. Subsequently, the unregulated voltage may damage the load.

In Figure 3.25, if the DC input varies between 10 V and 15 V while the zener diode's breakdown is 6.8 V and its maximum current is 250 mA, determine the following: maximum voltage drop across source resistor; R_s(min); minimum load if I(min) of zener diode equals 25 mA.

See how a zener diode can affect ripple voltage.

If we know the internal impedance of the zener diode, we can determine the effect the zener has on input ripple. If the zener diode has a maximum small signal breakdown impedance of 17 Ω, we can find the amount of ripple reduction by using the following equation.

$$V_z = \frac{R_z}{R_s} \times V_s$$

where V_z = output ripple
$\quad\quad V_s$ = input ripple
$\quad\quad R_s$ = series resistance
$\quad\quad R_z$ = zener impedance in the breakdown region

If the capacitor input filter delivered 8 V with a 400-mV ripple into the zener diode regulator, the output ripple would equal

$$V_s = \frac{17\,\Omega}{56\,\Omega} \times 400\ \text{mV} = 121.4\ \text{mV}$$

The output of the zener diode would equal 5.2 V with a 121.4-mV ripple.

Referring to Figure 3.26, determine the output ripple voltage.

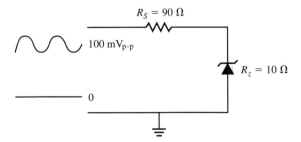

FIGURE 3.26

Progress Check
*You have now completed
objective 5.*

3.5 VOLTAGE MULTIPLIERS

Figure 3.27 shows an AC to DC power supply. An AC to DC power supply has
the disadvantage of having its maximum output voltage limited to about the
peak value of the AC line voltage. Using diodes, however, we can obtain voltages higher than the peak of the AC line voltage. With the addition of diodes
and capacitors to the AC to DC power supply in Figure 3.28, we have a **voltage multiplier** circuit called a **voltage doubler.**

As the name suggests, the circuit provides a DC output voltage twice the
peak of the AC line voltage. We could also call this circuit a peak-to-peak detector. Two half-wave rectifier circuits work together to give the combined
voltage. In Figure 3.28, point A is positive with respect to point B. With that

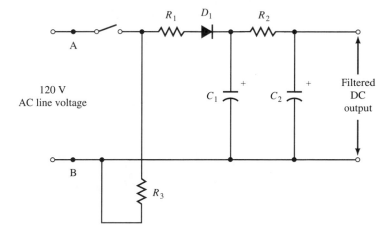

FIGURE 3.27 **AC to DC power supply**

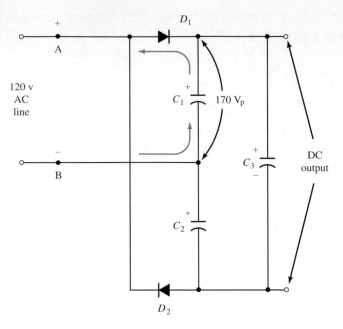

FIGURE 3.28 Converting the AC/DC power supply into a voltage multiplier. First alternation.

condition, the anode of diode D_1 goes positive with respect to the cathode. Simultaneously, the cathode of D_2 goes positive with respect to the anode. As we know, only diode D_1 can conduct. With D_1 conducting, capacitor C_1 charges through D_1 to the peak of the AC line voltage.

On the other alternation, opposite conditions exist. As shown in Figure 3.29, point A is now negative with respect to point B. While the anode of diode D_1 goes negative with respect to its cathode, the anode of diode D_2

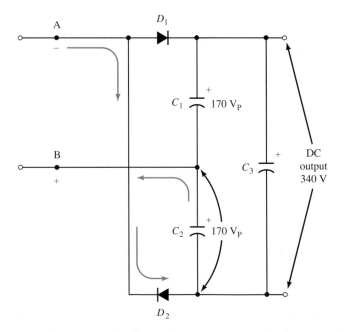

FIGURE 3.29 The voltage multiplier during the opposite (second) AC voltage alternation

goes positive with respect to its cathode. Only diode D_2 conducts. With D_2 conducting, capacitor C_2 charges through D_2 to the peak of AC line voltage.

Diodes D_1 and D_2 conduct alternately and charge both capacitors C_1 and C_2 to the peak value of the AC line voltage. The polarities of the voltages across the capacitors C_1 and C_2 show that the voltages are series-aiding. Shown as the output across the filter capacitor C_3, the sum of the voltages equals two times the peak of the AC line voltage.

With the circuit arrangement of Figures 3.28 and 3.29, one side of the AC line is not the common output. This will affect equipment operation by allowing stray 60-Hz signals to appear in the equipment. Figure 3.30 shows a voltage doubler circuit that overcomes this problem. Again, two half-wave rectifier circuits make up the entire circuit, and we see similar circuit action.

Let's examine the circuit action in Figure 3.30a. Point A is negative with respect to point B. Under these conditions, the cathode of D_1 goes negative with respect to its anode, while the anode of diode D_2 goes negative with respect to the D_2 cathode. Only diode D_1 conducts. Capacitor C_1 charges to the peak of AC line voltage through D_1.

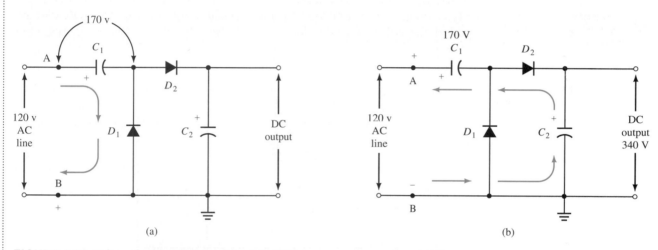

FIGURE 3.30 Voltage doubler with one side of the AC line as a common output. (a) First alternation. (b) Second alternation.

Figure 3.30b shows the circuit action during the other alternation of line voltage. With these conditions, only diode D_2 conducts. Since capacitor C_1 lies in series with diode D_2 and the AC line voltage, the AC line voltage plus the voltage across C_1 becomes applied across diode D_2. The voltage across capacitor C_1 series-aids the line voltage during this alternation. Capacitor C_2 charges to the peak of the line voltage plus the voltage across capacitor C_1. This sum of voltages equals twice the peak of the line voltage.

VOLTAGE TRIPLERS AND QUADRUPLERS

Interconnecting additional diode-capacitor sections will form voltage multipliers that produce a DC voltage three, four, or more times greater than the peak of the AC input voltage. Figure 3.31 shows a voltage tripler circuit. With this circuit we see a half-wave rectifier circuit, D_3-C_3, added to the half-wave voltage doubler circuit from Figure 3.30. Output voltages from the two circuits sum together and become applied across the load resistance.

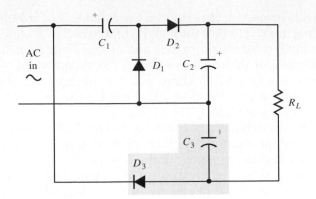

FIGURE 3.31 Voltage tripler circuit

Figure 3.32 illustrates a voltage quadrupler circuit. Compare Figures 3.32 and 3.30. Notice that the quadrupler circuit connects the inputs of two voltage multiplier circuits in parallel with the outputs connected in series. Typically, voltage tripler and quadrupler circuits generate the high DC voltage needed for color cathode ray tubes at the expense of lowering available current.

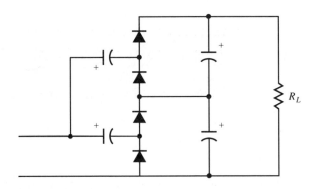

FIGURE 3.32 Voltage quadrupler circuit

✓ **Progress Check**
You have now completed objective 6.

REVIEW SECTION 3.5

1. A voltage doubler provides a DC output voltage _____ times the peak of the AC line voltage, without the need of _____ .
2. What is another name for a voltage doubler?
3. In Figure 3.28, C_2 is in parallel with C_1 and in series with C_3. True or false?
4. In Figure 3.28 C_1 and C_2 voltages are series-aiding; their combined voltage equals the voltage of C_3. True or false?
5. In Figure 3.28, if V_{in} = 240 V, what is the DC output voltage?
6. Which circuit will cause more 60-Hz interference, Figure 3.28 or Figure 3.29?
7. What is the voltage of C_1 in Figure 3.28?
8. In Figure 3.28, can D_1 and D_2 both be conducting at the same time?
9. What is the major disadvantage of voltage multipliers?

10. A voltage doubler has _____ diodes. A voltage tripler has _____ diodes. A voltage quadrupler has _____ diodes.
11. In Figure 3.31, if the input voltage is 120 V_P, what is the DC output voltage?
12. In Figure 3.32, if the input voltage is 120 V_P, what is the DC output voltage?
13. What is a typical application of voltage triplers and quadruplers?

3.6 DIODES USED AS SIGNAL CLIPPERS

Diodes also work within circuits called **signal clippers.** A signal clipper eliminates part of a waveform and passes only the signal that occurs above or below a given voltage or current level. Applications include limiting excessive amplitude, waveshaping, and controlling the amount of power delivered to a load.

Earlier, we studied the half-wave rectifier circuit. A half-wave rectifier works as a signal clipper by eliminating an entire alternation. Figure 3.33 shows two half-wave rectifiers configured as signal clippers. In Figure 3.33a, the diode allows only the positive half of the sine wave to reach the output. Therefore, the circuit functions as a **negative peak clipper.** Reversing the diode polarity, as seen in Figure 3.33b, allows only the negative half of the sine wave to reach the output. With this in mind, we can refer to the circuit shown in Figure 3.33b as a **positive peak clipper.** In each case, zero is the clipping reference point. For the first circuit, any level more positive than zero passes to the output. Any level more negative than zero passes to the output in the second circuit.

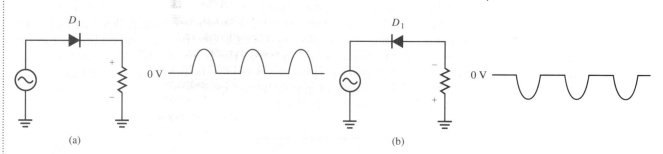

FIGURE 3.33 (a) Negative peak signal clipper. (b) Positive peak signal clipper.

We can reconfigure the circuits shown in Figure 3.33 so that the diode works in parallel with the load resistance. The diodes in Figure 3.34, are **shunt clippers.** A shunt clipper requires an extra resistor to protect the diode from excessive current flow during forward biasing. Along with the additional resistor, another difference between signal and shunt clippers remains.

When the diodes in Figure 3.34 conduct, no appreciable difference in the output appears if the load in the shunt clipper is at least 100 times larger than the series resistance. Indeed, because of the voltage divider action that occurs when a diode is reverse-biased, a larger ratio of R_s to R_c gives better results. With a forward bias applied to the diode, the diode action shorts out the load and eliminates the positive alternation. Again looking at Figure 3.34,

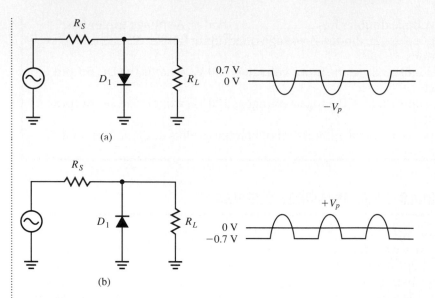

FIGURE 3.34 (a) Positive shunt clipper. (b) Negative shunt clipper.

we can see a small portion of the positive cycle at the output of the shunt clipper. The barrier potential of the diode provides this small, positive voltage. In effect, a forward-biased silicon diode appears as a 0.7-VDC source placed across the load, while a germanium diode works as a 0.3-VDC source.

On the other hand, a series clipper eliminates the positive alternation by preventing any current flow during the clipping. With no positive alternation, no current can reach the load. Consequently, the entire positive alternation is clipped. In most circuit applications, the clipping of the positive alternation can work as an advantage.

In Figure 3.35, a forward-biased diode allows current flow through the load and produces the negative half-cycle. During this half-cycle, the barrier potential of the diode series opposes the potential of the source voltage. Thus, the voltage across the load equals the peak voltage of the negative alternation minus the barrier potential voltage.

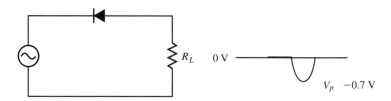

FIGURE 3.35 Series clipper producing a negative half-cycle

BIASED CLIPPERS

Figure 3.36 depicts a DC source placed in series with a diode shunt clipper. In Figure 3.36a, the DC source reverse biases the diode. Conduction will occur only if the AC signal can overcome both the DC source voltage and the diode barrier potential voltage. The diode stays in an off condition for a greater period of time; a larger amount of input signal reaches the output.

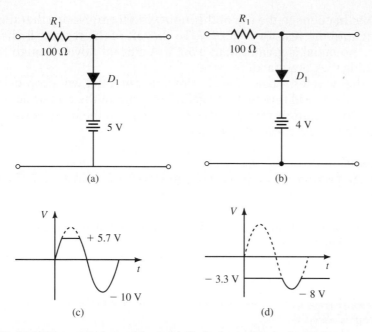

FIGURE 3.36 (a) A DC source reverse biases the diode shunt clipper. (b) A DC source forward biases the diode shunt clipper. (c) Oscilloscope waveform showing negative clipping. (d) Oscilloscope waveform showing positive clipping.

The diode is forward-biased in Figure 3.36b. As depicted in Figure 3.36d, the diode stays on longer and less of the input signal reaches the load.

Apply an input signal to Figure 3.36a. As the signal goes positive, the diode remains reverse-biased until the input signal voltage is great enough to overcome the barrier potential and battery. As long as the diode is reverse-biased, the input signal will be applied to the load. However, as soon as the diode is forward-biased, the input signal is clipped to the sum of the battery and barrier potentials. Figure 3.36c illustrates the resulting output signal. All of the negative alternation is passed to the load and part of the positive alternation. Only that part of the signal whose potential is greater than the battery and barrier potentials is clipped.

EXAMPLE 3.12

Determine the output from Figure 3.36a when the DC source voltage connected to the diode is 5 V and the AC input voltage is 10 V_p.

With no signal applied, the diode is reverse-biased by the 5-V battery. As the AC input waveform goes positive, the diode remains reverse-biased until the input waveform can cause the anode to be positive with respect to the cathode. During the time the diode is reverse-biased, the input waveform is felt across the output terminals. The diode is forward-biased when the amplitude of the input waveform is +5.7 V. To overcome the barrier potential, 0.7 V is required plus 5 V to overcome the battery. When the diode is forward-biased, the diode acts like a closed switch. Notice that the diode voltage drop and the battery voltage are across the load terminals. The output voltage will be held to their +5.7-V potential. As long as the input waveform voltage is greater than +5.7 V, the diode will be forward-biased.

What happens to the top of the input waveform? Notice that the voltage to the right of the resistor is +5.7 V. To the left of the resistor is the waveform whose potential is greater than +5.7 V. A current flows through the resistor, dropping the potential difference.

As the waveform continues in time, its potential will drop below +5.7 V, and the diode will reverse bias again. Throughout the remainder of the input waveform, the diode remains reverse-biased and the input waveform voltage will be felt across the load terminals.

Solution
Only the portion of the input voltage waveform above +5.7 V was clipped. Figure 3.36c shows the output waveform.

Figure 3.36b, d shows a DC source that places the diode in forward bias. Now, we need to find when the AC input voltage will shut off the diode. Let's analyze Figure 3.36d.

EXAMPLE 3.13

Let the DC voltage connected to the diode in Figure 3.36b equal 4 V. Let the AC input voltage equal 8 V_p. Determine the output.

With no signal applied to the input, the diode is forward-biased by the battery. At this time, the DC source voltage and the diode barrier voltage are series-opposing, which places −3.3 V across the output. This voltage remains across the output until the diode becomes reverse-biased and the DC voltage becomes electrically removed from the circuit. To stop the diode from conducting, the forward bias of the DC voltage must be removed. Removing the bias occurs when the input voltage waveform goes below 3.3 V. Figure 3.36d shows how the waveform would appear.

PRACTICE PROBLEM 3.6

Determine the output waveforms of Figures 3.37a and b. Label voltage values at maximum and minimum.

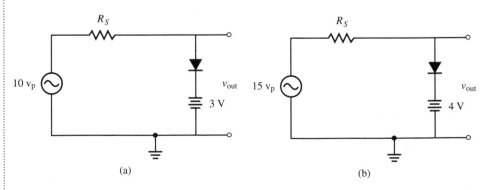

(a)   (b)

FIGURE 3.37

Although the figures in this section depict circuits that have sinewave inputs, clippers will work with any type of input waveform. If a waveform, such as a square wave, changes very fast, the stray capacitance within the circuit will distort the waveform.

NOTE: *In the diode discussions, there have been times when the diode voltage drop was a factor in circuit analysis and times when it was not. When should a technician take the diode drop into consideration? If the biasing potential is 10 or more times greater than the barrier potential, the technician may ignore the diode voltage drop. If, however, the biasing potential is less than 10 times, the technician should consider the diode drop. Why 10 times? Because of the tolerances of the components. Resistors commonly have 5% and 10% tolerances. The diode barrier potential may be from 0.6 to 1.0 V. We will find that tolerances in transistors will also vary greatly. Ten times works well with component tolerances.*

ADDING ANOTHER SHUNT ARM

If we add a negative shunt clipper arm to a positive shunt clipper arm, the circuit will clip on both the positive and negative signal peaks. Figure 3.38 shows both the circuit and the resulting waveform. Voltage sources V_{R1} and V_{R2} provide biasing for their respective diodes. D_1 will begin clipping the input waveform when the input waveform voltage exceeds V_{R1} and D_1 barrier potential. This part of the circuit is the positive clipping circuit. D_2 and V_{R2} will clip the incoming waveform when their combined potential is exceeded in a negative direction. D_2 and V_{R2} form the negative clipping circuit. The resulting output waveform will be clipped on both alternations. By changing the bias battery voltages, we can alter the clipping level on either the positive or negative alternations.

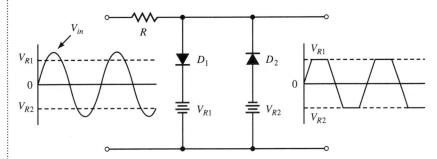

FIGURE 3.38 Adding another shunt arm to the shunt circuit

REVIEW SECTION 3.6
1. What is the name of a circuit that eliminates part of a waveform?
2. Draw a negative peak clipper, and explain how it operates. Also show the output waveform.
3. Draw a positive peak clipper, and explain how it operates. Also show the output waveform.
4. Predict the output waveform and voltage in the following circuit.

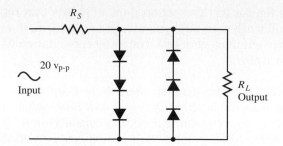

5. What is the name of the circuit in Figure 3.36a?
6. In Figure 3.36b, if the DC supply increased to 6 V what would the new waveform look like?
7. In a shunt clipper, when the diode is conducting (forward-biased), the voltage across the load is assumed to be approximately 0.7 V for silicon and 0.3 V for germanium. True or false?

3.7 BREAKDOWN VOLTAGE LIMITING

We know that a reverse-biased diode exhibits a high resistance until it reaches the breakdown voltage. At the breakdown point, the diode has high conductivity. Because of this characteristic, the breakdown voltage can provide very effective limiting action.

Although any diode could work for breakdown voltage limiting, zener diodes give the best results because their breakdown is controllable. Once a signal diode enters breakdown, its slope is so steep that it very quickly ex-

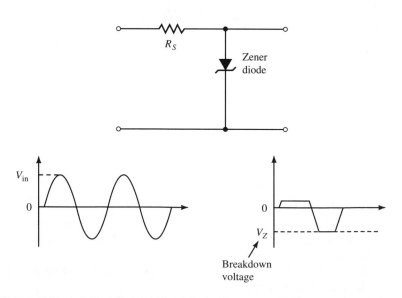

FIGURE 3.39 Breakdown voltage limiting circuit based on a zener diode

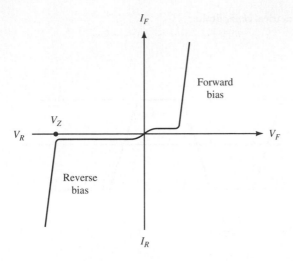

FIGURE 3.40 Voltage-current characteristic curve of a zener diode

ceeds P_{Dmax} due to a very sudden increase in I_R. Ordinary diodes have desirable forward-bias characteristics and high reverse-bias breakdown voltage levels. Zener diodes function in the reverse-biased condition. Figure 3.39 shows a simple limiting circuit based on a zener diode.

Viewing Figure 3.40, we see the voltage-current characteristic curve for a typical zener diode. During the positive voltage swing, the zener diode becomes forward-biased and conducts as an ordinary diode would. On the negative voltage swing, though, the zener diode becomes reverse-biased until the input level reaches the breakdown voltage point. Then, the diode conducts heavily in the reverse region and clips the negative alternation at the breakdown voltage point.

Figure 3.41 shows another variation of using zener diodes for clipping. Instead of using zener diodes only in the reverse-biased condition, the **symmetrical limiter** employs back-to-back zener diodes. In the circuit, diode D_2 clips the positive portion of the input wave, while diode D_1 clips the negative portion.

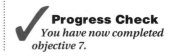

Progress Check
You have now completed objective 7.

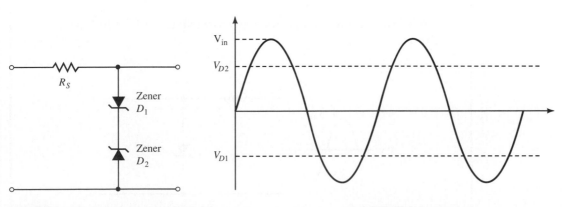

FIGURE 3.41 Symmetrical zener limiter

(continued on next page)

FIGURE 3.41 Symmetrical zener limiter (continued)

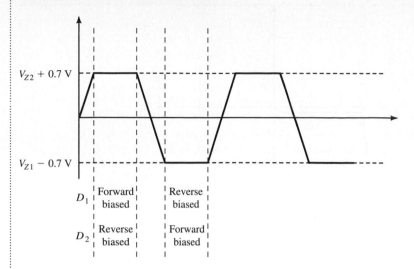

$V_{Z2} + 0.7\ V$

$V_{Z1} - 0.7\ V$

D_1 | Forward biased | Reverse biased

D_2 | Reverse biased | Forward biased

REVIEW SECTION 3.7

1. When a zener diode is forward-biased, what is its voltage drop?
2. When a zener diode is reverse-biased, but before breakdown, it acts like a(n) _____ (open, short).
3. Predict the output voltage and wave shapes of the following circuit.

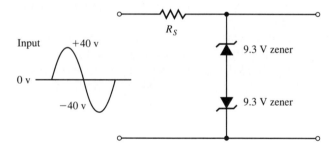

Input

+40 v

0 v

−40 v

R_S

9.3 V zener

9.3 V zener

4. Predict the output voltage and waveform shapes of the following circuit.

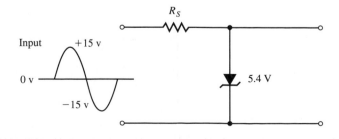

R_S

Input

+15 v

0 v

−15 v

5.4 V

Sometimes, a circuit design requires the combining of an AC voltage with a DC voltage. Instead of changing the shape of the waveform, the operation changes the level of the DC reference voltage. We call this process **clamping,** or **DC restoration.** Clamping circuits are also called **base-line stabilizers, DC filters,** or **DC inserters.**

Figure 3.42 shows a 4-V_{p-p} sine wave centered on a 2-VDC level. In this illustration, the center value of the sine wave has shifted in the positive direction by several volts. We call this a **clamped positive sine wave.** A circuit that shifts waveforms is called a **signal-clamping circuit.** In this section, we'll study several types of signal-clamping circuits.

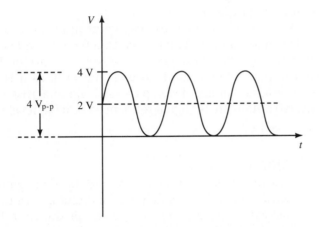

FIGURE 3.42 Four-volt peak-to-peak sine wave riding on a 2-VDC level

POSITIVE CLAMPER

Figure 3.43 shows a positive clamper. Note the position of the capacitor. In circuits that use diodes as peak detectors, the capacitor works in parallel with the load. In the positive clamper circuit, the capacitor lies in series with the source voltage. As the diode is forward-biased, the capacitor charges and the voltage has the illustrated polarity.

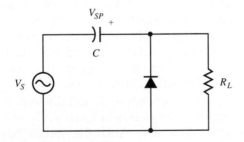

FIGURE 3.43 A positive clamper

After the capacitor has quickly charged, the diode is reverse-biased. Electrically it is removed from the circuit. It remains electrically removed from the circuit except during the short time near the peak voltage of the negative

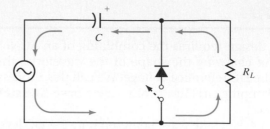

FIGURE 3.44 Diode electrically off

alternation. At that time, the AC voltage overcomes the DC voltage of the capacitor and the barrier potential of the diode. With a fully charged capacitor, the circuit resembles Figure 3.44.

From the perspective of the load, the capacitor appears as a DC source voltage equal to the peak voltage. With the load showing the marked polarity, current flows in a counterclockwise direction around the circuit. Electrically, the diode placed in parallel with the load does not exist as it is reverse-biased. Also, the AC source voltage develops an AC signal across the load. As the figure shows, we obtain an AC signal referenced to a DC level equal to the peak voltage.

NEGATIVE CLAMPER

If we reversed the terminals of the diode, we would have a negative clamper circuit. Voltage polarity across the capacitor would change. In turn, current would flow in a clockwise direction around the circuit shown in Figure 3.43. The top of the load would become negative with respect to ground. As a result, the entire sine wave would shift below ground reference.

CLAMPING AT OTHER THAN PEAK VOLTAGES

We can also clamp waveforms to voltages other than the peak voltage of the input waveform. Figure 3.45a illustrates a circuit that will clamp the input waveform to +15 V. Figure 3.45a shows the charge path for the capacitor. Observe that the negative alternation and the battery are series opposing. That is, they are pushing current in opposite directions. Using Kirchhoff's voltage loop, the capacitor charges to 15 V. Once the capacitor is fully charged, the diode is reverse-biased, which effectively opens that leg of the circuit. Switches are shown in Figure 3.45b and c to illustrate that no current flows in the diode leg. The switches are symbolic and not physically located in the circuit. During the positive alternation, Figure 3.45b shows that the source and the capacitor are series aiding. That is, they each are pushing current the same way. Their respective voltages add, and the load resistor drops +35 V. During the next negative alternation, Figure 3.45c shows that the source and capacitor are series opposing. Since the source has more push (the greater potential), the current flows through the resistor in the opposite direction. The load resistor drops −5 V. Figure 3.45d illustrates the resulting waveform. The output waveform peaks at +35 V and then −5 V. The peak-to-peak amplitude of the output waveform is the same as the input waveform. However the DC reference moved from 0 V on the input to +15 V on the output.

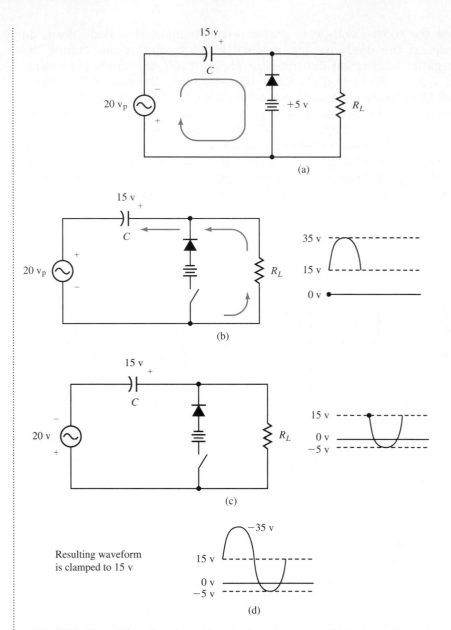

FIGURE 3.45 (a) Charging of capacitor in clamping circuit. (b) Positive alternation of clamping circuit. (c) Negative alternation of clamping circuit. (d) Clamped waveform.

Once the capacitor is effectively charged, the diode is reverse-biased, which effectively opens that leg of the circuit. Switches are shown in Figure 3.45b and c to illustrate that no current flows in the diode leg. The switches are symbolic and not physically located in the circuit.

EXAMPLE 3.14

Using Figure 3.46a, determine the output voltage, if the source waveform voltage is 20 V peak.

Since the source voltage is greater than 10 times the diode drop, do not consider the diode barrier potential in your calculations. During the first negative alternation illustrated in Figure 3.46b, the diode is forward-biased. Current flows in a clockwise direction from the top of the source, through the capacitor, then through the diode, and back to the bottom of the source. The capacitor will charge to the source peak voltage of 20 V. The left plate will be negative in respect to the right plate of the capacitor. With +20 V on the cathode of the diode, the diode is reverse-biased. The load resistor must be large so that the capacitor, which cannot discharge through the diode, will discharge through the resistor very slowly. Since the capacitor is able to hold its charge, it appears as a DC voltage source.

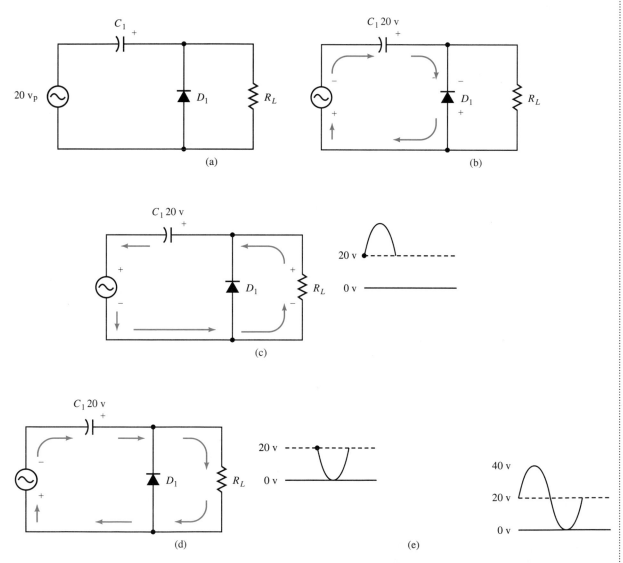

FIGURE 3.46 (a) Positive clamping circuit. (b) Charging of the capacitor in the $+V_P$ clamping circuit. (c) Positive alternation of the $+V_P$ clamping circuit. (d) Negative alternation of $+V_P$ clamping circuit. (e) $+V_P$ clamped waveform.

Look at Figure 3.46c. With the capacitor fully charged, it acts like a DC voltage source. The capacitor's charge keeps the diode reverse-biased most of the time. The diode will conduct only if the capacitor partially discharges. The AC source will now be a series-aiding source during its positive alternations and a series-opposing source during its negative alternations. When the source is at the peak of the positive alternation, Kirchhoff's voltage loop can help show what is happening. Observe that the polarities of both the source and the capacitor are pushing current in the same way. They are series aiding. Their voltages add and are dropped by the load. Since the capacitor is at 20 V and the source's $+V_p$ is also 20 V, the voltage dropped by the load resistor is 40 V.

Now use Figure 3.46d. During the negative alternation of the source, observe that the polarities of the capacitor and source are pushing current against each other. That is called series opposing. Since both have the same but opposing potentials, the resulting voltage felt by the load resistor is zero.

Of course, the capacitor cannot hold its charge indefinitely. During each negative alternation of the source, the diode will be forward-biased for a short time, recharging the capacitor to 20 V.

Solution
The output waveform, Figure 3.46e, is an AC waveform whose $+V_p$ is 40 V and whose $-V_p$ is 0 V.

PRACTICE PROBLEM 3.7

Determine the output waveform of the figures below. Label voltage values at minimum and maximum.

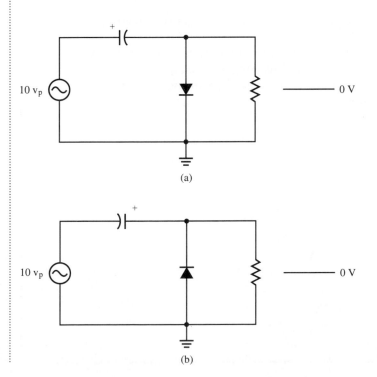

(a)

(b)

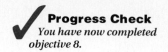
Diode clamping circuits provide simple and inexpensive DC level control. Simply changing the diode orientation or the bias battery potential gives a wide range of clamping levels. However, diode clamping circuits are susceptible to loading and will not work with nonperiodic waveforms.

REVIEW SECTION 3.8

1. What are three other names for a clamping circuit?
2. What is the difference between a clipper and a clamper?
3. Draw a clamped positive sine wave of 6 V_{p-p}, riding on a +4 VDC level.
4. A circuit that shifts waveforms is called a _____ _____ circuit.
5. In a clamper circuit, the capacitor is in _____ (series, parallel) with the voltage source.
6. Draw a positive clamper, and explain how it operates.
7. Is the following circuit a positive or a negative clamper?

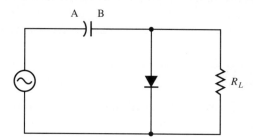

8. In the circuit above, which side of the capacitor will charge negative, A or B?
9. If the anode is connected to the capacitor in a clamper, it is a _____ (positive, negative) clamper.
10. Label the following waveforms as a clipper, positive clamper, or negative clamper:

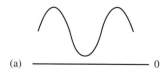

(a)

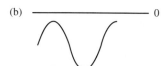

(b)

(c)

3.9 UNBALANCED LOADS AND THE DC RETURN

A diode and capacitor combination can shut down circuit operation if the load is unbalanced. When the resistance of a load is greater for one half-cycle than the resistance during the other half-cycle, an **unbalanced load** occurs. Figure 3.47 shows an unbalanced load.

In this circuit, unwanted charging occurs when using a capacitively coupled source. The internal capacitor within the source charges during positive alternations. Unfortunately, during negative alternations the diode is reversed-biased and no discharge path for the capacitor exists. During the positive alternation, the diode becomes forward-biased, and the load appears resistive at the value of the load resistor. When reverse biasing of the diode occurs, the load seems as if it is an infinite resistance, and the capacitor cannot discharge. When the capacitor charges to the peak voltage, it series-opposes the source. From the perspective of the diode, the net effective voltage is zero. The diode is then open for both alternations, and no output signal reaches the load.

We can overcome the unwanted effect of the unbalanced load by inserting a DC return resistor. Figure 3.48 shows a circuit with a DC return resistor. This resistor allows the capacitor to discharge while the diode is reverse-biased. To be effective, the return resistor must be small enough to allow the capacitor to discharge. Since the capacitor charges through resistor R_L in Figure 3.48, the DC return resistor must have a smaller value than resistor R_L.

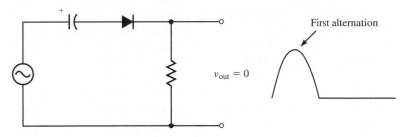

FIGURE 3.47 The effects of an unbalanced load on coupling capacitors

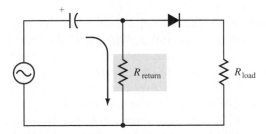

FIGURE 3.48 An unbalanced load with a DC return resistor. The arrow shows the discharge path.

REVIEW SECTION 3.9
1. What is meant by the term *unbalanced load?*
2. What is the function of R_{return} in Figure 3.48?

Progress Check
You have now completed objective 9.

Caution: Checking voltage and current levels in any circuit can be dangerous. This danger exists especially when testing the power supply section of an electronic device. Use extreme care, and follow these safety procedures at all times:

1. Remove all jewelry.

2. Do not touch any exposed circuit.

3. Insert test probes with one hand only. Do not insert probes with both hands. Keeping one hand behind your back or under the test bench will minimize the chances of current flowing through your heart should you accidentally touch a "live" circuit.

4. Take your time.

5. Be certain about what you do. If you are unsure about a procedure, *ask for help.*

Whenever an electronic system malfunctions, good troubleshooting techniques involve first checking the power supply. Every electronic system—whether a television receiver, microcomputer, or robotic controller—relies on its power supply for a given level of voltage and current. Cutting off or disrupting that level of voltage and current may cause symptoms that will lead a technician to the wrong conclusions. Many times, checking the power supply voltage, current, and signal levels will save a technician valuable time.

A fundamental step when troubleshooting a power supply is measuring the DC output voltage. If the DC output voltage check gives an abnormal measurement, the technician can narrow the search to two basic areas. One device in the power supply may have malfunctioned, or circuitry attached to the power supply may have developed a problem that causes the power supply to act differently.

If the output voltage of the power supply does not approximate a level specified in the manufacturer's schematic, disconnect the load circuitry from the power supply. Then, connect a dummy load, a resistance that would equal the Thevenin resistance seen by the power supply when connected to the original load circuitry. After attaching the dummy load, remeasure the output voltage. If the voltage returns to normal, then the search leads to the attached circuitry. Otherwise, the power supply has a faulty part. When we look at the following symptoms, we'll assume that our tests have taken us back to the power supply.

Progress Check
You have now completed objective 10.

TYPICAL FAULTS IN POWER SUPPLY CIRCUITS

Symptoms: No output voltage; blown fuse

Several possible causes—such as a leaky or shorted filter capacitor, leaky or shorted diodes, or a shorted transformer turn—may exist. Using a simple analog ohmmeter, we can check the filter capacitor, diodes, and fuse.

Using the voltage from the meter battery to charge, a capacitor will move the meter needle toward the infinite resistance mark. A shorted or leaky capacitor will not move the needle to infinity. In addition to using the ohmmeter, we can also use a capacitance, or Z, meter to check the capacitor for both value and leakage. Always replace a filter capacitor with the same value of farads and working voltage.

Unfortunately, finding a shorted transformer turn with an ohmmeter can be tricky. First, test to see if there is a short between the winding and the case of the transformer. If a short is found, replace the transformer. If no short is found, then apply a low-voltage AC signal to the primary of the transformer. Use an AC voltmeter to determine if there is an AC signal across the secondary. If not, you have a bad transformer and it needs to be replaced.

Symptom: No output; fuse is OK

If the transformer secondary voltage measures within the normal range, check for open rectifier diodes. A zero secondary voltage may show up at or before the primary windings of the transformer. If a check at the transformer primary also gives a zero reading, disconnect the power to the electronic equipment. Basics such as checking for a defective line switch, broken AC line cord, or defective outlet can provide answers.

Symptom: Low voltage and/or excessive ripple

As we saw in this chapter, ripple can be caused by poor filtering. Check for open diodes or a bad filter capacitor. When filter capacitors age, the electrolyte inside the capacitor dries. If checks of the diodes and filter capacitor display no problems, disconnect the original circuitry and attach a dummy load. If the ripple disappears, look for shorts or excessive loading in the regular load.

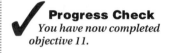

Progress Check
You have now completed objective 11.

TROUBLESHOOTING QUESTIONS
1. List five safety procedures to follow when working on a power supply.
2. If the DC voltage of a power supply is abnormal, in what two major sections could the problem be?
3. What are some possible causes for a blown fuse?
4. What does a circuit with a good fuse but no output indicate?
5. What symptom would a power supply with an open filter capacitor offer?
6. Draw a block diagram of a power supply, and explain the function of each stage, with expected input and output waveforms.

WHAT'S WRONG WITH THIS CIRCUIT?

The following circuit is a bridge rectifier power supply with a capacitive input filter and a zener regulator. The normal readings are:

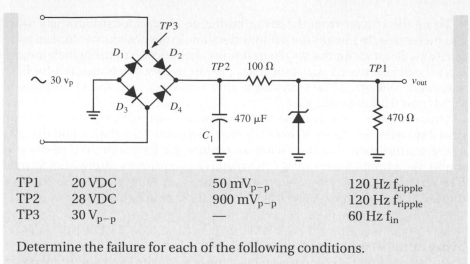

TP1	20 VDC	50 mV$_{p-p}$	120 Hz f$_{ripple}$
TP2	28 VDC	900 mV$_{p-p}$	120 Hz f$_{ripple}$
TP3	30 V$_{p-p}$	—	60 Hz f$_{in}$

Determine the failure for each of the following conditions.

Problem 1

TP1	19 VDC	130 mV$_{p-p}$	60 Hz f$_{ripple}$
TP2	25.9 VDC	1.7 V$_{p-p}$	60 Hz f$_{ripple}$
TP3	32 V$_{p-p}$	—	60 Hz f$_{in}$

Problem 2

TP1	0 VDC	0 mV$_{p-p}$	0 Hz f$_{ripple}$
TP2	30 VDC	Not measurable	60 Hz f$_{ripple}$
TP3	30 V$_{p-p}$	—	60 Hz f$_{in}$

Problem 3

TP1	13.3 VDC	18 V$_{p-p}$ (full wave)	120 Hz f$_{ripple}$
TP2	17.7 VDC	30 V$_{p-p}$	120 Hz f$_{ripple}$
TP3	30 V$_{p-p}$ (half wave)	—	60 Hz f$_{in}$

Problem 4

TP1	22.7 VDC	440 mV$_{p-p}$	120 Hz f$_{ripple}$
TP2	27.5 VDC	560 mV$_{p-p}$	120 Hz f$_{ripple}$
TP3	30 V$_{p-p}$	—	60 Hz f$_{in}$

SUMMARY

Most of today's electronic equipment requires DC operating voltages that may range below and above the AC line voltage. Because of this, modern power supplies have many different components. Some of these components are transformers, diodes, and capacitors.

Half-wave and full-wave rectifier circuits rectify the AC voltage to a pulsating DC voltage. A half-wave rectifier circuit allows current to flow to the load for only one alternation of the AC voltage. Full-wave rectifier circuits allow current to flow to the load for the full duration of each AC cycle.

Two types of full-wave rectifier circuits are used in electronic equipment. With the center-tapped full-wave rectifier circuit, one side of the load is connected directly to the center-tap of the secondary of the transformer, while the other side connects to the rectifiers. Each rectifier diode conducts on opposite AC alternations.

With the full-wave bridge rectifier circuit, four rectifier diodes provide rectification for the complete AC cycle. Instead of requiring a tapped transformer secondary, the full-wave bridge circuit rectifies the full secondary voltage on each alternation. The full-wave bridge rectifier has the advantage of using all the amplitude of the transformer secondary rather than only half the secondary amplitude as in a full-wave rectifier.

Even though diodes convert the AC voltage to pulsating DC voltage, the "rough" output is not suitable for most modern electronic applications. Capacitors and resistors make up filters. Each type of component works to minimize the size of the DC voltage output pulses.

Modern power supplies also require regulation or a constant voltage under varying load conditions. Because of several characteristics, zener diodes work well for power supply regulation.

Other circuits use diodes to multiply voltage. Rather than have the maximum output voltage limited by the peak value of the AC line voltage, a designer may use diodes to obtain higher voltage values. Voltage multiplier circuits may double, triple, or quadruple output voltages. Common uses for voltage multipliers include supplying the high DC voltage needed for color cathode ray tubes.

Other circuits use diodes as clippers. A clipper passes only the signal occurring above or below a given voltage or current. Negative and positive peak clippers are used in electronic circuits. One important use for clipping circuits is in an FM receiver. In this instance the clippper is referred to as a limiter. The circuit eliminates unwanted electrical noise picked up from outside sources such as motors, ignition systems, and bad weather conditions. Zener diodes, because of their special characteristics, also work well as limiters.

Some circuit designs require the combining of an AC voltage with a DC voltage. Diodes "clamp" or change the level of the DC reference voltage. Like clippers, signal clampers work on both positive and negative voltage pulses.

Troubleshooting electronic equipment often involves checking the power supply. Because of the voltages, safety should be a primary concern. Finding a fault in the power supply requires the technician to use a systematic approach.

ABOUT DATA SHEETS

Data sheets serve two purposes. First, they are the manufacturer's sales tool. The manufacturer will technically demonstrate how the product will meet your needs Second, they provide engineers and technicians with a necessary source for devide performance characteristics.

As you will observe in Appendix B of this book, there is no standard data sheet. Each manufacturer is free to publish the data in whatever format it chooses. However, in the competitive market of electronic devices, there are several common features found in most data sheets.

Data sheets are usually combined into book form. For example, transistors are usually found in discrete semi conductor data books. Included in a

MAXIMUM RATINGS

Rating	Symbol	Value	Unit
Drain-Source Voltage	V_{DS}	25	VDC
Drain-Gate Voltage	V_{DG}	25	VDC
Gate-Source Voltage	V_{GS}	−25	VDC
Gate Current	I_G	10	mADC
Total Device Dissipation @ $T_A = 25°C$ Derate above 25°C	P_D	200 2	mW mW/°C
Junction Temperature Range	T_J	125	°C
Storage Temperature Range	T_{stg}	−65 to +150	°C

MPF102

CASE 29-04, STYLE 5
TO-92 (TO-226AA)

JFET
VHF AMPLIFIER

N-CHANNEL — DEPLETION

Refer to 2N4416 for graphs.

ELECTRICAL CHARACTERISTICS ($T_A = 25°C$ unless otherwise noted.)

Characteristic	Symbol	Min	Max	Unit
OFF CHARACTERISTICS				
Gate-Source Breakdown Voltage ($I_G = -10$ μADC, $V_{DS} = 0$)	$V_{(BR)GSS}$	−25	—	VDC
Gate Reverse Current ($V_{GS} = -15$ VDC, $V_{DS} = 0$) ($V_{GS} = -15$ VDC, $V_{DS} = 0$, $T_A = 100°C$)	I_{GSS}	— —	−2.0 −2.0	nADC μADC
Gate-Source Cutoff Voltage ($V_{DS} = 15$ VDC, $I_D = 2.0$ nADC)	$V_{GS(off)}$	—	−8.0	VDC
Gate-Source Voltage ($V_{DS} = 15$ VDC, $I_D = 0.2$ mADC)	V_{GS}	−0.5	−7.5	VDC
ON CHARACTERISTICS				
Zero-Gate-Voltage Drain Current* ($V_{DS} = 15$ VDC, $V_{GS} = 0$ VDC)	I_{DSS}	2.0	20	mADC
SMALL-SIGNAL CHARACTERISTICS				
Forward Transfer Admittance* ($V_{DS} = 15$ VDC, $V_{GS} = 0$, f = 1.0 kHz) ($V_{DS} = 15$ VDC, $V_{GS} = 0$, f = 100 MHz)	$|y_{fs}|$	 2000 1600	 7500 —	μmhos
Input Admittance ($V_{DS} = 15$ VDC, $V_{GS} = 0$, f = 100 MHz)	$Re(y_{is})$	—	800	μmhos
Output Conductance ($V_{DS} = 15$ VDC, $V_{GS} = 0$, f = 100 MHz)	$Re(y_{os})$	—	200	μmhos
Input Capacitance ($V_{DS} = 15$ VDC, $V_{GS} = 0$, f = 1.0 MHz)	C_{iss}	—	7.0	pF
Reverse Transfer Capacitance ($V_{DS} = 15$ VDC, $V_{GS} = 0$, f = 1.0 MHz)	C_{rss}	—	3.0	pF

*Pulse Test: Pulse Width ≤ 630 ms; Duty Cycle ≤ 10%.

semiconductor data book would be devices such as diodes, bipolar transistors, JFET transistors, and MOSFET transistors. Devices such as op-amps, comparators, and special analog ICs are usually found in linear device data books. There are also data books for digital devices, for memory chips, and many other categories.

Many data books provide glossaries, indexes, and tables of contents. For example, in order to find the data sheet for a 2N3904, which is an NPN-type general purpose amplifier, a selection guide is usually provided in the first section. The selection guide tells you the page number where you can find the transistor. Also, data books may include a glossary of terms. For example, the term V_{CBO} means the minimum voltage applied to collector-base junction with the emitter open that resulted in damage to the junction. Hence, V_{CBO} should not be exceeded. Another common term in transistor data books h_{FE}. A technician would consider h_{FE} to be the transistor's current gain (β) H_{FE} is usually stated in terms of minimum and maximum at a give collector current (I_C) and a give collector-emitter voltage (V_{CE}). For example, the 2N3904 should gave and H between 100 and 300 when I_C is 10 mA to V_{CE} is 1 V.

Case style is also important to a technician. Cast style provides the physical dimensions of the transistor. A give type transistor may be packaged in different cases. One popular case for general purpose amplifiers t TO-92, which is a molded plastic case with set physical dimensions. A data book will provide a drawing of the TO-92, and the terminals will be labeled *emitter*, *base*, and *collector*.

Data sheets provide a technician with operating characteristics, physical dimensions, and terminal arrangements. Reading data book introductions, glossaries, and tables of contents will provide a technician with the knowledge of how the manufacturer laid out the product information.

PROBLEMS

1. What is the V_{TH} for the diode in Figure 3.49? What is the R_{TH}?

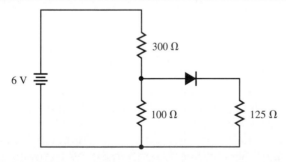

FIGURE 3.49

2. Treating the diode of Figure 3.49 as a piece of wire, determine how much current is flowing through it.
3. Your answer to problem 2 will change when you account for the 0.7-V barrier potential of the diode in Figure 3.49. Recalculate how much current is flowing through the diode.

Use the circuits in Figure 3.50 for problems 4–14.

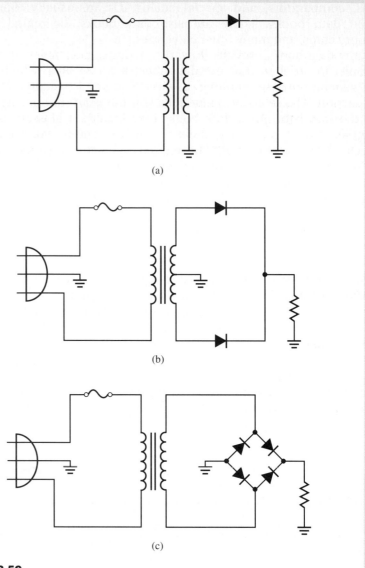

FIGURE 3.50

4. If all three circuits have the same turns ratio, which circuit has the highest DC voltage out?
5. Which circuit has the lowest output frequency?
6. If there is 15 V_{rms} across the secondary of Figure 3.50b, what is the peak inverse voltage across the top diode?
7. Identify the half-wave and full-wave rectifiers in Figure 3.50.
8. How is a bridge rectifier better than a center-tapped full-wave rectifier?
9. The secondary of circuit 3.50b has 22 V_p across it. Determine the peak load voltage.
10. There is an input of 22 V_p across the secondary of Figure 3.50b. Determine the peak current through the load when the load is a

2.2-kΩ resistor. Determine the average DC current through the load.

11. If the voltage across the secondary in Figure 3.50c is 18 V_p, what is the maximum peak inverse voltage felt across the diodes?

12. With 10 V_p across the secondary of Figure 3.50c, determine the average DC current through a load resistor of 820 Ω. Treat the diodes as pieces of wire when forward biased.

13. Repeat problem 12, but this time account for the diode voltage drops.

14. Which of the circuits in Figure 3.50 has an unbalanced load?

Use the circuit in Figure 3.51 for problems 15–17.

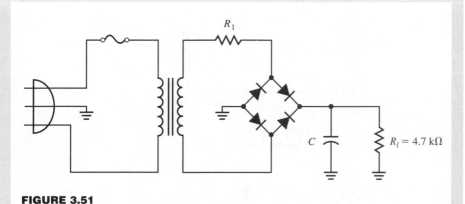

FIGURE 3.51

15. What is the function of R_1 in the circuit?
16. What is the function of the capacitor in the circuit?
17. If the rms voltage across the secondary is 20 V, determine the ripple amplitude when C = 10 μF. Ignore R_1 in calculating your answer.

The zener diode characteristics in Figure 3.52 are:
$$I_z(max) = 70 \text{ mA}, \quad I_z(min) = 7 \text{ mA}$$
Use this information for problems 18 and 19.

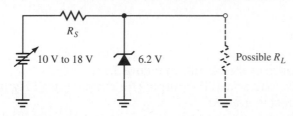

FIGURE 3.52

18. If the input voltage fluctuates between 10 and 18 V, what is the minimum value of R_s?
19. Using R_s from problem 18, what is the smallest load that can be placed across the zener diode without losing regulation?

Use the circuits in Figure 3.53 for problems 20–22.

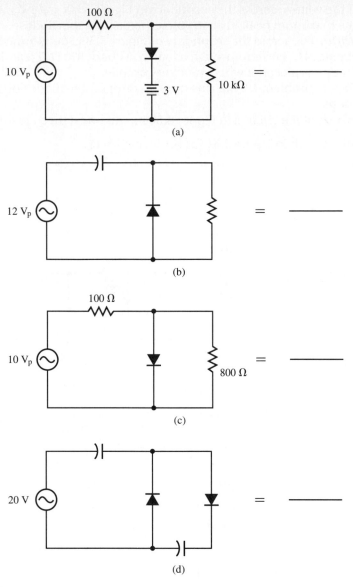

FIGURE 3.53

20. Identify each of the circuits in Figure 3.53.
21. Determine and sketch the outputs for the circuits of Figure 3.53. with respect to ground.
22. What happens to circuit operation if you reverse the diode in the circuit in Figure 3.53c?
23. What electrical parameter is sacrificed when you use a capacitive-diode voltage tripler?
24. Discuss the pros and cons of shunt clippers, comparing them with series clippers.

Use the circuit in Figure 3.54 for problems 25–27.

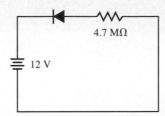

FIGURE 3.54

25. The diode in Figure 3.54 has a reverse resistance of 1 MΩ. How much current is flowing through it?
26. How much voltage is dropped across the 4.7-MΩ resistor in Figure 3.54 if the diode has a reverse resistance of 2 MΩ?
27. Predict what voltage you would read if you placed a DVM with an input impedance of 1 MΩ across the diode. The diode's reverse resistance is 2 MΩ.
28. A diode conducts 1 mA of current at 1.1 V and 21 mA of current at 1.9 V. Determine its dynamic AC resistance.
29. What is the function of a DC return circuit? In what types of diode circuits should a DC return be used?
30. What guideline is used to determine if a diode is working with large signals?

BIPOLAR JUNCTION TRANSISTORS

OBJECTIVES

✓ **As you read this chapter, concentrate on learning how to:**

1. Describe the interior regions of a bipolar transistor
2. Identify NPN and PNP transistors, and tell how they differ
3. Identify the junctions of a transistor and tell how they function
4. Properly bias NPN and PNP transistors for linear operations
5. Define and compute alpha and beta
6. Understand the damaging effects of thermal runaway and how to prevent them
7. Define how a transistor controls current
8. Define cutoff, saturation, and breakdown
9. Describe how to use a beta box to solve transistor circuits
10. Describe the effects of feedback on beta
11. Describe the different types of biasing configurations along with their advantages and disadvantages

INTRODUCTION

Transistors are remarkable devices that can amplify AC signals or switch a device from an on state to an off state and back. Modern electronic equipment uses transistors for signal transmission, video and audio signal reproduction, and voltage regulation. Like the diodes we studied in the previous chapters, transistors begin as a single crystal of silicon or germanium. Also, the transistor features the same techniques used in diode construction—the doping and joining of P- and N-types of semiconductor material. However, the transistor contains *three* sections of semiconductor material and therefore has two P-N junctions and three terminals. Because it has two junctions, it is often called a **bipolar junction transistor** (or BJT).

Two fundamental types of bipolar junction transistors exist. One type, called the **NPN transistor,** consists of a P-type region sandwiched between two N-type regions. Conversely, a **PNP transistor** consists of an N-type region sandwiched between two P-type regions.

Like diodes, transistors will not function without a source of energy. In most cases, this source is an applied DC voltage. Various methods of applying the DC voltages can be used, depending on the task given the transistor. In this chapter, we will look at DC voltage bias arrangements. **DC biasing** is a method of applying voltages to a transistor so that it has the correct DC voltage levels at the various terminals. Without the appropriate bias voltage, a transistor cannot properly amplify AC signals.

As we progress through the chapter, we will discuss the strengths, weaknesses, and limitations of emitter, base, voltage divider, and collector feedback bias arrangements. In addition, we will investigate the DC conditions, DC currents, and DC voltages that must be present before the transistor can amplify AC signals. We will study AC signals and transistor operation in Chapter 5.

4.1   WHAT'S INSIDE THE BIPOLAR JUNCTION TRANSISTOR?

In Figure 4.1, the three regions of a bipolar PNP transistor are labeled as the emitter, the base, and the collector. The two junctions of the transistor are called the **collector-base junction,** or **collector junction,** and the **emitter-base,** or **emitter, junction.** The base section is common to the two junctions.

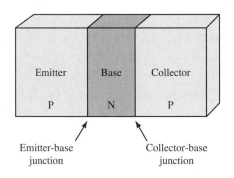

FIGURE 4.1   Block diagram showing the three regions of a bipolar junction PNP transistor

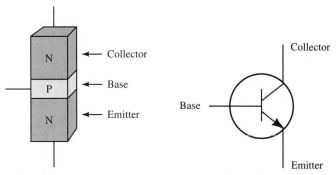

FIGURE 4.2   (a) Cross section of an NPN bipolar junction transistor and (b) its schematic symbol

Figure 4.2 shows the cross section of an NPN transistor and its schematic symbol. In this three-region device a layer of N-type material is the emitter, a layer of P-type material is the base, and a layer of N-type material is the collector. The emitter is very heavily doped and has many free electrons available for current flow. The thinner base region, which is very lightly doped, controls the number of charge carriers that leave the emitter. Depending on whether the transistor is an NPN or PNP type, the charge carrier will be either holes (in the PNP type) or electrons (in the NPN type). The largest region, the collector, often works as the output section of the transistor. The collector is not doped as heavily as the emitter section. It is designed to dissipate heat quickly.

Figure 4.3 shows a PNP transistor and its schematic. Again, the emitter is heavily doped; many free holes work as the charge carriers. The polarities of NPN and PNP transistors are simply reversed and will react oppositely to any voltage that would be applied to them.

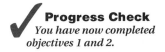

Progress Check
You have now completed objectives 1 and 2.

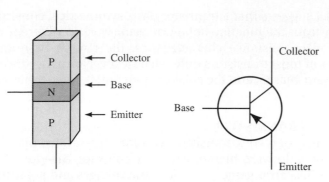

FIGURE 4.3 (a) Cross section of a PNP bipolar junction transistor and (b) its schematic symbol

---

REVIEW SECTION 4.1

1. List three applications for bipolar transistors.
2. The three sections of a bipolar transistor are the _____ , the _____, and the _____ .
3. Electrons are the charge carriers in the _____ (N, P) type of material, and holes are the charge carriers in the _____ (N, P) type of material.
4. The section of the bipolar transistor that is the largest and dissipates the most heat is the _____ .
5. The section of the bipolar transistor that has the heaviest doping is the _____ .
6. The _____ section of a bipolar transistor is very thin.
7. The base of an NPN bipolar junction transistor is _____ (N, P) material.
8. Draw the schematic symbols for the NPN and PNP bipolar transistors. Label the base, collector, and emitter leads.

4.2 THE TRANSISTOR AS TWO BACK-TO-BACK DIODES

A transistor works like two diodes connected back to back. If we considered only simple resistance measurements, Figure 4.4 would be an accurate representation. We cannot explain actual transistor operation in this manner, however, because two diodes connected back to back will not amplify a signal.

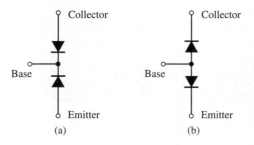

FIGURE 4.4 The transistor as a symmetrical device. (a) A PNP transistor. (b) An NPN transistor.

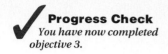
Figure 4.4 suggests that a transistor has a symmetrical construction, but it doesn't. Instead, the junctions of a transistor work in different bias regions. Despite similar resistance characteristics, the emitter-base and collector-base diodes of most transistors differ electrically. Normally, the emitter junction is forward-biased, and the collector junction is reverse-biased.

BIASING A TRANSISTOR

Figure 4.5 shows an NPN transistor with forward-reverse bias. The emitter-base junction is forward biased, while the collector-base junction is reverse biased. This bias arrangement provides the voltages and polarities necessary if the transistor is to be used as an amplifier.

The forward bias across the emitter-base junction reduces the size of the depletion region. This reduces the strength of the electrical field, which makes it easier for electrons to move into the base region of the transistor.

Because the base region of the transistor is very thin and very lightly doped most electrons do not recombine with holes in the P-type material of the base and flow out of the base lead. So, if most electrons leaving the emitter cannot exit from the base of the transistor, they must exit from the collector.

Since the collector-base junction is reverse biased, it expands. This further reduces the "effective" base area, as the collector-base depletion region expands into the base

Therefore, as V_{EE} increases, the number of electrons entering the base region increases. Due to the thinness of the base and the light doping, they "pile up." If enough negative charges "pile up" in the base, they repel electrons that had already reached the base. This effect pushes electrons towards the negative ion barrier near the collector-base junction. Eventually, the opposition of the negative ion barrier next to the collector-base junction is overcome. Electrons flow out of the collector, attracted by the positive terminal of the collector battery, V_{cc}.

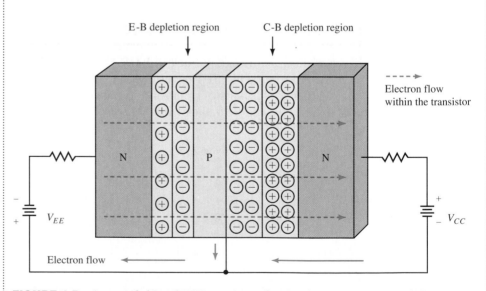

FIGURE 4.5   A properly biased NPN transistor, showing depletion regions and electron flow through the transistor

It is important to remember that the emitter current equals the sum of the base and collector currents:

$$I_E = I_B + I_C$$

The emitter supplies current to the collector, and the base controls the current through the transistor.

Figure 4.6 illustrates the biasing arrangement for a PNP transistor. One supply voltage connected between the emitter and base provides the proper polarity for forward-bias conditions. Another voltage source supplies collector-base bias and ties between the emitter and collector.

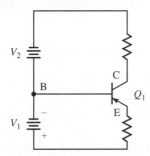

FIGURE 4.6   **A properly biased PNP transistor**

With the supply voltages connected to the junctions, the base is at about 0.7 V negative with respect to the emitter. The collector is biased at 10 V negative with respect to the emitter. With the collector 9.2 V more negative than the base, the base-collector junction is reverse-biased.

We can properly bias transistors in more than one way. Figure 4.7 shows an NPN transistor biased with only one supply voltage. Remember that in an NPN transistor the base is positive with respect to the emitter, and the collector is positive with respect to the base.

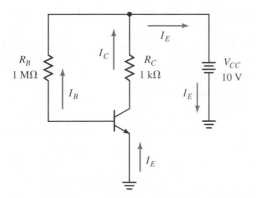

FIGURE 4.7   **An NPN transistor biased with only one supply voltage**

In Figure 4.7, the emitter of the transistor and the negative terminal of supply voltage tie directly to ground. Adding a 1-MΩ resistor, R_B, between the transistor base and the positive terminal of the supply voltage forward biases the base-emitter junction. Although voltage is dropped across the 1-MΩ base resistor, enough potential remains at the base to create the necessary 0.7-V bias condition.

Another resistor, the 1-kΩ R_C, allows the biasing of the base-collector junction. Resistor R_C connects between the collector and positive terminal of the supply voltage.

HOW THE TRANSISTOR CONTROLS CURRENT

One important feature of a transistor is that it controls the current that flows through it. Most of the current flows from the emitter to the collector, with the base acting as the controlling element. With a forward bias applied to the emitter-base junction, the base begins to draw a small current. Once the base current has started, most of the electrons pass through the thin base region and come under the control of the highly positive (in an NPN transistor), reverse-biased collector region.

As we have seen, most of the current flows directly from the emitter to the collector. A small fraction of this current takes a detour and flows into the base circuit. This small base current "turns on" the transistor. With no base current, the transistor cannot turn on.

As we progress through the different biasing methods, remember that the bias voltages connect to the transistor for the generation of a base current. The small base current determines the emitter current, which, in turn, determines the collector current.

Progress Check
You have now completed objective 4.

REVIEW SECTION 4.2

1. For a bipolar transistor to operate as an amplifier, the base-emitter junction must be (forward-, reverse-) biased.
2. Which current in a bipolar transistor is larger, the collector current or the base current?
3. Which current is smaller, the collector current or the emitter current?
4. The emitter current is the sum of the _____ current and the _____ current.
5. Draw a properly biased PNP transistor.
6. In a bipolar junction transistor, the small _____ current determines the _____ current.

4.3   ALPHA AND BETA

ALPHA

In most cases, 95 to 99% of a transistor's current from the emitter flows to the collector. Some transistor specification sheets include a measure of this percentage and express it as the ratio of collector current to emitter current. We call this ratio **alpha,** represented by the Greek letter α, and described by the equation

$$\alpha = \frac{I_C}{I_E}$$

In most of our work, we will assume that alpha equals unity, or 1. In other words, 100% of the emitter current reaches the collector. While this may not be true in all cases, the approximation of using unity for alpha makes the analysis of transistor circuits easier. By assuming that alpha equals 1, we assume that the emitter current value equals the collector current value. So when we solve circuits we can simply use the equation

$$I_E = I_C$$

BETA

Beta is a transistor specification that relates collector current to base current. It is expressed mathematically by the ratio

$$\frac{I_C}{I_B}$$

In transistor specification sheets, it is represented most often by the designation H_{fe} and sometimes by the Greek letter β (beta).

In Figure 4.8, we see that a small base current flows out the base terminal toward the positive plate of V_{BB}. If we knew beta, we could then predict how much base current we would need to control a larger collector current.

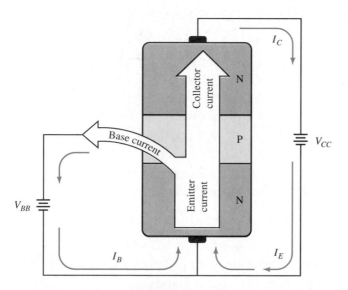

FIGURE 4.8   **A small base current controls the large collector current. Most current flows between the emitter and the collector.**

Values for beta can vary dramatically, typically ranging from 50 to 300. Beta values depend on the operating point of the transistor, the transistor type, the transistor application, and the temperature.

EXAMPLE 4.1

Using the circuit shown in Figure 4.9 as a reference, examine the relationships among currents, alpha, and beta. Solve for an unknown value of emitter, collector, or base current using the known currents, alpha, and beta. Calculate the beta and alpha of the circuit.

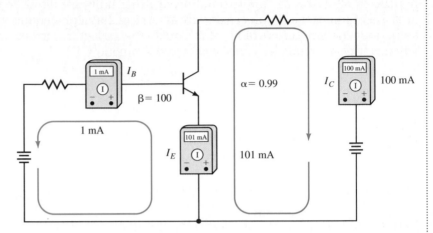

FIGURE 4.9

Kirchhoff's current law states that the emitter current (I_E) equals the sum of the base current (I_B) and the collector current (I_C). Thus, the collector current equals the emitter current minus the base current, and the base current equals the emitter current minus the collector current.

Solution

$$I_E = I_C + I_B = 100 \text{ mA} + 1 \text{ mA} = 101 \text{ mA}$$
$$I_C = I_E - I_B = 101 \text{ mA} - 1 \text{ mA} = 100 \text{ mA}$$
$$I_B = I_E - I_C = 101 \text{ mA} - 100 \text{ mA} = 1 \text{ mA}$$

Beta equals the collector current divided by the base current:

$$\beta = \frac{I_C}{I_B} = \frac{100 \text{ mA}}{1 \text{ mA}} = 100$$

Because beta is a function of two current values, when we know the beta value we can use it to determine values of current:

$$I_B = \frac{I_C}{\beta} = \frac{100 \text{ mA}}{100} = 1 \text{ mA}$$
$$I_C = I_B \times \beta = 1 \text{ mA} \times 100 = 100 \text{ mA}$$

Alpha equals the collector current divided by the emitter current:

$$\alpha = \frac{I_C}{I_E} = \frac{100 \text{ mA}}{101 \text{ mA}} = 0.99$$

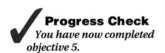

Progress Check
You have now completed objective 5.

THERMAL RUNAWAY

Beta will change as temperature changes. Changes in temperature may result from the temperature of the air, from the material that makes up the transistor, or from electrical conduction, or **self-heating.** As temperature increases, beta increases. Thus, for a fixed base current, the collector current will increase. Since we now have more collector current, there will be more self-heating. This in turn increases beta; thus the cycle repeats, with the transistor getting hotter and hotter. This cycle is known as **thermal runaway.** Unchecked, it can damage the transistor. Properly designed circuits protect against thermal runaway by reducing the effects that beta has on circuit operation.

✓ **Progress Check**
*You have now completed
objectives 6 and 7.*

REVIEW SECTION 4.3

1. Match the following formulas or statements to either alpha (α) or beta (β).

$\dfrac{I_C}{I_B}$ _____

Less than 1 _____

$\dfrac{I_C}{I_E}$ _____

95 to 99% _____

Range from 5 to 300 _____

2. Calculate the values for the emitter current, alpha, and beta in the following circuit.

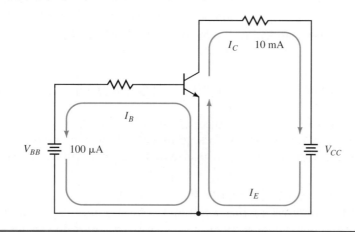

4.4 MORE ABOUT TRANSISTOR CHARACTERISTICS

POWER DISSIPATION IN TRANSISTOR CIRCUITS

In Figure 4.10, each transistor terminal shows an appropriate voltage. Let us use Kirchhoff's voltage law to calculate the voltage difference between the collector and the emitter (V_{CE}):

$$V_{CE} = V_C - V_E = 9.26\,V - 7.41\,V = 1.85\,V$$

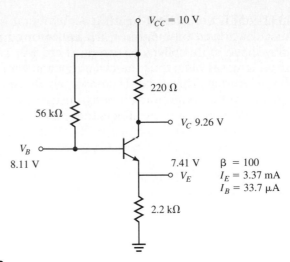

FIGURE 4.10

V_{CE} is the voltage drop across the transistor. We can use the voltage drop along with the amount of current flowing through the transistor (I_E) to determine the power dissipated by the transistor. Watt's law says that power (P) equals current times voltage; therefore:

$$P = I_E \times V_{CE} = 3.37 \text{ mA} \times 1.85 \text{ V} = 6.23 \text{ mW}$$

The 6.23 mW of power dropped across the transistor in the above calculation is due to the effects of DC biasing. You will often find it labeled as P_{DQ}:

$$P_{DQ} = 6.23 \text{ mW}$$

The Q in the abbreviation P_{DQ} represents quiescent. In this situation, quiescent refers to circuit operation in which no AC signal is present.

THE MINIMUM VOLTAGE NEEDED TO REVERSE BIAS THE COLLECTOR-BASE JUNCTION

The voltage measured from the base to the collector (V_{BC}) is another important parameter. We obtain this voltage by subtracting the collector voltage, V_C, from the base voltage, V_B. For the circuit shown in Figure 4.10, we have:

$$V_{BC} = V_B - V_C$$
$$8.11 \text{ V} - 9.26 \text{ V} = -1.15 \text{ V}$$

Never attempt to measure this voltage directly. Doing so places the voltmeter's internal resistance across the reverse-biased junction. Since both the junction and the voltmeter have high internal resistances, circuit operation may be seriously altered, possibly damaging the transistor. Instead, measure the base voltage to ground and the collector voltage to ground and then find the difference. Circuits that feature an NPN transistor should have a negative base-collector voltage with a value of at least 1 V. If the voltage does not appear as expected, the collector-base junction has lost its reverse bias.

VOLTAGES SEEN FROM THE BASE TO THE EMITTER

We can label the voltage from base to emitter as V_{BE}. This voltage equals the base voltage (V_B) minus the emitter voltage (V_E). For a silicon NPN transistor, the voltage is approximately 0.7 V. When measured, it may vary by ±0.1 V.

DC CUTOFF, SATURATION, AND BREAKDOWN

Figure 4.11 shows how the base and collector currents flow in the emitter region. Even though the base current makes up only a small fraction of the entire transistor current, it plays a major role in transistor operation. With no base current, collector current cannot exist. Inside the transistor, the resistance from emitter to collector increases to a very high level. While the full supply voltage appears across the transistor, no voltage appears across the load resistance. In effect, the transistor becomes "turned off," or enters a **cutoff** condition. We label the cutoff current I_{CBO} (collector current with base open).

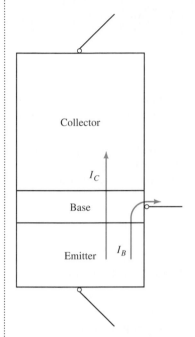

FIGURE 4.11 Electron current flow in the emitter region in an NPN transistor

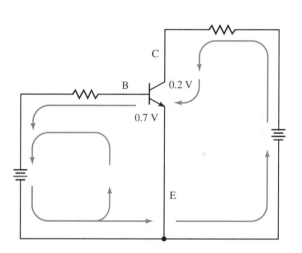

FIGURE 4.12 Electron (current) flow for a saturated NPN transistor

Sometimes, another condition can draw off enough base current so that the emitter-collector resistance decreases. We call this condition **saturation.** During saturation, any further increase in base current will not cause an increase in collector current. Figure 4.12 shows the electron flow for a saturated NPN transistor. Little or no voltage appears across the collector-to-emitter terminals. The amount of base current required for saturation depends on the beta of the transistor.

When the reverse-biased collector-base junction begins to conduct, the voltage from collector to emitter exceeds a value labeled V_{CEO}. We call this point the **breakdown voltage.** The breakdown voltage is the maximum voltage value seen from collector to emitter with the base open. Exceeding the breakdown voltage can permanently fuse the emitter to the collector.

Another breakdown voltage is the V_{EBO} (emitter to base, collector open), or the maximum reverse bias voltage that can be applied to the emitter-base junction. The junction will not sustain a large reverse bias voltage. Large negative voltages may appear at the base in the form of an input signal. If these negative voltages exceed V_{EBO}, the transistor may be damaged.

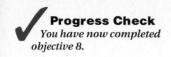

Progress Check
You have now completed objective 8.

Still another reverse bias voltage, called the V_{CBO} (collector to base, emitter open) is a maximum collector-to-base voltage. Exceeding this voltage will cause the breakdown, or failure, of the transistor in one of two ways. First, the transistor could start a reverse bias avalanche. Second, the depletion zone widens as the reverse bias increases. If the reverse bias reaches a high level, the depletion zone will widen until it touches the emitter region of the transistor. This is a shorted condition between the emitter and collector.

THE BETA BOX

Transistor circuit operation can be predicted by performing advanced calculations based on a transistor's solid state characteristics. Fortunately, predictions obtained by using advanced equations are rarely needed by a technician. Simpler, easier-to-use methods can predict circuit behavior, allowing for quick analysis and troubleshooting of a circuit.

In this text, we will use a method of circuit analysis based on a visual representation of equations derived from Kirchhoff's voltage law: Around any closed loop, the sum of voltages must be zero.

With this visual representation, we can solve for circuit values using Ohm's law. We call this method the **Beta Box Model**, Figure 4.13a.

Figure 4.13(b) shows a flowchart that you can use to help you remember the steps in using the model.

BETA BOX RULES

1. The beta box does not drop any voltage. The internal resistance of the beta box is zero.

2. Currents entering and leaving the beta box are not the same value. Although current values may change, certain constant relationships are observed:

    I_C is beta times larger than I_B
    I_C equals alpha times I_E and, since we assume alpha equals 1, $I_C = I_E$, so
    I_E is beta times larger than I_B

3. A node within the beta box obeys Kirchhoff's current law:

$$I_E = I_C + I_B$$

    and

$$I_C = I_E - I_B$$

4. When resistors are in the base and emitter sides of the base-emitter loop, the beta box must be equalized. This means we adjust resistance values by a factor of β to simulate equal currents flowing in all parts of the emitter-base loop. By equalizing the beta box, we can develop a simple Ohm's law equation for the base-emitter loop.

5. If a transistor circuit contains only a base resistor in the control loop, solve for the base current. Then, find I_E by multiplying I_B times beta.

To see how the beta box model can be used in transistor circuit analysis, we'll take a quick look at two examples, then move on to a more detailed examination of biasing in which the beta box model will continue to be used.

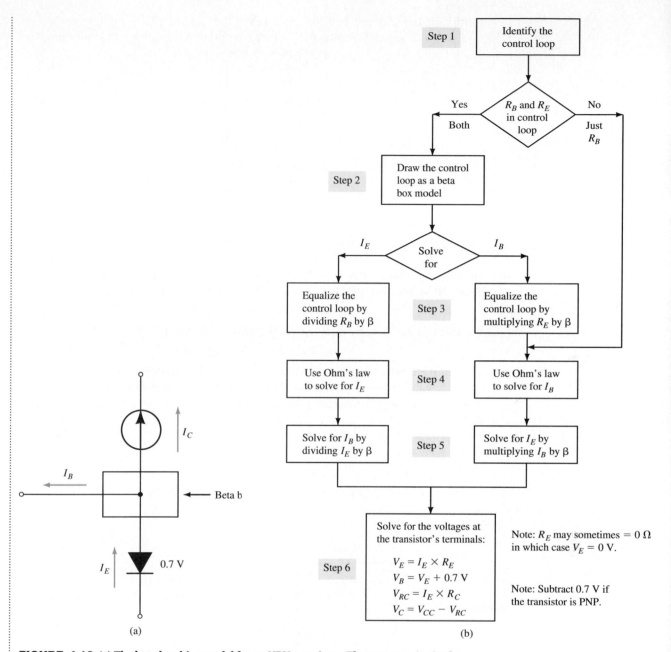

FIGURE 4.13 (a) The beta box bias model for an NPN transistor. The arrows point in the direction of electron flow. (b) Flowchart for beta box model.

The following figure shows a transistor circuit that has a base resistor and a collector resistor. The first step in using the Beta Box Model has already been performed. The base-emitter loop, which is highlighted in the figure, is the control loop for this circuit.

Since the control loop contains only one resistor, we can jump directly to step 5 of the Beta Box Model. Before we can calculate the base current we need to determine V_{RB}, the voltage across the base resistor,

$$V_{RB} = 12\,V - 0.7\,V = 11.3\,V$$

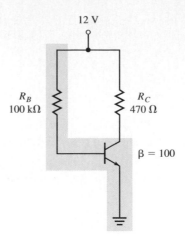

where 12 V equals the supply voltage, and 0.7 V is the base-emitter voltage drop. Next, we can use Ohm's law to determine the base current

$$I_B = 11.3 \text{ V}/100 \text{ k}\Omega = 113 \text{ μA}$$

We then multiply I_b by β to obtain the emitter current

$$I_E = 113 \text{ μA} \times 100 = 11.3 \text{ mA}$$

The figure below shows a transistor circuit with two power supplies; however, only one of these power supplies is in the base-emitter loop. Since the base-emitter loop is the control loop for the circuit, it has been highlighted in part a .

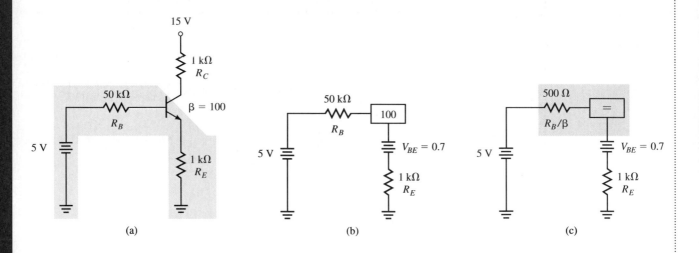

(a)   (b)   (c)

The highlighted loop is redrawn in part b as a beta box loop. The beta box loop looks like a simple series circuit. We could apply Ohm's law to this circuit, *except* the current is not the same everywhere. The current flowing though R_B is β times smaller than the current flowing though R_E . We will *simulate* equal currents by "equalizing" the loop as represented by the equal sign inside of the beta box in part c. The base resistance is divided by beta, so that we can treat the current flowing through this part of the loop as being equal to I_E. This technique works because if R_B became β times smaller, the current flowing through it, I_B, must become β times bigger

Since

$$I_E = \beta \times I_B$$

we have a situation in which the simulated current is the same everywhere, that is, it is equal to I_E everywhere within the beta box loop. We can now apply Ohm's law to the beta box loop.

The 5-V supply is opposed by V_{BE}, therefore the total effective voltage is

$$5\,V - 0.7\,V = 4.3\,V$$

The total effective resistance is the sum of the two resistances within the beta box loop

$$R_T = 500\,\Omega + 1\,k\Omega = 1.5\,k\Omega$$

I_E and I_B are then

$$I_E = 4.3\,V / 1.5\,k\Omega = 2.866\,mA$$
$$I_B = I_E / \beta = 2.866\,mA / 100 = 28.66\,\mu A$$

The voltages at the terminals are

$$V_E = I_E \times R_E = 2.866\,mA \times 1\,k\Omega = 2.866\,V$$
$$V_B = V_E + 0.7 = 3.566\,V$$
$$V_{RC} = I_E \times R_C = 2.866\,mA \times 1\,k\Omega = 2.866\,V$$
$$V_C = V_{CC} - V_{RC} = 15\,V - 2.866\,V = 12.134\,V$$

To summarize, the beta box model allows you to:

1. Use Ohm's law to solve a transistor circuit.

2. Minimize the number of formulas that you need to memorize.

3. Have a visual representation of your calculations, a representation that allows you to "see" how equations based on Kirchhoff's voltage law are used in transistor circuit calculations

Figure 4.13b shows a flowchart that outlines how to use the beta box. Every time we analyze a transistor circuit, we will follow the procedures shown in the flowchart.

Step 1.   Identify the control loop

Step 2.   Draw the control loop as a beta box model.

Step 3.   Equalize the control loop by dividing R_B by β or multiplying R_E times β.

Step 4.   Solve for either I_E or I_B. Each solution uses a beta box loop.

Step 5.   Knowing I_E, divide by β to find I_B. Knowing I_B, multiply by β to find I_E.

Step 6.   Use the current value to predict the terminal voltages of the transistor.

REVIEW SECTION 4.4

1. The power dissipated by a transistor is calculated by the _____ _____ across the collector and emitter multiplied by the _____ flowing through a transistor.
2. When measuring the base-collector voltage you should measure across the transistor terminals directly. True or false?
3. The measured base-collector voltage should be more than 1 V. True or false?
4. The base-emitter voltage is generally equal to _____ volts.
5. With no base current, the transistor operates in its normal range. True or false?
6. When saturation occurs, increasing the _____ _____ has no effect on the collector current.
7. If the collector-emitter voltage exceeds breakdown voltage, no damage will be done to the transistor. True or false?
8. The voltage drop and internal resistance of the beta box are equal to _____ .
9. What are the rules governing current levels entering and leaving a beta box?
10. _____ current law controls the current flowing to and from a node in the beta box.

4.5   USING BASE BIAS

There are two ways of establishing base bias. One method uses two power supplies, while another uses only one power supply. Figure 4.14 shows both configurations. In Figure 4.14a, two batteries establish the bias voltages. One battery, labeled V_{BB}, forward biases the emitter-base junction. The other battery, labeled V_{CC}, reverse biases the collector-base junction.

In Figure 4.14b, one battery, labeled V_{CC}, connects to the top of the circuit and provides voltages for both sides of the transistor. Forward bias current for the emitter-base junction flows through the base resistor, R_B. Reverse bias voltage is applied to the collector-base junction via R_C.

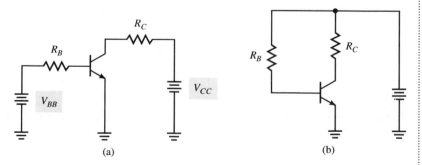

FIGURE 4.14 (a) Using two batteries to establish base bias. (b) Using one battery to establish base bias.

EXAMPLE 4.2

Solve for the value of the emitter, base, and collector currents and the emitter, base, and collector voltages for the circuit shown in Figure 4.15. Use the same techniques used in example 4.1.

The circuit shown in Figure 4.15 is a base-bias circuit using two power supplies and an emitter resistor.

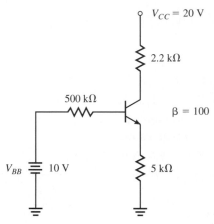

FIGURE 4.15 Base-bias transistor circuit using two power supplies

Solution

Figure 4.16a shows the model for the control loop, and Figure 4.16b shows the equalized model. The equivalent series circuit is shown in Figure 4.16c.

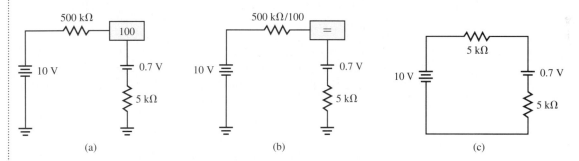

FIGURE 4.16 (a) Control loop model for the circuit shown in Figure 4.10. (b) Equalized model. (c) Equivalent series circuit.

We can use this equivalent series circuit to solve for the emitter current. We also know that the effective voltage is

$$10\,V - 0.7\,V = 9.3\,V$$

and that the effective resistance is

$$5\,k\Omega + 5\,k\Omega = 10\,k\Omega$$

Using Ohm's law to solve for the emitter (and, because they're equal, the collector) current, we find that:

$$I_E = I_C = \frac{9.3\ V}{10\ k\Omega} = 0.93\ mA$$

With beta equaling 100, the base current is

$$I_B = \frac{0.93\ mA}{100} = 9.3\ \mu A$$

The voltage on the emitter is

$$V_E = I_E \times R_E = 0.93\ mA \times 5\ k\Omega = 4.65\ V$$

The voltage on the base is

$$V_B = 4.65\ V + 0.7\ V = 5.35\ V$$

The collector resistor voltage drop is

$$V_{RC} = 0.93\ mA \times 2.2\ k\Omega = 2.05\ V$$

Subtracting the collector resistor voltage value from the collector source voltage value gives us the collector voltage value:

$$V_C = 20\ V - 2.05\ V = 17.95\ V$$

PRACTICE PROBLEM 4.1

Use the technique from example 4.2 to determine the collector-to-emitter voltage. Then calculate the power dissipation, or P_{DQ}, for the transistor shown in Figure 4.17.

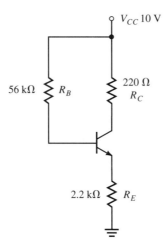

FIGURE 4.17

EXAMPLE 4.3

For the circuit shown in Figure 4.18, determine the values of: the voltage dropped across the base resistor, the base current, the collector current, the voltage dropped across the collector resistor, the collector voltage, and the emitter voltage.

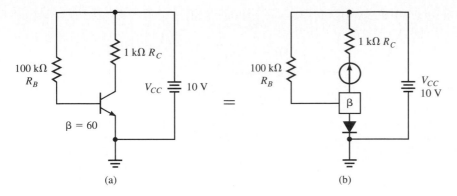

FIGURE 4.18

Figure 4.18a shows a base-biased circuit using a single supply voltage. Figure 4.18b shows the beta box equivalent of the circuit. Remember that beta is the link between the base circuit and the collector circuit.

Solution
The voltage drop across the base resistor (V_{RB}) equals the collector supply voltage minus the voltage to overcome the barrier potential:

$$V_{RB} = V_{CC} - 0.7\,V = 10\,V - 0.7\,V = 9.3\,V$$

We can use this value to find the value of the base current. Using Ohm's law, the base current equals

$$I_B = \frac{V_{RB}}{R_B} = \frac{9.3\,V}{100\,k\Omega} = 93\,\mu A$$

The collector current equals the base current multiplied by the beta value. This equation shows how a transistor amplifies current. The collector current value is:

$$I_C = I_B \times \beta = 93\,\mu A \times 60 = 5.58\,mA$$

We use this value and Ohm's law to find the amount of voltage dropped across the collector resistor:

$$V_{RC} = I_C \times R_C = 5.58\,mA \times 1\,k\Omega = 5.58\,V$$

Using Kirchhoff's voltage law, we can solve for the collector voltage. The collector voltage equals the collector supply voltage minus the voltage dropped across the collector resistor, or:

$$V_C = V_{CC} - V_{RC} = 10\,V - 5.58\,V = 4.42\,V$$

Looking at the schematic diagram, we can see that the emitter connects to ground. Thus, the emitter voltage equals zero.

EXAMPLE 4.4

Find the base, emitter, and collector currents for the NPN circuit shown in Figure 4.19. Then use the current values to determine the voltage values at the three transistor terminals. Designate these voltages—all with respect to ground—as the emitter voltage (V_E), the base voltage (V_B), and the collector voltage (V_C).

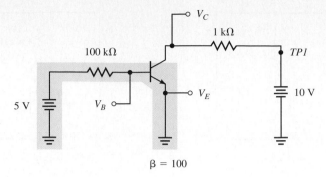

FIGURE 4.19

In addition to the information provided in the schematic, we also see that beta is given as 100. The shaded portion of the figure represents the base-emitter loop of the circuit. Whatever happens within this loop controls the remainder of the transistor circuit. Before proceeding, we must decide if the loop is closed or open. In other words, we need to know if current flows through the loop. If current flows, we have a closed loop. Whether or not current flows depends on whether the emitter-base junction is forward-biased or reverse-biased. Forward biasing the junction with a voltage supply, as shown in Figure 4.20, allows current to flow.

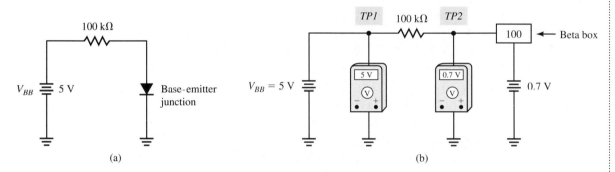

FIGURE 4.20 (a) Forward biasing the circuit shown in Figure 4.19. (b) Using the beta box to solve the circuit shown in Figure 4.19.

In Figure 4.20a, we see how the emitter-base loop would appear if we replaced the emitter-base junction with a diode. The bias voltage of 5 V is large enough to overcome the 0.7-V barrier potential, so the battery voltage establishes forward bias for the diode, or the emitter-base junction, and the loop is closed.

Now that we have established that the emitter-base loop is closed, we can develop an equation for the loop based on Kirchhoff's voltage law. Kirchhoff's voltage law states that the sum of voltages around any loop must be zero. Even with the base and emitter currents flowing in the circuit and having different values, we can still apply Kirchhoff's voltage law for closed loops.

Solution

With the beta box ground rules in mind, we can solve the circuit shown in Figure 4.20a. Let us think about those rules while we look at Figure 4.20b. First, we notice that only the base side of the beta box has a resistor; we do not need to equalize the beta box. Only when *both* sides of a base-emitter loop have resistors do we need to equalize the beta box. Remembering that the beta box drops no voltage, we can determine the voltage dropped across the 100-kΩ resistor.

In Figure 4.20b, test-point 1 (TP1) measures 5 V to ground. Test-point 2 measures 0.7 V to ground because of the semiconductor junction barrier potential. From this information, we can calculate the voltage across the 100-kΩ resistor.

$$5\,V - 0.7\,V = 4.3\,V$$

Now, we can apply Ohm's law to find the base current:

$$I_B = \frac{4.3\,V}{100\,k\Omega} = 43\,\mu A$$

Because beta = 100, we can determine the collector and emitter currents:

$$I_C = I_E = 43\,\mu A \times 100 = 4.3\,mA$$

These three solutions give the part of the answers needed for the example objective. From here, we can find the terminal voltages.

The emitter terminal in Figure 4.19 ties directly to ground in Figure 4.20. Therefore V_E, or the emitter voltage, equals zero. Remember that one of the rules of silicon bipolar transistors is that the base terminal voltage will be 0.7 V higher than the emitter voltage. The emitter voltage equals zero, so V_B (the base voltage) equals 0.7 V. Using a two-step process, we can now calculate the collector voltage. Applying Ohm's law, we see that the collector current multiplied by the collector resistor value will supply the value for voltage dropped across the resistor:

$$V_{RC} = 4.3\,mA \times 1\,k\Omega = 4.3\,V$$

We can subtract that voltage from the collector supply voltage to arrive at the collector terminal voltage:

$$V_C = 10\,V - 4.3\,V = 5.7\,V$$

We can apply the bias model, beta box, and the base-emitter loop as tools for solving other transistor circuits. Each concept is crucial to transistor operation. Frequent use of these tools will increase your understanding of both transistor and circuit functions. As you become familiar with these tools you will develop the skills needed to solve transistor circuits without memorizing many formulas. Be patient; keep at it.

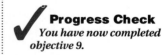

Progress Check
You have now completed objective 9.

EXAMPLE 4.5

Solve the circuit shown in Figure 4.21 for the emitter, base, and collector currents and the emitter, base, and collector voltages. The transistor is a silicon NPN transistor with a beta of 150.

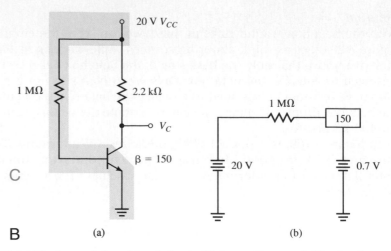

FIGURE 4.21 (a) Single supply base-biased circuit. (b) Control loop model for circuit shown in Figure 4.21.

The circuit shown in Figure 4.21a is a base-biased NPN configuration using a single power supply. Shading highlights the emitter-base loop. This loop controls transistor operation in the circuit.

Solution

Figure 4.21b shows a separate drawing of the control loop. We have a closed loop with a beta box that scales the emitter current to 150 times the value of the base current. The effective voltage of the loop is

$$20\,V - 0.7\,V = 19.3\,V$$

The beta box drops no voltage. Therefore, the entire 19.3 V must drop across the 1-MΩ resistor. We can use Ohm's law to determine the base current:

$$I_B = \frac{19.3\,V}{1\,M\Omega} = 19.3\,\mu A$$

With a beta of 150, we can calculate the collector current:

$$I_C = 150 \times 19.3\,\mu A = 2.895\,mA$$

The collector current value equals the emitter current value, so the emitter current is 2.895 mA. The emitter voltage is at ground potential and has a value of 0 V. Because the base voltage is 0.7 V higher than the emitter voltage, the base voltage value equals 0.7 V. To determine the collector voltage, we must first see how much voltage drops across the collector resistor:

$$V_{RC} = I_C \times R_C = 2.895\,mA \times 2.2\,k\Omega = 6.369\,V$$

As in the preceding example, we subtract this voltage value from the collector source voltage value to find the collector voltage:

$$V_C = 20\,V - 6.369\,V = 13.631\,V$$

Although it may seem that we employed a number of equations to find the solutions, the equations are nothing more than Ohm's law. Take another look at the equations. Notice that the $E = I \times R$ form of Ohm's law becomes apparent.

Using the beta box model and the circuit shown in Figure 4.22, find the values for the emitter, base, and collector currents and the emitter, base, and collector voltages using a 2.2-kΩ resistor for R_C.

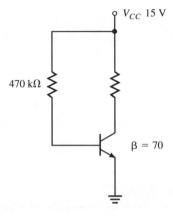

FIGURE 4.22

IMPROVING BASE BIAS

So far, we have discussed relatively simple transistor circuits. Those circuits contained a minimum of parts—two resistors, the battery supplies, and a transistor. Because there were so few parts it was easy to find the circuit values such as V_E and V_C. Unfortunately, though, base-bias circuits have problems.

Some of those problems exist because beta values may vary from transistor to transistor. Let us suppose that a circuit designer requires a steady base current of 20 μA. If the specified transistor has a beta value that ranges from 100 to 300, the collector current could vary by as much as a factor of 3.

REVIEW SECTION 4.5
1. V_B, V_C, and V_E are designated voltages with respect to _____.
2. The barrier potential of a forward-biased emitter-base junction is _____ volts for a silicon transistor.
3. The collector current is _____ times larger than the base current.
4. Many of the equations used in this section are forms of _____ law and _____ law.
5. Draw a beta box model, and explain procedures used to calculate the base, emitter, and collector voltages and the base and collector currents.

6. Looking at the following circuit, trace the current paths and calculate the base, emitter, and collector voltage values and the base, emitter, and collector current values.

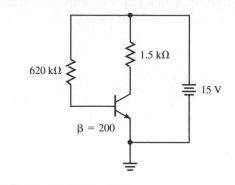

4.6 MINIMIZING THE EFFECTS OF BETA

Given that beta can change dramatically from transistor to transistor, a circuit designer must minimize the effects of beta as it varies. When constructing circuits, designers look for methods that guarantee consistent circuit performance. One method, adding an emitter resistor to the circuit, swamps out the effects of β by making R_E larger than R_B/β.

ADDING R_E IN A BASE-BIASED CIRCUIT AS A BIAS METHOD

In Figure 4.23a, a resistor has been added to the emitter section of the circuit. This emitter resistor, labeled R_E, becomes part of the base-bias arrangement and the control loop consisting of the collector voltage, the base resistor, and the emitter-base barrier potential. Figure 4.23b shows a model that represents the control loop with the beta box arrangement.

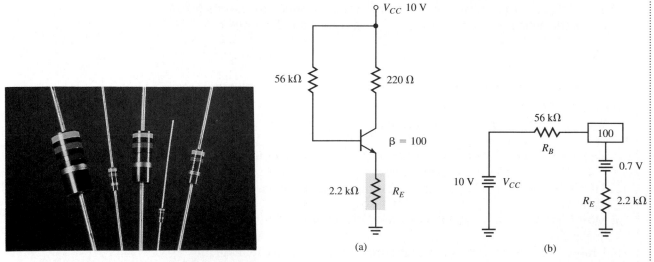

FIGURE 4.23 (a) Adding a feedback resistor (R_E) to a circuit to minimize variations in beta. *(Photograph courtesy of Allen-Bradley)* (b) Control loop model for the circuit.

In this diagram, resistors work on both the emitter and the base sides of the beta box. For a second, let us review one beta box rule that affects this arrangement: When resistors are in the base and emitter sides of the base-emitter loop, the beta box must be equalized so that Ohm's law can be used to solve the base-emitter loop.

In Example 4.6, we will see how we can apply the rule. Later, we will discuss how the emitter resistor minimizes change caused by beta variations.

EXAMPLE 4.6

Use the model and beta box from Figure 4.13 to solve for the emitter, base, and collector currents and the emitter, base, and collector voltages for the circuit shown in Figure 4.23a. The transistor is a doped silicon NPN device.

Because the circuit features an emitter resistor, the emitter terminal is no longer at ground potential, and the emitter voltage no longer equals 0 V. Therefore the base voltage no longer equals only 0.7 V.

All these changes make it more difficult to calculate the base current—we cannot easily determine the voltage drop across the base resistor. One side of the resistor connects to the collector source voltage, but the other side no longer is at 0.7 V. Earlier, we found the emitter current value by multiplying beta times the base current value. Because the emitter resistor is in the control loop, we cannot use that simple equation to find the emitter current.

Solution

To account for the effects of the emitter resistor and to calculate the emitter current, we must equalize the beta box. At present, a 100-to-1 difference exists between the emitter current and the base current. We must change the model of the control loop so that the base current appears equal to the emitter current. Understand that we are not changing the value of the base current. Instead, we are changing only the representation of the base current in the model so that we can easily calculate the emitter current value.

If we pretend that the base current becomes beta times larger, we must account for the change in our model. First, let us state that we will keep the supply voltage constant in our model. If supply voltage is constant, and current changes, then resistance must change. Ohm's law states that voltage equals current times resistance. If we show the voltage as a fixed value and if current increases by a factor of beta, then voltage can remain stationary only if the resistance decreases by a factor of beta.

When we equalize the model, we adjust the resistance value. Examine Figure 4.24a. In this figure, an equal sign replaces the value of beta inside the beta box and represents the equalization of the model. Furthermore, the base resistance is now beta times smaller than its original value, and equals 560 Ω. Figure 4.24b depicts this change.

In Figure 4.24b, we see a simple series circuit that has equal current everywhere. This current is the emitter current. All the work that we have gone through pays off. We can use Ohm's law to solve for the emitter current value:

$$I = \frac{V}{R}$$

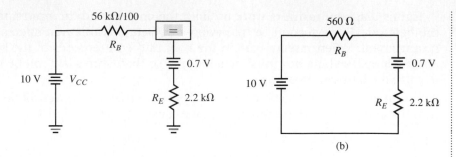

FIGURE 4.24 (a) Equalized model for the circuit shown in Figure 4.23a. (b) How the base resistance changes through equalization.

The effective voltage is

$$10\text{ V} - 0.7\text{ V} = 9.3\text{ V}$$

The effective resistance is

$$2.2\text{ k}\Omega + 560\text{ }\Omega = 2.76\text{ k}\Omega$$

With those values, the emitter current equals

$$I_E = I_C = \frac{9.3\text{ V}}{2.76\text{ k}\Omega} = 3.37\text{ mA}$$

Dividing the emitter current value by beta, the base current equals

$$\frac{3.37\text{ mA}}{100} = 33.7\text{ }\mu\text{A}$$

To determine the voltage at the emitter terminal, we use Ohm's law:

$$V_E = I_E \times R_E = 3.37\text{ mA} \times 2.2\text{ k}\Omega = 7.41\text{ V}$$

Adding 0.7 V to the emitter voltage value, we find that the base voltage equals

$$7.41\text{ V} + 0.7\text{ V} = 8.11\text{ V}$$

To calculate the collector voltage value, we need to find the amount of voltage dropped across the collector resistor:

$$V_{RC} = I_C \times R_C = 3.37\text{ mA} \times 220\text{ }\Omega = 0.741\text{ V}$$

Subtracting this value from the collector source voltage, we find that the collector voltage equals

$$10\text{ V} - 0.741\text{ V} = 9.26\text{ V}$$

We can use the methods from the example to solve many different transistor circuits. With these methods, we can apply Ohm's law to solve the circuit and use circuit models to see how the circuit operates. In addition, we have a method that frees us from memorizing equations. This method increases problem-solving abilities and allows us to troubleshoot many transistor circuits.

EXAMPLE 4.7

Find the values of the voltage dropped across the base and emitter resistors, the emitter current and voltage, the base voltage, the voltage dropped across the collector resistor, and the collector voltage.

Figure 4.25a shows a schematic of a base-bias circuit with an emitter resistor. Figure 4.25b shows the beta-equalized control circuit, and Figure 4.25c shows the equivalent control circuit.

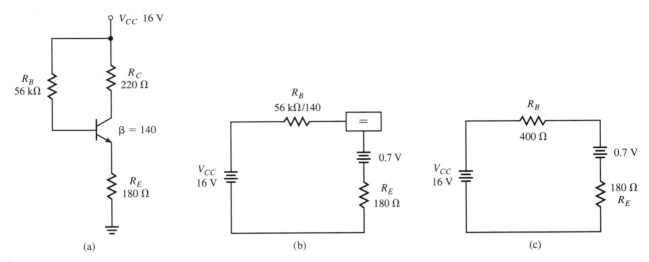

(a)   (b)   (c)

FIGURE 4.25 (a) Base-bias circuit with emitter resistor. (b) Beta equalizing control circuit. (c) Equivalent control circuit.

Solution

From the perspective of the emitter current, the base resistor has an effective value of

$$R_B \text{ (effective)} = \frac{R_B}{\beta} = \frac{56 \text{ k}\Omega}{140} = 400 \text{ }\Omega$$

Using Kirchhoff's series circuit law, we find that the voltage dropped across the base and emitter resistors equals

$$V_{CC} - 0.7 \text{ V} = V_{RB} + V_{RE} = 15.3 \text{ V}$$

By using Ohm's law and Kirchhoff's series circuit law, we can solve for the value of the emitter current:

$$I_E = \frac{V_{RB} + V_{RE}}{(R_B + R_E) \text{ effective values}} = \frac{15.3 \text{ V}}{400 \text{ }\Omega + 180 \text{ }\Omega} = 26.4 \text{ mA}$$

The emitter voltage is

$$V_E = I_E \times R_E = 26.4 \text{ mA} \times 180 \text{ }\Omega = 4.75 \text{ V}$$

The base voltage equals the emitter voltage plus the barrier potential:

$$V_B = V_E + 0.7 \text{ V} = 4.75 \text{ V} + 0.7 \text{ V} = 5.45 \text{ V}$$

Because the emitter current nearly equals the collector current, we can use the emitter current value and Ohm's law to estimate the value of the voltage dropped across the collector resistor:

$$V_{RC} = I_E \times R_C = 26.4 \text{ mA} \times 220 \text{ }\Omega = 5.81 \text{ V}$$

The collector voltage equals the collector supply voltage minus the voltage dropped across the collector resistor:

$$V_C = V_{CC} - V_{RC} = 16 \text{ V} - 5.81 \text{ V} = 10.2 \text{ V}$$

Using the circuit shown in Figure 4.26, determine the effective value of the base resistor with respect to the emitter current. Determine the values of the emitter current and voltage, the base voltage, and the collector voltage. Trace and label all current paths for the circuit.

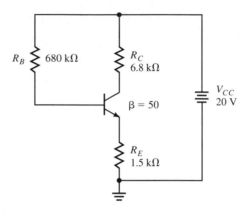

FIGURE 4.26

A DETAILED LOOK AT HOW THE EMITTER RESISTOR MINIMIZES THE EFFECTS OF BETA CHANGES

Let us return now to our hypothetical circuit that has a specified steady base current of 20 μA and a transistor with a beta value that ranges from 100 to 300. Beta in this circuit could increase from 100 to 300. As a result, the emitter and collector currents would also change by a factor of 3. With the emitter resistor in the circuit, this large a change cannot occur.

Figure 4.27a is the control loop first seen in Figure 4-24a. In this instance $\beta = 300$. We can equalize the 56-kΩ resistor by dividing its value by 300, or the new value for beta. Solving for I_E in Figure 4.27b results in a new emitter current of 3.896 mA. Compared with the previous value of 3.37 mA, the emitter current changes by a factor of only 1.16.

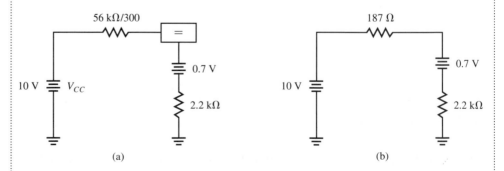

(a)   (b)

FIGURE 4.27 (a) Equalizing the emitter resistor. (b) Increasing the size of the emitter resistor. Beta changed by a factor of 3 (300 to 100), but I_E is now 3.896 mA compared with 3.37 mA, changing by a factor of only 1.16.

EXAMPLE 4.8

Determine the effects of increasing the value of the emitter resistance.

Figure 4.28a illustrates what happens if we increase the size of the emitter resistor in Figure 4.27 and beta changes. In the circuit, the value for the emitter resistor increases from the 2.2 kΩ of Figure 4.27a to 3.3 kΩ. As before, we will model the circuit with a beta value of 100, which produces an emitter current of 2.41 mA.

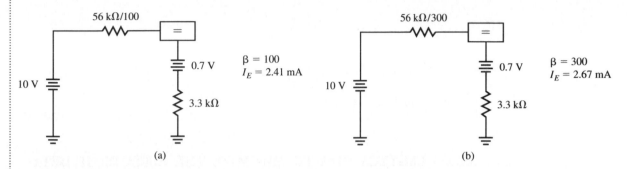

FIGURE 4.28 (a) Model of example circuit, showing the effects of increasing R_E on minimizing changes in beta. (b) Beta changes by a factor of 3, and I_E changes by a factor of 1.11.

Solution

In Figure 4.28b, the value of beta increases to 300, and the emitter current increases to 2.67 mA. Dividing 2.67 mA by 2.41 mA gives a change of 1.11. In Figure 4.27, a change of 1.16 existed. Ideally, we would like to reduce this factor to 1.00, or no change. It is difficult to reduce the factor to 1.00. However, designers attempt to construct circuits with an emitter current that does not change appreciably with drastic changes in beta.

Increasing the emitter resistance to minimize the effects of a changing beta is not always a practical solution. Each increase in emitter resistance reduces the amount of current delivered by the transistor to resistors connected at the collector. If the emitter current decreases, the collector current also decreases.

We could compensate for increasing the emitter resistance by changing the supply voltage level, but that is not always possible. The circuit may connect to a piece of equipment that features a fixed power supply. Or the power supply driving the transistor circuit may provide energy for other circuits as well. Changing the supply voltage to compensate for changes in one circuit could affect other circuit operations. Finally, there are limits on the voltages that can be applied to a transistor.

Looking at the circuit shown in Figure 4.29, solve for the emitter, collector, and base current values and the emitter, collector, and base voltage values.

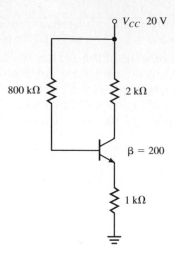

FIGURE 4.29

USING EMITTER BIAS TO MINIMIZE THE EFFECTS OF BETA

Many circuit designers use some sort of **emitter bias** to minimize beta and to stabilize circuit operation. Emitter biasing utilizes two voltage sources—one in the collector, another in the emitter. Figure 4.30 shows a typical emitter-biased circuit.

In the figure, the voltage connected to the emitter is labeled V_{EE}. Although most schematic drawings will not show the voltage source, it will be connected as shown in this figure. The collector voltage source will show in the same way. At the point where the voltage source connects to the emitter resistor, the polarity is negative with respect to ground. Figure 4.30 displays this point as $-V_{EE}$.

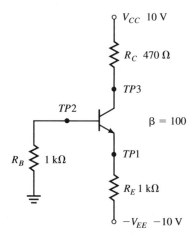

FIGURE 4.30 A typical emitter-bias NPN transistor circuit. The control loop is highlighted.

Even though Figure 4.30 does not show the difference in values, the base resistor in an emitter-bias circuit has a smaller value than the base resistors found in base-bias circuits. This difference is important. If the base resistor in an emitter-bias circuit has a small value, it will have a minimal effect on the DC operation of the circuit. A control loop model will show the small effect that the base resistor value has on circuit calculations.

In Figure 4.31, we see the control loop of a beta box model for an emitter-bias circuit. Because there are resistors on each side of the beta box, we must equalize the model. We have already solved for the emitter current in base-bias circuits that have an emitter resistor. We can likewise solve for the emitter current in an emitter-bias circuit. In other words, we are learning to use a consistent procedure for solving transistor circuits. When we equalize the model, we will reduce the base resistance by dividing it by beta.

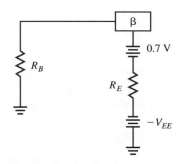

FIGURE 4.31 Beta box model of the control loop for the emitter-bias circuit shown in Figure 4.30

EXAMPLE 4.9

Using the circuit shown in Figure 4.30, solve for the emitter, collector, and base currents and the emitter, collector, and base voltages.

Figure 4.32a shows the control loop taken from the circuit in Figure 4.30 and Figure 4.32b illustrates the equalized model. Notice how the effective base resistance is reduced. As shown in the equivalent series circuit in Figure 4.32c, the effective base resistance is insignificant when compared with the value of the emitter resistance. In this case, the emitter resistance is

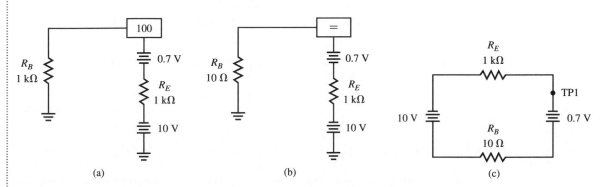

FIGURE 4.32 (a) Control loop model for the circuit shown in Figure 4.30. (b) Equalized model of the circuit. (c) Equivalent series circuit.

1000 Ω (1 kΩ). Since the part of the model affected by beta has no real impact on the value found for the emitter current, we can safely assume that beta has little or no effect on emitter-bias circuits. A properly designed emitter-bias circuit produces stable circuit operation with consistent operating points.

Solution

Using the equivalent series circuit in Figure 4.32c, we can apply Ohm's law when solving for the emitter current in Figure 4.30:

$$I_E = I_C = \frac{(10\,V - 0.7\,V)}{1010\,\Omega} = 9.21\,mA$$

To determine the voltage at the emitter terminal, or TP1, we must find the amount of voltage used by the emitter resistor:

$$V_{RE} = 9.21\,mA \times 1\,k\Omega = 9.21\,V$$

If we start at the ground connection of the emitter bias voltage and move up the emitter circuit, we first drop the 10-V potential of the bias voltage. At the point in the circuit marked $-V_{EE}$, we see $-10\,V$ to ground at the $-V_{EE}$, or emitter bias voltage point, in the circuit. Proceeding through the emitter resistor, we gain 9.21 V, which yields

$$-10\,V + 9.21\,V = -0.79\,V \text{ at TP1}$$

or the voltage at the emitter terminal equals $-0.79\,V$ with respect to ground.

To obtain the base voltage, we can add 0.7 V to the emitter voltage. This gives:

$$V_B = V_E + 0.7\,V = -0.79\,V + 0.7\,V = -0.09\,V$$

With all the calculated voltage values, we can now find the base current. We can use two methods to calculate the value of base current:

1. Divide the emitter current by the beta value.

$$I_B = \frac{I_E}{\beta} = \frac{9.21\,mA}{100} = 92.1\,\mu A$$

2. Use Ohm's law, and divide the voltage drop across the base resistor by the resistance.

$$I_B = \frac{V_{RB}}{R_B} = \frac{0.09\,V}{1000\,\Omega} = 90\,\mu A$$

Next, we can find how much voltage is dropped by the collector resistor:

$$V_{RC} = 9.21\,mA \times 470\,\Omega = 4.33\,V$$

As before, we can subtract the voltage drop of the collector bias or supply voltage to find the voltage at the collector terminal:

$$V_C = V_{CC} - V_{RC} = 10\,V - 4.33\,V = 5.67\,V$$

When testing emitter-biased circuits, you can expect to find the base voltage close to ground potential. In the example above, the base voltage had a value of 90 mV. In addition, the emitter voltage should equal about $-0.7\,V$ below ground potential.

PRACTICE PROBLEM 4.5

Determine the collector-to-emitter voltage for the transistor shown in Figure 4.30. Use this information to find the amount of power dissipated by the transistor.

PRACTICE PROBLEM 4.6

Solve for values of the emitter current, base current, collector current, base voltage, collector voltage, and emitter voltage for the circuit shown in Figure 4.33. Trace the current paths.

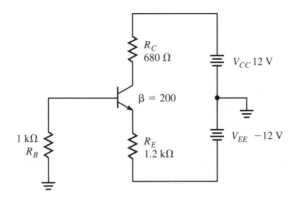

FIGURE 4.33

EXAMPLE 4.10

Solve for values of the emitter, base, and collector current and the emitter, base, and collector voltage using the circuit shown in Figure 4.34.

The control loop model, equalized model, and equivalent series circuit are diagrammed in Figure 4.35. Notice how insignificant the value of the base resistor divided by beta appears when compared with the value of the emitter resistor.

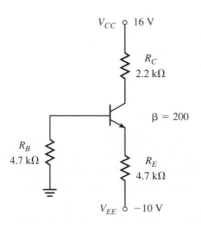

FIGURE 4.34

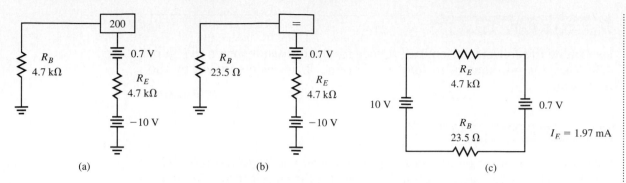

(a)   (b)   (c)

FIGURE 4.35 (a) Control loop model, (b) equalized model, and (c) equivalent series circuit for the circuit shown in Figure 4.34.

Solution

Using the equivalent series circuit, we solve for the emitter current value by using Ohm's law:

$$I_E = \frac{\left(10\,V - 0.7\,V\right)}{\left(4.7\,k\Omega + 23.5\,\Omega\right)} = 1.97\,mA$$

The emitter resistor has a voltage drop of

$$V_{RE} = 1.97\,mA \times 4.7\,k\Omega = 9.26\,V$$

To find the voltage at the emitter, we add the voltage drop value to the emitter source voltage value:

$$V_E = -10\,V + 9.26\,V = -0.74\,V$$

To determine the base voltage, we add 0.7 V to the emitter voltage:

$$V_B = -0.74\,V + 0.7\,V = -0.04\,V$$

As we know, the base voltage should be close to ground potential. To find the collector voltage, we need to know how much voltage is dropped by the collector resistor:

$$V_{RC} = 1.97\,mA \times 2.2\,k\Omega = 4.33\,V$$

Next, we subtract the collector resistor voltage value from the collector source voltage value to find a value for the collector voltage:

$$V_C = V_{CC} - V_{RC} = 16\,V - 4.33\,V = 11.67\,V$$

Emitter bias is an excellent means of making a circuit independent of beta and making it stable. However, using emitter bias requires two voltage sources. In the next sections, we will look at two bias arrangements that require only one voltage supply.

COLLECTOR FEEDBACK BIAS

Another bias method partially offsets changes in beta caused by transistor variations and by temperature changes. **Collector feedback bias** offers the economy of using only one power supply and two resistors. Figure 4.36 shows a collector feedback bias arrangement. The highlighted control loop uses part of the collector circuit. The use of the collector circuit makes the collector feedback bias arrangement different from the bias arrangements that we have already studied.

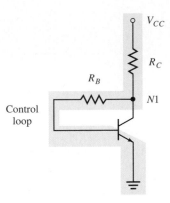

FIGURE 4.36 A collector feedback bias circuit. N1 is a feedback node. The control loop is highlighted.

Figure 4.37 shows the control loop model. Notice that resistors are on only one side of the beta box. However, we must equalize the model because of node N1 in the collector. We can refer to this node as a **feedback node.** Before studying how this node affects circuit operation, let us solve the control loop model.

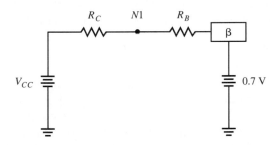

FIGURE 4.37 Control loop model of the circuit shown in Figure 4.36. Note that N1 is a feedback node.

Solving the control loop model for collector feedback bias requires a procedure different from that used to solve other bias arrangements. Figure 4.37 shows how this procedure differs from the others and combined with Figure 4.38 illustrates the beta box bias model of the NPN transistor. This model connects to the resistors and makes up the complete collector bias circuit.

The emitter current enters the beta box from the emitter terminal and splits into two parts. One part—the base current—flows out the base lead, while the other part—the collector current—flows out the collector. These two currents recombine at the feedback node, N1. Emitter current flows through the collector resistor. Even though the model resembles a series circuit, different currents flow in the loop. Before solving the simple series circuit and using Ohm's law to find the emitter current, we must equalize the model. A special rule exists for this situation: The resistance between the feedback node and the beta box is equalized by dividing the resistive value by beta.

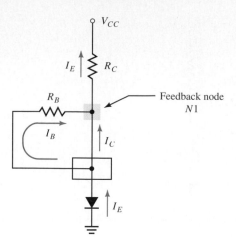

FIGURE 4.38 Beta box bias model of the NPN transistor, with external resistors shown in collector feedback bias arrangement

EXAMPLE 4.11

Solve for the values of the emitter, collector, and base current and the emitter, collector, and base voltage for the circuit shown in Figure 4.39.

Because the emitter terminal connects to ground, the emitter voltage equals 0 V. In turn, the base voltage equals 0.7 V. Unfortunately, we cannot calculate the base current directly. Thus, we will need to find the amount of emitter current that flows.

Solution

The control loop model for the circuit in Figure 4.39 is shown in Figure 4.40a. It contains a feedback node, so we may apply our special-case rule and equalize the base resistor. This is necessary because the current through the control loop is not the same everywhere within the loop. Figure 4.40b shows the equalized model. This leaves us with a series circuit where the currents equal the emitter current. We can apply Ohm's law and find that

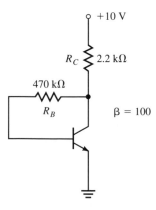

FIGURE 4.39

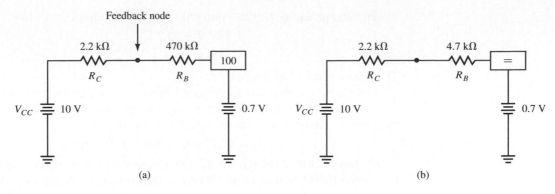

FIGURE 4.40 (a) Control loop model for the circuit shown in Figure 4.39. (b) Equalized model.

$$I_E = \frac{V_{CC} - V_{BE}}{R_C + R_{B(equalized)}}$$

$$I_E = \frac{10\text{ V} - 0.7\text{ V}}{2.2\text{ k}\Omega + 4.7\text{ k}\Omega} = 1.35\text{ mA}$$

The base current can be found by dividing the emitter current value by the beta value:

$$I_B = \frac{1.35\text{ mA}}{100} = 13.5\ \mu\text{A}$$

Finally, we can determine the collector voltage by first finding a value for the voltage dropped across the collector resistor:

$$V_{RC} = 1.35\text{ mA} \times 2.2\text{ k}\Omega = 2.97\text{ V}$$

and then subtracting that value from the collector supply voltage value:

$$V_C = V_{CC} - V_{RC} = 10\text{ V} - 2.97\text{ V} = 7.03\text{ V}$$

THE EFFECTS OF FEEDBACK

From our discussion of thermal runaway (on page 146), we know that beta is temperature-dependent. As temperature increases, the beta value increases. As beta increases, the amount of current flow increases, and the increased current causes temperature to rise again.

Feedback breaks the temperature-beta-current flow circle. In Figure 4.40, the voltage at the feedback node or collector terminal decreases. To see why the voltage decreases, let us look at the figure and solve for voltage and current values. For the sake of argument, we will say that the emitter current rises to 1.5 mA from the earlier level of 1.35 mA, giving us a 0.15-mA change.

With this information, we find that the voltage drop across the collector resistor becomes

$$1.5\text{ mA} \times 2.2\text{ k}\Omega = 3.3\text{ V}$$

Subtracting the voltage drop from the collector supply voltage, we find that the collector terminal voltage has lowered to

$$10\text{ V} - 3.3\text{ V} = 6.7\text{ V}$$

from the earlier level of 7.03 V.

Since the collector voltage has decreased, the voltage drop across the base resistor also decreases.

We can calculate the base current directly by using Ohm's law:

$$I_B = \frac{(6.7 \text{ V} - 0.7 \text{ V})}{470 \text{ k}\Omega} = 12.77 \text{ }\mu\text{A}$$

The base current has decreased.

If the base current has decreased, then the emitter current must also decrease, because the emitter current equals the base current multiplied by the beta value. This decrease is reflected in the formula

$$I_E = 12.77 \text{ }\mu\text{A} \times 100 = 1.277 \text{ mA}$$

We began our investigation by assuming that the emitter current had increased 0.15 mA to a level of 1.5 mA. Moreover, our calculations showed that the emitter current actually dropped 0.73 mA to a level of 1.277 mA. With feedback the change in circuit parameters caused by variations in β is reduced.

PRACTICE PROBLEM 4.7

Solve for the values of the emitter, base, and collector current and the emitter, base, and collector voltage for the circuit shown in Figure 4.41. Determine the collector-to-emitter voltage and the power dissipation of the transistor.

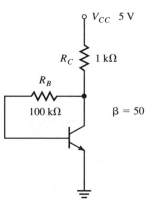

FIGURE 4.41

REVIEW SECTION 4.6

1. An emitter bias circuit has two voltage sources, one in the _____ and the other in the _____ .
2. In emitter bias, the base resistor has a _____ ohmic value.
3. In an emitter-bias circuit, the voltage on the base with respect to ground is approximately _____ volts.
4. In an emitter-bias NPN transistor circuit, the emitter voltage is approximately _____ volts.
5. Draw an emitter-bias circuit, and explain how to calculate values for the base, emitter, and collector voltages and currents.
6. Explain why the beta value can be ignored when using emitter bias.
7. Compare and contrast the characteristics of emitter bias and base bias.

8. The emitter resistor is added to the circuit to keep the emitter current value constant. True or false?

9. V_{CE} represents the voltage from collector to base. True or false?

10. $P_{DQ} = V_{CE} \times I_E$. True or false?

11. Collector feedback bias uses only two resistors. True or false?

12. With collector feedback bias, the base current flows only through the base resistor. True or false?

13. With collector feedback bias, the emitter voltage equals zero, and the base voltage is approximately 0.7 V. True or false?

14. To find the emitter current value in a collector feedback bias circuit, we must divide the collector resistance value by beta. True or false?

15. Draw the beta box model for a collector feedback bias circuit that uses an NPN transistor. Trace the current paths.

16. Explain how to calculate the values for the emitter, base, and collector voltages and currents for a collector feedback bias circuit.

17. For the following circuit, determine I_C, I_E, V_E, V_B, R_B, and beta.

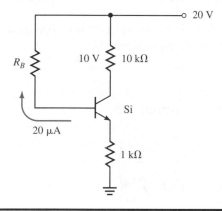

4.7   USING VOLTAGE DIVIDER BIAS TO MINIMIZE EFFECTS OF BETA

Voltage divider bias is perhaps one of the best methods for stabilizing a circuit. This type of bias works well for minimizing variations in beta. Designers

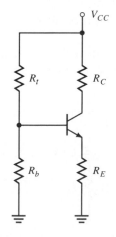

FIGURE 4.42 A typical voltage divider bias circuit

prefer voltage divider bias because it is more stable than either the base bias or emitter feedback bias methods seen earlier. As Figure 4.42 shows, voltage divider bias circuits require only one voltage supply.

In the figure, one power supply, labeled V_{CC}, the transistor, and four resistors make up the voltage divider bias circuit. Of the four resistors, one works as a collector resistor (R_C), another as an emitter resistor (R_E), and the remaining two resistors form the voltage divider. We will refer to the divider resistors as R_t for top resistor and R_b for bottom resistor. The emitter resistor works with the voltage divider to stabilize the circuit and minimizes changes in beta.

EXAMPLE 4.12

Figure 4.43 shows the model for the voltage divider control loop. Using the values shown in Figure 4.43, solve for the values of the emitter, base, and collector currents and the emitter, base, and collector voltages for a silicon NPN transistor.

To work with the model, we must thevenize the base section to reduce all the bias resistors to a single equivalent R_b.

Solution
Referring to Figure 4.44a, we have disconnected the beta box, the emitter resistor, and the 0.7-V barrier potential from the circuit in Figure 4.43b. Now, we can thevenize for terminals 1 and 2. Since terminal 2 connects to ground, we know that its voltage (V_2) is 0 V. Terminal 1 is at the center of the voltage divider; we can calculate its voltage (V_1) to ground as

$$V_1 = \frac{10 \text{ k}\Omega}{20 \text{ k}\Omega} \times 20 \text{ V} = 10 \text{ V}$$

Thus, the thevenized voltage equals $V_1 - V_2$, or 10 V.

To find the thevenized resistance, we reduce the collector supply voltage to 0 V and draw the diagram in Figure 4.44b. From the perspective of the thevenized terminals, the voltage divider resistors parallel each other. Using the product over the sum method, we have:

$$R_{TH} = \frac{10 \text{ k}\Omega \times 10 \text{ k}\Omega}{10 \text{ k}\Omega + 10 \text{ k}\Omega} = \frac{100 \text{ M}\Omega}{20 \text{ k}\Omega} = 5 \text{ k}\Omega$$

We can use this information to draw the Thevenin equivalent circuit shown in Figure 4.44c.

When we reconnect the emitter side of the control loop to the Thevenin equivalent circuit, we get Figure 4.44d. In effect, we have changed an unfa-

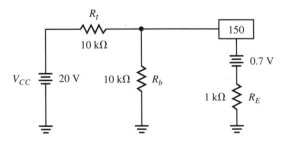

FIGURE 4.43 Control loop model for a voltage divider circuit.

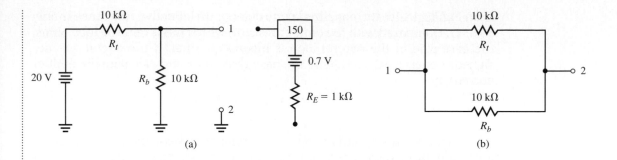

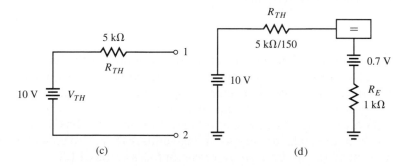

FIGURE 4.44 (a) Separating the voltage divider from the beta box. (b) Finding R_{TH}. (c) Thevenin equivalent circuit. (d) Thevenized model connected to the beta box.

miliar circuit into a familiar circuit that is easier to solve. Using Figure 4.43b, we find that the emitter current equals:

$$I_E = I_C = \frac{9.3 \text{ V}}{1033 \text{ } \Omega} = 9 \text{ mA}$$

The base current is:

$$I_B = \frac{9 \text{ mA}}{150} = 60 \text{ } \mu A$$

At the emitter terminal, the voltage is:

$$V_E = 9 \text{ mA} \times 1 \text{ k}\Omega = 9 \text{ V}$$

while the base voltage is:

$$V_B = 9 \text{ V} + 0.7 \text{ V} = 9.7 \text{ V}$$

With the collector resistor dropping

$$9 \text{ mA} \times 470 \text{ } \Omega = 4.23 \text{ V}$$

we can find the value for the collector terminal. Subtracting the voltage dropped across the resistor from the collector supply voltage, we have

$$20 \text{ V} - 4.23 \text{ V} = 15.77 \text{ V}$$

Look again at Figure 4.44c. The value of the thevenized resistance equals the resistance of the paralleled top and bottom resistors. The effective resistance, or the base side of the control loop, is smaller than either of the two voltage divider resistors and helps to minimize the effects of the control loop on the circuit.

In Figure 4.44d, the equalized control loop, the effective resistance is only 33.33 Ω. Compared with the emitter resistor, which has a value of 1000 ohms, the base side of the control loop is inconsequential. If the circuit was designed properly, we can ignore the base resistance and calculate the emitter current by

$$I_E = \frac{V_{TH} - 0.7\ V}{R_E}$$

As you can see, the equation makes no reference to beta; the circuit is almost independent of beta. Therefore the voltage divider bias configuration provides a stable form of bias.

EXAMPLE 4.13

Using Figure 4.45, solve for the Thevenin values for the base and emitter voltages. Determine the thevenized base resistance. Also, calculate the values for the emitter, base, and collector voltages and currents for the circuit.

Figure 4.45 shows a voltage divider bias circuit. The top resistor in the divider has a value of 12 kΩ, and the bottom resistor has a value of 2.2 kΩ. In addition, the collector supply voltage equals 15 V. The collector resistor has a value of 470 Ω, and the emitter resistor has a value of 120 Ω.

Solution
Figure 4.46a shows the base voltage divider. The Thevenin value of the base voltage equals:

$$V_{TH}\left(base\right) = \frac{R_B}{R_b + R_t} \times V_{CC} = \left(\frac{2.2\ k\Omega}{2.2\ k\Omega + 12\ k\Omega}\right) \times 15\ V = 2.32\ V$$

Figure 4.46b shows a diagram of the top and bottom resistors paralleling one another, and Figure 4.46c shows the thevenized resistance. The Thevenin resistance of the base equals:

$$R_{TH}\ (base) = R_T \parallel R_b = 12\ k\Omega \parallel 2.2\ k\Omega = 1.86\ k\Omega$$

Using those values, we can determine the resistance of the base. That value is:

$$\frac{R_{TH}\left(base\right)}{\beta} = \frac{1.86\ k\Omega}{150} = 12.4\ \Omega$$

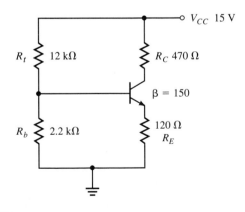

FIGURE 4.45

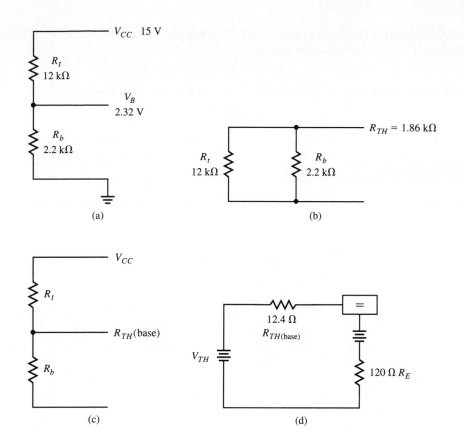

FIGURE 4.46

Moving to the beta box shown in Figure 4.46d, we can solve for the other circuit values. The voltage across the base and emitter resistors equals the thevenized voltage minus the base-to-emitter voltage, or the barrier potential:

$$2.32\,V - 0.7\,V = 1.62\,V$$

Using that value, we can solve for the emitter current with Ohm's law. Dividing the voltage dropped across the base and emitter resistors by the combined values of the voltage divider resistors gives us:

$$I_E = \frac{1.62\,V}{12.4\,\Omega + 120\,\Omega} = 13.36\,mA$$

Again using Ohm's law, we solve for the emitter voltage:

$$V_E = I_E \times R_E = 13.36\,mA \times 120\,\Omega = 1.60\,V$$

The base voltage equals the emitter voltage plus the barrier potential or:

$$V_B = V_E + 0.7\,V = 1.60\,V + 0.7\,V = 2.30\,V$$

The collector current equals the emitter current or:

$$I_C = I_E = 13.36\,mA$$

Applying Ohm's law and the values that we now have, we can solve for the voltage across the collector resistor:

$$V_{RC} = I_C \times R_C = 13.36\,mA \times 470\,\Omega = 6.28\,V$$

The collector voltage equals the collector supply voltage minus the voltage dropped across the collector resistor or:

$$V_C = V_{CC} - V_{RC} = 15\,V - 6.28\,V = 8.72\,V$$

The base current equals the collector current divided by the value of beta:

$$I_B = \frac{I_C}{\beta} = \frac{13.36\,mA}{150} = 89.1\,\mu A$$

PRACTICE PROBLEM 4.8

Using Figure 4.47, calculate the Thevenin values for the base voltage and the base resistance. Calculate the values for the base, emitter, and collector voltages and currents. Trace the current paths.

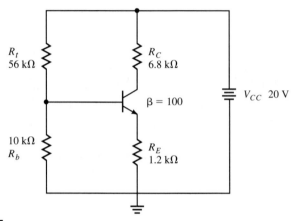

FIGURE 4.47

PRACTICE PROBLEM 4.9

Using the component values shown in Figure 4.48, determine values for the emitter, collector, and base currents and voltages.

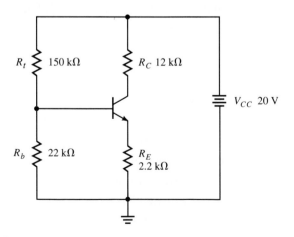

FIGURE 4.48

HOW TO SET UP THE VOLTAGE DIVIDER

Stiff and Firm Voltage Dividers

By definition, a voltage divider is considered **stiff** if the thevenized resistance divided by the beta value is less than or equal to 0.01 multiplied by the emitter resistance or:

$$\frac{R_{TH}}{\beta} \leq 0.01 \times R_E$$

When using this guideline, choose the lowest value of beta listed on the transistor specification sheet. Meeting the guideline allows the dropping of the thevenized resistance and beta values from circuit calculations with less than a 1% error resulting.

We can consider the voltage divider as **firm** if the thevenized resistance divided by the beta value is less than or equal to 0.1 multiplied by the emitter resistance or:

$$\frac{R_{TH}}{\beta} \leq 0.1 \times R_E$$

Again using the transistor specification sheet, choose the lowest value for beta. If the equation meets the guidelines, then the thevenized resistance and beta values may be dropped from circuit calculations with less than a 10% error.

A quick calculation also gives rough values for the emitter current and the terminal voltages. If the divider is at least firm, the value predictions will be off by no more than 10%.

EXAMPLE 4.14

Solve for the DC parameters of the circuit shown in Figure 4.49.

Assuming that the voltage divider is firm, we will use a quick method for computation. If we picture the voltage divider from Figure 4.49 as a stand-alone circuit, Figure 4.50a, we can apply the voltage divider theorem:

$$V_{TH} = \frac{2 \text{ k}\Omega}{6 \text{ k}\Omega} \times 18 \text{ V} = 6 \text{ V}$$

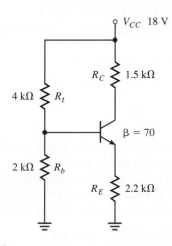

FIGURE 4.49

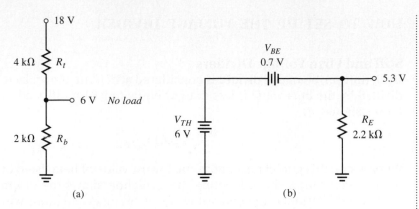

FIGURE 4.50 (a) Using the quick method for solving a firm voltage divider. (b) Using thevenized voltage to calculate V_E

This Thevenin voltage represents the voltage at the node between the top and bottom resistors when the divider is not loaded or connected to the transistor. Next, we can subtract 0.7 V to account for the voltage at the base terminal and the base-emitter drop from the thevenized voltage:

$$6\,V - 0.7\,V = 5.3\,V$$

Assigning this voltage to the emitter terminal, we can solve the rest of the circuit as shown in Figure 4.50b.

Applying Ohm's law to the emitter resistor, we find that:

$$I_E = I_C = \frac{5.3\,V}{2.2\,k\Omega} = 2.41\,mA$$

To find the base current, we divide the emitter current by the beta value and have:

$$\frac{2.41\,mA}{70} = 34.43\,\mu A$$

At this point, we know the values for all three currents and the base and emitter voltages. We can move to the collector. The collector resistor has a voltage drop of:

$$V_{RC} = I_C \times R_C = 2.41\,mA \times 1.5\,k\Omega = 3.62\,V$$

Subtracting this value from the collector supply voltage, we find that the voltage at the collector terminal equals

$$18\,V - 3.62\,V = 14.38\,V$$

An Alternative Method

Using beta and the effective base resistance, we will again solve the circuit shown in Figure 4.49. Figure 4.51 shows this procedure.

First, as Figure 4.51a shows, we thevenize the base part of the control loop to obtain the Thevenin equivalent circuit for the base. The Thevenin voltage equals approximately 6 V, and the Thevenin resistance is the result of $R_t \parallel R_b$:

$$R_{TH} = \frac{2\,k\Omega \times 4\,k\Omega}{2\,k\Omega + 4\,k\Omega} = \frac{8\,k\Omega}{6\,k\Omega} = 1.33\,k\Omega$$

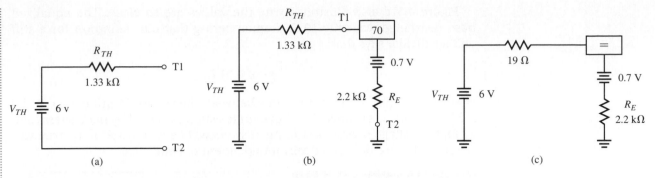

FIGURE 4.51 (a) Thevenizing the base part of the control loop in the circuit shown in Figure 4.49. (b) Attaching a control loop to the thevenized circuit. (c) Equalized control loop model of the circuit.

In Figure 4.51b, the emitter side of the control loop model attaches to the Thevenin equivalent circuit. Equalizing the control loop model gives us the circuit shown in Figure 4.51c. From this circuit, we can calculate the emitter current:

$$I_E = I_C = \frac{\left(6\,V - 0.7\,V\right)}{2219} = 2.39\,mA$$

With this value, we can also calculate the emitter voltage:

$$2.39\,mA \times 2.2\,k\Omega = 5.26\,V$$

Adding 0.7 V to the emitter voltage value, we arrive at the base voltage value:

$$5.26\,V + 0.7\,V = 5.96\,V$$

Next, we determine the amount of voltage dropped across the collector resistor:

$$2.39\,mA \times 1.5\,k\Omega = 3.59\,V$$

Subtracting this value from the collector supply voltage value, we have a value for the collector voltage:

$$18\,V - 3.9\,V = 14.10\,V$$

As we can see, the quick solution provided answers close to the answers obtained when we accounted for the effective base resistance and beta. Table 4.1 lists the values given by the two methods.

Table 4.1		
Parameter	Quick	Thorough
Emitter current	2.41 mA	2.39 mA
Base current	34.43 μA	34.14 μA
Collector current	2.41 mA	2.39 mA
Emitter voltage	5.3 V	5.26 V
Base voltage	6 V	5.96 V
Collector voltage	14.38 V	14.10 V

Figure 4.51c also illustrates why the values are so close. The equalized base resistance was only 19 Ω. Remembering that our definition for a stiff voltage divider says that

$$\frac{R_{TH}}{\beta} \leq 0.01\,R_E$$

sheds some light on the similarities between the values. Multiplying 0.01 by 2.2 kΩ, we get 22 Ω. Since the Thevenin resistance divided by beta is less than 22 Ω, the voltage divider is stiff. All answers will be within 1% of the true values, and the circuit is practically independent of beta.

PRACTICE PROBLEM 4.10

For the circuits shown in Figure 4.52, determine the values of the emitter, base, and collector currents and voltages using both the quick and thorough methods. Compare the results. Tell if the circuits are stiff, firm, or neither.

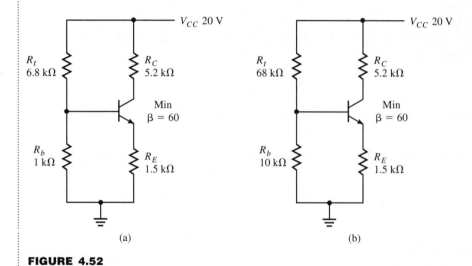

FIGURE 4.52

PRACTICE PROBLEM 4.11

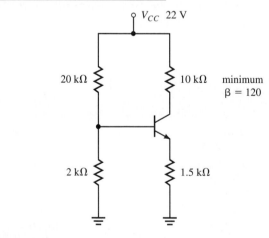

FIGURE 4.53

Determine the values of the emitter, collector, and base currents and the emitter, base, and collector voltages for the voltage divider circuit shown in Figure 4.53. Determine if the voltage divider is stiff or firm.

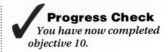

REVIEW SECTION 4.7

1. Of base bias, collector feedback bias, and voltage divider bias, which method provides the most circuit stability?
2. R_t and R_b are used as a _____ _____.
3. When we thevenize to find the thevenized resistance, R_t and R_b are in (series, parallel).
4. If beta varies from 50 to 300, what value of beta should be used to determine whether a voltage divider is stiff or firm?
5. Explain how to calculate the values of the base, emitter, and collector voltages and currents for a stiff voltage divider circuit.
6. Compare and contrast voltage divider bias and the other bias circuits that we have studied.

4.8 THE BIASING OF THE PNP TRANSISTOR

The voltages used to bias PNP transistors have the opposite polarity from those used to bias NPN transistors. Even though current flows in the opposite direction, we can still use the same skills and procedures that we used to solve NPN transistor circuits.

Figure 4.54 represents the beta box model for a PNP transistor. Notice that the collector and base currents flow into the node inside the beta box. These two currents combine to form the emitter current, which flows out the emitter. Also, note the polarity of the diode representing the base-emitter junction. The cathode, or negative side, of the diode attaches to the node within the beta box. To forward bias the transistor junction, the base must be negative with respect to the emitter.

In Figure 4.55, we see the base bias and emitter bias configurations for PNP transistors. In the base bias circuit, the collector supply voltage is now negative with respect to ground. To see how we can solve this circuit, consider the following example.

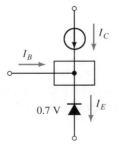

FIGURE 4.54 A beta box bias model for a PNP transistor. The arrows point in the direction of electron flow.

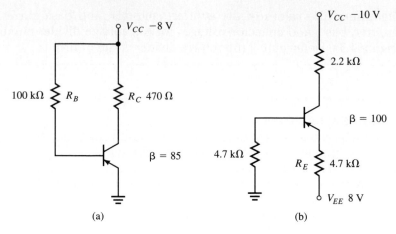

(a)   (b)

FIGURE 4.55 (a) Base-bias configuration for a PNP transistor. (b) Emitter-bias configuration for a PNP transistor.

EXAMPLE 4.15

Solve for the values of the emitter, base, and collector currents and the emitter, base, and collector voltages for the circuit depicted in Figure 4.55a.

Figure 4.56 shows the control loop model. Again, a similarity exists between this model and the model used for NPN base-bias circuits. Polarities for both the collector and the base-emitter bias, or supply, voltages reverse. Consequently, the base current flows into the base lead. As this is a simple base-biased circuit with no emitter resistor, we can directly solve for the base current.

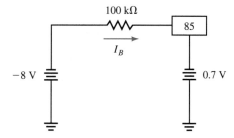

FIGURE 4.56 Control loop model for the circuit shown in Figure 4.55a. No equalization is needed.

Solution

The effective voltage of the control loop is 7.3 V. Since the beta box has only one resistor, we do not need to worry about equalization. We can use the 100-kΩ value in our calculations:

$$I_B = \frac{7.3 \text{ V}}{100 \text{ k}\Omega} = 73 \text{ μA}$$
$$I_E = 73 \text{ μA} \times 85 = 6.21 \text{ mA}$$

The voltage used by the collector resistor equals

$$6.21 \text{ mA} \times 470 \ \Omega = 2.92 \text{ V}$$

Because the side of the collector resistor connected to the collector terminal is more positive than the side attached to the collector supply voltage, we add V_{CC} to V_{RC}:

$$V_C = -8 \text{ V} + 2.92 \text{ V} = -5.08 \text{ V}$$

The collector terminal voltage *is negative* with respect to ground. Since the emitter terminal connects directly to ground, the emitter voltage equals zero. The base voltage is -0.7 V.

Given the negative values of the base and collector voltages, the P-type material of the collector is more negative with respect to ground than the N-type material of the base. This defines reverse bias.

EXAMPLE 4.16

Solve the emitter-biased PNP circuit shown in Figure 4.55b.

Again, except for the reversed polarities of the emitter and base-emitter bias, or supply, voltages, the control loop model for the PNP emitter bias closely resembles the model used for the NPN example. With resistors on both sides of the beta box, we must equalize the model. Figure 4.57 shows the entire process. Using the equalized model depicted in Figure 4.57b, we can calculate the emitter current:

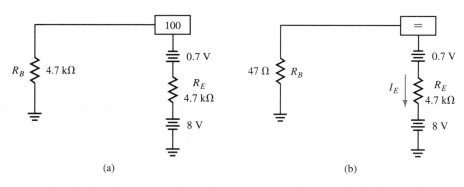

(a)   (b)

FIGURE 4.57 (a) Control loop model for the circuit shown in Figure 4.55b. (b) Equalized model of the circuit.

$$I_E = I_C = \frac{7.3 \text{ V}}{\left(4.7 \text{ k}\Omega + 47 \ \Omega\right)} = 1.54 \text{ mA}$$

The base current is the emitter current divided by beta, so

$$\frac{1.54 \text{ mA}}{100} = 15.4 \ \mu\text{A}$$

The emitter resistor uses:

$$1.54 \text{ mA} \times 4.7 \text{ k}\Omega = 7.24 \text{ V}$$

Because the side of the emitter resistor connected to the emitter terminal has a less positive voltage than the side connected to the emitter voltage supply, we can subtract V_{RE} from V_{EE}:

$$8\,V - 7.24\,V = 0.76\,V$$

Even though the voltage at the PNP emitter has the same magnitude as the NPN emitter, it has the opposite polarity. To obtain the base voltage, we subtract:

$$0.76\,V - 0.7\,V = 0.06\,V$$

As with the NPN emitter-biased circuit, the base terminal of the PNP emitter-biased circuit is almost 0 V.

One set of calculations remains. We can determine the voltage across the collector resistor with the following simple formula:

$$1.54\,mA \times 2.2\,k\Omega = 3.3\,V$$

The side of the collector resistor connected to the collector is more positive than the side connected to the collector voltage source. We add the voltage dropped across the collector resistor to the collector supply voltage to find the collector terminal voltage:

$$-10\,V + 3.39\,V = -6.61\,V$$

The collector voltage is negative with respect to ground.

COLLECTOR FEEDBACK BIAS FOR THE PNP TRANSISTOR

Let us study Figure 4.58. Here, we see a collector feedback bias arrangement with a PNP transistor. As the figure shows, the collector supply voltage is -12 V with respect to ground. Even though the earth ground symbols represent a point in the circuit that would have zero volts, we can use another point as the zero reference point. In Figure 4.58b, TP1 ties directly to the negative terminal of the collector voltage supply. Instead of using earth ground as the zero reference point, we can use TP1.

Our perspective, or reference, does change with our choice to use TP1 as the zero reference point. If we connect the voltmeter negative test lead to TP1 and the positive test lead to the earth ground, the ground point will read $+12$ V higher than the zero reference. Going one step further, we should look

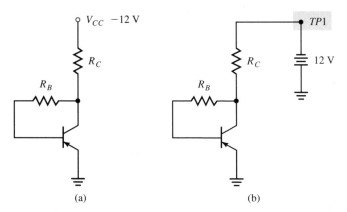

FIGURE 4.58 (a) Collector feedback bias arrangement with a PNP transistor. (b) Another view of the same circuit.

at Figure 4.59a. In Figure 4.59a, we have redrawn the circuit from the per-spective of the new reference point. The emitter, the "old ground," now con-nects to a positive 12 V. At the bottom of the collector resistor, TP1 works as the ground reference. To keep the ground symbol at the bottom of the dia-gram, we then turn the PNP transistor symbol upside down.

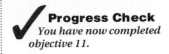
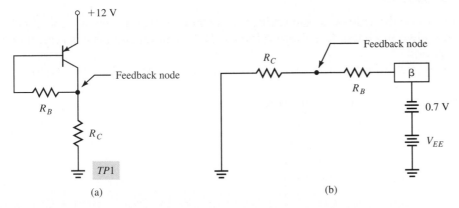

FIGURE 4.59 (a) The circuit shown in Figure 4.58a redrawn, upside down. (b) Beta box model of the circuit with the control loop and feedback node.

WHY DRAW THE PNP TRANSISTORS UPSIDE DOWN?

Disregarding the schematic location of the transistors, the model of the con-trol loop functions as it should. Figure 4.59b shows the beta box model of a control loop with a feedback node. The resistance between the beta box and the feedback node must be equalized by dividing the resistance by beta. We can solve the circuit by finding the series, or emitter, current.

Figure 4.60, should help to explain why the schematic shows the transis-tors upside down. On the left, an NPN transistor works with base bias. To the right, a PNP transistor uses collector feedback bias. The two share a common voltage supply, which is $+12$ V with respect to ground. Electron flow through both transistors goes up from ground, through the transistors, and to the voltage supply. If the schematic showed a separate reference and supply

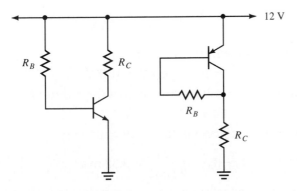

FIGURE 4.60 Schematic illustrating need for upside-down PNP transistors. Here an NPN (on the left) and a PNP (on the right) transistor share a common power supply.

voltage for the PNP transistor, the drawing would be cluttered. Instead, the drawing shows that the circuit will function normally by connecting the PNP as diagrammed.

___

EXAMPLE 4.17

Determine the values for the emitter, base, and collector currents for the circuit shown in Figure 4.61. Determine the DC terminal voltages as referenced to ground.

   The circuit featured in Figure 4.61 is an upside-down PNP transistor circuit using voltage divider bias. The divider is stiff, so we can use the quick method of finding the DC parameters.

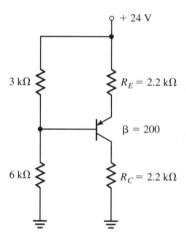

FIGURE 4.61 **Upside-down PNP transistor using voltage divider bias**

Solution
Using the voltage divider theorem, we will determine the voltage at the base to ground:

$$V_B = \frac{6 \text{ k}\Omega}{9 \text{ k}\Omega} \times 24 \text{ V} = 16 \text{ V}$$

and add 0.7 V to obtain the emitter voltage:

$$V_E = 16 \text{ V} + 0.7 \text{ V} = 16.7 \text{ V}$$

Then, we can determine the voltage drop across the emitter resistor:

$$24 \text{ V} - 16.7 \text{ V} = 7.3 \text{ V}$$

Using this voltage, we can calculate the emitter current:

$$I_E = I_C = \frac{7.3 \text{ V}}{2.2 \text{ k}\Omega} = 3.32 \text{ mA}$$

Using the value for the collector current, we can determine the voltage used by the collector resistor:

$$3.32 \text{ mA} \times 2.2 \text{ k}\Omega = 7.3 \text{ V}$$

Since the resistor connects to ground, the collector voltage also equals 7.3 V. Last, we can find the base current by dividing the collector current by beta:

$$I_B = \frac{3.32 \text{ mA}}{200} = 16.6 \, \mu A$$

PRACTICE PROBLEM 4.12

Redraw the circuit shown in Figure 4.61 rightside up and shift the ground voltage reference to the +24-V supply. The ground connections will change to a negative 24 V, and the positive 24-V connection will convert to ground. Determine the voltages and polarities at the emitter, base, and collector with respect to the new ground.

REVIEW SECTION 4.8
1. Why are many PNP transistor schematics drawn upside down?
2. What comparisons can be made between the voltage polarities required to bias NPN transistors and the voltage polarities required to bias PNP transistors?
3. Do electrons flow into or out of the emitter terminal of a PNP transistor?

4.9 COLLECTOR CURVES

It is possible to graph the response of the collector current to variations in base current and changes in voltage from collector to emitter. Such a graph becomes useful for providing information about transistor characteristics. Technicians rely either on test circuits that generate these curves (Figure 4.62) or on a curve tracer, a piece of test equipment that generates and displays collector curves.

Figure 4.63 shows a typical collector curve. The curve has three regions—saturation, breakdown, and active.

To refresh your memory, the breakdown region starts when the voltage from collector to emitter exceeds a value known as V_{CEO}. This voltage is the

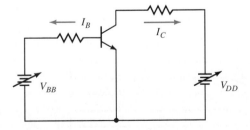

Note: Changes in V_{DD} affect V_{CE} for fixed values of base current.
Changes in V_{BB} produce collector curves for different values of I_B.

FIGURE 4.62 Test circuit for obtaining collector curve(s)

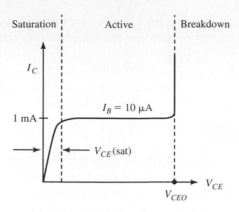

FIGURE 4.63 A typical collector curve of a transistor

voltage at which conduction in the reverse-biased collector-base junction begins a very rapid increase. The junction is not designed to conduct very large currents, so the rapid increase at the junction is usually very bad for the transistor. As you can see, the curve goes almost straight up upon reaching V_{CEO}. A vertical line on this type of graph indicates a short.

The active region lies between the voltage that just reverse biases the collector-base junction and V_{CEO}. During this span, the collector current remains almost constant. In this range of voltages, changes in V_{CEO} have almost no effect on collector current, so the collector is acting like a constant current source. The active region is the region that is used when we want to use the transistor as an amplifier.

A family of collector curves can be generated for each transistor by varying the base current. For each change in base current a new collector curve will be plotted, as in Figure 4.64. The bottom curve in the graph represents the condition in which $I_B = 0 \ \mu A$. At that point the transistor is essentially cut off. There is a small amount of current running through the transistor. This cur-

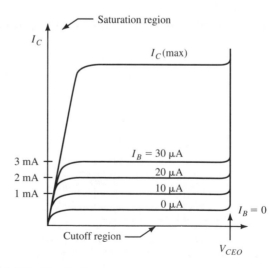

FIGURE 4.64 A family of collector curves

rent is due to leakage effects and other solid state imperfections in the transistor. Specification sheets will list this current as I_{CEX}.

Each time the base current is changed, a new collector curve results. Beta can be determined by dividing I_B into the resultant collector current. In Figure 4.64, beta is 100. The topmost curve is most often the maximum collector current that the transistor can provide. Exceeding this value can damage the transistor.

Collector curves are a function of the transistor itself. These values are not dependent on how the transistor is used in a circuit; rather, the curves can be used to predict how the transistor will react in a circuit.

DC LOAD LINES

To use collector curves, we also need to take a look at the DC load line. Before, our analysis of the DC operation of bipolar junction transistors focused on the control loop. This loop was usually the base-emitter loop, and, in the special case of the collector feedback bias configuration, included the collector circuit.

EXAMPLE 4.18

Determine the value of the current that flows through a circuit with a saturated transistor.

Now, we will focus on the emitter-collector loop and see how it limits circuit operation. Figure 4.65 shows a base-biased NPN circuit. When the transistor conducts, the emitter and collector sections form a closed loop. This section is highlighted in the figure.

When the transistor goes into saturation, it acts as a closed switch. Ideally, a closed switch should drop no volts, but the saturation voltage of a transistor is normally 0.2 to 0.3 V. Figure 4.65b shows the saturation voltage drop of the transistor. As a result, the circuit is a closed loop.

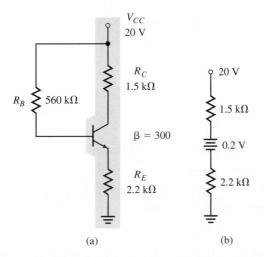

(a)   (b)

FIGURE 4.65 (a) Base-biased NPN transistor circuit. (b) Saturation loop of the transistor.

Solution

We can use Ohm's law to calculate the current that flows through a circuit with a saturated transistor:

$$I_C \text{ (sat)} = \frac{(20\text{ V} - 0.2\text{ V})}{3.7\text{ k}\Omega} = 5.35\text{ mA}$$

This value of current at saturation marks one end of the DC load line.

The other end of the DC load line marks the point where the transistor no longer conducts, or is cut off. We can designate this point as V_{CE} (cutoff). In Figure 4.66, the cutoff voltage is 20 V. To understand this concept, remember that no current flows. With no current flowing, the top of the emitter resistor is at ground potential, and the emitter voltage measures at 0 V. Therefore the bottom of the collector resistor measures at 20 V. The full supply potential is exerted across the transistor.

Because we know both end points of the DC load line, we can plot the line on a graph. Using the set of collector curves in Figure 4.67, we have plotted the load line. While the collector curves represent the characteristics of a transistor, the load line represents the limits of the circuit. The intersection of the load line and a collector curve yields the collector current and collector-to-emitter voltage for a given base current. This allows us to predict how

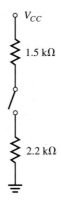

FIGURE 4.66 Illustrating cutoff. Transistor shown as an open switch.

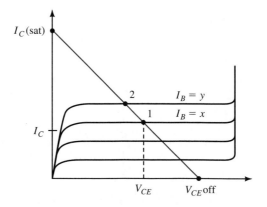

FIGURE 4.67 DC load line superimposed on collector curves

various levels of base current will affect circuit operation. If we set the value of the base current at *x*, then we arrive at the conditions associated with point 1. Changing the value of the base current to *y*, we find the conditions associated with point 2. From this partial graph, it becomes apparent that the base current decreases as the transistor moves toward cutoff. As the base current becomes larger, the transistor moves toward saturation.

EMITTER BIAS AND DC LOAD LINES

An emitter-biased transistor circuit has two power supplies in the emitter-collector loop. This does not make it any more difficult to calculate the ends of the load line. Let us take a look at Figure 4.68.

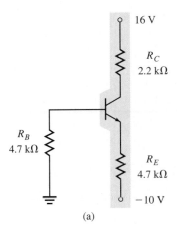

(a)

FIGURE 4.68 (a) Calculate the ends of the DC load line. (b) Control loop for the circuit.

EXAMPLE 4.19

Calculate the ends of the DC load line for the circuit shown in Figure 4.68a.

The ends of the DC load line are determined by the emitter and collector sections of the transistor circuit and the power supplies. Because emitter-biased circuits have two power supplies, we must account for both.

Solution

Figure 4.68b shows this highlighted portion of Figure 4.68a and shows it as if it were a separate circuit. A 0.2-V supply, the value of the saturation voltage, represents the transistor. To find the collector current at saturation, we can substitute the value of the saturation voltage into Ohm's law:

$$I_C(\text{sat}) = \frac{(26 \text{ V} - 0.2 \text{ V})}{(2.2 \text{ k}\Omega + 4.7 \text{ k}\Omega)} = 3.74 \text{ mA}$$

Because the absence of current sends the emitter terminal to -10 V and the collector terminal to $+16$ V, the cutoff voltage equals the sum of the two power supplies:

$$V_{CE} \text{ (cutoff)} = 16 \text{ V} - (-10 \text{ V}) = 26 \text{ V}$$

We have now determined the two end points of the DC load line.

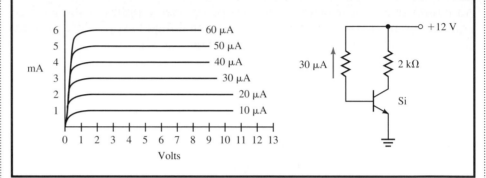

TROUBLESHOOTING

In the next chapter, we will investigate how transistors amplify AC signals. Often, the failure of a transistor to amplify properly occurs because some part of the DC circuit operation has broken down. In this section, we will discuss some of the failures that can happen in the DC circuit operation. Common practice dictates that we begin any troubleshooting process by measuring DC values and checking those measurements against the values that we predict. From there, we can move our problem solving to the AC operation of the circuit.

DON'T OVERLOOK THE POWER SUPPLY

Check the power supply first. Make sure that the voltage level coming out of the power supply is correct. Then check for proper distribution of the voltages around the circuit. If the voltages are missing, turn off the power, and recheck the power supply. Blown fuses may indicate a short circuit in the rectifier circuits. If the rectifiers check out, use an ohmmeter and test from the collector voltage point to ground for a short circuit. A short between the collector voltage supply and ground can cause fuses to blow.

USE YOUR EYES AND NOSE

A technician should complete the repair properly and quickly. A careful visual inspection of the circuitry may show which part has failed and may save the time needed for calculations or tests. Look for broken leads, bad solder joints, cracks in the printed circuit board, and burnt parts. Use your sense of smell to determine whether a part is running hot. Many solid state devices give a distinctive odor when running abnormally hot.

DON'T ASSUME ANYTHING

Often a circuit will malfunction or a part will run hot because it has been installed backwards. Always check the orientation of the transistor, capacitor, and diode leads. Sometimes, a wrong part has been installed. If you have access to the schematics or semiconductor substitution guides, check the part numbers and parameters. Assume nothing, and check all the possibilities.

CHECKING THE TRANSISTOR

As you check the circuitry, it may become necessary to check the transistor. An ohmmeter and a simple test can disclose problems within the transistor. While the ohmmeter test will not show if the beta is too low or if a transistor is partially shorted, it will show opens and shorts. More complex tests require the use of a curve tracer or transistor checker.

Both the curve tracer and transistor checker place the transistor under test while applying voltages to terminals. In most cases, the technician can select voltages that closely match the circuit conditions for the transistor. Using the transistor checker, a technician can use meter indications to show how the transistor acts with applied voltages. With the curve tracer, the technician can view the operating curves of the transistor while the transistor functions with the applied voltages. Then the technician can compare the actual transistor curves against the manufacturer's data sheets. An ohmmeter should detect very high resistance from emitter to collector in both directions with the base lead open. If the ohmmeter does not show a high resistance between the two leads, then something may be wrong within the transistor. The ohmmeter may not indicate an infinite open because a small amount of current can flow from collector to emitter with the base open.

Next, test the emitter-base junction. In one direction, the junction is forward-biased. To perform this test, you must make sure the voltage source inside the ohmmeter is large enough to overcome the barrier potential. Select ranges that do not exceed the maximum current or voltage ratings of a device. With forward bias, the resistance should measure less than 1000 Ω. Reversing the bias to the junction through reversing the meter leads should give an indication of over 100 kΩ. If the meter indicates a high or low resistance measurement in both directions, replace the transistor.

Remember also that the DC voltmeter can complete a circuit path. Figure 4.69 illustrates this potential problem. The boxed information contains the normal operating values for the circuit. If the emitter resistor opens, then the emitter current should measure 0 amps. However, if we set up the voltmeter as shown in the figure, the internal resistance of the meter completes the control loop path.

The equalized model for this situation can be seen in Figure 4.70. With the meter in this circuit, approximately 134 nA flow through the meter resistance, giving:

$$134 \text{ nA} \times 10 \text{ M}\Omega = 1.34 \text{ V}$$

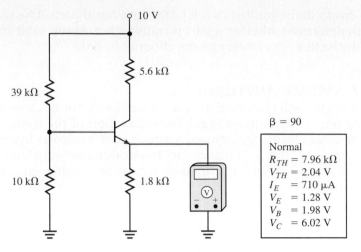

FIGURE 4.69 How a DC voltmeter can complete a circuit path

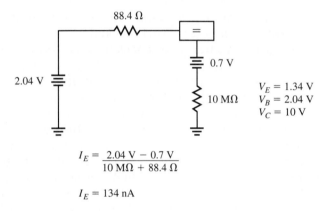

$$I_E = \frac{2.04 \text{ V} - 0.7 \text{ V}}{10 \text{ M}\Omega + 88.4 \text{ }\Omega}$$

$$I_E = 134 \text{ nA}$$

FIGURE 4.70 Equalized model of the circuit shown in Figure 4.69

a slightly higher than normal voltage reading. When we add 0.7 V for the base voltage value, the base voltage value shows as the Thevenin voltage, or 2.04 V. With the 10 MΩ from the meter in the emitter circuit, the transistor will not load down the voltage divider.

Problems may also occur when measuring voltages at the emitter terminal of an emitter-biased circuit. In Figure 4.71a, the meter is positioned improp-

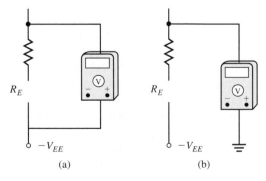

(a)   (b)

FIGURE 4.71 (a) Improper positioning of meter while measuring voltages at the emitter terminal of an emitter-biased transistor. (b) Proper positioning of the meter.

erly and completes the control loop. Positioned as shown in Figure 4.71b, the meter can test the emitter voltage without completing the control loop.

HOW TO CHECK THE FOUR BIAS PROTOTYPES

The Base-Bias Prototype
For any test of a transistor, start at the base. Figure 4.72 is a schematic of the base-bias prototype. It includes an emitter resistor. As we progress through the text, we will look at switching transistors and biasing. For now, though, let us concentrate on conventional biasing.

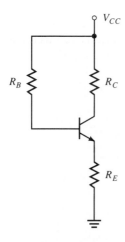

FIGURE 4.72 Base-bias prototype

Symptom	Probable cause
1. Base voltage equals 0 V.	Open base resistor
2. Lower than expected base voltage along with a lower than expected emitter voltage and 0 V at the collector voltage.	Open collector resistor
3. Lower than expected base voltage along with a lower than expected emitter voltage, and the collector voltage equals the collector supply voltage.	Faulty transistor
4. Lower than expected base voltage along with an emitter voltage that equals zero.	Emitter terminal shorted to ground
5. Higher than expected base voltage with a higher than expected emitter voltage and a lower than expected collector voltage.	Faulty transistor
6. Higher than expected base voltage with a higher than expected emitter voltage and a higher than expected collector voltage.	R_B too small

7. Higher than expected base voltage with a normal emitter voltage.	Faulty transistor
8. Base voltage equals the collector supply voltage with the emitter voltage nearly equaling the collector supply voltage.	Short in the base resistor circuit or an open emitter resistor
9. Base voltage equals the collector supply voltage with the emitter voltage at 0 V.	Faulty transistor

The Emitter-Bias Prototype

Figure 4.73 shows a schematic of the emitter-bias prototype. Remember that the base voltage of an emitter-biased NPN transistor circuit is close to 0 V, and the emitter voltage will measure at a negative 0.7 V.

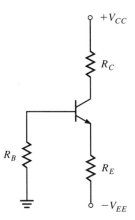

FIGURE 4.73 Emitter-bias prototype

Symptom	Probable cause
1. Base voltage is negative, but the emitter voltage remains 0.7 V lower than the base voltage.	Open collector resistor
2. Both the base and emitter voltages are at 0 V while the collector voltage nearly equals the collector supply voltage.	Open emitter resistor
3. Base voltage equals zero, emitter voltage equals the negative emitter supply voltage, and the collector voltage equals the collector supply voltage.	Open transistor. Check for shorts in the emitter and collector circuits before installing a new transistor.

Notes on shorts: If the emitter resistor circuit shorts, the negative emitter supply voltage will show at the emitter. If the collector resistor circuit shorts, the collector supply voltage will appear at the collector. Both of these conditions will substantially shorten the life of the transistor. If the base resistor circuit shorts, little or no difference will appear with the DC operation. This symptom will affect the AC circuit operation.

The Voltage Divider Prototype

Figure 4.74 shows the schematic of a voltage divider bias prototype.

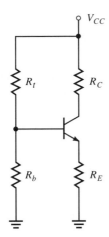

FIGURE 4.74 Voltage divider prototype

Symptom	Probable cause
1. Base voltage is 0 V.	R_t has opened, or the base resistor circuit has shortened.
2. Base voltage too low, and the voltage across the emitter resistor equals zero.	Short in the emitter resistor circuit
3. Base voltage too low, and the voltage across the emitter resistor is lower than normal but does not equal zero. Collector voltage equals the emitter voltage.	Open collector resistor
4. Base and emitter voltages are too high, with the collector voltage measuring too low.	Base resistor is open.
5. Base, emitter, and collector voltages are too high.	Faulty transistor
6. Base voltage equals the collector supply voltage; both the emitter and collector voltages are near the collector supply voltage value.	Short in the R_T circuit
7. Correct base voltage. Collector voltage equals the collector supply voltage, and the emitter resistance measures higher than normal.	Open emitter resistor

Symptom	Probable cause
8. Correct base voltage. Collector voltage equals the collector supply voltage, and the emitter voltage is normal.	Short in collector resistor circuit
9. Correct base voltage. Collector voltage equals the collector supply voltage, and the emitter voltage equals the base voltage.	Faulty transistor

The Collector Feedback Prototype

Figure 4.75 shows the prototype circuit for collector feedback bias.

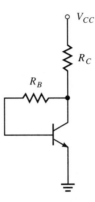

FIGURE 4.75 Collector feedback prototype

Symptom	Probable cause
1. Base and collector voltages near zero	Open collector resistor
2. Base voltage is zero; collector voltage equals the collector supply voltage.	Open base resistor
3. Base and collector voltages equal 0.7 V.	Short in base resistor circuit
4. Base voltage equals 0.7 V, and collector voltage equals the collector supply voltage.	Short in the collector resistor circuit
5. Base and collector values equal the collector supply voltage, and the emitter voltage equals zero.	Faulty transistor
6. Base, emitter, and collector supply voltages equal 0.7 V.	Open in emitter circuit, collector-base junction shorted

Answer questions 1–5 using the following circuit.

1. If R_{B1} was reduced in value, would the operating point on the DC load line move toward saturation or cutoff?

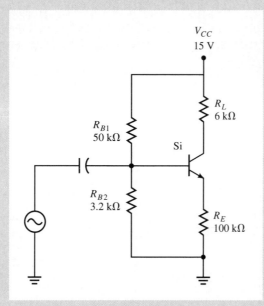

2. If R_E opened, V_{CE} would equal approximately _____.
3. If R_{B1} opened, V_C would equal approximately _____.
4. What resistor(s) compensate(s) for beta?
5. What resistor(s) compensate(s) for thermal runaway?

Determine the cause of the failure of the following base-biased circuit with an emitter resistor for each of the readings in 6–12. The normal readings are M1 = 0.54 V, M2 = 1.23 V, and M3 = 6.64 V.

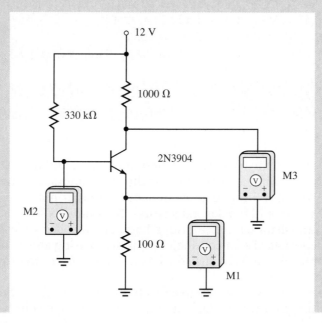

6. M1 = 11.65 V, M2 = 11.91 V, and M3 = 12.11 V
7. M1 = 0.0 V, M2 = 0.0 V, and M3 = 12.11 V
8. M1 = 0.0 V, M2 = 0.58 V, and M3 = 0.03 V
9. M1 = 1.03 V, M2 = 1.75 V, and M3 = 1.75 V
10. M1 = 0.0 V, M2 = 0 V, and M3 = 12.11 V
11. M1 = 0 V, M2 = 0.7 V, and M3 = 6.35 V

Determine the cause of the failure of the following voltage divider circuit for each of the readings in 12–16. The normal readings are M1 = 2.13 V, M2 = 2.8 V, and M3 = 9.1 V.

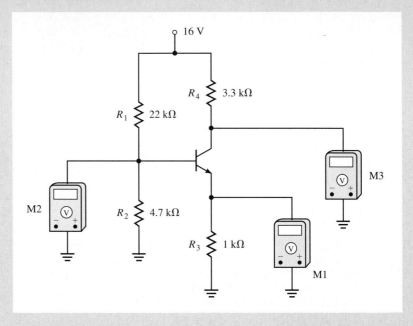

12. M1 = 0 V, M2 = 0 V, and M3 = 16.05 V
13. M1 = 4.14 V, M2 = 4.84 V, and M3 = 4.20 V
14. M1 = 0.45 V, M2 = 1.1 V, and M3 = 0.45 V
15. M1 = 0.45 V, M2 = 4.04 V, and M3 = 4.04 V
16. M1 = 0 V, M2 = 2.85 V, and M3 = 16.05 V

SUMMARY

Transistors can amplify signals or switch devices on and off. A bipolar transistor features the same construction as the semiconductor diode. However, a transistor contains three sections of semiconductor material and has two junctions. Two types of bipolar transistors—NPN and PNP—exist.

A bipolar transistor has an emitter, a base, and a collector region. Because each region consists of semiconductor material, biasing affects transistor operation. A properly biased transistor has a small base current controlling a large collector current.

Like diodes, transistors have specific characteristics. The ratio of collector current to emitter current is called alpha, and the ratio of collector current to base current is called beta. Both ratios affect the performance of a transistor. In addition, temperature can change the beta ratio. A transistor without some method of minimizing the effects of temperature on beta might be damaged by thermal runaway.

Other set transistor characteristics are cutoff, saturation, and breakdown. Cutoff occurs with the absence of base and collector current. The full supply voltage appears across the transistor instead of the load. With saturation, any further increase in base current will not cause an increase in collector current. Little or no voltage appears across the transistor. Some transistors use cutoff and saturation and act as switches. Breakdown in a transistor may occur when the collector-to-emitter junction or the base-to-collector junction is excessively reverse-biased. The collector-to-emitter voltage reaches a maximum point with the base open. Breakdown can damage a transistor.

A beta box is a model that illustrates current action within a transistor. We use the beta box to simplify solving transistor circuits. Using the beta box breaks the circuit down into equivalent circuits and control loops.

Several methods exist for biasing transistors. Base bias applies a bias voltage to the emitter-base and collector-base junctions. Feedback returns a portion of the output to the input side of the transistor. Collector feedback and emitter feedback bias use negative feedback. Voltage divider bias provides a stable method of minimizing the effects of variations in beta.

By using collector curves, a technician can graph the response of collector current to changes in base current and collector-emitter voltage. Using collector curves also allows the appplication of DC load lines.

Troubleshooting transistor circuits requires a consistent approach of checking power supply voltages and voltage distribution. A technician should also look for burnt resistors, improperly installed components, and blown fuses.

After checking the circuit for obvious malfunctions, a technician should look for defective transistors. Technicians can use an ohmmeter and a simple test to check transistors. More complex equipment such as curve tracers and transistor checkers also exists.

PROBLEMS

1. Determine the following for each of the circuits in Figures 4.76 through 4.80:
   a. Is this an NPN or a PNP transistor?
   b. What is the bias configuration (base bias, emitter bias, voltage divider bias, or collector feedback bias)?
   c. Calculate the DC emitter current.
   d. Calculate the DC base current.
   e. Calculate the DC voltages.
   f. Calculate the power dissipated by the transistor.
   g. Calculate the DC saturation current.
   h. Calculate the DC cutoff voltage.

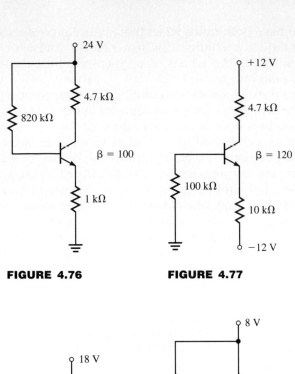

FIGURE 4.76

FIGURE 4.77

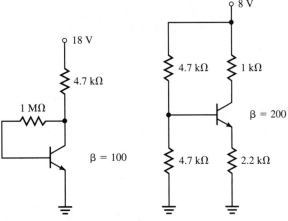

FIGURE 4.78

FIGURE 4.79

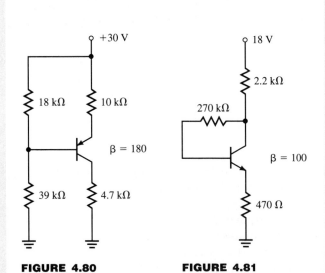

FIGURE 4.80

FIGURE 4.81

2. Develop and draw the beta box control loop model for the circuit in Figure 4.81.
3. Use Figure 4.82 for the remaining problems.
   a. What is V_{CE}?
   b. Plot the DC load line, and label the end points.
   c. Determine the Q point and label I_C and V_{CE}.

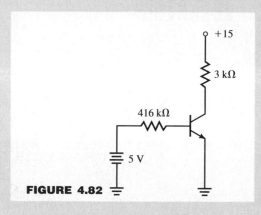

FIGURE 4.82

TRANSISTORS WORKING AS SMALL-SIGNAL AMPLIFIERS

Courtesy of Sprague-Goodman Electronics

OBJECTIVES

✓ **As you read this chapter, concentrate on learning how to:**

1. Define the concept of gain
2. Draw the DC equivalent circuit for an amplifier
3. Draw the AC equivalent circuit for an amplifier
4. Calculate AC emitter resistance (r'_e)
5. Describe the effect of using a swamping resistor in an amplifier circuit
6. Describe the common-emitter amplifier characteristics
7. Determine the output voltage, gain, input impedance, and output impedance of the common-emitter amplifier circuit, using the voltage divider bias
8. Determine the output voltage, gain, input impedance, and output impedance of the common-emitter amplifier circuit, using the collector feedback bias
9. Describe common-collector amplifier characteristics
10. Determine the output voltage, gain, input impedance, and output impedance of the common-collector amplifier circuit, using either voltage divider bias or collector feedback bias
11. Describe common-base amplifier characteristics
12. Determine the output voltage, gain, input impedance, and output impedance of the common-base amplifier circuit, using the voltage divider bias
13. Determine the gain of multistage small-signal amplifiers

INTRODUCTION

An **amplifier** increases the magnitude of some quantity. In electronic applications, the amplifier circuit increases the current or voltage of a small signal before applying the signal to an output device such as a speaker or to the input of another circuit. Transistor amplifier circuits may be found in radios, stereos, and many other electronic devices.

This chapter introduces **small-signal amplifiers,** or amplifiers that have input voltages in the millivolt range. In many cases, small signal amplifiers are found at the output of antennae or industrial sensing devices. This chapter also will build our knowledge about the DC characteristics of transistor circuits.

In this chapter, we will study the AC operation of bipolar junction transistors and how the AC operation interacts with the DC operations. In particular, we will look at how the DC emitter current affects the AC input resistance. Also, we will discover how to calculate voltage gain, how to determine input and output impedances, and how one transistor stage works with another stage. We will also learn to identify the three basic AC configurations—common emitter, common collector, and common base.

5.1 SOME FUNDAMENTAL CONCEPTS

WHAT IS GAIN?

A transistor circuit may provide current, voltage, or power gain. We define **gain** as the measure of how much larger an output signal is than an input signal. It is a measure of the amplifying ability of a transistor. **Voltage gain** is represented as A_v and is defined by the formula

$$A_v = \frac{v_{out}}{v_{in}}$$

Current gain is represented as A_i and is defined by the formula

$$A_i = \frac{i_{out}}{i_{in}}$$

Power gain is represented as A_p and is defined by the formula

$$A_p = A_v \times A_i$$

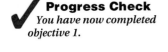

Progress Check
You have now completed objective 1.

EXAMPLE 5.1

Calculate voltage, power gain, and output current.

The input and output voltage of a transistor amplifier were experimentally measured and are shown in Figure 5.1. Also, the input current was measured as 200 μA (p-p), and the current gain $A_i = 100$.

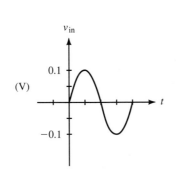

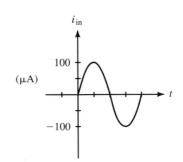

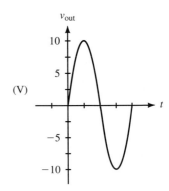

FIGURE 5.1

Solution

With the data given, it is most convenient to use peak-to-peak values read from the graph.

Using the definitions

$$A_v = \frac{V_{out\,(p-p)}}{V_{in\,(p-p)}} = \frac{20\ V}{0.2\ V} = 100$$

$$A_i = \frac{I_{out\,(p-p)}}{I_{in\,(p-p)}}$$

or

$$I_{out\,(p-p)} = A_i \times I_{in\,(p-p)} = 100 \times 200\ \mu A = 20\ mA$$

Using the calculated voltage gain and given current gain

$$A_p = A_v \times A_i = 100 \times 100 = 10{,}000$$

PRACTICE PROBLEM 5.1

For the waveshapes and gains given in Figure 5.2, calculate the output voltage (p-p), current gain, and power gain when $A_v = 10$.

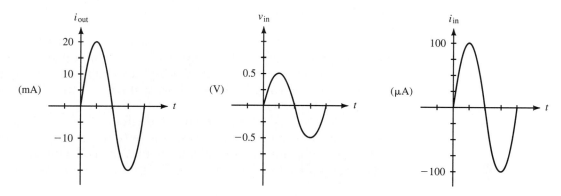

FIGURE 5.2

THREE TRANSISTOR AMPLIFIER CIRCUIT CONFIGURATIONS

Three basic amplifier circuit configurations—common emitter, common collector, and common base—exist. Since a transistor has only three terminals, one must be common to the other two for input and output connections. Figure 5.3 shows the three configurations. For the time being, each circuit has only the basic components. The diagrams do not show the values of bias resistors or capacitors.

In Figure 5.3a, we see a common-emitter circuit, the most widely used configuration. The input signal is applied between the base and the emitter, while the output signal is taken from between the collector and ground. Since the emitter connects to AC ground, it functions as a common reference point for the input and output. Two things to remember about common-emitter circuits are:

1. A common-emitter circuit can provide current, voltage, and power gain.

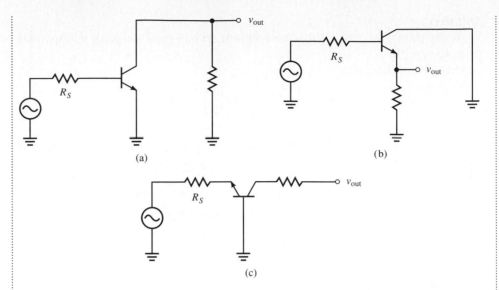

FIGURE 5.3   (a) Common-emitter configuration— AC equivalent circuit. (b) Common-collector configuration— AC equivalent circuit. (c) Common-base configuration— AC equivalent circuit.

2. Common-emitter circuits have a medium input impedance and a high output impedance.

   Since this chapter discusses amplifiers at low frequencies, assume that impedance and resistance are synonymous. In truth, impedance is the phasor sum of resistance and reactance. During this chapter we will ignore the effects of reactance. Later in this book we will address the effects of reactance on amplifying circuits.
   Figure 5.3b shows a common-collector circuit. The input signal is applied between the base and collector, while the output signal appears between the emitter and collector. For this configuration, the collector ties to AC ground and works as the reference point for AC signals. A common-collector circuit:

1. provides high current gain and a voltage gain of approximately one.

2. has a high input impedance and a low output impedance.

   The common-collector circuit functions well as an impedance-matching circuit. By placing a common-collector circuit between a circuit whose output has a high impedance and a circuit whose input impedance is low, the maximum signal power is transferred from the first circuit through the common collector to the second circuit. Remember: matching impedances provides for maximum power transfer.
   Figure 5.3c shows the common-base circuit. An input is applied between the emitter and the base, while the output appears between the collector and base. The base is common to both the input and output. A common-base circuit:

1. has a low input resistance with a very high output resistance, and

2. provides a high voltage gain and a current gain of approximately one.

COUPLING CAPACITORS

Later we will discuss multistage amplifiers. For now, we need to consider how those stages will connect together. For the purposes of this chapter, we will use **capacitor coupling.** Coupling capacitors are shown in Figure 5.4. Other methods of connecting multistage amplifiers are direct and transformer coupling.

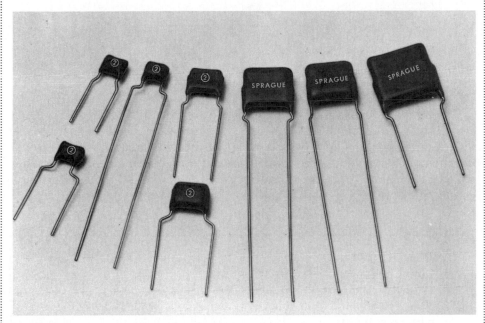

FIGURE 5.4 **Coupling capacitors.** *(Courtesy of Sprague Electric Co.)*

Coupling allows a signal to pass from one stage to the next stage or stages. Along with being easy and inexpensive, using capacitors for coupling allows the isolation of DC and bias voltages. As a result, those voltages cannot affect other stages. However, coupling capacitors will pass AC signals from one stage to another. In addition, coupling capacitors must pass the lowest frequency that needs to be amplified.

SINGLE-STAGE AND MULTISTAGE AMPLIFIERS

Small-signal amplifiers can exist as either a single transistor or as several transistors connected in series. Figure 5.5 shows the generic symbol for a **single-stage amplifier.** Realistic amplifier circuits require more than one stage. As a rule, single transistor stages cannot provide the required gain with acceptable distortion levels. Secondly, amplification creates heat. Rather than all the heat being dissipated by a single transistor, several transistors dissipate the heat, allowing each transistor to operate at cooler temperatures. Excessive heat is a major cause of component failure in most electronic devices. Figure 5.6 shows the generic symbols used in a **multistage amplifier.**

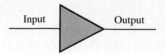

FIGURE 5.5 The schematic symbol for a single-stage amplifier. The symbol may be modified for specific types of amplifiers.

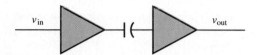

FIGURE 5.6 A multistage amplifier with capacitive coupling

The amplified signal may be a complex waveform such as an audio signal carried by a radio frequency. A single-stage amplifier could not handle the task of amplifying the high-frequency signals from the antenna and amplifying the audio signal for reproduction. In addition, the single-stage amplifier could not separate the audio from the radio frequency and match input and output impedances.

REVIEW SECTION 5.1

1. Define gain.
2. Write the equations for voltage, current, and power gain.
3. The current gain is 150, and the power gain is 6000. The voltage gain must be:
    (a) 15,000
    (b) 0.04
    (c) 40
    (d) 900,000
4. The most widely used transistor amplifier configuration is the common emitter. True or false?
5. What are the characteristics of the common-emitter circuit? What gains can the common-emitter circuit provide?
6. The common-collector circuit can be used for _____.
7. The common-base circuit provides _____ input resistance and _____ output resistance.
8. What is the typical voltage gain range of the common-collector circuit?
9. What is the major advantage of capacitor-coupled amplifiers?

5.2 EQUIVALENT CIRCUITS

To make our analysis of transistor circuits easier, we can use the superposition theorem to break the entire circuit down into easily recognizable simple circuits. Since both DC and AC signals drive the amplifier circuits, we can break the large circuit down into either DC or AC equivalent circuits. From there, we can break the circuit down even more into simple series or parallel circuits. This method works well both to further our understanding of amplifier circuits and as a troubleshooting tool.

Figure 5.7 shows a two-stage amplifier. An **amplifier stage** may be identified as an active device and its related components. Q_1 is the active component of the first stage, and Q_2 is the second stage's active component. C_1 and C_3 are the coupling capacitors that separate the two stages from the source, each other, and the output device. C_2 separates the two stages. To which stage does C_2 belong? It belongs to both stages. When considering the first stage by itself, we would draw C_2 as the output capacitor of the first stage. Likewise we would draw C_2 as the input capacitor when considering the second stage.

DC EQUIVALENT CIRCUITS

Using Figure 5.7, consider the first stage only. R_1 and R_2 set the bias for Q_1.

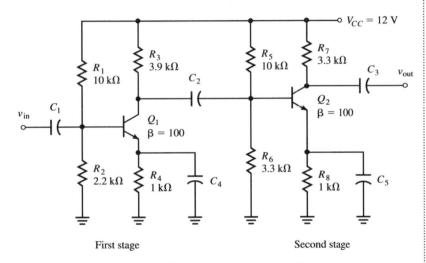

FIGURE 5.7 A two-stage transistor amplifier using capacitive coupling

---

EXAMPLE 5.2

Draw a DC equivalent circuit for the first stage of the circuit shown in Figure 5.7. Including the connection to the DC power supply, draw the stage for transistor Q_1. Stop drawing at all open points.

For our analysis, drawing the DC equivalent circuit is not difficult. Essentially, we concentrate on how the different parts of the circuit behave under DC signal conditions. We can treat all capacitors in the circuit as if they are open because capacitors block DC current. Then, we can trace the DC signal from the DC supply voltage through the bias resistors and to the transistor. With the simplified circuit, finding current, voltage, and resistance values becomes a matter of applying either Ohm's law or Kirchhoff's voltage and current laws.

Solution

Shown in Figure 5.8a is the DC equivalent circuit for stage 1 of the circuit.

To solve the DC equivalent circuit, first solve the DC control loop. Find the Thevenin voltage at TP1 using the voltage divider method.

$$V_{TH} = V_{CC}\left(\frac{R_2}{R_1 + R_2}\right)$$

$$= 12 \text{ V}\left(\frac{2.2 \text{ k}\Omega}{10 \text{ k}\Omega + 2.2 \text{ k}\Omega}\right) = 2.2 \text{ V}$$

Then solve for the Thevenin resistance. In Figure 5.8b the DC supply is shorted. Looking back into the bias resistor from TP1, R_1 and R_2 are in parallel.

$$R_{TH} = R_1 \parallel R_2 = 10 \text{ k}\Omega \parallel 2.2 \text{ k}\Omega = 1.8 \text{ k}\Omega$$

Now in Figure 5.8c use the beta box to equalize the circuit. To equalize the circuit divide R_{TH} by beta (beta is equal to 100).

$$\frac{R_{TH}}{\beta} = \frac{1.8 \text{ k}\Omega}{100} = 18 \text{ }\Omega$$

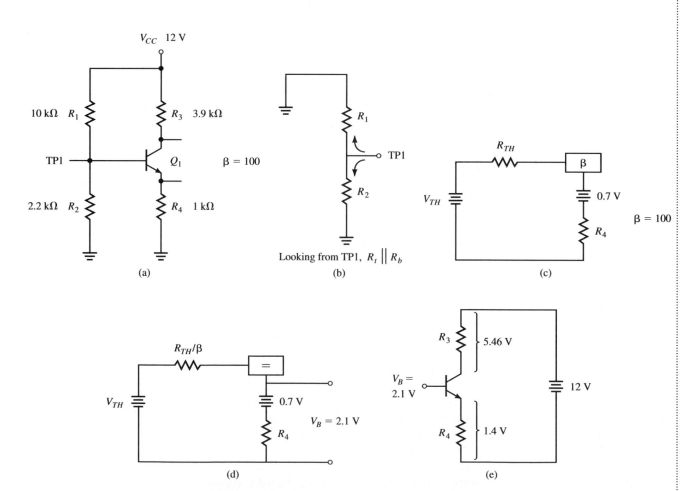

FIGURE 5.8 DC analysis of the first stage of the circuit shown in Figure 5.7

In Figure 5.8d the DC control loop has been reduced to a simple series circuit. Since the voltage at the base is loaded down less than 10%, the circuit is firm.

Secondly, we need to find the emitter current, I_E (Figure 5.8e). The emitter voltage is one diode drop less than V_B.

$$V_B - 0.7\,V = V_E$$
$$2.1\,V - 0.7\,V = 1.4\,V$$

We can use Ohm's law to determine the emitter current. It will be the emitter voltage divided by the emitter resistance.

$$I_E = \frac{V_E}{R_E} = \frac{1.4\,V}{1\,k\Omega} = 1.4\,mA$$

Next we need to find the collector voltage (V_C). The collector current is the same as the emitter current, that is

$$I_C = I_E$$

V_C may be found by subtracting the collector resistor voltage (V_{RC}) drop from V_{CC}. To find V_{RC} we use Ohm's law.

$$I_C \times R_C = V_{RC}$$
$$1.4\,mA \times 3.9\,k\Omega = 5.46\,V$$

Now, we have enough information to solve for V_C:

$$V_{CC} - V_{RC} = V_C$$
$$12\,V - 5.46\,V = 6.54\,V$$

The DC equivalent circuit has allowed us to solve for all the DC transistor voltages and currents.

From here, let us move to the AC analysis of the stage.

Progress Check
You have now completed objective 2.

Draw the DC equivalent circuit for the *second* stage of the circuit shown in Figure 5.7. Figure 5.9 shows the second stage as a separate amplifier. Determine the DC voltage at the base, emitter, and collector.

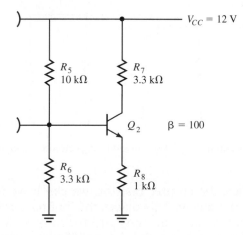

FIGURE 5.9 DC equivalent circuit for the second stage of the circuit shown in Figure 5.7

AC EQUIVALENT CIRCUITS

For the second part of our analysis, we will break the transistor amplifier circuits into AC equivalent circuits. Figure 5.10 is the first stage of Figure 5.7.

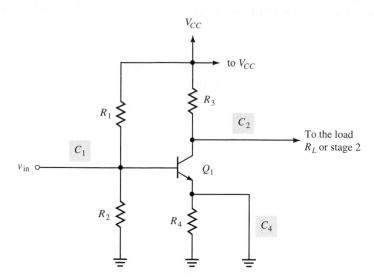

FIGURE 5.10 Beginning AC analysis of the first stage from Figure 5.7. Note: highlighted capacitors are being "treated" as if they were pieces of wire.

For the purposes of our AC circuit analysis, we can treat capacitors as direct signal paths. While the capacitor blocks DC signals, it will pass AC signals. Now, we must think in different terms than we did when drawing the DC equivalent circuit. The AC equivalent circuit is drawn from the perspective of the AC input voltage.

Figure 5.11 shows this AC equivalent circuit. The AC signal source symbol represents the input voltage. Looking from the perspective of the AC input voltage, *all DC sources would appear as paths to ground.* Both resistors R_1 and R_2 have one side tied to AC ground with the other sides connected to the transistor base.

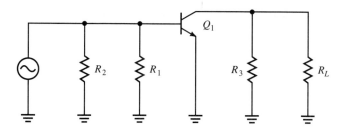

FIGURE 5.11 AC equivalent circuit for stage 1 of the multistage circuit shown in Figure 5.7

Since a capacitor parallels the emitter resistor, we can leave the resistor out of the equivalent circuit drawing. This places the emitter terminal of the transistor at AC ground. At the collector, the collector resistor, R_3, also connects to AC ground, because it ties to a DC source. The load resistance, R_L, represents the input impedance of the second stage and is in parallel with the collector resistor.

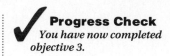
PRACTICE PROBLEM 5.3

Draw the *AC* equivalent circuit for the second stage of the circuit shown in Figure 5.7.

AC EMITTER RESISTANCE

The **AC emitter resistance** is an internal characteristic of the transistor. When we use the beta box, the symbol for AC emitter resistance, r'_e, will go inside the box, just below the emitter-base node. Figure 5.12 shows the positioning of the AC emitter resistance symbol.

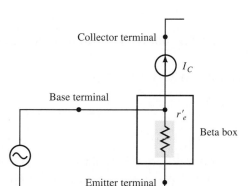

FIGURE 5.12 AC base-emitter loop showing the beta box model. The highlighted areas are the location of r'_e, the AC emitter resistance.

EXAMPLE 5.3

Calculate the AC emitter resistance value from the base-emitter curve. Obtain a plot of the base-emitter junction either from specification sheets or from a curve tracer. Figure 5.13 shows a sample plot.

This plot is the forward part of a diode curve. Choose two points on the curve that you believe represent the operating conditions of the transistor in the circuit. Next, find the current associated with each point. Using a form of Ohm's law, we can determine the AC emitter resistance:

$$r'_e = \frac{\Delta V}{\Delta I}$$

Solution
Point 1 is at 1.3 V. The current at this point is 11 mA. Point 2 is at 0.8 V and has 1 mA of current.

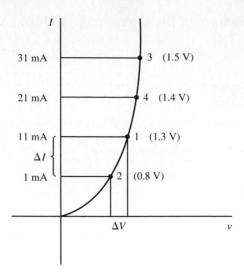

FIGURE 5.13 Typical base-emitter diode curve

With these values, we can use the following formula to arrive at a value for the AC emitter resistance:

$$r'_e = \frac{\left(1.3 \text{ V} - 0.8 \text{ V}\right)}{\left(11 \text{ mA} - 1 \text{ mA}\right)} = \frac{0.5 \text{ V}}{10 \text{ mA}} = 50 \ \Omega$$

Before we discuss the second method of calculating the AC emitter resistance, take a look at points 3 and 4 in Figure 5.13. Graphically, ΔV becomes smaller as the selected voltages move well above the 0.7-V barrier potential. In turn, the AC emitter resistance has a smaller value. We call the AC emitter resistance a **dynamic resistance** because its value changes as circuit conditions change. Since AC emitter resistance changes, we will approximate its value as we analyze the transistor circuits.

PRACTICE PROBLEM 5.4

Recalculate the AC emitter resistance, r'_e, for the transistor with characteristics shown in Figure 5.13 using points 3 and 4, and compare with the result of example 5.3.

The second method for calculating r'_e is easier to do. Although it is less accurate, the answer will be close enough to use for our purposes. Use the following relationship:

$$r'_e = \frac{25 \text{ mV}}{I_E}$$

Explaining why 25 mV is used is beyond the scope of this text. I_E is the DC emitter current.

EXAMPLE 5.4

Determine the r'_e of transistor Q_1 using the approximation

$$r'_e = \frac{25\ mV}{I_E}$$

Solution

Using the voltage divider bias circuit shown in Figure 5.14, we can find a value for the AC emitter resistance. The DC equivalent circuit looks like the circuit shown in Figure 5.8. Since the divider is firm, we can use the quick calculation method to find the Thevenin voltage:

$$V_{TH} = \frac{2.2\ k\Omega}{12.2\ k\Omega} \times 12\ V = 2.16\ V$$

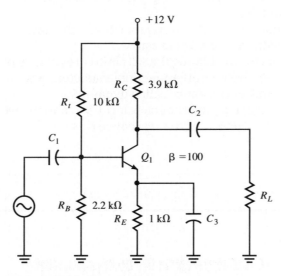

FIGURE 5.14 Voltage divider bias circuit

Subtracting 0.7 V from the Thevenin voltage, we get the value for the emitter voltage:

$$V_E = 2.16\ V - 0.7\ V = 1.46\ V$$

Now, let us use Ohm's law to calculate the emitter current value:

$$I_E = \frac{1.46\ V}{1\ k\Omega} = 1.46\ mA$$

Last, we can find the value for the AC emitter resistance:

$$r'_e = \frac{25\ mV}{1.46\ mA} = 17.1\ \Omega$$

Progress Check
You have now completed objective 4.

As you gain experience, you will find out that the AC emitter resistance will have a small value. Values below 25 Ω are common, with some values measuring as low as 5 Ω.

Determine the AC emitter resistance for the following DC emitter currents: 10 mA, 5 mA, 2.5 mA.

REVIEW SECTION 5.2
1. What are the three types of AC equivalent circuits used to analyze transistor circuits?
2. Coupling capacitors block _____.
3. In DC equivalent circuits, capacitors act as _____ circuits.
4. In AC equivalent circuits the emitter bypass capacitor shorts the emitter to ground, and the DC supply is replaced by a short circuit to ground. True or false?
5. The load resistance in an AC equivalent circuit appears in _____ with the collector resistor.
6. Explain the steps for calculating the AC emitter resistance of a transistor from the base-emitter junction characteristics.
7. Why is the AC emitter resistance called dynamic?
8. The emitter voltage in a transistor circuit is 1.8 V, and the DC emitter resistance is 2 kΩ. The AC emitter resistance r'_e is:
   (a) 12.5 Ω
   (b) 27.8 kΩ
   (c) 1.4 Ω
   (d) 2.78 Ω

5.3 EXPLORING THE COMMON-EMITTER CONFIGURATION

CALCULATING TRANSISTOR GAIN

Using Figure 5.15, we can analyze the circuit shown in Figure 5.14. Figure 5.15a shows the DC model for the control loop before equalization. Figure 5.15b shows the AC control loop for the same circuit. The internal resistance of the source is represented by the symbol R_S. From the perspective of the AC signal, R_t and R_B parallel each other. Each connects to the base terminal, with the remaining side connected to the AC ground. Although the internal resistance is an internal transistor characteristic, it is drawn below the beta box to indicate that it is a part of the emitter circuit. The beta for the transistor is given inside the beta box.

Even though we know that the transistor has a base-emitter junction barrier potential of 0.7 V, Figure 5.15b does not show this voltage. AC signals see DC voltages as shorts. Because a capacitor bypasses the DC emitter resistor and places the emitter terminal at AC ground, the DC emitter resistance is also missing.

Figure 5.15c further simplifies the AC control loop. Here, two bias resistors combine into one equivalent resistance. The AC signal will reach the base lead and the beta box with no loss of signal strength. This occurs because the

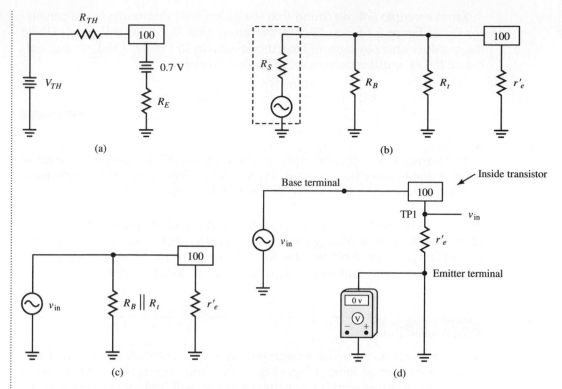

FIGURE 5.15 (a) DC model for the control loop of the circuit shown in Figure 5.14. (b) AC control loop for the circuit. (c) Simplification of the AC control loop. (d) Signal flow of the circuit.

bias resistors lie in parallel with the AC source. From the AC source, a direct path leads to the base.

For a moment, let us review two rules about the beta box model:

1. The beta box model has no internal resistance.

2. The beta box model drops no voltage.

Using these rules, we can state that the entire AC input signal will reach the top of the AC emitter resistor shown in Figure 5.15c. Figure 5.15d depicts the signal flow. Because the bias resistors have no effect on the AC signal as it reaches the base, they no longer show in the figure.

Remembering that TP1 actually resides within the transistor, we cannot measure voltage at this point. If we placed an AC voltmeter on the emitter terminal of the transistor, the scale would read 0 V. The bypass capacitor places the emitter terminal at AC ground. Despite our inability to measure voltage at TP1, we can calculate the AC emitter current with the formula:

$$i_e = \frac{v_{in}}{r'_e}$$

EXAMPLE 5.5

Using Figure 5.14, give the AC input voltage a value of 10 mV (p-p). Then determine the emitter current.

From example 5.4, we found that the AC emitter resistance (r'_e) of transistor Q_1 in figure 5.14 was 17.1 Ω. We know that V_{in} = 10 mV (p-p). Using these values and our control loop model shown in Figure 5.15d, we can calculate the AC emitter current value for the transistor:

$$i_e(\text{p-p}) = \frac{10\text{ mV (p-p)}}{17.1\ \Omega} = 585\ \mu\text{A (p-p)}$$

In Chapter 4, we defined alpha as the measure of the amount of emitter current that reaches the collector. Typical values were above 0.95, with many transistors having an alpha of 0.99. Since alpha is close to 1, or unity, we decided to assume that alpha is 100%. With this assumption, we could approximate the collector current by saying that it equaled the emitter current. Alpha also applies to AC currents. Consequently, we can use the emitter current of 585 μA to predict the value for collector current. Knowing this, we can proceed to the next example, which shows the gain loop model.

EXAMPLE 5.6

Now, we can determine the voltage gain for the circuit shown in Figure 5.14.

To do this, let us look at Figure 5.16, the gain loop model that will work with most common-emitter circuits. Later, we will look at variations of the same model. Note that the beta box in the model features an equality sign. The loop has been equalized. In fact, because the AC emitter current equals the AC collector current, the loop is self-equalizing.

Also, notice that the collector resistance shows that the collector supply voltage is an AC ground. The AC output is developed off the collector terminal. In the figure, the AC output lies just above the constant current source symbol. When we investigated Figure 5.15d, we explained the presence of the input voltage at the top of the AC emitter resistance. Using this gain loop model, we can calculate the AC output and determine the gain of the transistor.

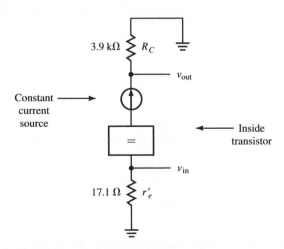

FIGURE 5.16 Gain loop that will work with common-emitter circuits

Solution

From example 5.5 and the discussion, we know that the AC collector current equals 585 µA. Use Ohm's law to determine the voltage drop across the collector resistance:

$$v_{rc} = 585\ \mu A \times 3.9\ k\Omega = 2.28\ V$$

Since the collector resistance connects to ground, the AC output to ground must equal 2.28 V (p-p). Given that we originally applied 10 mV and increased the signal size to 2.28 V (p-p), we have a gain of:

$$A_v = \frac{v_{out}}{v_{in}} = \frac{2.28\ V_p}{10\ mV_p} = 228$$

This gain figure represents the gain with no load.

We can also determine the AC voltage gain by dividing the collector resistance by the internal emitter resistance. The reason is that r'_e and R_C share a common current. V_{out} is i_c multiplied by R_C. v_{in} is $i_e \times r'_e$. Since i_c and i_e are the same current, i_c may be substituted for i_e. We know that

$$A_v = \frac{v_{out}}{v_{in}}$$

We can rewrite the gain relationship as

$$A_v = \frac{i_c R_c}{i_c r'_e}$$

Since i_c is in both the numerator and the denominator, it cancels out, leaving

$$A_v = \frac{r_c}{r'_e} = \frac{3.9\ k\Omega}{17.1\ \Omega} = 228$$

In Figure 5.17, we redraw the gain loop model from the perspective of the constant current source. For clarity, we can drop the beta box from the drawing; it has been equalized, drops no voltage, and has no internal resistance. Now, we have reduced the gain loop model to a series circuit. We can apply the voltage divider theorem, which says that voltages in a series circuit are proportional to the resistances in that series circuit. The voltage across the collector resistor is proportional to the voltage across the internal resistance of the emitter. Given this rule, the following formula becomes apparent:

$$A_v = \frac{v_{out}}{v_{in}} = \frac{r_c}{r'_e}$$

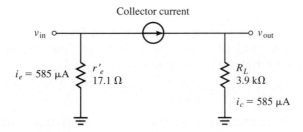

FIGURE 5.17 Equivalent series circuit for the gain loop model drawn from the perspective of the constant current source

USING DC EMITTER BIAS

Because DC emitter bias provides good circuit stability and reduces the effects of beta, emitter bias has gained wide usage.

EXAMPLE 5.7

Use the skills that we have developed to analyze the emitter-biased common-emitter circuit shown in Figure 5.18.

Figure 5.18 shows a common-emitter circuit that utilizes emitter bias. Of the three capacitors in the circuit, C_1 and C_2 act as coupling capacitors and isolate DC voltage levels. Connected as an output coupler, capacitor C_2 prevents the DC level of the collector from reaching the output. The bypass capacitor, C_3, places the emitter terminal at AC ground.

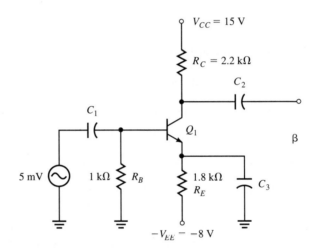

FIGURE 5.18 Emitter-biased common-emitter circuit

Realizing that the coupling capacitor "sets the boundaries" of the DC circuit operation, we develop the DC control loop model shown in Figure 5.19. This model is the same as the model used for the emitter-biased circuit in Chapter 4. The superposition theorem tells us that we can analyze the DC operation of the circuit, analyze the AC effects, and then combine the two answers to solve the circuit.

Solution

When we use the equalized control loop model of Figure 5.19b, we can ignore the base resistance of 10 Ω because R_E is more than 100 times greater than the 10 Ω. We find that the DC emitter current equals:

$$I_E = \frac{7.3\,V}{1.8\,k\Omega} = 4.06\,mA$$

We can use this information to predict the internal resistance of the emitter:

$$r'_e = \frac{25\,mV}{4.06\,mA} = 6.2\,\Omega$$

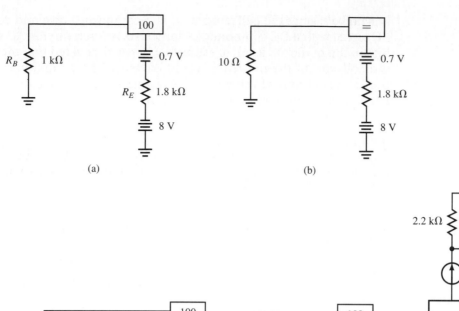

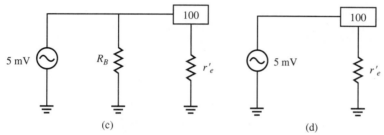

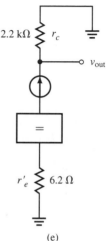

FIGURE 5.19 (a) Gain loop model of the circuit shown in Figure 5.18. (b) Equalized control model of the circuit. (c) AC control loop of the circuit. (d) Simplification of the AC control loop. (e) Common emitter gain loop model.

Figure 5.19c shows the AC control loop, and Figure 5.19d shows its simplified representation. The entire AC signal reaches the base through a direct path and loses no signal strength. Since the beta box has a resistor on only one side, we do not need to equalize the box. We can determine the emitter current with the formula:

$$I_E = \frac{5 \text{ mV (peak)}}{6.2 \ \Omega} = 806 \ \mu A \ (\text{peak})$$

The gain loop model is shown in Figure 5.19e. Using the information we developed earlier, we can calculate the gain of the circuit:

$$A_v = \frac{r_c}{r'_e} = 355$$

Using the voltage gain value, we can predict the size of the output voltage:

$$v_{out} = A_v \times v_{in} = 355 \times 5 \text{ mV}_p = 1.78 \text{ V}_p$$

EXAMPLE 5.8

Combine the DC and AC voltages to find the voltage level at the collector of transistor Q_1 in the circuit shown in Figure 5.18.

The voltage at the collector of Q_1 is a composite wave and consists of an AC signal with a DC component. Many texts will refer to the AC signal as riding on top of the DC level. In example 5.7, we discovered that the AC signal at the collector of the transistor would have a 1.78-V_p amplitude. To see how the DC conditions of the circuit affect this signal, we need to complete the DC analysis of the collector in Figure 5.18.

Solution

The DC emitter current equals 4.06 mA. This current flows into the transistor and out the collector. Thus, the collector current also equals 4.06 mA. The DC voltage used by the collector resistor is:

$$V_C = 4.06 \text{ mA} \times 2.2 \text{ k}\Omega = 8.9 \text{ V}$$

Subtracting this value from the collector source voltage value, we can arrive at a value for the collector voltage:

$$V_C = V_{CC} - V_{RC} = 15 \text{ V} - 8.9 \text{ V} = 6.1 \text{ V}$$

The AC signal sits astride this DC level, producing the waveform pictured in Figure 5.20.

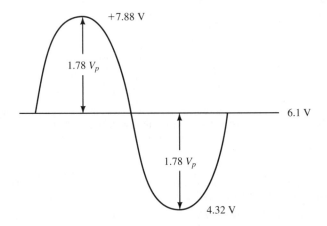

FIGURE 5.20 Waveform showing an AC signal sitting astride a DC level

If we tested the collector output with an oscilloscope, we would need to set the oscilloscope for DC coupling. If we set the oscilloscope for AC coupling, we would see the 1.78-V_p signal, but not the DC shift.

HOW A LOAD CAN AFFECT TRANSISTOR GAIN

In Figure 5.21, the collector portion of the circuit has a load resistor attached to capacitor C_2. Because the capacitor prevents any DC voltages from reaching the load or any DC voltages present on the load from reaching the transistor, the load resistor has no effect on DC operation. Capacitor C_2 acts like a short for AC operation; the load resistor will affect AC operation. To see how, let us proceed to the next example.

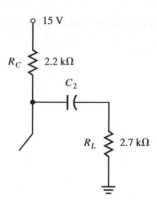

FIGURE 5.21 Collector circuit with an attached load

EXAMPLE 5.9

Determine how a load affects the AC operation of a common-emitter amplifier and its voltage output.

The collector circuit is redrawn in the AC configuration shown in Figure 5.22a. The collector resistor and load resistor parallel each other during AC operation. These two resistances combine into an equivalent AC collector resistance, or r_c. Remember that the AC collector resistance is different from

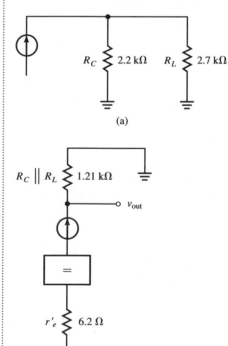

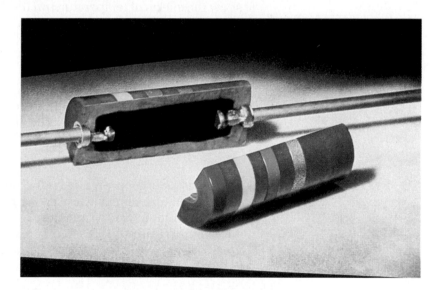

FIGURE 5.22 (a) Redrawing of the circuit shown in Figure 5.21. *(Photograph courtesy of Allen-Bradley)* (b) Gain loop model of the circuit, incorporating changes caused by a load.

the DC collector resistance. After combining the resistances, the gain loop model resembles Figure 5.22b.

Solution

Using the gain loop model of Figure 5.22b and the techniques that we developed earlier, we can conclude that the AC voltage gain is:

$$A_v = \frac{r_c}{r_e'} = \frac{1.21 \text{ k}\Omega}{6.2 \text{ } \Omega} = 195$$

Because of the load, the gain has dropped significantly. Unloaded, the transistor produced a gain of 355. Loaded, the gain dropped to 195. However, the loaded circuit is the more realistic since it duplicates actual circuit conditions. Most of the time, we will calculate gain under loaded conditions.

STABILIZING GAIN BY SWAMPING OUT THE AC EMITTER RESISTANCE

In Chapter 4, we discovered that transistor characteristics vary from transistor to transistor. In this chapter, we found that the AC emitter resistance is influenced by the operating conditions of the circuit. In addition, the AC emitter resistance is temperature-dependent. A look at example 5.10 will show the effects of temperature on the AC emitter resistance.

EXAMPLE 5.10

Find the effects of temperature change on the r_e' for the circuit shown in Figure 5.18. Assume that the temperature has risen.

Previously, we calculated a value of 6.2 Ω for r_e'.

$$r_e' = \frac{25 \text{ mV}}{I_E} = \frac{25 \text{ mV}}{4.06 \text{ mA}} = 6.2 \text{ } \Omega$$

This is a fairly good predictor of r_e' at a temperature of 18°C. To account for temperature changes, we could multiply r_e' by the temperature factor. The temperature factor is the ratio of the transistor temperature to the reference temperature of 18°C. Temperature units must be in kelvins. Zero degrees Celsius is equal to 273 degrees kelvin.

$$0°C = 273°K$$

$$\text{Temperature factor} = \frac{\text{transistor temperature } (\text{in}° \text{ C}) + 273°\text{K}}{\text{room temperature } (291°\text{K})}$$

When the temperature equals 18°C, the new term equals 1 and does not change our original answer.

Solution

Let us suppose that the transistor heats to a temperature of 50°C. The temperature factor becomes:

$$\frac{\left(50°C + 273° \text{ K}\right)}{291° \text{ K}} = 1.11$$

The value for r_e' changes to:

$$r_e' = 1.11 \times 6.2\,\Omega = 6.88\,\Omega$$

In turn, the loaded gain changes to:

$$A_v = \frac{1.21\,k\Omega}{6.88\,\Omega} = 176$$

ADDING A SWAMPING RESISTOR

Working through the examples, we saw that the gain value dropped from 195 to 176. To reduce the effect of temperature variations on transistor operation the circuit designer may place a **swamping resistor** in the circuit. To swamp something is to dampen its effect. Since the circuit designer wants to reduce the effect of temperature variations he can use a swamping resistor.

EXAMPLE 5.11

Determine the effects of the swamping resistor on the operation of the circuit shown in Figure 5.23.

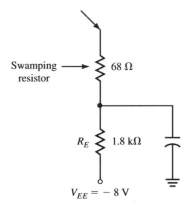

FIGURE 5.23 Circuit with a swamping resistor in place

The 68-Ω resistor is placed into the emitter circuit in series with the emitter resistor. The 68-Ω value is the nearest standard value resistor that is approximately ten times r_e'. This will alter the DC operation of the circuit. First, we will determine the effects the swamping resistor has on the DC emitter current. As you may have guessed, a change in the emitter current will lead to changes in the AC emitter resistance.

Solution

We can calculate a new value for the emitter current by adding the swamping resistor value to the value of the emitter resistor. Our calculation should look like this:

$$I_E = \frac{7.3\,V}{\left(1.8\,k\Omega + 68\,\Omega\right)} = 3.91\,mA$$

This formula is based on the modified control loop shown in Figure 5.24a. Since we ignored the resistance value of the base before, we will continue to ignore it. The new emitter current value will change the AC emitter resistance to:

$$r'_e = \frac{25 \text{ mV}}{3.91 \text{ mA}} = 6.39 \ \Omega$$

This may not seem like a drastic change until we look at the updated control loop model shown in Figure 5.24b. Because the swamping resistor, R_{sw}, is not bypassed, we can include it in the control loop model. Our gain calculation becomes:

$$\frac{r_e}{\left(r_{sw} + r'_e\right)}$$

Substituting values into the equation, we arrive at:

$$A_v = \frac{1.21 \text{ k}\Omega}{68 \ \Omega + 6.39 \ \Omega} = 16.3$$

as a new voltage gain value.

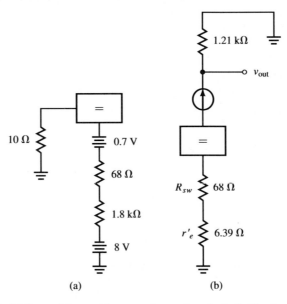

(a)   (b)

FIGURE 5.24 (a) Control loop with swamping resistor included in the analysis. (b) Gain loop with swamping resistor included in the analysis.

Comparatively, when looking at the previous values of 195 and 176, we have seen a tremendous loss in voltage gain. Nevertheless, the circuit design benefits from stability and consistent performance. If we use the same temperature factor that we used in example 5.10 and recalculate the AC emitter resistance for Figure 5.23, we find that the AC emitter resistance equals:

$$1.11 \times 6.39 \ \Omega = 7.09 \ \Omega$$

In turn, this value changes the gain value to:

$$\frac{1.21 \text{ k}\Omega}{\left(68 \ \Omega + 7.09 \ \Omega\right)} = 16.11$$

A change in temperature from 18°C to 50°C—a 32-degree increase—only changed the gain value from 16.2 to 16.11. The circuit has achieved good stability. Any circuit designer must weigh the additional cost of swamping against product reliability, performance, and reputation.

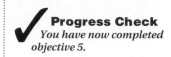

Progress Check
You have now completed objective 5.

WHY WORRY ABOUT DESIGN FACTORS?

Generally, technicians do not actively participate in circuit design. Yet, it helps us diagnose problems if we know why parts are in a circuit and what function they perform. Even though it is unlikely that technicians will make calculations involving temperature changes, it is important to know about the effects of temperature on circuit operations.

PHASE INVERSION

Of the three basic transistor circuit configurations, only the common-emitter configuration produces an AC output signal that is 180 degrees out of phase with the input signal. If the AC signal on the base is at its positive peak, the AC signal at the collector will be at its negative peak. When the AC signal at the base reaches its negative peak, the output signal at the collector goes to its positive peak.

Progress Check
You have now completed objective 6.

Consider the circuit action that occurs when the AC input reaches its positive peak. Temporarily, the base voltage increases. If the base voltage increases, the instantaneous bias level also increases and causes an increase in the emitter current. As we have seen, an increase in emitter current also causes an increase in collector current. When the collector current increases, the voltage dropped by the collector resistor also increases:

$$V_{RC} = I_C R_C$$

As a result of the voltage drop increase, the collector voltage must decrease:

$$V_C = V_{CC} - V_{RC}$$

Any action by the base voltage will have the opposite effect on the collector voltage and will produce a 180-degree phase shift.

REVIEW SECTION 5.3

1. AC signals see DC voltage sources as (opens, shorts).
2. To high frequencies a capacitor acts like a(n) (short, open).
3. If a 1.5-kΩ emitter resistor is bypassed with an emitter bypass capacitor, the resistor appears as a short to the signal. True or false?
4. Draw a common-emitter amplifier. Draw the common-emitter AC and DC equivalent control circuits.
5. What is the formula for I_E in a common-emitter amplifier?
6. $i_e = i_c$ in a common-emitter amplifier. True or false?
7. If $I_C = 2$ mA, and $R_C = 4.7$ kΩ, what is the value of V_{RC}?
8. If $V_{out} = 2$ V_{p-p} and $V_{in} = 10$ mV_{p-p}, what is the value of A_v?
9. R_C/r'_e is a possible formula for A_v in a common-emitter amplifier. True or false?
10. Refer to Figure 5.14. Which capacitors are coupling capacitors? Which are bypass capacitors?
11. Refer to Figure 5.14. Assume $R_B = 22$ kΩ, $R_C = 4.7$ kΩ, $R_E = 3.3$ kΩ. Calculate the following: V_B, V_C, V_E, r'_e, A_v.

12. Explain why when R_L is added to Figure 5.14 the DC conditions do not change.
13. When R_L is connected to a common-emitter amplifier, what happens to the voltage gain of that amplifier? Explain.
14. What is the difference between R_C and r_c?
15. What is the function of a swamping resistor?
16. What is the formula for voltage gain when a swamping resistor is added to the circuit?
17. In a common emitter circuit, what is the phase relationship between V_{in} and V_{out}?

5.4 ANALYZING THE INPUT IMPEDANCE OF A COMMON-EMITTER CIRCUIT THAT USES VOLTAGE DIVIDER BIAS

Figure 5.25 illustrates a common-emitter circuit that uses voltage divider bias. Because the voltage divider is stiff, we can utilize the quick method of finding the emitter current when we solve the following example.

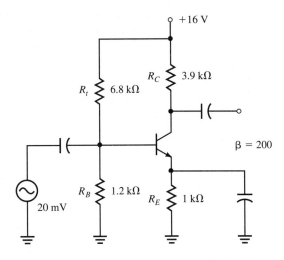

FIGURE 5.25 Common-emitter circuit using voltage divider bias

EXAMPLE 5.12

Determine the input impedance of the base and the input impedance of the stage.

First, we will use the quick method of finding the emitter current. Then, we will use the emitter current value to find r_e'. With the value for r_e', we can solve for the input impedance of the base and the circuit. Note that the emitter resistor is bypassed, and no AC swamping resistor exists.

Solution
To solve the circuit, we need to find the voltage at the base:

$$V_B = V_{TH} = \frac{1.2\ k\Omega}{8\ k\Omega} \times 16\ V = 2.4\ V$$

We can subtract 0.7 V from the base voltage value to find the emitter voltage:

$$V_E = 2.4\text{ V} - 0.7\text{ V} = 1.7\text{ V}$$

Now, we can use these values and Ohm's law to find the value for the emitter current:

$$I_E = \frac{1.7\text{ V}}{1\text{ k}\Omega} = 1.7\text{ mA}$$

With that value, we can determine the value for the AC emitter resistance:

$$r'_e = \frac{25\text{ mV}}{1.7\text{ mA}} = 14.7\ \Omega$$

Figure 5.26a shows the AC control loop using this information. The question mark in the box shows that we do not know the value for the internal impedance of the source. For now, we will assume that the source approximates an ideal voltage source and has little or no internal resistance. Thus, we do not need to equalize the loop to find the emitter current; our equation looks like this:

$$i_e = \frac{20\text{ mV}\,(\text{p-p})}{14.7\ \Omega} = 1.36\text{ mA}\,(\text{p-p})$$

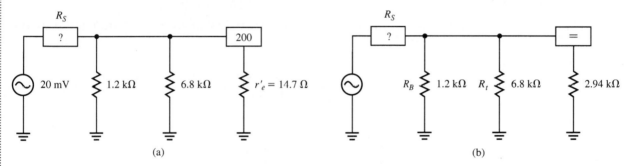

(a)   (b)

FIGURE 5.26 (a) AC control loop model of the circuit shown in Figure 5.25. (b) AC control loop model from the perspective of the base current.

Let us consider the control loop from the perspective of the base current. Figure 5.26b shows this perspective. The emitter resistance has been equalized by multiplying it by the beta value. In the figure, the 2.94-kΩ value represents the input impedance of the base. Since we have bypassed the emitter resistor, the emitter resistance equals zero. This allows our model and the formula Z_{in} (base) $= \beta r'_e$ to agree. We can establish the following rule: The input impedance of the circuit is calculated by combining the bias resistances with Z_{in} (base).

The beta box has no internal resistance and does not drop voltage, so we can remove it from the equalized model. By showing the beta box as a direct path, the diagram of Figure 5.26b also shows that $R_t \parallel R_b \parallel Z_{in}$ (base). Performing the calculation and using the reciprocal method of calculating resistances in parallel, we find:

$$Z_{in}\text{ (stage)} = 1.2\text{ k}\Omega \parallel 6.8\text{ k}\Omega \parallel 2.94\text{ k}\Omega = 757\ \Omega$$

When we studied the base bias circuit, we ignored the base resistor for AC operation because the resistor is larger than the base input impedance. In a voltage divider bias circuit, however, one of the bias resistors may have a smaller value than the base input impedance. We must include the bias resistor values in our analysis of the AC operation of the circuit.

EXAMPLE 5.13

Determine the effects of the voltage divider bias resistors on AC operation when the input source has a low internal impedance. Then, compare this with the effects the base resistors have when the input source has a high internal resistance.

To see why we must include the bias resistors in our analysis, we need to perform some additional calculations. Figure 5.27 shows an AC control loop equalized for the base current. The internal impedance of the source is listed as 50 Ω. Using Ohm's law, we determine that the base current equals:

$$i_b = \frac{20 \text{ mV}}{(2.9 \text{ k}\Omega + 50 \text{ }\Omega)} = 6.78 \text{ }\mu\text{A}$$

When we multiply the base current value by the beta value, we find the emitter current value:

$$i_e = 200 \times 6.78 \text{ }\mu\text{A} = 1.36 \text{ mA}$$

This value compares nicely with our original answer of 1.36 mA. Little change has occurred.

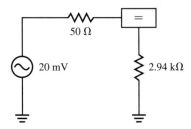

50 Ω

20 mV

2.94 kΩ

FIGURE 5.27 AC control loop equalized for the base current

Solution

Moving to Figure 5.28a, we can begin to analyze how the bias resistors affect AC circuit operation. By using Z_{in} (stage), we determine the total current drawn out of the source. This total current includes the AC currents flowing through R_t, R_b, and Z_{in} (base) and has a value of 24.8 μA. Flowing through the internal resistance of the source, R_S, the total current produces a voltage drop of:

$$24.8 \text{ }\mu\text{A} \times 50 \text{ }\Omega = 1.24 \text{ mV}$$

When we subtract the voltage drop value from input voltage value, we can predict the voltage across the combination of R_t, R_b, and Z_{in} (base). Looking at Figure 5.28b, this voltage equals the voltage across the base.

Calculating the base current, we find that:

$$i_b = \frac{18.76 \text{ mV}}{2.94 \text{ k}\Omega} = 6.38 \text{ }\mu\text{A}$$

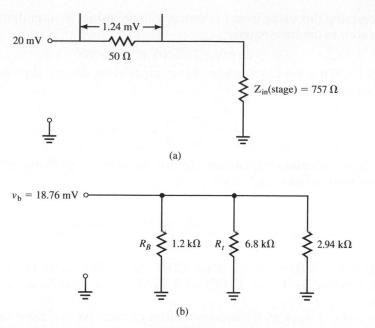

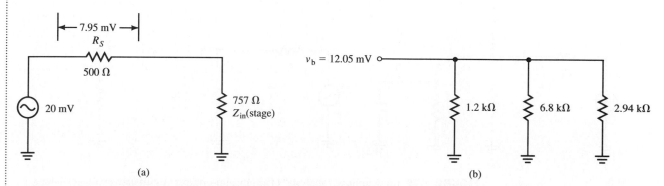

FIGURE 5.28 (a) Analysis of how bias resistors affect AC circuit operation. (b) Reduced effective AC voltage due to internal impedance of the source.

With this answer, we can find the emitter current:

$$i_e = 200 \times 6.38 \ \mu A = 1.28 \ mA$$

This value still compares favorably with our original answer of 1.36 mA. All these equations work well for a small source internal resistance value. However, if the source has a larger value for its internal resistance, we would need to change our calculations.

In Figure 5.29a, the internal resistance of the source is 500 Ω. Changing the value of the source resistance, we can use the same procedure used to solve the circuit shown in Figure 5.29b. In Figure 5.29a, the total current draw from the source equals 15.9 μA. This current produces a voltage drop across the source internal resistance of:

$$15.9 \ \mu A \times 500 \ \Omega = 7.95 \ mV$$

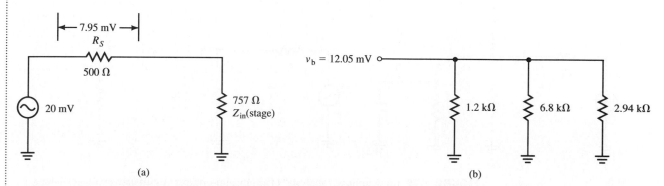

FIGURE 5.29 (a) Simplified AC control loop. (b) Reduced effective AC voltage due to internal impedance of the source.

Subtracting this value from the source voltage value, we find that the voltage delivered to the base equals:

$$20 \text{ mV} - 7.95 \text{ mV} = 12.05 \text{ mV}$$

With 12.05 mV feeding the base input impedance, we calculate that the base current equals:

$$i_b = \frac{12.05 \text{ mV}}{2.94 \text{ k}\Omega} = 4.10 \text{ μA}$$

When we calculate the emitter current, we find a significant difference between the new and original answers:

$$i_e(\text{original}) = 1.36 \text{ mA}$$
$$i_e(\text{new}) = 200 \times 4.10 \text{ μA} = 820 \text{ μA}$$

SOME OBSERVATIONS AND CONCLUSIONS ABOUT ANALYZING COMMON-EMITTER CIRCUITS THAT USE VOLTAGE DIVIDER BIAS

From all our work with common emitter circuits, we can draw some conclusions about how to analyze AC circuit operation. If we know that the internal resistance of the source has a low value, we can ignore the bias resistors of a voltage divider circuit. Otherwise, we must include those values in our analysis.

For a common-emitter circuit using voltage divider bias:

$$Z_{in}(\text{stage}) = [R_t \parallel R_b \parallel Z_{in}(\text{base})]$$

If the source has a large internal resistance or if an actual resistance is present, as Figure 5.30a shows, determine the value for Z_{in} (stage) and proceed to Figure 5.30b. We can use the voltage divider theorem to find the base voltage, which is the voltage remaining after the drop across the source internal resistance. After finding this value, we can use Ohm's law and Figure 5.30c to calculate the base current: $I_b = v_b/Z_{in}(\text{base})$. Multiplying the base current by the beta value will give the emitter current.

We can continue to calculate the base input impedance as beta multiplied by the emitter resistance plus the AC emitter resistance. If the emitter resis-

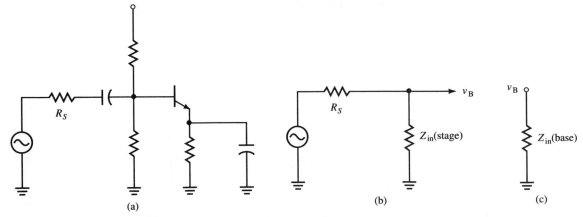

FIGURE 5.30 (a) Graphical "analysis" of a common-emitter circuit using voltage divider bias. Note: In this case R_s is at least 10% of Z_{in} (stage). (b) Input control loop. (c) Equivalent input base impedance.

tor is bypassed, we can use zero for the emitter resistance value. If the circuit features a swamping resistor, then the emitter resistance equals the value of the swamping resistor.

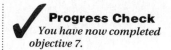

Progress Check
You have now completed objective 7.

PRACTICE PROBLEM 5.6

Assuming the transistor in Figure 5.31 is silicon and has a beta of 100, determine the following values for stiff coupling:
(a) r'_e
(b) Z_{in} (stage)
(c) DC base voltage
(d) DC emitter voltage
(e) DC collector voltage
(f) v_{out}

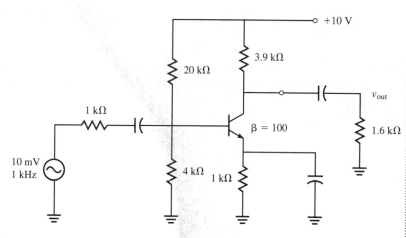

FIGURE 5.31

REVIEW SECTION 5.4

1. What is meant by stiff voltage divider bias?
2. Calculate V_B, V_E, V_C, I_E, and r'_e for the following figure.

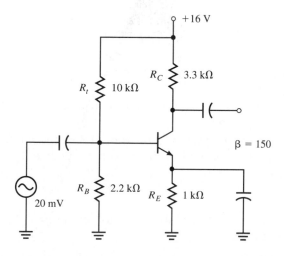

3. Refer to question 2. Draw the AC control loop (no R_s).
4. Refer to questions 2 and 3. If beta = 150 and V_{in} = 10 mV$_{p\text{-}p}$, determine the following: Z_{in}(base), Z_{in}, I_B, I_E (no R_s).
5. In the figure in question 2, if R_s = 300 Ω, what are the values of V_B, I_B, I_E?
6. In the figure in question 2, if the emitter bypass capacitor opens, what would happen to the Z_{in}(base) and Z_{in}?
7. Explain why Z_{in}(base) equals $r'_e \times$ beta.
8. Z_{in} must be less than the value of R_B. True or false?
9. What is the formula for Z_{in}(base) if a swamping resistor is added?
10. To DC, R_t and R_b are in (series, parallel).
   To AC, R_t and R_b are in (series, parallel).

5.5 ANALYZING A COMMON-EMITTER CIRCUIT THAT USES COLLECTOR FEEDBACK BIAS

In Figure 5.32, we see a common-emitter circuit that uses collector feedback bias to set up the DC operating voltages. We will use this circuit in the next example.

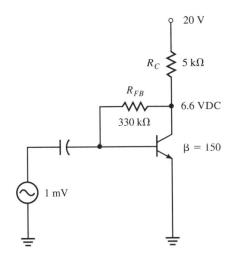

FIGURE 5.32 A common-emitter circuit using collector feedback bias

---

EXAMPLE 5.14

Find the DC emitter current, the AC emitter resistance, and the gain for the circuit shown in Figure 5.32.

Figure 5.33a shows the control loop model for this circuit. Because the control loop contains a feedback node, we must equalize the control loop for the emitter current.

Solution
To equalize the control loop, use the beta box in Figure 5.33a. The feedback resistor is divided by beta.

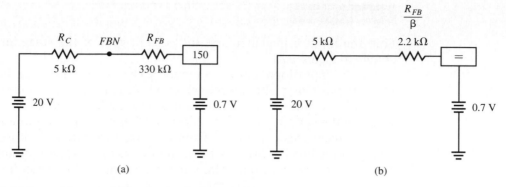

(a)   (b)

FIGURE 5.33 (a) Control loop model for the circuit shown in Figure 5.32. FBN is the feedback node. (b) Equalized control loop model.

$$R\left(equ\right) = R_{FB} = \frac{330 \text{ k}\Omega}{150} = 2.2 \text{ k}\Omega$$

Figure 5.33b shows the equalized control loop as a simple series circuit. Now, we can calculate the DC emitter current:

$$I_E = \frac{\left(20 \text{ V} - 0.7 \text{ V}\right)}{7.2 \text{ k}\Omega} = 2.68 \text{ mA}$$

With the DC emitter current, we can find the AC emitter resistance:

$$r'_e = \frac{25 \text{ mV}}{2.68 \text{ mA}} = 9.33 \text{ }\Omega$$

In turn, this leads to a gain calculation:

$$\frac{5 \text{ k}\Omega}{9.33 \text{ }\Omega} = 536$$

Figure 5.34 diagrams the AC input loop for the circuit. The feedback resistor complicates computations and also makes it difficult to see relationships in simple series or series-parallel combinations. Nevertheless, we can use our knowledge about this type of circuit to construct a procedure for simplifying the model.

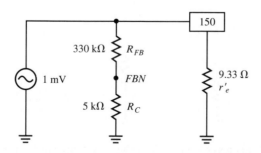

FIGURE 5.34 AC control loop for the circuit shown in Figure 5.32

EXAMPLE 5.15

Find the input impedance and the current gain of the stage for the circuit shown in Figure 5.31.

The AC input loop shown in Figure 5.34 has resistances on both sides of the beta box along with a feedback node. To completely equalize the model, we must account for both the beta box and the feedback node. First, we will equalize the beta box with the objective of showing the amount of base current. To reach this objective, we must realize two things. Because the source voltage signal has a direct path to the transistor base, all the input voltage reaches the base terminal. The emitter current on the opposite side of the beta box is beta times larger than the base current.

Solution

Figure 5.35 shows the first step in analyzing the model. On the emitter side of the beta box, the resistance changes to 1.4 kΩ. We can calculate this by multiplying beta by the AC emitter resistance:

$$150 \times 9.33 \ \Omega = 1.4 \ k\Omega$$

Not surprisingly, this equals the input impedance of the base. Once we know the input impedance of the transistor base, we can calculate the current drawn from the source by the base:

$$I_b = \frac{1 \ mV}{1.4 \ k\Omega} = 714 \ nA$$

This is not the entire current pulled from the source. Some AC current flows through the other parallel branch shown in Figure 5.32. To find the amount of the branch current, we will need to compute the AC voltage at the feedback node.

Earlier, we did a preliminary DC analysis of the circuit and found values for the emitter current, the AC emitter resistance, and the voltage gain. The voltage gain was 536. On the basis of this information, the AC voltage at the feedback node must be:

$$v_{fbn} = 536 \times 1 \ mV = 536 \ mV$$

Since the common-emitter circuit has a 180-degree phase shift, the output is −536 mV when the input is +1 mV. A 537-mV drop appears across the feedback resistor. Knowing the value of the voltage drop across the feedback re-

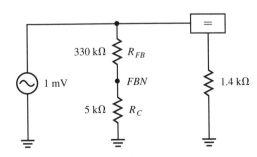

FIGURE 5.35 AC control loop analysis of a circuit using collector feedback bias

sistor and the value of the feedback resistor, we can apply Ohm's law to find the current through the feedback resistor:

$$i_f = \frac{537 \text{ mV}}{330 \text{ k}\Omega} = 1.62 \text{ } \mu\text{A}$$

This current merges with the collector current and flows through the collector resistor. Now, we can find the total current drawn by adding the values of the two branch currents:

$$1.62 \text{ } \mu\text{A} + 714 \text{ nA} = 2.34 \text{ } \mu\text{A}$$

We know the total current value and the input voltage value, so we can use Ohm's law to find the circuit impedance (Z).

$$Z = \frac{E}{I} = \frac{1 \text{ mV}}{2.34 \text{ } \mu\text{A}} = 427 \text{ } \Omega$$

The source voltage sees the stage as a 427-Ω resistor. This value is Z_{in}(stage).

As a bit of a review, let us see how we could have used the model to predict the same result. Using the model, we must

1. compute Z_{in}(base) as normal;

2. eliminate R_C from the input model—the source does not see R_C because of the AC collector voltage;

3. equalize R_{FB} to account for the current flow through it. To equalize the feedback resistor, we can divide the resistor value by the voltage gain value.

4. calculate Z_{in}(stage) by using product over sum for the model of Figure 5.36.

In addition, we must determine the current gain.

To find the current gain, we must find a value for the output current by:

$$i_{out} = \frac{536 \text{ mV}}{5 \text{ k}\Omega} = 107 \text{ } \mu\text{A}$$

Using the current pulled out of the source, 1.62 μA, we find that the current gain equals:

$$A_i = \frac{107 \text{ } \mu\text{A}}{1.62 \text{ } \mu\text{A}} = 66$$

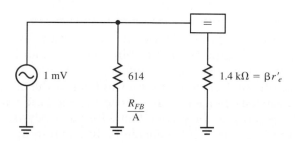

FIGURE 5.36 Equalized feedback resistor in a collector feedback circuit

Progress Check
You have now completed objective 8.

Determine the AC output voltage, AC output current, current gain, and input impedance of the stage for the circuit shown in Figure 5.37.

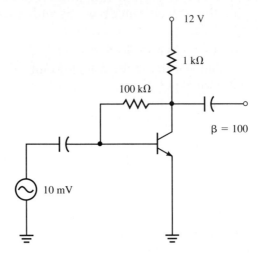

FIGURE 5.37

REVIEW SECTION 5.5
1. Refer to Figure 5.32. If $R_{FB} = 180$ kΩ, $R_C = 1.2$ kΩ, and beta = 75, determine the value of I_E, r'_e, and A_v.
2. Refer to question 1. What are the values of Z_{in} and A_i?
3. What is the formula for A_v in a collector feedback bias circuit?
4. Refer to questions 1, 2, and 3. If $V_{out} = 10$ V_{p-p}, what is the value of V_{in}?
5. Refer to Figure 5.32. If $R_{FB} = 330$ kΩ, $V_{CC} = 10$ V, $R_C = 1.5$ kΩ, and beta = 110, determine the following: V_B, V_C, V_E.

5.6 EXPLORING THE COMMON-COLLECTOR CONFIGURATION

The common-collector circuit, also known as the **emitter-follower circuit,** provides current gain and impedance matching. Voltage gain is slightly less than one. Figure 5.38a shows two common-collector circuits. In Figure 5.38a, voltage divider bias sets up the DC conditions that allow the circuit to work. Figure 5.38b uses base bias. Since we have already studied the effects of voltage divider bias, let us move to the second figure for our analysis.

Capacitors C_1 and C_2 are coupling capacitors. In a common-collector circuit, the AC signal couples into the base and is taken off the emitter junction. With no collector resistor, the collector is at AC ground and works as the common terminal. The output signal flows to the 100-Ω resistor connected

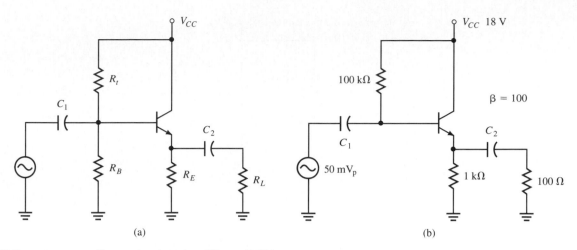

FIGURE 5.38 Two common-collector circuits using different DC bias arrangements

to the right side of capacitor C_2. Since the capacitor blocks DC voltages, the load resistor has no effect on the DC operation of the circuit.

For the same reason, we can ignore the source voltage and its internal resistance. Capacitor C_1 isolates the source from the DC conditions set by the bias resistors. As usual, we will begin our analysis by determining the DC voltages established by the bias configuration. Then, we will look at the AC operating conditions and draw the AC equivalent circuit.

EXAMPLE 5.16

Analyze the DC and AC resistances, voltages, and currents for the circuit shown in Figure 5.38b.

Figure 5.39a displays the DC equivalent circuit. The DC control loop model (Figure 5.39b) allows us to calculate the emitter current. First, equalize R_B. Then we can proceed through the following equations:

$$R_{B(equalized)} = \frac{R_B}{\beta} = \frac{100\text{ k}\Omega}{100} = 1\text{ k}\Omega$$

$$I_E = \frac{17.3\text{ V}}{\left(1\text{ k}\Omega + 1\text{ k}\Omega\right)} = 8.65\text{ mA}$$

$$V_E = 8.65\text{ mA} \times 1\text{ k}\Omega = 8.65\text{ V}$$

$$V_B = 8.65\text{ V} + 0.7\text{ V} = 9.35\text{ V}$$

$$I_B = \frac{8.65\text{ mA}}{100} = 86.5\text{ }\mu\text{A}$$

$$V_C = V_{CC} = 18\text{ V}$$

$$r'_e = \frac{25\text{ mV}}{8.65\text{ mA}} = 2.89\text{ }\Omega$$

When we draw the AC equivalent circuit in Figure 5.40a, we can treat the capacitors as direct paths for current. Since the AC signal has a direct path to the base, the AC signal does not decrease before reaching the base terminal. Figure 5.40b shows the control loop model for the circuit. As the model shows, no resistance exists between the signal and the base. The emitter circuit consists of the AC emitter resistance in series with the parallel combina-

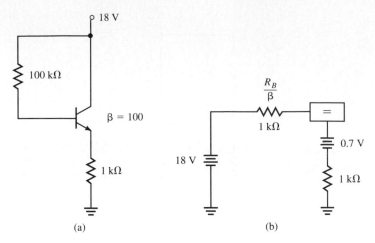

FIGURE 5.39 (a) DC equivalent circuit of Figure 5.38b. (b) DC control loop for the circuit of Figure 5.38b.

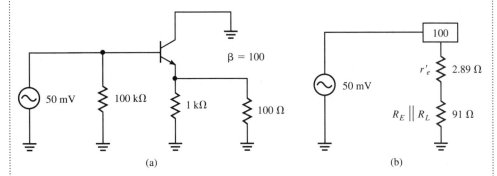

FIGURE 5.40 (a) AC equivalent circuit for Figure 5.38b. (b) AC control loop for the circuit shown in Figure 5.38b.

tion of the emitter resistor and load resistor. From here, we can calculate the AC emitter current, i_e.

Solution

With resistance on only one side of the beta box, we do not need to equalize the AC control loop. Thus, the AC emitter current equals:

$$\frac{50\text{ mV}_p}{93.89\ \Omega} = 533\ \mu A_p$$

To find the output voltage, let us consider the gain loop model for the common-collector circuit shown in Figure 5.41. Since the beta box drops no voltage, all of the source input voltage arrives at test point 1. As before, we cannot measure the voltage at TP1 because the test point is inside the transistor. We can use the model to predict what will happen at the output voltage tap.

Using the value for the input signal, we know that 50 mV arrives at the test point. The AC emitter circuit forms a voltage divider driven by the 50 mV. Therefore, we can predict the output voltage using the voltage divider relationship:

$$v_{out} = \frac{50\text{ mV} \times 91\ \Omega}{93.89} = 48.46\text{ mV}$$

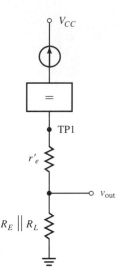

FIGURE 5.41 Gain loop model for the circuit shown in Figure 5.38b

The output voltage is less than the input voltage in a common-collector circuit. Also, the emitter resistor in parallel with the load resistor determines the gain of the circuit:

$$A_v = \frac{\left(R_E \| R_L\right)}{R_E \| R_L + r'_e}$$

In most common collector circuits, the value for the AC emitter resistance will be small compared with the parallel resistance value. When we apply the above formula to Figure 5.41 and use the values from the circuit shown in 5.40, we get:

$$A_v = \frac{91\,\Omega}{93.89\,\Omega} = 0.969$$

Because the gain is almost 1, we can define the common-collector circuit as a "unity gain" amplifier for voltages.

DETERMINING THE INPUT IMPEDANCE OF A COMMON-COLLECTOR TRANSISTOR CIRCUIT

Even though the common-collector circuit does not provide voltage gain, it has merit in that it can match impedance between circuits. The common-collector circuit can also be used to provide current gain, where $A_i = \beta$. In this section, we will learn how to determine the input impedance of a common-collector transistor.

EXAMPLE 5.17

Determine the input impedance of the circuit shown in Figure 5.38b.

The AC control loop, including the internal resistance of the source, for the circuit has been redrawn in Figure 5.42. Since the internal resistance lies in series with the output voltage of the source, the signal no longer has a direct path to the base.

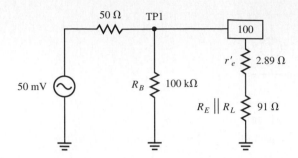

FIGURE 5.42 AC control loop for Figure 5.38b including internal impedance of the source

Starting our analysis in the base part of the AC control loop, we see that the 100-kΩ and the 50-Ω resistors form a voltage divider. The voltage at TP1 is:

$$v_{TP1} = \frac{100 \text{ k}\Omega}{\left(100 \text{ k}\Omega + 50 \text{ }\Omega\right)} \times V_{in} = 49.98 \text{ mV}$$

Since we lost less than 1/10 of one percent of the signal, we can say that the voltage at the test point equals the input voltage. If we thevenize the test point and look back toward the input, we see that the 50-Ω resistance parallels the 100-kΩ resistor. Figure 5.43a illustrates that:

$$50 \text{ }\Omega \parallel 100 \text{ k}\Omega = 49.98 \text{ }\Omega$$

Because the 100-kΩ bias resistor has little effect on AC operation, we can remove it from the AC control loop model shown in Figure 5.43b. Effectively, we have reduced the control loop model to a simple series circuit. There are resistances on both sides of the beta box, so we should equalize the model. Remember that we must solve for the base current, not the emitter current. To do this, we convert the control loop so that the AC emitter current looks like the base current. *Equalizing the AC control model for the base current requires that the emitter resistor appears beta times larger.*

Figure 5.44 shows the equalized control loop model. Using the equalized model, we can calculate the base current value:

$$i_b = \frac{50 \text{ mV}}{50 \text{ }\Omega + 9.4 \text{ k}\Omega} = 5.29 \text{ }\mu\text{A}$$

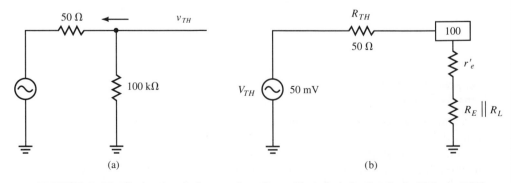

FIGURE 5.43 Eliminating the base resistor from AC analysis for the ciruit of Figure 5.38b

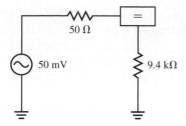

FIGURE 5.44 Equalized AC control loop for the circuiit of Figure 5.38b

The $9.4 = k\Omega$ value represents the input impedance of the base as "seen" by the AC source. This current flows through the source voltage, drops some voltage across the internal resistance, and equals:

$$5.29\ \mu A \times 50\ \Omega = 265\ \mu V$$

By using the common collector as a buffer between the load and the source, we can preserve almost all the original signal. As the current draw increases, the amount of voltage lost across the internal resistance also must increase. The amount of current drawn from the source equals the base current.

OBSERVATIONS AND CONCLUSIONS ABOUT THE COMMON-COLLECTOR CIRCUIT

In a common-collector circuit, the effective AC emitter resistance equals the value of the emitter resistor in parallel with the load resistor:

$$r_e = R_E \parallel R_L$$

Also, from the perspective of the AC signal driving the base, the input impedance of the transistor equals beta multiplied by the sum of the DC emitter resistance and the AC emitter resistance:

$$B(r_e + r'_e) = Z_{in}(base)$$

The base resistor in a base bias resistor network is almost always large enough that it has a minimal effect on AC operation.

Because the common-collector configuration presents a high impedance to the AC signal, it does not load down the source. This makes the common collector a good choice for coupling signals into low-impedance loads. Although the common-collector circuit does not provide voltage gain, the circuit does provide good current gain. Finally, the output voltage of the common collector is in phase with the input voltage.

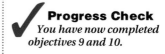

Progress Check
You have now completed objectives 9 and 10.

REVIEW SECTION 5.6
1. What is another name for a common-collector amplifier?
2. A common collector provides voltage gain. True or false?
3. The input on a common collector is between which two leads? The output is between which two leads?
4. For the following circuit diagram, determine V_B, V_C, V_E, I_B, r'_e.
5. If R_L in question 3 is 220 Ω, what is the value of i_e, if $V_{in} = 10\ V_{p\text{-}p}$?

6. Explain why V_{out} is less than V_{in} in a common-collector amplifier.
7. A common collector is characterized by a _____ input impedance and a _____ output impedance.
8. What is an application of a common-collector amplifier?
9. What is the phase relationship between V_{in} and V_{out} in a common-collector amplifier?
10. Compare and contrast the characteristics of a common collector and any common-emitter amplifier.

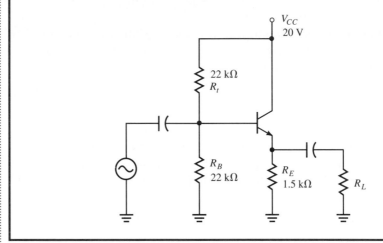

5.7 EXPLORING THE COMMON-BASE CONFIGURATION

With the common-base configuration, we have typically a low input impedance and good voltage gain, but no current gain. In effect, the input current equals the output current, and the emitter current equals the collector current. We will approximate the current gain of a common-base transistor circuit to equal 1. No phase shift from input to output exists in the common-base circuit. The common-base circuit is generally used in high-frequency applications.

EXAMPLE 5.18

Figure 5.45a shows a common-base circuit. The figure shows the transistor on its side, which is typical for common-base schematic drawings. For this figure, emitter bias sets the DC characteristics. Determine the AC input impedance of the circuit.

Figure 5.45b shows the DC control loop for the circuit. A close study of the control loop shows the absence of the base resistor. With the AC input signal placed on the emitter, the base resistor becomes unnecessary. With no AC input to the base, we do not need to worry about shorting the signal.

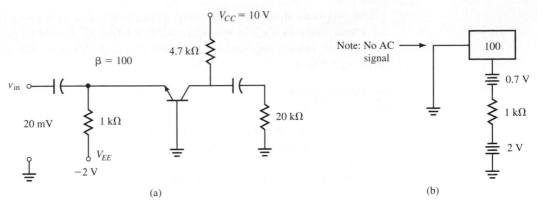

FIGURE 5.45 (a) A common-base circuit. (b) DC control loop model of the circuit.

Solution

With no base resistor, our analysis becomes easier. Because the loop does not need equalizing, we can calculate the emitter current directly:

$$I_e = \frac{(2\text{ V} - 0.7\text{ V})}{1\text{ k}\Omega} = \frac{1.3\text{ V}}{1\text{ k}\Omega} = 1.3\text{ mA}$$

That answer allows us to calculate the AC emitter resistance:

$$r'_e = \frac{25\text{ mV}}{1.3\text{ mA}} = 19.2\ \Omega$$

Figure 5.46a depicts the AC input loop. Again, the lack of resistance on the base side of the beta box allows us to proceed with equalizing the loop. Consequently, we can remove the beta box. Figure 5.46b shows the redrawn control loop without the beta box. This view makes it easy to see that the emitter resistor is in parallel with the AC emitter resistance. Seen by the AC source, the input impedance is:

$$R_E \,\|\, r'_e = 1\text{ k}\Omega \,\|\, 19.2\ \Omega = 18.8\ \Omega$$

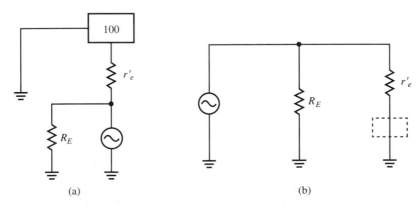

FIGURE 5.46 (a) AC input loop of the circuit shown in Figure 5.45a. (b) Redrawn control loop model.

We can now draw an important conclusion. For common-base circuits, the input impedance closely approximates the AC emitter resistance. With such a low input impedance, the internal resistance of the source causes some signal loss.

EXAMPLE 5.19

Find the amount of signal reaching the emitter terminal for the circuit shown in Figure 5.45a. For this example, we can give the internal resistance a value of 10 Ω.

To determine the amount of signal that will reach the emitter terminal for the circuit shown in Figure 5.45a, we will need to set up the voltage divider shown in Figure 5.47.

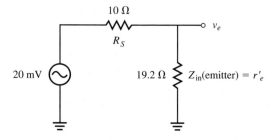

FIGURE 5.47 Voltage divider effect on the input signal of a common-base circuit

Solution
The voltage divider drops the input voltage to 13.15 mV.

$$\frac{19.2 \ \Omega}{29.2 \ \Omega} = 0.658$$

$$v_e = 0.658 \times V_{in} = 13.15 \text{ mV}$$

This is the voltage that will be amplified.

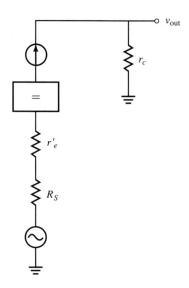

FIGURE 5.48 Gain loop model for a common-base circuit

Using the gain loop model shown in Figure 5.48, we can determine the gain of the transistor. Because the input impedance of the transistor has a small value, we can calculate the stage gain with:

$$A_v = \frac{r_c}{r'_e + r_s}$$

EXAMPLE 5.20

Determine the transistor gain, the stage gain, and the AC output voltage for the circuit shown in Figure 5.45a.

Solution
In the example, the transistor gain is:

$$\frac{\left(4.7 \text{ k}\Omega \parallel 20 \text{ k}\Omega\right)}{r'_e} = \frac{3805 \ \Omega}{19.2 \ \Omega} = 198$$

With a signal level of 13.15 mV reaching the emitter, the AC output equals:

$$v_{out} = 13.15 \text{ mV} \times 198 = 2.60 \text{ V}$$

Using the numbers for the stage gain, we arrive at the same answer. With stage gain equaling

$$\frac{3805 \ \Omega}{29.2 \ \Omega} = 130$$

We can multiply the stage gain by the input signal voltage to arrive at:

$$v_{out} = 20 \text{ mV} \times 130 = 2.6 \text{ V}$$

USING VOLTAGE DIVIDER BIAS WITH THE COMMON-BASE CIRCUIT

Figure 5.49 shows a common-base circuit biased with a voltage divider. A bypass capacitor at the voltage divider node places the base at AC ground. The

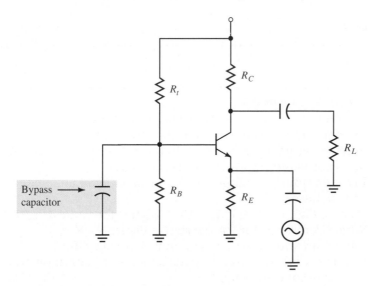

FIGURE 5.49 Common-base circuit using voltage divider bias

AC signal is injected into the emitter through a coupling capacitor. We can analyze the configuration by stepping through the following outline:

1. Set up the DC control loop model, and solve for the emitter current.

2. Use the emitter current value to calculate the AC emitter resistance. This value equals the input impedance of the circuit.

3. Use the gain loop model of Figure 5.48 to predict circuit gain.

4. Predict the AC output voltage by multiplying the stage gain by the input voltage signal.

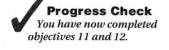

Progress Check
You have now completed objectives 11 and 12.

PRACTICE PROBLEM 5.8

For Figure 5.50, determine the input impedance, the gain, and the AC output voltage.

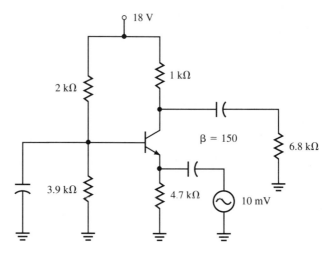

FIGURE 5.50

REVIEW SECTION 5.7
1. List four characteristics of a common-base amplifier.
2. Refer to Figure 5.45a. Trace the DC current paths.
3. Refer to Figure 5.45a. If $R_E = 470\ \Omega$ and $R_C = 2.2\ k\Omega$, determine the following values: V_B, V_C, V_E, I_C, r'_e, and Z_{in}.
4. In a common-base amplifier we can conclude that Z_{in} equals r'_e. True or false?
5. Refer to Figure 5.45a. If $R_S = 10\ \Omega$, what is the value of v_e?
6. What is a formula for voltage gain in Figure 5.49?
7. What is an application of a common-base amplifier?
8. Compare and contrast the characteristics of a common-base and a common-collector amplifier.

5.8 STAGE GAIN VERSUS TRANSISTOR GAIN

As we have looked at single-stage amplifiers, we have discussed transistor gain and stage gain. Even though the labels for transistor gain and stage gain may seem similar, the two characteristics have different definitions. Moving toward multistage amplifiers, the differences become more important.

EXAMPLE 5.21

Determine the transistor and stage gains and the AC output voltage for the circuit shown in Figure 5.51.

Working through the circuit shown in Figure 5.51, we find that the transistor gain can be calculated with the formula:

$$A_v = \frac{r_c}{r'_e}$$

That is, transistor gain equals the DC collector resistance divided by the AC emitter resistance.

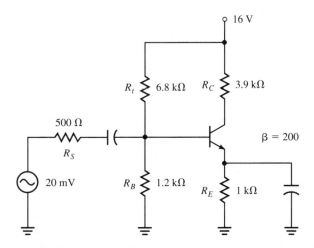

FIGURE 5.51 Common-emitter circuit—AC source with a large internal impedance

Solution
Since the circuit contains no load resistor, the DC collector resistance equals the value of the collector resistor and the gain equals:

$$\frac{r_c}{r'_e} = \frac{3.9 \text{ k}\Omega}{14.7 \text{ }\Omega} = 265$$

In circuits driven by large internal impedances, not all of the input signal reaches the base of the transistor. Some of the signal drops across the internal impedance of the source. Obviously, the internal impedance of the source can reduce the effectiveness of the amplifier. For the circuit shown in Figure 5.51, the stage gain equals the output voltage of the transistor divided by the input signal of the circuit. Using the transistor gain and the input signal reaching the base, we find that the transistor output voltage is:

$$265 \times 12 \text{ mV} = 3.2 \text{ V}_p$$

Thus, the stage gain is:

$$\frac{3.2 \text{ V}}{20 \text{ mV}} = 160$$

Even with a transistor that supplies a gain of 265, the circuit provides a stage gain of only 160.

We can calculate stage gain directly without first determining the output voltage through other techniques. To see how, let us look at Figure 5.51 one more time. In this diagram, we see a voltage divider driven by a 20-mV source. This source "sees" its own internal resistance of 500 Ω and the input impedance of the stage. The input impedance was found to equal:

$$Z_{in} = 757 \text{ }\Omega$$

Setting up a voltage divider relationship for the circuit, we find that the percentage of applied input voltage reaching the base equals:

$$\frac{757 \text{ }\Omega}{1257 \text{ }\Omega} = 0.6$$

Multiplying this value by the transistor gain value, we get:

$$0.6 \times 265 = 159$$

This agrees with the 160 stage gain value that we calculated in the example. Again, rounding off causes the small discrepancy.

REVIEW SECTION 5.8
1. The internal resistance of a source driving a transistor amplifier increases the overall stage gain. True or false?
2. Stage gain is a measure of the output voltage of the transistor divided by the _____ _____ of the circuit.
3. The input signal on the base of a transistor is always less than or at best equal to the input signal of the circuit. True or false?
4. The internal impedance of the signal source and the input impedance of the transistor stage form a _____.
5. Overall stage gain can be determined by multiplying the gain of the transistor by the ratio of [z_{in} stage] /R_s, where [z_{in} stage] is the input impedance into the transistor amplifier *circuit* and R_s is the internal impedance of the source. True or false?

5.9 MULTISTAGE AMPLIFIERS

A single transistor amplifier usually cannot provide the amount of gain often needed by modern electronic circuits. Higher gain can be achieved by cascading, which means connecting the output of one amplifier stage to the input of another amplifier stage. When amplifier stages are **cascaded** the voltage gains of individual stages multiply. A three-stage amplifier that has individual stage gains equaling 30 would have an overall gain of 27,000.

Figure 5.52 shows several amplifier stages cascaded, or connected in series together. Each cascaded stage works with the other stages with the overall function of the circuit influenced by the individual actions. To calculate the overall gain of the entire circuit, the individual stage gains can be multiplied together.

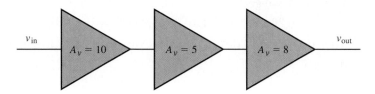

FIGURE 5.52 Several amplifier stages cascaded together

EXAMPLE 5.22

Calculate the output signal of the circuit shown in Figure 5.52.

The first stage has a gain of 10; the second-stage gain is 5; and the gain of the third stage is 8.

Solution

Multiplying those numbers, we find that the entire circuit has a voltage gain of:

$$A_v = 10 \times 5 \times 8 = 400$$

Since we have arrived at 400 as an answer, we know that the input signal will be 400 times larger by the time it reaches the output. An input signal of 1 mV would produce an output signal of 400 mV.

Manufacturers use another word for gain, the **decibel** (dB). The decibel is one-tenth of a bel. Using the term *decibel* relates sound intensity to the physiology of the human ear. Sounds differing by tenths of a decibel can be separated by the human ear. For the present, decibels, shown as A'_v will represent changes in voltage gain. When we progress to power amplifiers, the decibel will also represent power gain.

If we have a value for the voltage gain, we can calculate the decibel rating with the formula:

$$A'_v = 20 \log A_v$$

If we translate the gain of each stage into decibels, we can add the decibels together to find the overall voltage gain.

EXAMPLE 5.23

Determine the overall gain for the circuit shown in Figure 5.52 by converting each stage gain to decibels and then adding the decibel gain of the stages together.

Solution
First, we will convert each stage gain from the circuit to decibels using the formula: $A'_v = 20 \log A_v$

Stage 1: $A'_v = 20 \log 10 = 20$ dB
Stage 2: $A'_v = 20 \log 5 = 13.98$ dB
Stage 3: $A'_v = 20 \log 8 = 18.06$ dB

Adding the results, we get:

$$20 \text{ dB} + 13.98 \text{ dB} + 18.06 \text{ dB} = 52.04 \text{ dB}$$

which is the equivalent of the answer that we found in example 5.22.

PRACTICE PROBLEM 5.9

Convert the following values to decibel equivalents:

156, 34, 89, 908

If we need to convert a decibel value to a standard value, we could use a calculator to follow this sequence:

Enter the decibel value.
Divide it by 20.
Press INV or SECOND FUNCTION then the log button.

Now that we know that 400 equals 52.04 decibels, let us convert 52.04 dB back to standard notation.

EXAMPLE 5.24

Convert a decibel value to standard notation.

Solution
Using your calculator, enter 52.04, and then divide it by 20. You should obtain 2.602 as an answer. Now, first press the INV key and then the LOG key. The calculator should display 399.94. Since the 52.04 was truncated to two decimal places, a slight discrepancy will appear because of rounding off.

PRACTICE PROBLEM 5.10

Convert these decibel values back to standard form:

34 dB, 16.5 dB, 22 dB, 9 dB

DETAILED MULTISTAGE ANALYSIS
The circuit shown in Figure 5.53 is a two-stage transistor amplifier circuit. While the first stage features a common-emitter configuration with voltage divider bias, the second stage includes a common-collector configuration

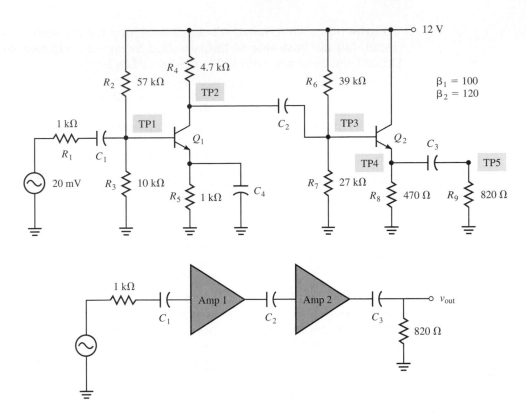

FIGURE 5.53 Two-stage transistor amplifier

that again uses voltage divider bias. Capacitor C_1 couples the input into the first stage. Because some resistance exists between the base terminal and the input source, the overall gain of the circuit will decrease.

Capacitor C_2 couples the output from the first stage into the base of the second stage. Another capacitor, C_3, links the output signal to the output load resistor. Capacitor C_4 works as a bypass capacitor and places the emitter terminal of transistor Q_1 at AC ground.

EXAMPLE 5.25

Find the DC and AC voltages at the five designated test points in the circuit shown in Figure 5.53.

Solution: The Second Stage
We begin our analysis with the output stage, or the common-collector circuit. First, we need to determine the Thevenin voltage at test point TP3. We can calculate this value by using the voltage divider theorem:

$$V_{TH} = \frac{27\ k\Omega}{\left(27\ k\Omega + 39\ k\Omega\right)} \times 12\ V = 4.9\ V$$

Since the base of transistor Q_2 loads the divider and lowers the base voltage, this is not the DC voltage for the test point. The base voltage is less than the Thevenin voltage.

Figure 5.54 shows the equalized control loop for the second stage of the circuit. On the base side of the beta box, we see an effective resistance of 133 Ω. This resistance, with 120 as the beta of Q_2 is:

$$\frac{\left(39\text{ k}\Omega \parallel 27\text{ k}\Omega\right)}{120} = 133\ \Omega$$

From the model of the control loop, we can find the emitter current for Q_2:

$$\frac{4.2\text{ V}}{603\ \Omega} = 6.97\text{ mA}$$

Using this value, we can find the value for the emitter voltage for Q_2 and voltage for test point TP4:

$$6.98\text{ mA} \times 470\ \Omega = 3.3\text{ V}$$

Adding 0.7 V to the emitter voltage gives the base voltage for Q_2 and voltage for TP3:

$$3.28\text{ V} + 0.7\text{ V} = 4.0\text{ V}$$

Test point 5 has a DC voltage of zero volts because capacitor C_3 provides DC isolation between test point 5 and test point 4.

Next, we use the emitter current value to find the AC emitter resistance (r'_e) for transistor Q_2. This value is:

$$\frac{25\text{ mV}}{6.98\text{ mA}} = 3.58\ \Omega$$

This resistance, r'_e, works in series with the emitter resistance consisting of the emitter resistor in parallel with the load resistor. With these values, we can calculate the input impedance of the base of transistor Q_2:

$$Z_{in}(\text{base}) = \text{beta}(r'_e + r_e) = 120(3.58\ \Omega + 299\ \Omega) = 36.3\text{ k}\Omega.$$

This input resistance is in parallel with resistors R_5 and R_6, which means that the stage impedance equals:

$$Z_{in}(\text{stage}) = 36.3\text{ k}\Omega \parallel 27\text{ k}\Omega \parallel 39\text{ k}\Omega = 11.08\text{ k}\Omega$$

Since a common-collector configuration makes up the second stage, we can set the voltage gain to unity.

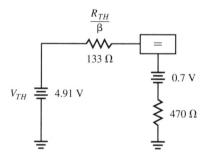

FIGURE 5.54 Equalized control loop for the second stage of the circuit shown in Figure 5.53

Solution: The First Stage

We can begin our DC analysis of the first amplifier stage by again finding the Thevenin voltage:

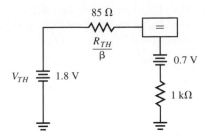

FIGURE 5.55 Equalized control loop for the first stage of the circuit shown in Figure 5.53

$$V_{TH} = \frac{10 \text{ k}\Omega}{67 \text{ k}\Omega} \times 12 \text{ V} = 1.8 \text{ V}$$

Since the first stage has a firm voltage divider, we can assume that the base voltage of transistor Q_1 equals the Thevenin voltage; see Figure 5.55. We find this voltage at test point 1. To find the emitter voltage of Q_1, we subtract 0.7 V from the base voltage:

$$V_E = 1.8 \text{ V} - 0.7 \text{ V} = 1.1 \text{ V}$$

Next, we can determine the emitter current of the transistor by using Ohm's law:

$$I_E = I_C = \frac{1.1 \text{ V}}{1 \text{ k}\Omega} = 1.1 \text{ mA}$$

With this value, we can find the voltage drop across resistor R_4:

$$V_{R4} = 1.1 \text{ mA} \times 4.7 \text{ k}\Omega = 5.2 \text{ V}$$

To obtain the collector voltage, we subtract the resistor R_4 voltage drop from the collector source voltage, giving us the voltage at TP2:

$$V_{TP2} = 12 \text{ V} - 5.12 \text{ V} = 6.9 \text{ V}$$

To begin our AC analysis of the first stage, we calculate the AC emitter resistance of the transistor:

$$r'_e = \frac{25 \text{ mV}}{1.1 \text{ mA}} = 22.7 \text{ } \Omega$$

The transistor voltage gain of the first stage is:

$$A_{v_1} = \frac{r_c}{r'_e} = \frac{\left(4.7 \text{ k}\Omega \parallel 11 \text{ k}\Omega\right)}{22.9 \text{ } \Omega} = \frac{\left(3.3 \text{ k}\Omega\right)}{22.9 \text{ } \Omega} = 144$$

Of the values used for the gain equation, the 11 kΩ is the input impedance of the second stage. Figure 5.56 diagrams the relationship of the impedance with the collector resistor. Since the DC voltage supply is an AC ground, the collector resistor works in parallel with the second-stage output impedance.

Next, we move to the input side of transistor Q_1. We start by determining the input impedance found at the base:

$$Z_{in}(base) = \beta(r'_e) = 100 \times 22.9 \text{ } \Omega$$

The bypass capacitor, C_4, places the emitter at AC ground. Since the emitter resistance equals zero, it does not show in the calculation. Three resistances—the input impedance of the base paralleling resistors R_2 and R_3—make up the input impedance of the first stage:

$$Z_{in}(stage 1) = 10 \text{ k}\Omega \parallel 57 \text{ k}\Omega \parallel 2.29 \text{ k}\Omega = 1.8 \text{ k}\Omega$$

This input resistance also works in series with the source resistance, R_1.

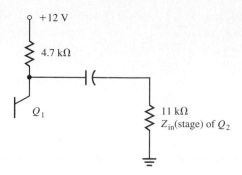

FIGURE 5.56 Relationship of the input impedance of Q_2 and the collector of Q_1 from the circuit in Figure 5.53

Because of the source resistance, only 12.9 mV of AC voltage reaches the transistor base. Figure 5.57 shows the relationship between the resistors and how the voltage divider principle gives the 12.9-mV value. To find the AC voltage at test point 2, we can multiply the base voltage by the transistor gain:

$$12.9 \text{ mV} \times 144 = 1.86 \text{ vAC}$$

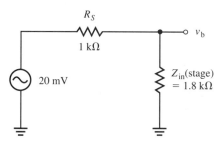

FIGURE 5.57 Input impedance of stage 1 from Figure 5.53 and its relationship to the internal impedance of the source

This AC voltage couples directly to the base of transistor Q_2 through capacitor C_2. The AC voltage at test point 3 equals 1.86 vAC. Since the common collector found in the amplifier second stage has a voltage gain of 1, the voltages at test points 4 and 5 also equal 1.86 vAC. With the gain of the first transistor at 144 and that of the second transistor at 1, the overall gain of the circuit equals 144:

$$144 \times 1 = 144$$

Progress Check
You have now completed objective 13.

REVIEW SECTION 5.9

1. In a multistage amplifier, if $A_{v1} = 20$, $A_{v2} = 5$, and $A_{v3} = 15$, what is the value of A_v total?
2. A human ear can detect a change of _____ decibels.
3. If $A_v = 250$, what is the value of A_v' (gain)?
4. Overall gain in a multistage amplifier is (added, multiplied) when using decibels.

5. Find the decibel equivalent of the following numbers: 210, 5860, 18, 14.
6. Refer to Figure 5.53. Change V_{CC} to 15 V, and determine A'_{v1}, A'_{v2}.
7. Explain why a resistor that opens in stage 2 does not affect Q_1 DC conditions.

TROUBLESHOOTING

Troubleshooting a multistage amplifier demands an organized approach to finding the fault. Before looking at the individual amplifier circuits, check the power supply voltages. In any device that has multiple printed circuit boards or connections, check for proper power supply distribution. Keep a log of the failures and record symptoms and involved parts. If possible, obtain schematic drawings from the manufacturer. If no schematic is available, draw a simple schematic of the circuit.

EXAMPLE 5.26

Determine the results of an open bypass capacitor for the circuit shown in Figure 5.58a.

With the open bypass capacitor, the control loop becomes modified as shown in figure 5.58b. The emitter resistor is part of the control loop and works as a path for AC current. Using Ohm's law we can calculate a new value for the AC current:

$$i_e = \frac{5 \text{ mV}}{1.8 \text{ k}\Omega} = 2.78 \text{ } \mu\text{A}$$

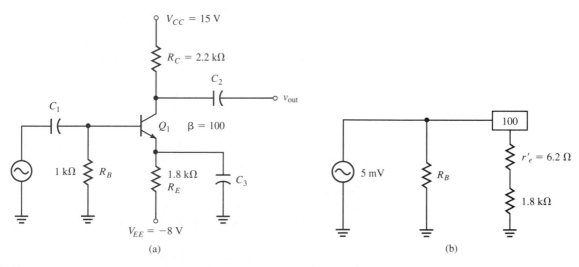

FIGURE 5.58 (a) Circuit with an open bypass capacitor. (b) Control loop model.

The new current has decreased to a minuscule amount, and the amplifier is almost useless.

Solution

Figure 5.59 shows the modified gain loop model. With the failure of the by-pass capacitor, the emitter resistor becomes a part of the gain loop. Gain now falls to:

$$\frac{R_c}{r'_e + R_e} = \frac{2.2 \text{ k}\Omega}{1.8 \text{ k}\Omega} - 1.22$$

Since the gain has dropped from 355 to 1.22, the output signal reduces to:

$$5 \text{ mV} \times 1.22 = 6.1 \text{ mV}$$

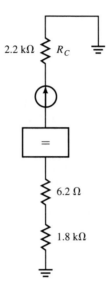

FIGURE 5.59 Modified gain loop for a circuit with an open bypass capacitor

ISOLATING FAULTS IN A MULTISTAGE CIRCUIT

Figure 5.60 shows a multistage amplifier circuit in block diagram form. To test the circuit, supply the input signal. A signal generator can produce the voltages and currents normally present at the input of the circuit. When applying the input signal, make sure that the applied signal fits the circuit needs. An audio generator will not work for a radio-frequency circuit.

After making a visual inspection and checking the supply voltages, start troubleshooting by checking the output of the last stage. Use circuit predictions or schematic information to verify the output values. If the value is not correct, check the output of the preceding stage. Moving from the last stage to the first stage will quickly lead to the problem area.

When troubleshooting the bad stage, check the DC bias levels by using the guidelines from the troubleshooting section at the end of Chapter 4. If no DC problems surface, check the AC part of the circuit. As we have seen, a faulty capacitor can "kill" the amplifier output.

Although small signal amplifiers will not normally distort a signal, a check of the waveform shape may lead to the problem area. A distorted signal is

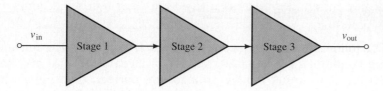

FIGURE 5.60 Block diagram of a multistage amplifier circuit

one that does not retain the proper shape. Figure 5.61 shows several wave-form examples of distorted signals such as clipping. A small-signal amplifier with a clipped output may have too much gain. The input signal overdrives the amplifier or bias components may have changed values, causing level shifts.

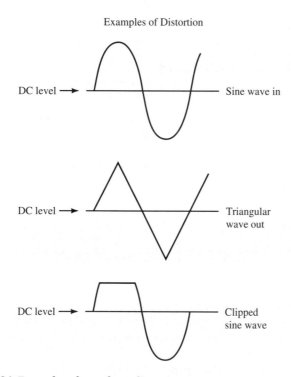

FIGURE 5.61 Examples of waveform distortion

TROUBLESHOOTING QUESTIONS
1. Explain how to check a bypass capacitor and a coupling capacitor using an oscilloscope and a signal generator.
2. When a bypass capacitor opens, what happens to the voltage gain of that stage?
3. Explain how to troubleshoot a multistage circuit.
4. Explain what clipping is and what causes it.

The following circuit is a common-emitter amplifier. The normal operating conditions are V_{in} = 20 mV, TP1 = 20 mV, TP2 = 0 mV, and TP3 = 840 mV. Assuming the DC voltages are correct, determine which coupling or bypass capacitor is faulty for each of the following conditions:

1. V_{in} = 20 mV, TP1 = 0 V, TP2 = 0 V, and TP3 = 0 V
2. V_{tn} = 20 mV, TP1 = 20 mV, TP2 = 20 mV, and TP3 = 4 mV
3. V_{in} = 20 mV, TP1 = 20 mV, TP2 = 0 mV, and TP3 = 1.69 $V_{p\text{-}p}$

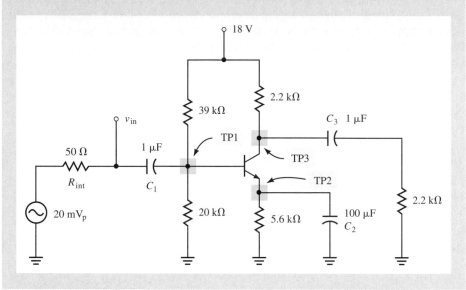

SUMMARY

Transistor amplifiers increase some quantity in a circuit. Electronic circuitry uses many types of transistor amplifiers. Those amplifier types come in three basic configurations called the common emitter, common collector, and common base.

When transistors amplify, they provide either voltage, current, or power gain. Gain is the measure of the output signal compared with the input signal.

In each of the three basic transistor configurations, one section is common to the other two for both input and output connections. Each name—common emitter, common base, and common collector—names the section that is common to the other two sections.

Signals must pass from one stage to another in a multistage amplifier. One method of coupling the stages together is using capacitors. Capacitor coupling gives an easy and inexpensive means of passing signals while isolating DC and bias voltages.

A multistage amplifier dissipates heat better than a one-stage amplifier. In addition, the multistage amplifier can link different frequencies and impedances. Each stage may use a different type of bias.

To make circuit analysis easier, we can break amplifier circuits down into equivalent circuits. Because amplifier circuits use both DC and AC voltages, we use AC and DC equivalent circuits. Using AC source voltages creates dynamic AC signal factors such as an AC emitter resistance. From there, we can break the circuit down even further into simple series and parallel circuits. This allows us to use Ohm's law, Kirchhoff's laws, and the beta box theory.

All three transistor configurations use some type of bias such as voltage divider bias, emitter bias, collector feedback bias, or base bias. Each method influences characteristics such as gain, impedance, and performance under varying thermal conditions.

Almost every piece of electronic equipment uses multistage amplifier circuits. Instead of relying on one stage for all the requirements, designers will cascade, or link, transistor stages together. When measuring gain, technicians can measure either transistor gain or stage gain. A unit called the decibel is used for the measurement of gain. Adding the gains of the stages together gives the total gain for the multistage amplifier circuit.

As does all other electronic troubleshooting, working with multistage amplifiers requires an organized approach. This approach involves not only checking for proper voltages but also signal tracing. Technicians use equipment such as multimeters, signal generators, and oscilloscopes when working with multistage amplifiers. Using the proper techniques will allow a technician to narrow the problem from the complete circuit to a single transistor stage.

PROBLEMS

1. Determine the following for the circuits shown in Figures 5.62, 5.63, 5.64, and 5.65:
   (a) The AC emitter current
   (b) The AC base current
   (c) The AC emitter–base diode resistance, r'_e
   (d) The unloaded voltage gain of the transistor
   (e) The unloaded voltage gain of the transistor in decibels
   (f) The input impedance of the base, $Z_{in}(base)$
   (g) The input impedance of the circuit, $Z_{in}(stage)$
   (h) The voltage gain of the circuit. Include losses due to source resistance, loading, and any swamping resistor, if present

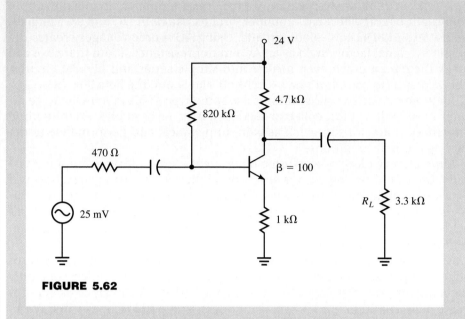

FIGURE 5.62

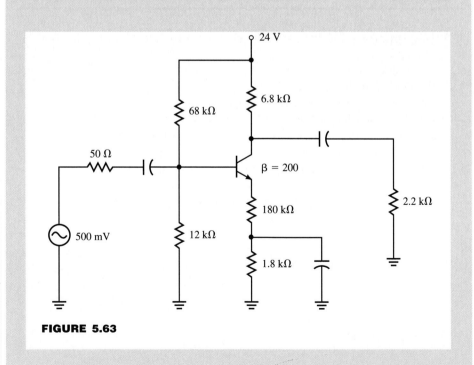

FIGURE 5.63

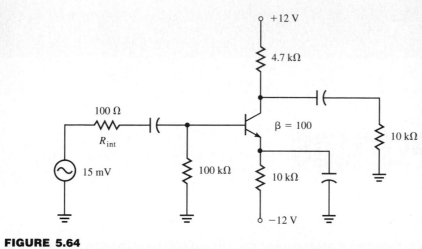

FIGURE 5.64

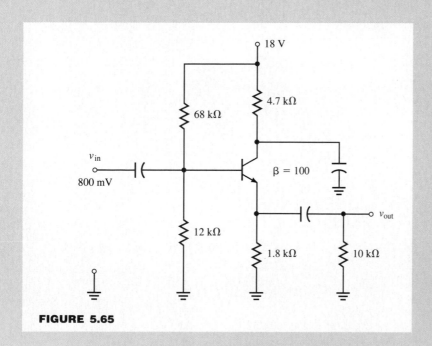

FIGURE 5.65

6

POWER AMPLIFIERS

OBJECTIVES

✓ **As you read this chapter, concentrate on learning how to:**

1. Plot the ends of an AC load line
2. Plot the Q point on a load line
3. Determine the maximum, usable AC output swing
4. Determine the maximum amount of power that a transistor can deliver to a load
5. Determine the operating frequency of a tuned class C amplifier
6. Identify whether a circuit is operating class A, class AB, class B, or class C
7. Describe the different methods used to couple amplifier stages

INTRODUCTION

At the input of an amplifier, the signal is usually small, with an output swing less than 10% of the AC load line. Measuring the output swing and comparing it with the load line provides a way to define small-signal operation. At the final stage of an amplifying system, the large-signal amplifier delivers a large power gain and the required output power for the system.

In this chapter, we will look at how bipolar junction transistors amplify large signals. Four basic large-signal amplification signal circuit classes will be studied—class A, class B, class AB, and class C. We will compare the pros and cons of each type and how amplifier efficiency varies with operating conditions. We will also see how those larger signals require transistors to dissipate larger amounts of power and heat.

Also, we will look at another type of load line, called the AC load line. As we discuss the AC load line, we will discover how the parameters shown by the line limit circuit operation and the amplitude of the output signal.

With large signal amplification, the circuit will have more distortion. In this chapter, we will discuss different types of distortion and how to minimize them.

THE QUIESCENT, OR Q, POINT

The DC voltages at the transistor terminals represent circuit operation with no AC input. We can define this condition as **quiescent.** The term *quiescent* literally means quiet. The circuit can be quiescent under two different conditions: when there is no AC input and when there is an AC input at the zero crossing point. Thus, the quiescent, or Q, point will show on the DC and AC load lines. Unless the DC and AC resistances are equal, the load lines will only intersect at the **Q point.**

Figure 6.1 shows DC voltages marked on a sine wave. We call these points the **zero crossing points.** Note that the zero crossing points occur at 0°, 180°, and 360° on the sine wave. When we apply an AC sine wave to the circuit, the DC voltages will be present only when the AC signal crosses the zero axis. When the AC signal crosses the zero axis, the transistor circuit passes through the Q point.

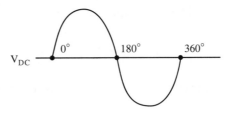

FIGURE 6.1 DC voltages marked on a sine wave. At the zero crossing point, the circuit is quiescent.

LOAD LINES

In Chapter 4, we used DC load lines to calculate the points where a transistor would enter cutoff and saturation. Used with the characteristic curves of the transistor, the load line shows us the operating points that limit transistor operation. Along with showing the Q point for the circuit, the DC load line also shows the DC cutoff and saturation points.

Figure 6.2a depicts a DC load line for a circuit having a cutoff voltage of 22 V and a saturation current of 6.875 mA. Working between cutoff and saturation, a transistor will amplify signals and conduct current. Entering saturation, the transistor acts as a closed switch. The output voltage drops across the resistor. Going into cutoff, the transistor no longer conducts current.

In this chapter, we will use both DC and AC load lines. Since both DC and AC voltages influence circuit operation, we use two load lines to find the operating points for the circuit. Figure 6.2b shows an AC load line and an AC signal representing the output voltage. As with the DC load line, we can establish the end points of the AC load line by finding values for saturation and cutoff.

AC CUTOFF AND SATURATION VOLTAGES

Figure 6.2c is a voltage divider biased common-emitter amplifier. Figure 6.2d is the DC and AC load lines for Figure 6.2c.

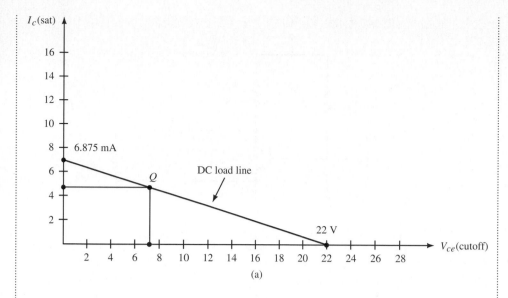

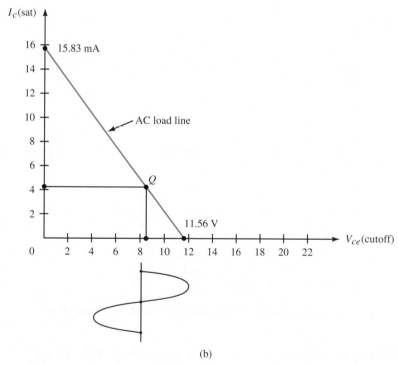

(b)

FIGURE 6.2 (a) DC load line for a circuit having a cutoff voltage of 22 V and a saturation current of 6.875 mA. (b) AC load line. (Continued on next page.)

One end point of an AC load line shown in Figure 6.2d represents the voltage across the transistor when the AC signal had reduced current flow to zero. It is the AC cutoff voltage. The other end of the AC load line is the AC saturation point. For AC saturation to occur, the AC signal must drive the voltage across the transistor to zero.

To determine the AC saturation current of an amplifier, we first find the collector-to-emitter quiescent voltage and the DC quiescent current. From a spec sheet, we can assume that the voltage across the transistor will be 0.2

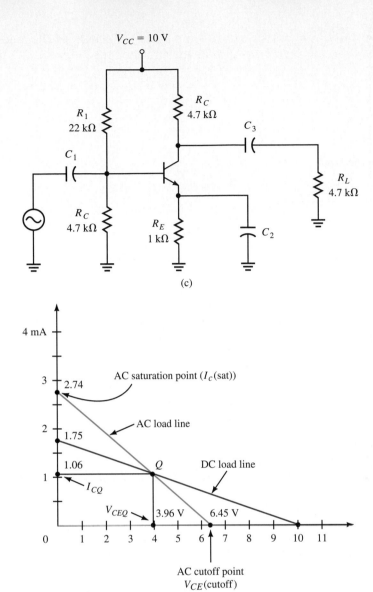

(c)

(d)

FIGURE 6.2   **(c) common emitter circuit. (d) The DC and AC lead lines for the circuit 6.2 (c).**

V when the transistor enters saturation. This means that the collector-to-emitter voltage, or V_{CE}, will drop to zero and transfer all of the voltage drop to the AC collector resistance. To solve for AC saturation current, we use the formula

$$I_C(\text{sat}) = I_{CQ} + \frac{V_{CEQ}}{r_C}$$

where

I_{CQ} is DC quiescent collector current

V_{CEQ} is the DC quiescent collector-to-emitter voltage

r_C is the AC loaded collector resistance

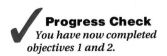

Progress Check
You have now completed objectives 1 and 2.

This formula says that AC saturation current is the sum of DC quiescent current and the maximum instantaneous current caused by the AC signal. The maximum instantaneous current is equal to the quiescent collector-to-emitter voltage being dropped across the AC collector resistance.

The following procedure should illustrate how to solve for AC saturation current.

First, we need to find base bias voltage with no AC signal applied.

$$V_B = \frac{V_{CC} \times R2}{R1 + R2} = \frac{10V \times 4.7\,k\Omega}{22\,k\Omega + 4.7\,k\Omega} = 1.76V$$

Now we can find the emitter voltage.

$$V_E = V_B - V_{BE} = 1.76\,V - 0.7\,V = 1.06\,V$$

Using Ohm's law we can solve for the quiescent emitter current.

$$I_E = \frac{V_E}{R_E} = \frac{1.06\,V}{1\,k\Omega} = 1.06\,mA$$

We will also use the assumption that I_E is equal to I_C.

We now know the DC quiescent current (I_{CQ}) is equal to 1.06 mA.

Next we need to solve for the quiescent collector-to-emitter voltage.

$$V_{CEQ} = V_{CC} - (V_{RC} + V_E)$$

Since we know V_{CC} and V_E, we only need V_{RC} to find V_{CEQ}.

$$V_{RC} = I_C \times R_C = 1.06\,mA \times 4.7\,k\Omega = 4.98V$$

Substituting the known voltages, we can solve for V_{CEQ}.

$$V_{CEQ} = V_{CC} - (V_{RC} + V_E)$$
$$V_{CEQ} = 10\,V - (4.98\,V + 1.06\,V) = 3.96\,V$$

The last value we need to know in order to solve for AC saturation current is the AC collector resistance. AC collector resistance is the parallel resistance of R_C and R_L.

$$r_c = \frac{R_C \times R_L}{R_C + R_L} = \frac{4.7\,k\Omega \times 4.7\,k\Omega}{4.7\,k\Omega + 4.7\,k\Omega} = 2.35\,k\Omega$$

Now we can substitute the calculated values into the AC saturation formula.

$$I_{C(sat)} = I_{CQ} + \frac{V_{CEQ}}{r_c} = 1.06\,mA + \frac{3.96\,V}{2.35\,k\Omega} = 2.74\,mA$$

On Figure 6-2d, the AC saturation point is drawn in where V_{CE} is zero volts and I_C is 2.74 mA. This is the top of the AC load line.

Now we are ready to solve for the AC collector-to-emitter cutoff voltage, $V_{CE(cutoff)}$. We will use the formula:

$$V_{CE(cutoff)} = V_{CEQ} + (I_{CQ} \times r_C)$$

The formula tells us that the AC collector-to-emitter cutoff voltage is the sum of the collector-to-emitter quiescent voltage and the voltage drop caused by maximum AC current change across the AC collector resistance.

We have already solved for all of the unknown values. By substituting we can solve for the AC collector-to-emitter cutoff voltage.

$$V_{CE(cutoff)} = V_{CEQ} + (I_{CQ} \times r_C)$$
$$V_{CE(cutoff)} = 3.98\,V + (106\,mA \times 2.35\,k\Omega)$$
$$V_{CE(cutoff)} = 6.47\,V$$

On Figure 6-2d, the AC collector-to-emitter cutoff point is drawn in where I_C is 0 mA and V_{CE} is 6.47 volts. This is the bottom of the AC load line.

Also observe that both the DC and AC load lines pass through the Q point. If they do not, then an error has been made in the calculations.

FINDING THE MAXIMUM AMPLITUDE OF AN AC OUTPUT VOLTAGE

As the input signal alters the transistor bias, the output swings from one peak to the other peak. The maximum AC output voltage is determined by the position of the Q point on the AC load line. The largest possible AC signal is achieved when the Q point centers on the AC load line. If the Q point does not center on the load line, the smaller distance from the Q point to one of the end points of the load line determines the maximum AC output signal without distortion.

In Figure 6.2b, the Q point is closer to the AC cutoff point than to the AC saturation point and marks the maximum amplitude of:

$$11.6\,V - 8.5\,V = 3.1\,V$$

The peak amplitude of the circuit output voltage may equal no more than $\pm 3.1\,V$. This gives us a value for the largest possible AC output voltage without saturation or cutoff clipping. If the input signal forced the output to have a greater peak amplitude, we would encounter distortion.

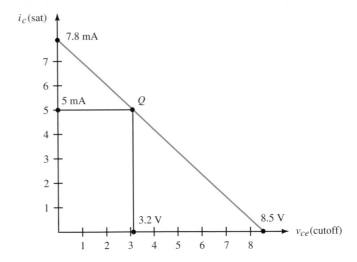

✔ **Progress Check**
You have now completed objective 3.

PRACTICE PROBLEM 6.1

Determine the maximum AC output swing without distortion in Figure 6.3.

FIGURE 6.3

CLIPPING

Figure 6.4 shows how positive and negative current swings of the output become clipped. These waveforms show the effects of an input signal forcing the AC output signal to cause either saturation or cutoff. The input signal overdrives and distorts the output signal.

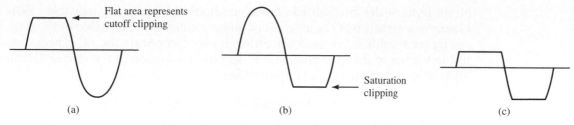

FIGURE 6.4 (a) Cutoff clipping of an output waveform. (b) Saturation clipping. (c) Distortion caused by cutoff and saturation clipping. Waveforms apply to commn emitter circuit.

With a clipped output, the output signal is no longer present for the full input cycle. Figure 6.4a shows cutoff clipping, Figure 6.4b shows the effect of saturation clipping, and Figure 6.4c shows the distortion caused by both.

REVIEW SECTION 6.1
1. A synonym for *quiescent* is _____.
2. A transistor is at its quiescent point when no _____ signal is applied or at the _____ points.
3. Zero crossing points occur at _____, _____, and _____ degrees on the sine wave.
4. When the transistor is working between cutoff and saturation, it does not amplify signals or conduct current. True or false?
5. The transistor acts as a closed switch upon entering _____, and upon entering _____ it no longer conducts current.
6. The end points of an AC load line denote _____ and _____ voltages.
7. The AC output voltage is at a maximum when the Q point is centered on the AC load line. True or false?
8. If the Q point isn't centered on the load line the largest distance from the Q point to an end point determines the maximum AC output without distortion. True or false?
9. When clipping occurs, the output signal is still present for the full cycle. True or false?

6.2 CLASSIFYING AMPLIFIERS

We can separate transistor amplifiers into either **current, voltage,** or **power amplifiers.** In addition to these three general types, we can also divide transistor amplifiers by the type of bias and by the relative time during which the amplifiers produce an output for a given cycle of input signal. The four classes of amplifiers studied in this text are A, B, AB, and C.

Several differences among the classes of amplifiers occur that concern the distortion and efficiency of the AC signal. **Distortion** is an unwanted change in the shape of an AC signal waveform. Although some distortion occurs in all amplifiers, the amount of distortion should be as low as possible. **Efficiency** is a measure of the amount of power converted to a usable output signal by an amplifier. We measure efficiency by comparing the AC power output delivered to the load against the DC power supplied to the circuit. In the form of an equation, efficiency looks like:

$$\text{efficiency} = \eta = \frac{P_L}{P_{DC}}$$

where power is measured in RMS watts. A high rate of efficiency means less power is wasted in the form of heat and more power is delivered to the load.

REVIEW SECTION 6.2

1. Transistor amplifiers can be used as _____,
   _____, _____ amplifiers.
2. The time transistors produce an output further classifies amplifiers into classes _____, _____, _____, and
   _____.
3. _____ changes the shape of an AC signal waveform undesirably.
4. Efficiency measures the amount of power that is converted into a usable output signal by an amp. True or false?
5. If there is a higher efficiency rate, then more power will be converted into heat and wasted. True or false?

6.3 CLASS A AMPLIFIERS

A transistor circuit in which the transistor conducts the entire input cycle is a **class A amplifier.** The base bias and input signal allow the conduction of collector current for the entire time of one complete input cycle. Because the transistor conducts for the full input cycle, it does not enter either cutoff or saturation. Class A amplifiers have the lowest amount of distortion, but they also have the lowest level of efficiency. The amplifiers we have studied so far have all been class A amplifiers.

Figure 6.5 shows an emitter-biased common-emitter circuit. Nothing in the schematic drawing for the circuit sets the circuit apart from the small-signal circuits that we saw in earlier chapters. However, detailed schematics will indicate the size of the peak-to-peak AC output voltage. If the output voltage is greater than or equal to 10% of the power supply voltage, the circuit may work as a power amplifier. For the purposes of this chapter, we will assume that all the example transistors can dissipate the power required for power amplification.

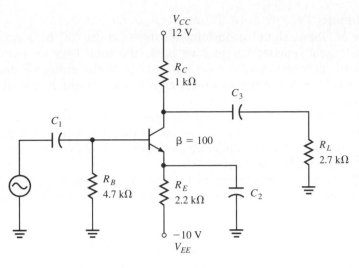

FIGURE 6.5 Emitter-biased, common-emitter circuit

EXAMPLE 6.1

Determine the DC and AC operating limits of the circuit shown in Figure 6.5.

   We will begin our analysis of the circuit by examining the DC operating conditions. Without the correct DC operating condition, the AC characteristics of the circuit will not be correct.

Solution
DC Analysis

   As mentioned, the circuit shown in Figure 6.5 uses emitter bias to set the DC operating conditions. We begin by determining the DC voltages at the transistor terminals. Since we have emitter bias, the voltage at the emitter terminal would equal approximately -0.7 V. The base voltage would be approximately 0 V with respect to ground.

   From previous work, we know that the circuit is independent of beta. Therefore, we can ignore the base resistor in our DC calculations. To obtain the emitter current, we calculate that:

$$I_E = \frac{10\,V - 0.7\,V}{2.2\,k\Omega} = 4.23\,mA$$

Using Ohm's law, we can calculate the voltage dropped across the collector resistor:

$$V_{RC} = I_E \times R_C = 4.23\,mA \times 1000\,\Omega = 4.23\,V$$

Then we can subtract that dropped voltage from the collector supply voltage to find the DC collector voltage:

$$V_C = V_{CC} - V_{RC} = 12\,V - 4.23\,V = 7.77\,V$$

Subtracting the emitter voltage from the collector voltage will give a value for the collector-to-emitter voltage:

$$V_{CE} = 7.77\,V - (-0.7\,V) = 8.47\,V$$

AC Analysis

The AC input signal superimposes itself on the DC bias voltages. When the AC signal reaches its positive peak, the total bias on the base of the transistor increases, and the output voltage at the collector decreases. Because we have a common-emitter circuit, the output is out of phase with the input.

When the AC signal into the base goes negative, the bias at the base decreases and lowers the emitter current. If the emitter current decreases, the voltage drop across the collector resistors also decreases, and the output voltage at the collector rises.

In the circuit shown in Figure 6.5, the presence of a bypass capacitor ties the emitter terminal to AC ground. Using the value for the emitter current, we can find the AC emitter resistance:

$$r'_e = \frac{25 \text{ mV}}{4.23 \text{ mA}} = 5.91 \ \Omega$$

The AC collector resistance equals:

$$r_c = 1 \text{ k}\Omega \parallel 2.7 \text{ k}\Omega = 730 \ \Omega$$

Because the AC collector resistance value is not the same as the DC collector resistance, we know that we will have two different load lines. Figure 6.2b shows the AC load line. The AC input signal forces a change in the quiescent operating conditions. If the AC input on the base goes negative, the bias at the base becomes lower than the quiescent operating point. In turn, the current through the transistor will decrease. If the AC signal becomes large enough to turn the transistor off, no current will flow through the transistor. We will have a net change of:

$$\Delta I_C = I_C - I_C \text{(cutoff)} = 4.23 \text{ mA} - 0 \text{ mA} = 4.23 \text{ mA}$$

Because the current changed, we can treat it as an AC value. Multiplying the change in current by the AC collector resistance will show the change in output voltage of the transistor:

$$\Delta V_C = \Delta I_C + r_C = 4.23 \text{ mA} \times 730 \ \Omega = 3.09 \text{ V}$$

The voltage drop across the AC collector resistance transfers to the collector terminal. Before, the transistor had 8.47 V at the collector terminal. Now, the terminal shows:

$$V_{CE} \text{(cutoff)} = V_{CEQ} + \Delta V_C = 8.47 \text{ V} + 3.09 \text{ V} = 11.56 \text{ V}$$

This value represents the voltage across the transistor when the AC signal had reduced current flow to zero. It is the r_{CE}(cutoff) voltage and one end point of the AC load line shown in Figure 6.2b.

Let us find the other end point of the AC load line—the AC saturation point. For saturation to occur, the AC signal must drive the voltage across the transistor to zero. We will see a change in the 8.47-V output for Figure 6.5. This voltage transfers to the AC collector resistance. Using Ohm's law, we get:

$$i_c = \frac{V_{CE}}{R_C} = \frac{8.47 \text{ V}}{730 \text{ k}\Omega} = 11.6 \text{ mA}$$

Since the current has increased, we can add this value to the quiescent current value to obtain the total current at saturation:

$$i_{c(sat)} = I_{CQ} + i_C = 11.6 \text{ mA} + 4.23 \text{ mA} = 15.83 \text{ mA}$$

We can mark this point on the AC load line as the saturation point. Once we have the two end points, we can plot the load line on the graph.

We can find the AC voltage gain by dividing the AC collector resistance by the AC emitter resistance:

$$A_v = \frac{r_c}{r'_e} = \frac{730\ \Omega}{5.91\ \Omega} = 123.5$$

PRACTICE PROBLEM 6.2

Compute the DC and AC load lines for Figure 6.6.

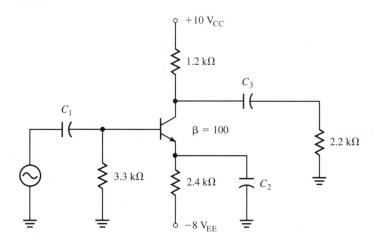

FIGURE 6.6

EXAMPLE 6.2

Plot the load line for the circuit shown in Figure 6.7.

The circuit shown in Figure 6.7 is a common-emitter configuration with a swamping resistor placed in the emitter terminal circuit. A voltage divider bias arrangement makes up the DC configuration for the circuit.

Solution
DC Analysis

We will begin our analysis by looking at the DC operation of the circuit. The DC voltage at the base can be determined with the voltage divider principle:

$$V_B = \frac{V_{CC} \times R_2}{R_1 + R_2} = \frac{18\ V \times 470\ \Omega}{1.5\ k\Omega + 470\ \Omega} = 4.3\ V$$

The divider is stiff, so we can ignore the Thevenin resistance of the base and the loading effect of the transistor. To calculate the emitter voltage, we have:

$$V_E = 4.3\ V - 0.7\ V = 3.6\ V$$

With this value, we can find a value for the emitter current:

$$I_E = \frac{V_E}{r_E + R_E} = \frac{3.6\ V}{100\ \Omega} = 36\ mA$$

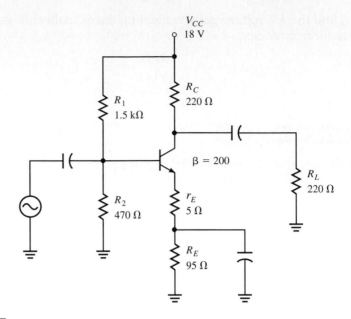

FIGURE 6.7

The voltage drop across the DC collector resistance equals:

$$V_{RC} = 36 \text{ mA} \times 220 \text{ } \Omega = 7.92 \text{ V}$$

Subtracting this value from the collector supply voltage, we find the collector voltage:

$$V_C = V_{CC} - V_{RC} = 18 \text{ V} - 7.92 \text{ V} = 10.08 \text{ V}$$

Using these values, we can determine the voltage across the transistor with no AC input signal:

$$V_{CE} = V_C - V_E = 10.08 \text{ V} - 3.6 \text{ V} = 6.48 \text{ V}$$

With the additional letter Q representing a quiescent value, the DC quiescent operating points are:

$$I_{EQ} = 36 \text{ mA} \qquad \text{and} \qquad V_{CEQ} = 6.48 \text{ V}$$

AC Analysis

First, we will find the total AC resistance, r_t, of the circuit. The collector terminal sees:

$$r_C = R_C \parallel R_L = 220 \text{ } \Omega \parallel 220 \text{ } \Omega = 110 \text{ } \Omega$$

Then add the swamping resistor, r_E.

$$5 \text{ } \Omega = \text{AC emitter resistance}$$
$$r_t = r_c + r_E = 115 \text{ } \Omega$$

If the AC signal reduces the current flowing through the transistor to zero, we see a voltage change of:

$$36 \text{ mA} \times 115 \text{ } \Omega = 4.14 \text{ V}$$

When this voltage transfers from the AC resistance of the circuit to the transistor, the AC signal places the transistor into cutoff. The AC cutoff voltage is:

$$V_{Ce(cutoff)} = 6.48 \text{ V} + 4.14 \text{ V} = 10.62 \text{ V}$$

If the voltage across the transistor reduces to zero because of saturation, it introduces a change of 6.48 V into the circuit. This voltage transfers to the AC resistance and produces a current change equaling:

$$i_{C(sat)} = \frac{6.48 \text{ V}}{115 \text{ }\Omega} = 56.3 \text{ mA}$$

Adding the current change value to the quiescent change value gives a value for saturation current:

$$i_{C(sat)} = 56.3 \text{ mA} + 36 \text{ mA} = 92.3 \text{ mA}$$

Using these values, we can plot the AC load line for the circuit. Figure 6.8 shows the load line. As you can see, the Q point is closer to cutoff than to saturation. The maximum output swing without clipping equals:

$$v_{out \text{ (pk)}} = v_{ce}(\text{cutoff}) - V_{CEQ} = 10.62 \text{ V} - 6.48 \text{ V} = 4.14 \text{ V}_p$$

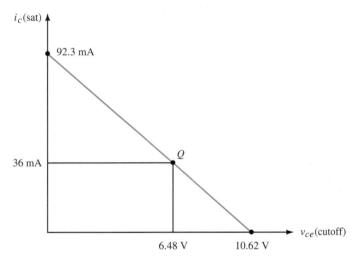

FIGURE 6.8   AC load line for the circuit shown in Figure 6.7

PRACTICE PROBLEM 6.3

Determine the DC load line for Figure 6.7 when the 470-Ω resistor is changed to a 680-Ω resistor.

EXAMPLE 6.3

Using values from the previous example, determine the maximum AC input that will not drive the output into clipping.

Using the information from the previous example, we have 36 mA of current flowing without the presence of an AC signal.

Solution
With 36 mA of current the AC emitter resistance equals:

$$r'_e = \frac{25 \text{ mV}}{36 \text{ mA}} = 0.694 \text{ }\Omega$$

Since the circuit has a swamping resistor, we add the value of the resistor to the AC emitter resistance value to find the total emitter resistance:

$$r_e = r_e + r_e' = 5.694 \ \Omega$$

To find the gain of the circuit, we divide the AC collector resistance by the emitter resistance:

$$A_v = \frac{r_c}{r_e} = \frac{110 \ \Omega}{5.694 \ \Omega} = 19.3$$

If the maximum output voltage can have a peak of 4.14 V, the maximum input peak voltage must equal:

$$V_{in(pk)} = \frac{V_{out(pk)}}{A_V} = \frac{4.14 \ V}{19.3} = 214.5 \ mV$$

If the input signal exceeds 214.5 mV, the gain of the circuit will cause the output to exceed 4.14 V. Clipping will result.

PRACTICE PROBLEM 6.4

Determine the AC load line for the circuit of Figure 6.7 when the swamping resistor is changed to a 22-Ω value.

Next, we will examine a common-collector circuit, shown in Figure 6.9. With this type of circuit, the output will be in phase with the input.

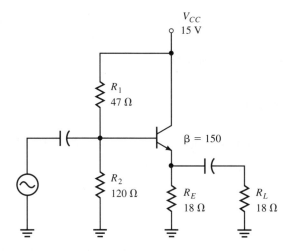

FIGURE 6.9 Common-collector circuit

EXAMPLE 6.4

Determine the maximum AC operating conditions for the circuit shown in Figure 6.9.

Since the transistor is biased with a voltage divider, we can use the voltage divider theorem to determine the base voltage. Also, with a stiff voltage divider, we can ignore the Thevenin resistance and the loading effects of the transistor.

Solution

DC Analysis

The base voltage equals:

$$V_B = \frac{120\ \Omega}{47\ \Omega + 120\Omega} \times 15\ V = 10.78\ V$$

To obtain the emitter voltage, we subtract 0.7 V from the base voltage value:

$$V_E = 10.77\ V - 0.7\ V = 10.08\ V$$

We can use this value to find the emitter current:

$$I_E = \frac{10.07\ V}{18\ \Omega} = 559\ mA$$

Since the collector connects directly to the power supply, we can easily find the collector voltage:

$$V_{CC} = V_C = 15\ V$$

With all this information, we know the DC voltage values at the quiescent point. Now, we can find the voltage across the transistor:

$$V_{CE} = V_C - V_E = 15\ V - 10.07\ V = 4.93\ V$$

AC Analysis

To determine the AC load line, we need to calculate what happens when the AC signal drives the transistor into cutoff and saturation. If the circuit is driven into cutoff, we will have a net drop in current equal to 559 mA. This current change transfers voltage from the AC emitter resistance to the transistor. For Figure 6.9, this equals:

$$18\ \Omega \parallel 18\ \Omega = 9\ \Omega$$

Using Ohm's law, we find the amount of voltage transferred to the transistor:

$$559\ mA \times 9\ \Omega = 5.031\ V$$

The collector-to-emitter AC cutoff voltage equals:

$$v_{ce}(\text{cutoff}) = 5.031\ V + 4.93\ V = 9.961\ V$$

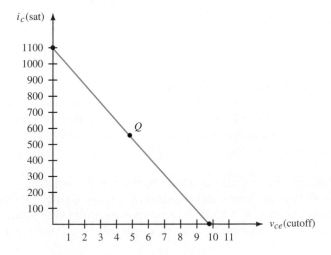

FIGURE 6.10 AC load line for the circuit shown in Figure 6.9

Next, we will determine the saturation point. In this case, the voltage across the transistor reduces to zero at saturation and introduces a 4.93-V change. This change transfers voltage to the AC emitter resistance and produces a current change equal to:

$$\frac{4.93 \text{ V}}{9 \text{ }\Omega} = 548 \text{ mA}$$

Adding this value to the quiescent current value determines the total current flowing during saturation:

$$548 \text{ mA} + 559 \text{ mA} = 1.1 \text{ A}$$

Figure 6.10 shows the AC load line for Figure 6.9. As you can see, the Q point is very close to the center of the AC load line. This is the ideal place for the Q point of a class A amplifier.

PRACTICE PROBLEM 6.5

Determine the AC load line for Figure 6.11. Locate the Q point, and determine the maximum AC output (without clipping).

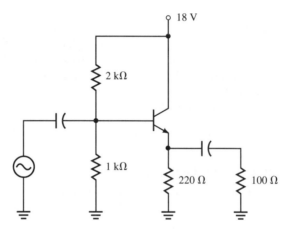

FIGURE 6.11

CLASS A AMPLIFIER POWER DISSIPATION

In Figure 6.12, we see the same AC load line drawn in three different ways. The difference in the lines is the position of the Q point. In Figure 6.12a, the Q point is closer to the saturation end point. Positioned in this way, it limits the maximum voltage swing at the output to 2 V before saturation clipping begins.

In Figure 6.12b, the Q point is closer to the cutoff end point. In this position, the cutoff voltage limits the maximum output swing. As diagrammed, the maximum swing is 2 V. Figure 6.12c shows a centered Q point. In this position, the output can change by a peak amplitude of 5 V before any clipping will occur.

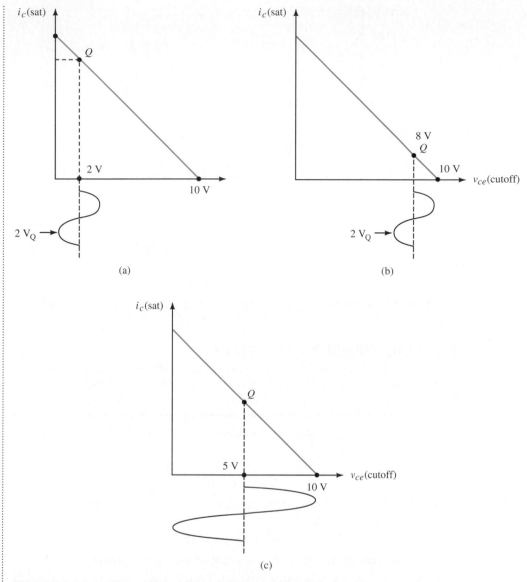

FIGURE 6.12 (a) AC load line with Q point closer to the saturation end point. (b) AC load line with Q point closer to the cutoff end point. (c) AC load line with Q point centered.

A simple inspection shows that the operating condition shown in Figure 6.12c allows the maximum voltage from a class A amplifier without clipping. If we move the Q point so that the collector-to-emitter voltage equals 5.1 V, the distance to the other end point becomes 4.9 V. The lower voltage defines the maximum output of the circuit without clipping.

Let us set up a practice AC load line like the line shown in Figure 6.13. On this load line, AC cutoff occurs at 15 V, and the AC saturation current equals 30 mA. If the Q point centers on the load line, the quiescent collector-to-emitter voltage across the transistor must equal 7.5 V. At the same time, the current flowing through the transistor should equal 15 mA. Using those qui-

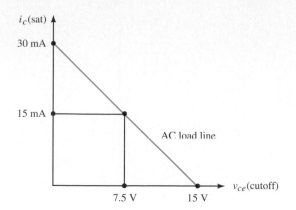

FIGURE 6.13 Practice AC load line

escent values, we can calculate the power dissipated by the transistor at quiescence using the P = I × E formula:

$$P_Q = 15\text{ mA} \times 7.5\text{ V} = 112.5\text{ mW}$$

The transistor dissipates this power without the presence of an AC signal.

DELIVERING POWER TO THE LOAD

Our practice circuit can deliver a maximum peak signal of 7.5 V to an output load with no clipping. Relating this value to the DC or quiescent values requires the conversion of the peak voltage and current values to RMS or effective AC voltage. An **effective** or **root mean square value** of an AC voltage or current is the voltage value that causes the circuit resistance to expend the same amount of power that it would for an equivalent DC voltage. The RMS, or effective, voltage value of our practice circuit would equal:

$$V_{RMS} = 7.5\text{ V}_p \times 0.707 = 5.3\text{ V}$$

where 0.707 is the conversion factor between peak and RMS power.
The current value would equal:

$$I_{RMS} = 15\text{ mA} \times 0.707 = 10.6\text{ mA}\cdot$$

Multiplying the two values gives the amount of dissipated power:

$$P_D = 5.3\text{ V} \times 10.6\text{ mA} = 56.2\text{ mW}$$

Comparing this dissipation value with the quiescent power dissipation value gives:

$$\eta = \frac{P}{P_Q} = \frac{56.2\text{ mW}}{112.5\text{ mW}} = 0.499 \times 100 = 49.9\%$$

The circuit delivered just under half of the power dissipated by the transistor. Under optimum conditions, the load will receive no more than 50% of the power dissipated by the transistor when no signal is present. This is the best that class A can provide.

Progress Check
You have now completed objective 4.

PRACTICE PROBLEM 6.6

Using Figure 6.14 and the load line from Figure 6.15, determine the amount of power used by the transistor and the amount of power delivered to the load with no signal present.

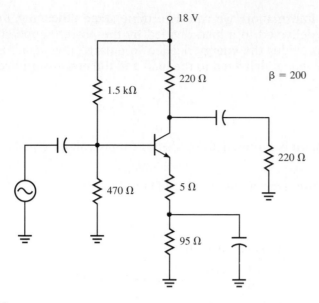

FIGURE 6.14

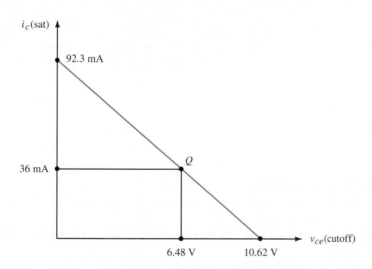

FIGURE 6.15

STAGE EFFICIENCY AND THE CLASS A AMPLIFIER

Under optimum conditions, the class A amplifier circuit can deliver half as much power to a load as that dissipated by the transistor. Since we can calculate the power dissipated by the transistor by multiplying the quiescent collector-to-emitter voltage by the quiescent collector current, we can express the maximum power delivered to a load as one-half of the quiescent dissipated power. The formulas would appear as:

$$P_Q = V_{CEQ} \times I_{CQ}$$
$$P_L = 0.5 \times V_{CEQ} \times I_{CQ}$$

With a centered quiescent point and a symmetrical output signal, the collector supply voltage is double the quiescent collector-to-emitter voltage, and the average supply current equals the quiescent collector current.

With that information, we can determine **stage efficiency,** or the maximum power delivered to a load divided by the power supplied by the DC source that provides the energy needed to enlarge the signal. Substituting values for the power delivered to the load and the power supplied by the DC source gives:

$$\frac{P_L}{P_{DC}} = \frac{(0.5 \times V_{CEQ} \times I_{CQ})}{I_{CQ} \times V_{CC}}$$

Since the Q point is centered, the collector supply voltage equals:

$$V_{CC} = 2 \times V_{CEQ}$$

and changes the stage efficiency formula to:

$$\eta = \frac{P_L}{P_{DC}} = \frac{(0.5 \times V_{CEQ} \times I_{CQ})}{(I_{CQ} \times 2 \times V_{CEQ})}$$

Canceling like terms leaves us with:

$$\frac{P_L}{P_{DC}} = \frac{0.5}{2} = 0.25 \times 100 = 25\%$$

At best, the load receives 25% of the power delivered by the DC source. If the Q point does not center on the load line, the efficiency of the class A amplifier suffers even more. When we include all the current drains, such as the quiescent collector current and the currents through the bias resistors, that drain through the voltage source, the efficiency of class A power circuits drops even lower. Compared with other classes of amplifiers, the class A amplifier has poor efficiency.

REVIEW SECTION 6.3

1. Class A amplifier circuits have transistors conducting for the entire input cycle. True or false?
2. What factor keeps the class A amplifier from entering saturation or cutoff?
3. Class A amplifiers have the lowest amount of _____ and _____.
4. The position of the Q point on the load line determines the maximum output voltage that can be achieved without clipping. True or false?
5. The RMS voltage of an AC signal equals the voltage necessary to expend the same power with a DC voltage. True or false?
6. Class A amplifier circuits can deliver only half as much power to the load than that which is dissipated by the transistor. True or false?
7. Write the formulas necessary to find the maximum power delivered to a load, and describe each variable.
8. The voltage dropped across the collector resistor will decrease the output signal strength. True or false?
9. With the Q point centered on the load line the class A amplifier will operate at its lowest efficiency. True or false?

6.4 CLASS B AMPLIFIERS

As we have seen, class A amplifiers have some disadvantages. With class A amplification, the transistor will dissipate twice as much power as it delivers to a load. Although additional circuitry adds stability, it also draws additional current. Also, as the power to the load increases, the power rating of the transistor must also increase.

To offset these disadvantages, many circuit designs implement bias arrangements that cause the transistor to operate as something other than a class A amplifier. A **class B amplifier** transistor has the base biased to collector current cutoff. Collector current is present for only one alternation of an applied sine wave input signal.

Figure 6.16 shows two output waveforms from amplifier circuits. In Figure 6.16a, a sine wave applied to a class A amplifier causes the entire input signal to be transferred to the output with no loss of signal. Figure 6.16b shows how class B amplifier operation affects the output waveform. In class B operation, a transistor shuts down for half the input cycle. Only half the input signal reaches the output. Either the positive or negative alternation will be missing from the transistor waveform.

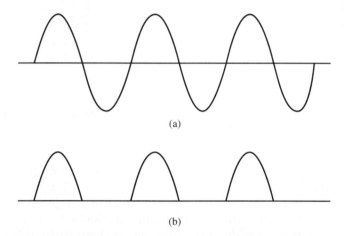

(a)

(b)

FIGURE 6.16 (a) Sine wave output from a class A amplifier (360°). (b) Sine wave output from a class B amplifier (180°).

COMPLEMENTARY CLASS B AMPLIFIER OPERATION

During class B operation, the transistor is off without the presence of a signal and will not conduct current. By itself, this makes the class B amplifier more efficient than the class A amplifier. With class A operation, the transistor conducts even without the presence of a signal. Figure 6.17 illustrates class B operation through an AC load line. Note the position of the Q point on the load line.

Figure 6.18 shows an NPN and a PNP transistor amplifier circuit drawn to represent the AC configuration. Each silicon transistor has an output taken across the AC emitter resistance with the collector at AC ground. The two transistors have an identical AC input signal. The NPN and PNP transistors form a complementary pair.

If the circuitry attached to the base has no DC voltage, the transistors will shut off because of the absence of a signal. Because the transistors are con-

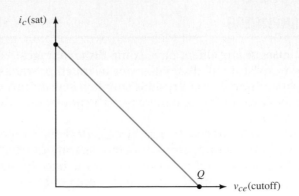

FIGURE 6.17 AC load line for class B operation

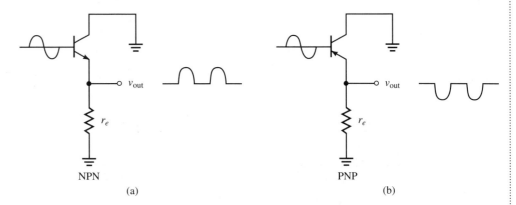

FIGURE 6.18 (a) Class B amplifier providing positive output signal. (b) Class B amplifier providing negative output signal.

structed from silicon, 0.7 V must appear across the base-emitter junction before conduction can occur. Even with the absence of a DC voltage at the base, the AC signal into the base, if greater than 0.7 V, provides the needed voltage to bias the base-emitter junction for conduction. In Figure 6.18a, the positive alternation provides the necessary voltage for conduction of the NPN transistor. In Figure 6.18b, the negative alternation provides the necessary voltage for conduction of the PNP transistor.

When conduction begins, current flows through the AC emitter resistor. In turn, a voltage develops across the emitter resistance and becomes the output voltage. Looking at Figure 6.18, we see two transistor circuits capable of providing an output for one-half of the input cycle. Combining the two circuits will provide a class B with an output for the entire input cycle.

CLASS B PUSH-PULL CIRCUITS

Class B amplifiers offer greater efficiency than class A amplifiers, but they also have more distortion. Some circuits utilize two transistors as a push-pull arrangement in which the two transistors amplify alternate halves of the input signal cycle. We call the arrangement **push-pull** because the transistors share amplification. With a common load, the two amplifiers provide an output signal that appears as an amplified copy of the input. This arrangement cuts the distortion.

Figure 6.19 shows the two circuits from Figure 6.18 connected together and forming a class B push-pull amplifier circuit. In this circuit, the AC signal couples to both transistor bases. When the AC signal goes positive, the NPN transistor conducts. Current flows through the emitter resistor shared by both transistors. The positive alternation appears across the resistor. Since the positive alternation is the wrong polarity for biasing the PNP transistor, the PNP transistor will not conduct.

When the AC signal changes polarity, both transistor bases become negative with respect to ground. The PNP transistor conducts, while the NPN

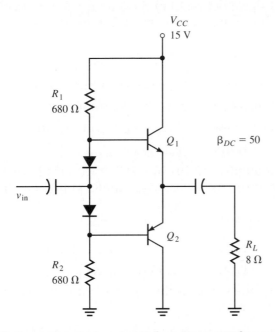

FIGURE 6.19 Connecting the two circuits from Figure 6.18 forms a push-pull class B amplifier.

transistor is turned off. As the PNP transistor conducts, current flows into the collector lead of the PNP transistor and out the emitter. Then, the current flows into the emitter resistor and produces a negative voltage with respect to ground. The negative alternation of the AC signal appears across the emitter resistor. Combined, the two transistors conduct for almost the entire AC input cycle.

DISSIPATING AND DELIVERING POWER WITH THE CLASS B AMPLIFIER

Most circuit designs will drive a class B push-pull amplifier from the cutoff point to the saturation point. Using Figure 6.19, we can determine the maximum power delivered to the load by the push-pull amplifier.

EXAMPLE 6.5

Determine the efficiency of the class B push-pull amplifier circuit shown in Figure 6.19.

When both transistors operate in the cutoff state, they split the power supply voltage because the transistors have the same characteristics.

Solution
In Figure 6.19, the quiescent collector-to-emitter voltage equals:

$$V_{CEQ} = \frac{15\,V}{2} = 7.5\,V$$

When either transistor saturates, the 7.5 V transfers to the load. Thus, the maximum unclipped signal delivered to the load is 7.5 V. Using Ohm's law, we can determine the maximum peak current through the load:

$$I_C(sat) = \frac{7.5\,V}{8\,\Omega} = 937.5\,mA$$

Converting both the current and voltage values to rms values:

$$P = 0.707(7.5\,V) \times 0.707(937.5\,mA) = 3.52\,W$$

Our calculations tell us that for any class B push-pull circuit:

$$P_L(max) = 0.5\,V_{CEQ} \times I_{C(sat)}$$

The 0.5 results from multiplying 0.707 by 0.707. If we substitute the collector supply voltage for the quiescent collector-to-emitter voltage, we have:

$$P_L\,(max) - \frac{0.5(V_{CC})}{2} \times I_{C(sat)}$$

which equals:

$$P_L(max) = 0.25\,V_{CC} \times I_{C(sat)}$$

The power drawn from the supply is the collector supply voltage divided by the collector supply current or:

$$P_{DC} = V_{CC} \times I_{CC}$$

While the collector supply current equals the collector current at saturation, the collector supply current flows only for the half-cycle that the NPN transistor conducts. To determine the amount of current, we can find the average amount that would flow for a half-cycle pulse:

$$0.318 \times I_{C(sat)}$$

The power supply provides:

$$P_{DC} = V_{CC} \times 0.318[I_{C(sat)}]$$

We can determine the maximum stage efficiency with the formula:

$$\frac{P_L\,(max)}{P_{DC}} = \frac{[0.25(V_{CC}) \times I_C(sat)]}{[V_{CC} \times 0.318(I_C(sat))]}$$

Canceling like terms breaks the formula down to:

$$\frac{P_{L(max)}}{P_{DC}} = \frac{0.25}{0.318} = 0.786 \times 100 = 78.6\%$$

This is the maximum stage efficiency of a class B push-pull amplifier. This value compares well against the maximum stage efficiency of 25% provided by the class A amplifier.

BIASING THE CLASS B AMPLIFIER

Figure 6.20 shows a popular method, called **diode bias,** used for biasing push-pull class B amplifiers. In Figure 6.20, diodes provide the forward bias voltage for the base-emitter junction of the transistor. The biasing diodes match the electrical and thermal characteristics of the base-emitter junctions. To understand how the diode bias arrangement works, let us take a look at the current mirror concept.

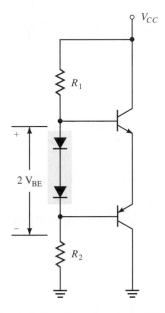

FIGURE 6.20 Diode bias for a class B amplifier

THE CURRENT MIRROR CONCEPT

Figure 6.21a shows the **current mirror.** In this circuit, a discrete diode functions in series with a current-limiting resistor; the entire series combination is placed between the collector supply voltage and ground. Figure 6.21b shows a transistor configured as the biasing diode. In this way, the circuit designer exactly matches the characteristics of the biasing diode to the transistor.

With a 20-V power supply attached to the top of a 2.2-kΩ resistor labeled as R_1 in Figure 6.21a, we can calculate the current through the bias diode. Approximately

$$\frac{20 \text{ V} \times 0.7 \text{ V}}{2.2 \text{ k}\Omega} = 8.77 \text{ mA}$$

of biasing current flows through the diode.

Moving to Figure 6.21c, we see the diode curves for the biasing diode and the base-emitter junction of the amplifier. Because the curves match, the bias current flowing through the diode sets up the base-emitter voltage for the amplifier. The voltage across the diode equals the voltage across the base-emitter junction of the transistor, which makes the current through the transistor equal to the current through the diode. This is the current mirror effect.

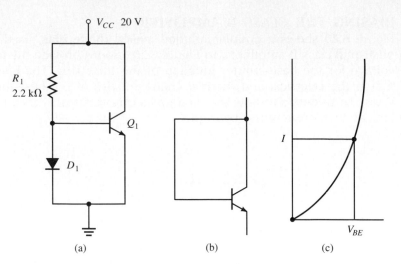

FIGURE 6.21 (a) Current mirror. (b) Transistor configured as the biasing diode. (c) Diode curves for the biasing diode and the base-emitter junction of the amplifier.

PRACTICE PROBLEM 6.7

In Figure 6.21, let V_{CC} = 15 V and R_1 = 1 kΩ. Determine the amount of current flowing through the transistor.

PRACTICE PROBLEM 6.8

In Figure 6.21, let V_{CC} = 12 V and R_1 = 2.7 kΩ. Determine the current flowing through the transistor.

TEMPERATURE STABILITY AND THE CLASS B AMPLIFIER

Returning to Figure 6.20, we can determine the current through the bias section of the circuit by using this formula:

$$I(bias) = \frac{[V_{CC} - 2(V_D)]}{2R}$$

where $2V_D$ accounts for the diode drops and 2R represents the total resistance of the voltage divider.

As temperature changes, the voltage across the diode will change. This affects the bias current. If the circuit designer makes V_{CC} significantly higher than 2 × V_D, small changes in the diode voltage will have minimal effect on the bias current. Therefore, larger values of V_{CC} minimize temperature effects.

OPERATING A CLASS B AMPLIFIER WITH AC AND DC CONDITIONS

The circuit shown in Figure 6.22 shows a push-pull arrangement using diode compensation bias and an 8-Ω load. Many power amplifiers will have low-impedance loads. Let us take a closer look at the circuit shown in Figure 6.22.

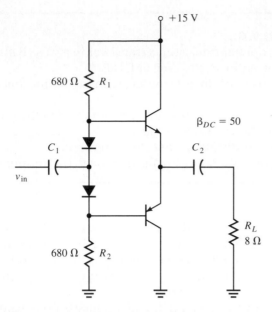

FIGURE 6.22 Push-pull amplifier with diode bias

EXAMPLE 6.6

Determine the current gain through the amplifier transistor.
  We start our analysis by finding the amount of DC bias current. The circuit uses the current mirror concept.

Solution
We find the DC bias current by dividing the sum of the collector supply voltage and the diode voltage drop by the resistance in series with the diodes.

$$\frac{15 \text{ V} - 1.4 \text{ V}}{2 \times 680 \text{ }\Omega} = 10 \text{ mA}$$

Since the bias arrangement uses the current mirror principle, the emitter current flowing through the transistors equals 10 mA.
  Applying an AC signal to capacitor C_1 produces push-pull action and a complete sine wave across the 8-Ω emitter resistor. This gives a gain of approximately ≤ 1, a common collector characteristic, and a current gain of beta = 50.

PRACTICE PROBLEM 6.9

Using Figure 6.22, change V_{CC} to 12 V. Let each 680-Ω resistor equal 470 Ω. Determine the amount of current flowing through the transistors. Do the component value and supply voltage value changes affect voltage gain? Do the changes affect current gain?

6.5   CLASS AB AMPLIFIERS

Operating a class B push-pull circuit with no DC bias causes a type of distortion called **crossover distortion.** Figure 6.23 illustrates a waveform with crossover distortion. With no DC bias voltage, the input signal will fall below 0.7 V for a short period of time. Neither transistor will become forward biased, and the output voltage equals zero.

To minimize crossover distortion, we can apply a small amount of DC bias voltage to the push-pull configurations. This changes the class B amplifier into a **class AB amplifier.** With a class AB amplifier, base biasing allows collector current to flow for more than one alternation of the applied signal but not for the time of one complete cycle. Class AB amplifiers compromise be-

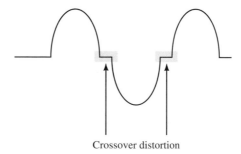

Crossover distortion

FIGURE 6.23 **Waveform with crossover distortion**

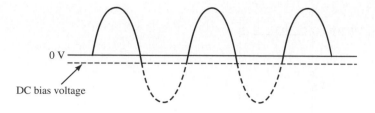

0 V

DC bias voltage

Note: Dashed lines, transistor is off
Solid lines, transistor is on

FIGURE 6.24 **Sine wave from a class AB amplifier. More than 180°, less than 360°.**

tween the higher efficiency of the class B amplifiers and the lower distortion of the class A amplifiers (Figure 6.24).

BIASING THE CLASS AB AMPLIFIER

Figure 6.25 shows the effect of adding DC bias on the load line. The Q point moves slightly up from the end point of the load line. Let us investigate the biasing that produces class AB operation with the circuit shown in Figure 6.26.

The circuit features two transistors connected together in a push-pull configuration. For now, we will concentrate on how DC signals affect circuit action. Note that no resistors are in the transistor paths between the collector supply voltage and ground. Only the transistors limit current in this path.

A class AB biasing arrangement causes one transistor to conduct before the other transistor stops conducting. For a short period of time, a path between power and ground exists through the transistors. Because of this factor, the bias arrangement must be carefully constructed to avoid runaway current through the transistors.

To ease the runaway problem, we must use a matched pair of transistors. A matched pair of NPN and PNP transistors must have the same electrical and thermal characteristics. Because the transistors form a series circuit between the collector supply voltage and ground, the same DC current flows through both transistors—$I_{CQ1} = I_{CQ2}$. In addition, the collector supply voltage splits evenly between the two transistors. Approximately 0.7 V appears across each resistor marked R_2.

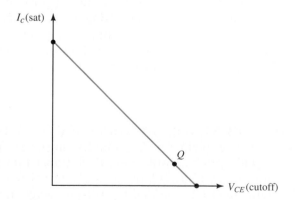

FIGURE 6.25 **Effect of adding DC bias shown through a load line**

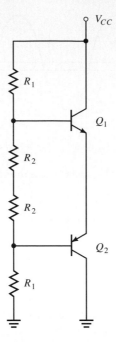

FIGURE 6.26 Class AB push-pull amplifier

Unfortunately, the voltage divider bias configuration can make the transistor section susceptible to temperature changes. As the transistors heat because of current flow, the base-emitter junction characteristics change. However, the values of the bias resistors do not change. Since the values of the resistors do not change the bias voltages remain the same while the barrier potential of the base-emitter junction drops. As a result, the amount of current flowing through the transistors sharply increases and damages the components. For this reason, most circuits utilize the diode biasing arrangement that we saw during our study of class B amplifier circuits.

QUASICOMPLEMENTARY-SYMMETRY AMPLIFIERS

Higher-power amplifiers that produce 10 or more watts of power will use either transformer coupling or a **quasicomplementary-symmetry** configuration. The term *quasi* indicates that the complementary-symmetry section of the amplifier appears before the output stage. Using a matched pair of high-power amplifier transistors adds cost to the circuit. Putting the complementary-symmetry section before the output stage provides the phase inversion required to drive PNP transistors at a reduced cost without sacrificing performance.

QUASICOMPLEMENTARY-SYMMETRY AMPLIFIER OPERATION

Figure 6.27 shows a quasicomplementary-symmetry amplifier. Transistors Q_3 and Q_2 form a push-pull complementary pair. Each transistor operates for half the input cycle, and the two have identical characteristics. Driver transistor Q_1 boosts the signal before it enters the push-pull stage. Even though bias resistors R_1 and R_2 look like a voltage-bias arrangement, neither con-

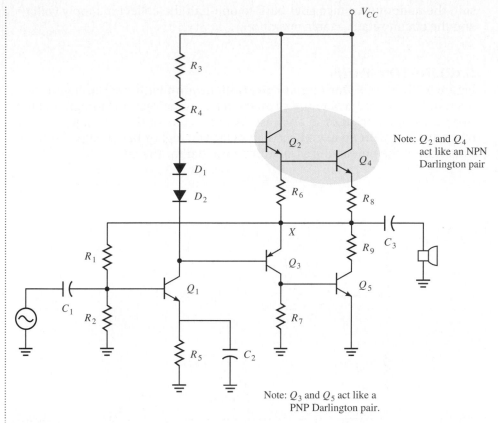

FIGURE 6.27 Schematic of a quasicomplementary-symmetry amplifier

nects to Q_1 collector supply voltage. Instead, these resistors tie to the emitters of transistors Q_2 and Q_3. In the circuit shown in Figure 6.27, the voltage at test point X is one-half of the collector supply voltage. The voltage divider ratio (R_1 and R_2) sets the voltage at the Q_1 base to 0.7 V higher than V_{R5}.

STABILIZING THE QUASICOMPLEMENTARY-SYMMETRY AMPLIFIER

Negative feedback through the closed loop helps to stabilize the circuit. As the voltage at test point X increases, the bias of Q_1 also increases. Q_1 then produces more current. Flowing through the diodes and the voltage divider resistors, this current produces a larger voltage drop. The voltage drop lowers the base voltages of transistors Q_2 and Q_3, which lowers the voltage seen at test point X.

Transistors Q_4 and Q_5 form the power output stage. With no signal applied to the circuit, the voltage developed across resistors R_6 and R_7 is small; therefore not enough voltage develops across resistors R_6 and R_7 to forward bias transistors Q_4 and Q_5.

When a positive signal is applied to the circuit, transistor Q_3 has less voltage, while transistors Q_1 and Q_2 have more voltage. At transistor Q_3, the collector voltage rises. With less voltage, the emitter voltage at Q_2 falls, which also provides less bias voltage for transistors Q_5 and more bias for Q_4. As a result, the midpoint voltage goes back to one-half the collector supply voltage, and the circuit retains stable operation.

DARLINGTON PAIRS

Figure 6.28 shows a **Darlington pair configuration** with the Darlington connection highlighted inside the shaded circle. Sometimes a Darlington pair is two transistors packaged in one case. We can think of the Darlington pair as one transistor with an extremely high beta. The higher beta results from one transistor of the pair acting as an input amplifier for the other.

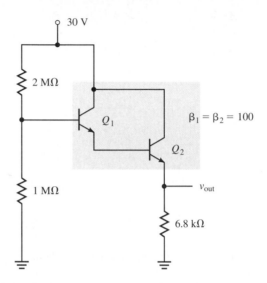

FIGURE 6.28 Darlington pair, common-collector configuration

Base current from the first transistor appears beta times larger at its emitter. Because the emitter of the first transistor connects to the base of the second transistor, the current appears beta times larger at the emitter of the second transistor. In the form of an equation, the total beta would appear as:

$$B_T = B_1 B_2 = 200 \times 200 = 40,000$$

If each transistor had a beta of 200, the total beta would equal 40,000. A high beta produces better circuit stability, higher current gain, and a higher input impedance. In the following example, we will learn more about how the Darlington pair works.

EXAMPLE 6.7

Determine the emitter current flowing through transistor Q_2 in Figure 6.28.

We start our analysis by calculating the amount of voltage on the base of transistor Q_1. Using the voltage divider principle, we have:

$$V_{B2} = \frac{1\,M\Omega}{3\,M\Omega} \times 30\,V = 10\,V$$

Dropping 0.7 V for the base-emitter barrier potential of transistor Q_1 leaves:

$$10\,V - 0.7\,V = 9.3\,V$$

The 9.3 V is also the voltage on the base of transistor Q_2. Dropping another 0.7 V for the base-emitter barrier potential leaves:

$$9.3\,V - 0.7\,V = 8.6\,V$$

The 8.6 V is also the voltage at the emitter of transistor Q_2.

We can now calculate the emitter current for transistor Q_2 by using Ohm's law:

$$I_{E2} = \frac{8.6\,V}{6.8\,k\Omega} = 1.265\,mA$$

Dividing that amount by 100 gives us the base current for Q_2:

$$I_{B2} = \frac{1.265\,mA}{100} = 12.65\,\mu A$$

and the emitter current for Q_1. Now, we can determine the base current for transistor Q_1:

$$I_{B1} = \frac{12.65\,\mu A}{100} = 126.5\,nA$$

This base current can now be compared with the current flowing through the voltage divider. First, we will find the value of the current flowing through the divider:

$$\frac{30\,V}{3\,M\Omega} = 10\,\mu A$$

Then, we will divide that amount by the amount of base current flowing through transistor Q_1:

$$\frac{10\,\mu A}{126.5\,nA} = 79$$

The voltage divider is definitely firm, almost stiff, and gives a stable bias point.

Two important reasons exist for the stability of the bias point. Current flowing through the emitter or load resistor equals 1.265 mA, while the base current at the input to the circuit equals 126.5 nA. Thus, the overall current gain equals:

$$A_I = \frac{1.265\,mA}{126.5\,nA} = 10,000$$

This is a hefty current gain. In addition, the beta of transistor Q_1 multiplied by the beta of transistor Q_2 also equals 10,000. The current gain equals the product of the transistor betas. When the two betas have the same value, the current gain equals beta squared.

Beta also helps to raise the input impedance of the base:

$$Z_{in}(base) = \beta_1 \times \beta_2 \times r_e$$

As we work through the next example, we will see the importance of the impedance formula.

EXAMPLE 6.8

Determine the input impedance of the circuit drawn in Figure 6.28.

From the preceding discussion, we know that the input impedance of the base will equal:

$$Z_{in(base)} = \beta^2 r'_e = 10,000 \times 6.8 \text{ k}\Omega = 68 \text{ M}\Omega$$

Because this figure is very high, the impedance of the reverse-biased collector-base junction comes into play:

$$Z_{in(base)} = \beta_2 \times r'_e \| r'_c$$

To determine r'_c, find h_{oe} or h_{ob} (h parameter) on a transistor specification sheet. Then

$$r'_c = \frac{1}{h_o}$$

For example, a 2N2405 transistor has an h_{ob} of 0.5 umhos, so

$$r'_c = \frac{1}{0.5 \text{ umhos}} = 2 \text{ M}\Omega$$

In every other calculation involving the base impedance, we did not consider the AC collector resistance because the impedance of the reverse-biased collector-base junction is typically 2 MΩ.

In this case, though, we cannot ignore the AC collector resistance because it is the smaller of the two values:

$$68 \text{ M}\Omega \| 2 \text{ M}\Omega = 1.994 \text{ M}\Omega$$

The value of the AC collector resistance sets the upper limit of the base input impedance for Darlington pair configurations with very high impedances. Since the 1.94 MΩ of the base input impedance works in parallel with the resistors of the voltage divider, the input impedance of the stage equals:

$$Z_{in}(stage) = 1.994 \text{ M}\Omega \| 10 \text{ M}\Omega \| 20 \text{ M}\Omega = 1.5 \text{ M}\Omega$$

PRACTICE PROBLEM 6.10

Rework Figure 6.28 with the value of beta for each of the two transistors equaling 170.

Because the basic AC configuration uses the common collector, the voltage gain of the circuit shown in Figure 6.28 is approximately one. In the next example, we will calculate the gain of a Darlington pair connected in the common-emitter configuration.

EXAMPLE 6.9

A voltage divider biases the circuit shown in Figure 6.29. Determine the voltage gain of the circuit.

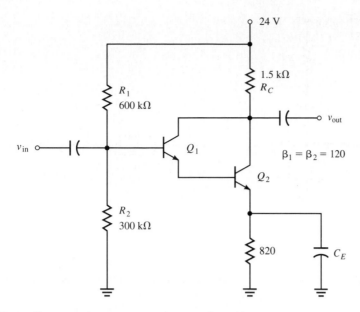

FIGURE 6.29 Darlington pair, common-emitter configuration

In the figure:
- 24 V
- R_1 600 kΩ
- 1.5 kΩ R_C
- v_{in}
- Q_1
- v_{out}
- $\beta_1 = \beta_2 = 120$
- Q_2
- R_2 300 kΩ
- 820
- C_E

Solution

We start by finding the voltage at the base of transistor Q_1:

$$V_{BQ1} = \frac{300\text{ k}\Omega}{900\text{ k}\Omega} \times 24\text{ V} = 8\text{ V}$$

The voltage at the emitter of Q_1 equals:

$$V_{EQ1} = 8\text{ V} - 0.7\text{ V} = 7.3\text{ V}$$

which is also the voltage at the base of transistor Q_2. From this, we can determine the voltage at the emitter of Q_2:

$$V_{EQ2} = 7.3\text{ V} - 0.7\text{ V} = 6.6\text{ V}$$

By applying Ohm's law, we can determine the current flowing through Q_2:

$$I_{Q2} = \frac{6.6\text{ V}}{820\text{ }\Omega} = 8.05\text{ mA}$$

We can ignore the AC emitter resistance of the circuit because the emitter resistance is bypassed. Next, we will calculate the AC emitter resistance for Q_2:

$$r'_{e2} = \frac{25\text{ mV}}{8.05\text{ mA}} = 3.1\text{ }\Omega$$

By looking at the AC beta box model shown in Figure 6.30a, we can see that the AC input signal develops across the AC emitter resistance of both transistors. Because of this, we need to determine the value of the AC emitter resistance for Q_1.

To find that value, we find the amount of base current flowing out of Q_2:

$$I_{BQ2} = \frac{8.05\text{ mA}}{120} = 67.1\text{ }\mu\text{A}$$

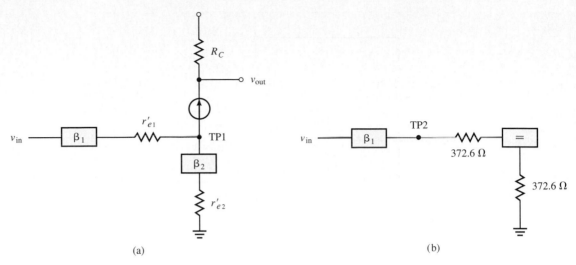

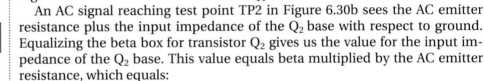

FIGURE 6.30 (a) AC beta box model of circuit shown in Figure 6.29. (b) Redrawing of (a).

This current is also the emitter current for Q1. Using that value, we can determine the AC emitter resistance for Q1

$$r'_{e_1} = \frac{25\,\text{mV}}{67.1\,\mu\text{A}} = 372.6\,\Omega$$

Figure 6.30b includes the value for the Q_1 AC emitter resistance.

An AC signal reaching test point TP2 in Figure 6.30b sees the AC emitter resistance plus the input impedance of the Q_2 base with respect to ground. Equalizing the beta box for transistor Q_2 gives us the value for the input impedance of the Q_2 base. This value equals beta multiplied by the AC emitter resistance, which equals:

$$r'_{e2} = 120 \times 3.105\,\Omega = 372.6\,\Omega$$

It is important to note that the AC signal at TP2 is cut in half by the time it reaches test point TP1 in Figure 6.30a. Since the gain of transistor Q_2 equals the collector resistance divided by the AC emitter resistance, the overall gain of the circuit equals:

$$A_V = \frac{1}{2} \times \frac{r_c}{r'_{e_2}} = \frac{r_c}{2r'_{e_2}}$$

Using this value, we can determine the overall gain of the circuit as:

$$A_V = \frac{1.5\,\text{k}\Omega}{6.2\,\Omega} = 241.9$$

PRACTICE PROBLEM 6.11

Calculate the gain of the circuit shown in Figure 6.29. Account for a 4.7-kΩ load resistor connected between the output tap and ground.

EXAMPLE 6.10

Determine the input impedance of the Darlington pair shown in Figure 6.29.

Using the analysis shown in Figure 6.29, let the AC resistance seen on the right side of the Q_1 beta equal 745.2 Ω.

Solution

To an AC signal applied to the left side of the beta box for Q_1, the resistance is:

$$\beta_1 \times 745.2\ \Omega = 89.2\ k\Omega$$

PRACTICE PROBLEM 6.12

Determine the input impedance for the circuit shown in Figure 6.29. Include the effects on the bias resistors. Without changing the beta values, how might the input impedance of the circuit be improved while maintaining the same gain?

REVIEW SECTION 6.5

1. Applying small amounts of DC bias voltage to push-pull amplifier configurations eliminates _____, changing a class B amp into a class AB amp.
2. What does base biasing of a class AB amp allow the collector current to do?
3. Class AB amps combine the higher _____ of a class B amp and the lower _____ of a class A amp.
4. To bias the class AB amp connect two _____ in a push-pull configuration.
5. Class AB biasing causes one transistor to conduct before the other stops conducting. True or false?
6. A class AB amp must be biased carefully to avoid runaway current through the transistors. True or false?
7. To defeat runaway current a pair of PNP and NPN transistors with unmatched electrical and thermal characteristics is used. True or false?
8. Describe what can happen when a voltage divider bias is used in a class AB amp, and give a solution to this problem.
9. Describe what the term *quasi* refers to in a quasicomplementary-symmetry amplifier.
10. Describe how the quasicomplementary-symmetry amp operates and how it is stabilized. Use circuit diagrams.
11. A Darlington pair refers to two _____ often packed in one case.

Class C amplifiers have the base biased so that collector current flows for less than one alternation of the applied sine wave input signal. Many class C amplifiers conduct for much less than 180° of the input cycle. Figure 6.31 shows typical voltage and current waveforms taken from the collector lead of a class C amplifier. Note that the output voltage from the collector is not a sine wave, although the circuit has an AC sine wave input.

To reproduce a sine wave at its output, the class C amplifier requires an LC circuit designed to resonate at a particular frequency. The range of frequencies into the circuit must be restricted to ensure proper amplifier action. Class C operation is usually limited to radio frequency (RF) amplifiers that have tuned LC circuits in order to keep the physical size of the inductors and capacitors small. Those amplifiers normally utilize frequencies above 20 kHz.

Because the collector current flows for a short time during each input cycle, distortion goes to a high level. However, the tuned circuit allows the amplifier to provide sufficient energy needed for resonant action. Class C amplifiers have the greatest efficiency of the four classes and give high power amplification.

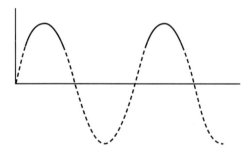

Note: Dashed lines, transistor is off
Solid lines, transistor is on

FIGURE 6.31 **Typical current waveform taken from a class C amplifier collector lead**

CLASS C CIRCUIT OPERATION

Figure 6.32 shows a class C amplifier circuit. Note that the circuit does not use DC bias. With a quiescent input signal, the transistor will not conduct. Applying an AC signal to the base of the transistor causes the input section of the circuit to act as a negative **clamper.** Figure 6.33 diagrams the clamper action.

When the top of the source in Figure 6.33 goes positive with respect to ground, the base-emitter junction of the transistor becomes forward-biased. A large amount of current flows in the base-emitter loop and charges the capacitor in the base section of the circuit. Figure 6.33 shows the polarity of the capacitor.

In Figure 6.34, the top of the source goes negative with respect to ground. Under these conditions, the base-emitter junction becomes reverse-biased and appears as an open circuit. Since the current from the source cannot flow through the high impedance of the reverse-biased transistor, it must

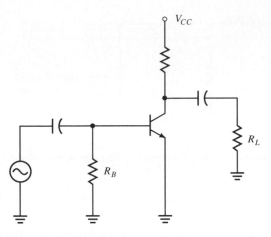

FIGURE 6.32 Class C amplifier circuit

flow through the base resistor. With a larger-value base resistor, the capacitor slowly loses the charge gained during the positive alternation.

By selecting the proper size of base resistor and paying attention to the RC time constant of the base loop when the base-emitter junction is reverse-biased, we can either prevent the quick discharge of the coupling capacitor or keep the charge constant. The positive alternation replaces any voltage returned by the coupling capacitor during the reverse biasing of the transistor junction. If the charge across the capacitor remains constant, the input coupling capacitor can substitute for a DC source.

Because of the polarity of the substitute DC source, the transistor goes into cutoff. The process of charging the coupling capacitor takes only a few positive alternations. After reaching full charge, the capacitor voltage allows the transistor to conduct during only the positive peaks.

To see why this occurs, let us review how clampers work. In a diode clamper circuit, the voltage across the capacitor will equal approximately the peak input voltage minus the diode barrier potential, or $V_p - 0.7$ V. In Figure 6.34, a positive AC input signal would cause the voltage across the capacitor to series-oppose the input signal. Substituting the input voltage value for V_p, the capacitor voltage for $-(V_p - 0.7$ V), and -0.7 V for the barrier potential of the diode, we have:

$$V_p - V_p + 0.7 \text{ V} - 0.7 \text{ V} = 0$$

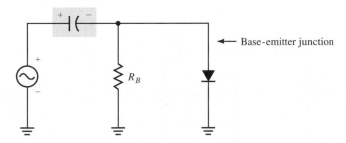

FIGURE 6.33 Diagram of clamper action in a class C amplifier, with base-emitter diode closed

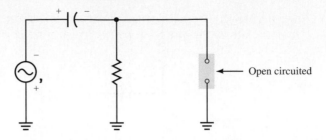

FIGURE 6.34 Clamper action in a class C amplifier with base-emitter diode open

With a net effective voltage equaling zero, no current flows. As we know, the capacitor does return some of its charge through the base resistor during the negative alternation. For a brief amount of time, the input voltage exceeds the voltage opposing it. With the small amount of current to flow, the transistor conducts and produces class C operation.

CLASS C OPERATING CONDITIONS

Because the class C amplifier circuit has no independent DC bias source, the Q point for the circuit is at cutoff. No current flows through the transistor without an AC signal into the base. Figure 6.35 shows the Q point at cutoff, thus $V_{CEQ} = V_{CC}$. To saturate the transistor, the voltage across the transistor must be reduced to zero. Figure 6.36 shows the AC circuit for a class C ampli-

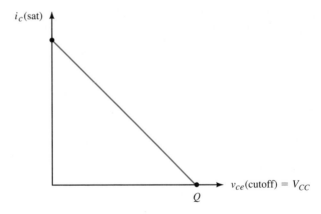

FIGURE 6.35 AC load line of a class C amplifier

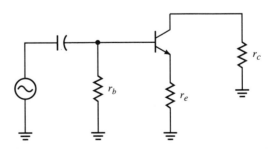

FIGURE 6.36 AC configuration of a class C amplifier

fier. As the voltage across the transistor reduces to zero, it transfers to the collector and emitter resistances. Thus, we can calculate the saturation current for an untuned class C amplifier as:

$$i_C(\text{sat}) = \frac{V_{CC}}{r_c + r_e}$$

EXAMPLE 6.11

Determine the cutoff voltage and the saturation current for the untuned class C amplifier shown in Figure 6.37.

Since the circuit operates as a class C amplifier, it has no DC bias. Thus, the transistor is off, and the entire supply voltage appears across the transistor.

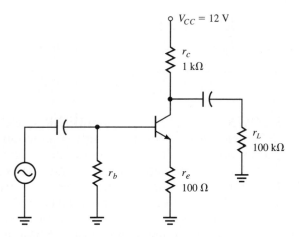

FIGURE 6.37 **Untuned class C amplifier**

Solution

The collector-to-emitter voltage at cutoff equals the collector supply voltage:

$$V_{CE}(\text{cutoff}) = V_{CC} = 12\,\text{V}$$

To determine the AC saturation current, we first find the value for the AC collector resistance:

$$r_c = 1\,\text{k}\Omega \,\|\, 100\,\text{k}\Omega = 990\,\Omega$$

The total AC resistance seen between ground and the collector supply voltage equals:

$$990\,\Omega + 100\,\text{k}\Omega = 1090\,\Omega$$

With those values, we can find the saturation current value:

$$I_{(\text{sat})} = \frac{12\,\text{V}}{1090\,\Omega} = 11\,\text{mA}$$

Now, we have both end points of the AC load line for an untuned class C amplifier.

Determine AC cutoff and saturation for the circuit shown in Figure 6.38.

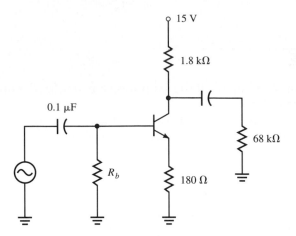

FIGURE 6.38

EMITTER BREAKDOWN IN CLASS C AMPLIFIER CIRCUITS

Looking at Figure 6.39, we see that the base-emitter junction is reverse-biased and appears as an open circuit. With the junction acting as an open circuit, the source and the coupling capacitor series-aid one another. Two voltages affect the base-emitter junction. In Figure 6.39, we see a worst-case condition. The input cycle has reached its negative peak, while the voltage across the base-emitter junction reaches a negative value twice the input peak voltage. This combination of voltages can cause the base-emitter junction of the transistor to go into reverse breakdown.

Many RF power transistors have a small reverse breakdown rating, such as 5 V. Transistor specification sheets will label this rating as BV_{ebo}. In Figure 6.40, we see a method used by many designers to prevent reverse breakdown. A small-signal diode with a breakdown rating of 25 V or higher works in series with the base-emitter junction. Placing the small-signal diode in series with the junction keeps the base-emitter loop open because the diode remains open.

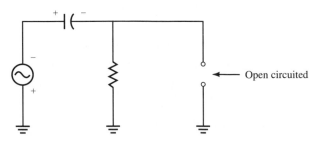

FIGURE 6.39 Worst-case condition of a class C amplifier

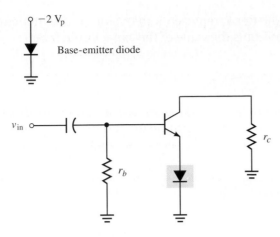

FIGURE 6.40 Protecting against reverse breakdown in a class C amplifier

TUNED CLASS C AMPLIFIERS

Although some applications use the narrow output pulses from an untuned class C amplifier, many circuits require a sine wave output. The narrow output pulses of the class C amplifier are converted into a sine wave through the use of a resonant circuit. The resonant circuit functions as part of the collector section of the amplifier. Figure 6.41 shows this type of circuit.

In the circuit, a parallel resonant tank circuit works in parallel with the load resistor and becomes a part of the collector circuit. Very low frequencies see the inductor as a short, while very high frequencies see the capacitor as a short. Thus, only frequencies near the resonant frequency of the tank develop across the load resistor.

We can calculate the resonant frequency (f) of the tank circuit by using the formula

$$f = \frac{1}{2\pi\sqrt{LC}}$$

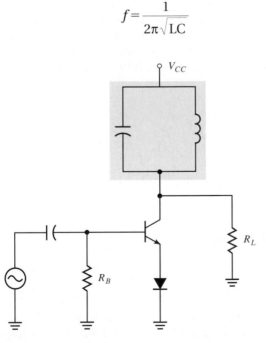

FIGURE 6.41 Tuned class C amplifier

In the formula, the letter L represents the value of the inductor in henries and the letter C represents the value of the capacitor in farads.

PRACTICE PROBLEM 6.14

In a tuned class C amplifier, the capacitor has a value of 0.001 µF and the inductor has a value of 100 µH. Determine the resonant frequency.

PRACTICE PROBLEM 6.15

In a tuned class C amplifier, C = 100 pF and L = 1 mH. Determine the resonant frequency.

HOW THE TUNED CIRCUIT ACTION BEGINS

With a short-duration, high-frequency input pulse into the tank, the inductor has a high reactance, while the capacitor acts as a short. This pulse shows as the output of the class C transistor collector. Consequently, most of the collector current flows into the capacitive branch of the tank circuit and charges the capacitor.

When the transistor goes into cutoff, the pulse is removed from the tank circuit. The capacitor discharges through the inductor. As the capacitor discharges, the discharge current builds up a magnetic field around the inductor. Once the capacitor has fully discharged, the magnetic field collapses. This collapsing magnetic field causes current to flow in the circuit, which recharges the capacitor. With energy repeatedly passing between the capacitor and inductor, the energy transfer produces a sine wave.

Each time energy transfers between the capacitor and inductor, some of it dissipates as heat because of circuit effects or the effects of the magnetic field. Instead of producing a perfect sine wave, the resonant tank circuit produces a damped sine wave. Figure 6.42 shows a damped sine waveform.

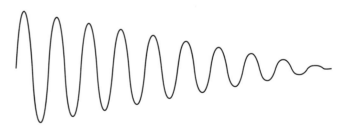

FIGURE 6.42 Damped sine wave

EXAMPLE 6.12

Determine the value of a base resistor in Figure 6.41.

Let the capacitor have a value of 1 µF. The lowest input frequency (*f*) to a class C amplifier is 20 kHz.

Solution

At 20 kHz, the time (t) for one complete cycle equals

$$t = \frac{1}{f} = \frac{1}{20 \text{ kHz}} = 50 \text{ μs}$$

We divide this value in half to find the time for the negative half-cycle: 25 μs. This is the time that the input coupling capacitor has for discharging. To prevent the capacitor from discharging too much, we set the RC time constant to 100 times this duration:

$$100 \times 25 \text{ μs} = 2.5 \text{ ms} \leq \text{the target RC time}$$

If the value of the capacitor is 1 μF, we can determine the value of the base resistor:

$$\frac{2.5 \text{ ms}}{1 \text{ μF}} = 2.5 \text{ k}\Omega$$

This should be a minimum value for the base resistor. Lower values allow the capacitor to lose too much charge.

PRACTICE PROBLEM 6.16

The lowest input frequency into the circuit for Figure 6.38 is 50 kHz. Determine the size of R_b.

MAINTAINING THE SINE WAVE

If the circuit could replace the dissipated energy, the sine wave would not damp out. Since the class C amplifier can produce a series of output pulses, the energy dissipated in the tank circuit can be recovered. For effective circuit action, the output rate of the pulses should match the resonant frequency of the tank circuit.

It is possible to set the resonant frequency of the tank circuit to a multiple of the pulse rate. For example, a current pulse produced as the output of a transistor every 20 μs would have a frequency of 50 kHz. We can find this conversion with the formula:

$$f = \frac{1}{t} = \frac{1}{20 \times 10^{-6}} = 50 \text{ kHz}$$

This frequency is the base or fundamental of the resonant tank circuit.

Working with multiples of the fundamental frequency allows the tank circuit to be tuned to 100 kHz, 150 kHz, or 200 kHz and still produce a sine wave. Any tuned frequency that is a multiple of 50 kHz would produce a sine wave. We refer to a circuit that uses multiples of a resonant frequency as a **frequency multiplier.** The output voltage delivered to the load is a multiple of the frequency delivered to the base of the transistor.

Frequency multipliers have the disadvantage of dissipating power while tuning to a multiple of a resonant frequency. Let us look at Figure 6.43. In this figure, we see three drawings of tank circuits. Figure 6.43a shows a tank circuit tuned to the base frequency. With this circuit, every positive peak restores energy to the tank. Figure 6.43b shows a tank circuit tuned to the second **harmonic,** or multiple of the fundamental frequency. Energy becomes

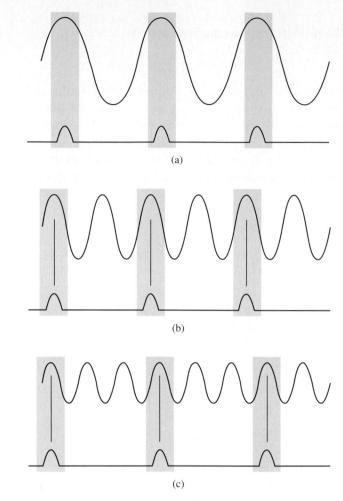

FIGURE 6.43 (a) Tank circuit tuned to the base frequency. Pulses every cycle. (b) Tank circuit tuned to the second harmonic. A pulse every other cycle. (c) Tank circuit tuned to the third harmonic. A pulse every third cycle.

restored to the circuit every other cycle. Consequently, the overall amplitude of the sine wave drops.

In Figure 6.43c, the tank circuit is tuned to the third harmonic of the fundamental frequency. Once again, the output amplitude of the sine wave drops. At some point, a designer will decide how much power to sacrifice while obtaining higher frequencies. If the power loss is too high, the designer may consider other types of frequency multipliers.

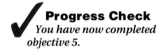

Progress Check
You have now completed objective 5.

DISSIPATING AND DELIVERING POWER WITH THE CLASS C AMPLIFIER

We know that no current will flow through a class C amplifier without the presence of an AC signal. With no current flowing, the DC collector voltage is dropped across the collector and emitter terminals. Look again at Figure 6.41. The path from the collector supply voltage to the collector terminal goes through the inductive section of the tank circuit. From the perspective of the collector supply voltage, the capacitor acts as an open circuit, while the inductor acts as a short circuit.

Since the emitter is at ground, the quiescent voltage across the transistor equals the collector supply voltage value. Driving the transistor into saturation will produce the maximum unclipped sine wave shown in Figure 6.44. The largest undistorted peak-to-peak wave doubles the collector supply voltage amount. To determine the power delivered by a saturated transistor, we need to find the **quality factor,** or Q, of the tank circuit. Then, we need to convert the Q value into an effective resistance and then determine the effective AC resistance. The following example illustrates the procedure for finding the effective AC resistance.

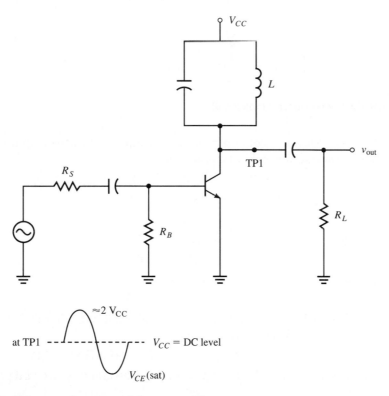

FIGURE 6.44 Sine wave of a saturated class C amplifier

EXAMPLE 6.13

Determine the resonant frequency and quality factor of the tank circuit shown in Figure 6.45. Find the effective AC resistance of the circuit.

When the Q of a coil in a tank circuit is above 10, we can use that value to approximate the Q of the circuit. We can also use the Q of the coil to determine the effective AC resistance created by the DC winding resistance of the coil.

Solution
First, we will use the formula

$$f = \frac{1}{2\pi\sqrt{LC}}$$

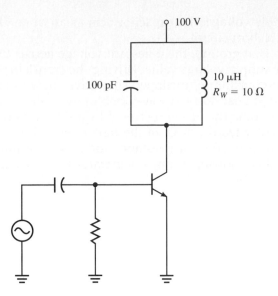

FIGURE 6.45 Collector tank circuit

to find the resonant frequency of the tank circuit. Multiplying the inductance value by the capacitance value, we get:

$$L \times C = 100 \text{ pF} \times 10 \text{ }\mu\text{H} = 1 \times 10^{-15}$$

Taking the square root of this answer, we obtain:

$$31.6 \times 10^{-9}$$

Multiplying this value by 2π we get:

$$198.6 \times 10^{-9}$$

Dividing one by this number gives us:

$$5.03 \text{ MHz, or } 5 \text{ MHz}$$

Next, we will find the quality of the coil. First, we will determine the reactance of the circuit at the resonant frequency:

$$X_L = 2\pi fL = 2\pi \times 5 \text{ MHz} \times 10 \text{ }\mu\text{H} = 314 \text{ }\Omega$$

Dividing this number by the DC winding resistance of the coil gives the Q of the coil:

$$Q(\text{coil}) = \frac{314}{10} = 31.4$$

From here, we will use the Q of the coil to determine the effective AC resistance created by the DC winding resistance of the coil:

$$R(\text{AC parallel}) = Q(\text{coil}) \times X_L = 31.4 \times 314 \text{ }\Omega = 9.86 \text{ k}\Omega$$

Figure 6.46 shows the results of our calculations.

In Figure 6.46, we pretend that the tank circuit has an infinite resistance at the resonant frequency. Any circuit deficiencies for this "perfect" circuit result from the presence of the 9.86-kΩ effective resistance of the tank circuit. We can use the effective resistance to calculate the power delivered to the load. From Figure 6.44, the maximum peak AC output equals the collector supply voltage. Thus, we have:

$$P_L = \frac{V^2}{R} = \frac{(0.707 \times V_{CC})^2}{9.86 \text{ k}\Omega} = 506 \text{ mW}$$

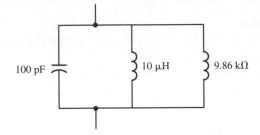

100 pF   10 µH   9.86 kΩ

FIGURE 6.46 Equivalent AC circuit for the tank in Figure 6.45

PRACTICE PROBLEM 6.17

Change the DC winding resistance of the coil in Figure 6.45 to 25 Ω. Recompute the effective AC resistance of the tank. Determine the power delivered to the load under these conditions.

CLASS C AMPLIFIERS AND EFFICIENCY

Class C amplifiers have an efficiency of almost 100%. We can use the next example to continue our analysis of Figure 6.45.

Progress Check
You have now completed objective 6.

EXAMPLE 6.14

Determine the efficiency of the class C circuit shown in Figure 6.45.

From example 6.13, we know that the output power to the load is 506 mW. The power dissipated by the transistor has a relationship with the amount of time the transistor conducts during 10% of the input cycle. Most power transistors have a saturation voltage equaling approximately 0.2 V.

Solution
When the transistor conducts, the saturation current equals:

$$i_C(\text{sat}) = \frac{100 \text{ V}}{9.86 \text{ k}\Omega} = 10.14 \text{ mA}$$

Since we know the average value of the saturation voltage for a power transistor, we can find the amount of power dissipated by the transistor:

$$P = 10.14 \text{ mA} \times 0.2 \text{ V} = 2.03 \text{ mW}$$

Because the transistor conducts for only 10% of the input cycle, it consumes power for only 10% of the cycle. This means that the average power dissipated for the entire cycle is:

$$P_{avg} = 2.03 \text{ mW} \times 0.1 = 0.2 \text{ mW}$$

The total power used in the circuit equals:

$$P_T = P_L + P_{avg} = 506 \text{ mW} + 0.2 \text{ mW} = 506.2 \text{ mW}$$

with a circuit efficiency that is:

$$\eta = \frac{P_L}{P_T} = \frac{506 \text{ mW}}{506.2 \text{ mW}} \times 100 = 99.9\%$$

1. Class C amplifiers conduct for more than one alternation of the applied AC input signal. True or false?
2. To get a sine wave output with a class C amp you must tune the circuit. True or false?
3. Class C amps, when tuned correctly, give the greatest _____ of the four classes and give high _____ amplification.
4. Describe how the class C amplifier circuit operates.
5. In a class C amp, current flows through the transistor without an AC signal into the base. True or false?
6. Give the equation for saturation current in an untuned class C circuit. Describe the variables.
7. When a class C amp is tuned, the resonant circuit functions as part of the _____ section of the amp.
8. In a parallel resonant tank, with high frequencies the _____ will act as a short, while with low frequencies the _____ will act as a short.
9. Describe the tuned circuit action of a class C amp.
10. A frequency multiplier describes a circuit that uses a multiple of a _____ _____.
11. The efficiency of class C amps is equal to almost _____ percent.

6.7 COUPLING AMPLIFIER STAGES TOGETHER

As we have studied different amplifier circuit types, we have also seen different ways to couple the amplifier stages together. High voltage gains can be attained through the use of cascaded amplifier stages. In this context, **cascade** means connecting the output of one amplifier to the input of another amplifier. The input signal is amplified by one stage and then passed to the next stage for further amplification. Amplification continues until the needed gain is acquired. In a cascaded amplifier, the stage gains multiply together to provide a total amplifier gain.

Most methods of coupling use some method to prevent DC voltages from one stage from upsetting the bias of the next stage. Four basic methods of coupling exist—RC coupling, transformer coupling, direct coupling, and tuned coupling.

RC COUPLING

Resistor-capacitor (RC) coupling provides the least expensive method of connecting two amplifier stages. A resistor loads the stage, while a capacitor couples the stages together. Figure 6.47 shows an example of RC coupling.

Because amplifiers work with frequencies, the value of the coupling capacitor must show a low reactance to the frequency of operation. In effect, the coupling capacitor acts as a short circuit for AC signals and as an open

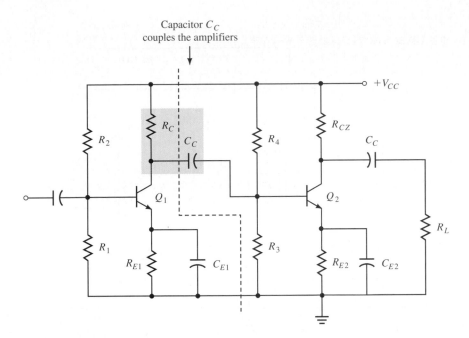

Capacitor C_C
couples the amplifiers

FIGURE 6.47 RC coupling

circuit for DC bias voltages. Without the blocking action of the coupling capacitor, the high collector voltage of a transistor could change the bias voltages of the next stage.

Although RC coupling provides low-cost cascading, the method has some limitations. If the AC signal goes to a low frequency, the capacitive reactance increases. The capacitor will no longer pass the complete AC signal. Some of the AC signal voltage drops across the capacitor, and the circuit loses low-frequency response.

TRANSFORMER COUPLING

Class A and B amplifier circuits will often employ **transformer coupling.** Figure 6.48 shows a common-emitter driver stage transformer-coupled to a push-pull amplifier stage. This method works well when the circuit requires phase-splitting or impedance matching. Transformer coupling exhibits good signal transfer with no DC interaction between stages. Transformer coupling has the disadvantages of size, added cost, and poor frequency response.

DIRECT COUPLING

Direct coupling, shown in Figure 6.49, offers good low-frequency response. Output from one amplifier directly connects to the input of another amplifier. With the removal of any capacitive or inductive properties, the frequency response of the circuit improves. Circuits that require good low-frequency response use direct coupling. Let us use the following example to learn more about direct-coupled amplifiers.

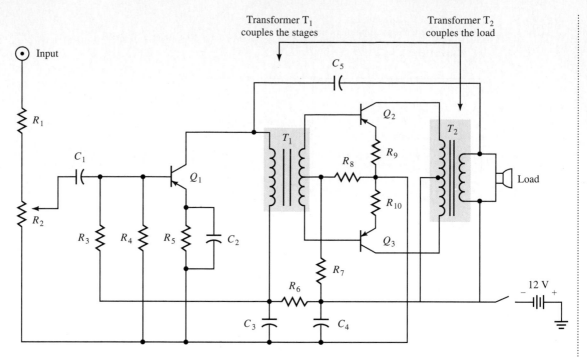

FIGURE 6.48 Transformer coupling

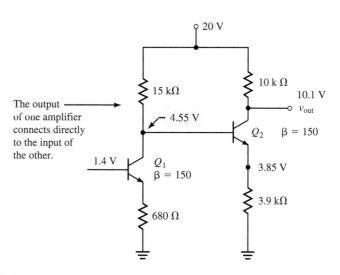

FIGURE 6.49 Direct coupling

EXAMPLE 6.15

Determine the overall gain of the two-stage direct-coupled amplifier shown in Figure 6.49.

There is 1.4 V at the base of transistor Q_1. This voltage biases the transistor into conduction and produces 0.7 V at the emitter of Q_1.

Solution

With the base and emitter voltages for transistor Q_1, we can calculate the current flowing through the transistor:

$$I_E = \frac{0.7 \text{ V}}{680 \, \Omega} = 1.03 \text{ mA}$$

This current flows out the collector and produces a voltage drop across the 15-kΩ load resistor. The remaining 4.5 V drops across the collector, feeds the base voltage of transistor Q_2, and forward biases the second transistor.

Since the emitter-base junction has a voltage drop of 0.7 V, we can predict a value for the voltage at the emitter of Q_2:

$$4.55 \text{ V} - 0.7 \text{ V} = 3.85 \text{ V}$$

This voltage is applied to the 3.9-kΩ resistor connected to transistor Q_2. We can now compute the emitter current for Q_2:

$$I_E(2) = \frac{3.85 \text{ V}}{3.9 \text{ k}\Omega} = 0.987 \text{ mA}$$

The emitter current flows out the collector and through the 10-kΩ collector resistor. When the circuit is quiescent, the collector voltage equals 10.1 VDC.

The gain of the second stage is:

$$\frac{r_c}{\left(r_e + r'_e\right)}$$

We can ignore the AC emitter resistance because the emitter resistance has not been bypassed. Therefore, the second-stage gain equals:

$$\frac{10 \text{ k}\Omega}{3.9 \text{ k}\Omega} = 2.56$$

The gain of the first stage also is

$$\frac{r_c}{\left(r_e + r'_e\right)}$$

Again, we can ignore the AC emitter resistance.

In addition, we can ignore the loading effects that the second stage may have on the first stage, because the input impedance of the Q_2 base equals beta multiplied by the sum of the emitter resistance and the AC emitter resistance or:

$$Z_{in}(\text{base})_2 = \text{beta}(r_e + r'_e) = 150(3.9 \text{ k}\Omega) = 585 \text{ k}\Omega$$

The base impedance of the second transistor is large enough to prevent serious loading of the first stage. We can approximate the gain of the first stage as

$$\frac{15 \text{ k}\Omega}{680 \, \Omega} = 22$$

Finally, the overall gain of the two-stage amplifier equals

$$22 \times 2.56 = 56.4$$

REFERENCING THE DIRECT-COUPLED AC SIGNAL TO GROUND

If we injected an AC signal into the circuit shown in Figure 6.49, it would be amplified by a factor of 56.4. However, if the circuit is to work as designed,

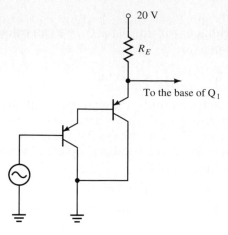

FIGURE 6.50 Darlington pair amplifier with the AC signal referenced to ground

the DC voltage at the base of transistor Q_1 must remain at 1.4 V. This means that the AC signal into the base of Q_1 will ride on a 1.4-VDC level.

Some amplifier circuits require that the AC signal reference to ground instead of a DC voltage level. In this case, the AC input signal must be isolated from the DC bias at the base of Q_1. Figure 6.50 shows a Darlington pair PNP combination with a ground-referenced input stage. The emitter-base junction voltage drops of the two PNP transistors raise the DC output voltage to 1.4 V. This voltage equals the DC bias voltage required to make the circuit of Figure 6.49 work as designed.

As you may suspect, direct coupling poses problems with DC voltage interaction and stabilization. Any circuit using direct coupling must have amplifier output voltages that are correct for the input of the next stage.

TUNED COUPLING

Even though transformer coupling offers poor frequency response, some designs use that limitation for better circuit performance. Figure 6.51 shows

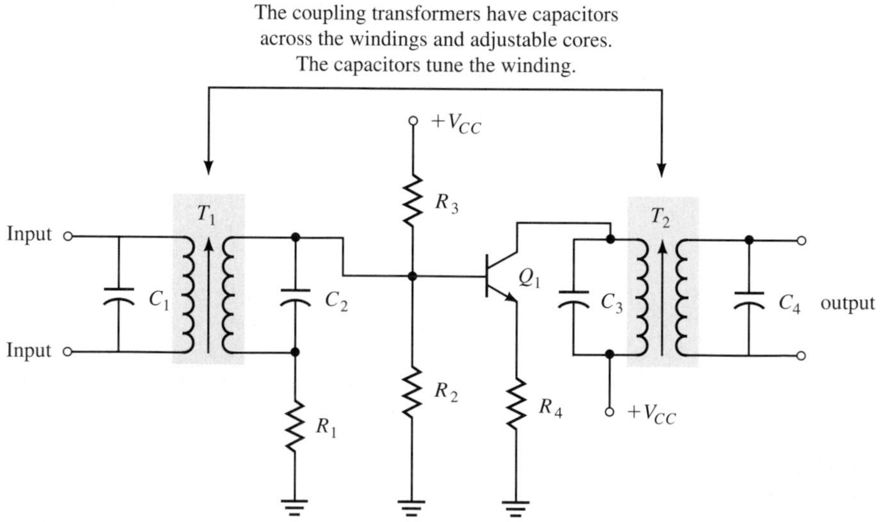

FIGURE 6.51 Tuned coupling

how **tuned coupling** transformers can limit the frequency response of an amplifier circuit. The coupling transformers tune to a specific frequency through the use of a capacitor across the winding and an adjustable core. Tuned coupling circuits are often found in the intermediate-frequency amplifiers used in radio receivers.

✓ **Progress Check**
You have now completed objective 7.

REVIEW SECTION 6.7

1. High voltage gains can be achieved through _____ amplifier stages.
2. In a cascaded amp configuration, the stage gains _____ together for the total amplifier gain.
3. The four basic methods of coupling amplifiers are _____, _____, _____, and _____ coupling.
4. In RC coupling, a _____ loads the stage, while a _____ couples the stages together.
5. One problem with RC coupling is that when the AC signal goes too low in frequency the _____ _____ increases, causing a loss in AC signal voltage.
6. _____ coupling is most often used in class A and B amplifiers.
7. Direct coupling is used with no _____ or _____ properties, giving good frequency response to the circuit.
8. Describe how the direct-coupled AC signal is referenced to ground.
9. The _____ _____ circuit gives better circuit performance than transformer coupling.

TROUBLESHOOTING

TROUBLESHOOTING LARGE-SIGNAL AMPLIFIERS

Many of the troubleshooting techniques that we used when discussing small-signal amplifiers will also work with large-signal amplifiers. Any troubleshooting procedure should begin with an overall look for burnt, broken, or discolored parts. Shorted output transistors often result in burnt emitter resistors. Also, the procedure should include checks of the power supply voltages.

When checking the amplifier stages, examine the input signal into the individual stages. This type of check should include DC voltage measurements at the emitter, base, and collector terminals of the amplifier transistor. Because we are working with AC signals, we also need to use an oscilloscope to check input and output waveforms. An oscilloscope becomes an especially useful tool when looking for distortion problems.

Because the operating points of the individual transistors in a direct-coupled amplifier hinge on the operation of all the transistors in the amplifier, a technician cannot isolate a defective stage by measuring voltages. Instead, checking each transistor in the amplifier becomes a priority. In many cases, direct-coupled amplifiers will have multiple transistor failures.

The midpoint voltage is a useful reference tool for checking the operation of a direct-coupled amplifier. A properly biased direct-coupled amplifier will have a midpoint voltage that equals one-half the supply voltage. Large variations from this point indicate trouble in the circuit.

We know that the first transistor in a direct-coupled amplifier controls the operation of the remaining transistors. If we force the first transistor into cutoff by shorting the base and emitter leads and then check the midpoint voltage, we can see whether the problem exists in the predriver or output stages. With the first transistor cut off, the midpoint voltage should either drop to 0 V or become equal to the supply voltage. This occurs because one output transistor will saturate, while the other goes into cutoff. No voltage change indicates a problem within the first stage. A voltage change moves our troubleshooting to the predriver or output stages.

An abnormal midpoint voltage also indicates a shorted or saturated transistor in a complementary-symmetry pair. A zero midpoint voltage indicates a shorted transistor. If the midpoint voltage equals the supply voltage, one transistor has become saturated.

TROUBLESHOOTING QUESTIONS

1. When troubleshooting, you should first look for
_____, _____, or _____
parts.
2. Checking the amplifier stages should be done by examining the
_____ signal to the individual stages.
3. Oscilloscopes are used to look at the AC signals and are extremely useful when looking for _____.
4. Often, direct-coupled amps will have multiple _____ failures.
5. The _____ voltage is a useful tool in troubleshooting because a properly biased direct-coupled amplifier will have a _____ voltage equal to one-half of the supply voltage.
6. How is the midpoint voltage a helpful tool in troubleshooting a complementary-symmetry amplifier?

WHAT'S WRONG WITH THESE CIRCUITS?

The following circuit is a class B push-pull amplifier. The normal measured readings are

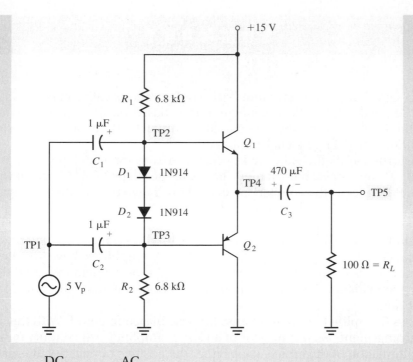

	DC	AC
TP1	0 V	5 v_p
TP2	8.41 V	5 v_p
TP3	7.26 V	5 v_p
TP4	8.75 V	4.7 v_p
TP5	0 V	4.7 v_p

Problem 1

	DC	AC
TP1	0 V	5 v_p
TP2	3.5 V	5 v_p
TP3	2.85 V	5 v_p
TP4	3.5 V	Severely distorted sine wave
TP5	0 V	Severely distorted sine wave

Problem 2

	DC	AC
TP1	0 V	5 v_p
TP2	11.62 V	5 v_p
TP3	0 V	5 v_p
TP4	11.34 V	Severely distorted sine wave
TP5	0 V	Severely distorted sine wave

Problem 3

	DC	AC
TP1	0 V	5 v_p
TP2	10.02 V	5 v_p
TP3	8.86 V	Ground noise
TP4	13.25 V	500-mV peaks
TP5	0 V	500-mV peaks

S U M M A R Y

Large-signal amplifiers have four basic configurations called class A, class B, class AB, and class C. Each configuration is based on the type of bias used and the length of time that the amplifier conducts. Both DC and AC signal conditions affect large-signal amplifiers.

Distortion and efficiency are key characteristics for each type of amplifier circuit. Distortion is an unwanted change in the shape of an AC signal waveform. Efficiency is the comparison of AC power delivered to the load against the DC power delivered to the circuit.

Class A amplifiers conduct for the entire input cycle. Because collector current flows for the entire cycle, the transistor does not enter cutoff or saturation. Of the four amplifier classes, class A amplifiers have the lowest amount of distortion and efficiency. Under the best operating conditions, class A amplifiers will deliver no more than 25% of the total power dissipated to the load.

Class B amplifiers have the transistor base biased to cutoff. With the biasing arrangement, each transistor of a class B amplifier will conduct for only one alternation of the input signal. Either the positive or negative alternation will be missing from the transistor waveform. Class B amplifiers are more efficient than class A amplifiers because the transistor will not conduct current without the presence of a signal. Unfortunately, class B amplifiers also have more distortion. Crossover distortion occurs when a class B amplifier operates with no DC bias.

A class AB amplifier uses base biasing that allows the transistor to conduct for slightly more than one alternation of the input signal. Class AB amplifiers work between the higher efficiency of class B amplifiers and the low distortion of class A amplifiers. Biasing for a class AB amplifier causes one transistor in a push-pull arrangement to conduct before the other transistor has stopped conducting.

Quasicomplementary-symmetry amplifiers produce output power greater than 10 W. We call the amplifier configuration quasicomplementary-symmetry because the complementary-symmetry portion of the circuit appears before the output stage of the circuit. The quasicomplementary-symmetry configuration offers phase inversion for driving NPN transistors without a loss of performance or a cost increase.

A Darlington pair amplifier configuration is two transistors usually housed in one case. One transistor acts as an input amplifier for the other. Darlington pairs exhibit an extremely high beta and a high base impedance.

Class C amplifiers are biased so that collector current flows for less than one alternation of the applied input signal. In practice, many class C amplifiers conduct for much less than one alternation. Of the four amplifier classes, class C amplifiers have the greatest efficiency. Also, class C amplifiers give high power amplification. Because class C amplifiers conduct for only a portion of the input signal, distortion can go to a high level. However, class C amplifiers work with tuned circuits and amplify only resonant frequencies or harmonics. This cuts the distortion problem.

Four basic methods for coupling amplifier stages exist. They are RC coupling, transformer coupling, direct coupling, and tuned coupling. Coupling or cascading stages together gives high voltage gains. To find the total gain of cascaded stages, we multiply the stage gains together. In addition, the coupling methods must not change the biasing of the individual stages.

Many of the troubleshooting methods used for working with small-signal amplifiers will also work with large-signal amplifiers. In many cases, a technician can look for physical evidence such as burnt or discolored wiring and components. Checking amplifier stages involves measuring voltages and waveforms.

PROBLEMS

For problems 1–10, refer to Figure 6.52.
  1. Determine the DC voltages at the three transistor terminals.

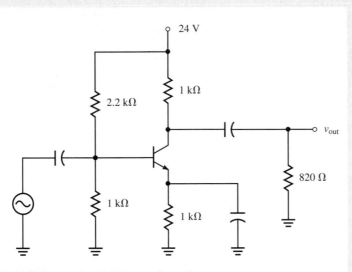

FIGURE 6.52 Circuit for problems 1 through 10

  2. What is I_{CQ} for this circuit?
  3. Determine and plot the DC load line.
  4. Determine and plot the AC load line.
  5. Determine and plot the Q point on the AC load line.
  6. What is the maximum AC output signal that does not cause clipping?
  7. What is the power dissipated by the transistor?
  8. Assuming that the maximum AC output signal referred to in question 6 is applied to the load, how much power is delivered to the load?
  9. Ignoring bias currents, what is the power efficiency of the circuit?
  10. Including biased currents, what is the power efficiency of the circuit?

Problems 11–20: Repeat the questions for 1–10, but this time analyze Figure 6.53.

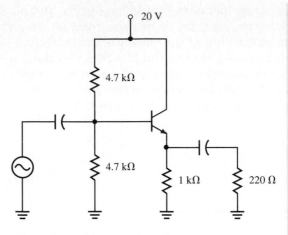

FIGURE 6.53 Circuit for problems 11 through 20

For problems 21–30, refer to Figure 6.54.

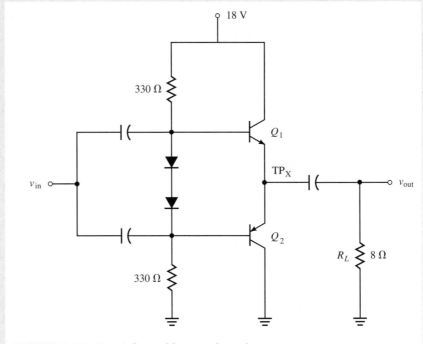

FIGURE 6.54 Circuit for problems 21 through 30

21. Determine the current flowing through the voltage divider.
22. What is the quiescent current through Q_1 and Q_2?
23. What is the DC voltage at TP_x?
24. What is the DC voltage at the base of Q_1?
25. What is the DC voltage at the base of Q_2?
26. What is the maximum AC peak output voltage?
27. What is the maximum peak current that can be delivered to the load?
28. What is the maximum power that can be delivered to the load?
29. What is the efficiency of the circuit?
30. What is the function of the two diodes in the circuit?

For problems 31 and 32, refer to the circuit in Figure 6.55.

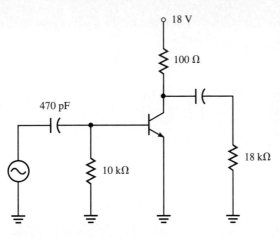

FIGURE 6.55 Circuit for problems 31 and 32

31. What is the cutoff voltage?
32. What is the saturation current?

For problems 33–38, refer to Figure 6.56.

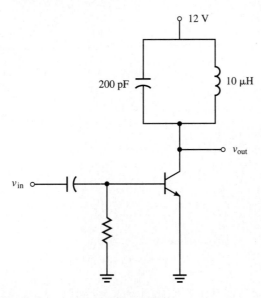

FIGURE 6.56 Circuit for problems 33 through 38

33. What is the resonant frequency of the tank?
34. What is the Q of the coil at resonance?
35. What is the effective parallel resistance of the tank?
36. What is the maximum peak-to-peak output of the tank?
37. What is the power delivered to the load?
38. If the transistor is on 20% of the time, what is the power efficiency of the circuit?

JUNCTION FIELD EFFECT TRANSISTORS

Tom Way, IBM

✓ **As you read this chapter, concentrate on learning how to:**

1. Define what a JFET does
2. Describe the differences between N-channel and P-channel JFETs
3. Describe the sections of a JFET
4. Describe the differences between depletion and enhancement mode operation
5. Use a normalized universal JFET drain curve to predict JFET circuit operation
6. Determine the DC voltages at the terminals of a JFET
7. Calculate the DC current flowing through a JFET for each of the JFET bias arrangements discussed in this chapter
8. Define transconductance
9. Determine the gain of a JFET amplifier
10. Use a JFET as a common-drain amplifier
11. Use a JFET as a common-gate amplifier
12. Troubleshoot JFET circuits

INTRODUCTION

Until this chapter, we have discussed either semiconductor diodes or bipolar junction transistors. We have seen how transistors can amplify small and large signals. In this chapter, we will begin to look at other types of semiconductor devices. Field effect transistors have different properties than the bipolar junction transistors (BJTs) that we have studied. As the name implies, a **field effect transistor (FET)** controls current through an electrical field produced by an applied voltage. Bipolar junction transistors control current with a base current.

Since the controlling portion of a FET draws very little current, the FET has a very high input impedance. Circuits that require input impedances higher than those found in bipolar semiconductor devices often use **junction field effect transistors,** or **JFET**s. These uses include analog switches, ultrahigh impedance amplifiers, voltage-controlled resistors, and current sources.

Aside from working as a voltage-controlled device, JFETs differ from bipolar junction transistors in other ways. The JFET is a unipolar device. Unlike the bipolar junction transistor, the JFET does not require two different types of charge carriers to operate. A JFET will use either negative or positive charge carriers, but not both types, during normal operation. Although JFETs are voltage-controlled devices, they do not provide the amount of voltage gain seen with current-controlled bipolar junction transistors.

7.1   BASIC JFET OPERATION

Figure 7.1 shows cross-sectional views of typical JFETs. Small, heavily doped regions of one type of semiconductor extend into a bar of the opposite type of semiconductor material. These diagrams show the two basic types of JFETs: an N-channel JFET (Figure 7.1a) and a P-channel JFET (Figure 7.1b). In an **N-channel JFET,** the bar of semiconductor material is N-type silicon; the majority carrier is electrons. A **P-channel JFET** has a bar made of P-type silicon; the majority current carrier is holes.

The three sections of each JFET shown in Figure 7.1 are the source, drain, and gate of the JFET. Comparatively, the **drain, source,** and **gate** in a JFET work the same as the collector, emitter, and base in a bipolar junction transistor. Let us use an N-channel JFET as an example. One end of the N-type bar is the source and the other end of the bar is the drain, with the drain more positive than the source. Two small P-type regions connected internally make up the gate. We can define the **channel** of an N-channel JFET as the region between the two P-type regions and as the region of a JFET through which current flows. We would see opposite polarities in a P-channel JFET. Electrons carry current in an N-channel JFET, while holes carry current in a P-channel JFET.

In the N-channel JFET, electrons enter at the source, flow through the channel, and exit at the drain. The P-type material gate is the control terminal. The amount of voltage at the gate determines the amount of current that will flow through the JFET. Current can flow through a JFET even if the controlling voltage is zero.

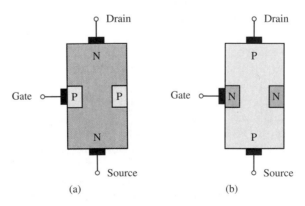

FIGURE 7.1   **(a) Block diagram of an N-channel JFET. (b) Block diagram of a P-channel JFET.**

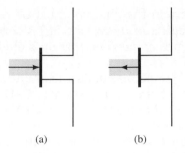

(a)   (b)

FIGURE 7.2   **(a) Schematic symbol for an N-channel JFET. (b) Schematic symbol for a P-channel JFET.**

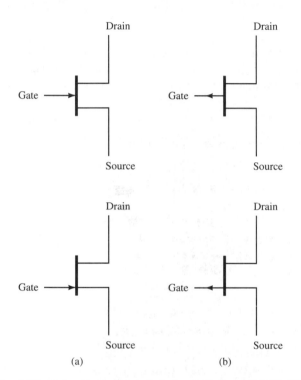

(a)   (b)

FIGURE 7.3   **(a) Generic schematic symbols for an N-channel JFET. (b) Generic schematic symbols for a P-channel JFET.**

Figure 7.2 shows the schematic symbols for the two types of JFETs. *As in the BJT schematic drawings, the arrowhead points to the N-type material.* When the arrowhead of the gate terminal points into the device, it has an N-type channel. A schematic symbol with the arrow pointing out of the device has a P-type channel. Figure 7.3 shows "generic" schematic figures for JFETs.

Progress Check
You have now completed objectives 2 and 3.

CONNECTING SUPPLY VOLTAGES TO THE JFET

In an N-channel JFET, current flows from the source to the drain with the gate grounded. The gate junction must be reverse-biased so that its voltage can control the drain current. Except for a small amount of leakage current, no current flows in the gate circuit.

The JFET will conduct at its maximum value when its gate and source are tied together. In order to control current flow through the channel, the

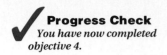
amount of majority current carriers in the channel must be reduced. This mode of operation is called the **depletion mode.** If a FET's current is being controlled by depleting majority current carriers, then it is operating in the depletion mode.

During normal operation, the JFET operates at its maximum value of drain current when the gate voltage equals the source voltage. CAUTION: Never allow the gate to be more than 0.5 V more positive than the source. If the difference rises above the 0.5-V level, the gate-source diode will become forward-biased and will begin to conduct, which will damage the JFET.

In Figure 7.4 an N-channel JFET is connected to two supply voltages. One supply voltage is connected between the drain and source (V_{DD}), while the other is connected between the gate and source (V_{GG}). V_{DD} and V_{GG} are supply voltages. Any voltage developed between the drain and the source can be called the drain-to-source voltage (V_{DS}) and any voltage developed between the gate and source may be called the gate-to-source voltage. Although in Figure 7.4 V_{DD} and V_{DS} may appear to be the same, later on we will differentiate between them. First, we will examine how the drain-to-source voltage can affect the JFET.

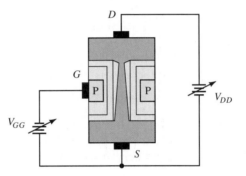

FIGURE 7.4 How the depletion zone grows together when the JFET has gate and drain bias

THE SOURCE-TO-DRAIN SUPPLY VOLTAGE

Figure 7.5 graphs the action of the drain current (I_D) versus the drain-to-source voltage (V_{DS}). As the drain-to-source voltage level slowly rises toward the knee, current rises linearly with the voltage. With small values of current,

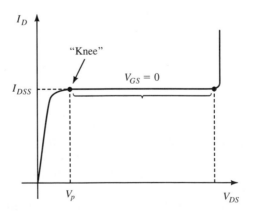

FIGURE 7.5 A JFET drain curve

the JFET channel acts as a linear resistor. Increasing the V_{DS} causes some dramatic changes. The areas of the channel near the junctions become positive with respect to the source terminal. Since the gate regions connect externally to the source, the junctions are reverse-biased. In turn, the areas of the depletion regions increase so that the depletion regions extend further into the channel region.

PINCH-OFF VOLTAGE

As the depletion region increases, the resistance of the channel increases and its conductance decreases. Further increasing the source-to-drain voltage causes the depletion regions to extend into the channel until they nearly meet. The amount of drain voltage needed to make the depletion regions nearly meet is called the **"pinch-off" voltage** (V_P).

Measuring the pinch-off voltage is not difficult. Shorting the gate to the source connection causes the gate-to-source voltage (V_{GS}) to equal 0 V and produces the maximum drain current (I_{DSS}). Slowly increasing the drain-to-source voltage while monitoring the drain current will stabilize the drain current. Once the drain current becomes constant, V_{DS} is at the pinch-off voltage point.

In Figure 7.5, the pinch-off voltage is at the "knee" of the curve. For a JFET, we will find that the pinch-off voltage has the same value as, but opposite polarity to, the gate-source cutoff voltage or $V_{GS(off)}$. Increasing the source-to-drain voltage beyond the pinch-off voltage will cause little additional change in the depletion regions and the drain current will have reached saturation.

Saturation for a JFET has a different meaning than for a bipolar junction transistor. With the BJT, saturation is the state of having a small collector-to-emitter voltage. For the junction field effect transistor, **saturation** means supplying the maximum amount of drain current at a given gate voltage.

THE GATE-TO-SOURCE SUPPLY VOLTAGE

Applying the gate-to-source voltage to the JFET provides a fixed reverse-bias voltage. Figure 7.6a shows a JFET with an applied gate-to-source voltage but zero drain-to-source voltage. With zero drain-to-source voltage, the depletion regions extend uniformly into the channel. Adding the drain-to-source voltage causes drain current to flow and the depletion region to become wedge-shaped. Figure 7.6b shows the results of applying the drain-to-source voltage.

In Figure 7.6b, we can see that the wedge-shaped portion of the depletion region grows because of current flow. With no drain current, the depletion region extends into the channel because of the applied gate-to-source voltage. This tells us that less drain current is required to reach the pinch-off voltage with an increased gate-to-source (V_{GS}) bias voltage. Since the PN junctions have a reverse bias, the gate-to-source region has a high resistance. Also, the gate current has a small value because it exists as the leakage current of a reverse-biased PN junction.

It is difficult to manufacture JFETs like those shown in Figure 7.6a, b. One method of manufacture is illustrated in Figure 7.6c, d. Notice that the JFET is constructed on a silicon base called a substrate. First, the N-type doping is introduced and then the P-type. The P-type material, which is the gate, forms a continuous loop. Current must flow from the source under the gate and out the drain. The channel is formed around the P-type gate.

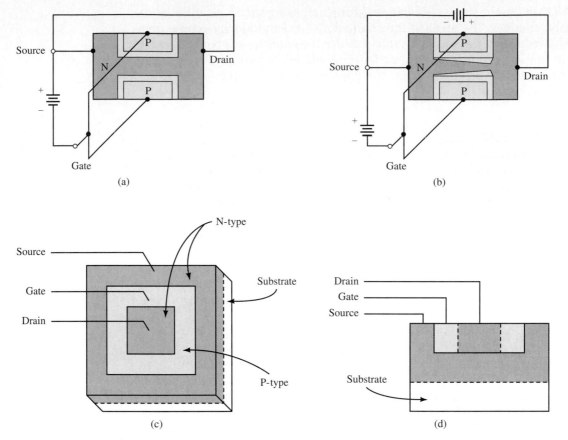

FIGURE 7.6 (a) Block diagram of a JFET using only source bias. (b) How the depletion zone grows together when the JFET has source and drain bias. (c) Top of JFET. (d) Cross-sectional view of JFET.

CHARACTERISTICS OF THE JFET DRAIN

Junction field effect transistors have good transconductance. That is, for a constant gate-source voltage, the drain current is nearly constant. Figure 7.7 shows a set of characteristic curves for the drain current of a JFET. From the graphs, you may see that a JFET will work as a constant current source by applying a constant gate-to-source voltage.

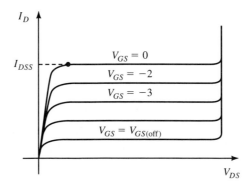

FIGURE 7.7 Set of characteristic curves for the drain current of a JFET. Each drain curve represents a different V_{GS}.

If the JFET reaches a state where the gate has shorted to the source, the drain current reaches its maximum value. On the vertical axis we label this point as I_{DSS}. This is the point on the horizontal axis where the gate-to-source voltage reaches cutoff, or the pinch-off voltage. The I_{DSS} point would be one end point of the load line.

DRAIN CURVES

Figure 7.8 shows a typical drain curve, which has three regions of JFET operation. Increases in the drain-to-source voltage produce proportional increases of I_D in the **ohmic region.** In the ohmic region, the JFET acts like a variable resistor. After the pinch-off voltage has been reached, the drain current remains constant and the JFET operates as a constant current source. The last region of the drain curves is the **breakdown region.** When the drain-to-source voltage becomes too large, the JFET loses its ability to control current and breaks down, acting like a short.

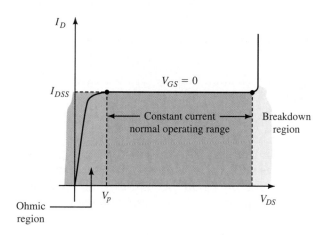

FIGURE 7.8 The regions of a JFET drain curve

Figure 7.9 plots the transconductance curve using I_D and V_{GS}. Since this curve is the same for all FETs, it may also be called a universal curve. The curve has been normalized. To normalize I_D, the value of I_D is divided by I_{DSS}.

$$I_{D(N)} = \frac{I_D}{I_{DSS}}$$

Notice that we plot the ratio of drain current to maximum drain current. That means I_{DSS} is plotted on the vertical axis at the 1.00 point. $V_{GS(cutoff)}$ is plotted on the horizontal axis at the 1.00 point also.

In order to use information we have found on a normalized graph, we have to denormalize. That is, we change the normalized ratio back to its real value. If we found V_{GS} by graphing, we can find its real value by multiplying the ratio by $V_{GS(cutoff)}$.

$$V_{GS} = V_{GS(N)} \times V_{GS(cutoff)}$$

In example 7.1, we will use a normalized JFET graph and I_D to find V_{GS}.

Using example 7.1, we will take a closer look at the characteristic curve for a JFET.

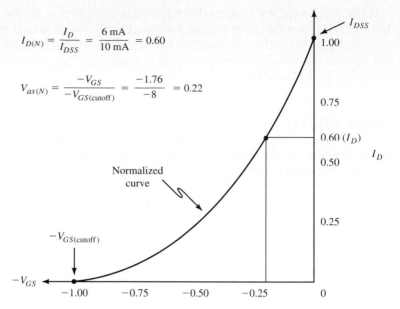

$$I_{D(N)} = \frac{I_D}{I_{DSS}} = \frac{6\text{ mA}}{10\text{ mA}} = 0.60$$

$$V_{as(N)} = \frac{-V_{GS}}{-V_{GS(\text{cutoff})}} = \frac{-1.76}{-8} = 0.22$$

FIGURE 7.9   Normalized transconductance curve for a JFET

EXAMPLE 7.1

Using Figure 7.9, find the gate-to-source voltage needed to produce 6 mA of drain current.

    If we use a JFET that has an I_{DSS} of 10 mA, we can plot one point of the load line at 1.00 on the vertical axis of the universal curve. For this example, the JFET will stop conducting when the gate-to-source voltage reaches -8 V. The -8 V represents the gate-to-source cutoff voltage $V_{GS\,(\text{cutoff})}$ and is plotted at the -1.00 point of the horizontal axis.

Solution

Since we know that the JFET has an I_{DSS} that equals 10 mA, we can also plot the desired drain current on the load line. Sixty percent of 10 mA gives us the desired 6 mA of drain current. Then, we draw a perpendicular line from the 0.60 mark on the vertical axis until it intersects the JFET curve.

    From the intersect point, we draw another line down to the horizontal axis. This line should intersect the horizontal axis at either 0.22 or 0.23 and represents the percentage of the gate-to-source cutoff voltage applied to the JFET that would produce 6 mA of drain current. On the graph, we have

$$0.22 \times -8\text{ V} = -1.76\text{ V}$$

To have 6 mA of drain current flowing from the JFET, we must apply -1.76 V at the gate.

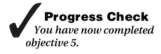

Progress Check
You have now completed objective 5.

PRACTICE PROBLEM 7.1

Using the graph in Figure 7.9, determine V_{GS} if $I_D = 0.40\ I_{DSS}$.

THE JFET IS A SQUARE LAW DEVICE

The JFET characteristic curve forms a parabola. From the graphs that we have seen, we know that the drain current does not suddenly cut off. Instead, it continues to drop. To find out how the JFET functions, we should look at the equation for the drain current (I_D):

$$I_D = I_{DSS}\left(1 - \frac{V_{GS}}{V_{GS(cutoff)}}\right)^2$$

The equation shows that the square root of the drain current value equals a fixed point of the gate-to-source voltage multiplied by the saturation current value. The JFET is a square law device.

If we require linear outputs from a JFET, we must restrict the device to a small portion of its characteristic curve because the curve does not have a linear portion. Since the curve is continually changing, any signal applied will reflect the curve change in the FET's output. Any change other than amplitude in the output signal when compared with the input signal is called **nonlinear distortion.** In small-signal applications the distortion is insignificant. However, in large-signal applications the distortion would be unacceptable. As an example, nonlinear distortion could change a sine wave input signal into an output signal that would resemble a triangular wave.

REVIEW SECTION 7.1

1. A device in which P-type semiconductor material extends into a bar of N-type material is called a P-channel JFET. True or false?
2. The source, drain, and gate of a JFET can be compared to the _____, _____, and _____ of a bipolar junction transistor.
3. _____ carry current in an N-channel JFET; while _____ carry current in a P-channel JFET.
4. Unlike BJTs, JFETs operate with _____ control voltage.
   (a) Forward-biased
   (b) Reverse-biased
   (c) Neutral
5. JFETs always conduct with the gate tied to the source, or in depletion mode. True or false?
6. If the source-to-drain voltage is increased past the pinch-off voltage there is a _____ in the depletion regions.
   (a) Large change
   (b) Decrease
   (c) Small change
7. The area where a JFET loses its ability to control current occurs in a region referred to as:
   (a) Ohmic
   (b) Breakdown
   (c) Normal operation

8. The JFET is a square law device because the square root of the drain current is equal to a fixed point of the gate-to-source voltage multiplied by the _____ _____ _____.

9. A JFET must be restricted to a small part of its characteristic curve, if we want linear output; otherwise the _____ will have too much of a change.

7.2 BIASING THE JFET

MATCHING JFET CHARACTERISTICS

Before looking at biasing configurations for junction field effect transistors, we need to look at how the characteristics of JFETs may vary because of manufacturing processes.

Some JFET characteristics exhibit more variation than the corresponding characteristics of bipolar junction transistors. A JFET with a grounded source terminal may work well as a current source but the amount of current may not be predictable. Also, the gate-to-source voltage may vary significantly with a given value of drain current. Figure 7.10 indicates how variations within a batch of JFETs with the same part number can make predicting the current through the device more difficult.

In Figure 7.10, we see two curves showing how a particular JFET may react in a circuit. The top curve represents the maximum operating characteristics of the device. This line is marked with the end points of -8 V for $V_{GS(off)}$ and 16 mA of I_{DSS}. The lower curve represents the minimum operating characteristics of the JFET. The end points for this line are at -2 V and 4 mA. An actual JFET may function anywhere between these two lines.

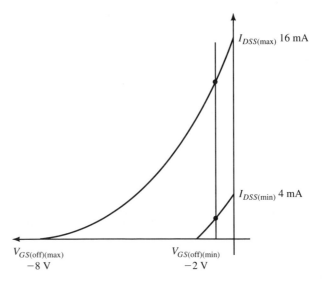

FIGURE 7.10 How variations within a batch of JFETs with the same part number can make predicting current difficult

USING GATE BIAS FOR THE JFET

Figure 7.11a shows gate bias using two supply voltages. As shown in Figure 7.11b, using one bias voltage would not provide the correct bias voltages for the gate. At first glance, gate bias resembles the base biasing arrangement used for bipolar junction transistors. However, some significant differences exist. Foremost is that the PN junction of a bipolar transistor is forward biased. By design, the JFET PN junction would be destroyed when trying to conduct current.

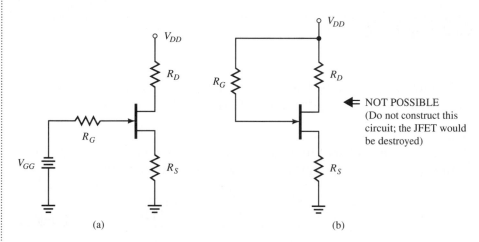

FIGURE 7.11 (a) Gate bias with two source voltages. (b) Incorrect method of gate biasing.

Some of those differences become apparent when we attempt to draw the control loop for the gate circuit. Figure 7.12a shows the control loop as it might appear. As drawn, the gate-to-source voltage box represents the voltage across the reverse-biased gate diode. We see no beta parameter because no relationship between the current flowing through the gate resistor and the current flowing through the source resistor exists. Remember that the only current flowing through the gate resistor is a small amount of leakage current.

For an MPF102 JFET, this leakage current would have a value of approximately 2 nA. Specification sheets label the leakage current as I_{GSS}. Usually, we can ignore the leakage current. For instance, if we have a gate resistor with a large value of 1 MΩ, the voltage across the gate resistor will equal:

$$2 \text{ nA} \times 1 \text{ M}\Omega = 2 \text{ mV}$$

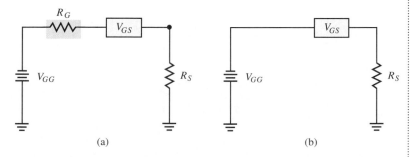

FIGURE 7.12 (a) Gate bias control loop. (b) Gate bias control loop with the gate resistor eliminated.

Such a small voltage does not affect the JFET operation. Figure 7.12b shows a redrawn JFET circuit without the gate resistor.

Looking at the redrawn figure, we see a loop with two voltages and a single resistor. If we knew the value of the gate-to-source voltage, we could apply Ohm's law to the loop. Yet, that voltage does not represent the voltage across a forward-biased diode. Thus, we cannot automatically use 0.7 V as a value for the gate-to-source voltage and for solving the control loop. Example 7.2 can show how to solve the control loop.

EXAMPLE 7.2

For the circuit shown in Figure 7.12b, solve for the current flowing through the source resistor.

Use the following values when calculating the amount of current flowing through the source resistor: $V_{GS} = -2\,V$; $V_{GG} = -10\,V$; $R_S = 1k\Omega$.

Solution
We can use Ohm's law to find the amount of current flowing through the resistor (I_{RS}). In equation form, our solution would look like this:

$$I_{RS} = \frac{V_{GG} - V_{GS}}{R_S} = \frac{-10\,V\,2\,(22\,V)}{1\,k\Omega} = \frac{-8\,V}{1\,k\Omega} = 8\,mA$$

PRACTICE PROBLEM 7.2

For the circuit shown in Figure 7.12b, change the value of V_{GS} to -14 V and solve for the current flowing through the source resistor.

Although the answer to example 7.2 provides an approximation of the amount of current flowing through the source resistor, the method may not always prove satisfactory. For this method to work, V_{GG} would need a value large enough to swamp the value given for V_{GS}.

Figure 7.13 shows another attempt at reducing the control loop for the JFET. In the figure, the gate-to-source voltage equals one-quarter of the gate-to-source cutoff voltage. Using example 7.3, let us see how this method would affect our solution.

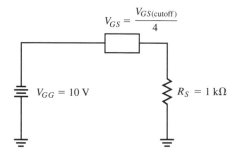

FIGURE 7.13 Gate-bias control loop with R_G eliminated

EXAMPLE 7.3

Calculate the current for the circuit shown in Figure 7.13.
   For the example, we assume that $V_{GS(cutoff)}$ equals -4 V.

Solution
If the gate-to-source voltage value is one-quarter of the -4 V, it has a value of
-1 V. Using that value in our equation, we have:

$$I_{RS} = \frac{-10\ V - \left(-1\ V\right)}{1\ k\Omega} = -9\ mA$$

PRACTICE PROBLEM 7.3

Use a value for $V_{GS(off)}$ of -6 V and the one-quarter approximation method
for V_{GS} to determine the current flowing in Figure 7.14.

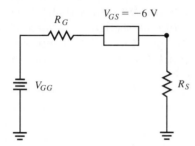

FIGURE 7.14

   As far as percentages are concerned, the answer found in example 7.3 differs a great deal from the answer found in example 7.2. To account for the
difference, we would need to change some values in every manufactured circuit. This is not a realistic solution.
   Going back to figure 7.10, we can plot a bias line on top of the characteristic curves. This straight, vertical line represents the fixed voltage placed on
the gate by the gate-to-source voltage. As you can see, it can produce a wide
range of drain current values. Because of the wide range, most circuit designers will reject the use of gate bias.

USING SELF-BIAS FOR A JFET
The self-biased circuit shown in Figure 7.15 is an improvement over the gate
bias method. In a **self-biased circuit,** the gate ties to DC ground through a
gate resistor. Because gate current is 0 A, the gate voltage is at 0 V. Also, the
self-biased circuit will always employ a source resistor. As current flows
through the JFET channel, a positive voltage develops across the source resistor (R_S). In an N-channel JFET, the positive voltage at the source and the
zero voltage at the gate produce a negative gate-to-source voltage.

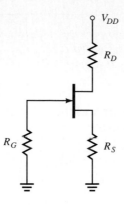

FIGURE 7.15 Self-biased circuit

EXAMPLE 7.4

Analyze the circuit shown in Figure 7.15 given the following values: $R_G = 100$ $k\Omega$, $R_D = 2\ k\Omega$, $R_S = 1\ k\Omega$, $V_{DD} = 12\ V$, $V_{GS(cutoff)} = -8\ V$, and $I_{DSS} = 10\ mA$.

We can plot current versus voltage for the source resistor by using Ohm's law. A plot for a resistor should produce a straight line since Ohm's law is a linear equation. With no current through the resistor, zero voltage is developed across the resistor. We can mark this point at the origin of the graph shown in Figure 7.16 and label it as P_1.

We can use any current value less than the value given for I_{DSS}. In this case, we will choose 2 mA. On the graph, this would place P_2 at 0.2 on the vertical axis because

$$I_{D(N)} = \frac{I_D}{I_{DSS}} = \frac{2\ mA}{10\ mA} = 0.2$$

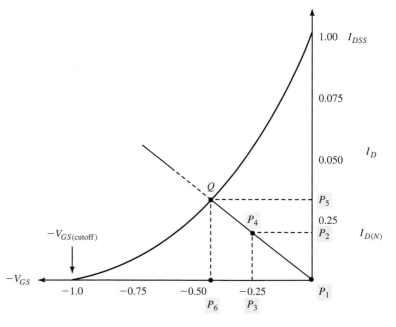

FIGURE 7.16 Self-bias analysis graph for example 7.4

Solution

First, we will determine the amount of voltage that our chosen current would develop when it flows through the source resistor:

$$V_S = 2\,\text{mA} \times 1\,\text{k}\Omega = 2\,\text{V}$$

Then, we will find this voltage on the graph. Dividing 2 V by 8 V gives the point for this voltage: 0.25. The 0.25 represents 25% of -8 V. We mark this point on the horizontal axis and label it as P_3.

Drawing perpendicular lines from the axis of P_2 and P_3 gives us point P4 at the intersection. Now draw a bias line from P_1 to P_4. If necessary, extend the line until it crosses the JFET curve. The point where the bias line and the JFET curve intersect is the Q point for the circuit. In this example, the Q point is at 0.33 along the I_{DSS} axis, which is marked as P_5. The current flowing through the JFET channel is:

$$I_D = 0.33 \times I_{DSS} = 0.33 \times 10\,\text{mA} = 3.3\,\text{mA}$$

Marked as P_6, the voltage at the Q point equals:

$$V_{GS} = 0.42 \times V_{GS(off)} = 0.42 \times -8\,\text{V} = -3.36\,\text{V}$$

The JFET has 0 V at the gate and $+3.36$ V at the source, with 3.3 mA of current flowing through the channel.

PRACTICE PROBLEM 7.4

Analyze the circuit used for example 7.4. Use a value of 820 Ω for the source resistor.

EXAMPLE 7.5

Solve the self-biased circuit shown in Figure 7.17.

The JFET has an I_{DSS} of 16 mA and a $V_{GS(cutoff)}$ of -12 V. The source resistor is 2.2 kΩ. Figure 7.18 shows the graph, and Figure 7.17 shows the circuit.

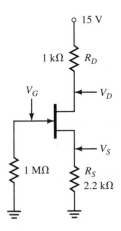

FIGURE 7.17 JFET circuit

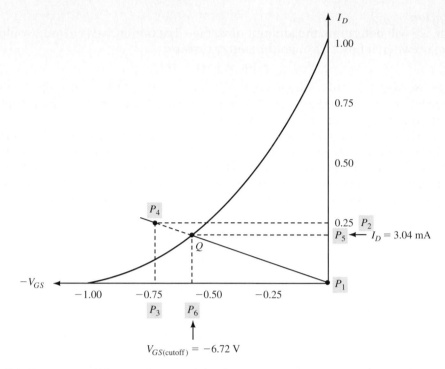

FIGURE 7.18 Self-bias analysis graph for the JFET circuit in Figure 7.17

Solution

Since we are working with a self-biased circuit, P_1 is marked at the origin. The current chosen to plot the self-bias line is 4 mA. This means that we assume that:

$$\frac{4 \text{ mA}}{16 \text{ mA}} = 0.25$$

We locate 0.25 along the I_D axis and mark it as P_2. A current of 4 mA through a 2.2-kΩ resistor will produce:

$$4 \text{ mA} \times 2.2 \text{ k}\Omega = 8.8 \text{ V}$$

The 8.8 V is:

$$\frac{8.8 \text{ V}}{12 \text{ V}} = 0.73 \text{ of } V_{GS(\text{cutoff})}$$

This point is marked as P_3. Lines drawn from P_2 and P_3 intersect at P_4, which is marked on the graph. A bias line is drawn from the origin to P_4. Since it intersects the JFET curve, we do not need to extend the line.

The bias line and the JFET curve intersect at the Q point. The Q point has coordinates of P_5 and P_6, which are marked on the graph. P_5 is 0.19 up the I_D axis. This tells us that:

$$I_D = 0.19 \times 16 \text{ mA} = 3.04 \text{ mA}$$

P_6 is 0.56 out to $V_{GS(\text{cutoff})}$, which gives us:

$$V_{GS} = 0.56 \times -12 \text{ V} = -6.72 \text{ V}$$

EXAMPLE 7.6

Calculate the voltage value for the drain (V_D) in Figure 7.17.

The current flowing through the source resistor equals the drain current. We determined that the current flowing through the source resistor equals 3.04 mA in the last example. This current also flows through the drain resistor. In Figure 7.17, the drain resistor is 1 kΩ.

Solution
The voltage drop across the resistor is:

$$V_{RD} = 3.04 \text{ mA} \times 1000 \text{ } \Omega = 3.04 \text{ V}$$

Subtracting this value from the drain supply voltage value gives us the voltage at the drain terminal:

$$V_D = 15 \text{ V} - 3.04 \text{ V} = 11.96 \text{ V}$$

PRACTICE PROBLEM 7.5

For the circuit shown in Figure 7.17, change the value of R_S to 1.5 kΩ, and calculate the voltage value for the drain.

PRACTICE PROBLEM 7.6

Determine the drain voltage for Figure 7.17 if the drain resistor has a value of 1.5 kΩ. Does this change in the drain resistor value affect the drain current?

✔ **Progress Check**
You have now completed objective 6.

Let's review the steps used to find the Q point of a self-biased JFET circuit.

1. Construct the transconductance curve.

2. Choose any value of I_D that is less than I_{DSS}.

3. Using I_D and R_S, find V_S.

4. Since V_G is zero, the value of V_{GS} must be the same as V_S but with the opposite polarity.

5. From the 0,0 point on the graph through the intersection point for I_D and V_{GS}, draw the bias line that will cut across the transconductance curve.

6. The intersection point of the bias line and the transconductance curve is the Q point of the FET.

Before we leave this section, let us look at Figure 7.19. This figure shows a graphical comparison of gate bias and self-bias. You can see that the vertical spread between Q_1 and Q_2 for the gate bias line is larger than the vertical spread between Q_3 and Q_4 for the self-bias line. That is, the possible values of

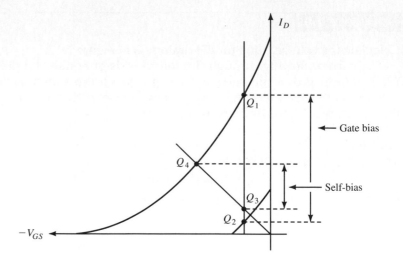

FIGURE 7.19 Comparing the effects of gate bias and self-bias

I_D for a given V_{GS} are less when using self-bias. Self-bias improves predictable JFET performance.

BIASING THE JFET WITH A VOLTAGE DIVIDER

Figure 7.20 shows a JFET biased with a voltage divider. At first glance, this configuration appears to apply a positive voltage to the gate. However, the voltage applied to the source resistor is more positive than the voltage at the gate. As a result, the amount of current flowing through the channel is less than the maximum available current, and the JFET becomes reverse-biased.

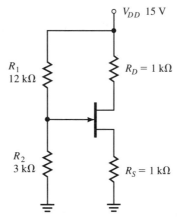

FIGURE 7.20 Voltage divider bias

By looking at Figure 7.21, we can see how voltage divider bias affects the bias line. Placing a Thevenin voltage on the gate of the JFET tips the slope of the line closer to horizontal. In itself, this enhances the predictability of the JFET in a circuit.

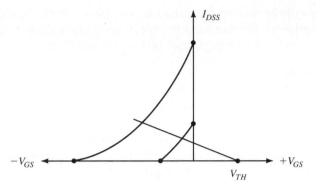

FIGURE 7.21 Bias line for voltage divider bias

Moving back to the circuit shown in Figure 7.20, we can calculate the voltages and currents flowing in the circuit in several different ways. Let us start our analysis of the circuit with example 7.7.

EXAMPLE 7.7

Predict the voltages at the three JFET terminals.

The JFET will not load down the voltage divider because the JFET biasing will reverse its gate diode. This will remain true as long as the amount of current flowing through the device remains less than the I_{DSS}. If the expected current exceeds I_{DSS}, then the circuit will not work as expected. We start our analysis by finding the voltage at the gate lead of the JFET.

Solution
The voltage at the gate lead equals:

$$V_{TH} = \frac{V_{DD} \times R_2}{R_1 + R_2}$$
$$V_{TH} = \frac{15\,V \times 3\,k\Omega}{12\,k\Omega + 3\,k\Omega}$$
$$= 3\,V$$

We will assume a gate-to-source voltage of -2 V so that we can predict the voltage at the top of the source resistor:

$$3\,V - (-2\,V) = 5\,V$$

With 5 V at the top of the source resistor, we can find the corresponding value for the drain current:

$$I_D = I_S = \frac{5\,V}{1\,k\Omega} = 5\ mA$$

Using this information, we can now predict the voltage at the drain terminal:

$$15\,V - (I_D R_D) = 15\,V - 5\,V = 10\,V$$

PRACTICE PROBLEM 7.7

For the circuit shown in Figure 7.20, change the value of the 3-kΩ resistor to 4.7 kΩ, and predict the voltages at the three JFET terminals.

Knowing the three terminal voltages allows us to check for the correct DC operating conditions. We can obtain a more accurate answer by determining the gate-to-source cutoff voltage and the I_{DSS} for the JFET.

EXAMPLE 7.8

Predict the terminal voltages and the drain current using the JFET curve shown in Figure 7.22.

The circuit shown in Figure 7.20 has an I_{DSS} of 20 mA and a $V_{GS(off)}$ of 8 V. Because V_G is not zero, the bias line does not start at the origin. Instead, the line begins along the x axis at V_{TH} in Figure 7.22.

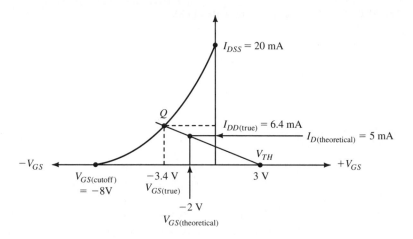

FIGURE 7.22 Analysis of voltage divider bias using a nonnormalized curve

Solution

Step 1: Use the characteristic curve for the JFET. I_{DSS} is the top vertical point of the curve. $V_{GS(cutoff)}$ is at the bottom horizontal point of the curve.

Step 2: Find V_{TH}, which will also be V_G, using the voltage divider method. Since V_{TH} is a positive voltage, plot it to the right of the intersection of the vertical and horizontal axis.

$$V_{TH} = \frac{R_2 \times V_{DD}}{R_1 + R_2}$$

$$= \frac{3\ k\Omega \times 15\ V}{12\ k\Omega + 3\ k\Omega}$$

$$V_G = 3\ V$$

Step 3: Find theoretical V_S by adding 2 V_{GS} to V_{TH}.

$$V_S = V_{TH} + V_{GS}$$

$$= 3\ V + 2\ V$$

$$V_S = 5\ V$$

Step 4: Find the corresponding theoretical I_D by using Ohm's law.

$$I_D = \frac{V_S}{R_S}$$

$$= \frac{5\ V}{1\ k\Omega}$$

$$I_D = 5\ mA$$

Step 5: Plot a bias line from V_{TH} to the intersection of I_D and V_{GS} so that the bias line also intersects the curve.

Step 6: Find the true I_D by drawing a horizontal line to the I_D axis. Notice that the line intersects at 6.2 mA.

Step 7: Find the true V_{GS} by drawing a vertical line down to the V_{GS} axis. Notice that the line intersects at -3.4 V.

Step 8: Find the true V_S.

$$V_S = I_D \times R_S$$
$$V_S = 6.4 \text{ V}$$

Step 9: Find the true V_D.

$$V_D = V_{DD} - I_D R_D$$
$$V_D = 15 \text{ V} - 6.4 \text{ mA} \times 1 \text{ k}\Omega$$
$$V_D = 8.6 \text{ V}$$

EXAMPLE 7.9

Analyze the circuit shown in Figure 7.23 using the graph in Figure 7.24.

The circuit shown in Figure 7.23 uses voltage divider bias. As usual, we will use -2 V for a theoretical V_{GS}.

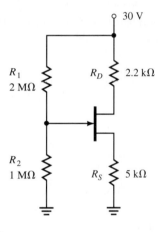

FIGURE 7.23

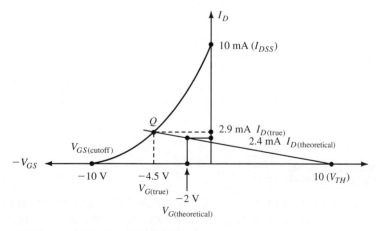

FIGURE 7.24 Graphic analysis of voltage divider bias

Solution

Step 1: Use the characteristic curve for the JFET. I_{DSS} is equal to 10 mA and $V_{GS(cutoff)}$ is equal to -10 V.

Step 2: Find V_{TH} (which will also be V_G) and plot it.

$$V_{TH} = \frac{R_2 \times V_{DD}}{R_1 + R_2}$$

$$= \frac{1\ M\Omega \times 30\ V}{2\ M\Omega + 1\ M\Omega}$$

$$V_G = 10\ V$$

Step 3: Find theoretical V_S by adding 2 V_{GS} to V_{TH}.

$$V_S = V_{TH} + V_{GS}$$
$$= 10\ V + 2\ V$$
$$= 12\ V$$

Step 4: Find the corresponding theoretical I_D by using Ohm's law.

$$I_D = \frac{V_S}{R_S}$$

$$= \frac{12\ V}{5\ k\Omega}$$

$$= 2.4\ mA$$

Step 5: Plot a bias line from V_{TH} through the intersection of theoretical I_D and V_{GS} so that the bias line also intersects the curve.

Step 6: Find the true I_D by drawing a horizontal line to the I_D axis. Notice that the line intersects at 2.9 mA.

Step 7: Find the true V_{GS} by drawing a vertical line down to the V_{GS} axis. Notice that the line intersects at -4.5 V.

Step 8: Find the true V_S.

$$V_S = I_D \times R_S$$
$$= 2.9\ mA \times 5\ k\Omega$$
$$= 14.5\ V$$

Step 9: Determine V_D by subtracting the drain resistor voltage (V_{RD}) from V_{DD}.

$$V_D = V_{DD} - (I_D \times R_D)$$
$$= 30\ V - (2.9\ mA \times 2.2\ k\Omega)$$
$$= 23.6\ V$$

PRACTICE PROBLEM 7.8

Let $V_{GS(off)}$ equal -8 V and I_{DSS} equal 12 mA for the JFET seen in Figure 7.23. Recompute and redraw the bias line. Determine the new values for V_{GS} and I_D.

EXAMPLE 7.10

Determine the power dissipated (P_D) by the JFET shown in Figure 7.23.

From the previous example, we determined that 2.9 mA of current flows through the JFET.

Solution

The voltage dropped by the JFET can be determined by subtracting the source voltage from the drain voltage:

$$V_D - V_S = 23.62\,V - 14.5\,V = 9.12\,V$$

Using Watt's law, we find that the amount of dissipated power is:

$$9.12\,V \times 2.9\,mA = 26.45\,mW$$

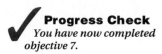

Progress Check
You have now completed objective 7.

REVIEW SECTION 7.2

1. The actual operating characteristics of a JFET are unpredictable because of the process of:
   (a) Manufacturing
   (b) Hookup
   (c) Biasing
2. There are no variations observed in a batch of JFETs with the same part numbers. True or false?
3. Gate biasing of a JFET is similar to _____ biasing for BJTs.
4. Leakage current is normally referred to as:
   (a) I_{GS}
   (b) I_{GSS}
   (c) I_{GG}
5. When a JFET is self-biased, the gate ties to a DC ground through a:
   (a) Resistor
   (b) Capacitor
   (c) Diode
6. In a self-biased circuit the gate voltage is equal to zero. True or false?
7. With the bias line closer to the horizontal, as with the voltage divider bias, a JFET circuit becomes less predictable. True or false?
8. When the voltage applied to the source resistor is more positive than the voltage at the gate, the JFET is:
   (a) Neutral-biased
   (b) Forward-biased
   (c) Reverse-biased

7.3 CURRENT SOURCE BIAS

Current source bias will produce a stable I_D, which will be independent of the JFET used and any changes in specifications. A circuit diagram of a JFET biased with a current source is shown in Figure 7.25.

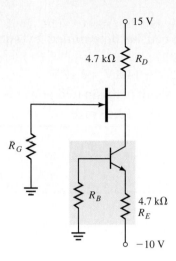

FIGURE 7.25 A JFET circuit using current source bias

EXAMPLE 7.11

In this circuit, we see a bipolar transistor being used as a current source. The first step in analyzing this circuit is to determine what the BJT is doing. The BJT in Figure 7.25 is emitter-biased. This means that to a first approximation, the base is at 0 V with respect to ground, and the emitter is at -0.7 V with respect to ground.

Solution
We use Ohm's law to determine the current flowing in the emitter:

$$-10\text{ V} - \left(-0.7\text{ V}\right) = \frac{-9.3\text{ V}}{4.7\text{ k}\Omega} = 1.98\text{ mA}$$

This current flows out the collector and into the JFET. Therefore,

$$I_E = I_S = I_D$$

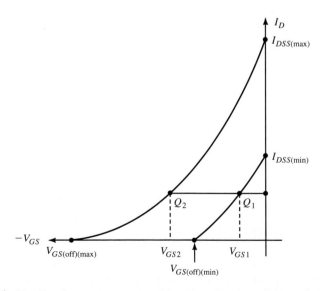

FIGURE 7.26 Bias line for a current source bias. Note that Q_1 and Q_2 produce the same I_D.

This current will establish a V_{GS} from gate to source in the JFET. The bias line established by a current source can be seen in Figure 7.26. Note that it does not matter whether the maximum or minimum JFET curve is used. The drain current will be the same, and only V_{GS} will change. This is the best possible bias for a JFET.

When establishing current source bias, a circuit designer needs to keep two things in mind. I_E must be less than I_{DSS}. To ensure that this is so, a circuit designer must build the circuit around the minimum guaranteed I_{DSS}; hence the lower curve is used. (This should be done for the other bias arrangements as well.)

Second, the bipolar must remain in forward-reverse bias for it to act like a constant current source. Let us suppose that the lower curve has been used, as it should be, in Figure 7.26. The amount of current through the JFET causes a V_{GS} of -2 V to be developed.

Since the gate of the JFET is at ground, the source terminal of the JFET is now at $+2$ V to ground. This voltage is also the voltage at the collector of the bipolar transistor. With the base of the bipolar transistor at ground, V_B is equal to 0 V, and V_C is equal to $+2$ V as already discussed. The more positive voltage is on the collector terminal. Consequently, the collector-base diode is reverse-biased and will act as a current source.

REVIEW SECTION 7.3
1. _____ _____ bias will produce a stable I_D.
2. In order to serve as a constant current source, the BJT must be _____ biased.
3. Describe the bias line on a JFET using current source bias.
4. When using current source biasing, the JFET gate is tied to (V_{DD}, ground).

7.4   TRANSCONDUCTANCE

Throughout this chapter, we have relied on the JFET curve to solve problems and analyze circuits. We define this curve as the transconductance curve. When working with JFETs, **transconductance** is the measure of how the gate-to-source voltage affects the amount of drain current flowing through the device. Generally, the transconductance of a JFET will equal a few thousand micromhos at a few milliamps.

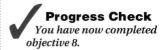

Progress Check
You have now completed objective 8.

NOTE: *Mhos is a measure of conductance, as is siemens. 1 mho = 1 siemen. Siemen is the newer designation. However, the data books currently on the market use the term mhos. It is the author's decision to use mhos as a measure of conductance because students will probably find mhos used in their data books.*

Two different measurements of transconductance in JFETs exist. The transconductance of the device at the Q point is called g_m, while g_{mo} is the transconductance of the JFET when the drain current reaches saturation (I_{DSS}). Transconductance is always specified when the gate-to-source voltage equals zero or when the drain current saturates. The value of g_m depends on the amount of drain current flowing at a given time. Some variation in gain will occur as the amount of drain current begins to change.

Mathematically, we can express the relationship of g_{mo} and g_m as:

$$g_m = g_{mo}\left(1 - \frac{V_{GS}}{V_{GS(off)}}\right)$$

We use this calculation when we find the voltage gain and analyze the AC properties of the JFET. Because the values of transconductance can vary over a wide range, we will always use the minimum values for transconductance when we analyze JFET circuits.

Let us break transconductance down even more. **Conductance (G)** is a measure of how easily current flows through a resistance. It is the inverse of resistance and can be expressed as:

$$G = \frac{1}{R}$$

In this equation, G represents the conductance measured in mhos.

If we have a 100-Ω resistor, the conductance of the resistor equals:

$$G = \frac{1}{100} = 10 \text{ millimhos}$$

The value of the resistance fixes the value of the conductance. No control element exists within the resistor that could change its conductance. When we move to the JFETs, however, the gate-to-source voltage does act as the control element.

DETERMINING g_{mo} OR TRANSCONDUCTANCE AT I_{DSS}

We can find the value for transconductance at the saturated drain current with two different methods. Sometimes the manufacturer of the JFET will list a value for g_{mo} on the device specification sheet. Unfortunately, not every manufacturer will use g_{mo} to represent transconductance at the maximum drain current.

Other manufacturers may substitute a value called **forward transfer admittance,** designated as **yfs,** for the transconductance value. In this instance, admittance is the reciprocal of impedance. As an example, the data sheet for an MPF102 JFET lists a value between 2000 μmhos and 7500 μmhos for the forward transfer admittance.

If you cannot find or recognize a value for g_{mo} on a specification sheet, a method exists for calculating the value. The formula:

$$g_{mo} = \frac{2(I_{DSS})}{V_{GS(off)}}$$

resembles an inverted version of Ohm's law. R = E/I, inverted, is 1/R = I/E.

EXAMPLE 7.12

Determine the transconductance for a JFET that has the minimum values of 2 mA for I_{DSS} and -2 V for $V_{GS(off)}$.

Drop the negative sign for $V_{GS(off)}$ when using the formula to find the transconductance.

Solution

Applying the g_{mo} formula, we find that:

$$\frac{2 \times I_{DSS}}{V_{GS(off)}} = \frac{4 \text{ mA}}{2 \text{ V}} = 2 \text{ millimhos} = 2000 \text{ μmhos}$$

PRACTICE PROBLEM 7.9

Determine the g_{mo} for a JFET that has the minimum values of 4 mA for I_{DSS} and -5 V for $V_{GS(off)}$.

FINDING g_m OR TRANSCONDUCTANCE AT THE Q POINT

Once g_{mo} is known, then g_m, which is the transconductance at the Q point, may be found. In example 7.13, we will find g_m for the circuit in Figure 7.23.

EXAMPLE 7.13

Determine quiescent transconductance using the information provided in Figures 7.23 and 7.24.

The minimum values for the JFET are: $I_{DSS} = 10$ mA and $V_{GS} = -10$ V.

Solution

First, we find g_{mo} by using the g_{mo} equation.

$$g_{mo} = \frac{2(I_{DSS})}{V_{GS}}$$
$$= \frac{2(10 \text{ mA})}{10 \text{ V}}$$
$$= \frac{20 \text{ mA}}{10 \text{ V}}$$
$$= 2000 \text{ μmhos}$$

Second, we determine g_m using the g_m formula.

$$g_m = g_{mo}\left[1 - \left(\frac{V_{GS}}{V_{GS(cutoff)}}\right)\right]$$
$$= 2000 \text{ μmhos} \times \left[1 - \left(\frac{4.5 \text{ V}}{10 \text{ V}}\right)\right]$$
$$= 1100 \text{ μmhos}$$

Thus, the quiescent transconductance of the circuit is 1100 μmhos.

As we move into the next section of this chapter, we will learn that we can find a value for AC voltage gain by applying the quiescent transconductance value.

7.5   USING JFETS AS AMPLIFIERS

High input impedance also makes the JFET valuable as an amplifier. In addition to the high input impedance, the JFET also exhibits low feedback capacitance. Having a low input capacitance becomes important at high frequencies. A JFET amplifier circuit can provide voltage gain for a low cost and at low-drain currents. However, the circuit also has the disadvantages of having a high output impedance, poor linearity, and sometimes unpredictable gain.

Figure 7.27 shows a simple AC model of a JFET. No current flows from the gate terminal. Remembering that the drain terminal works much like the collector of a bipolar junction transistor, we have a constant current source. JFETs configured as common-source amplifiers or source followers work like common-emitter and common-collector transistor amplifiers, respectively.

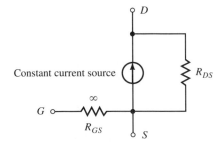

FIGURE 7.27 Equivalent JFET circuit with an infinite gate-to-source resistance

COMPARING THE VOLTAGE GAINS OF BJTs AND JFETs

A JFET amplifier has one major shortcoming when compared with the bipolar transistor amplifier. The transconductance of a JFET amplifier is much lower than the transconductance of a transistor amplifier at the same current. Table 7.1 shows how the two devices would compare given a quiescent current of 1 mA and a 15-V power supply.

Table 7.1		
	BJT Amplifier	**JFET Amplifier**
Voltage gain	-300	-7.5
Transconductance	40,000 μmhos	1000 μmhos

For a further illustration of the gain difference, let us calculate and compare voltage gains of a bipolar transistor and junction field effect transistor. Voltage gain in a bipolar transistor is calculated as:

$$A_V = \frac{r_c}{r'_e}$$

Converting the AC resistance of the emitter junction to a conductance value will allow us to begin our comparison. The AC transconductance of a bipolar transistor equals 1 divided by the AC emitter resistance. If we use a value of 25 Ω for the AC emitter resistance, we have a transconductance of:

$$g_{m'e} - \frac{1}{r'_e} = \frac{1}{25\,\Omega} = 40,000\,\mu\text{mhos}$$

Rewriting this formula based on the AC transconductance of the emitter gives us:

$$A_V = g_{m'e} \times r_c$$

For the JFET, the voltage gain equals the transconductance multiplied by the drain resistance or:

$$A_V = g_m \times r_d$$

Assuming that the drain resistance equals the collector resistance, we can establish a ratio of gains. Most JFETs will have an average transconductance of 5000 to 7000 μmhos. For this example the ratio would show:

$$\frac{g_m \text{ for bipolar transistor}}{g_m \text{ for JFET}} = \frac{40,000}{7000}$$

The bipolar transistor produces 5.7 times as much gain as the JFET.

In many cases, the AC emitter resistance is in the 5-Ω range. At 5 Ω, the AC transconductance of the transistor emitter equals:

$$g_{m'e} = \frac{1}{r'_e} = 200,000\,\mu\text{mhos}$$

The ratio then becomes:

$$\frac{200,000\,\mu\text{mhos}}{7000\,\mu\text{mhos}} = 28.6$$

Again, the transistor has a superior gain factor. For this reason, circuit designs will not usually employ JFETs as simple amplifiers unless the design requires a high input impedance, low input current, or low feedback capacitance.

Because single JFET amplifiers produce lower gain than bipolar transistors, some designs will use multiple stages of JFET amplifiers. However, most designs will employ a combination of JFET stages and bipolar transistor stages. These designs will feature the JFET as the input stage and the bipolar transistor as the output stage. While the JFET gives a high input impedance with a minimum amount of noise, the bipolar transistor provides higher gain and a lower output impedance.

FINDING THE VOLTAGE GAIN OF A JFET AMPLIFIER

Although we are working with a different type of semiconductor device, the method for calculating voltage gain has not changed. Voltage gain equals the output voltage divided by the input voltage. With a common-source JFET amplifier, the input voltage is the AC input voltage at the gate.

To find the output voltage, we need to find a value for the AC drain resistance and the AC drain current. We can find the relationship between input voltage and the AC drain current by using the quiescent transconductance value:

$$g_m = \frac{i_d}{V_{in}}$$

Rearranging the equation, we find that:

$$V_{in} = \frac{i_d}{g_m}$$

Substituting the voltage gain equation for the gate-to-source value, we have:

$$A_V = \frac{V_{(out)}}{V_{(in)}} = \frac{(i_d \times r_d)}{\dfrac{i_d}{g_m}}$$

Inverting and multiplying, we get:

$$A_V = (i_d \times r_d) \times \frac{g_m}{i_d}$$

The values for i_d cancel, leaving us with:

$$A_V = g_m \times r_d = \text{voltage gain}$$

Voltage gain for a JFET amplifier equals the quiescent transconductance of the amplifier multiplied by its AC drain resistance.

USING THE JFET AS A COMMON-SOURCE AMPLIFIER

Figure 7.28 shows a **common-source JFET amplifier** using self-bias. The input voltage is applied through a coupling capacitor to the gate, while the output signal is passed across the load resistor. When we consider AC operation, the load resistor parallels the drain resistor. Since the source resistor is bypassed, the source of the JFET is at AC ground. During AC operation, this makes the source common to the gate and the drain.

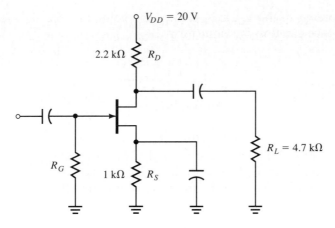

FIGURE 7.28 A common-source JFET amplifier stage

Find the unloaded and loaded gain for the JFET shown in Figure 7.28.

Figure 7.29 shows the transconductance curve for the JFET shown in Figure 7.28; $I_{DSS} = 16$ mA and $V_{GS(cutoff)} = -12$ V.

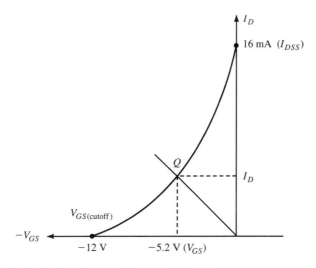

FIGURE 7.29 Analysis of the JFET used in Figure 7.28

Solution
We start by calculating g_{mo}:

$$g_{mo} = 2 \times \frac{I_{DSS}}{V_{GS(cutoff)}}$$
$$= 2 \times \frac{16 \text{ mA}}{-12 \text{ V}}$$
$$= 2667 \text{ } \mu\text{mhos}$$

Figure 7.29 shows the self-bias line for the JFET and indicates the intersection of the line with the letter Q. From this intersection point, a perpendicular line is drawn to the V_{GS} axis, giving us a value of -5.2 V. Once we know g_{mo} and V_{GS} we can find the value for g_m:

$$g_m = g_{mo}\left[1 - \left(\frac{V_{GS}}{V_{GS(cutoff)}}\right)\right]$$

$$= 2667\,\mu\text{mhos}\left[1 - \left(\frac{-5.2\text{ V}}{-12\text{ V}}\right)\right]$$

$$= 1511\,\mu\text{mhos}$$

We need to know the loaded and unloaded drain resistance as well as g_m to find voltage gain (A_v). Unloaded drain resistance is simply the drain resistor (R_D). The loaded resistance is the drain resistor in parallel with the load resistor.

$$r_d = R_D \parallel R_L$$

$$r_d = 2.2\text{ k}\Omega \parallel 4.7\text{ k}\Omega$$

$$r_d = 1.5\text{ k}\Omega$$

Unloaded voltage gain is g_m multiplied by R_D.

$$A_v = g_m \times R_D$$
$$= 1511\,\mu\text{mhos} \times 2.2\text{ k}\Omega$$
$$= 3.32$$

Loaded voltage gain is g_m multiplied by r_d.

$$A_v = g_m \times r_d$$
$$= 1511\,\mu\text{mhos} \times 1.5\text{ k}\Omega$$
$$= 2.27$$

Even though the JFET does not supply a large amount of gain, the circuit would still work well as an input stage of an amplifier. Since the input impedance of the gate is almost infinite, we can set the input impedance of the stage by selecting the gate resistor that would give the desired match. The output impedance of the circuit equals the value given when the drain resistor parallels the AC drain-to-source resistance.

PRACTICE PROBLEM 7.10

Let I_{DSS} equal 10 mA and $V_{GS(off)}$ equal -10 V. Using the procedure seen in example 7.14, recalculate the gain for the circuit shown in Figure 7.28.

EXAMPLE 7.15

Determine the effects the drain-to-source resistance has on the gain calculated in example 7.14 if the resistance equals 10 kΩ.

Turning to Figure 7.30, we see that the drain-to-source resistance parallels the drain resistance. In example 7.14, we found that the drain resistance equals 1.5 kΩ.

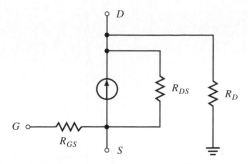

FIGURE 7.30 Relationship of drain-to-source resistance to an external drain resistance

Solution
Since the drain-to-source resistance of 10 kΩ parallels the 1.5-kΩ resistance, we have:

$$10\,k\Omega \parallel 1.5\,k\Omega = 1.30\,k\Omega$$

The gain of the circuit drops to:

$$A_V = 1511\ \mu\text{mhos} \times 1.3\,k\Omega = 1.96$$

EXAMPLE 7.16

Determine the gain of the circuit diagrammed in Figure 7.31, given:

$$I_{DSS} = 20\ \text{mA}$$
$$V_{GS(\text{cutoff})} = -10\ \text{V}$$

The circuit shown in Figure 7.31 uses current source bias. First, we need to find the emitter current of the bipolar transistor that acts as a current source:

$$I_E = \frac{-10\,\text{V} - (-0.7\,\text{V})}{3.3\,k\Omega} = 2.82\ \text{mA}$$

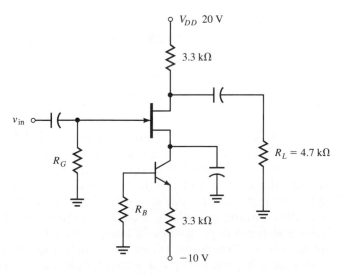

FIGURE 7.31 Common-source JFET amplifier with current source bias

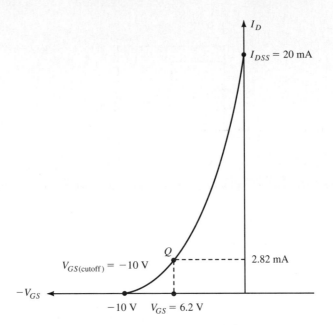

FIGURE 7.32 Graphic analysis of common-source amplifier

The JFET will adjust the gate-to-source voltage so that it produces an equal amount of current.

We draw the bias line representing the 2.82 mA as shown in Figure 7.32. The intersection of the bias line and the JFET curve is the Q point. Draw a line straight down from the Q point to V_{GS}. V_{GS} is equal to 6.2 V.

Next, we will determine a value for g_{mo}:

$$g_{mo} = 2 \times \frac{I_{DSS}}{V_{GS(cutoff)}} = \frac{40\ mA}{10\ V} = 4000\ \mu mhos$$

With that value, we find a value for g_m:

$$g_m = g_{mo}\left(1 - \frac{V_{GS}}{V_{GS(cutoff)}}\right) = 4000\ \mu mhos\left(1 - \frac{6.2\ V}{10\ V}\right) = 1520\ \mu hmos$$

The unloaded voltage gain equals:

$$A_{v(unloaded)} = g_m \times R_D = 1520\ \mu mhos \times 3.3\ k\Omega = 5.0$$

while the loaded voltage gain is:

$$A_{v(loaded)} = g_m \times r_d = 1520\ \mu mhos \times 3.3\ k\Omega \,\|\, 4.7\ k\Omega = 2.95$$

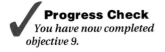

Progress Check
You have now completed objective 9.

USING THE JFET AS A COMMON-DRAIN AMPLIFIER

Figure 7.33 shows an amplifier configuration called the **common drain.** Once again, the input signal is applied to the gate. This time, though, the output is taken from the source, and the drain connects directly to the drain voltage supply. Since the drain voltage supply acts as an AC ground, the drain becomes the common point for the other two terminals. Because of its configuration, we also call this a **source follower.** As in the common collector or emitter follower configuration, the output gain is less than unity.

When the output voltage is measured across the source resistor, it equals the source voltage. Thus, the input voltage measures higher than the output

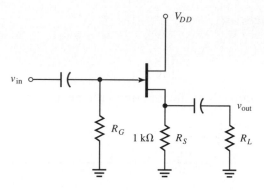

FIGURE 7.33 Common-drain JFET amplifier

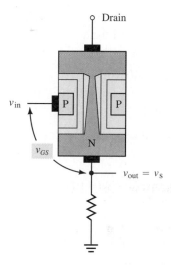

FIGURE 7.34 Block diagram of common-drain operation. Note: N_{GS} across the reverse-biased gate diode.

voltage by the amount of gate-to-source voltage. Figure 7.34 shows this relationship and the lower part of the source follower circuit.

Since voltage gain equals the output voltage divided by the input voltage, we have:

$$A_V = \frac{V_S}{V_S + V_{GS}}$$

You may see that the best gain figures occur at unity. Changing the last equation, we find that:

$$A_V = \frac{(i_d \times r_s)}{(i_d \times r_s) + \left(\dfrac{i_d}{g_m}\right)}$$

Factoring out the i_d, we arrive at:

$$A_V = \frac{(i_d \times r_s)}{i_d\left(r_s + \dfrac{1}{g_m}\right)}$$

Canceling out the i_d gives us:

$$A_V = \frac{r_s}{\left(r_s + \dfrac{1}{g_m}\right)}$$

Let us see how we can apply this information to the next example.

EXAMPLE 7.17

Determine the AC output voltage if the input voltage equals 100 mV.
The g_m of the JFET shown in Figure 7.33 equals 5000 μmhos.

Solution
Using the equation that we developed, we find that:

$$A_V = \frac{1\text{ k}\Omega}{\left(1\text{ k}\Omega + 1/5000\text{ μmhos}\right)}$$

$$A_V = \frac{1\text{ k}\Omega}{\left(1\text{ k}\Omega + 200\right)}$$

$$A_V = \frac{1\text{ k}\Omega}{\left(1.2\text{ k}\Omega\right)} = 0.833$$

The output voltage equals:

$$V_O = 0.833 \times 100\text{ mV} = 83.3\text{ mV}$$

PRACTICE PROBLEM 7.11

For the circuit shown in Figure 7.33, change the value of g_m to 3000 μmhos, and determine the AC output voltage if the input voltage equals 100 mV.

Progress Check
You have now completed objective 10.

USING THE JFET AS A COMMON-GATE AMPLIFIER

Figure 7.35 shows another JFET amplifier configuration called the **common gate.** With this circuit, the input injects into the source, while the output measures at the drain. Voltage gain for a common-gate amplifier equals:

$$A_v = g_m \times r_d$$

Since the gate terminal is bypassed and becomes an AC ground, it also works as a common terminal for AC signals. With a common-gate amplifier, we no longer have an infinite input impedance.

The input impedance equals the gate-to-source voltage, or the AC voltage measured across the reverse-biased gate diode divided by the input, or source, current. In equation form, this would appear as:

$$Z_{in(gate)} = \frac{V_{in}}{i_{in}}$$

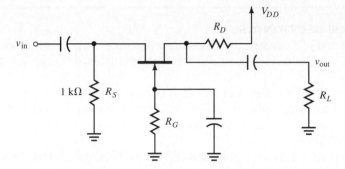

FIGURE 7.35 Common-gate JFET circuit and the lower part of the source follower circuit

or

$$Z_{in(gate)} = \frac{V_{GS}}{i_s}$$

Since the source current equals the transconductance multiplied by the gate-to-source voltage, we can change the equation to:

$$Z_{in(gate)} = \frac{V_{GS}}{g_m \times V_{GS}}$$

which reduces to:

$$Z_{in(gate)} = \frac{1}{g_m}$$

EXAMPLE 7.18

Determine the input impedance of the JFET if the transconductance equals 5000 μmhos.

Solution

$$Z_{in(gate)} = \frac{1}{g_m} = \frac{1}{5000 \text{ μmhos}} = 200 \text{ } \Omega$$

EXAMPLE 7.19

Determine the input impedance to the stage shown in Figure 7.35 if the source resistor has a value of 1 kΩ.

The source resistor parallels the input impedance.

Solution

$$Z_{in} = \frac{R_S \times Z_{in(gate)}}{R_S + Z_{in(gate)}}$$

$$\frac{1 \text{ k}\Omega \times 200 \text{ } \Omega}{1 \text{ k}\Omega + 200 \text{ } \Omega} = 0.167 \text{ } \Omega$$

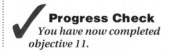

Progress Check
You have now completed objective 11.

1. When a JFET is set up as a common-source amplifier it works like a BJT configured as a common-collector transistor amplifier. True or false?
2. When a JFET rather than a BJT is used as an amplifier, a major shortcoming is observed; the problem is that a JFET amplifier has a _____ that is much lower than that of a transistor amplifier.
3. A way around the problem of JFET amplifiers producing less gain than BJT amplifiers is to use a _____ circuit design.
4. An advantage to designing an amplifier circuit with a JFET at the source and a BJT at the output is:
   (a) High input impedance and low output impedance
   (b) Equal input and output impedance
   (c) Low input impedance and high output impedance
5. Because we are using a different semiconductor device, we must use a different way to compute the voltage gain of a JFET. True or false?
6. The voltage gain of a JFET amplifier is equal to the _____ voltage divided by the _____ voltage.
7. When a JFET is self-biased the input signal passes through a coupling capacitor connected to the _____, while the output voltage can be measured across the load resistor.
8. Another name for a JFET used as a common-drain amplifier is a _____ _____.

7.6 OTHER APPLICATIONS FOR JFETs

SOURCE FOLLOWERS

Because of their high input impedance, JFETs work well when configured as the source follower shown in Figure 7.36. In many cases, an electronic device will require a following stage that does not draw a large amount of current.

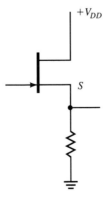

FIGURE 7.36 A JFET configured as a source follower

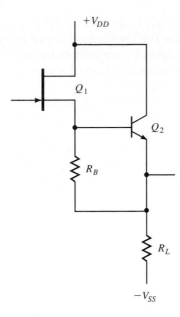

FIGURE 7.37 Adding a bipolar transistor as the current source

Often, the source follower configuration will appear in the input stages of electronic test equipment such as oscilloscopes. In other situations, the source follower also works as an input buffer.

JFET source followers have a high output impedance that may degrade the performance of a stage. Because the output impedance can form a voltage divider with a load resistor, the output voltage may be even less as part of the output voltage is dropped across the output impedance. As a result, the JFET source follower has an uncontrollable DC offset voltage. This condition will produce distortion within the circuit.

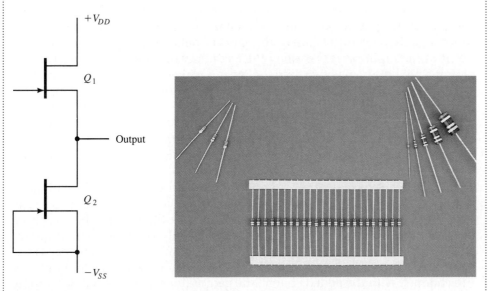

FIGURE 7.38 Matched pair of JFETs. (*Photograph Courtesy of Allen-Bradley*)

Replacing the load resistor with a semiconductor device such as a diode, bipolar junction transistor, or another JFET reduces the voltage offset. Figure 7.37 shows the addition of the improved current source in the form of a transistor. Using a transistor as the current source provides a fairly constant current source along with a low output impedance.

However, the offset voltage does not completely reduce to zero and can still pose problems. Using a matched JFET pair, as shown in Figure 7.38, cuts the offset voltage to zero. When the JFET operates at a current less than the drain saturation current, it provides better linearity and acts as a good current source. Matched JFET source followers often appear as input buffers because of their low offset voltage and high input impedance.

VARIABLE RESISTORS

In a graph of the drain current of the JFET versus its drain source voltage, the curves nearly resemble straight, sloped lines when the voltage value is less than the gate-to-source voltage minus the threshold voltage. We can define the **threshold voltage** as the amount of gate-to-source voltage present when drain current begins to flow. Figure 7.7 shows a set of graphed curves. In addition to appearing as straight and sloped lines, they indicate a resistive characteristic. This means that a JFET will function as a voltage-controlled resistor for small signals of either polarity. The JFET behaves much like a linear resistor.

Resistance values of JFETs may vary from fractions of ohms to nearly infinity. JFETs used for power applications may have an internal resistance of 0.1 Ω. One common application for a JFET variable resistor is the automatic gain control (AGC) in electronic stereo or television equipment. An AGC control automatically adjusts the gain of an amplifier.

LINEAR AND LOGIC SWITCHES

JFETs work well as linear switches. However, most switching designs allow for the possibility of gate conduction within the JFET. Without some safeguards, signals at the gate will allow conduction to occur at the gate. Keeping the gate below ground potential will keep the JFET in the ohmic region. If the input signals become negative, the gate is held below the most negative point of the input cycle. To allow JFET conduction, the control input operates at a point more positive than the most positive point of the input cycle.

JFETs also work well in switching a load to ground. In this case, the JFET functions as a logic switch. Effectively, the device is either on or off, that is, either saturated or cutoff. Since the JFET draws no current from the logic device at its input, the JFET works well as an interface between logic devices and high-current or high-voltage loads. As an example, a single JFET driven by a logic device can switch loads of 10 A or more.

JFET switches have several limitations. In the on state, a JFET has an internal resistance of 25 to 200 Ω. Along with stray capacitance values within the device, the resistance forms a low-pass filter. The filter limits the switching speed of the JFET to frequencies of a few megahertz or less. In addition, JFETs can add transients to a signal during the off and on sequence. **Transients** are produced when a control signal at the gate capacitively couples to the channel. With the added charge, the gate suddenly steps from one supply voltage to another and transfers the charge to the channel.

TROUBLESHOOTING

Usually, you will find troubleshooting JFET circuits easier than troubleshooting bipolar transistor circuits. As with all the electronic circuits that we have studied, check the power supply voltages and the input signals to the stage. After establishing the quality of those voltages and signals, begin to troubleshoot the JFET stage.

NORMAL JFET OPERATING CONDITIONS

Figure 7.39 shows a JFET circuit using voltage divider bias. Under normal operating conditions, the current through resistor R_1 equals the amount of current through resistor R_2. In addition, the drain current equals the source current. Also, no current flows out of the JFET gate.

CONDITIONS WITH A SHORTED JFET JUNCTION

If the JFET junction shorted, current would flow at the gate terminal. In turn, the current flowing through resistor R_1 would be a different amount than the amount of current flowing through resistor R_2. Also, the source current would rise to a level close to saturation or I_{DSS}. Finally, the drain current would not equal the source current because a portion of the source current would flow at the shorted gate.

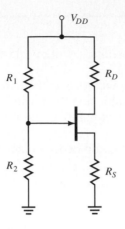

FIGURE 7.39 JFET circuit using voltage divider bias

CONDITIONS WITH AN OPEN JFET JUNCTION

When a JFET junction opens, the depletion region cannot form. Without the depletion region, the JFET cannot control current. As a result, the gate-to-source voltage will not equal 0 V. Nevertheless, the drain current will reach the saturation level.

BIAS RESISTOR PROBLEMS

If resistor R_1 opens, the bias changes from voltage divider bias to self-bias. Unfortunately, the circuit will seem to work. In this instance, compare your predicted voltages with the actual voltages. With an open bias resistor R_1, no current will flow through resistor R_2, and the DC voltage on the gate will equal 0 V. The DC voltage on the drain will rise to a higher than expected level.

If resistor R_2 opens, forward bias will damage the JFET. The voltage on the gate will equal the supply voltage since no complete path for current to flow through resistor R_1 exists. When resistor R_2 opens, the JFET junction will open also.

If the drain resistor opens, we lose our source and drain currents. Voltage at the drain will fall to zero. If the source resistor opens, we again lose the source and drain currents. However, the voltage on the drain will equal the power supply voltage.

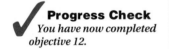

Progress Check
You have now completed objective 12.

TROUBLESHOOTING QUESTIONS

1. With a shorted JFET, current flows at its maximum rate at the source. True or false?
2. The depletion region cannot form with an open JFET junction. True or false?
3. When the depletion region is not formed the JFET works as an excellent current controller. True or false?
4. If the top bias resistor opens, in a voltage divider bias JFET, the circuit will appear to work normally. True or false?
5. When the drain resistor opens, voltage at the drain will go to the maximum value. True or false?

WHAT'S WRONG WITH THIS CIRCUIT?

The following circuit is a JFET source follower amplifier. The normal readings are:

	DC	AC
TP1	0 V	10 mV
TP2	0 V	10 mV
TP3	2.5 V	8.4 mV
TP4	0 V	8.4 mV

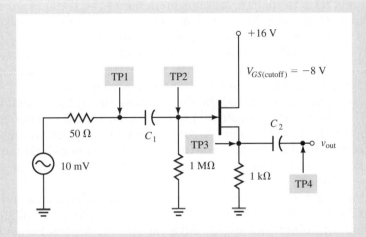

Determine the faults for the following readings:

1.

	DC	AC
TP1	0 V	10 mV
TP2	0 V	0 mV
TP3	2.5 V	0 mV
TP4	0 V	0 mV

2.

	DC	AC
TP1	0 V	0 mV
TP2	0 V	0 mV
TP3	2.5 V	0 mV
TP4	0 V	0 mV

3.

	DC	AC
TP1	0 V	10 mV
TP2	0 V	10 mV
TP3	8 V	0 mV
TP4	0 V	0 mV

SUMMARY

Junction field effect transistors are semiconductor devices that differ from bipolar junction transistors in several ways. Bipolar transistors are current-controlled devices; JFETs control drain current through an electrical field with an applied voltage. Also, JFETs have higher input impedances than traditional transistors. In addition, and perhaps most important, the junction field effect transistor is a unipolar device that does not require two different types of charge carriers to operate.

Two basic types of JFETs—called N-channel and P-channel—exist. The designations refer to the construction material of the JFET. Electrons carry current in an N-channel JFET, while holes carry current in the P-channel device. Every JFET has three sections, called the source, drain, and gate, which correspond to the collector, emitter, and base of a BJT. Charge carriers enter at the source, flow through the channel, and exit at the drain. Voltage at the gate determines the amount of current flowing through the JFET.

JFETs work in a mode called the depletion mode. In the depletion mode, the JFET conducts with the gate tied to the source. Cutoff is produced by reverse-biasing the gate. Since the N-channel JFET is a depletion mode device and its diode should not be forward-biased, the gate cannot be more than 0.5 V more positive than the source. A P-channel JFET would have the opposite polarity voltages.

Junction field effect transistors exhibit a quality called transconductance. Transconductance is the measure of how the gate-to-source voltage affects the amount of current flowing through the device. A constant gate-source voltage provides a nearly constant drain current at most levels. Since the JFET has stable transconductance it works well as a constant current source. However, the JFET has a lower transconductance than a bipolar junction transistor.

Not all JFETs of the same type have the same operating characteristics. Because of the manufacturing processes, some characteristics may show more variation than others. These characteristics include the gate-to-source voltage and the amount of current that the JFET will supply.

Like bipolar transistors, JFETs use different methods of bias. These bias methods include gate bias, self-bias, and voltage divider bias.

JFETs configured as common-source amplifiers and source followers operate like transistors configured as common-emitter and common-collector amplifiers, respectively. JFET amplifier configurations include the common-drain and common-source amplifiers. In addition, JFETs can operate as amplifiers, input buffers, variable resistors, and switches.

Troubleshooting JFET circuits may seem easier than troubleshooting bipolar transistor circuits. Conditions such as a shorted or open junction and bias resistor problems cause standard symptoms to arise.

1. A JFET has 100 μA of leakage current out of the gate at a test voltage of 12 V. What is the reverse resistance of the JFET junction?
2. A JFET has an I_{DSS} of 8 mA and a $V_{GS(off)} = -4$ V. What is the I_D when $V_{GS} = -1.5$ V?
3. Determine the g_{mo} for the JFET used in problem 2.

Refer to Figure 7.40 for problems 4 through 7.

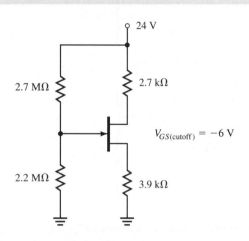

FIGURE 7.40 Circuit for problems 4 through 7

4. Assume $V_{GS} = -2$ V. Determine I_D.
5. Assume $V_{GS} = -2$ V. Determine the power dissipated by the JFET.
6. Assume $V_{GS} = \frac{1}{4} V_{GS(off)}$. Determine I_D.
7. Assume $V_{GS} = \frac{1}{4} V_{GS(off)}$. Determine the power dissipated by the JFET.

Refer to Figure 7.41 for problems 8 through 13.

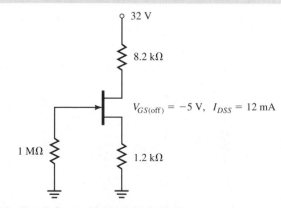

FIGURE 7.41 Circuit for problems 8 through 13

8. Assume $V_{GS} = -2$ V. Determine I_D.
9. Assume $V_{GS} = -2$ V. Determine P_D.
10. Assume $V_{GS} = \frac{1}{4} V_{GS(off)}$. Determine I_D.
11. Assume $V_{GS} = \frac{1}{4} V_{GS(off)}$. Determine P_D.
12. Using the listed values of $V_{GS(off)}$ and I_{DSS}, determine I_D using the universal JFET curve.
13. Using the listed values of $V_{GS(off)}$ and I_{DSS}, determine P_D using the universal JFET curve to predict I_D and V_{DS}.

Refer to Figure 7.42 for problems 14 through 17.

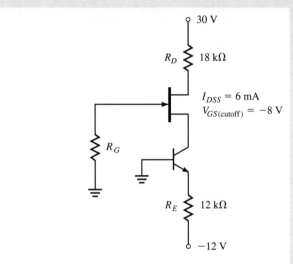

FIGURE 7.42 Circuit for problems 14 through 17

14. What is the value of I_D?
15. What is the voltage at the source of the JFET with respect to ground?
16. What is the P_D of the JFET?
17. What is the P_D of the bipolar?

Refer to Figure 7.43 for problems 18 through 22.

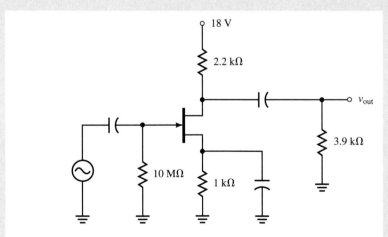

FIGURE 7.43 Circuit for problems 18 through 22.

$I_{DSS} = 10$ mA; $V_{GS(cutoff)} = 28$ V.

18. Determine the g_{mo} for the JFET.
19. Determine I_D for the JFET.
20. Using the answer from problem 19, determine g_m at the Q point.
21. Determine the voltage gain, A_v, for the JFET.
22. Does the 10-MΩ gate resistor have any effect on DC operation of the circuit? On AC operation of the circuit?

Refer to Figure 7.44 for problems 23 through 28.

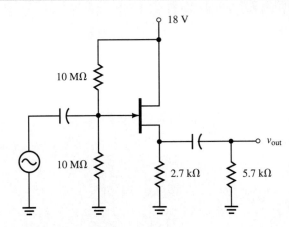

FIGURE 7.44 Circuit for problems 23 through 28.
$I_{DSS} = 10$ mA; $V_{GS(cutoff)} = 28$ V.

23. What is the input impedance into the circuit?
24. Determine I_D.
25. Determine the DC voltage at the source terminal.
26. Determine g_m at the Q point.
27. A common-gate JFET amplifier has a g_m of 5000 μmhos. What is the input impedance looking into the gate?
28. The common-gate circuit mentioned in problem 27 has an AC drain resistance of 4.7 kΩ. What is the voltage gain of the circuit?

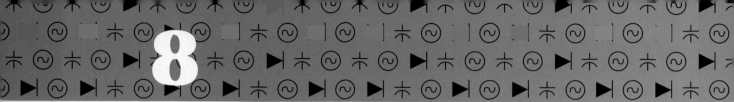

8

MOSFETS AND MOSFET CIRCUITS

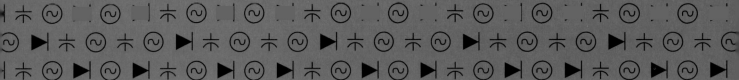

✓ **As you read this chapter, concentrate on learning how to:**

1. Describe the advantages of MOSFETs over JFETs or bipolar devices
2. Recognize the schematic symbols for the depletion MOSFET (D MOSFET)
3. Describe D MOSFET operation
4. Recognize the schematic symbols for the enhancement MOSFET (E MOSFET).
5. Describe E MOSFET operation
6. Recognize various D MOSFET bias arrangements
7. Recognize various E MOSFET bias arrangements
8. Recognize the cascode amplifier and be able to state its purpose
9. Describe a MOSFET voltage-controlled switch
10. Recognize a CMOS circuit and its application
11. Handle MOSFETs properly to minimize the possibility of electrostatic discharge damage

INTRODUCTION

In this chapter, we will take a look at another type of field effect transistor called the **metal oxide semiconductor field effect transistor,** or **MOSFET.** Some service manuals may refer to the MOSFET as an **insulated gate field effect transistor,** or **IGFET.** As the last descriptive name clearly describes, the MOSFET has a gate that is insulated from the remainder of the device. Unlike the JFET, the polarity of the bias on the gate will not affect MOSFET operation.

Because of their construction, MOSFETs appear as two distinct types called depletion and enhancement MOSFETs. **Depletion MOSFET**s will conduct with a forward, reverse, or zero gate bias, while **enhancement MOSFET**s must have a forward bias. Depletion MOSFETs, while in the depletion mode, work the same as JFETs. Like JFETs, MOSFETs are both N-channel and P-channel devices. Consequently, four types of MOSFETs exist.

Several characteristics—such as a low "on" resistance, a high "off" resistance, low leakage currents, low capacitance, and good high-frequency response—make MOSFETs valuable for a number of electronic applications. MOSFETs are used as amplifiers, switches, and controlled-current sources.

8.1 ADVANTAGES OF THE MOSFET

MOSFETs provide the circuit designer with several important advantages. One advantage is lowered power dissipation, which occurs because of the small control circuit. Another advantage is that the insulated gate lead provides higher input impedances than those found in JFET circuits. A third advantage is that the MOSFET has a very high gate resistance, which is not affected by the polarity of bias on the gate. For a JFET transistor, high input resistance occurs only when the gate-to-source PN junction is reverse-biased.

Finally, the thin layer of glass (silicon dioxide) that separates the gate region from the conducting channel cuts the amount of leakage current within the device. Leakage current can limit the performance of a JFET by reducing the input impedance. The MOSFET insulated gate construction also prevents temperature from affecting leakage currents in the gate circuit. In the junction-type FET, the gate current or leakage current of the reverse-biased PN junction is temperature-sensitive. The diagram in Figure 8.1 shows the insulating layer enlarged for the purpose of clarity. Because the actual device has an extremely thin insulating layer, electrostatic charges can damage the MOSFET.

Table 8.1 lists some typical values and parameters for MOSFETs.

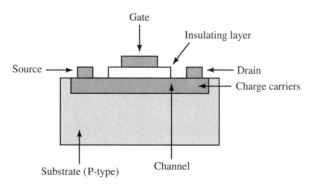

FIGURE 8.1   **Block diagram of a MOSFET**

Table 8.1 Typical values and parameters for MOSFETs

Input resistance	$= 10^{14}\ \Omega$
Transconductance	$= 10{,}000\ \mu\text{mhos}$
Output resistance	$= 200\ \Omega$
Input capacitance	$= 6\text{pF}$
Cutoff frequency	$= f_c = \dfrac{g_m}{2}$

REVIEW SECTION 8.1
1. Another name for a MOSFET is an IGFET. Describe the construction of this FET.
2. The two distinct types of MOSFETs are _____ and _____ types.
3. Enhancement-type MOSFETs must have a _____ bias to operate.
4. MOSFETs are generally used as _____, _____, and _____ _____ sources.
5. Lower power dissipation occurs in MOSFETs because of the small control circuit. True or false?
6. Due to a thin layer of glass separating the gate from the conducting channel, there is a small leakage current in MOSFETs. True or false?
7. Describe how the insulated gate helps with the operation of the MOSFET; include leakage current, temperature, and biasing.

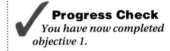

Progress Check
You have now completed objective 1.

8.2 TYPES OF MOSFETs

MOSFETS

Two terms are used to describe a MOSFET: mode and type. Type refers to the construction of a MOSFET. Mode describes how a MOSFET operates.

An enhancement-type MOSFET is constructed so that it will conduct only if its channel is enhanced with major current carriers. Only a forward bias will enhance its channel. When an enhancement-type MOSFET is biased to conduct, it operates in the enhancement mode. Forward bias for a silicon bipolar device is 0.7 V. But MOSFETs are unipolar devices, and each MOSFET has its own forward bias voltage.

A depletion-type MOSFET is constructed so that it will conduct with a reverse, zero, or a forward bias applied to the gate. It will stop conducting only when its channel is depleted of major charge carriers. When a reverse bias is being applied to the depletion-type MOSFET, its channel is being depleted of major charge carriers and its operation is called the depletion mode. A depletion-type MOSFET may also be forward-biased. When it is forward-biased, its channel is being enhanced with major charge carriers, and its operation is called the enhancement mode.

Remember, an enhancement-type MOSFET will operate only in the enhancement mode. A depletion-type MOSFET will operate in either the depletion mode or the enhancement mode.

THE DEPLETION MOSFET, OR D MOSFET

Figure 8.1 represents a depletion-type MOSFET. A metal control electrode or gate acts like the plate of a capacitor. Any charge placed on the gate induces an equal and opposite charge on the semiconductor channel underneath the insulating oxide layer. With the induced charge, the channel controls conductivity between two ohmic contacts at opposite ends of the channel. As with JFETs, we call those contacts the **source** and the **drain.**

When properly biased, the depletion MOSFET, or **D MOSFET,** reduces current flow by depleting the major charge carriers within the channel. Even without any voltage applied to the gate, charge carriers reside in the channel. For the sake of definition, we can label the bias voltage as the reverse gate voltage. However, the insulated-gate electrode of the MOSFET can also increase the conductance to the channel without increasing gate current.

MORE ABOUT D MOSFETS

Figure 8.2 shows a cross-sectional view of an N-channel D MOSFET. In most applications, the substrate lead and the source lead are connected together. The substrate connection is made for the largest possible reverse bias across the source and drain junctions.

In Figure 8.3, we see the schematic symbols for the depletion-type N-channel (Figure 8.3a) and P-channel (Figure 8.3b) MOSFETs. When we compare this schematic figure with the schematic figure for a JFET, we can see an extra terminal. The extra terminal is connected to the body or substrate of the MOSFET. In Figure 8.3a, the arrowhead on the substrate lead points to the channel N-type material. For the P-channel MOSFET shown in Figure 8.3b, the arrowhead points away from the channel. Since the channel for a depletion-type MOSFET normally conducts, it is shown as a solid vertical line.

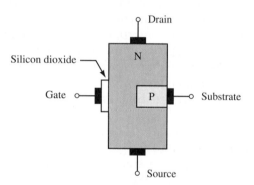

FIGURE 8.2   **Cross-sectional view of an N-channel D MOSFET**

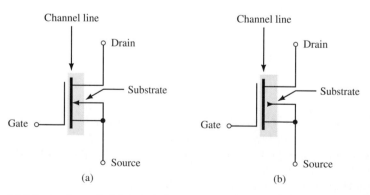

FIGURE 8.3   **(a) Schematic symbol for an N-channel D MOSFET. (b) Schematic symbol for a P-channel D MOSFET.**

In both cases, the channel area between the source and drain consists of the same material as the source and drain. That is, in a P-type depletion MOSFET, if the drain and source are P-type materials, the channel is also a P-type substance. This allows measurable drain current with a zero gate voltage. In addition, the D MOSFET is constructed so that the capacitance between the gate and source and between the gate and drain are reduced. Because of the reduced internal capacitance, the D MOSFET exhibits good high-frequency response.

Figure 8.4 shows the characteristic curves for an N-channel depletion-type MOSFET. Since we are working with a depletion-mode device, we may continue to call the value of the current with the gate shorted to the source the I_{DSS} current. Also, we may continue to label the gate-to-source voltage at which the drain current equals zero as $V_{GS(off)}$. When we worked with JFETs, we called this voltage the "pinch-off" voltage.

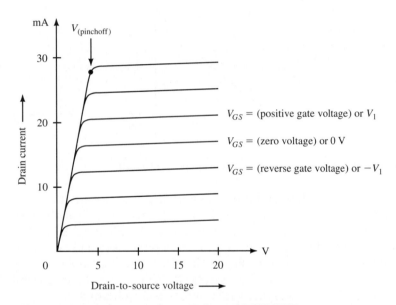

FIGURE 8.4   Characteristic drain curves for an N-channel D MOSFET

In Figure 8.4, with low drain-to-source voltages and no gate-to-source voltage, the channel resistance remains constant, while the drain current varies linearly with the drain voltage. A further increase in drain voltage causes a voltage drop in the channel. As a result, there is a greater voltage difference between the gate and points close to the drain. At a given drain-to-source voltage, the quantity of carriers at the drain becomes depleted. Any further increase in the drain-to-source voltage does not change the channel current (I_D) because the carriers have already been depleted. The drain current has reached the saturation point.

As curves in Figure 8.4 indicate, a D MOSFET can operate with forward, reverse, or zero gate bias. As the $-V_1$ curve also shows, a given reverse gate voltage will cause the channel to pinch off at a smaller drain voltage. Another curve, labeled V_1, shows that a given forward gate voltage will cause the channel to pinch off at a larger drain voltage.

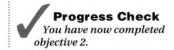

Progress Check
You have now completed objective 2.

D MOSFET OPERATION

The D MOSFET gate lead almost operates as an open circuit. Any leakage current is in the picoamp range. Comparatively, JFETs have leakage currents through the reverse-biased gate diode that range in the nanoamp area.

Interestingly, D MOSFETs can operate in either the depletion mode or the enhancement mode. Since the D MOSFET has an insulated gate lead, its operating range may extend into the first quadrant. In Figure 8.5, the parabolic curve extends past the maximum drain current (I_{DSS}) into the first quadrant of the graph, or into the enhancement mode. Operating in the second quadrant of the graph, the device functions in the depletion mode.

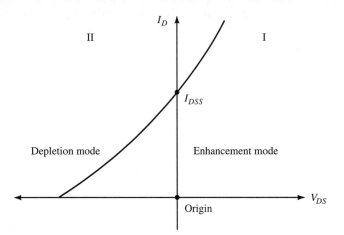

FIGURE 8.5   Transconductance curve for a D MOSFET

To better understand the depletion and enhancement modes, let us take a look at Figure 8.6. To begin with, the silicon dioxide has **dielectric** properties as it insulates the gate. It will support an electrical field. While the semiconductor material acts as one conductive plate, the lead acts as another plate. Effectively, the gate, the silicon dioxide insulating material, and the channel form a capacitor. Since the silicon dioxide supports an electrical field, the buildup of charges on the gate influences the channel.

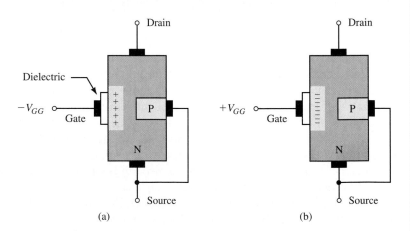

FIGURE 8.6   **(a) Depletion mode. Cross-sectional view of how an applied negative voltage affects D MOSFET operation. (b) Enhancement mode. Cross-sectional view of how an applied positive voltage affects D MOSFET operation.**

Figure 8.6a shows a cross-sectional view of how an applied negative voltage affects the carrier within the MOSFET. The negative voltage on the gate forces electrons away from the portion of the channel near the gate. In turn, a net positive charge remains in the channel and depletes the number of majority current carriers in the channel.

In Figure 8.6b, the N-channel MOSFET works with an applied positive voltage. Remember that a positive applied voltage would not work with JFETs because it would forward bias the gate diode. In the case of the MOSFET, though, the silicon dioxide acts as an insulator and prevents the flow of current out of the gate. The positive charge on the gate causes negative charges in the channel. These negative charges enhance, or add to, the negative charge carriers already present in the channel. With the increase of available charge carriers, current flow becomes easier.

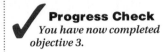

Progress Check
You have now completed objective 3.

8.3 THE ENHANCEMENT MOSFET, OR E MOSFET

A correctly biased enhancement MOSFET, or **E MOSFET**, will allow current flow only when the number of charge carriers within the channel is increased. The gate must be forward-biased to induce carriers into the channel and to permit conduction. Figure 8.7 shows a cross section of an enhancement-type MOSFET.

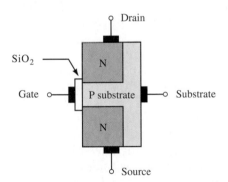

FIGURE 8.7   Cross-sectional view of an E MOSFET. (SiO_2 = silicon dioxide.)

No complete path for charge carriers to move from the source to the drain exists. In this case, a P-type substrate isolates two N-type layers from one another.

As with the D MOSFET, a layer of silicon dioxide insulates the gate from the substrate so that the device can use capacitance to control the flow of current. In Figure 8.8, we see the schematic symbols for N-channel and P-channel E MOSFETs. Look again at the schematic symbol for the D MOSFET in Figure 8.3. Note that the channel line in that figure for the D MOSFET is solid. In Figure 8.8, however, the channel line for the enhancement-type MOSFET is broken, indicating the channel is normally nonconducting and requires a forward-biased voltage to cause conduction.

Figure 8.9 shows how the capacitance of the gate creates a path for current to flow from the source to the drain. In the diagram, a positive voltage is placed on the gate. While the P-type material of the substrate serves as the

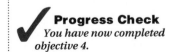

Progress Check
You have now completed objective 4.

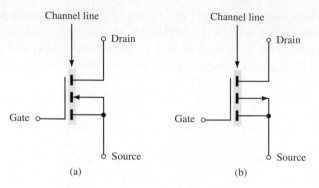

FIGURE 8.8 Schematic symbol (a) for an N-channel and (b) a P-channel E MOSFET

second plate of the capacitor, the silicon dioxide acts as the dielectric. The positive charges on the gate attract negative charges on the opposite side of the silicon dioxide. An N-type channel is formed through the P-type substrate.

We refer to this negatively charged area as an **inversion layer,** or an **induced channel** through which electrons emitted from the source can move to the drain. To establish the inversion layer, a certain amount of gate voltage is required. We define this voltage as the **threshold voltage,** abbreviated V(th). In a P-channel enhancement MOSFET, the application of a negative gate voltage pulls holes into the region and causes the channel to change from N-type to P-type. Again, the channel provides a source-to-drain conduction path.

In the E MOSFET, the gate electrode covers the entire area between the source and the drain. Since the electrode covers a larger area, the gate voltage can induce a conductive channel between the source and the drain. As opposed to the D MOSFET, which uses a reverse, a forward, or a zero gate voltage for conduction, the E MOSFET utilizes only a forward gate voltage. The forward gate voltage draws additional charge carriers into the channel to increase conductance. Little or no conductance exists in the channel with either zero or reverse bias. An E MOSFET will not conduct until a sufficient amount of positive voltage becomes applied to the gate.

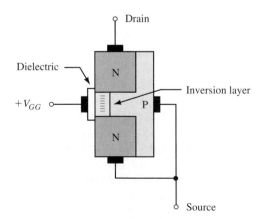

FIGURE 8.9 How the capacitance of the gate creates a path for current to flow from the source to the drain

Figure 8.10a shows a transconductance curve for an enhancement-type MOSFET. Since E MOSFETs operate differently from D MOSFETs, we must think of the maximum drain current in other terms. The maximum drain current ($I_{D(on)}$) no longer limits the amount of current that can flow through a device. For 0 V on the gate, the drain current will equal zero for all source-to-drain voltages less than the breakdown voltage.

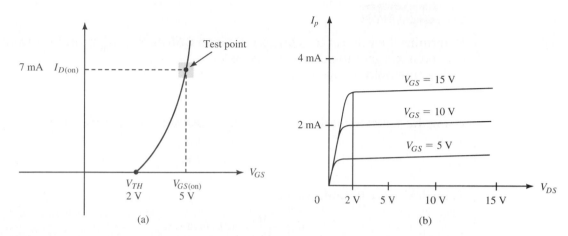

FIGURE 8.10 (a) Transconductance curve for an E MOSFET. (b) Characteristic curves for an E MOSFET.

As we have learned, the threshold voltage will induce charge carriers into the channel region and cause conduction. In E MOSFETs, the threshold, or $V_{GS(th)}$ voltage, becomes an important parameter. This is the value of gate-to-source voltage at which drain current begins to flow and the device begins to conduct.

Again, let us look at the characteristic curve in Figure 8.10b. If the forward gate voltage has a greater value than the threshold voltage and the drain-to-source voltages, the drain current will increase linearly with the drain-to-source voltage. The important value in the characteristic curve of E MOS-FETs is the value of $V_{GS} - V_{GS(th)}$, or the amount of gate-source voltage above the threshold. As the drain voltage increases, the voltage between the gate and points in the channel close to the drain becomes smaller. The drain current does not increase for further increases in drain voltage. For a forward gate voltage where V_{DS} is greater than V_{th}, more carriers become induced into the channel, and the drain current reaches saturation.

To further illustrate E MOSFET action, Figure 8.10a depicts the transconductance curve for a typical E MOSFET. Note that current cannot flow until V_{GS} exceeds the threshold voltage. In addition, a test point on the curve marks the point where both V_{GS} and I_D are in the on state. However, this point does not mark the only point along the curve where the E MOSFET conducts.

Instead, the test point gives us a reference point for calculating the constant, **K,** for a given type of E MOSFET. The constant is a value proportional to the width-length ratio of the channel and sets the "conduction factor" for the manufacturing process (K is equal to I/V_G^2). During the manufacturing process, the values of the constant and the threshold voltage set the parameters for the MOSFET. Knowing those values allows us to determine the curve of drain current versus gate-to-source voltage.

To solve for any drain current value other than the maximum drain current for this E MOSFET, we would use the following equation:

$$I_D = K[V_{GS} - V(th)]^2$$

with I_D equaling the drain current, V_{GS} equaling the applied voltage gate to source, and $V(th)$ equaling the threshold voltage.

EXAMPLE 8.1

Determine the value of the constant K for an E MOSFET with a drain current of 7 mA and a gate-to-source voltage of 5 V.
   The threshold voltage equals 2 V.

Solution
Using the equation for drain current and substituting values, we have:

$$I_D = K(V_{GS} - V(th))^2$$
$$7\ mA = K(5\ V - 2\ V)^2$$
$$7\ mA = K \times 9\ V^2$$

Transposing the formula, we have:

$$K = \frac{7\ mA}{9\ V^2} = 0.78\ mA/V^2$$

PRACTICE PROBLEM 8.1

An E MOSFET has an I_D of 8 mA and a V_{GS} of 6 V. The threshold voltage equals 1.5 V. Determine the value of K for this device.

PRACTICE PROBLEM 8.2

Another E MOSFET has an $I_{D(on)}$ equal to 6 mA. Determine K if V_{GS} is equal to 4 V and the threshold voltage is equal to 1 V.

EXAMPLE 8.2

Using the E MOSFET and the value of K from example 8.1, determine how an increase of the gate-to-source voltage to 7 V affects the device.

Solution
With a new value for the gate-to-source voltage, we will have a new value for the drain current. This value is:

$$I_D = \frac{0.28\ mA}{V^2} \times \left(7\ V - 2V\right)^2$$

$$I_D = \frac{0.78\ mA}{V^2} \times 25\ V^2$$

$$= 19.5\ mA$$

Remember that the values obtained in this example work only with an E MOSFET that has a curve that matches the parameters given.

The gate-to-source voltage of the E MOSFET in practice problem 8.2 is increased to 5.5 V. Determine the drain current under these conditions.

REVIEW SECTIONS 8.2 AND 8.3

1. The depletion-type MOSFET, when in depletion mode, conducts with the gate tied to the source, which allows it to operate in a manner similar to that of the JFET. True or false?
2. Describe the capacitive action of the D MOSFET.
3. Generally, when a MOSFET is used, the substrate lead and the drain are connected to create the largest possible reverse bias. True or false?
4. Because of its high internal capacitance, the D MOSFET exhibits good frequency response. True or false?
5. Describe the operation of a D MOSFET.
6. As with a D MOSFET, the E MOSFET controls current through the channel by _____ caused by the silicon dioxide (glass) dielectric.
7. In an N-channel E MOSFET, the negatively charged area caused by a positive voltage at the gate is referred to as an _____ _____.
8. The path created by the capacitance causes a DC current flow from the gate to the drain. True or false?
9. An E MOSFET won't conduct until a sufficient amount of _____ voltage is applied at the gate.
10. Describe the differences between D and E MOSFET operations.
11. $V_{GS(th)}$ denotes the value of the _____ -to- _____ voltage when the drain current flows and the device conducts.

✓ **Progress Check**
You have now completed objective 5.

8.4   D MOSFET BIASING AND CIRCUIT ANALYSIS

DEPLETION BIAS

For the D MOSFET in depletion mode, biasing is similar to the biasing used for a JFET. It is called **depletion bias.** The D MOSFET continues to act as a square law device and operates over the universal curve of FETs. Nevertheless, some differences exist. From a practical point of view, either forward or reverse bias exists for the gate of a D MOSFET. We could apply voltages of either polarity to the gate. Or we could bias the gate by applying no signal to the gate. Instead, with the DC gate-to-source voltage equaling zero, an AC input signal applied directly to the gate would act as a bias voltage.

EXAMPLE 8.3

Solve for the DC values of the D MOSFET drain current and voltage and the gate-to-source voltage. Determine the gain of the circuit.

Figure 8.11 shows a D MOSFET using self-bias, while Figure 8.12 depicts the plot of the self-bias line for the 180-Ω source resistor. This line intersects the transconductance curve at the Q point marked on the graph.

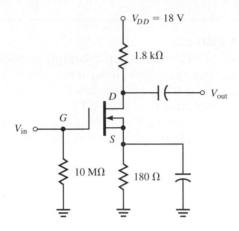

FIGURE 8.11 D MOSFET using self-bias

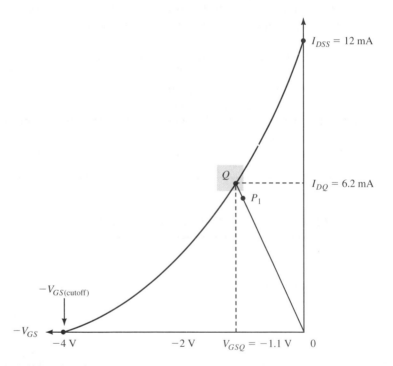

FIGURE 8.12 Self-bias line for a D MOSFET

Solution
DC Circuit Analysis

First, we will analyze the DC operation of the MOSFET circuit. The self-bias line is drawn by selecting a drain current that is less than I_{DSS}. We have selected 5 mA. Multiply I_D by the source resistor to find the source voltage.

$$V_S = I_D \times R_S$$
$$= 5\,\text{mA} \times 180\,\Omega = 0.9\,\text{V}$$

Since the gate voltage is 0 V, the gate-to-source voltage (V_{GS}) is -0.9 V. Place a point, P_1, at the intersection of the I_D and V_{GS}. Now draw a line from the 0 point on the graph through P_1 so that the transconductance curve is intersected. The point where the self-bias line and the transconductance curve intersect is the Q point. Drawing a line horizontally from the Q point to the I_D line provides the DC operating drain current (I_{DQ}), which is 6.2 mA. Drawing a line vertically from the Q point to the V_{GS} line provides the DC operating gate-to-source voltage (V_{GSQ}), which is -1.1 V.

AC Circuit Analysis

To find the gain of the circuit we need to analyze the AC circuit. First we need to find the maximum transconductance (g_{mo}).

$$g_{mo} = \frac{2 \times I_{DSS}}{V_{GS(cutoff)}}$$
$$= \frac{2 \times 12\,\text{mA}}{4\,\text{V}}$$
$$= 6000\,\mu\text{mhos}$$

Now we need to find the transconductance at the Q point (g_m).

$$g_m = g_{mo}\left[1 - \left(\frac{V_{GS}}{V_{GS(cutoff)}}\right)\right]$$
$$= 6000\,\text{mmhos}\left[1 - \left(\frac{-1.1\,\text{V}}{-4\,\text{V}}\right)\right]$$
$$= 6000\,\text{mmhos}\,(1 - 0.275) = 4350\,\text{mmhos}$$

Using the gain formula for FETs, we can find the gain of the circuit.

$$A_v = g_m \times R_D$$
$$= 4350\,\mu\text{mhos} \times 1.8\,\text{k}\Omega = 7.8$$

Since ohms and mhos are inverse to each other and cancel, gain has no units. Also since no load is present the gain is unloaded.

PRACTICE PROBLEM 8.4

Change the values on Figure 8.11 from a 180-Ω source resistor to a 210-Ω source resistor and recompute the answers for example 8.3.

USING VOLTAGE DIVIDER BIAS WITH A D MOSFET

Figure 8.13 shows a D MOSFET biased through a voltage divider. An important application difference surfaces between the JFET and the D MOSFET. With the D MOSFET, we can accomplish voltage divider biasing without a source resistor. In this case, the 0 V at the source and the positive gate Thevenin voltage cause V_{GS} to be positive. Remember that D MOSFETs can also work in the enhancement mode. This circuit operates in the enhancement mode.

Let us analyze the circuit action for Figure 8.13 through example 8.4.

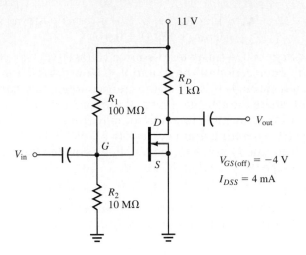

FIGURE 8.13 Voltage divider bias for the D MOSFET

EXAMPLE 8.4

Solve the voltage divider circuit shown in Figure 8.13. Find values for the D MOSFET terminal voltages, transconductance, and circuit gain.

To begin, we will determine the voltage at the gate of the MOSFET.

Solution

The value of gate voltage is formed by using the voltage divider method:

$$V_{GS} = \frac{10 \text{ M}\Omega}{110 \text{ M}\Omega} \times 11 \text{ V} = 1 \text{ V}$$

This value puts the MOSFET in the enhancement mode. We solve by using the formula:

$$I_D = I_{DSS}\left[1 - \left(\frac{V_{GS}}{V_{GS(off)}}\right)\right]^2$$

or

$$I_D = 4 \text{ mA}\left[1 - \left(\frac{+1 \text{ V}}{-4 \text{ V}}\right)\right]^2 = 4 \text{ mA}\left(1 + 0.25\right)^2 = 4 \text{ mA}\left(1.5625\right)$$

This gives a value of 6.25 mA for I_D. This value is greater than the value for I_{DSS}. Any value within the brackets greater than 1 also answers a requirement for enhancement-mode operation.

From here, we can use the drain current value to determine the voltage at the drain with respect to ground:

$$V_{RD} = 6.25 \text{ mA} \times 1 \text{ k}\Omega = 6.25 \text{ V}$$

Subtracting this value from V_{DD}, we find that V_D equals:

$$V_D = 11 \text{ V} - 6.25 \text{ V} = 4.75 \text{ V}$$

We now know the DC voltage values for each of the three MOSFET terminals.

AC analysis of the circuit follows the tried-and-true pattern. Calculating g_{mo} , we get:

$$g_{mo} = \frac{8\,mA}{4\,V} = 2000\,\mu mhos$$

Since we are in enhancement mode, g_m is greater than g_{mo}. Using the formula for g_m, we find that:

$$g_m = 2000\,\mu mhos\left[1-\left(\frac{+1\,V}{-4\,V}\right)\right]$$
$$= 2000\,\mu mhos\,(1+0.25) = 2500\,\mu mhos$$

As you can see, g_m may equal more than 100% of g_{mo} when the device operates in the enhancement mode. Circuit gain equals:

$$A_v = g_m \times R_D = 2500\,\mu mhos \times 1\,k\Omega = 2.5$$

PRACTICE PROBLEM 8.5

Change the value of the 10-MΩ resistor shown in Figure 8.13 to 8.2 MΩ and rework example 8.4.

USING ZERO BIAS WITH THE D MOSFET

So far, we have looked at depletion and enhancement bias with a D MOSFET. Another type of bias unique to the D MOSFET, called **zero bias,** also exists. Figure 8.14 shows a D MOSFET circuit using zero bias. Both the gate and the source have the same DC voltage, which is zero volts. The applied gate-to-source voltage is zero.

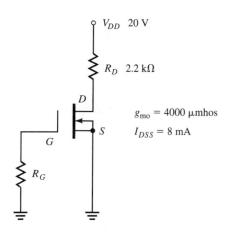

FIGURE 8.14 D MOSFET using zero bias

EXAMPLE 8.5

Determine the circuit parameters for the D MOSFET in the circuit in Figure 8.14.

We have already established that the circuit uses zero bias and that the applied gate-to-source voltage equals 0 V.

Solution
Because of the zero bias $V_{GS} = 0$ V and $I_D = I_{DSS}$. So $I_D = 8$ mA.
The voltage drop across the drain resistor equals:

$$V_{RD} = 8 \text{ mA} \times 1.5 \text{ k}\Omega = 12 \text{ V}$$

Subtracting this from V_{DD}, we find that:

$$V_D = 20 \text{ V} - 12 \text{ V} = 8 \text{ V}$$

Now, we know the DC voltage values for the source, the gate, and the drain. We also know how much DC current flows through the MOSFET. Next, we will turn our attention to the AC analysis. Since $V_{GS} = 0$ V,

$$g_m = g_{mo} = 4000 \text{ }\mu\text{mhos}$$

Consequently, the gain is:

$$A_v = g_m \times R_D = 4000 \text{ }\mu\text{mhos} \times 1.5 \text{ k}\Omega = 6$$

PRACTICE PROBLEM 8.6

Change the value for the drain resistor in Figure 8.14 to 1 kΩ and rework example 8.5.

> **REVIEW SECTION 8.4**
> 1. Biasing the D MOSFET, in depletion mode, is similar to biasing a
> _____.
> 2. Voltages of only positive polarity can be applied to bias a D MOSFET. True or false?
> 3. The difference in voltage divider biasing between JFETs and D MOSFETs is that D MOSFET biasing can be accomplished without a
> _____ _____.
> 4. When using zero bias the gate and the source must be at the
> _____ DC voltage.

Progress Check
You have now completed objective 6.

8.5 E MOSFET BIASING AND CIRCUIT ANALYSIS

USING VOLTAGE DIVIDER BIAS WITH AN E MOSFET
Since the E MOSFET must operate in the enhancement mode, biasing techniques such as self-bias, current source bias, and zero bias will not work. Each of these bias types produces depletion-mode operation and will not "turn on" the E MOSFET. Providing that the source stays at DC ground, voltage divider bias and gate bias can produce enhancement-mode operation. Let us look at voltage divider bias with the E MOSFET.

EXAMPLE 8.6

Find the circuit parameters for the circuit shown in Figure 8.15.

In this case, the on-state drain current ($I_{D(on)}$) for the E MOSFET equals 10 mA at an applied source-to-gate voltage of 10 V. The threshold voltage equals 2 V.

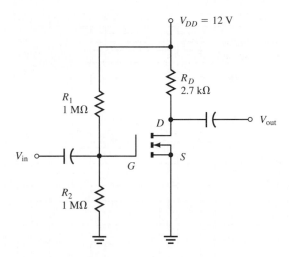

FIGURE 8.15 E MOSFET using voltage divider bias

Solution

The first step is to determine the constant for the E MOSFET:

$$10 \, \text{mA} = K(10 \, \text{V} - 2 \, \text{V})^2 = K \times 64 \, \text{V}^2$$

$$K = \frac{10 \, \text{mA}}{64 \, \text{V}^2} = 0.156 \, \frac{\text{mA}}{\text{V}^2}$$

DC Analysis

Next, we will determine the voltage on the gate:

$$V_G = \frac{1 \, \text{M}\Omega}{2 \, \text{M}\Omega} \times 12 \, \text{V} = 6 \, \text{V}$$

Note that there is no source resistor. With no source resistor, the applied gate-to-source voltage will remain positive. Since we know V_{GS} is a constant value, we can determine the drain current:

$$I_D = K(V_{GS} - V_{TH})^2$$

$$I_D = 0.156 \, \frac{\text{mA}}{\text{V}^2} (6 \text{V} 22 \text{V})^2 = 2.5 \, \text{mA}$$

With the drain current value, we can determine the drain voltage:

$$V_{RD} = I_D \times R_D$$

$$V_{RD} = 2.5 \, \text{mA} \times 2.7 \, \text{k}\Omega = 6.75 \, \text{V}$$

Subtracting the drain voltage value from V_{DD}, we obtain the V_D:

$$V_D = V_{DD} - V_{RD} = 12 \, \text{V} - 6.75 \, \text{V} = 5.25 \, \text{V}$$

AC Analysis

Reading the g_m from a chart that compares it with I_D is the easiest way to determine g_m for a given E MOSFET. We can use a graph, such as the one shown in Figure 8.16, to estimate a value for transconductance. This graph is called a **forward transfer admittance graph.** In this instance, the plot on the graph shows that for an I_D of 2.5 mA g_m equals 1000 μmhos. The gain of the circuit is:

$$A_v = g_m \times R_D = 1000 \text{ μmhos} \times 2.7 \text{ kΩ} = 2.7$$

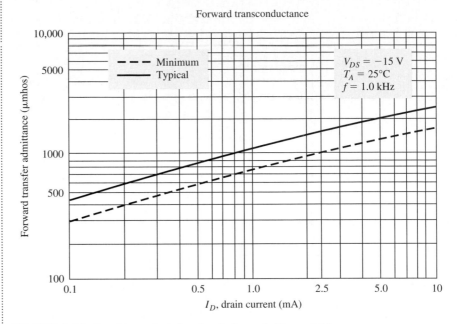

FIGURE 8.16 Graphic analysis for circuit shown in Figure 8.15

PRACTICE PROBLEM 8.7

Change the 1-MΩ resistor in Figure 8.15 to an 820-kΩ resistor and rework example 8.6.

DRAIN FEEDBACK BIAS

Another kind of bias, called **drain feedback bias,** also triggers the E MOSFET. Figure 8.17 shows an example of this bias type. In this circuit, a portion of the output is brought back to the gate to establish V_{GS}. Since no gate current

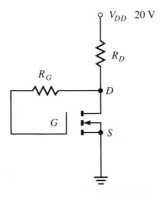

FIGURE 8.17 Drain feedback bias for an E MOSFET

flows, no voltage drops across the gate resistor. Because of this, the drain-to-source voltage equals the gate-to-source voltage.

EXAMPLE 8.7

Using the circuit in Figure 8.17, calculate a value for the drain resistor. Determine the circuit gain.

Solution

In Figure 8.17, we find a circuit that has the same transconductance curve as the circuit shown in Figure 8.10. Part of the drain feedback circuit design centers around $I_{D(on)}$. If the design dictates a drain-to-source voltage of 5 V, the gate-to-source voltage will equal 5 V, and the drain current will equal 7 mA.

If 7 mA of current flows through the drain resistor, we can use Ohm's law to calculate the value of the drain resistor. From the schematic, we know that the value of V_{DD} is 20 V. With the voltage values in hand, we can solve for the value of R_D:

$$R_D = \frac{15 \text{ V}}{7 \text{ mA}} = 2.14 \text{ k}\Omega$$

The closest stand and resistor would be 2.2 kΩ. As in the previous example, we can obtain g_m from the forward transfer admittance graph. Then, we can find the gain by multiplying the transconductance value by the value of the drain resistor. At 7 mA, g_m is typically 2400 μmhos.

$$A_v = g_m \times R_D$$
$$= 2400 \text{ μmhos} \times 2.2 \text{ k}\Omega = 5.3$$

PRACTICE PROBLEM 8.8

An E MOSFET with a V_{DD} of 25 V has a drain current that equals 8 mA. Determine the size of the drain resistor in a drain feedback circuit. The drain voltage should equal 16 V.

Example 8.8 will show how we can predict circuit behavior for drain feedback bias when the drain current does not equal the drain current during the on state.

EXAMPLE 8.8

Using the specifications given in Figure 8.18 for drain current, gate-to-source voltage, and threshold voltage, determine the constant of an E MOSFET. Plot the transconductance curve for the E MOSFET.

Beginning our analysis, we use the value for a drain current during the on state even while we suspect that the circuit cannot produce this current. From the specifications for the MOSFET, we know that the drain current and applied gate-to-source voltage during the on state equal 8 mA and 4 V, respectively. The threshold voltage equals 2 V.

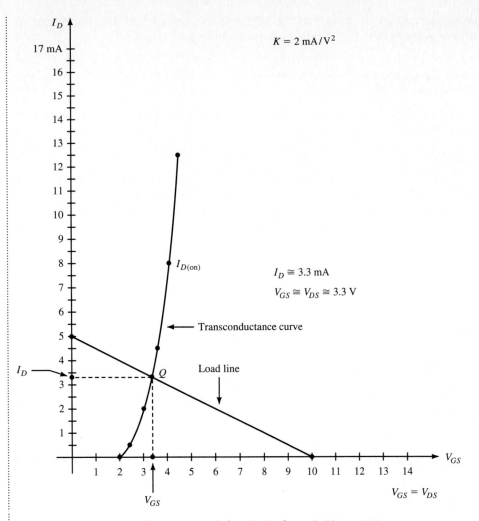

FIGURE 8.18 Transconductance graph for circuit shown in Figure 8.17

Solution

We can use these values to determine the constant:

$$I_{D(on)} = K\left(V_{GS} - V(th)\right)^2$$

$$K = \frac{I_{D(on)}}{\left(V_{GS} - V(th)\right)^2} = \frac{8 \text{ mA}}{\left(4 \text{ V} - 2 \text{ V}\right)^2} = \frac{8 \text{ mA}}{4 \text{ V}^2} = 2\frac{\text{mA}}{\text{V}^2}$$

Rearranging the values, we find that the constant is 0.002. Now that we have a value for the constant, we can plot the drain current for several different applied gate-to-source voltages and draw the transconductance curve for this E MOSFET on a piece of graph paper. Figure 8.18 shows how the graph would appear.

PRACTICE PROBLEM 8.9

An E MOSFET has an $I_{D(on)}$ equal to 6 mA at a V_{GS} of 4 V. Determine K, and graph the transconductance curve.

EXAMPLE 8.9

Figure 8.19 shows an E MOSFET circuit using drain feedback bias. Determine a value for the saturation current. Also, find a value for the drain-to-source voltage at cutoff. Using those values, draw a load line for the circuit and plot the Q point. Determine the actual value of I_D and V_{GS} of the circuit.

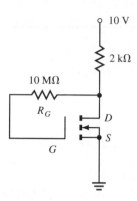

FIGURE 8.19 E MOSFET circuit using drain feedback bias

Solution

First, we will determine the saturation current. Assuming V_{DS} equals 0 V at saturation, we have:

$$I_{D(sat)} = \frac{V_{DD}}{R_D} = \frac{10\,V}{2\,k\Omega} = 5\,mA$$

and the first point for a load line. When the E MOSFET is open, no current flows through the device. Therefore, the entire supply voltage is across the E MOSFET. Consequently, the drain-to-source voltage at cutoff equals the drain supply voltage, or 10 V:

$$V_{DS(cutoff)} = V_{DD} = 10\,V$$

This value gives us the second point for the load line. We can plot this load line on the same graph used for the transconductance curve of Figure 8.18. The intersection of the load line and the transconductance curve is the Q point for the circuit. Using this point, we can determine the drain current and the applied gate-to-source voltage for the circuit shown in Figure 8.19. These values are marked on the graph depicted in Figure 8.18.

Progress Check
You have now completed objective 7.

Change the 2–kΩ drain resistor on Figure 8.19 to a 1.2–kΩ resistor, and re-compute the saturation current and cutoff voltage. Draw the load line, and determine the new Q point.

8.6 MOSFET APPLICATIONS

THE CASCODE AMPLIFIER

Applying a voltage to the gate is the controlling mechanism of the MOSFET. Remember, the gate is capacitive and its capacitive reactance may cause problems. As you may recall from the equation $X_c = 1/(2\pi fc)$, capacitive reactance is inversely proportional to frequency. At higher frequencies, the capacitive reactance of the insulated gate will decrease. If the reactance becomes too low, the insulated gate will appear as a short to the input signal. The device will not act properly.

Using a cascode amplifier will help to overcome this problem. In a cascode amplifier circuit, two MOSFETs function in series. Figure 8.20 illustrates a cascode amplifier circuit. In the figure, the top transistor, Q_1, presents an impedance to the output of the bottom transistor, Q_2.

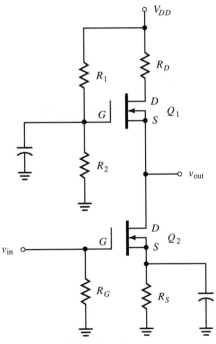

FIGURE 8.20 Cascode amplifier circuit

The transistor Q_1 is configured as a common-gate circuit and has an input impedance of $1/g_m$. Transistor Q_2 is configured as a common-source amplifier circuit. The gain for Q_1 is:

$$A_v = g_m \times \frac{1}{g_m} = 1, \text{ or unity}$$

With a unity gain, transistor Q_2 isolates the top transistor from the input signal and reduces the overall capacitance of the circuit. Why the capacitance is reduced will be discussed in Chapter 9 when discussing the Miller Effect. Given its arrangement in the common-gate circuit, transistor Q_1 takes the signal from transistor Q_2 and amplifies it with a gain of:

$$A_v = g_m \times R_D$$

Overall, the entire circuit produces the desired gain of a common source amplifier and increases higher frequency capabilities by reducing input capacitance.

Figure 8.21 shows an interesting device called the **dual-gate MOSFET.** Comparatively, the MOSFET in Figure 8.21 works much like the circuit shown in Figure 8.20. Both are cascode amplifiers. But because the MOSFET in Figure 8.21 has a dual gate, a circuit designer can build a cascode amplifier using only one transistor.

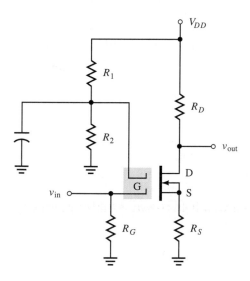

FIGURE 8.21 Dual-gate MOSFET circuit

Dual-gate MOSFETs seem like an almost custom-made device for cascode amplifiers. Figure 8.22 uses a dual-gate MOSFET. To understand how the dual-gate MOSFET operates, consider that the drain of the lower FET connects to the source of the upper FET. The circuit shown in Figure 8.22 has tuned input and output circuits that limit the band of amplified frequencies. Dual-gate MOSFETs work as high-frequency amplifiers and as mixers.

Progress Check
You have now completed objective 8.

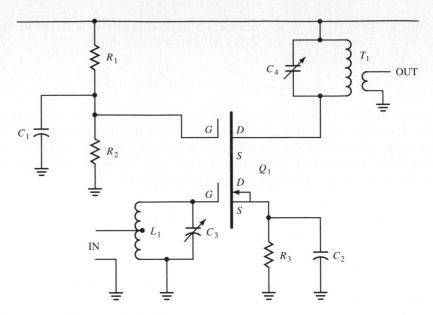

FIGURE 8.22 Dual-gate MOSFET circuit

CMOS ANALOG AND DIGITAL APPLICATIONS

Analog Applications

MOSFETs have many uses in analog circuitry. Since MOSFETs have a combination of a low "on" resistance, an extremely high "off" resistance, low leakage currents, and low capacitance, they have become widely used as voltage-controlled switches. An ideal analog switch acts like an ideal mechanical switch in that it will pass a signal to a load without affecting the quality of the signal.

Figure 8.23 shows an E MOSFET used as an analog switch. The device does not conduct with its gate grounded. With the gate grounded, the drain-to-source resistance will equal approximately 10,000 MΩ. No signal can pass through. Applying the supply voltage to the gate allows conduction in the MOSFET channel. With the E MOSFET saturated the drain-to-source resis-

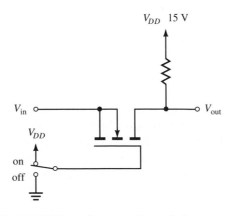

FIGURE 8.23 E MOSFET used as an analog switch

tance drops to a range of 25–100 Ω. Like mechanical switches, the MOSFET switch will allow signals to pass in either direction.

The circuit shown in Figure 8.23 has several limitations. Although the circuit will work for positive signals up to 10 V, larger signals will cause the drain-to-source resistance to rise. In turn, the MOSFET will go back to its nonconductive state. In addition, the circuit will only switch signals that are in the positive range. Negative signals will drive the MOSFET to turn on with a grounded gate. Tying the gate to a negative 15 V along with the positive 15 V will allow the circuit to switch signals in range from a negative 10 V to a positive 10 V.

Figure 8.24 shows another analog application for MOSFETs. Because MOSFETs have no gate current, they are excellent choices as current sources for op-amps. In the figure, the E MOSFET acts as a current sink. Current is sampled through a resistor and compared with the noninverting input of the op-amp. With no gate current, the sampling can occur with fewer errors. Op-amps will be covered in detail in Chapter 11.

A device such as a bipolar junction transistor with its base current would introduce errors into the sample.

Progress Check
You have now completed objective 9.

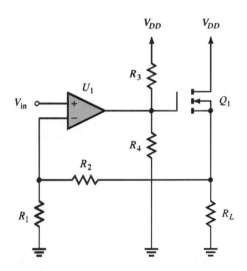

FIGURE 8.24 E MOSFET used as a current source for an op-amp

Digital Applications

In addition to the analog applications, MOSFETs also work well for digital applications such as logic circuits. In Figure 8.25a, we see two E MOSFETs stacked one on top of the other. The two transistors make up a **CMOS circuit,** or a *c*omplementary pair of *MOS*FET transistors. While the top transistor is a P-channel device, the bottom transistor is an N-channel device. Essentially, the two MOSFETs function much like a push-pull switch.

When an input signal goes positive and exceeds the threshold voltage of the bottom transistor, the bottom transistor will turn on and provide a path to ground for the output. Essentially, the output is pulled to ground. If we labeled the positive input voltage as a one, then a zero would represent the input at ground. A zero input signal would shut off the bottom transistor and turn on the top transistor. This action provides a path to power for the output. Thus, a zero input signal causes the output to rise to V_{DD}, or a logic 1.

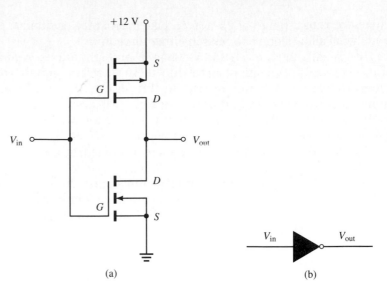

+12 V

V_{in}

V_{out}

(a)

V_{in} ▶ V_{out}

(b)

FIGURE 8.25 (a) E MOSFETs are connected together to produce an inverted output. (b) Logic symbol for an inverter.

We can describe this action as an inverter, one of the basic structures of the logic family. Figure 8.25b shows the logic symbol for an inverter. An **inverter** changes a high input to a low output or a low input to a high output. Unfortunately, inverters have one bad characteristic. As the input jumps between supply voltage and ground, the MOSFETs can conduct. This conduction can produce large current spikes.

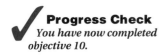

Progress Check
You have now completed objective 10.

REVIEW SECTION 8.6
1. Capacitive reactance disrupts the normal operation in a MOSFET because when the frequency of the applied signal gets too high, the insulated gate will appear as a _____ to the input signal.
2. Describe how the cascode amplifier overcomes the problem caused by the capacitive reactance.
3. _____-_____ MOSFETs act as high-frequency amplifiers and as mixers.
4. Why do MOSFETs work well in analog circuits?
5. How are MOSFETs used as switches with both positive and negative input signals?
6. Describe the construction of a push-pull switch.
7. Show how MOSFETs are used in digital applications.

TROUBLESHOOTING

Electrostatic damage may occur through the simple handling of a MOSFET. Here are a few things you should know about handling static-sensitive devices.

1. Metal oxide silicon devices should always arrive packed in antistatic, conductive foam. Never remove the MOSFET from the container until you are ready to install it.

2. Sometimes the MOSFET package will include a metal ring or foil that surrounds the four leads. This ring or foil shorts out the leads and helps to prevent static buildup. Always remove the ring before installing the MOSFET.

3. Connect all test instruments, metal test benches, and soldering irons to earth ground.

4. Wear an antistatic wrist strap that places the body at earth ground.

5. Do not inject signals into these devices with no applied DC voltages.

6. Never remove the device from the circuit when the power is on.

7. Use antistatic floor mats.

✓ **Progress Check**
You have now completed objective 11.

TROUBLESHOOTING QUESTIONS
1. What precautions do manufacturers take to protect MOSFETs during shipping?
2. A metal ring is sometimes found surrounding the leads of a MOSFET. What purpose does this ring serve?
3. All of the tools, as well as yourself, should be connected to what, when handling MOSFETs?
4. What shouldn't be applied to MOSFETs when no DC voltages are present?
5. With the power on, what shouldn't be done with a MOSFET device?

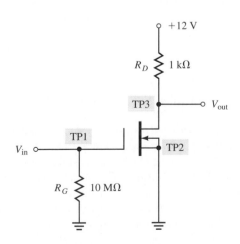

6. Given the figure above, determine if R_G, R_D, or the MOSFET has failed. Is the failed part open or bad? The DC voltages at all three test points are equal to 0 V; V_{GS}(cutoff) = 28 V; I_{DSS} = 6 mA.

SUMMARY

MOSFETs are another member of the field effect transistor family. A MOSFET, or metal oxide semiconductor field effect transistor, has a gate that is insulated from the device and uses an electrical field across a gate capacitance to control current. MOSFETs have several advantages. The polarity of the bias voltage and changes in ambient temperature do not adversely affect MOSFET operation. In addition, MOSFETs have low leakage currents, low input capacitance, and good high-frequency response.

There are two types of MOSFETs and two modes of operation. The two types of MOSFETs are the depletion, or D, MOSFET, and the enhancement, or E, MOSFET. The two modes of operation are the depletion mode and the enhancement mode.

The D MOSFET can operate in the depletion or the enhancement mode. When operating in the depletion mode, the D MOSFET operates similarly to the JFET.

Characteristics seen with JFETs, such as I_{DSS} and V_{GS}, also apply to the D MOSFET. In the depletion mode, current control is done by removing major charge carriers from the conduction channel. The D MOSFET can also operate in the enhancement mode. By applying a forward bias, the conduction channel is enhanced with additional major charge carriers.

D MOSFETs will conduct with forward bias, reverse bias, or zero bias, while E MOSFETs will conduct only with a forward bias. For the D MOSFET, several types of bias methods exist. Depletion bias is similar to the bias used for a JFET. This type of bias consists of an AC input signal applied directly to the gate. D MOSFETs also use voltage divider bias. However, MOSFETs do not require a source resistor. Since the D MOSFET has unique bias qualities, the device will also operate with zero bias or no gate-to-source voltage.

The E MOSFET will operate only in the enhancement mode. Major charge carriers must be introduced into the conduction channel before the E MOSFET will conduct. This is called the enhancement mode. It is done by using forward-biasing arrangements, such as voltage divider bias and drain feedback bias.

E MOSFETs cannot use self-bias, current source bias, or zero bias since those bias types would produce depletion mode operation. Voltage divider bias will work with the E MOSFET. In addition, the E MOSFET will conduct with drain feedback bias. With this type of bias, a portion of the output is brought back to the gate and establishes the gate-to-source voltage.

MOSFETs are excellent devices for both analog and digital applications. Analog applications include voltage-controlled switches and current sources for op-amps. Digital applications include inverters.

Simply handling a MOSFET may introduce electrostatic damage. As with all static-sensitive devices, precautions must be observed. Some of these precautions include packing the devices in antistatic, nonconductive foam, grounding all test equipment and benches, wearing an antistatic wrist strap, and using antistatic floor mats.

PROBLEMS

Refer to Figure 8.26 for problems 1 through 4.
1. Determine I_D for the circuit.
2. Determine g_{mo} for the MOSFET used in the circuit.
3. Determine g_m at the Q point.
4. Determine the unloaded voltage gain.

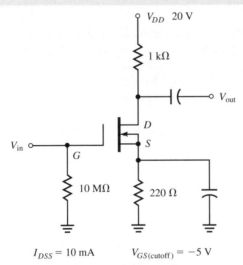

FIGURE 8.26 Circuit for problems 1 through 4

Refer to Figure 8.27 for problems 5 through 12.
5. Determine the input impedance of the circuit.
6. Determine the mode of operation.
7. Determine V_{GS}.
8. Determine g_{mo}.
9. Determine I_D.
10. Determine g_m at the Q point.
11. What is the DC voltage at the drain with respect to ground?
12. What is the unloaded voltage gain of the circuit?
13. A D MOSFET has the following specifications: $I_{DSS} = 10$ mA and $V_{GS(off)} = -8$ V. A positive 4 V is applied to the gate. Determine I_D.
14. The D MOSFET mentioned in problem 13 is used in a circuit with zero bias. Determine I_D.
15. An E MOSFET has the following specifications: $I_{D(on)} = 12$ mA at 10 V. The threshold voltage is $+3$ V. What is the value of K for this particular device?

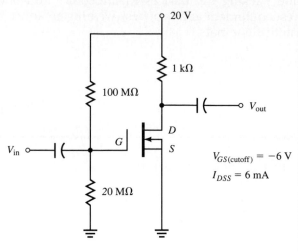

FIGURE 8.27 Circuit for problems 5 through 12

Refer to Figure 8.28 for problems 16 through 18.
16. The E MOSFET has an $I_{D(on)}$ of 8 mA at a V_{GS} of 12 V. Determine the R_D that sets up this condition.
17. Determine the value of K for the E MOSFET. Use this value to plot out a transconductance curve for the device. Plot out at least 5 points.
18. Determine the load line for the circuit if $R_D = 1.2$ kΩ.

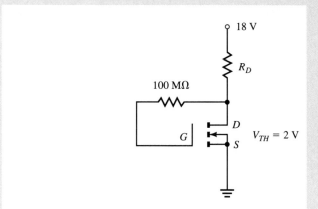

FIGURE 8.28 Circuit for problems 16 through 18

9

TRANSISTOR SWITCHING

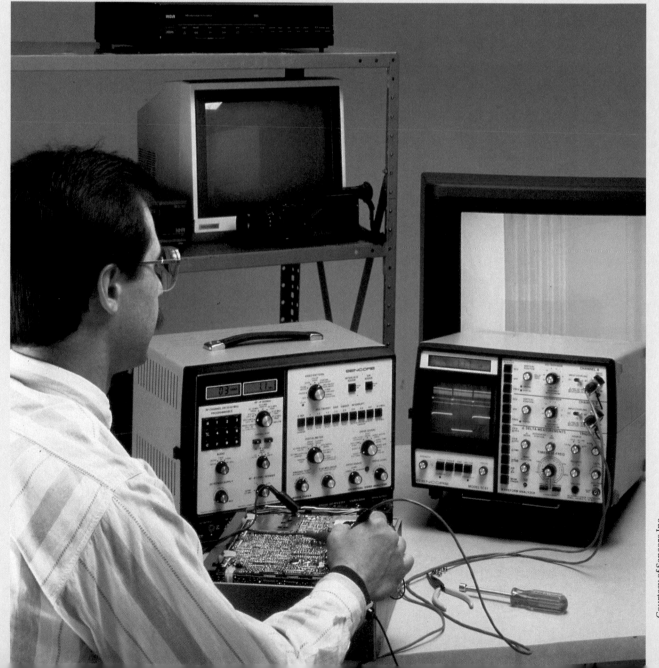

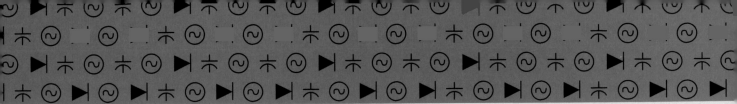

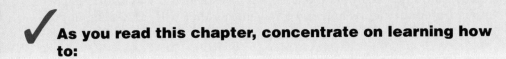

✓ **As you read this chapter, concentrate on learning how to:**

1. Determine if a transistor switch is operating correctly
2. Determine the effects of leakage current in transistor switching circuits
3. Determine the effects of saturation voltage on the transistor switch and the voltage delivered to a load
4. Determine the minimum base current required to saturate a transistor switch
5. Determine the maximum input frequency to a transistor switch
6. Improve the switching speeds of transistor switches using speedup capacitors
7. Analyze Schmitt triggers using discrete transistors as the switching elements
8. Analyze a transistor switch in which the transistor is a FET
9. Set up and analyze unijunction transistor relaxation oscillators
10. Recognize different types of signals
11. Analyze the operation of ramp generators

INTRODUCTION

In this chapter, we will look at the transistors we have studied in a different way. Rather than concentrate on the transistor as an amplifier, we will study transistors that function as solid state switches. Transistor switches allow us to switch devices off and on rapidly. In addition to rapid switching, transistor switches can be used to switch a number of control circuits with one control signal. Although the need for switching an attached device on and off is clear, we will also look at some special switching circuits in this chapter. For the most part, we will consider transistor operation as the device moves from saturation to cutoff. The active region, the region used for amplification, will have little impact in this chapter.

Among the circuits that we will see, the Schmitt trigger circuit minimizes erratic switching operation due to electrical noise at the input of the circuit. We will also look at several different types of ramp generators, a circuit that produces a ramp waveform at its output. Along with other applications, ramp waveforms are useful in timing applications and in controlling the sweep rate of oscilloscopes.

In addition, we will examine another type of transistor—the unijunction transistor, or UJT. This device has many uses in industrial electronic applications. As we study UJT operation, we will investigate how this device may be used to produce ramps and spike waveforms that work in timing applications as well as switching applications.

9.1   TRANSISTOR SWITCHING OPERATIONS

Any electronic switching function requires a device that will not conduct under certain conditions and will conduct under other conditions. Figure 9.1 shows a transistor in a common-emitter configuration attached to a load resistor. Of the three possible transistor configurations, the common emitter provides both the current and voltage gain needed for switching. In the common-emitter configuration, the transistor controls an output signal with a small input signal. In this particular diagram, the transistor acts like an ideal switch. An ideal switch would show as either a complete open or a complete short. In addition, an ideal switch does not dissipate energy and changes conditions instantaneously. Also, an ideal switch requires no power to stay in a condition or to switch to a condition.

However, the transistor is neither a mechanical device nor an ideal switch. A solid state device will exhibit small variations from the ideal because of internal resistance, leakage resistance, and capacitance. While the internal resistance of a device prevents it from becoming a true short circuit, the leakage resistance allows current to flow within the device when it is open. Capacitance, caused by the terminals of a semiconductor acting as plates, keeps the device from switching instantaneously.

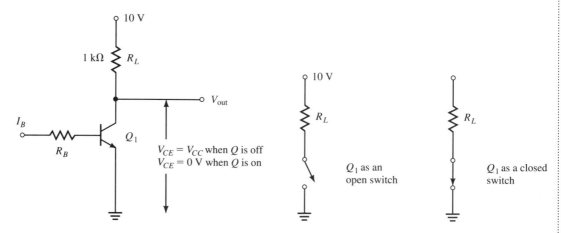

FIGURE 9.1   Transistor switch using the common-emitter configuration and illustrating the concept of an ideal switch

In a transistor switch, a positive voltage placed on the base of the transistor causes the transistor to enter saturation. The collector-to-emitter voltage of the transistor, or V_{CE}, drops to near zero. During this time, current flows through the load. When voltage is removed from the base, the transistor enters cutoff, and the current drops to zero. During cutoff, the collector-to-emitter voltage equals the collector supply voltage.

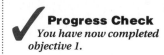

Progress Check
You have now completed objective 1.

TRANSISTOR SWITCHING CHARACTERISTICS

Leakage Current

Figure 9.2 shows the collector curves for a typical NPN transistor. Note the two shaded areas on this diagram. The area labeled as the cutoff region lies under the collector curve where the base current equals zero. Cutoff occurs when both transistor junctions are reverse-biased. During cutoff, the transistor has a high output impedance. Even though no base current flows, a small amount of collector current continues to flow. Leakage current causes the flow of collector current.

Transistor specification sheets may show leakage current as I_{CBO}, I_{CO}, or even I_{CEX}. I_{CBO} is a common designation, meaning that collector current, or I_C, combines with the base open, or BO, condition. As we know, leakage current flows even when the transistor is not conducting and does not allow the device to work as a true circuit open. Therefore, a small amount of current flows through the load at all times. In addition, this current is highly sensitive to temperature changes.

Let us take a look at how leakage current affects the operation of a transistor switch.

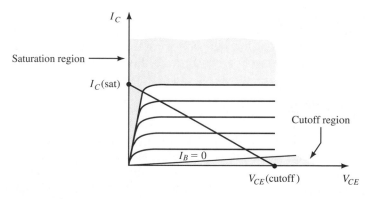

FIGURE 9.2 Characteristic curves of a transistor showing the switching regions

---

EXAMPLE 9.1

Determine the voltage at the output of the circuit shown in Figure 9.1.

From a specification sheet, we know that the leakage current for a 2N3904 NPN transistor is 50 nA.

Solution

In the circuit of Figure 9.1, this leakage current will produce a voltage drop across the load of

$$50 \text{ nA} \times 1 \text{ k}\Omega = 50 \text{ μV}$$

The voltage at the output would then be

$$V_{out} = 10 \text{ V} - 50 \text{ μV} = 9.99995 \text{ V}$$

PRACTICE PROBLEM 9.1

A transistor has a leakage current of 100 nA. If this current flows through an 8-kΩ load, how much voltage will be dropped because of the leakage current?

Progress Check
You have now completed objective 2.

As you can see from example 9.1, the amount of leakage current flowing through the load when the transistor is off has a minimal effect on circuit operation.

Saturation Voltages

Now let us study the second shaded region in Figure 9.2, the saturation region. During saturation, both junctions of the transistor are forward-biased. Also, a saturated transistor has a larger-than-normal base current and a smaller-than-normal collector-to-emitter voltage. When the transistor reaches saturation, the collector does not act as a constant current source. Internally, the transistor has only a small amount of resistance.

The saturation voltage, or V_{CE}(sat), will usually equal 0.2 to 0.3 V. However, as the current drawn from the device begins to increase, the saturation voltage also increases. For further illustration of the increase, refer to the specification sheet of the 2N3904 shown in Figure 9.3. The saturation voltage for the transistor equals 0.2 V when the collector current equals 1 mA and rises to 0.3 V with 50 mA of current flowing through the transistor.

EXAMPLE 9.2

Using the specification of V_{CE}(sat) for a 2N3904, determine the effects the saturation voltage has on the circuit operation for Figure 9.1.

If an ideal switch were connected to the load and closed, then the load voltage would be V_{CC}. However, a transistor switch, which is saturated, would drop approximately 0.2 V. The load voltage would then be less than ideal.

Solution

The first equation depicts the reduction in the collector-to-emitter voltage drop:

$$V(\text{load}) = V_{CC} - V_{CE}(\text{sat}) = 9.8 \text{ V}$$

This loss of voltage affects the current through the load:

$$I(\text{load}) = \frac{9.8 \text{ V}}{1 \text{ k}\Omega} = 9.8 \text{ mA}$$

MAXIMUM RATINGS

Rating	Symbol	Value	Unit
Collector-Emitter Voltage	V_{CEO}	40	Vdc
Collector-Base Voltge	V_{CBO}	60	Vdc
Emitter-Base Voltage	V_{EBO}	6.0	Vdc
Collector Current — Continuous	I_C	200	mAdc
Total Device Dissipation @ T_A = 25°C Derate above 25°C	P_D	625 5.0	mW mW/°C
*Total Device Dissipation @ T_C = 25°C Derate above 25°C	P_D	1.5 12	Watts mW/°C
Operating and Storage Junction Temperature Range	T_J, T_{stg}	−55 to +150	°C

*THERMAL CHARACTERISTICS

Characteristic	Symbol	Max	Unit
Thermal Resistance, Junction to Case	$R_{\theta JC}$	83.3	°C/W
Thermal Resistance, Junction to Ambient	$R_{\theta JA}$	200	°C/W

*Indicates Data in addition to JEDEC Requirements.

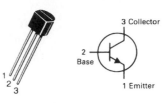

2N3903
2N3904

CASE 29-04, STYLE 1
TO-92 (TO-226AA)

GENERAL PURPOSE TRANSISTOR

NPN SILICON

ELECTRICAL CHARACTERISTICS (T_A = 25°C unless otherwise noted.)

Characteristic		Symbol	Min	Max	Unit
OFF CHARACTERISTICS					
Collector-Emitter Breakdown Voltage(1) (I_C = 1.0 mAdc, I_B = 0)		$V_{(BR)CEO}$	40	—	Vdc
Collector-Base Breakdown Voltage (I_C = 10 μAdc, I_E − 0)		$V_{(BR)CBO}$	60	—	Vdc
Emitter-Base Breakdown Voltage (I_E = 10 μAdc, I_C = 0)		$V_{(BR)EBO}$	6.0	—	Vdc
Base Cutoff Current (V_{CE} = 30 Vdc, V_{EB} = 3.0 Vdc)		I_{BL}	—	50	nAdc
Collector Cutoff Current (V_{CE} = 30 Vdc, V_{EB} = 3.0 Vdc)		I_{CEX}	—	50	nAdc
ON CHARACTERISTICS					
DC Current Gain(1)		h_{FE}			—
(I_C = 0.1 mAdc, V_{CE} = 1.0 Vdc)	2N3903 2N3904		20 40	— —	
(I_C = 1.0 mAdc, V_{CE} = 1.0 Vdc)	2N3903 2N3904		35 70	— —	
(I_C = 10 mAdc, V_{CE} = 1.0 Vdc)	2N3903 2N3904		50 100	150 300	
(I_C = 50 mAdc, V_{CE} = 1.0 Vdc)	2N3903 2N3904		30 60	— —	
(I_C = 100 mAdc, V_{CE} = 1.0 Vdc)	2N3903 2N3904		15 30	— —	
Collector-Emitter Saturation Voltage(1) (I_C = 10 mAdc, I_B = 1.0 mAdc) (I_C = 50 mAdc, I_B = 5.0 mAdc)		$V_{CE(sat)}$	— —	0.2 0.3	Vdc
Base-Emitter Saturation Voltage(1) (I_C = 10 mAdc, I_B = 1.0 mAdc) (I_C = 50 mAdc, I_B = 5.0 mAdc)		$V_{BE(sat)}$	0.65 —	0.85 0.95	Vdc
SMALL-SIGNAL CHARACTERISTICS					
Current-Gain — Bandwidth Product		f_T			MHz
(I_C = 10 mAdc, V_{CE} = 20 Vdc, f = 100 MHz)	2N3903 2N3904		250 300	— —	

FIGURE 9.3 **Specification sheet for 2N3903 and 2N3904 transistors.** (*Courtesy of Motorola, Inc.*)

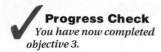

Progress Check
You have now completed objective 3.

A 4.7-kΩ load is designed to operate with 5 V across it. How much current should flow through the load? Since 0.3 V is lost during saturation, the load receives only 4.7 V. How much did the current through the load drop because of $V_{CE}(sat)$?

If we operate the transistor of Figure 9.1 correctly, we must make sure that enough base current is present for the transistor to reach saturation. In the next example, we will investigate how to keep the base current at the required level.

Determine the minimum base current required to saturate the transistor of Figure 9.1.

In example 9.2, we determined that 9.8 mA of current would reduce the saturation voltage to 0.2 V. Again turning to the transistor specification sheet, we will check the amount of DC current gain, or beta. With collector currents near 10 mA, beta will be in the range of 100 to 300. To ensure that we will have sufficient base current, we will use the minimum value for beta.

Solution
Dividing the amount of current flowing through the load by the minimum value of beta will give us the minimum value of base current:

$$I_B(min) = \frac{9.8 \text{ mA}}{100} = 98 \text{ μA}$$

If we provide a minimum of 98 μA of current into the base, we can be sure that the transistor of Figure 9.1 will saturate when it is turned on.

Progress Check
You have now completed objective 4.

Transistor Switching Speeds

In this section, we will consider transistor switching speeds and the characteristics associated with them. The speed at which a transistor can switch from on to off is important in digital applications where analog signals must keep pace with digital signals. Since a transistor cannot act as an ideal switch and switch instantaneously from one condition to another, a time delay develops. Obviously, keeping the time delay as short as possible is an important part of the switching circuit. When we study transistor switching speeds, we will also analyze waveforms and time intervals.

Figure 9.4 shows three varieties of step voltage waveforms that might be used to switch transistors on and off. Figure 9.4a is a pulse waveform, where the on time is considerably less than the off time. The relationship between on time and the waveform period is called **duty cycle.**

$$\text{Duty cycle} = \frac{\text{on time}}{\text{waveform time}} \times 100$$
$$= \frac{t(\text{on})}{t(\text{on}) + t(\text{off})} \times 100$$

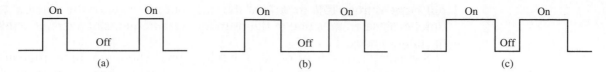

FIGURE 9.4   Step voltage waveforms. (a) Pulse waveform. (b) Square waveform.
(c) Rectangular waveform.

Figure 9.4b depicts a square wave that has an equal on time and off time. This waveform has a 50% duty cycle. The last waveform, shown in Figure 9.4c, is a rectangular waveform that has a long on time and a short off time. Duty cycle is more than 50%.

All three waveforms exhibit **step voltage change.** Voltage changes levels in sharp, rapid transitions, or steps. In theory, this transition takes place in zero seconds. In practice, that is not possible. Let us look at how a transistor handles step voltages. In addition, we will see how to optimize a switching circuit so that the circuit will use less time when switching from one level to the other.

We can compare input waveforms and output waveforms in Figure 9.5. The input waveform in Figure 9.5a is the theoretical ideal for purposes of comparison and shows how the transistor reacts. In Figure 9.5, we see that the input to the base is an ideal step change. As the base current goes positive, it creates a voltage across the base resistor that forward biases the emitter-base diode. The output waveform in the figure is drawn just below the input. As the two waveforms show, there is a delay before the output collector current begins to rise. We define this delay as **delay time.**

A more precise definition for delay time is the time required for the collector current to reach 10% of its final value after the base current begins to flow. Delay time will usually be indicated in symbols as t_d. The timing diagram of Figure 9.5b shows the next interval of time, called **rise time.** We define rise time as the amount of time needed by a transistor to change its out-

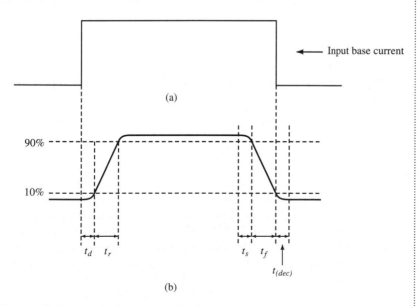

FIGURE 9.5   (a) Ideal input. (b) Time intervals of output waveform.

put current from 10% to 90% of the final output value of the current. Rise time, designated t_r, is one of the primary measurements of a circuit's ability to change states.

Switching a transistor off also requires time. There are three intervals of time associated with turn-off. We can define **storage time** as the amount of time required by a transistor output to drop from 100% of the output level to 90% of the output level after the removal of the input base current. It is designated t_s.

Before the transistor can shut off, the charge carriers crossing the collector-base diode must be removed. These charge carriers become stored or trapped in the collector-base region when the junction reverse biases. Removing the charge carriers from the depletion zone requires an interval of time called the storage time. The length of storage time depends on whether the transistor is saturated and the hardness of the saturation. If you go back to example 9.3, you should see that we needed a minimum of 98 μA of current to saturate the transistor. More base current would overdrive the transistor. This affects switching speed and storage time.

Fall time is defined as the time required to drop from 90% of the output level to 10% of the output level. It is designated t_f.

Figure 9.5b shows another time interval, called **decay time** *(t(dec))*, or the time required for the transistor to switch completely off. Since the transistor effectively turns off by the time the current has dropped to 10% of the output level, this time is usually not included in calculating switching speeds.

EXAMPLE 9.4

Determine the time required to turn on and off a 2N3904 transistor. Assume that the transistor will be in saturation when turned on. What are the implications of this information?

The specification sheet for the transistor lists switching times required for going from one state to another. From the specification sheet, we have:

$$t_d = 35 \text{ ns} \qquad\qquad t_r = 35 \text{ ns}$$
$$t_s = 200 \text{ ns} \qquad\qquad t_f = 50 \text{ ns}$$

Solution
Adding the times together gives us the amount of time needed to switch this transistor on and off:

$$35 \text{ ns} + 35 \text{ ns} + 200 \text{ ns} + 50 \text{ ns} = 320 \text{ ns}$$

If a transistor cannot keep up with the rate of switching, the output waveform becomes distorted. Look at Figure 9.6A. In this diagram, a step voltage turns on a transistor. Immediately, though, the transistor turns off. Since the transistor has just reached 100% of the final output current, no discernible pulse width exists. When the transistor attempts to turn off immediately, the output waveform resembles a voltage spike instead of the input step voltage.

As a general guideline, a reasonable reproduction of the input voltage at the output requires a pulse width at least 10 times larger than the total switching time. You can see the pulse width diagrammed in Figure 9.6b. Fol-

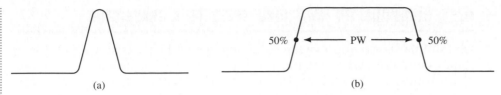

(a)   (b)

FIGURE 9.6   (a) Output current and voltages are not allowed to stabilize. (b) Allowing
output to stabilize. PW = pulse width.

lowing this guideline means that we should have a pulse width of at least
3200 ns for the circuit shown in Figure 9.1. Similar arguments also exist for
the total time off (T). Subsequently, the minimum time off should equal 3200
ns, and the total time for a complete cycle equals the sum of time on plus
time off, or 6400 ns.

Converting this to a frequency yields:

$$f = \frac{1}{T} = 156 \text{ kHz}$$

This information tells us that if a frequency greater than 156 kHz is applied to
a type 2N3904 transistor, the output will be distorted and start to resemble
Figure 9.6a rather than Figure 9.6b.

PRACTICE PROBLEM 9.3

A transistor has the following specifications:

$t_d = 15$ ns $t_r = 30$ ns
$t_s = 150$ ns $t_f = 50$ ns

Using the guidelines discussed in this section, determine the minimum pulse
width time.

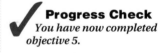

Progress Check
*You have now completed
objective 5.*

REVIEW SECTION 9.1
1. An electronic switch is a device that will not _____
   under certain conditions.
2. List some properties of an ideal switch.
3. What properties make a transistor switch different from an ideal
   switch?
4. Leakage current prevents the transistor from acting like a true short
   circuit. True or false?
5. What does $V_{CE}(\text{sat})$ denote, and what is this normally equal to?
6. Quick switching speeds are preferable for digital applications. True
   or false?
7. What do these parameters mean: delay time, rise time, fall time,
   storage time, and decay time?
8. What is required of the input voltage pulse in order to give a
   reasonable output reproduction, and what is it compared to?

9.2 IMPROVING THE SWITCHING SPEED OF A TRANSISTOR

One simple way to improve the switching rate of the transistor is not driving it into saturation. When the transistor does not reach saturation, the storage time reduces to almost zero. In the case of a 2N3904 transistor, this eliminates up to 200 ns of switching time. On the down side of the time gain, though, the nonsaturated, conducting transistor will have a voltage V_{CE}(sat) across the collector-to-emitter terminals. The switch consumes power. So, when deciding which transistor switch to use in a particular application, a designer often has to make tradeoffs.

ADDING A SPEED-UP CAPACITOR TO THE SWITCHING CIRCUIT

Placing a **speed-up capacitor** into a transistor switching circuit also improves switching performance. This is diagramed in Figure 9.7. As Figure 9.7 shows, the speed-up, or **commutating, capacitor** is placed across the base resistance. Before we discuss how to calculate the value of the speed-up capacitor, let us consider how it affects circuit operation.

First let us see what happens to the base-emitter diode without a speed-up capacitor. When the transistor turns on and the diode has become forward-biased, the charge carriers stored in the previously reverse-biased depletion region must be removed. This consumes time. Both delay time and rise time can improve if the voltage across the base-emitter diode goes to zero or close to zero just before the transistor turns on. In addition, the junction capacitance of the base-emitter diode also charges and discharges faster if the transistor is overdriven, which means driving the transistor into hard saturation.

FIGURE 9.7 Adding a speed-up capacitor to a transistor switching circuit. *(Photograph courtesy of AVX Corporation)*

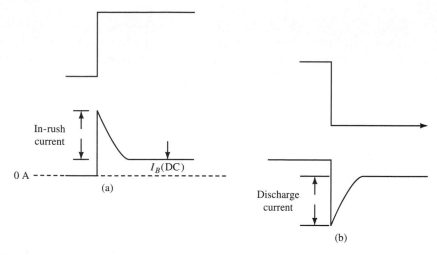

FIGURE 9.8   (a) Step voltage waveform at turn-on. (b) Waveform at turn-off.

Unfortunately, overdriving the transistor affects storage time. We have found a way to turn on the transistor faster at the expense of turning it off more slowly. Turning the transistor off faster requires a large negative input voltage that depletes the stored charge carriers of the collector-base diode faster. However this reverse biases the base-to-emitter voltage and extends the time required to turn on the transistor.

Seemingly, a "no-win" situation exists. Yet, the speed-up capacitor offers a solution. Look at the waveforms in Figure 9.8. In Figure 9.8a, we see what happens at turn-on. A positive-going step voltage is applied to the input of Figure 9.7. This changing voltage recognizes the speed-up capacitor as a short across the base resistor. For a small amount of time, while the capacitor charges up, there is a large in-rush of current. Consequently, this over-drives the transistor and turns the device on faster.

After the capacitor charges, the current into the base drops to the value given by dividing the input voltage value by the value of the base resistor. The heavy saturation of turn-on is reduced, and the amount of current flowing through the collector-base diode drops. Because an excessive amount of charge carriers do not reside in the forward-biased collector-base diode, the amount of storage time does not lengthen.

In Figure 9.8b, we see what happens at turn-off. When the input voltage is reduced to zero the speed-up capacitor discharges through the base resistor and produces a negative voltage. This negative voltage removes charge carriers quickly from the collector-base diode and decreases the storage time. The speed-up capacitor improves turn-on time as well as turn-off time. Examples 9.5 and 9.6 depict the proper selection and calculation of the speed-up capacitor.

EXAMPLE 9.5

Set up the transistor switch of Figure 9.9 to ensure saturation.

Figure 9.9 shows a transistor switch set up with a 1-kΩ load and a 47-kΩ base resistor. The transistor was tested and found to have a beta of 173.

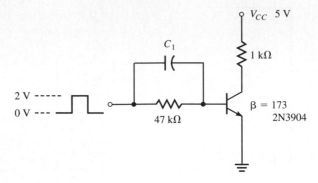

FIGURE 9.9 Transistor switch

Solution

To saturate the transistor, the voltage from collector to emitter must drop to 0.2 V. This means the collector resistor must drop 4.8 V. With 4.8 V across the collector resistor, the collector current must equal:

$$I_C = \frac{4.8\text{ V}}{1000\ \Omega} = 4.8\text{ mA}$$

Dividing the collector current by the beta value, we get:

$$I_B = \frac{4.8\text{ mA}}{173} = 27.75\ \mu\text{A}$$

Generally, use the value of beta at the saturation specification given for the transistor for the calculations. This should ensure that the transistor will reach saturation.

This current flows through the 47-kΩ base resistor producing a voltage drop of:

$$V_{RB} = 27.75\ \mu\text{A} \times 47\text{ k}\Omega = 1.3\text{ V}$$

The base-emitter diode adds another 0.7 V to the base-emitter loop. To produce the desired base current, we must input the sum of the base voltage and base-to-emitter voltage values, or

$$V_{BB} = V_{RB} + V_{BE}$$
$$= 1.3\text{ V} + 0.7\text{ V} = 2\text{ V}$$

PRACTICE PROBLEM 9.4

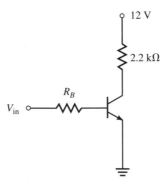

FIGURE 9.10

A transistor with a beta of 150 is connected to a 2.2-kΩ load as shown in Figure 9.10. Determine the base current needed to saturate the transistor. Assume $V_{CE} = 0.2$ V.

EXAMPLE 9.6

Determine the size of the speed-up capacitor when the input signal is 200 kHz.

The signal was applied to the circuit in Figure 9.9 without the speed-up capacitor, C_1. The result is shown in Figure 9.11a. In this display, the top waveform is the input voltage to the base. The horizontal sensitivity setting for

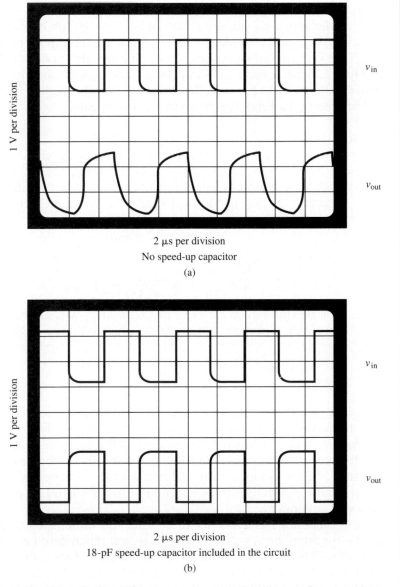

2 µs per division
No speed-up capacitor

(a)

2 µs per division
18-pF speed-up capacitor included in the circuit

(b)

FIGURE 9.11 (a) Waveform display without a speed-up capacitor. (b) Speed-up capacitor placed across the base resistor.

both the waveforms shown in Figure 9.11 is 2 μs per division, and the vertical sensitivity for the top waveform is 1 V per division.

The bottom waveform, the output voltage measured at the collector, has a vertical sensitivity of 2 V per division. Since we are working with a common-emitter configuration that produces a 180° phase shift, the output is inverted. The second waveform in Figure 9.11a indicates that the transistor reaches saturation. As you can see from the figure, the waveform is distorted.

In Figure 9.11b, a speed-up capacitor is placed across the base resistor. The shape of the output waveform has significantly improved. The following calculations show how the speed-up capacitor was selected. Figure 9.11a shows that the total time for the pulse will be 5 μs.

Half this time, or 2.5 μs, is the transistor's off time. This time is often called the **recovery time,** or **settling time.** It is the time required for the speed-up capacitor to return to its discharged condition.

When the speed-up capacitor discharges through the base resistor, it produces a voltage spike similar in shape to the current spike shown in Figure 9.8b. This negative voltage spike forces charge carriers out of the collector-base junction faster than if the transistor had been left to switch off itself. To ensure that this spike has the effect we want, we should not allow the amplitude of the negative spike to return to 0 V too quickly.

Solution

Let us set the voltage remaining across the plates of the capacitor after it discharges to 5% of its starting value. A capacitor requires 3 RC time constants to discharge 95%. Normally, some residual voltage will remain on the plates of the capacitor so that it does not discharge completely before the input pulse is reapplied. Thus:

$$3RC = T = 2.5 \text{ μs}$$

To solve for C, rearrange.

$$C = \frac{T}{3R} = \frac{2.5 \text{ μs}}{3 \times 47 \text{ k}\Omega} = 17.7 \text{ pF}$$

The nearest standard value is 18 pF. This value was inserted into the circuit of Figure 9.9. It produced the results shown in Figure 9.11b.

As you can see, the collector output voltage waveform is less distorted than the waveform shown in Figure 9.11a. Unfortunately, there is a limit to how much we can improve the waveform. If the capacitor has a larger value than that calculated in example 9.6, it will not fully discharge during the settling time. The next input pulse will not see the capacitor as a short. Therefore, the large inrush of current needed to turn on the transistor of Figure 9.9 faster will not appear.

If the capacitor has a smaller value than the value calculated in example 9.6, the negative input voltage at the base will reach zero before the settling time ends. This affects the ability of the circuit to reduce turn-off time.

Let us take a look at one more example of a BJT switch before we turn our attention to some applications for transistor switches.

EXAMPLE 9.7

Determine the size of the base resistor and the speed-up capacitor when the input signal to the circuit in Figure 9.12 is 5 V at 140 kHz.

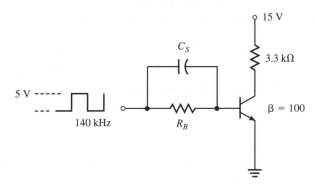

FIGURE 9.12 Switching circuit for example 9.7

Solution

We start the analysis by finding the collector current needed to drop the collector-to-emitter voltage to 0.2 V. This current is:

$$I_C = \frac{14.8\ V}{3.3\ k\Omega} = 4.48\ mA$$

If this amount of current flows, the collector resistor will drop 14.8 V and will leave 0.2 V on the collector. Using a beta value of 100, we can predict the base current needed:

$$I_B = \frac{I_C}{\beta}$$
$$= \frac{4.48\ mA}{100}$$
$$= 44.8\ \mu A$$

With a 5-V input, 4.3 V is dropped across the base resistor when the transistor conducts. Using the values that we have, we set a value for the base resistor:

$$R_B = \frac{4.3\ V}{I_B} = \frac{4.3\ V}{44.8\ \mu A} = 96\ k\Omega$$

The next lowest standard value is a 91-kΩ resistor. Selecting a lower-value resistor ensures that we will have enough base current to saturate the transistor. With the base resistor value set, we can turn our attention to finding a value for the speed-up capacitor.

At an input frequency of 140 kHz, we have a total time for one cycle equal to:

$$\frac{1}{140\ kHz} = 7.14\ \mu s$$

Dividing this in half gives an amount for the settling time:

$$\frac{7.14\ \mu s}{2} = 3.57\ \mu s$$

Leaving 5% of the charging voltage present means that the settling time will be equal to 3 RC time constants.

$$T = 3.57 \ \mu s = 3 \ RC = 3 \times 91 \ k\Omega \times C$$

$$C_s = \frac{3.57 \ \mu s}{(3 \times 91 \ k\Omega)} = 13 \ pF$$

Using the next available value, we can select a 15-pF capacitor and insert it into the circuit.

Progress Check
You have now completed objective 6.

REVIEW SECTION 9.2
1. What are the benefits of not driving a transistor into saturation? What are the drawbacks?
2. A commutating capacitor can be placed across the base resistor to speed up switching. True or false?
3. What happens to the base-emitter diode under normal switching conditions?
4. What effect does adding a speed-up capacitor have during turn-on?
5. Describe the effect of a speed-up capacitor during the turn-off cycle.
6. What happens to the value of beta if the transistor is overdriven?
7. What effects will a speed-up capacitor have on switching if it is too large or too small?

9.3 TRANSISTOR SWITCHING APPLICATIONS

SCHMITT TRIGGERS

A transistor switching configuration called the **Schmitt trigger** has become widely used in digital systems. This circuit configuration responds to a slow-changing input signal with an output signal that has fast transition times. Figure 9.13 shows a simple Schmitt trigger circuit.

Schmitt triggers are designed to detect when an input voltage crosses a voltage threshold. In most cases, these thresholds are called the **upper threshold,** or **upper trip point (UTP),** and the **lower threshold,** or **lower trip point (LTP).** When a voltage swing in the proper direction crosses either of these trip points, the outputs of the circuits switch. Almost any shape waveform can work as an input to the device. The output is designed to switch rapidly from one voltage level to another. This rapid switching produces a step voltage output. Schmitt triggers appear as squaring circuits and amplitude comparators.

Figure 9.14a shows two transistors configured as a Schmitt trigger. Note that transistors Q_1 and Q_2 share a common-emitter resistor. The resistor plays an important role in the circuit operation. With the input voltage applied to the base of Q_1, the output voltage comes from the collector of Q_2. Neither transistor acts as an amplifier; both function as switches.

With no input to the circuit, Q_1 is off. During this time, Q_2 conducts. The output voltage will be less than the collector supply voltage. When an input voltage large enough to turn on Q_1 is applied, Q_1 turns on and Q_2 turns off. At

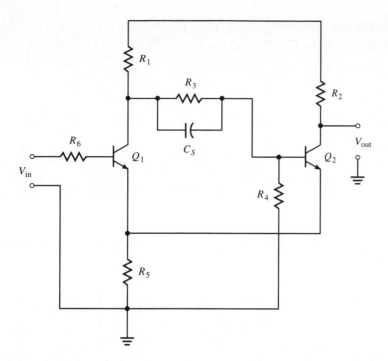

FIGURE 9.13 Simple Schmitt trigger circuit

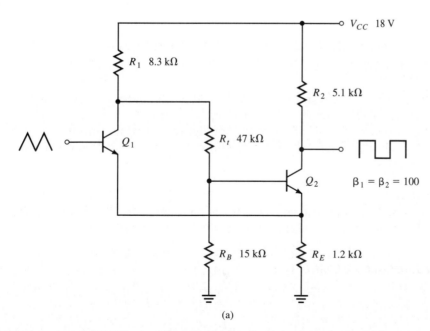

(a)

FIGURE 9.14 (a) A transistor-driven Schmitt trigger. (b) The Schmitt trigger when Q_1 is off and Q_2 is on. (c) Input waveforms at the base of Q_1 and output waveforms at the collector of Q_2. (continued on next page)

FIGURE 9.14 continued

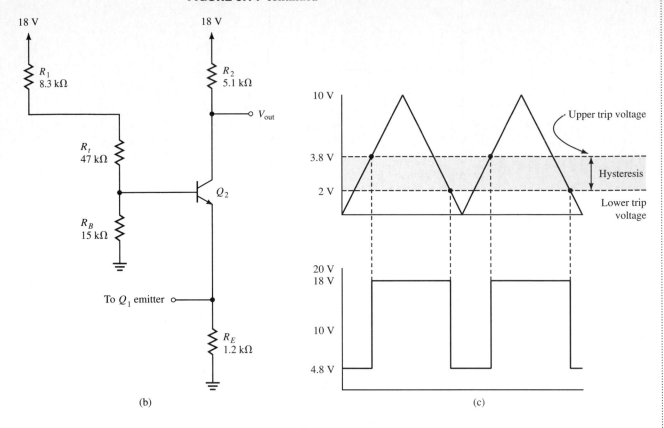

(b)

(c)

this time, the output taken from the collector of Q_2 will rise toward the collector supply voltage.

ANALYZING FOR THE UPPER THRESHOLD POINT

EXAMPLE 9.8

For the circuit shown in Figure 9.14a, predict the amount of voltage seen at the emitter of Q_2 when the input to Q_1 is zero volts.

We start our analysis of the circuit of Figure 9.14 by assuming that Q_1 is off and Q_2 is conducting. Figure 9.14b shows the equivalent circuit when Q_1 is off and Q_2 is on.

Solution
Q_2 is biased through the voltage divider that consists of resistors R_1, R_t, and R_b. The total resistance of this voltage divider in Figure 9.14 is:

$$R_{total} = (15 \text{ k}\Omega + 47 \text{ k}\Omega + 8.3 \text{ k}\Omega) = 70.3 \text{ k}\Omega$$

The voltage at the base of Q_2 is:

$$V_B = \frac{V_{CC} \times R_b}{R_{total}}$$

$$V_B = \frac{18V \times 15 \text{ k}\Omega}{70.3 \text{ k}\Omega}$$

$$V_B = 3.84V$$

Subtracting 0.7 V for the barrier potential voltage of Q_2, we can predict the amount of voltage seen at the emitter of Q_2:

$$V_E = 3.84\,V - 0.7\,V = 3.14\,V$$

Because of the shared emitter resistor, 3.14 V is dropped at the emitter of Q_1. To turn on Q_1, we must have at least 3.84 V applied to its base. This voltage is the upper threshold voltage for the circuit shown in Figure 9.14a.

The emitter resistor is important to the switching speed of the Schmitt trigger. When connected in parallel, bipolar transistors are "current hogs." That is to say, one of the transistors will try to take all the current and starve the other transistor. When the emitters of Q_1 and Q_2 are connected together, each transistor has the same emitter current source as the other. As one of the transistors starts to conduct, it will starve the other transistor of emitter current, causing a more rapid change in collector voltage.

When the input voltage at the base of Q_1 equals the upper threshold voltage, Q_1 begins to conduct. V_C of Q_1 begins to decrease, which decreases V_B of Q_2. As Q_2 begins to cut off, the Schmitt trigger output begins to rise. Since bipolars are current hogs and the emitter resistor limits the available current, Q_1 begins to take all the available emitter current and starves Q_2, driving Q_2 into cutoff faster. The result is a fast-rising output waveform even if the input waveform is slow-rising. When Q_2 is in cutoff, the output voltage should be very near V_{CC}.

ANALYZING FOR THE LOWER THRESHOLD POINT

With Q_1 on, we can look at what must happen to switch back or turn Q_2 on and Q_1 off. If the input voltage to Q_1 decreases, the emitter current flowing through the device will also drop. In turn, this lowers the amount of current flowing out of the collector of Q_2. Consequently, the voltage drop across R_1 goes down.

A smaller voltage drop across R_1 causes the voltage at the base of Q_2 to rise. At some point, Q_2 will switch on. The voltage at the base of Q_1 when Q_2 switches on is called the lower threshold point. When Q_2 begins to turn on, it begins to hog all the emitter current, starving Q_1 into cutoff. The upper threshold point and the lower threshold point cannot be the same amount of voltage. The difference between the amount of voltage seen at the two trip points is known as the **hysteresis voltage.**

The hysteresis voltage is illustrated in Figure 9.14c. The Schmitt trigger will change from a low value to a high value only on the rise side of the waveform and only if the upper trip point has been crossed. Both conditions must have been met. When the upper trip point is crossed on the falling side of the input waveform, the output does not change. The Schmitt trigger will change from a high value to a low value only on the fall side of the waveform and only if the lower trip point has been crossed. Notice that, although the input waveform has slow rising and falling sides, the output has fast rising and falling sides. The hysteresis voltage (V_H) is the difference between the upper trip point (UTP) and the lower trip point (LTP).

$$V_H = V_{UTP} - V_{LTP}$$
$$3.84\,V - 2\,V = 1.84\,V$$

To turn Q_2 on, the input voltage must cross the upper threshold point while going positive. After this occurs, any positive or negative voltage crossing the upper threshold point will have no effect on the circuit. To turn Q_1 back off, the input voltage must cross the lower threshold point during a negative swing. Once this occurs, any positive or negative voltage crossing the lower threshold point will have no further effect on circuit operation.

EXAMPLE 9.9

Calculate the lower threshold point for the circuit shown in Figure 9.14a. Most circuits have resistors and other components whose tolerances are at least 5%. When calculating circuit currents and voltages, you need not carry answers out to more than three significant digits. This example uses standard components whose tolerance is 5%, and calculations are rounded to three significant digits.

For the modified drawing in Figure 9.15, the lower threshold point is the voltage at which the input voltage equals the Q_2 base voltage. In order to solve for V_B of Q_2, we must first solve for the current (I_2) going through R_b when the LPT is reached. Then we use Ohm's law to solve for V_B.

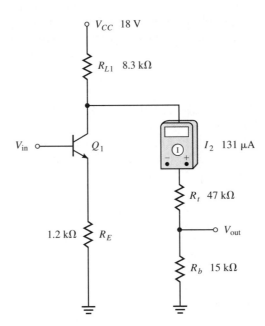

FIGURE 9.15 Schmitt trigger from Figure 9.14a when Q_1 is on and Q_2 is off

Solution
We start by placing the above statement about the lower threshold point into a simple equation. This equation shows that the current flowing through the base resistor produces a voltage drop equal to the input voltage:

$$I_2 \times R_b = V_{in}$$

Since $V_{in} = (I_E R_E) + 0.7\,V$, we have:

$$I_2 \times R_b = (I_E \times R_E) + 0.7\,V$$

With this information, we get:

$$I_2 \times R_b - 0.7\,V = I_E \times R_E \qquad \text{(Conclusion 1)}$$

Conclusion 1 may be rewritten as:

$$I_E = \frac{(I_2 \times R_b) - 0.7 \text{ V}}{R_E}$$

Next, we write an equation for the loop in Figure 9.15.

$$(I_2 \times R_b) + (I_2 \times R_t) + R_{L1}(I_2 + I_c) = 18 \text{ V}$$

Substituting known values into the equation gives us:

$$(I_2 \times 15 \text{ k}\Omega) + (I_2 \times 47 \text{ k}\Omega) + 8.3 \text{ k}\Omega(I_2 + I_c) = 18 \text{ V}$$

We can factor out I_2.

$$I_2(15 \text{ k}\Omega + 47 \text{ k}\Omega + 8.3 \text{ k}\Omega) + (8.3 \text{ k}\Omega \times I_c) = 18 \text{ V}$$

Now we can combine like terms.

$$I_2(70.3 \text{ k}\Omega) + (8.3 \text{ k}\Omega \times I_c) = 18 \text{ V}$$

Basic algebra tells us that we cannot solve a single equation with two unknowns. However, we can use conclusion 1 to eliminate I_C from the equation. Also remember that I_C is equal to I_E. Substitute I_E for I_C.

$$I_2(70.3 \text{ k}\Omega) + (8.3 \text{ k}\Omega \times I_E) = 18 \text{ V}$$

Now substitute conclusion 1 for I_E.

$$I_2\left(70.3 \text{ k}\Omega\right) + \left[8.3 \text{ k}\Omega \times \frac{(I_2 \times R_b) - 0.7 \text{ V}}{R_E}\right] = 18 \text{ V}$$

Once again we can substitute the known values.

$$I_2(70.3 \text{ k}\Omega) + \left[8.3 \text{ k}\Omega \times \frac{I_2 \times 15 \text{ k}\Omega - 0.7 \text{ V}}{1.2 \text{ k}\Omega}\right] = 18 \text{ V}$$

The next two steps will simplify the equation.

$$I_2\left(70.3 \text{ k}\Omega\right) + \left[8.3 \text{ k}\Omega \times \left(\frac{I_2 \times 15 \text{ k}\Omega}{1.2 \text{ k}\Omega} - \frac{0.7 \text{ V}}{1.2 \text{ k}\Omega}\right)\right] = 18 \text{ V}$$

$$I_2(70.3 \text{ k}\Omega) + [8.3 \text{ k}\Omega(12.5 \times I_2 - 583 \text{ }\mu\text{A})] = 18 \text{ V}$$

Then multiply the second term

$$I_2(70.3 \text{ k}\Omega) + [8.3 \text{ k}\Omega \times (12.5 \times I_2) - (8.3 \text{ k}\Omega \times 583 \text{ }\mu\text{A})] = 18 \text{ V}$$

Combine like terms

$$I_2(70.3 \text{ k}\Omega) + [(104 \text{ k}\Omega \times I_2) - 4.84 \text{ V}] = 18 \text{ V}$$

Discard unneeded brackets.

$$(I_2 \times 70.3 \text{ k}\Omega) + (104 \text{ k}\Omega \times I_2) - 4.84 \text{ V} = 18 \text{ V}$$

Move the voltage to the right side of the equation.

$$(I_2 \times 70.3 \text{ k}\Omega) + (104 \text{ k}\Omega \times I_2) = 18 \text{ V} + 4.84 \text{ V} = 22.8 \text{ V}$$

Combine the remaining terms.

$$I_2 \times 174.3 \text{ k}\Omega = 22.8 \text{ V}$$

Rearrange to solve for I_2.

$$I_2 = \frac{22.8 \text{ V}}{174.3 \text{ k}\Omega} = 131 \text{ }\mu\text{A}$$

When multiplied by the value of the base resistor, this gives us a value for the voltage at the base of Q_2 and the lower threshold voltage:

$$\begin{aligned}\text{LTP} &= I_2 \times R_b \\ &= 131 \text{ }\mu\text{A} \times 15 \text{ k}\Omega \\ &= 2.0 \text{ V}\end{aligned}$$

Change R_b on Figure 9.15 to 22 kΩ and recompute the LTP.

Although the number of equations seems intimidating, the equations developed to calculate the lower threshold point are based on Kirchhoff's voltage law and Ohm's law. Working through the equations allows you to analyze the circuit and see how the lower threshold point affects circuit action.

Input-Output Characteristics

Since Schmitt triggers provide fast transition times at the output, the configuration becomes particularly useful when converting a sine wave input signal to a series of pulses at the output. We call this series of rectangular wave pulses a **pulse train.** Pulses in a pulse train must appear at regular intervals. The quantity of pulses that appear within 1 s is defined as the **pulse repetition frequency** or **rate.** Figure 9.16 compares the input signal and output signal of a Schmitt trigger.

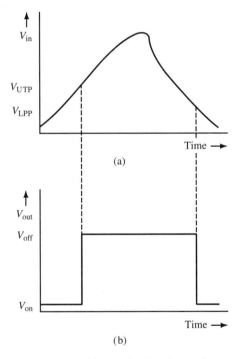

FIGURE 9.16 (a) Input signal for a Schmitt trigger. (b) Output signal for a Schmitt trigger.

EXAMPLE 9.10

Determine the output voltage of the circuit shown in Figure 9.14a when Q_2 is conducting.

With Q_2 conducting, Q_1 is off, and no collector current flows out of Q_2. The base of Q_2 is then driven by the voltage divider that consists of R_b, R_t, and R_{L1}. The total resistance of this divider is 70.3 kΩ.

Solution

Applying the voltage divider theorem to the divider, we get:

$$\frac{15 \text{ k}\Omega}{70.3 \text{ k}\Omega} \times 18 \text{ V} = 3.8 \text{ V}$$

or a voltage that we calculated earlier. Subtracting 0.7 V, we get an emitter voltage of:

$$3.8 \text{ V} - 0.7 \text{ V} = 3.1 \text{ V}$$

We can use this value to determine the amount of emitter current flowing through the emitter resistor:

$$\frac{3.1 \text{ V}}{1.2 \text{ k}\Omega} = 2.58 \text{ mA}$$

This current flows out the collector of Q_2 and through R_{L2} and produces a voltage drop of:

$$2.58 \text{ mA} \times 5.1 \text{ k}\Omega = 13.2 \text{ V}$$

When subtracted from the collector voltage supply, it leaves:

$$18 \text{ V} - 13.2 \text{ V} = 4.8 \text{ V}$$

when Q_2 conducts. This is plotted in Figure 9.14c

✓ Progress Check
You have now completed objective 7.

REVIEW SECTION 9.3

1. A Schmitt trigger has an _____ signal that is slow-changing, and an _____ that has a fast transition.
2. What happens when a voltage swing in the correct direction crosses a trip point of a Schmitt trigger?
3. Schmitt triggers appear as _____ circuits and _____ comparators.
4. Describe the action of the two transistors in a Schmitt trigger.
5. Describe the action of the transistors when an applied voltage is much greater than the upper threshold voltage.
6. What is the difference between the upper and lower threshold voltages known as?
7. What two conditions must be met for a Schmitt trigger to switch properly?
8. A Schmitt trigger is good for switching a sinewave input into what type of output?

9.4 FET SWITCHING

JFET SWITCHING APPLICATIONS

Both JFETs and MOSFETs work well in switching applications. When working with JFET switching circuits, we need to know two parameters. These parameters—$R_{DS}(on)$ and $I_D(cutoff)$—tell us how a JFET switch varies from the ideal condition. $R_{DS}(on)$ is the resistance between the source and the drain when the JFET is on. $I_D(cutoff)$ is the leakage current through the device when it is cut off.

To see how we can use these parameters in our analysis, let us proceed to example 9.11.

EXAMPLE 9.11

Using the circuit shown in Figure 9.17, calculate the amount of voltage dropped across the load resistance and the amount of current flowing through the load.

The circuit shown in Figure 9.17 shows a JFET switch. For this example, the device is an N-channel JFET designated as a 2N4856. The JFET has a maximum leakage current of 0.5 μA and maximum drain-to-source resistance of 25 Ω.

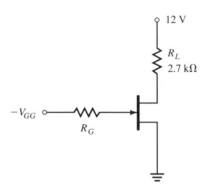

FIGURE 9.17 JFET switching circuit

Solution

First, let us look at the device when it is off. When $V_{GS(cutoff)}$ is applied to the device, the current will drop to no more than 0.5 μA. This current will flow through the load, producing:

$$5 \, \mu A \times 2.7 \, k\Omega = 13.5 \, mV$$

The output voltage when the device is off is:

$$12 \, V - 13.5 \, mV = 11.9865 \, V$$

Since we have seen an extremely small change in voltage from input to output, we can ignore the leakage current and treat the JFET as open when it is not conducting. This rule applies to nearly all JFETs.

When the JFET is on and saturated, the drain-to-source resistance will drop only a small voltage.

$$V_{DS(on)} = \frac{25 \, \Omega}{\left(25 \, \Omega + 2.7 \, k\Omega\right)} \times 12 \, V = 111 \, mV$$

We next determine the voltage dropped across the load and from this the current flowing through the load. These values equal:

$$V_2 = 12 \, V - 111 \, mV = 11.9 \, V$$

and

$$I_2 = \frac{11.9 \, V}{2.7 \, k\Omega} = 4.4 \, mA$$

Checking the specifications for the 2N4856 JFET, we find that the device can provide a minimum I_{DSS} of 50 mA. Therefore, the on-state current requirements are no problem.

FETs have switching characteristic parameters similar to those discussed earlier for the BJT. Both devices have turn-on times and turn-off times. The specifications for the 2N4856 JFET list: delay time = 6 ns; rise time = 3 ns; turn-off time = 25 ns; however, JFETS do not have storage times.

✓ **Progress Check**
You have now completed objective 8.

MOSFET SWITCHING APPLICATIONS

MOSFETs can also can work as switches. Figure 9.18 shows an E MOSFET configured as a switch. Generally, MOSFETs have specifications similar to those of JFETS—turn-on times, turn-off times, $R_{DS}(on)$, and $I_{DS}(cutoff)$. MOSFET switches have the advantage of low power consumption. Nevertheless, MOSFET switches also have the disadvantage of slower switching speeds. Manufacturers of MOS-based devices have continued to improve switching speeds of the devices they produce.

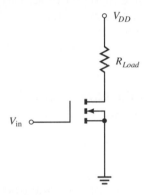

FIGURE 9.18 E MOSFET switch

REVIEW SECTION 9.4

1. Which parameters give the information on how a JFET switch varies from the ideal conditions?
2. What is an advantage to using MOSFETs instead of JFETs as switches?
3. Name a disadvantage to using MOSFETs instead of JFETs as switches.

9.5 UJT OPERATION

Figure 9.19 is a cross-sectional representation of the **unijunction transistor,** or UJT. From the appearance of the drawing, it may seem that the UJT and the JFET have a great deal in common. Although the two devices have a similar construction technique, they operate very differently. In Figure 9.19a the

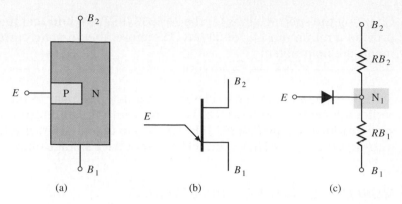

FIGURE 9.19 (a) Cross-sectional view of a UJT. (b) Schematic symbol of a UJT. (c) Electrical equivalent of a UJT.

P-type material is heavily doped. Also, note the designations of the UJT terminals. The three terminals are referred to as Base 1 (B_1), Base 2 (B_2), and the emitter (E).

Specifically, Figure 9.19 shows an N-channel UJT that has its emitter attached to the P-type material. This UJT has heavily doped P-type material with many excess holes available to conduct current and a lightly doped N-channel. In fact, the channel has a resistance that we can measure with an ohmmeter. Typical resistance values for the N-channel are in the range of 7 kΩ to 10 kΩ when the emitter is open. Specification sheets for a UJT list this value as the **interbase resistance** (R_{BB}).

Another important UJT specification is the **intrinsic standoff ratio** (ISR). This is simply the ratio of the lower base resistance to the total resistance of the channel before the emitter is turned on:

$$\eta = \frac{R_{B1}}{\left(R_{B1} + R_{B2}\right)}$$

Let us use the following example to see how the intrinsic standoff ratio affects UJT operation.

EXAMPLE 9.12

Using the representation of the UJT shown in Figure 9.19, determine how much voltage must be applied to the emitter to forward bias the PN junction. An 18-V supply attaches to B_2.

Let the interbase resistance equal 10,000 Ω and the intrinsic standoff ratio equal 0.7. With an intrinsic standoff ratio of 0.7, R_{B1} is 7000 Ω and R_{B2} is 3000 Ω. Since the ratio of voltages in a series circuit equals the ratio of resistances in a series circuit, we can apply the voltage divider method to N_1 of Figure 9.19c. The intrinsic standoff ratio is simply another form of the voltage divider method.

Solution
Using the interbase resistance and R_{B1} as the values for the voltage divider, we have:

$$\frac{7000 \ \Omega}{10,000 \ \Omega} = 0.7$$

which verifies the claim that the instrinsic standoff ratio is a form of voltage divider action. Then

$$0.7 \times 18\,V = 12.6\,V \text{ at } N_1$$

To forward bias the emitter, we must place 0.7 V more on the emitter terminal. This places the emitter 0.7 V higher than the N_1 node. So:

$$V_p = 12.6\,V + 0.7\,V = 13.3\,V$$

We label UJT emitter voltage as V_p because it signifies the peak emitter voltage.

PRACTICE PROBLEM 9.6

A UJT has the following specifications: $R_{BB} = 12\,k\Omega$; the intrinsic standoff ratio is 0.6. If the device is powered by $V_{BB} = 20\,V$, what voltage is required to turn on (fire) the emitter?

THE NEGATIVE RESISTANCE OF THE UJT

When the emitter diode begins to conduct, the excess holes present in the P-type material migrate across the forward-biased diode and then enter the lower base region of the device. These positively charged holes attract electrons. Since a large number of holes exist in the semiconductor substance, the electrons have an easier time moving through the lower base. In effect, the conduction or firing of the emitter decreases the resistance of the lower base. A negative resistance forms due to an increase in current while the voltage drops. This characteristic makes the UJT useful.

A plot of emitter voltage versus emitter current is shown in Figure 9.20. V_v and I_v represent the valley voltage and valley current, while V_p and I_p represent peak voltage and peak current. At first, as the voltage increases at the emitter terminal, the leakage current also increases. Ohm's law describes the UJT action.

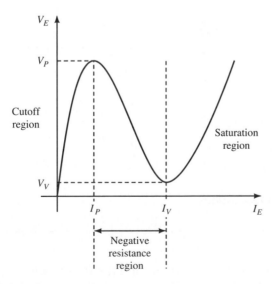

FIGURE 9.20 UJT characteristic curve

After the peak voltage is reached, the junction becomes forward biased and the voltage on the emitter decreases but the current continues to increase. This occurs because the resistance decreases faster than the voltage. The region of the characteristic curve between the peak voltage and the valley voltage is called the **negative resistance region.** After reaching the valley voltage, any further attempts to increase the current force the voltage to rise. Increases in current beyond the value of the valley current will saturate the junction.

UJT RELAXATION OSCILLATOR

A common circuit using a UJT is a **UJT relaxation oscillator.** Such a circuit produces a ramp-out waveform or a series of very-fast-rising high-amplitude voltage spikes. These spikes are taken from B_1 or B_2, while the ramp is taken from the emitter terminal. Industrial electronic applications use the voltage spikes to turn on semiconductor power devices such as SCRs or triacs.

A UJT relaxation oscillator is depicted in Figure 9.21. In this circuit, an RC network connects to the emitter. Before the emitter fires, the emitter capacitor will charge through the emitter resistor. The resistor ties to the 12-V supply. At some point, the voltage across the capacitor reaches the peak voltage needed to fire the emitter.

When this happens, the emitter turns on and presents the capacitor with a low-impedance path for discharging. Discharge occurs rapidly because the resistance of the lower base drops quickly. Also, a large amount of current flows around the emitter–base 1 loop. When the capacitor has discharged below the valley current, the current that it supplies will no longer be enough to keep the emitter in the valley region. The emitter reopens. When this happens, the capacitor begins to charge through the emitter resistor, and the entire process repeats.

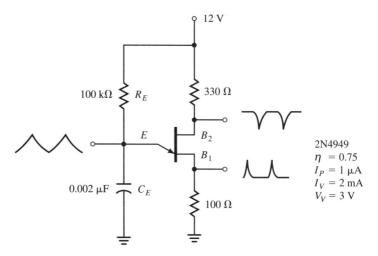

FIGURE 9.21 UJT relaxation oscillator

EXAMPLE 9.13

Analyze the circuit of Figure 9.21. Determine the frequency of operation and the output voltage across the capacitor.

According to specifications for the device, we need at least 1 μA of current to fire the device.

Solution

We start by calculating the peak voltage needed to fire the emitter:

$$0.75 \times 12\,\text{V} + 0.7\,\text{V} = 9.7\,\text{V}$$

At this time, the voltage across R_E equals:

$$12\,\text{V} - 9.7\,\text{V} = 2.3\,\text{V}$$

Since we know that 1 μA fires the emitter, the emitter resistor must have a value of:

$$R_{E(max)} = \frac{V_{RE}}{I_P}$$

$$R_{E(max)} = \frac{2.3\,\text{V}}{1\,\mu\text{A}} = 2.3\,\text{M}\Omega$$

Next, when the device discharges, the voltage drops to the valley voltage, or V_V. For the UJT used in Figure 9.21, this voltage equals 3 V. At this time, the voltage across the emitter resistor is:

$$V_{RE} = V_{CC} - V_V$$

$$V_{RE} = 12\,\text{V} - 3\,\text{V} = 9\,\text{V}$$

and the current associated with the condition is 2 mA.

Using those values, we can calculate the minimum size of the emitter resistor:

$$R_{E(min)} = \frac{V_{RE}}{I_V}$$

$$R_{E(min)} = \frac{9\,\text{V}}{2\,\text{mA}} = 4.5\,\text{k}\Omega$$

If R_E is less than 4.5 kΩ, we will get more than 2 mA through the emitter, and the UJT will enter the saturation region of the characteristic curve. When it does so, the emitter will not reopen, and the capacitor will not recharge. Circuit action stops. Thus, the acceptable value range for the emitter resistor is 4.5 kΩ to 2.3 MΩ. In this case, the circuit utilizes a 100-kΩ emitter resistor.

Next, we turn our attention to the frequency of operation. We can assume that the discharge time is very nearly instantaneous because of the large drop of resistance in the lower base. This eases our calculations. The emitter capacitor normally cycles from the valley voltage to the peak voltage. In this case, we have a change of:

$$V_{C(change)} = V_P - V_v$$

$$V_{C(change)} = 9.7\,\text{V} - 3\,\text{V} - 6.7\,\text{V}$$

The initial charging current would be:

$$I_{C(initial)} = \frac{V_{CC} - V_V}{R_E}$$

$$I_{C(initial)} = \frac{12\,\text{V} - 3\,\text{V}}{100\,\text{k}\Omega}$$

$$I_{C(initial)} = 90\,\mu\text{A}$$

However, this current will decay as the capacitor charges.

The exponential decay of the charging current forces us to use base e logs to determine the time it takes the capacitor to charge.

The following step-by-step procedure will help you find the oscillator frequency.

Step 1. Find V_{RE} when the UJT is saturated. In the solution we have found that to be 9 V.

Step 2. Find V_{RE} when the UJT is at its peak voltage ready to fire. In the solution we have found that to be 2.3 V.

Step 3. Divide the saturation voltage by the peak voltage to obtain 9 V/2.3 V = 3.91.

Step 4. Press the base e logarithm key. That key's script usually looks like this—LN, iN, in. If you are not sure, check your calculator manual. Your answer should be 1.36.

Step 5. Multiply the logrithmn by the RC time constant to obtain the waveform's period.

$$1.36 \times R_E \times C_E = 1.36 \times 100 \text{ k}\Omega \times 0.002 \text{ }\mu\text{F} = 272 \text{ }\mu\text{s}$$

Step 6. The oscillator's frequency is the inverse of the waveform's period.

$$f = \frac{1}{272 \text{ }\mu\text{s}} = 3.67 \text{ kHz}$$

EXAMPLE 9.14

Analyze the UJT relaxation oscillator shown in Figure 9.22. Determine the output frequency and the size of the voltage spikes across R_1.

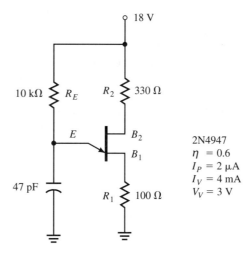

FIGURE 9.22 UJT relaxation oscillator

Solution
Starting with the peak voltage required to fire the emitter, we have:

$$V_P = (\eta \times V_{CC}) - V_{BE}$$
$$V_P = (0.6 \times 18 \text{ V}) + 0.7 \text{ V} = 11.5 \text{ V.}$$

Because the required peak current is 2 μA, the maximum value for R_E is

$$R_{E(max)} = \frac{V_{CC} - V_P}{I_P}$$

$$R_{E(max)} = \frac{(18\,V - 11.5\,V)}{2\,\mu A} = 3.25\,M\Omega$$

If the emitter resistor value is larger than 3.25 MΩ, the emitter will not fire. Next, we will determine the minimum value of emitter resistor by using the starting voltage across the emitter resistor, the valley current, and Ohm's law:

$$R_{E(min)} = \frac{V_{CC} - V_v}{I_v}$$

$$= \frac{18\,V - 3\,V}{4\,mA} = 3.75\,k\Omega$$

The circuit may use a resistor ranging from 3.75 kΩ to 3.25 MΩ. We have chosen to use a 10-kΩ resistor.

To find the output frequency, we follow the procedure shown in the preceding example:

Step 1. Find V_{RE} when the UJT is saturated.

$$V_{RE} = V_{CC} - V_V = 18\,V - 3\,V = 15\,V$$

Step 2. Find V_{RE} when the UJT is at its peak voltage ready to fire.

$$V_{RE} = V_{CC} - V_P = 18\,V - 11.5\,V = 6.5\,V$$

Step 3. Divide the saturation voltage by the peak voltage to obtain 15 V/6.5 V = 2.31.

Step 4. Press the base e logarithm key. Your answer should be 0.836.

Step 5. Multiply the logarithm by the RC time constant to obtain the waveform's period.

$$0.836 \times R_E \times C_E = 1.36 \times 10\,k\Omega \times 47\,\mu F = 393\,ns$$

Step 6. The oscillator's frequency is the inverse of the waveform's period.

$$f = \frac{1}{393\,ns} = 2.54\,MHz$$

Next, we determine the output amplitude of the voltage spikes across R_1. To understand what happens, take a look at Figure 9.23a. At the instant the peak voltage is reached, the emitter diode fires and becomes a low-imped-

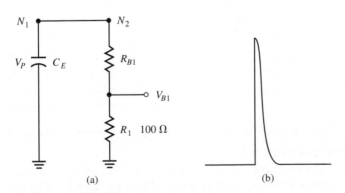

(a) (b)

FIGURE 9.23 (a) UJT oscillator for example. (b) Spike waveform.

ance path. In the figure, a wire between N_1 and N_2 represents the path. As you can see, the capacitor voltage is dropped across the combination of R_{B1}, the lower base of the UJT, and the external resistance attached to the B_1 lead. These two resistances form a voltage divider. Unfortunately, R_{B1} is a dynamic impedance and will change rapidly. Without more mathematical equations, we can only estimate the voltage across R_{B1} when the peak voltage is reached.

The valley voltage represents the estimation. At the moment that the UJT fires, the voltage across the external resistor is approximately:

$$V_p - V_v = 11.5\,V - 3\,V = 8.5\,V$$

The output across R_1 resembles the spike voltage waveform shown in Figure 9.23b. It will peak at no more than 8.5 V and quickly diminish as the resistance of the lower base drops.

We can calculate V_{B1} before the emitter fires, by using a simple voltage divider arrangement. The total resistance shown in Figure 9.24 is 6.43 kΩ. The voltage across the divider is:

$$\frac{100}{6.43\,k\Omega} \times 18V = 0.28\,V$$

Progress Check
You have now completed objective 9.

We can now conclude that the output from B_1 is a positive-going spike that starts at 0.28 V, rises to 8.5 V, and decays down to 0.28 V. The spikes appear at a rate of 2.54 MHz.

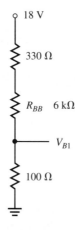

FIGURE 9.24 Equivalent circuit for Figure 9.23a

PRACTICE PROBLEM 9.7

Assume the standoff ratio of the UJT in Figure 9.22 has changed to 0.75. Determine the new peak voltage. Check to see if R_E is still within the acceptable range. If it is in the acceptable range, determine the new output frequency.

REVIEW SECTION 9.5
1. What do the following UJT specifications mean: I_p, V_p, V_v, and I_v?
2. Draw the schematic symbol for a UJT. How does it differ from the JFET symbol?

3. What is the intrinsic standoff ratio, and what equation is used to find this value?
4. What are the designations for the three UJT terminals?
5. What is meant by interbase resistance?
6. How does the negative resistance form in a P-type unijunction transistor?
7. What happens when the valley current is surpassed?
8. Describe the functional cycle of a UJT relaxation oscillator.
9. To determine the decay of current in a capacitor you would use the _____ _____ _____.

9.6 GENERATING SIGNALS

Throughout this text, we have discussed both DC and AC circuits. When we analyze DC circuits or the DC portion of a circuit, we can rely on familiar tools such as Ohm's law, Kirchhoff's law, and Thevenin equivalent circuits. When we analyze AC circuits or the AC portion of a circuit, we must also consider how voltages vary with time. In addition, we must remember how capacitors and resistors affect the amplitude and shape of waveforms. Before we begin to study advanced small-signal amplifiers, it may be wise to step back and take another look at signals.

SINE WAVES
Every time we work with small signals, we work with some amplitude of a sinusoidal or sine wave. Many texts will refer to a certain voltage at a certain frequency, such as a 5-μV signal at 3 MHz. That is a reference to a sine wave like the one pictured in Figure 9.25. An equation to illustrate the sine wave would appear as:

$$V = A \sin 2\pi ft$$

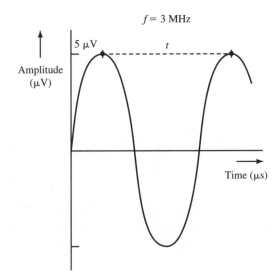

FIGURE 9.25 A sine wave

In the equation, A represents amplitude, *f* represents a frequency in hertz, and *t* represents some value of time.

If we discuss **frequency response,** we are defining the way that a circuit changes the amplitude of an applied sine wave as a function of frequency. Stereo equipment specifications will often list frequency responses for a range of 20 Hz to 20 kHz. Linear circuits driven by sine waves always respond with sine waves at their outputs. The sine waves may have a different amplitude and phase, but they still exist as sine waves.

SQUARE WAVES

Figure 9.26 shows a square wave or a signal that varies with time. Since the square wave works as a function of time, we consider both amplitude and frequency when working with square waves. Unlike the linear circuit driven by a sine wave, a linear circuit driven by a square wave will not respond with another square wave. As we saw earlier in this chapter, characteristics such as fall time, rise time, and decay time alter the appearance of the square wave.

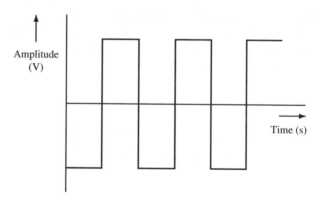

FIGURE 9.26 A square wave

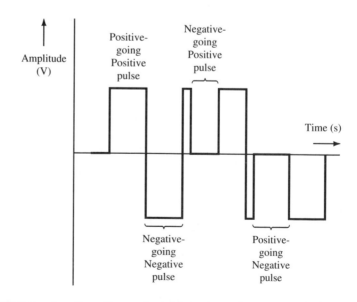

FIGURE 9.27 Examples of negative- and positive-going pulses

Pulses, like those shown in Figure 9.27 look similar to square waves. However, amplitude and one additional variable called pulse width define the pulse. When we discuss pulses, we talk about the pulse repetition rate and the ratio of the repetition rate to the repetition time period. Generally, we will see a "train" of equal pulses. Both square waves and pulses work as positive and negative polarities. In digital electronics, we see many applications for square waves and pulses.

RAMP AND TRIANGLE SIGNALS

A **ramp signal** has a slow linear rise and a rapid linear fall, Figure 9.28a. A **triangle signal** has a voltage that rises and falls at a constant rate as indicated by its symmetrical shape, Figure 9.28b. Transistor switching circuits also produce ramp voltages. This type of waveform is used in timing circuits and to control the horizontal sweep of CRT screens in television receivers, computer monitors, and oscilloscopes.

Progress Check
You have now completed objective 10.

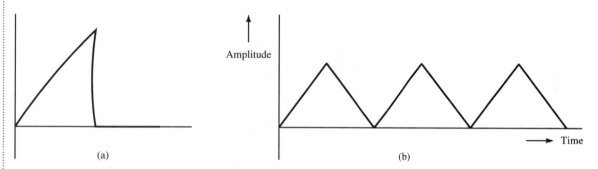

FIGURE 9.28 (a) Ramp waveform. (b) Triangle waveform.

SOME BASIC CIRCUITS

Knowing about the various signals is only one step in the process of understanding signal amplification. Some basic RC circuits affect the quality and amplitude of the signal as it travels from input to output. With that in mind, let us look at a differentiator, an integrator, a time delay circuit, and some ramp generators.

THE DIFFERENTIATOR CIRCUIT

The simple RC circuit shown in Figure 9.29 provides an output proportional to the rate of change of the input waveform. Given a square wave input signal, a **differentiator** will provide a spiked output waveform. As you may sus-

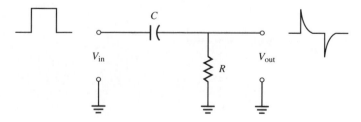

FIGURE 9.29 Differentiator circuit

pect, the differentiator is the same as a high-pass filter. An input signal produces a smaller output signal. Figure 9.30 illustrates the input and output waveforms. Applications utilize differentiators for detecting the leading and trailing edges of pulse trains.

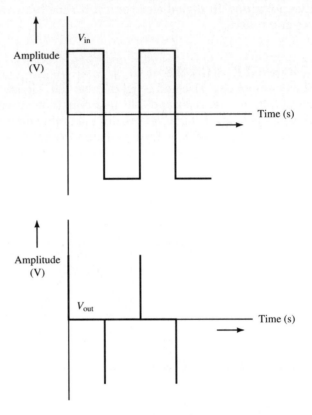

FIGURE 9.30 Input and output waveforms of a differentiator

THE INTEGRATOR CIRCUIT

The integral of something is the accumulation of a value over a period of time. An **integrator,** such as the one shown in Figure 9.31, accumulates volt-

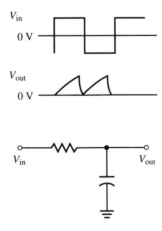

FIGURE 9.31 Integrator circuit

age over a period of time. A square input signal into an integrator produces an almost-perfect ramp signal at the output. As the input holds a steady voltage, the output rises or falls linearly. Just as the differentiator is a high-pass filter, the integrator is a low-pass filter. Integrators that have a large voltage across a large resistor work well as constant current sources.

THE TIME DELAY CIRCUIT

Since capacitors can affect the timing of a pulse, we can use a capacitor to produce a delayed pulse. Figure 9.32 shows two buffers symbolized by the triangles. The first buffer provides the input signal through a low source resistance. Because the RC circuit introduces decays, the output buffer switches slightly after the input buffer. Figure 9.33 illustrates how the time delay affects the output waveform.

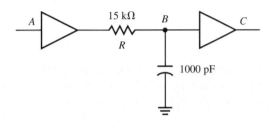

FIGURE 9.32 Time delay circuit

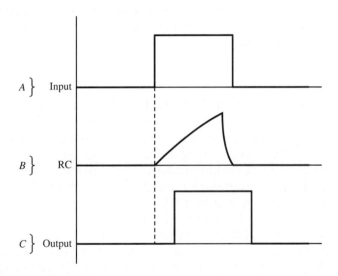

FIGURE 9.33 Input, RC, and output waveforms of a time delay circuit

GENERATING RAMP SIGNALS WITH AN RC CIRCUIT

The simplest **ramp generator** consists of a transistor switch in parallel with a capacitor. Such an arrangement is depicted in Figure 9.34. A voltage at the base resistor biases the transistor switch at saturation. About 0.2 V will appear across the transistor and the capacitor. The output will remain at this 0.2-V level until the switch is turned off when a negative input voltage is applied to the base of the transistor.

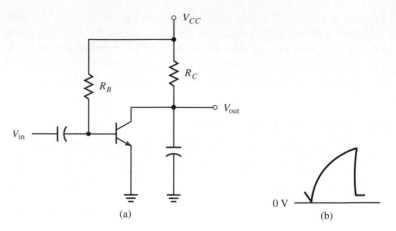

FIGURE 9.34 (a) RC ramp wave generator. (b) Typical ramp wave generator output.

Once the transistor is turned off, the capacitor can charge through the collector resistor. This charging voltage will have a shape similar to that diagrammed in Figure 9.34. Once the input pulse is removed from the circuit or its effect on the base of the transistor is no longer felt, the transistor will turn back on. When this happens, the transistor provides a low-resistance path for the capacitor to discharge and allows the rapid return to the lower output voltage, or $V_{CE}(sat)$. If the circuit of Figure 9.34b needed to produce a repetitive ramp output, a continuous stream of negative-going pulses would arrive at the base of the transistor.

RAMP GENERATORS AS CONSTANT CURRENT SOURCES

Using a constant current source improves the linearity of the output shown in Figure 9.34b. The circuit shown in Figure 9.35a is called a constant ramp generator. Its output is shown in Figure 9.35b. Although there are extra components in the circuit, the output ramp waveform nearly resembles a straight line. This advantage is important enough that it outweighs the additional cost of the extra parts shown in Figure 9.35.

For the most part, the circuit operation is similar to what we just discussed in the section on the RC ramp generator. Q_1 normally biases on through the base resistor. The base current is set so that Q_2 enters saturation, which keeps the voltage across the capacitor at 0.2 V. Q_2 also biases on through the voltage divider network. The normal path for electron flow is through the emitter of Q_1, out the collector of Q_1, into the collector of Q_2, out the emitter of Q_2, and into the power supply. Biasing Q_2 should allow the transistor to operate in the active region. In turn, its collector appears as a constant current source.

When a negative-going pulse becomes applied to the base of Q_1, the transistor shuts off. Capacitor C_2 begins to charge through Q_2. Now we can see the significant difference between this circuit and the circuit shown in Figure 9.34. The collector of Q_2 provides a constant current that the output capaci-

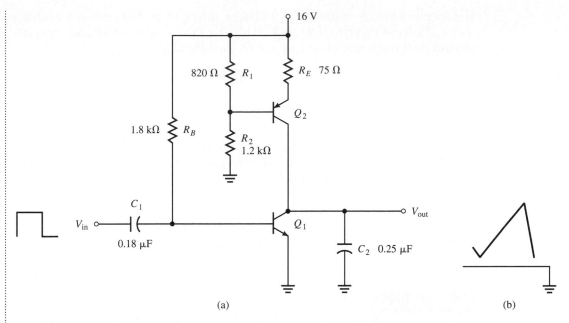

(a)

(b)

FIGURE 9.35 (a) Constant current ramp generator. (b) Typical constant current ramp generator output.

tor uses during charging. This feature is responsible for the linearity of the output ramp as shown in Figure 9.35b.

When Q_1 is allowed to turn back on, it provides a low-impedance path through which the output capacitor can discharge. This action also explains the rapid return to the saturation voltage. As before, if we want a continuous stream of output ramps, a continuous stream of input pulses must be applied to the circuit.

EXAMPLE 9.15

Analyze the constant current ramp generator of Figure 9.35a.

The input to the circuit will be a negative-going pulse of -2 V with a duration of 25 μs. Time between the pulses is 2.5 μs. The beta of each transistor is 100. We must check to see that the transistor is held off for the entire time the negative pulse is applied to the base.

Solution
When the pulse is applied to the base, the voltage across the base resistor is:

$$V_{RB} = V_{CC} - V_{BE} - (-2\,V)$$
$$= 16\,V - 0.7\,V + 2\,V$$
$$= 17.3\,V$$

Then:

$$I_{C1} = \frac{17.3\,V}{1.8\,k\Omega} = 9.6\,mA$$

The input coupling capacitor will charge through the base resistor with an initial charging current of 9.6 mA. With the following equation, we can approximate the amount of voltage across capacitor C_1:

$$V_{C1} = \frac{(I \times t)}{C}$$

$$V_{C1} = \frac{(9.6 \text{ mA} \times 25 \text{ μs})}{0.18 \text{ μF}} = 1.3 \text{ V}$$

C_1 has an exponential charge. However, the charging current will decay as C_1 charges. Since we used the initial charging current to test our answer, our answer will be larger than the measured amount. When the base is pulled down by -2 V from the pulse and the capacitor charges to 1.388 V, the voltage on the base will reach 0 V at the end of the input pulse. Even at the end of the input pulse, the input voltage to the base remains slightly negative. Therefore, the input transistor remains off during the entire negative input pulse.

EXAMPLE 9.16

Using the circuit shown in Figure 9.35, determine the charging current.

Solution

With Q_1 off, the output capacitor charges through the collector of Q_2. To find this charging current, we need to determine the emitter current of Q_2. Q_2 is biased with a voltage divider consisting of R_1 and R_2. Applying the voltage divider method, we get:

$$\frac{1.2 \text{ kΩ}}{(1.2 \text{ kΩ} + 820 \text{ Ω})} = 0.594$$

$$V_{BE} = 0.594 \times 16 \text{ V} = 9.5 \text{ V on the base of } Q_2$$

Since Q_2 is a PNP transistor, its emitter potential is 0.7 V higher than the base when it is turned on.

$$V_{E2} = 9.5 \text{ V} + 0.7 \text{ V} = 10.2 \text{ V}$$

We can now figure out the voltage across the emitter resistor:

$$16 \text{ V} - 10.2 \text{ V} = 5.8 \text{ V}$$

The current through the emitter resistor is:

$$\frac{5.8 \text{ V}}{75 \text{ Ω}} = 77.3 \text{ mA}$$

This is also the charging current for C_2. Now we can determine the output voltage across C_2 by using the equation

$$A_V = \frac{(I \times t)}{C}$$

While I represents the charging current, t is the time that Q_1 is off, and C is the value of C_2. Substituting values into the equation gives us:

$$\frac{(77.3 \text{ mA} \times 25 \text{ μs})}{0.250 \text{ μF}} = 7.73 \text{ V}$$

The output voltage will be a ramp that starts from 0.2 V or V_{CE}(sat) and rises to a peak of 7.73 V. This peak answer is accurate for one reason. The charging current through Q_2 will not decay because the collector of Q_2 is a constant current source.

To make sure that the collector of Q_2 remains a constant current source, we should check to see that Q_2 never enters saturation. The smallest collector-to-emitter voltage for Q_2 occurs when the output across C_2 reaches its peak. At that time we have:

$$V_{CC} = V_{C2} + V_{CE2} + V_{RE}$$

Substituting values gives us:

$$7.73\ V + V_{CE2} + 5.8\ V = 16\ V$$

Rearranging, we have:

$$V_{CE2} = 16\ V - 7.73\ V - 5.8\ V = 2.47\ V$$

Even at the peak output, the collector-to-emitter voltage of Q_2 is well away from saturation voltage. Consequently, the collector of Q_2 remains a constant current source for the entire charging period. Many circuit designs will have a minimum of 1 to 2 V for the collector-to-emitter voltage to ensure that the transistor remains in the active region.

We also need to check that Q_1 can discharge C_2 rapidly enough to bring the voltage across the output capacitor back to the starting voltage of 0.2 V. The time available for discharge is 2.5 μs. The voltage to be removed from C_2 is:

$$7.73\ V - 0.2\ V = 7.71\ V$$

Once again, we use the equation:

$$V = \frac{(I \times t)}{C}$$

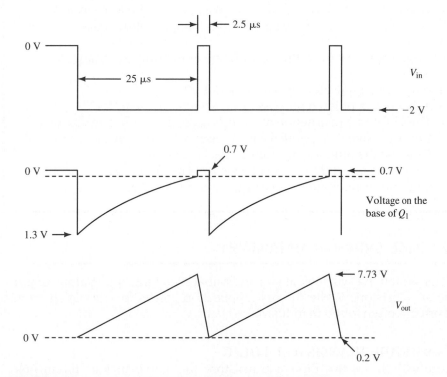

FIGURE 9.36 Typical waveforms from the circuit in Figure 9.35a

Substituting values, we find that:

$$7.71 \text{ V} = \frac{\left(I \times 2.5 \, \mu s\right)}{0.250 \, \mu F}$$

Rearranging those values, we have:

$$I_{CE1} = \frac{7.71 \text{ V} \times 0.250 \, \mu F}{2.5 \, \mu s} = 771 \text{ mA}$$

With a beta of 100, the base current would have to be:

$$\frac{771 \text{ mA}}{100} = 7.71 \text{ mA}$$

The base current flows through the base resistor when Q_1 is on and equals:

$$\frac{\left(16 \text{ V} - 0.7 \text{ V}\right)}{1.8 \text{ k}\Omega} = 8.5 \text{ mA}$$

This amount of current is more than enough to successfully discharge C_2. Figure 9.36 shows the waveforms for this circuit.

Progress Check
You have now completed objective 11.

PRACTICE PROBLEM 9.8

Change C_2 in Figure 9.35 to 0.33 μF. Recompute the new output peak of the ramp. Check to see if Q_2 stays out of saturation.

REVIEW SECTION 9.6
1. Write an equation for a general sine wave, and define its parts.
2. What is true about linear circuits if they have a sine wave input?
3. Square waves have many applications in digital electronics. True or false?
4. Describe, with a diagram, the differences in ramp and triangular waveforms.
5. What is an application of a differentiator?
6. A differentiator is a high-pass filter; what is an integrator?
7. The simplest construction of a ramp signal generator consists of a transistor switch in parallel with a resistor. True or false?
8. Describe the function of a ramp signal generator.
9. What are some differences between a normal ramp signal generator and a ramp generator as a constant current source?

9.7 MORE SWITCHING APPLICATIONS

In this section, we will look at two additional switching applications that use bipolar transistors. While the first application comes from a digital circuit, the other is often found in industrial systems.

TRANSISTOR-TRANSISTOR LOGIC

Figure 9.37 shows two bipolar transistors configured into a "totem pole." Generally, this configuration will appear as an output stage of a **transistor-**

transistor logic (TTL) integrated circuit. Because of the circuit design, the transistors do not turn on at the same time. Bias voltages control circuit action at points x and y. These points connect to the rest of the logic circuit that controls the totem pole. For the purposes of our discussion, we need only know that if point x biases transistor Q_2 into conduction, point y will bias transistor Q_1 off and vice versa.

When Q_1 conducts, it creates a path from the output to ground through the transistor. Since Q_1 is designed to saturate, the output voltage should equal approximately 0.2 V, or $V_{CE}(sat)$. In TTL circuits, any voltage up to 0.8 V is a logic zero. A logic zero represents either off or false. Since the output voltage is less than 0.8 V, any logic device connected to the output would interpret the output voltage as a logic zero.

To change the output logic level, point y must bias Q_1 off. An "off" Q_1 appears as an open switch. Since Q_1 is off, transistor Q_2 is biased to conduct and appears as a closed switch. Q_2 provides a path to the voltage source. The output voltage of transistor Q_2 appears as a logic one or a true. In a TTL circuit, a high output, which is also called one or on, is not less than 2.4 V. Any voltage that ranges between 2 V and 5 V equals a logic one. When Q_2 conducts, the maximum current flow is 400 μA. This current flows into the output and through the diode (D_1), Q_2, the 130-Ω resistor, and into the power supply. With 400 μA flowing through the 130-Ω resistor, the voltage drop across the resistor equals:

$$V_{RC} = I_C R_C$$
$$= 400 \text{ μA} \times 130 \text{ Ω}$$
$$= 52 \text{ mV}$$

Using the values that we have, we can determine the voltage across Q_2, or V_{CE2} during conduction. The voltage at the collector of Q_2 equals:

$$V_C = V_{CC} - (I_C R_C)$$
$$= 5 \text{ V} - 0.052 \text{ V} = 4.95 \text{ V}$$

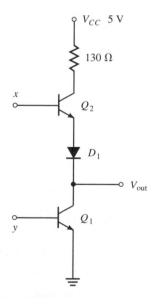

FIGURE 9.37 Two bipolar transistors configured into a "totem pole"

Since we want an output that is not less than 2.4 V, the emitter voltage equals:

$$V_{E2} = V_{out} + V_D$$
$$= 2.4\,V + 0.7\,V = 3.1\,V$$

V_D is the voltage drop across the barrier potential of the diode. Across the collector-emitter junction, the voltage equals:

$$V_C - V_E = 4.498\,V - 3.1\,V = 1.398\,V$$

By looking at those voltage values, we can conclude that transistor Q_2 will not reach saturation when it conducts. From our earlier discussions, we know that a transistor that does not reach saturation will switch on and off faster than one that does. In part, this explains why the switching times required to go from high to low logic levels differ from the switching times required to go from low to high logic levels.

PRACTICE PROBLEM 9.9

Determine if Q_2 saturates when the output of Figure 9.37 is 3.5 V. (NOTE: 3.5 V is a "typical" output for standard TTL circuits.)

INTERFACING BETWEEN VOLTAGE LEVELS

Figure 9.38a shows a transistor working as an interface between two electronic circuits that use different voltage and/or current levels. In the circuit, v_{in} represents the connection between the transistor switch, Q_1, and a logic circuit. In this case, the TTL device decides whether to turn attached equipment, such as a motor, on or off. Figure 9.38b shows the symbol for an electric motor. As we know, the TTL output will equal either 0.2 V for an off condition or a minimum of 2.4 V for an on condition. Obviously, neither of these voltages could power an electric motor running off the AC line voltage seen in Figure 9.38b.

Several not-so-obvious steps make the entire operation possible. When current flows through the coil windings, the relay connected to the collector of Q_1 energizes. More than likely, the amount of current will exceed the

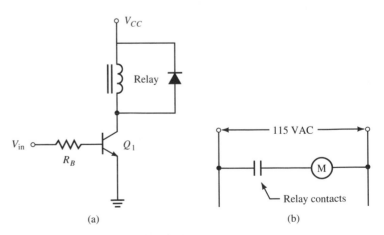

(a)   (b)

FIGURE 9.38 Transistor working as an interface between two electronic circuits that use different voltage and current levels amount of current provided by the TTL device. Q_1 provides a current path between ground and the power supply. Both the power supply voltage, or V_{CC}, and the current through Q_1 can be greater than the amount of voltage or current available from the TTL device. Input impedance from the base of Q_1 partially isolates the relay part of the circuit from the TTL part of the circuit.

Next, the current through the relay coil builds up a magnetic field. The field closes the normally open relay contacts shown in Figure 9.39b. As long as the relay is energized, the contacts remain closed and the motor runs. When the TTL output goes to a logic zero or off, it also removes the bias from Q_1 and the transistor stops conducting. Since the transistor no longer conducts, the relay loses its energizing current and the magnetic field around the coil collapses. Consequently, the relay contacts open and the motor stops.

The diode connected across the relay acts as a flyback diode. Any counter-electromotive force created when the magnetic field collapses is called **flyback** and discharges through the diode. Without a discharge path, the counter-emf voltages and currents could damage Q_1. You may see such a diode in any circuit that has inductive loads switched off and on.

REVIEW SECTION 9.7
1. Describe the action of a "totem pole" two-transistor configuration.
2. What voltages, in digital electronics, represent off, false, or logic zero?
3. In digital electronics, what is the voltage required for a logic one, on, or true?
4. What is one reason for the difference between switching times to go from low to high and switching times to go from high to low?
5. Briefly describe how a transistor operating on a small voltage can control a motor using high AC voltages.
6. What is the reason to include a flyback diode in high-voltage circuits where a transisor switch is used?

 TROUBLESHOOTING

The types of trouble that appear in switching circuits are the following: output does not switch when triggered, the output waveform does not have rapid rising and falling sides, the output rings, and the switching circuit loads down the previous circuit. All these problems (and others) can be solved using sound troubleshooting practices. First, identify the problem. Then, figure out what the output waveform should look like. Third, take waveform and voltage readings. Next, compare them with the calculated values. Finally, note the differences and ask why.

In Figure 9.35a Q_1 is a transistorized switch. The output should be a ramp waveform. If we find no ramp on the output but rather 0 V, we need to look for possible shorts from the output's vantage point. Looking from the output terminal, there are three possible paths for a short or a nonfunctional component. This discussion will use dynamic troubleshooting. Looking from the output, there are three possible malfunctioning components, Q_1, C_1, and the constant current source, Q_2.

First, check Q_1. Q_1 should be cutoff when the capacitor is charging, and saturated when the capacitor is discharging. Whether Q_1 is cutoff or saturated depends on the input signal. Check the input signal to ensure that its amplitude can switch the transistor. If it can, then we need to isolate the transistor from the circuit. If possible, lift the collector lead off the circuit board and connect it to the bottom of R_E via a jumper. This isolates Q_1 for the circuit and also provides a complete current path for transistor operation. With an oscilloscope, check the output of the collector of Q_1. If the oscilloscope reads near 0 V, then the emitter-collector of Q_1 is shorted. Replace Q_1.

If, however, a reasonable signal was seen on the collector of Q_1, then check C_2. After soldering Q_1 back into the circuit, lift the top lead of C_2, inject a low-frequency signal (about 100 Hz) on C_2, and observe with an oscilloscope. If the signal is what you expected, then C_2 is not defective. If, however, the signal is near 0 V, then C_2 may be shorted and should be replaced.

If C_2 functions properly, check Q_2. Calculate the emitter voltage of Q_2. The ideal V_E should be about 10.2 V. Now check the voltage at the emitter Q_2. Are the ideal and the measured within 1 V of each other? If they are, then Q_2 is functioning properly. If the emitter voltage is 16 V, then Q_2 is not conducting, and may be open. If the emitter voltage is near zero, then R_E may be open.

Common problems you may encounter are open and shorted transistors, shorted, leaky, and open capacitors, and open resistors.

WHAT'S WRONG WITH THIS CIRCUIT?

Determine the problem in the following circuit for each situation.

Situation 1: The input pulse is correct, but the transistor switch turns back on too soon. What can be wrong?

Situation 2: The transistor has been checked and it is good. Beta is accurate, yet the transistor does not saturate when it is on. What can be wrong?

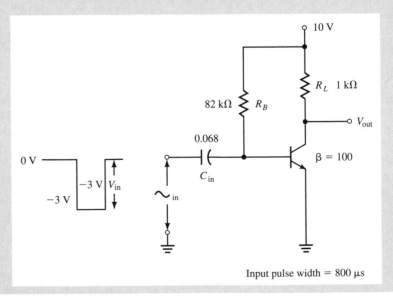

SUMMARY

In many ways, electronic switching devices work like mechanical switching devices. Bipolar junction transistors (BJTs), junction field effect transistors (JFETs), metal oxide semiconductor field effect transistors (MOSFETs), and unijunction transistors (UJTs) can function as electronic switches.

Even though all the listed semiconductor devices function sufficiently as switches, none of the devices are ideal switches. An ideal switch shows as either a complete open or a complete short, does not dissipate energy, and changes conditions instantaneously. Also, an ideal switch does not require power to stay in a condition or to switch to a condition.

Any solid state device is a nonideal switch. Solid state devices have internal resistances, leakage resistances, and capacitance. Internal resistance prevents the device from becoming a complete short. Leakage resistance permits leakage current to flow within the device and prevents it from becoming a complete open. Capacitance keeps the device from switching from one condition to another instantaneously. Because of these characteristics, solid state switches exhibit time factors such as delay time, rise time, storage time, fall time, and decay time.

For BJTs, the common-emitter configuration works best for electronic switching. The common emitter offers both voltage and current gain. In addition, the common-emitter configuration controls a large output signal with a small input signal. Transistor switches do not work in the active region. Instead, BJT switches use the cutoff and saturation regions to turn off and on. To speed switching, many transistor switching circuits have speed-up capacitors that reduce transition time.

Another transistor configuration is called the Schmitt trigger. A Schmitt trigger is a switch that detects voltage thresholds. The thresholds are called the lower threshold point and the upper threshold point. This configuration responds to a slow-changing input signal by having an output signal with fast transition times.

Along with JFETs and MOSFETs, the UJT also performs well for switching tasks that involve digital applications. When compared with FETs, the UJT has different quantities of doping. For example, an N-channel UJT has heavily doped P-type material with a large number of excess holes and lightly doped N-type material. UJTs have a common use as a relaxation oscillator. A relaxation oscillator produces a ramp-out waveform or a series of very-high-amplitude spikes. These spikes turn on other semiconductor devices such as SCRs and triacs.

When working with DC and AC signal amplification, you should consider several types of signals. Those signals vary because of time, amplitude, and frequency. Signal types are sine waves, square waves, ramp, and triangle waves and pulses.

Different bipolar junction transistor and junction field effect transistor signal amplifiers have different frequency responses or bandwidths. For example, a stereo amplifier may have a frequency response of 20 Hz to 20,000 Hz. Frequency response specifies the frequency range that the amplifier can adequately amplify.

Refer to Figure 9.39 for problems 1–9.

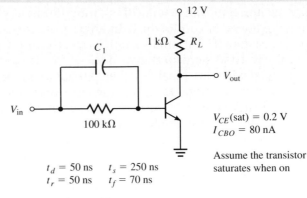

FIGURE 9.39 Circuit for problems 1–9

1. What is the output voltage when the transistor is off?
2. What is the output voltage when the transistor is on?
3. What is the current through the load when the transistor is off?
4. What is the current through the load when the transistor is on?
5. If β (min) = 50, what is the smallest value of V_{in} that will cause the transistor to saturate?
6. What is the total time required to turn on the transistor?
7. What is the total time required to turn off the transistor?
8. Using the guideline presented in this chapter that relates duration of pulse width (PW) to switching times, determine the upper frequency limit of the transistor switch. This problem assumes that the speed-up capacitor is not being used.
9. What value should C_1 be, if an input frequency of 350 kHz is applied to the circuit?

Refer to Figure 9.40 for problems 10–17.

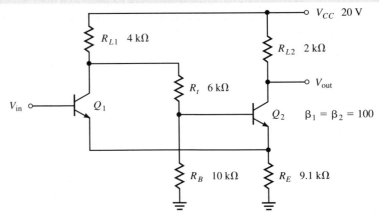

I_{CBO} = 50 nA for both transistors

FIGURE 9.40 Circuit for problems 10–17

10. When V_{in} is low, which transistor is normally on?
11. When V_{in} is low, what is the DC voltage at the base of Q_2?
12. When V_{in} is low, what is the V_{out}?
13. Determine the upper trip point.
14. Determine V_{out} when Q_1 is on.
15. What is the LTP of the circuit?
16. What is the hysteresis of the circuit?
17. Plot the hysteresis loop for the circuit on graph paper.

Refer to Figure 9.41 for problems 18–24.

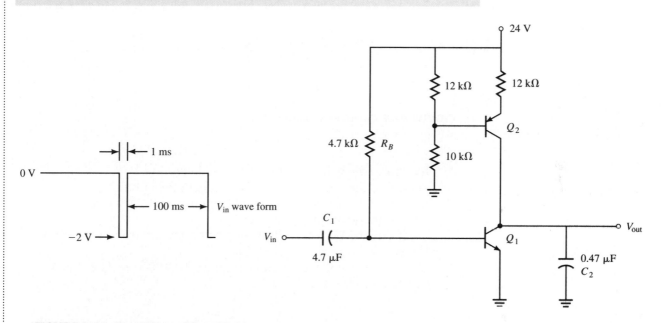

FIGURE 9.41 Circuit for problems 18–24

18. When the circuit is quiescent, is Q_1 biased on or off?
19. When the circuit is quiescent, is Q_2 biased on or off?
20. When Q_1 is on, what voltage do you expect to be across it?
21. What happens to the voltage across C_2 when Q_1 is off?
22. When the circuit is quiescent, what is the voltage at the base of Q_1?
23. Given the input shown, determine the amplitude of the ramp output.
24. If V_{CE2} is allowed to drop to a minimum of 2 V, what is the largest possible output ramp?

Refer to Figure 9.42 for problems 25–30.

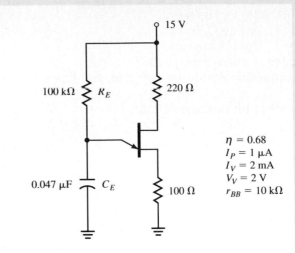

FIGURE 9.42 Circuit for problems 25–30

25. Determine the voltage required to fire the emitter.
26. Determine the largest value of R_E usable in the circuit.
27. Determine the minimum value of R_E usable with the circuit.
28. Determine the output frequency.
29. Determine the peak amplitude of the voltage spike across the 100-Ω resistor.
30. Determine the voltage amplitude across the 100-Ω resistor before the UJT emitter fires.

10

MORE SMALL-SIGNAL AMPLIFICATION

✓ **As you read this chapter, concentrate on learning how to:**

1. Recognize the impact of negative feedback on amplifier circuits

2. Interpret h parameters listed in transistor specification sheets

3. Analyze the low- and high-frequency limits of a bipolar transistor amplifier circuit

4. Recognize when the AC junction capacitance may have a bearing on transistor circuit waveform analysis

5. "Millerize" feedback capacitances

6. Analyze the low- and high-frequency limits of a junction field effect transistor amplifier circuit

7. Determine the gain of a differential amplifier

8. Calculate the common mode rejection ratio of a differential amplifier

INTRODUCTION

In this chapter, we will look at advanced methods of analyzing small-signal amplifiers. Essentially, we will use knowledge that we have accumulated throughout the text and some additional information about small-signal amplifiers. We will find that most amplifiers have distinct lower and upper frequency limits. As we discuss signals and waveform analysis, you may find your understanding of how transistors amplify signals increasing.

Moving from discrete devices such as transistors and diodes to linear integrated circuits, we will also talk about differential amplifiers. Since differential amplifiers work with both DC and AC signals, we will discuss the DC and AC analysis of the differential amplifier circuit. Also, we will look at biasing for the differential amplifier and at differential amplifier characteristics.

As we look at each device, we will also examine how frequencies affect the performance of the device. In addition, we will see how the individual devices shape waveforms and affect the bandwidth of a small-signal amplifier.

10.1 NEGATIVE FEEDBACK

Throughout our study of transistors, we have seen how feedback affects circuit action. Feeding a part of the output signal back into the input signal can significantly improve the performance of an amplifier. Improvement occurs when the quality of the feedback signal is compared with the quality of the input signal. Using the comparison, we can correct any circuit distortion problems.

Negative feedback opposes the amplification of signals. By reducing gain, negative feedback also increases input impedance and decreases output impedance. More important, negative feedback causes circuit gain to be independent of individual amplifier gain and improves circuit stability. In addition, negative feedback causes low-frequency response to remain at a constant level.

Along with those benefits, we know that negative feedback cuts amplifier distortion, drift, and noise. As previously discussed, even with the absence of an input signal, a nonideal amplifier may generate an output signal. The output signal may be due to characteristics such as drift and noise. A transistor is not an ideal linear amplifier. Different signal levels may cause distortion within an amplifier circuit. While a power amplifier may experience distortion at low signal levels, another amplifier may have harmonic distortion at large signal levels. Negative feedback reduces distortion levels to a minimum value. Because of negative feedback, the difference between the output and input signals remains small. Effectively, negative feedback cancels distortion by producing its own distortion in the opposite direction of the original unwanted distortion.

Negative feedback networks may depend on frequency or amplitude. If the network is frequency-dependent, we see an amplifier that has gain versus frequency characteristics. An amplitude-dependent feedback network can produce a nonlinear amplifier or a current source that has nearly infinite output impedance. By feeding back a signal proportional to the output source, the feedback network becomes a good current source.

Progress Check
You have now completed objective 1.

REVIEW SECTION 10.1

1. How does feedback correct distortion problems in an amplifier circuit?
2. Name some of the benefits of negative feedback.
3. What could happen, in a nonideal amplifier, even without an input signal?
4. Negative feedback increases distortion to a maximum value. True or false?

10.2 SMALL-SIGNAL TRANSISTOR PARAMETERS

We can consider a transistor as a device with two input terminals and two output terminals. One terminal is common to both the input and output. Figure 10.1 illustrates how our input-output device would appear. In Figure 10.1, an input voltage (v_1) to the network sets up a current labeled i_1. An out-

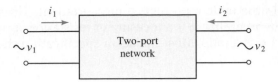

FIGURE 10.1 Two-port network

put voltage (v_2) across the terminals of the network sets up another current called i_2.

In Figure 10.2 the network is divided into two distinct equivalent circuits so that we can make circuit analysis easier.

Hybrid, or h, **parameters** constitute an advanced method of analyzing transistor performance under small-signal conditions. If we convert the input port into a voltage source in series with an impedance and the output port into a current source with a shunt impedance, we have an equivalent circuit similar to a transistor operating in its linear region. The output, or collector, of the transistor acts as a current source. Figure 10.2 represents the conversion of the ports into voltage and current sources.

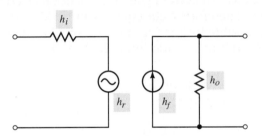

FIGURE 10.2 Two-port network with h parameters

For the circuit shown in Figure 10.2, the h parameters mark the small-signal behavior of the transistor at a given bias point. In addition, the h parameters can describe common-emitter, common-base, or common-collector configurations. Table 10.1 defines the h parameters and lists the parameter for each specific configuration:

Table 10.1				
Parameter	**Description**	**Common emitter**	**Common base**	**Common collector**
h_1	Input impedance with a shorted output	h_{ie}	h_{ib}	h_{ic}
h_f	Forward current gain with a shorted output	h_{re}	h_{rb}	h_{Fc}
h_r	Reverse voltage gain with an open input	h_{fe}	h_{fb}	h_{rc}
h_o	Output admittance with an open input	h_{oe}	h_{ob}	h_{oc}

Each parameter in the table represents a measurement. Throughout this text, you should remember that h_i is a measure of resistance, h_o is a measure of conductance, and h_f is a ratio. Most transistor specification sheets will use h_{fe} to represent beta.

EXAMPLE 10.1

Using the specifications for a 2N3904 transistor, calculate the AC emitter resistance.

The 2N3904 transistor has the following small-signal characteristics when tested at a current of 1 mA with a collector-to-emitter voltage of 10 V:

$$h_{fe} = 100\text{–}400$$
$$h_{ie} = 1\text{–}10 \text{ k}\Omega$$
$$h_{re} = 0.5 \times 10^{-4} \text{ to } 8 \times 10^{-4}$$
$$h_{oe} = 1\text{–}40 \text{ } \mu S$$

The test frequency is 10 kHz.

Solution

Since h_{fe} represents the forward current gain, we will use its value to approximate the value of beta. Therefore, beta equals 100 to 400. In addition, h_{ie} gives us an approximate value for the input impedance. If the input impedance of the base equals the beta value multiplied by the AC emitter resistance, we have:

$$Z_{in}(base) = beta(r'_e) = h_{ie} = h_{fe} (r'_e)$$

Rearranging the equation, we find that:

$$r'_e = \frac{h_{ie}}{h_{fe}}$$

Since h_{ie} and h_{fe} have a wide range of values, we should use the geometric averages for the values. For h_{fe}, the geometric average equals the square root of the product of the values:

$$h_{fe} = \sqrt{100 \times 400} = \sqrt{40,000} = 200$$

We can apply the same reasoning to find the geometric average of h_{ie}:

$$h_{ie} = \sqrt{1 \text{ k}\Omega \times 10 \text{ k}\Omega}$$
$$= \sqrt{10 \text{ M}\Omega} = 3.16 \text{ k}\Omega$$

Using those values, we can find a value for the AC emitter resistance:

$$r'_e = \frac{3162 \text{ }\Omega}{200} = 15.8 \text{ }\Omega$$

This value is different from, and more accurate than, the value approximated with the equation:

$$r'_e = \frac{25 \text{ mV}}{I_E} = \frac{25 \text{ mV}}{1 \text{ mA}} = 25 \text{ }\Omega$$

PRACTICE PROBLEM 10.1

The value of h_{ie} ranges from 1 kΩ to 5 kΩ, while the value of h_{fe} ranges from 100 to 300. Taking a geometric average, determine r'_e.

When we break an amplifier circuit down into an equivalent network, the network represents the small-signal electrical operation of the amplifier circuit at a given bias point. This allows us to use easily understood laws such as Kirchhoff's voltage and current laws and Ohm's law to analyze the circuit action. Manufacturers will show the hybrid parameters along with recommended bias points on the transistor specification sheets. Remember that the hybrid parameters are accurate only for a given bias point and for small-signal applications.

Progress Check
You have now completed objective 2.

REVIEW SECTION 10.2

1. What equations would be used for a bipolar junction transistor, when applying Thevenin's theorem?
2. How do we get an equivalent circuit that is similar to a transistor operating in its linear region, when the transistor is working under small-signal conditions?
3. List the hybrid, or *h*, parameters, and give the description of each.
4. What does breaking an amplifier into an equivalent network allow us to do?

10.3 FINDING THE BANDWIDTH OF BJT AND JFET AMPLIFIERS

For a given input amplitude, an ideal amplifier would have a consistent output amplitude across the frequency spectrum, Figure 10.3. However, a practical amplifier has output frequency limits. These limits are the frequencies

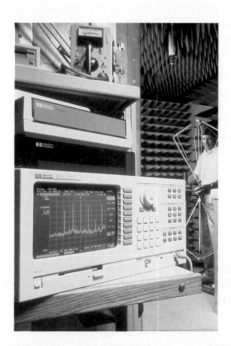

FIGURE 10.3 Frequency spans and bandwidths are analyzed and measured quickly by a spectrum analyzer. *(Photograph courtesy of Hewlett-Packard)*

where the output power has decreased to one-half the maximum output power. These points are also called the **half-power points** or -3 dB points. Look at Figure 10.4. The lower frequency point is called the low cutoff frequency (f_L) and the upper frequency point is called the high cutoff frequency (f_H). The frequencies between f_L and f_H are called the **bandwidth.** In equation form, bandwidth (BW) is defined as

$$BW = f_H - f_L$$

Output power changes as a function of frequency. The cause is capacitive reactance. Remember, capacitive reactance is inversely proportional to frequency.

$$X_c = \frac{1}{2\pi fC}$$

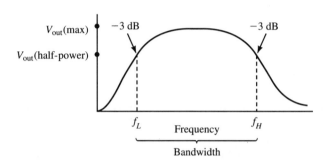

FIGURE 10.4 **Output amplitude over a spectrum of frequencies**

LEADING RC NETWORKS

In Figure 10.5 the capacitor and the resistor form a leading RC network. As the input frequency increases the capacitive reactance decreases, allowing more of the signal to be developed across the resistor. When the capacitive reactance equals the resistance, each component will drop the same amount of voltage. That voltage will be equal to 0.707 of the input voltage.

$$V_{out} = 0.707 \times V_{in}$$

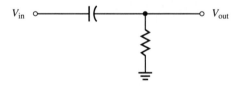

FIGURE 10.5 **The capacitor and the resistor form a leading RC network**

The circuit is called a **leading network** because the output voltage will lead the input voltage. In Figure 10.6 the output voltage is drawn at 0 degrees. The input voltage is at -45 degrees. The output voltage is ahead of, or leading, the input voltage by 45 degrees.

Because the output amplitude increases as frequency increases, the circuit is also called a **highpass circuit.**

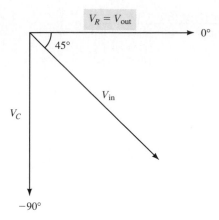

FIGURE 10.6 The output voltage is drawn at 0°.

The frequency where the output amplitude is 0.707 of the input voltage and the output voltage leads the input voltage is called the half-power or −3 dB point.

LAGGING RC NETWORKS

In Figure 10.7 the capacitor and the resistor form a lagging RC network. As the input frequency increases, the capacitive reactance decreases, allowing more of the signal to be developed across the resistor. When the capacitive reactance equals the resistance, each component will drop the same amount of voltage. That voltage will be equal to 0.707 of the input voltage.

$$V_{out} = 0.707 \times V_{in}$$

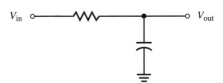

FIGURE 10.7 The capacitor and the resistor form a lagging RC network.

The circuit is called a **lagging network** because the output voltage will lag the input voltage. In Figure 10.8 the resistor voltage is drawn at 0 degrees. However, the output voltage is developed across the capacitor and is behind, or lagging, the input voltage by 45 degrees.

Because the output amplitude decreases as frequency increases, the circuit is also called a **lowpass circuit.**

The frequency where the output amplitude is 0.707 of the input voltage and the output voltage lags the input voltage is called the half-power or −3 dB point.

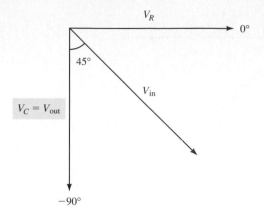

FIGURE 10.8 The resistor is drawn at 0°.

FINDING DOMINANT CRITICAL FREQUENCY OF A CIRCUIT

In both the leading and lagging RC networks, the critical frequency was determined when the capacitive reactance equaled the resistance. We can find the value of that frequency with a little algebra.

First we mathematically write out the definition for critical frequency.

$$R = X_c$$

Then we substitute the formula for capacitive reactance.

$$X_c = \frac{1}{2\pi f C}$$

$$R = \frac{1}{2\pi f C}$$

Last, we rearrange to solve for frequency (f).

$$f = \frac{1}{2\pi f C}$$

Now, anytime we know the capacitance and the resistance of an RC network, we can solve for its critical frequency. Because there is more than one RC network in an amplifier, there will be more than one frequency. However, only one of the critical frequencies will set the circuit's critical frequency which is called dominant critical frequency.

PRACTICE PROBLEM 10.2

Determine the critical frequency of an RC network when the total resistance equals 10 kΩ and the capacitance is 220 pF.

BIPOLAR JUNCTION TRANSISTOR LOW-FREQUENCY SIGNAL ANALYSIS

Figure 10.9 shows a common-emitter amplifier. As we analyze the circuit, we need to consider three capacitors: C_1, C_2, and C_3. While C_1 and C_3 act as coupling capacitors, C_2 is a bypass capacitor. During normal circuit operation, all three capacitors appear as shorts to the AC signal. As the frequency decreases, the reactance of the capacitor increases and allows less of the AC sig-

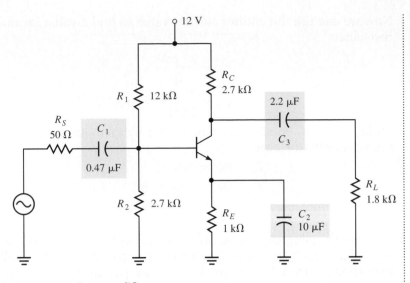

FIGURE 10.9 Common-emitter amplifier

nal to pass, or in the case of C_2, removes the short across R_E, reducing circuit gain.

Any DC signal treats the first coupling capacitor (C_1) as an open. The capacitor isolates the AC voltage source from its DC bias voltage. Because of the capacitors, we face the problem of retaining good circuit performance while decreasing the signal frequency. Example 10.2 may provide a solution to our problem.

EXAMPLE 10.2

Determine the critical low frequencies for the circuit shown in Figure 10.9.

We can begin our analysis by finding a value for the base voltage of the transistor.

Solution
With the equation:

$$V_B = \frac{R_2}{R_1 + R_2} \times V_{CC}$$

$$V_B = \frac{2.7 \text{ k}\Omega}{14.7 \text{ k}\Omega} \times 12 \text{ V} = 2.2 \text{ V}$$

we find that the base voltage equals 2.2 V. Subtracting the 0.7 V from the barrier voltage drop from the base voltage gives us a value for the emitter voltage:

$$V_E = V_B - V_{BE}$$

$$V_E = 2.2 \text{ V} - 0.7 \text{ V} = 1.5 \text{ V}$$

Using the emitter voltage value, we can find a value for the emitter current:

$$I_E = \frac{V_E}{R_E}$$

$$I_E = \frac{1.5 \text{ V}}{1 \text{ k}\Omega} = 1.5 \text{ mA}$$

Now we can use the emitter current value to find a value for the AC emitter resistance:

$$r'_e = \frac{25\text{ mV}}{I_E}$$

$$r'_e = \frac{25\text{ mV}}{1.5\text{ mA}} = 16.67\ \Omega$$

Multiply the AC emitter resistance value by the beta value to find the input impedance for the base:

$$Z_{in}(\text{base}) = \beta \times r'_e$$

$$Z_{in}(\text{base}) = 100 \times 16.67 = 1.67\text{ k}\Omega$$

We have completed our preliminary analysis.

Moving on, we can use the information that we have accumulated to see how the coupling capacitors affect the circuit. In Figure 10.10a, we see the equivalent amplifier network attached to capacitor C_1. Electrically removing the capacitor from the network leaves the resistances seen by the plate of the capacitor. As illustrated in Figure 10.10a, b, and c, the bias resistors combine with the base input impedance to give a resistance value of:

$$R_{in} = R_2 \parallel R_1 \parallel Z_n(\text{base})$$

$$R_{in} = 2.7\text{ k}\Omega \parallel 12\text{ k}\Omega \parallel 1.67\text{ k}\Omega = 950\ \Omega$$

This resistance is in series with the internal 50-Ω resistance of the source. The total Thevenin resistance seen by the plates of capacitor C_1 equals 1000 Ω ($R_{TS} = R_S + R_{in}$).

With the additional information, we can calculate the lower critical frequency for the network formed by the capacitor and its Thevenin resistance in Figure 10.10c:

$$f_l = \frac{1}{2\pi C_1 R_{TS}}$$

$$f_l = \frac{1}{2\pi \times 0.47\ \mu F \times 1000\ \Omega} = 339\text{ Hz}$$

Frequencies above the 339-Hz level will reach the base with little loss of signal strength. Frequencies below 339 Hz will be attentuated because part of the signal is dissipated across the reactance of the capacitor.

Figure 10.10d shows the Thevenin circuit for the output coupling capacitor (C_3). From the plates of the capacitor, the 2.7-kΩ and 1.8-kΩ resistances are in series. The collector of the transistor functions as a constant current source and has a near-infinite resistance. Since we have thevenized the circuit, we can treat the collector as an open circuit. Therefore, in Figure 10.10e, the total Thevenin resistance seen by the capacitor plates equals 4.5 kΩ and the critical frequency of the network formed by the output coupling capacitor equals:

$$f_l = \frac{1}{2\pi R_{TH} C_3}$$

$$f_l = \frac{1}{2\pi \times 4.5\text{ k}\Omega \times 2.2\ \mu F}$$

$$f_l = 16\text{ Hz}$$

The last capacitor in Figure 10.9 that can affect low-frequency operation is the bypass capacitor (C_2). The Thevenin circuit for this capacitor is dia-

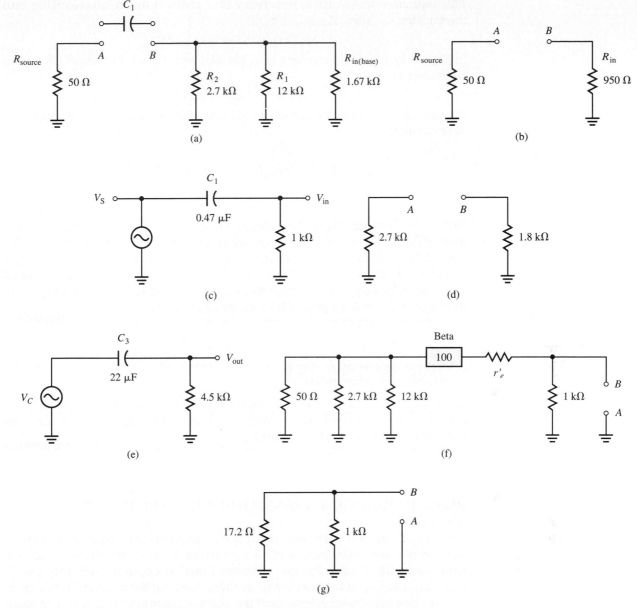

FIGURE 10.10 (a, b, and c) Bias resistors combine with the base input impedance. (d and e) Thevenizing for C3. (f and g) Thevenizing for C2.

grammed in Figure 10.10f. Note that it "sees" resistances on the other side of the beta box. These resistances must be equalized.

$$R_{TH} = R_{(source)} \parallel R_1 \parallel R_2$$
$$R_{TH} = 50\ \Omega \parallel 2.7\ k\Omega \parallel 12\ k\Omega = 48.9\ \Omega$$

Dividing by β to equalize the effect this resistance has on C_2, we get:

$$R_{effective} = \frac{R_{TH}}{\beta}$$
$$\frac{48.9\ \Omega}{100} = 0.489\ \Omega$$

This equalized resistance is in series with r'_e, which means that looking into the emitter, C_2 "sees" $R_{effective} + r'_e$:

$$16.77\ \Omega + 0.489\ \Omega = 17.2\ \Omega$$

In Figure 10.10g this resistance is in parallel with R_E, so the overall Thevenin resistance is:

$$17.2\ \Omega \parallel 1\ k\Omega = 16.9\ \Omega$$

We can now calculate the critical frequency of the network formed by the bypass capacitor:

$$f_l = \frac{1}{2\pi C_3 R}$$

$$\frac{1}{2\pi \times 10\ \mu F \times 16.9\ \Omega} = 942\ Hz$$

By finding these values, we have determined three critical frequencies associated with the lower limit. The frequencies are 338 Hz, 16 Hz, and 942 Hz. 942 Hz is the dominant critical low frequency for the circuit because injecting a signal frequency less than 942 Hz into the amplifier causes a loss of signal strength. A frequency lower than 942 Hz does not allow the bypass capacitor to properly shunt the signal, which causes gain to drop.

PRACTICE PROBLEM 10.3

An amplifier has three networks that limit low-frequency responses. The critical frequency of these three networks is 10 kHz, 5 kHz, and 8 kHz. Which frequency is the dominant low frequency?

BIPOLAR JUNCTION TRANSISTOR HIGH-FREQUENCY ANALYSIS

The high-frequency response of a bipolar junction transistor is limited by the emitter and collector junction capacitances. Most transistor specification sheets list a value for the collector junction capacitance and label it C_{OB}, C_{OBO}, or C_{CB}. If the specification sheet does not list a value for the emitter junction capacitance, you may find the approximate value with the equation:

$$C'_e = \frac{1}{2\pi\left(f_T \times r'_e\right)}$$

In the equation, f_T is the frequency that causes transistor gain to drop to unity. We refer to that frequency as the **gain-bandwidth product.**

EXAMPLE 10.3

Determine the emitter junction capacitance for the 2N3904 transistor shown in Figure 10.9.

Solution

In the circuit, the AC emitter resistance equals 16.7 Ω. The specification sheet shows that the gain-bandwidth product is 300 MHz. With those values, we can approximate a value for the emitter junction capacitance:

$$C'_e = \frac{1}{2\pi(f_T \times r'_e)} = \frac{1}{2\pi \times 300 \text{ MHz} \times 16.7 \text{ }\Omega} = 31.8 \text{ pF}$$

PRACTICE PROBLEM 10.4

The gain bandwidth product of a transistor is 100 MHz. The value of r'_e is equal to 25 Ω. Determine the value of C'_e.

Figure 10.11 shows the junction capacitances and their relationship to the rest of the AC circuit. In this diagram, R_C represents the AC collector resistance and R_{TH} represents the total AC resistance from the perspective of the base. Also, the transistor is boxed in to show that these two capacitances, which are drawn outside the transistor, are actually within the transistor.

Progress Check
You have now completed objectives 3 and 4.

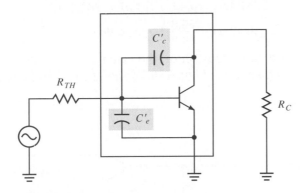

FIGURE 10.11 AC equivalent circuit for the circuit shown in Figure 10.9 showing the internal junction capacitors

MILLER'S THEOREM

On the specification sheet for a 2N3904 transistor, the collector junction capacitance has a value of 4 pF. High frequencies may cause some of the output signal to feed back through the capacitance as its reactance decreases. When the collector junction capacitance begins to act as a feedback capacitor, analysis becomes very difficult.

Miller's theorem reduces that difficulty by allowing us to split the effects of the junction capacitance into two components. One component affects the input circuit while the other affects the output circuit. In equation form, Miller's theorem appears as:

$$C_{in}\left(\text{Miller}\right) = \left(1 + A_v\right) C_{OB}$$

$$C_{out}\left(\text{Miller}\right) = \left(\frac{A_v + 1}{A_v}\right) C_{OB}$$

EXAMPLE 10.4

Determine the C_{in}(Miller) and the C_{out}(Miller) for the circuit shown in Figure 10.9.

Before finding the answers for the Miller theorem equations, we must determine the circuit gain.

Solution

From before, we know that the AC emitter resistance is 16.7 Ω.

The AC collector resistance is:

$$r_c = R_c \| R_L$$

$$r_c = 1.8 \text{ k}\Omega \| 2.7 \text{ k}\Omega = 1.08 \text{ k}\Omega$$

Using the AC resistance values, we can find the circuit gain:

$$A_V = \frac{r_c}{r'_e}$$

$$\frac{1.08 \text{ k}\Omega}{16.7 \text{ }\Omega} = 64.8$$

Now that we have a value for the circuit gain, we can "millerize" the collector junction capacitance:

$$C_{in}\left(\text{Miller}\right) = \left(64.8 + 1\right) 4 \text{ pF} = 263.2 \text{ pF}$$

$$C_{out}\left(\text{Miller}\right) = \frac{65.8}{64.8} 4 \text{ pF} = 4.06 \text{ pF}$$

If the circuit gain is greater than 10, set the value for C_{out}(Miller) equal to the collector junction capacitance. Figure 10.12 shows the results of the calculations.

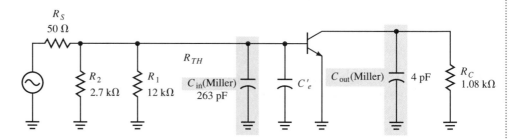

FIGURE 10.12 AC equivalent circuit for the circuit shown in Figure 10.9 after the collector junction capacitance has been "millerized"

A transistor circuit has a gain of 15. If the C_{OB} equals 40 pF, determine C_{in}(Miller) and C_{out}(Miller).

As Figure 10.12 illustrates, the Miller capacitances are in shunt with the signal. At high frequencies, the capacitances will short out a portion of the signal. We can calculate the critical high frequency for both the input and output terminals by using:

$$f_h = \frac{1}{2\pi RC}$$

EXAMPLE 10.5

Determine the upper critical frequency for the input circuit formed by the shunt capacitances pictured in Figure 10.12.

   We start our analysis by determining the total shunt capacitance for the input terminal.

Solution
C_{in} (Miller), as found in example 10.4, parallels the emitter junction capacitance, as found in example 10.3. The total capacitance for the input terminal equals:

$$C_T = C_{in}(Miller) + C'_e$$
$$C_T = 262.3 \text{ pF} + 31.8 \text{ pF} = 294.1 \text{ pF}$$

Next, we can find the total Thevenin resistance from the perspective of the base:

$$R_{TH} = R_S \parallel R_2 \parallel R_1$$
$$R_{TH} = 50 \ \Omega \parallel 2.7 \text{ k}\Omega \parallel 12 \text{ k}\Omega = 49 \ \Omega$$

Then we can calculate the critical frequency of the shunt capacitance that affects the input terminal:

$$f_h = \frac{1}{2\pi R_{TH}C_T}$$
$$f_h = \frac{1}{2\pi \times 49 \ \Omega \times 294.1 \text{ pF}} = 11,044,066 \text{ Hz}$$

This means that we can apply frequencies below 11 MHz to the input side of the circuit shown in Figure 10.9.

EXAMPLE 10.6

Determine the critical frequency of the network formed by the shunt capacitances affecting the output terminal of the circuit shown in Figure 10.9.

   We know that C_{out}(Miller) equals 4 pF and that the AC collector resistance equals 1.08 kΩ.

Solution
The critical high frequency of the output shunt network equals:

$$f_h = \frac{1}{2\pi R_{TH} \times C_{out}(Miller)}$$
$$\frac{1}{2\pi \times 1.08 \text{ k}\Omega \times 4 \text{ pF}} = 36.8 \text{ MHz}$$

The value of C_{out}(Miller) is 6 pF for the circuit of Figure 10.9. Determine the critical frequency of the output shunt network.

We have calculated the high cutoff frequency for the input and the output terminals. For high-frequency analysis, the dominant frequency is the lowest frequency. Thus, the two dominant frequencies are:

$$11,044,066 \text{ Hz} = f_h$$

and

$$942 \text{ Hz} = f_l$$

The bandwidth of the amplifier shown in Figure 10.9 is:

$$11,044,066 \text{ Hz} - 942 \text{ Hz} = 11,043,124 \text{ Hz}$$

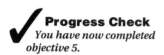

Progress Check
You have now completed objective 5.

JUNCTION FIELD EFFECT TRANSISTOR LOW-FREQUENCY SIGNAL ANALYSIS

Figure 10.13 shows a JFET amplifier and provides the transconductance value. The circuit uses self-bias and does not need a coupling capacitor on the gate. Even if a coupling capacitor existed in the circuit, the reverse biasing of the gate diode would make our analysis of the circuit easier. Because there is no input capacitor in Figure 10.8, the dominant low frequency is 0 Hz.

We can thevenize the output coupling capacitor using the same methods that we used for our bipolar junction transistor analysis. The Thevenin resistance seen by the output capacitance is:

$$R_{TH} = R_L + R_D$$
$$R_{TH} = 40 \text{ k}\Omega + 10 \text{ k}\Omega = 50 \text{ k}\Omega$$

Since the drain functions as a constant current source and electrically looks like an open, it does not affect the value of the thevenized resis-

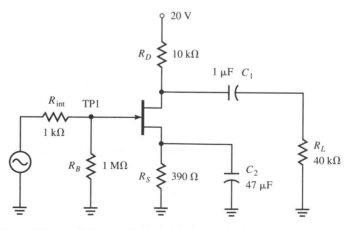

FIGURE 10.13 Self-biased JFET amplifier. $g_m = 5000 \mu$mhos.

tance. Calculating the critical frequency for the output portion of the circuit gives us:

$$f_l = \frac{1}{2\pi R_{TH} C_1}$$

$$f_l = \frac{1}{\left(6.28 \times 50 \text{ k}\Omega \times 1 \text{ }\mu F\right)} = 3.18 \text{ Hz}$$

Also the source bypass capacitor sees the source resistance and an impedance equal to 1 divided by the transconductance as part of the circuit. The thevenized resistance for the source bypass capacitor equals:

$$R_{TH} = \frac{1}{g_m} \parallel R_S$$

$$R_{TH} = \frac{1}{5000 \text{ }\mu mhos} \parallel 390 \text{ }\Omega$$

$$R_{TH} = 200 \text{ }\Omega \parallel 390 \text{ }\Omega = 132 \text{ }\Omega$$

Now, we can calculate the cutoff frequency for the source bypass network. The equation appears as:

$$f_l = \frac{1}{2\pi \times 132 \text{ }\Omega \times 47 \text{ }\mu F} = 25 \text{ Hz}$$

For the circuit shown in Figure 10.13, the three low cutoff frequencies are: 0 Hz, 3.18 Hz, and 25 Hz. Since 25 Hz is the highest low cutoff frequency, it is the dominant low frequency:

$$f_l = 25 \text{ Hz}$$

PRACTICE PROBLEM 10.7

Change the transconductance of the JFET in Figure 10.13 to 2000 μmhos. Recompute the break frequency of the source bypass capacitor network. Is it still the dominant low-frequency network?

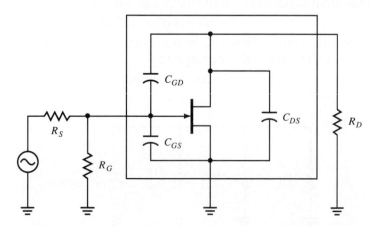

FIGURE 10.14   JFET circuit showing the terminal capacitances. The capacitors inside the box are not physically present, but are effects of the JFET.

JUNCTION FIELD EFFECT TRANSISTOR HIGH-FREQUENCY ANALYSIS

Figure 10.14 shows the AC configuration for a JFET amplifier. Like the bipolar junction transistor, the JFET has internal capacitances. These capacitances are between the gate and source, or C_{GS}; between the gate and drain, or C_{GD}; and between the drain and source, or C_{DS}. At high frequencies, the internal capacitances of the JFET limit circuit performance.

EXAMPLE 10.7

Determine the critical frequency of the input part of the circuit shown in Figure 10.13.

The JFET is an MPF102. Two capacitance values exist on the JFET specification sheet. The first is the input capacitance (C_{iss}), which is measured by the manufacturer by AC shorting the output and measuring the input capacitance. The JFET, MPF102, has a C_{iss} equal to 7 pF. The second specification sheet capacitance is C_{rss}, which is measured by the manufacturer and equals the feedback capacitance, C_{gd}. For the MPF102, C_{rss} equals 3 pF. We can use the values of C_{RSS} and C_{ISS} to determine the value of the gate-to-source capacitance.

Solution
Because C_{gs} and C_{gd} are in parallel, their summed capacitance is equal to C_{iss}. To find C_{gs} we need to rearrange our equation.

$$C_{iss} = C_{gs} + C_{gd}$$
$$C_{gs} = C_{iss} - C_{gd}$$
$$C_{gs} = 7\,\text{pF} - 3\,\text{pF} = 4\,\text{pF}$$

The gain of the circuit is:

$$A_v = g_m(r_d)$$
$$A_v = 5000\ \mu\text{mhos} \times 8\ \text{k}\Omega = 40$$

Using the gain value we can millerize the gate-to-drain capacitance:

$$C_{in}(\text{Miller}) = (A_v + 1)C_{GD} = 41 \times 3\,\text{pF} = 123\,\text{pF}$$

The C_{in}(Miller) value parallels the gate-to-source capacitance, which makes the total shunt capacitance of the input equal to:

$$123\,\text{pF} + 4\,\text{pF} = 127\,\text{pF}$$

Figure 10.15 shows the results of our calculations. The total shunt capacitance sees a Thevenin resistance equal to:

$$1\,\text{M}\Omega \,\|\, 1\,\text{k}\Omega = 999\,\Omega$$

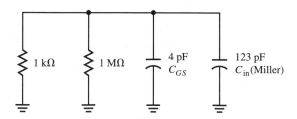

FIGURE 10.15   Lagging shunt network for the input of the circuit shown in Figure 10.13

Using the values for the shunt capacitance and the Thevenin resistance, we can find the critical frequency for the JFET amplifier circuit:

$$f_h = \frac{1}{(2\pi \times 999\ \Omega \times 127\ \text{pF})} = 1.25\ \text{MHz}$$

EXAMPLE 10.8

Determine the critical frequency of the output shunt network for the circuit shown in Figure 10.12.

Solution
We start our analysis by millerizing the gate-to-drain capacitance:

$$C_{out}(\text{Miller}) = \left(\frac{41}{40}\right) C_{GD} = 3.07\ \text{pF}$$

Generally, JFETs will have a drain-to-source capacitance of 2 pF.

Figure 10.16 shows the results of our calculations. Since C_{out}(Miller) parallels the drain-to-source capacitance, the entire shunt capacitance equals:

$$3\ \text{pF} + 2\ \text{pF} = 5\ \text{pF}$$

The Thevenin resistance seen by the shunt capacitance is the output resistance, R_D:

$$R_D = 10\ \text{k}\Omega \parallel 40\ \text{k}\Omega = 8\ \text{k}\Omega$$

Using the resistance and capacitance values, we can determine the critical frequency of the output shunt network:

$$f_C = \frac{1}{2\pi RC}$$

$$f_C = \frac{1}{2\pi \times 8\ \text{k}\Omega \times 5\ \text{pF}}$$

$$f_C = 3.98\ \text{MHz}$$

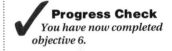

Progress Check
You have now completed objective 6.

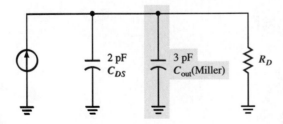

FIGURE 10.16   AC output shunt circuit showing the drain-to-source capacitance and C_{out}(Miller)

REVIEW SECTION 10.3
1. What is the definition of bandwidth, and what is its equation? Describe the terms used.
2. How is the critical frequency found in a capacitive resistive circuit?

3. What happens to the reactance of the capacitor if frequency decreases?
4. High-frequency response of a BJT is limited by what?
5. How is the approximate value of the emitter junction capacitance found?
6. What does Miller's theorem allow us to do?
7. In the analysis of a low-frequency signal, why can the gate diode in a JFET using self-bias be disregarded from the circuit?
8. Where are the internal capacitances found in a JFET?

10.4 DIFFERENTIAL AMPLIFIERS

So far, we have limited our study of semiconductor devices to diodes and transistors. Now, we will begin to broaden our knowledge by examining linear integrated circuits, Figure 10.17. Differential amplifiers are the "foot sol-

FIGURE 10.17 **Integrated circuits.** *(Photograph courtesy of International Rectifier, Semiconductor & IC Division)*

diers" of the linear integrated circuit family. Although we will study the basic form of transistors connected as a differential amplifier, our primary thrust takes us to the differential amplifier as a key element of most linear integrated circuits.

Since the differential amplifier works as the building block for many feedback and linear amplifiers, knowing about its characteristics is important for understanding the operation of other devices. A **differential amplifier**'s output is the difference between two inputs. Along with having good temperature stability, the differential amplifier direct couples and amplifies both DC and AC signals. In addition, the differential amplifier rejects unwanted noise common to both input lines.

DC AMPLIFICATION WITH DIFFERENTIAL AMPLIFIERS

Today's electronic applications range from communications and video equipment to computer systems. Many devices that either control other devices or require computing power rely on DC signals along with AC signals. As we know from our study of coupling capacitors in Chapter 6, a DC amplifier cannot use a capacitor or transformer to couple two stages together.

Still referring to Chapter 6, we can use direct coupling for a DC amplifier. Figure 10.18 shows a direct-coupled DC amplifier circuit with the common-emitter configuration. However, direct coupling introduces several problems. Without any type of coupling device, biasing for the amplifier becomes limited. Since the source voltage and the collector resistor must provide collector bias for the first transistor and base bias for the second transistor, the choice of biasing options narrows. Also, any change in the collector current of the first transistor becomes amplified by the second transistor. That is, any operating point change amplifies and makes maintaining a stable operating point difficult. This disadvantage is called **DC drift.**

Now that you have seen why a simple DC amplifier will not work for most modern electronics applications, let us move to a basic differential amplifier. Figure 10.19 shows an NPN differential amplifier. The configuration depicts two balanced transistors that provide good DC amplification. Because the amplifier exhibits symmetry, the effects of temperature on collector current are minimized.

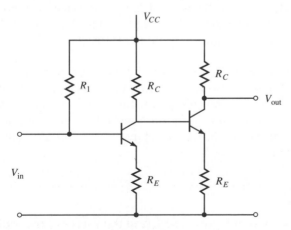

FIGURE 10.18   **Common-emitter DC amplifier**

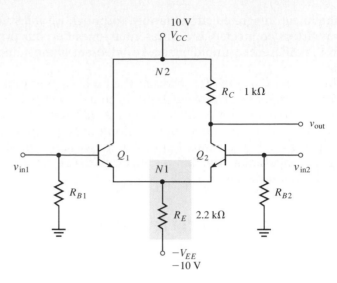

FIGURE 10.19 Basic differential amplifier

Instead of responding to temperature changes, the differential amplifier responds to differences between the two input signals. This difference is reflected in the output signal. Any variable that affects both input signals causes little or no output change. A differential amplifier is a stable DC amplifier because both inputs respond to the same change. Differential amplifier configurations may have the output taken from a single collector. Also, differential amplifiers may have one input grounded so that the difference voltage equals the single input voltage.

BIASING DIFFERENTIAL AMPLIFIERS

Figure 10.19 shows the most common form of a differential amplifier. In this form, the amplifier has an input for each transistor base and a single output taken from the collector of Q_2. Ideally, the matched set of transistors have an identical beta, an identical AC emitter resistance, and the same base-to-emitter voltages.

Biasing for the transistors is accomplished through a common-emitter resistor. For this form of differential amplifier, we can call the emitter resistor the "**tail resistor**" and the current flowing through the resistor the "**tail current.**" Current flowing through the matched transistors splits evenly for each transistor.

Even though the base resistors return the base current to ground, they actually have very little effect on the DC operation of the amplifier. As with an emitter-biased circuit, the DC voltage on the base of each transistor equals zero. With the bases at ground potential, the voltage at N1 equals -0.7 V. As you will see in example 10.19, we now have a way to calculate the tail and emitter current values.

EXAMPLE 10.9

Using the circuit shown in Figure 10.19, find the values of the emitter and tail currents.

The emitter source voltage equals -10 V, while the tail resistor has a value of 2.2 kΩ.

Solution
Applying Ohm's law to the tail resistor, we have:

$$\frac{-10\text{ V} - (-0.7\text{ V})}{2.2\text{ k}\Omega} = -4.22\text{ mA}$$

This current flows up through the tail resistor and splits evenly at node N1. The emitter currents equal:

$$I_{E1}\text{ and }I_{E2} = \frac{4.22\text{ mA}}{2} = 2.11\text{ mA}$$

PRACTICE PROBLEM 10.8

A differential amplifier has an emitter source voltage (V_{EE}) of -8 V and a tail resistance of 4.7 kΩ. How much current flows into the emitter of Q_2? What is the AC emitter resistance of Q_2?

AC ANALYSIS OF THE DIFFERENTIAL AMPLIFIER
To see the AC relationships of the circuit shown in Figure 10.19, we will redraw the circuit and use the superposition theorem. Let us start by analyzing the circuit with input voltage V_{in1} applied to the circuit and input voltage V_{in2} reduced to zero. Figure 10.20a shows how the AC circuit would look to the first input with the second input grounded.

In this situation, Q_1 acts as an emitter follower and Q_2 acts as a common-base amplifier for the V_{in1} signal. Since neither configuration produces a phase shift, the output at the collector of Q_2 is in phase with the V_{in1} input voltage. We can define this type of input signal as a **noninverting input.**

The AC signal into the base of Q_1 has several paths to ground. While one path leads through the AC and DC emitter resistance of Q_1, the other path

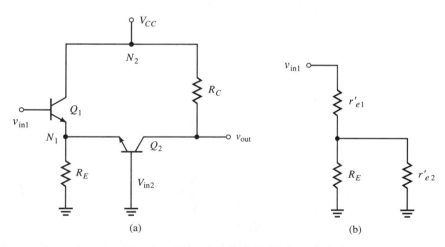

(a)   (b)

FIGURE 10.20   (a) AC equivalent circuit of a differential amplifier for N1 when N1 is grounded. (b) AC equivalent circuit paths to ground.

goes through the AC resistances of Q_1 and Q_2. Figure 10.20b illustrates this relationship. Looking at the figure, we can ignore the tail resistor because of its large value.

Considering the second path to ground, the total AC resistance of two matched transistors equals $2\,r'_e$.

The output is taken across the collector resistor, which causes the gain for a signal injected into the base of Q_1 to equal:

$$A_{v1} = \frac{R_C}{2\,r'_e}$$

Moving to Figure 10.21, we can see how the circuit would appear to an AC input signal on the base of Q_2 with the AC voltage at the base of Q_1 reduced to zero. This signal sees Q_2 as a common-emitter amplifier and Q_1 as a common-base amplifier. Because of the configuration, Q_2 has phase inversion and Q_1 has no phase inversion. Overall, though, the circuit has phase inversion. We define the input signal to Q_2 as an **inverting input.**

In this circuit, the AC emitter resistance of Q_1 shunts the emitter resistor. Otherwise, the various AC paths to ground mirror the paths seen in Figure 10.21.b.

The gain for the second input signal equals:

$$A_{v2} = \frac{-R_C}{2\,r'_e}$$

with the minus sign indicating that we are seeing a phase inversion. Using the superposition theorem dictates the use of the minus sign. From the superposition theorem, we find that:

$$V_{(out)} = V_{(out1)} + V_{(out2)}$$

Going back to our previous work, we know that:

$$V_{(out1)} = \frac{R_C}{2r'_e} \times V_{in1}$$

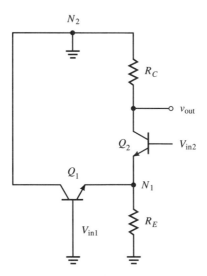

FIGURE 10.21 AC equivalent circuit of a differential amplifier from the perspective of N2 with N1 grounded

and

$$V_{(out2)} = \frac{-R_C}{2r'_e} \times V_{in2}$$

Therefore,

$$V_{(out)} = \frac{R_C}{2\,r'_e} \times v_{in1} + \frac{-R_C}{2\,r'_e}\,v_{in2}$$

$$V_{(out)} = \frac{R_C}{2\,r'_e}\left(v_{in1} - v_{in2}\right)$$

Substituting

$$A = \frac{R_C}{2\,r'_e}$$

we have

$$= v_{(out)} = A(v_{in1} - v_{in2})$$

As you may suspect, if the two input signals equaled one another, no AC output could exist. An output signal will exist only when the two input signals differ. We have now defined a differential amplifier.

Progress Check
You have now completed objective 7.

PRACTICE PROBLEM 10.9

In Figure 10.19, R_c = 4.7 kΩ, $-V_{EE}$ = 10 V, and R_E = 10 kΩ. Determine the single-ended gain of the circuit.

DIFFERENTIAL AMPLIFIER PARAMETERS

Input Offset Current

The **input offset current** is the difference between the two DC base currents. In equation form, this would look like:

$$I_{in}(off) = I_{B1} - I_{B2}$$

with $I_{in}(off)$ representing the input offset current. This specification tells us whether the two transistors match. An exact match will show that the two base currents equal one another and that the input offset current value is zero.

Input Bias Current

Although the input bias current parameter may seem similar to the input offset current, it is the average of the two base currents, or:

$$I_{in}\left(bias\right) = \frac{\left(I_{B1} + I_{B2}\right)}{2}$$

$I_{in}(bias)$ represents the input bias current. This parameter becomes particularly useful in determining the base current values. In equation form, the base current values equal:

$$I_{B1} = 2\,I_{in(bias)} - I_{b2}$$
$$I_{B2} = 2\,I_{in(bias)} - I_{b1}$$

Input Offset Voltage

With both transistor bases grounded as shown in Figure 10.19, the output signal at the Q_2 collector should be 0 V. Any deviation from this quiescent output voltage is the **output offset voltage.** An input offset voltage will return or null the output signal to the predicted quiescent value.

Common Mode Rejection Ratio

If we connected the bases of the two transistors from Figure 10.19 together, the amplifier would have a common-mode configuration. In this configuration, the same amplitude signal becomes applied to both bases. We define any voltage applied equally and simultaneously to the input terminals as a **common-mode signal.** Common-mode signals exist because of noise, drift, and differences in ground potential. Although common mode does not work in practical applications such as communications devices, it gives us a way to measure the ability of a differential amplifier to reject noise.

Common-mode noise, static, interference, and other forms of unwanted signals drive both sides of a differential amplifier evenly. These signals enter the circuit through the power supply connections. In Figure 10.19, common-mode noise sees the emitter resistor as a path to ground. Common-mode signal gain equals:

$$A_{cm} = \frac{R_C}{2R_E}$$

The value of the emitter resistor multiplied by two is the denominator because its current splits evenly at node 1 and sees two identical paths. This cuts the signal amplitude in half. Doubling the size of the emitter resistor accounts for the amplitude cut. Since the emitter resistor swamps out the AC emitter resistance for common-mode signals, we do not need to consider the effects of the AC resistance.

EXAMPLE 10.10

Determine the common-mode gain for the differential amplifier shown in Figure 10.19.

The collector resistor has a value of 1 kΩ, and the emitter resistor has a value of 2.2 kΩ.

Solution
Using the equation

$$A_{cm} = \frac{1\text{ k}\Omega}{2(2.2\text{ k}\Omega)} = 0.227$$

gives a value for the differential amplifier common-mode gain. Surprisingly, we see a loss of signal strength. Reducing the noise signals improves the signal-to-noise ratio of the amplifier.

Considering common-mode rejection, the differential amplifier stands as an excellent circuit. This becomes especially true when replacing the tail resistor with a constant current source. In equation form, the common-mode rejection ratio is:

$$\frac{A_d}{A_{cm}}$$

For the equation, A_d is the differential gain of the circuit while A_{cm} is the common-mode gain of the circuit. Differential gain is the amplifier gain when unequal signals are applied to the inputs. Most specification sheets list the value for common-mode rejection ratio (CMRR) in decibels. To convert the common-mode rejection ratio to decibels, use 20 log CMRR.

PRACTICE PROBLEM 10.10

A differential amplifier has a differential gain of 80 and a common-mode gain of 0.8. Determine the common-mode rejection ratio in decibels.

Progress Check
You have now completed objective 8

The Constant Current Source for a Differential Amplifier

In most practical applications, the differential amplifier has a constant current source connected to its emitters. Although we have used a large emitter resistor as a constant current source, in integrated circuit applications, a large resistance or capacitance can cause problems.

From our past studies of transistor behavior, we know that the collector current of a transistor acts like a constant current source. The collector current remains constant if the base-to-emitter voltage also remains constant. Not surprisingly, providing the constant base-to-emitter voltage is a high priority.

As Figure 10.22a shows, a transistor with a controlled reference voltage provides a constant current source. Moreover, we can use one or more PN junction diodes to supply the needed reference voltage. By using the 0.7-V forward voltage of diodes and resistances from a voltage divider, we can obtain a very precise reference voltage for the base-emitter junction of the transistor. Figure 10.22b shows a diode-stabilized constant current source. Other configurations may use a transistor with the collector shorted to the base or zener diodes. The shaded area of Figure 10.23 shows a differential amplifier with the addition of the constant current source.

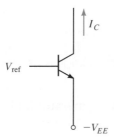

FIGURE 10.22 Transistor constant current source.

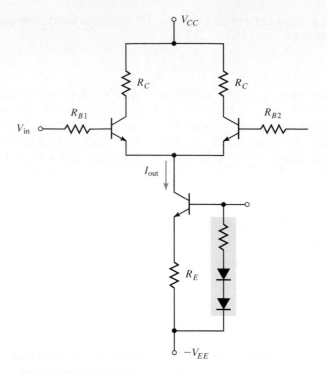

FIGURE 10.23 Differential amplifier with a constant current source. The shaded area shows diode stabilization.

REVIEW SECTION 10.4

1. What are some positive characteristics of a differential amp?
2. Biasing for transistors in a differential amp is accomplished by a common-emitter capacitor. True or false?
3. To what do the terms "tail resistor" and "tail current" refer?
4. In AC analysis of a differential amp, when will an output signal exist?
5. What happens to the input offset current if the two base currents are equal?
6. What term refers to a deviation from the quiescent output voltage?
7. To what are the common-mode signal gain and the common-mode rejection ratio equal?

 TROUBLESHOOTING

Figure 10.24 shows a test setup using a signal generator and an oscilloscope. We can use this setup to determine the bandwidth of an amplifier. Let us walk through testing a small-signal amplifier.

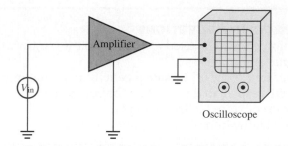

FIGURE 10.24 Test setup with a signal generator and an oscilloscope.

Before turning on the signal generator, adjust its signal amplitude to zero. Next, set the output of the generator so that it produces sine waves. We can determine the bandwidth of the amplifier either by using techniques learned in this chapter or by checking any available technical literature. Set the output frequency of the generator to the midpoint of the amplifier bandwidth.

Connect the oscilloscope to the output of the amplifier and then turn on the oscilloscope power. Slowly adjust the output amplitude of the generator so that the output is the maximum unclipped signal. After attaining the required amplitude, do not change the amplitude setting of the generator.

Adjust the image size on the oscilloscope screen so that it fits between the zero and 100% markings on the faceplate. In some instances, you may need to decalibrate the volts/division setting to adjust the size of the image. Generally, the decalibration knob is the inner knob of the volts/division assembly.

Slowly increase the signal frequency until the image size has dropped to 70% of its original amplitude. Check the adjustment of the seconds/division or the time base function of the oscilloscope so that a clear image is displayed. Measure the time for one cycle, and convert the time to frequency:

$$f = \frac{1}{t}$$

You have now measured the upper frequency limit (f_H) of the amplifier.

Now, readjust the midband frequency of the amplifier so that the image again fills out to 100%. After the image fills the screen, turn the generator frequency down until the amplitude drops to 70% of its midband amplitude. Adjust the seconds/division function for a clear image, and measure the time for one complete cycle. By converting this measurement to frequency, you will find a value for the low-frequency cutoff (f_L). The bandwidth (BW) of the amplifier equals:

$$BW = f_H - f_L$$

WHAT DO YOU CHECK IF THE BANDWIDTH IS WRONG?

If the low critical frequency is wrong, check the input and output coupling capacitors and the bypass capacitor. Sometimes, the reactances associated with these capacitors may vary from their listed tolerances. A resistance problem should show as a discrepancy between DC operation of the circuit and your DC predictions.

WHAT'S WRONG WITH THIS CIRCUIT?

Normal readings for the following circuit are:

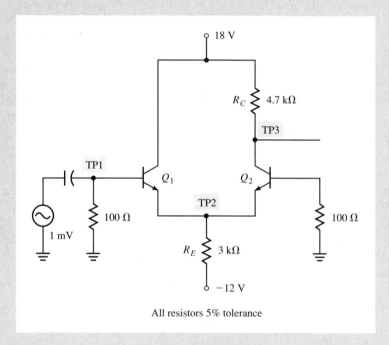

All resistors 5% tolerance

	DC		AC
TP1	0.08 V	TP1	1 mV
TP2	−0.6 V	TP2	500 μV
TP3	9 V	TP3	180 mV

Will the circuit function for the following readings in DC? If not, why not?

	DC
TP1	0.08 V
TP2	−0.6 V
TP3	8.12 V

Will the circuit function for the following readings in DC and AC? If not, why not?

	DC		AC
TP1	0.08 V	TP1	1 mV
TP2	−0.6 V	TP2	1 mV
TP3	18 V	TP3	0 V

SUMMARY

Negative feedback improves the performance of an amplifier. An amplifier circuit can eliminate distortion by feeding back and comparing the output signal with the input signal. In addition, negative feedback increases input impedance, decreases output impedance, and causes circuit gain to be independent of individual amplifier gain.

Small-signal transistor amplifiers have additional parameters called hybrid parameters. Hybrid, or h, parameters define the small-signal characteristics of a transistor at a given bias point. The small-signal characteristics are input impedance with a shorted output (h_i), forward current gain with a shorted output (h_f), reverse voltage gain with an open input (h_r), and output admittance with an open input (h_o).

Differential amplifiers work as the building blocks for many linear integrated circuit applications. A differential amplifier applies the input signal voltage between two symmetrical inputs. Differential amplifiers have good temperature stability and amplify both DC and AC signals. Like other semiconductor devices, differential amplifiers have specific characteristics. Those characteristics are input offset current, input bias current, input offset voltage, and common-mode rejection ratio.

Refer to Figure 10.25 for problems 1–9.

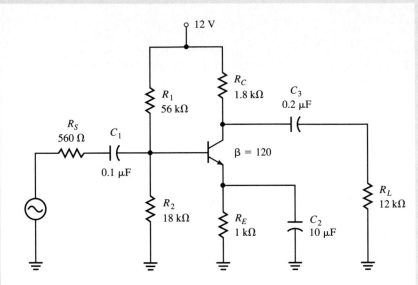

FIGURE 10.25 Circuit for problems 1–9. $\beta = 120$.

1. Determine the critical frequency for the input RC network.
2. Determine the critical frequency for the output RC network.
3. Determine the critical frequency for the bypass RC network.
4. What is the dominant low frequency for the circuit?
5. What is the gain for the circuit?
6. If C_{OB} is 3 pF, determine C_{in}(Miller). Determine C_{out}(Miller).
7. If C'_e is 22 pF, determine the critical frequency of the high-frequency input network.
8. Determine the critical frequency of the high-frequency output network.
9. What is the dominant high frequency of the circuit?

Refer to Figure 10.26 for problems 10–15.

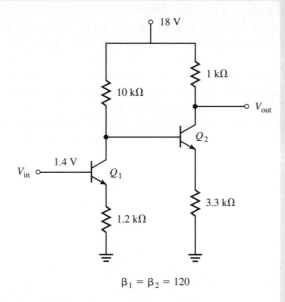

FIGURE 10.26 Circuit for problems 10–15. $\beta = 120$.

10. What is the low-frequency limit of the circuit?
11. What is the DC voltage at the base of Q_2?
12. What is the DC output voltage?
13. What is the gain of the first stage?
14. What is the gain of the second stage?
15. What is the overall gain of the circuit?

Refer to Figure 10.27 for problems 16–21.

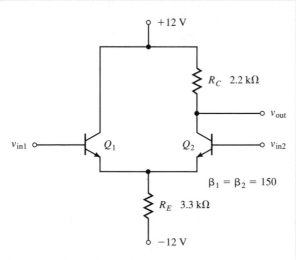

FIGURE 10.27 Circuit for problems 16–21. $\beta_1 - \beta_2 = 150$.

16. Determine the tail current for the circuit with both inputs grounded.
17. What is r'_{e1}? What is r'_{e2}?
18. What is the gain of the circuit?
19. The circuit has an A_{cm} of 0.33. Determine the common mode rejection ratio for the circuit. Express it in decibels; that is, find CMRR'.
20. Determine the input impedance as seen by an AC signal injected into the base of Q_1. Refer to Figure 10.5f to help you draw the beta box model for Z_{in}(base) for the circuit.
21. If the input at the base of Q_1 is 10 mV and the input to the base of Q_2 is 4 mV, what will the amplitude of the AC signal be at the output?

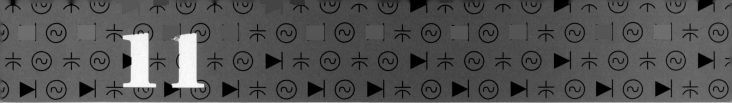

OPERATIONAL AMPLIFIER
FUNDAMENTALS

OBJECTIVES

As you read this chapter, concentrate on learning how to:

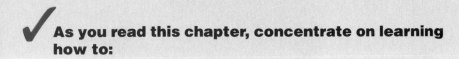

1. Identify and properly label the terminals in an operational amplifier schematic symbol
2. Identify whether an op-amp circuit is being operated open-loop or closed-loop
3. Determine the output voltage of an op-amp, using the equation $v_{out} = A(v_1 - v_2)$
4. Determine the effects of input offset current on op-amp circuits
5. Explain the need for nulling circuits
6. Find typical op-amp specifications on a sample operational amplifier specification sheet
7. Use the gain-bandwidth product to determine the gain of an op-amp at a particular frequency
8. Use the slew rate specification to determine how long it takes an output voltage to change from one level to another level
9. Calculate the power bandwidth of an op-amp circuit
10. Calculate the signal-to-noise ratio of an op-amp given the common mode rejection ratio of the op-amp

INTRODUCTION

When an electronics technician talks about op-amps, he or she is probably talking about a linear integrated circuit known as an operational amplifier. In this type of circuit, almost all the electronics needed for circuit operation are enclosed in a single package commonly called an integrated circuit, or IC. Types of IC packages include single in-line packages (SIPS), dual in-line packages (DIPS), flat packs, and surface mount devices shown in Figure 11.1.

It is most often understood that the op-amp IC is a general-purpose amplifier that can be used in many different applications, especially linear ap-

FIGURE 11.1 (a) Typical IC packages. (b) IC packages placed on circuitboard.

plications. In **linear applications,** the input signal is reproduced without distortion and at a larger amplitude. Most op-amps are designed to have very high gains under open-loop conditions (no feedback), with typical values ranging from 5000 to 100,000.

Op-amps with very high gain have a narrow bandwidth of about 0 to 15

Hz. This bandwidth can easily be extended up into the hundreds-of-kilohertz range by decreasing the gain of the op-amp circuit.

As you read this chapter, you will notice that many of the electrical characteristics of the operational amplifier are similar to the characteristics of the differential amplifier, which was discussed in Chapter 10. This similarity is not surprising because the input stage of an operational amplifier is a differential amp.

11.1 BEGINNING OP-AMP INFORMATION

The schematic symbol used to represent an operational amplifier is shown in Figure 11.2. Note that there are two inputs, shown on the left-hand side. These inputs are the inverting and noninverting inputs. The noninverting input is labeled with the plus sign, and the inverting terminal is marked with the minus sign.

When the voltage applied to the noninverting terminal is more positive with respect to ground than the voltage applied to the inverting terminal, the output will be positive with respect to ground, as shown in Figure 11.3.

When the input to the noninverting terminal is less positive with respect to ground than the voltage applied to the inverting terminal, the output will be negative with respect to ground, as shown in Figure 11.4.

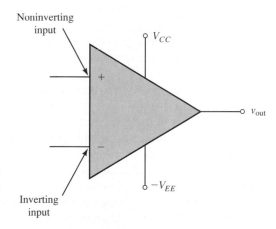

FIGURE 11.2 Operational amplifier schematic symbol

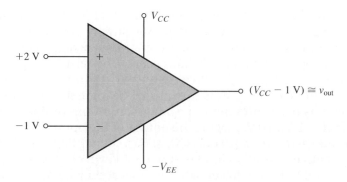

FIGURE 11.3 Positively saturated op-amp, open-loop operation

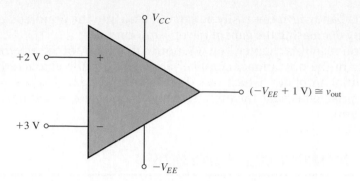

FIGURE 11.4 Negatively saturated op-amp, open-loop operation

OP-AMP POWER SUPPLY REQUIREMENTS AND LIMITATIONS

There are two power supply connections. The top connection is where the positive supply voltage is shown in a schematic. This is usually designated V_{CC}. The bottom connection is where the negative supply connection would be shown in a schematic. This negative voltage is often designated $-V_{EE}$.

These power supply levels set the limits of output voltage swing. When the op-amp output goes positive with respect to ground, it can reach a maximum of about 95 V_{CC}. This 5 "loss" is due to voltages dropped within the op-amp itself. In some cases this loss may be as much as 2 V.

When the output of the op-amp goes negative with respect to ground, it can reach a maximum of 95 V_{EE}. Once again, this 5 loss is due to voltage drops within the op-amp itself. As with the positive output, the loss may be as much as 2 V, in which case, a negative output would saturate the op-amp at $(-V_{EE} + 2 \text{ V})$.

Many commercially available op-amps have maximum power supply limits of ± 18 to ± 22 V. Minimum power supply limits for op-amps vary as well. The minimum power supply level of one op-amp, the LM4250C, can be as low as ± 1 V. However, unless you know for sure that you can go lower, use ± 5 V as a minimum supply voltage to ensure op-amp operation.

In most op-amp circuits, the power supplies are equal values and opposite polarities. This means that if the positive supply is set to $+15$ V, then the negative supply is set to -15 V. Similarly, if the negative supply is set to -6 V, then the positive supply is set to $+6$ V. In this arrangement, commonly referred to as a **dual polarity power supply,** the output of the op-amp can swing positive (above ground) and negative (below ground) in equal amounts.

In some special situations only one of the op-amp's power supply connections is wired to power; the other power supply connection is wired to ground. An example of this is when an op-amp is acting as an interface between an analog system and a digital system. The positive supply of the op-amp could be wired to $+5$ V, while the negative supply is wired to ground. The analog inputs to the op-amp are selected to force the op-amp's output to swing from positive saturation to negative saturation. In this case, the ideal swing is from $+5$ V to 0 V, making the output of the op-amp compatible with transistor-transistor logic (TTL) (74XX series, digital ICs) thresholds. Practically, the positive saturation level is about $+4$ V, which is an acceptable logic 1 in TTL circuits, but the lower saturation level may be as high as 1 or 2 V. This level is not an acceptable logic 0 for a TTL gate. Pulling the negative supply below ground forces the negative saturation level to drop below 0.8 V, the maximum voltage that is accepted as a logic 0.

In most cases, an op-amp has both a positive and a negative supply voltage applied to it. The ground for the op-amp is through a component or components that are wired between an op-amp terminal and ground potential. This ground is the center point between the two power supply levels. In this text, the power supply connections in circuits that have op-amps are not shown so that diagrams aren't cluttered.

THE OUTPUT TERMINAL

The single output is indicated on the right side of the schematic symbol. It is usually driven by a class B push-pull amplifier (discussed in Chapter 6).

Progress Check
You have now completed objective 1.

BASIC OP-AMP OPERATION

The input stage of an operational amplifier is a differential amplifier. This means that the difference between the two inputs is the voltage that is amplified by the device:

$$v_{out} = A(v_1 - v_2)$$

where A is the voltage gain of the circuit
   v_1 is the input to the noninverting terminal
   v_2 is the input to the inverting terminal

If the signals into the two input terminals are exactly the same voltage, amplitude, polarity, and phase, the output should be 0 V.

Many op-amps have very high open-loop gains. Figure 11.5 shows an op-amp operating in an open-loop configuration. In this circuit none of the output signal is returned to the input signal.

In a closed-loop circuit, part of the output signal is fed back to the input side of the circuit. An example is shown in Figure 11.6. In Figure 11.6, v_{out} drives a voltage divider consisting of R_f and R_b. The divider lowers the output voltage to the value of v_f, the voltage developed across R_b. It is this voltage, v_f, that is returned to the input of the circuit.

Progress Check
You have now completed objective 2.

You came across this idea of closed-loop operation in Chapter 4, when we studied the DC characteristics of collector feedback bias. In Chapter 5 we studied the AC characteristics of collector feedback bias.

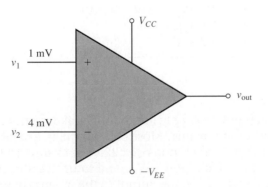

FIGURE 11.5 Differential inputs, open-loop operation

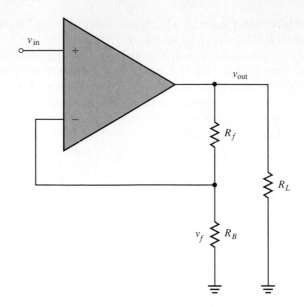

FIGURE 11.6 Op-amp circuit, closed-loop operation

Now, let's consider open-loop operation. Open-loop gain is often designated A_{OL} or A_{VOL}. Example 11.1 shows how to calculate the output voltage of the op-amp circuit shown in Figure 11.5.

EXAMPLE 11.1

Assume that the operational amplifier has an open-loop gain equal to 100,000. If the input at the inverting terminal is +4 mV and the input at the noninverting terminal is +1 mV, determine the size of the output voltage.

Solution
The differential input is −3 mV. It is this voltage that is amplified. Therefore the output voltage is

$$v_o = A(v_1 - v_2)$$
$$v_o = 100{,}000\ (1\text{ mV} - 4\text{ mV})$$
$$v_o = 100{,}000\ (-3\text{ mV})$$
$$v_o = -300\text{ V}$$

Unfortunately op-amp power supply limitations make it impossible to get this type of voltage out of an op-amp. Most commercially available op-amps have maximum power supply limitations of ±22 V. In example 11.1, the output swings towards the negative supply, $-V_{EE}$, and saturates the op-amp. If the op-amp power supplies were ±22 V, the output of the op-amp in example 11.1 would be approximately $(-V_{EE} + 1\text{ V}) = -21\text{ V}$.

Let's change the input voltages to the op-amp of Figure 11.5. The voltage at the noninverting terminal is 20 mV. The voltage at the inverting terminal is 18 mV. The op-amp has an open loop gain of 100,000, and the power supplies are at ±22 V. Determine the output voltage.

EXAMPLE 11.2

Determine the output voltage for the op-amp of Figure 11.7. The inverting voltage is +5 μV and the noninverting voltage is +3 μV. The open-loop gain is once again 100,000. The power supply limitations are ±18 V.

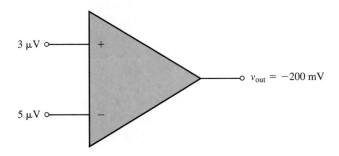

FIGURE 11.7 Differential inputs producing an output. $A_{OL} = 100,000$.

Solution
In this case, the differential input is 2 μV. This is amplified to obtain

$$v_o = A(v_1 - v_2)$$
$$v_o = 100{,}000(3\ \mu V - 5\ \mu V)$$
$$v_o = 100{,}000(-2\ \mu V)$$
$$v_o = -200\ mV$$

This is easily within the power supply limitations of this op-amp circuit. The −200 mV, not the saturation voltage, appears at the output.

EXAMPLE 11.3

The input to an op-amp with an A_{OL} of 5000 is +2 mV at the inverting terminal and 0 mV at the noninverting terminal. Determine the output of the device. The power supplies are at ±16 V.

Solution
The difference between the two inputs is 2 mV. So the output is

$$v_o = A(v_1 - v_2)$$
$$v_o = 5000(0\ mV - 2\ mV)$$
$$v_o = -10\ V$$

The polarity of the output is negative with respect to ground because the input to the noninverting terminal is more negative than the input at the inverting terminal.

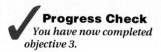
PRACTICE PROBLEM 11.2

An operational amplifier has an A_{OL} of 2000. The input at the noninverting terminal is +1 V and at the inverting terminal +1.01 V. Determine the output voltage and its polarity to ground. The power supplies are at ±22 V.

REVIEW SECTION 11.1

1. Draw the schematic symbol for an operational amplifier.
2. Which "connection point" in the op-amp schematic symbol is used to designate the output?
3. The positive and negative power supply voltages for an op-amp must be equal in magnitude. True or false?
4. What will be the polarity of the output of an op-amp with respect to ground if the inverting input is more positive than the noninverting input?
5. What is meant by the designation A_{OL}?
6. The negative power supply terminal of an op-amp can be grounded. True or false?
7. What is the ideal output voltage of an op-amp when the two input signals are exactly the same?
8. Typical op-amp open-loop gains range from 5000 to _____ .
9. Why do many op-amp circuits require the positive and negative power supply voltages to be of equal magnitude?
10. An op-amp has its output driven into positive saturation. This voltage is usually V_{CC} = _____ V.
11. Op-amps have no minimum power supply requirements. True or false?

11.2 ELECTRICAL PARAMETERS AND IMPEDANCES OF OP-AMPS

Although we are not going to investigate how the operational amplifier works internally, we cannot ignore the effects of the internal circuitry on the operation of circuitry connected to the terminals of the device. We need to understand these effects and how to compensate for them when they affect circuit operation.

INPUT IMPEDANCE

The input impedance of an op-amp (or the input resistance) is measured across the input terminals, as shown in Figure 11.8. While it is drawn as a single effective resistance, it is really the Thevenin resistance of the internal connections between the two input terminals. Typically, this resistance is very high. Typical values range from 500 kΩ to 2 MΩ. This op-amp specification is usually tested under open-loop conditions.

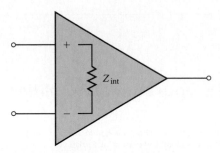

FIGURE 11.8 Input impedance of an operational amplifier

INPUT BIAS CURRENTS

In Figure 11.9a, two signals drive the inputs of the operational amplifier.

The voltage output is equal to the difference between the two input signals times the gain of the op-amp. This is the differential input discussed in Section 11.1.

Because op-amps are not ideal devices, the output may not equal A $(v_1 - v_2)$ exactly. This is because of differences in input bias currents and other effects.

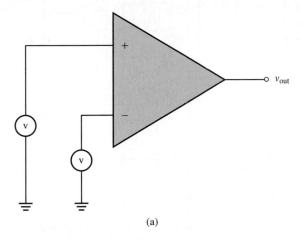

(a)

FIGURE 11.9a Op-amp with signals applied

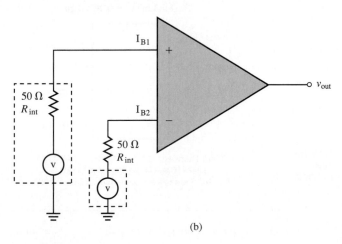

(b)

FIGURE 11.9b Op-amp bias currents

LM741/LM741A/LM741C/LM741E Operational Amplifier

General Description

The LM741 series are general purpose operational amplifiers which feature improved performance over industry standards like the LM709. They are direct, plug-in replacements for the 709C, LM201, MC1439 and 748 in most applications.

The amplifiers offer many features which make their application nearly foolproof: overload protection on the input and output, no latch-up when the common mode range is exceeded, as well as freedom from oscillations.

The LM741C/LM741E are identical to the LM741/LM741A except that the LM741C/LM741E have their performance guaranteed over a 0°C to +70°C temperature range, instead of −55°C to +125°C.

Schematic and Connection Diagrams (Top Views)

TL/H/9341–1

Metal Can Package

NC

OFFSET NULL — 1

INVERTING INPUT — 2

NON−INVERTING INPUT — 3

8 — NC

7 — V⁺

6 — OUTPUT

5 — OFFSET NULL

V⁻

TL/H/9341–2

**Order Number LM741H, LM741AH,
LM741CH or LM741EH
See NS Package Number H08C**

Dual-In-Line or S.O. Package

OFFSET NULL — 1 | 8 — NC

INVERTING INPUT — 2 | 7 — V⁺

NON−INVERTING INPUT — 3 | 6 — OUTPUT

V⁻ — 4 | 5 — OFFSET NULL

TL/H/9341–3

**Order Number LM741J, LM741AJ, LM741CJ,
LM741CM, LM741CN or LM741EN
See NS Package Number J08A, M08A or N08E**

FIGURE 11.10 741 op-amp specification sheet (partial)   (continued on next page)

FIGURE 11.10 (continued)

Absolute Maximum Ratings

If Military/Aerospace specified devices are required, please contact the National Semiconductor Sales Office/ Distributors for availability and specifications. (Note 5)

	LM741A	LM741E	LM741	LM741C
Supply Voltage	±22V	±22V	±22V	±18V
Power Dissipation (Note 1)	500 mW	500 mW	500 mW	500 mW
Differential Input Voltage	±30V	±30V	±30V	±30V
Input Voltage (Note 2)	±15V	±15V	±15V	±15V
Output Short Circuit Duration	Continuous	Continuous	Continuous	Continuous
Operating Temperature Range	−55°C to +125°C	0°C to +70°C	−55°C to +125°C	0°C to +70°C
Storage Temperature Range	−65°C to +150°C	−65°C to +150°C	−65°C to +150°C	−65°C to +150°C
Junction Temperature	150°C	100°C	150°C	100°C
Soldering Information				
N-Package (10 seconds)	260°C	260°C	260°C	260°C
J- or H-Package (10 seconds)	300°C	300°C	300°C	300°C
M-Package				
Vapor Phase (60 seconds)	215°C	215°C	215°C	215°C
Infrared (15 seconds)	215°C	215°C	215°C	215°C

See AN-450 "Surface Mounting Methods and Their Effect on Product Reliability" for other methods of soldering surface mount devices.

ESD Tolerance (Note 6)	400V	400V	400V	400V

Electrical Characteristics (Note 3)

Parameter	Conditions	LM741A/LM741E			LM741			LM741C			Units
		Min	Typ	Max	Min	Typ	Max	Min	Typ	Max	
Input Offset Voltage	T_A = 25°C R_S ≤ 10 kΩ R_S ≤ 50Ω		0.8	3.0		1.0	5.0		2.0	6.0	mV mV
	T_{AMIN} ≤ T_A ≤ T_{AMAX} R_S ≤ 50Ω R_S ≤ 10 kΩ			4.0			6.0			7.5	mV mV
Average Input Offset Voltage Drift				15							μV/°C
Input Offset Voltage Adjustment Range	T_A = 25°C, V_S = ±20V	±10				±15			±15		mV
Input Offset Current	T_A = 25°C		3.0	30		20	200		20	200	nA
	T_{AMIN} ≤ T_A ≤ T_{AMAX}			70		85	500			300	nA
Average Input Offset Current Drift				0.5							nA/°C
Input Bias Current	T_A = 25°C		30	80		80	500		80	500	nA
	T_{AMIN} ≤ T_A ≤ T_{AMAX}			0.210			1.5			0.8	μA
Input Resistance	T_A = 25°C, V_S = ±20V	1.0	6.0		0.3	2.0		0.3	2.0		MΩ
	T_{AMIN} ≤ T_A ≤ T_{AMAX}, V_S = ±20V	0.5									MΩ
Input Voltage Range	T_A = 25°C							±12	±13		V
	T_{AMIN} ≤ T_A ≤ T_{AMAX}				±12	±13					V
Large Signal Voltage Gain	T_A = 25°C, R_L ≥ 2 kΩ V_S = ±20V, V_O = ±15V V_S = ±15V, V_O = ±10V	50				50	200		20	200	V/mV V/mV
	T_{AMIN} ≤ T_A ≤ T_{AMAX}, R_L ≥ 2 kΩ, V_S = ±20V, V_O = ±15V V_S = ±15V, V_O = ±10V V_S = ±5V, V_O = ±2V	32 10				25			15			V/mV V/mV V/mV

(continued on next page)

FIGURE 11.10 (continued)

LM741/LM741A/LM741C/LM741E

Electrical Characteristics (Note 3) (Continued)

Parameter	Conditions	LM741A/LM741E			LM741			LM741C			Units
		Min	Typ	Max	Min	Typ	Max	Min	Typ	Max	
Output Voltage Swing	$V_S = \pm 20V$ $R_L \geq 10\ k\Omega$ $R_L \geq 2\ k\Omega$	± 16 ± 15									V V
	$V_S = \pm 15V$ $R_L \geq 10\ k\Omega$ $R_L \geq 2\ k\Omega$				± 12 ± 10	± 14 ± 13		± 12 ± 10	± 14 ± 13		V V
Output Short Circuit Current	$T_A = 25°C$ $T_{AMIN} \leq T_A \leq T_{AMAX}$	10 10	25	35 40		25			25		mA mA
Common-Mode Rejection Ratio	$T_{AMIN} \leq T_A \leq T_{AMAX}$ $R_S \leq 10\ k\Omega, V_{CM} = \pm 12V$ $R_S \leq 50\Omega, V_{CM} = \pm 12V$	80	95		70	90		70	90		dB dB
Supply Voltage Rejection Ratio	$T_{AMIN} \leq T_A \leq T_{AMAX}$, $V_S = \pm 20V$ to $V_S = \pm 5V$ $R_S \leq 50\Omega$ $R_S \leq 10\ k\Omega$	86	96		77	96		77	96		dB dB
Transient Response Rise Time Overshoot	$T_A = 25°C$, Unity Gain		0.25 6.0	0.8 20		0.3 5			0.3 5		μs %
Bandwidth (Note 4)	$T_A = 25°C$	0.437	1.5								MHz
Slew Rate	$T_A = 25°C$, Unity Gain	0.3	0.7			0.5			0.5		$V/\mu s$
Supply Current	$T_A = 25°C$					1.7	2.8		1.7	2.8	mA
Power Consumption	$T_A = 25°C$ $V_S = \pm 20V$ $V_S = \pm 15V$		80	150		50	85		50	85	mW mW
LM741A	$V_S = \pm 20V$ $T_A = T_{AMIN}$ $T_A = T_{AMAX}$			165 135							mW mW
LM741E	$V_S = \pm 20V$ $T_A = T_{AMIN}$ $T_A = T_{AMAX}$			150 150							mW mW
LM741	$V_S = \pm 15V$ $T_A = T_{AMIN}$ $T_A = T_{AMAX}$					60 45	100 75				mW mW

Note 1: For operation at elevated temperatures, these devices must be derated based on thermal resistance, and T_j max. (listed under "Absolute Maximum Ratings"). $T_j = T_A + (\theta_{jA} P_D)$.

Thermal Resistance	Cerdip (J)	DIP (N)	HO8 (H)	SO-8 (M)
θ_{jA} (Junction to Ambient)	100°C/W	100°C/W	170°C/W	195°C/W
θ_{jC} (Junction to Case)	N/A	N/A	25°C/W	N/A

Note 2: For supply voltages less than $\pm 15V$, the absolute maximum input voltage is equal to the supply voltage.

Note 3: Unless otherwise specified, these specifications apply for $V_S = \pm 15V$, $-55°C \leq T_A \leq +125°C$ (LM741/LM741A). For the LM741C/LM741E, these specifications are limited to $0°C \leq T_A \leq +70°C$.

Note 4: Calculated value from: BW (MHz) = 0.35/Rise Time(μs).

Note 5: For military specifications see RETS741X for LM741 and RETS741AX for LM741A.

Note 6: Human body model, 1.5 kΩ in series with 100 pF.

Keep in mind during the following discussion Figure 11.9, particularly 11.9b, where the internal resistances of the two sources are shown to be 50 Ω.

The input circuit of an operational amplifier, in which the first stage is a differential amplifier built around bipolar transistors, requires bias currents to function properly. Specification sheets for op-amps list the input stage current as **input bias current.** On Figure 11.10, a specification sheet for a 741 operational amplifier, the input bias current is listed as typically 80 nA, while the maximum value is given as 500 nA. Other op-amps have different input bias currents. For example, a μA709, general-purpose op-amp has an input bias current of 1500 nA. This is the maximum value possible. The actual value is probably a great deal less.

The input bias current (I_B) is the average of the two bias currents. If we designate the bias current associated with the noninverting terminal as I_{B1} and the bias current associated with the inverting terminal as I_{B2}, then

$$I_B = \frac{(I_{B1} + I_{B2})}{2}$$

EXAMPLE 11.4

This example uses specifications for the μA709 op-amp. Assume the worst-case bias current: $I_B = 1500$ nA. The minimum open-loop gain for a μA709 op-amp is 15,000. In addition, assume that the bias currents have split evenly. How will this affect the operation of the op-amp in Figure 11.9a?

Solution
We know that

$$I_B = \frac{I_{B1} + I_{B2}}{2}$$

If we assume that $I_{B1} = I_{B2}$, we can substitute I_{B1} for I_{B2}

$$I_B = \frac{I_{B1} + I_{B1}}{2}$$
$$I_B = \frac{2\,I_{B1}}{2} = I_{B1}$$

We have found that

$$I_B = I_{B1} = I_{B2} = 1500 \text{ nA}$$

These currents flow through the 50 Ω resistances. Consequently, we can calculate a DC voltage drop:

$$V_{source} = 1500 \text{ nA} \times 50 \text{ }\Omega$$
$$V_{source} = 75 \text{ }\mu\text{V}$$

This DC voltage is present at each input. Because the input of the op-amp amplifies only differences, the 75 μVDC at each input does not affect circuit operation because

$$V_{out} = A(v_1 - v_2)$$
$$V_{out} = A(75 \text{ }\mu\text{V} - 75 \text{ }\mu\text{V})$$
$$V_{out} = A(0 \text{ V})$$
$$V_{out} = 0 \text{ V}$$

Therefore, the DC component of the input voltage is not amplified.

Using the specification sheet in Figure 11.10, determine I_{B1} and I_{B2} for a 741 op-amp when the bias currents split evenly. Do this for the typical value and for the maximum value.

In Figure 11.11 an op-amp is driven by a single signal. The input voltage is applied to the noninverting terminal, and the inverting terminal is grounded. Therefore, we have an output that is equal to

$$v_{out} = A(v_1 - v_2)$$
$$v_{out} = A(v_1 - 0\,V)$$
$$v_{out} = Av_1$$

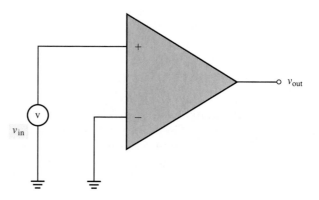

FIGURE 11.11 Imbalances due to input bias currents

EXAMPLE 11.5

Use the specifications for a μA709 op-amp given in example 11.4 and proceed under the assumption that $I_{B1} = I_{B2} = 1500$ nA. Determine the DC output voltage that results from the unbalanced input circuitry.

Solution
If the source providing the signal to the noninverting terminal of the op-amp in Figure 11.11 has 50 Ω of internal impedance, then we know, from example 11.4, that there is 75 μVDC at the noninverting terminal of the op-amp.

There is no DC voltage at the inverting terminal of Figure 11.11. Consequently, there is a difference in DC potential, and this difference is amplified.

$$v_{out} = A(75\ \mu V - 0\ V)$$
$$v_{out} = 15,000\ (75\ \mu V)$$
$$v_{out} = 1.125\ V$$

The DC voltage is present because the input circuitry was not balanced. It is why you will see op-amp circuits with circuit balancing resistors, as shown in Figure 11.9.

In Figure 11.9b a 50-Ω resistor is connected between the inverting terminal and ground. This 50 Ω balances the resistance "seen" by the noninverting terminal, in this case the internal resistance of the source. As a result, the DC levels at the two input terminals are again equal and have no effect on the output of the circuit.

There is another concern relating to input bias currents: Suppose the bias currents into the input terminals aren't equal. There is an op-amp specification that we can use to examine this situation. It is listed as **input offset current.** In the 741 specification sheet in Figure 11.10 the typical value is 20 nA and the maximum value is 200 nA.

The input offset current for the μA709 op-amp is 500 nA. This is the maximum value, and it represents a worst-case condition. That is, there could be as much as a 500-nA difference between I_{B1} and I_{B2}. Example 11.6 demonstrates how this might affect the operation of the op-amp.

EXAMPLE 11.6

Using the specifications for an μA709 op-amp, we have an input bias current of 1500 nA and an input offset current of 500 nA. Determine how this will affect the operation of the op-amp circuit in Figure 11.12.

Solution
Let's rearrange the formula for input bias current:

$$I_B = \frac{I_{B1} + I_{B2}}{2}$$
$$2I_B = I_{B1} + I_{B2}$$

From this and the specifications, we conclude that the maximum current is

$$I_{B1} + I_{B2} = 3000 \text{ nA}$$

and, at most, there can be a 500-nA difference.

So we have either

$$I_{B1} = 1750 \text{ nA} \quad \text{and} \quad I_{B2} = 1250 \text{ nA}$$

or

$$I_{B1} = 1250 \text{ nA} \quad \text{and} \quad I_{B2} = 1750 \text{ nA}$$

We can now determine voltage drops across the 50-Ω resistors connected to the input terminals of Figure 11.12.

$$v_1 = 1750 \text{ nA} \times 50 \text{ }\Omega$$
$$v_1 = 87.5 \text{ }\mu\text{V}$$

And at the other terminal

$$v_2 = 1250 \text{ nA} \times 50 \text{ }\Omega$$
$$v_2 = 62.5 \text{ }\mu\text{V}$$

So, even though the circuit designer balanced the inputs resistively, there might be a DC differential due to an imbalance of bias currents. This difference is amplified, producing an output of

$$A(v_1 - v_2) = 15,000(87.5 \text{ }\mu\text{V} - 62.5 \text{ }\mu\text{V})$$
$$A(v_1 - v_2) = 375 \text{ mV}$$

even when the AC signal is quiescent.

If this DC offset voltage at the output terminal is unacceptable, the output will have to be nulled.

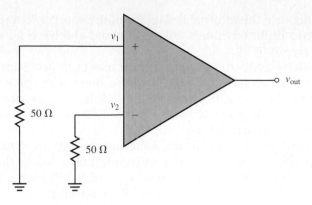

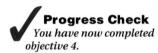

FIGURE 11.12 Balanced input impedances

PRACTICE PROBLEM 11.4

The μA741 op-amp has a maximum input bias current of 500 nA and a maximum input offset current of 200 nA. Determine the maximum output offset voltage that could be caused by the bias currents.

INPUT OFFSET VOLTAGE

Operational amplifiers, although amazing devices, are not perfect. Small imbalances in the internal circuitry cause imbalances at the terminals of the device. In this section we will discuss input offset voltages and how a nulling circuit can be used to correct for unwanted input voltages.

The small voltages that cause input offset voltages are distributed throughout the internal circuitry of an op-amp. However, we can ignore the internal circuitry of the op-amp and approximate the effects of input offset voltages by modeling the internal circuitry.

In Figure 11.13, the input offset voltage (V_{io}) is shown as a small DC battery at the input of the op-amp. In this case, V_{io} was connected to the noninverting input of the op-amp. It could just as easily have been connected to the inverting input of the op-amp.

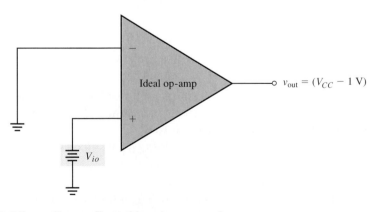

FIGURE 11.13 Effects of input offset voltage on output voltage

Remember that the representation of Figure 11.13 is just a model. These voltages are distributed throughout the internal circuitry of the op-amp. We cannot measure them directly, but the model allows us to predict how input offset voltage will affect op-amp behavior.

The specification sheet for the 741 op-amp lists the typical input offset voltage as 1 mV and the maximum as 6 mV. Either of these input levels will cause the output of the op-amp to saturate in open-loop conditions.

EXAMPLE 11.7

Determine the output voltage of a 741 op-amp with a V_{io} of $+1$ mV at the noninverting terminal. The power supplies are at ± 22 V.

Solution

$$v_{out} = A(v_1 - v_2)$$
$$v_{out} = 100,000 \times 1 \text{ mV}$$
$$v_{out} = +100 \text{ V}$$

This exceeds the power supply limit, so the output is $+21$ V.

EXAMPLE 11.8

Determine the output voltage of a 741 op-amp with a V_{io} of -6 mV at the noninverting terminal. The power supplies are at ± 22 V.

Solution

$$v_{out} = A(v_1 - v_2)$$
$$v_{out} = 100,000 \times -6 \text{ mV}$$
$$v_{out} = -600 \text{ V}$$

Because the output voltage value is greater than the negative saturation value, the output is -21 V.

PRACTICE PROBLEM 11.5

An op-amp has a V_{io} of -500 μV. The power supply levels are ± 12 V. Determine the output voltage caused by this situation.

PRACTICE PROBLEM 11.6

An op-amp has a V_{io} of $+20$ μV. The power supply levels are at ± 18 V. Determine the output voltage of the op-amp.

The direction in which op-amps saturate varies. One op-amp may saturate positively, another op-amp may saturate negatively. **Nulling circuits** are used to eliminate the effects of input offset voltage on the output voltage.

OUTPUT OFFSET VOLTAGE

If the two inputs into an op-amp have exactly the same voltage with respect to ground, then the differential input is 0 V. Under these conditions, the output of the op-amp should be 0 V as well. Because of slight internal imperfections in the internal circuitry of the op-amp, there may be a voltage at the output. This voltage is referred to as **output offset voltage** and is designated V_{oo}. Output offset voltage can be caused by input offset voltage.

There are two ways to reduce this problem. First, most circuits are closed-loop. In Chapter 12, we will see that negative feedback can reduce V_{oo} to insignificant values.

Second, many op-amps have a way of adjusting this unwanted output voltage to 0 V. The process of adjusting the output of an op-amp to 0 V when the differential input is 0 V is known as **nulling.**

Figure 11.14 shows a nulling circuit for a single 741 op-amp. Many dual and quad 741 packages do not come with nulling inputs. In Figure 11.14, a 10-kΩ potentiometer is placed across pins 3 and 9. The wiper arm terminal of

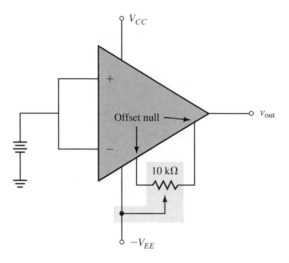

FIGURE 11.14 741 op-amp nulling circuit

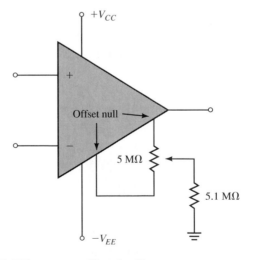

FIGURE 11.15 301 op-amp nulling circuit

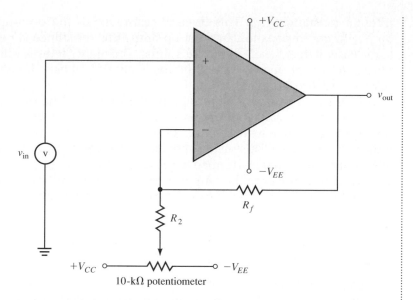

FIGURE 11.16 Nulling circuit for op-amps without nulling inputs

the potentiometer is connected to the negative power supply. The wiper arm is then moved back and forth, while the output is monitored with a voltage meter. The output is zeroed, or nulled, when it reaches 0 V with a differential input of 0 V.

Other types of op-amps also have terminals available for nulling the output. Figure 11.15 shows a nulling circuit for a 301 op-amp. Refer to the technical literature for the op-amp you are using to determine the type of nulling circuit that you should use if the output offset voltage becomes a problem.

Figure 11.16 shows a typical arrangement for nulling an op-amp that doesn't have nulling inputs. In this case the 10-kΩ potentiometer can be adjusted to apply either a positive or a negative voltage to the inverting input as needed. R_f and R_2 are part of a feedback loop. These components will be discussed in Chapter 12.

✓ **Progress Check**
You have now completed objective 5.

OUTPUT IMPEDANCE

Figure 11.17 shows how the output terminal of an op-amp appears to most loads. Looking back into the output terminal, we see a voltage source with an

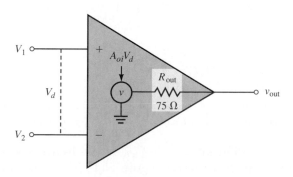

$R_{int} = R_{out}$ = output impedance of the op-amp

FIGURE 11.17 Output impedance of an op-amp

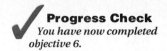

internal resistance, R_{int}. This internal resistance of the op-amp is the op-amp's output impedance. In most op-amps this resistance is nearly 0 Ω. In some cases it may be a few hundred ohms. The manufacturer will specify this parameter under open-loop conditions. Figure 11.10 lists the typical output resistance of the 741 as 75 Ω.

It is important to remember that, in most circuits, the output of the op-amp appears as a voltage source with an internal resistance of 0 Ω. Remember, a voltage source with an internal resistance of 0 Ω is considered to be an ideal voltage source. This means that all the voltage provided by the op-amp at its output terminal is dropped across the load. Practically, you can attach very small resistances to the output of most op-amp circuits without any loss of signal strength, provided the op-amp can deliver the current required.

REVIEW SECTION 11.2

1. If an op-amp with power supply voltages of ±22 V saturates positively, its output voltage will most likely be _____.
2. What should be the output of an op-amp when the differential input is 0 V?
3. What does V_{oo} mean?
4. What does "nulling" mean when discussing op-amp circuits?
5. What is the typical output impedance of most op-amps?
6. Input bias current is the sum of _____ and _____ .
7. Input offset current is the difference between _____ and
   _____ .

11.3 FREQUENCY-DEPENDENT CHARACTERISTICS OF AMPS

In this section we will look at the specifications of op-amps when they are operated with AC input signals. We will investigate the relationship between gain and bandwidth, the slew rate of an op-amp, and how it affects the power bandwidth of an operational amplifier and the ability of an op-amp to suppress electrical noise, a specification given as common mode rejection ratio.

GAIN-BANDWIDTH PRODUCT

The effects of signal frequency on circuit operation were discussed in Chapter 10. Like other circuits, op-amps also have frequency limitations. If you increase the frequency above the bandwidth of the device, the gain begins to drop.

The op-amp specification known as f_{unity} is the frequency at which the gain of the op-amp drops to 1 (unity). Figure 11.18 is a graph that plots gain versus frequency for the 741 op-amp. This type of graph is known as a **Bode plot** and can be used to find the critical frequency (f_c) and the f_{unity} of an op-amp. In Figure 11.18, when the frequency of operation is at 10 Hz, the open-

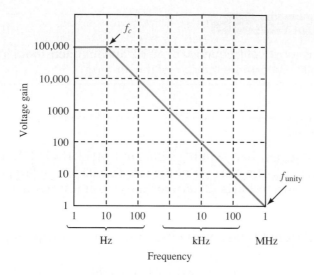

FIGURE 11.18 Bode plot of differential voltage gain versus frequency

loop gain is given as 10^5, which is a gain of 100,000. Multiplying the 10 Hz times the gain, we get

$$f_{unity} = 10\,\text{Hz} \times 100,000$$
$$f_{unity} = 1\,\text{MHz}$$

One megahertz is the gain-bandwidth product of the 741 op-amp. The **gain-bandwidth product** represents the f_{unity} of the op-amp. You can verify this by finding 1 MHz on the Bode plot in Figure 11.18 and crossing over to the vertical axis, where you will see that the gain is 1. The gain-bandwidth product for a given op-amp is a constant value. Once the 10-Hz level is passed, the output gain of the op-amp starts to drop. This drop is known as **roll off.** The 10 Hz is the critical frequency of the op-amp when tested under open-loop conditions.

EXAMPLE 11.9

An op-amp has a critical frequency of 15 Hz. The f_{unity} is 1.2 MHz. Determine the typical DC open-loop gain (A) of the amplifier.

Solution

$$f_{unity} = A \times \text{bandwidth}$$

Rearranging, we get

$$A = \frac{f_{unity}}{\text{bandwidth}}$$
$$A = \frac{1.2\,\text{MHz}}{15\,\text{Hz}}$$
$$A = 80,000$$

PRACTICE PROBLEM 11.7

An op-amp has a critical frequency of 12 Hz and an open-loop, low-frequency gain of 50,000. Determine the f_{unity} of the op-amp.

EXAMPLE 11.10

An op-amp has a gain-bandwidth product of 1 MHz. If we drop the frequency down below f_{unity} the gain (A) of the circuit will go up. Suppose we drop the frequency down to 1 kHz. Determine the gain at this frequency.

Solution
The gain-bandwidth product is a constant for any given op-amp, so we now have:

$$f_{unity} = gain \times bandwidth$$
$$1 \text{ MHz} = A \times f$$
$$1 \text{ MHz} = A \times 1 \text{ kHz}$$

Moving terms around, we get

$$A = \frac{f_{unity}}{f}$$
$$A = \frac{1 \text{ MHz}}{1 \text{ kHz}}$$
$$A = 1000$$

PRACTICE PROBLEM 11.8

An op-amp has an f_{unity} of 2 MHz. The critical frequency is 40 Hz. First, using the concept of gain-bandwidth product, determine the low-frequency, open-loop gain of the op-amp. Next, determine the gain of the op-amp when it is operated at a frequency of 500 kHz.

Example 11.11 shows why most op-amps are not operated open-loop.

EXAMPLE 11.11

An op-amp has an f_{unity} of 1 MHz and an A_{ol} of 100,000. Determine the bandwidth.

$$f_{unity} = A \times f$$
$$1 \text{ MHz} = 100,000 \times f$$

Rearranging the terms we get

$$f = \frac{1 \text{ MHz}}{100,000}$$
$$f = 10 \text{ Hz}$$

The bandwidth of this op-amp starts at DC and rises to only 10 Hz.

This is the same information we obtained from the Bode plot for a 741 op-amp (Figure 11.18). The numbers used in this example are typical of a 741 op-amp.

As you can see, if we want a wider bandwidth, gain has to be reduced. This means some type of negative feedback configuration will need to be wired to the op-amp. We will learn how this is done in Chapter 12.

PRACTICE PROBLEM 11.9

An op-amp has an f_{unity} of 10 MHz. Determine the gain of the op-amp at 2 MHz.

Progress Check
You have now completed objective 7.

SLEW RATE

Slew rate is a measure of how quickly the output voltage of an op-amp can change in response to a changing input voltage. It is usually specified in op-amp documentation by the amount of voltage swing per microsecond. A typical value for the slew rate of a 741 op-amp is 0.5 V/μs. An op-amp that can change states more quickly has a higher slew rate. For example, a 709, which responds more quickly than a 741, has a typical slew rate of 3 V/μs.

Some op-amp specification sheets list several slew rates. These slew rate values are tested under closed-loop conditions for various levels of gain. The 709 will slew at 0.3 V/μs when the closed-loop gain is unity (1), but will slew at a rate of 3 V/μs when the closed-loop gain is increased to 10.

EXAMPLE 11.12

An op-amp has a slew rate of 0.7 V/μs. The output of the op-amp is 0 V. A change at the input causes the op-amp output to rise to 10 V. How long will it take the output to reach this new level?

Solution
The new level is 10 V. Because the starting voltage was 0 V, we have a 10-V change. If we divide this change by the slew rate, we can determine the time (*t*) it takes the output to reach the new level:

$$\text{slew rate} = \frac{\Delta V}{t}$$

$$t = \frac{\Delta V}{\text{slew rate}}$$

$$t = \frac{10 \text{ V}}{0.7 \text{ V} / \mu s}$$

$$t = \frac{10 \text{ V} \times 1 \mu s}{0.7 \text{ V}}$$

$$t = 14.3 \ \mu s$$

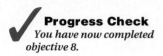
PRACTICE PROBLEM 11.10

An op-amp has a slew rate of 70 V/μs. It is set up so that the output changes from positive saturation to negative saturation. Because this particular op-amp has power supply levels at ±18 V, the output swings from +17 V to −17 V (or the other way around) for a total change of 34 V. How long will this take?

POWER BANDWIDTH

There is a phenomenon associated with slew rate known as the power bandwidth. The **power bandwidth** is the range of frequencies over which an op-amp can amplify large signals without introducing slew rate distortion.

Figure 11.19 illustrates how exceeding the power bandwidth distorts a sinewave input. The type of distortion shown in Figure 11.19 is known as **slew rate distortion.**

In symbols, power bandwidth can be expressed as

$$f_{max} = \frac{\text{slew rate}}{2\pi \times V_p}$$

where V_p is the peak output voltage and f_{max} is the upper frequency limit. Below this frequency slew rate distortion will not occur.

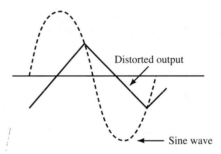

FIGURE 11.19 Slew rate distortion

EXAMPLE 11.13

A 8-V peak signal is at the output of an op-amp. The slew rate of the op-amp is 0.5 V/μs. Determine the power bandwidth of the op-amp.

Solution
Using the equation above, we have

$$f_{max} = \frac{0.5 \, \text{V} / \mu s}{2\pi \times 8 \, V_p}$$

$$f_{max} = 9.95 \, \text{kHz}$$

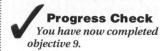

Progress Check
You have now completed objective 9.

PRACTICE PROBLEM 11.11

Determine the power bandwidth of an op-amp that has a slew rate of 3 V/μs. The output signal is 10 V_p.

COMMON MODE REJECTION RATIO

Common-mode noise enters circuitry via some common point, often the power supply (as discussed in Chapter 10). In a differential amplifier, common-mode noise can enter through the tail resistor. When amplified, this noise causes unwanted "signals" to appear at the output of an amplifier. The **common mode rejection ratio** (CMRR) is a measure of an op-amp's ability to minimize this noise.

The common mode rejection ratio is listed in op-amp specification sheets in decibels (dB). When we use the prime symbol (') with common mode rejection ratio (CMRR'), our solution will be in decibels. When the common mode rejection ratio is expressed without the prime symbol (CMRR), our solution will be in gain, a unitless number. The minimum CMRR' for a μA709 op-amp is 65 dB, and for the μA741 it is 70 dB (see Figure 11.10).

We convert these numbers into a ratio by the formula

$$CMRR' = 20 \log CMRR$$

where

$$CMRR = \frac{A}{A_{cm}}$$

$$A = \text{differential mode gain}$$

$$A_{cm} = \text{common-mode gain}$$

This is a comparison of wanted signal strength (differential mode gain) versus unwanted signal strength (common-mode gain).

EXAMPLE 11.14

What is the ratio of A to A_{cm} if the CMRR' of an op-amp is 65 dB?

$$CMRR' = 20 \log CMRR$$

$$\frac{CMRR'}{20} = \log \frac{A}{A_{cm}}$$

$$\frac{65 \, dB}{20} = \log \frac{A}{A_{cm}}$$

$$3.25 \, dB = \log \frac{A}{A_{cm}}$$

Using the antilog function on a scientific calculator we get

$$1778.3 = \frac{A}{A_{cm}}$$

which means that the op-amp will amplify desired signals 1778.3 more times than it will amplify unwanted common-mode noise.

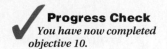
PRACTICE PROBLEM 11.12

Determine the ratio of A to A_{cm} if the CMRR' is 70 dB.

REVIEW SECTION 11.3

1. The frequency at which the gain of an op-amp drops to 1 is known as _____ .
2. The upper frequency limit of most op-amps operated open-loop is typically 10–15 Hz. True or False?
3. Once the critical frequency of an op-amp has been exceeded, the open-loop gain of the op-amp begins to rise. True or false?
4. Define slew rate.
5. Give the formula for power bandwidth.
6. Power bandwidth is a small-signal phenomenon. True or false?
7. Common mode rejection ratios are most often listed in _____ in an op-amp specification sheet.
8. The common mode rejection ratio is a comparison of _____ with _____ .

TROUBLESHOOTING

As with transistors, there is a DC setup that must be established before a signal can be applied to an op-amp. If a signal applied to an op-amp is distorted or nonexistent on the output, a check of the DC setup may aid in troubleshooting the problem.

First, inspect the op-amp and associated circuitry. Look for signs of overheating, such as burned resistors, "oozing" capacitors, IC packages that are cracked and discolored, or raised or broken PCB runs. If a component has suffered heat damage, replace it.

Second, check the supply voltages applied to the op-amp. The schematic should indicate the value of the voltages. If the voltages are not in the tolerance required, the power supply should be troubleshot.

Third, check the DC voltages at the inverting and noninverting inputs. These voltages should be identical. Only a difference of a few microvolts should be observed. If there is a significant difference between the two inputs, then the op-amp is not bootstrapping and is defective.

There are other problems that can occur after a board has been assembled or repaired. One is improperly installed components (ICs). ICs may be installed upside down. Components of the wrong value can be installed, for example, using a 10-kΩ resistor rather than a 1-kΩ resistor. The difference to look for is a orange band rather than a red band. Electrolytic capacitors can also be installed backwards. Improper solder techniques can introduce shorts into a circuit. Too much solder can splash across two leads of an IC. These kinds of problems only occur after a board has been repaired. Most of them can be detected by observation.

WHAT'S WRONG WITH THIS CIRCUIT?

The expected output of the figure below is +10 V, the actual output is +8 V. $V_{TP1} = 1$ mV, $V_{TP2} = 0.8$ mV, $A_{ol} = 50{,}000$, and $V_{CC} = \pm12$ V. Determine why the actual output voltage is less than expected.

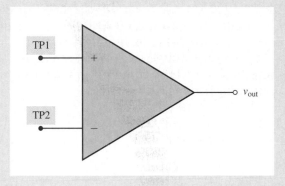

SUMMARY

The input stage to an op-amp is a differential amplifier. The difference between the voltages of the noninverting and the inverting input terminals is amplified.

Many operational amplifiers have large open-loop gains. Values of 5000 to 100,000 can be found in commercially available op-amps. Open-loop gain is often designated A_{ol}.

Most op-amps are powered by two voltage supplies. The positive supply commonly is equal to the negative supply so that the output of the op-amp can swing in equal amounts both above and below ground.

Input bias current is a measure of the average of the bias currents flowing at the input terminals of an op-amp. Input offset current is a measure of how far apart the two input bias currents are. Differences between them can produce output offset voltage.

The high values of open-loop gain for most op-amps can present problems in circuits built around these devices. The high open-loop gain often results in outputs that are saturated.

Slight imperfections in the internal circuitry of an op-amp may cause an output offset voltage even when the differential inputs are equal. Nulling circuits can be used to reduce output offset voltage to 0 V.

The input impedance of most op-amps is quite high. Values in the megohms are not uncommon. Input impedance is often measured under open-loop conditions.

The output impedance of most op-amps is very close to 0 Ω. In many cases the output of an op-amp can act like an ideal voltage source.

The gain-bandwidth product of an op-amp allows us to see how reducing gain can increase bandwidth and vice versa. The frequency at which open-loop gain drops to 1 is f_{unity}.

Slew rate is a measure of how quickly the output voltage of an op-amp can change in response to changes in the input voltage of the circuit. Slew rate is often specified in volts per microsecond.

Power bandwidth is a measure of the maximum frequency for large-signal operations. Exceeding power bandwidth causes slew rate distortion.

The common mode rejection ratio (CMRR) is a measure of how well an op-amp suppresses common-mode noise. This ratio is a comparison of wanted signal strength with unwanted signal strength. It is often listed in decibels. Op-amps have very high common-mode rejection ratios; thus they are good at suppressing common-mode noise.

PROBLEMS

1. An op-amp has an open-loop gain of 200,000. If the differential input to the op-amp is 5 μV, determine the output voltage.
2. An op-amp has power supply levels set to ± 18 V. The differential input is 30 mV. There is 50 mV on the noninverting terminal and 20 mV on the inverting terminal. The open-loop gain is 50,000. Determine the output voltage.
3. An op-amp has an output of 10 V and a gain of 100,000. What is the differential input voltage?
4. An op-amp has an output of 5 V and a differential input of 2 mV. What is the open-loop gain of the op-amp?
5. The voltage at the noninverting terminal is 5 V. The voltage at the inverting terminal is 6 V. Will the output voltage be positive or negative with respect to ground?
6. An op-amp is powered by a positive voltage of $+15$ V and a negative voltage of -15 V. When the op-amp saturates, its output voltage will be either $+13.5$ V or -13.5 V. If the op-amp has an open-loop gain of 50,000, what is the largest differential input that will just saturate the output?
7. What process can be used to eliminate output offset voltage?
8. The input bias current of an op-amp is 500 nA. What are the values of I_{B1} and I_{B2} if these two values are balanced?

9. The input bias current of an op-amp is 1000 nA. The input offset current of the same op-amp is 300 nA. If I_{B1} is larger than I_{B2}, what are the values of I_{B1} and I_{B2}?

10. An op-amp has an input offset voltage of 100 μV. The open-loop gain of the device is 50,000. What is the output offset voltage?

11. An op-amp has an f_{unity} of 2 MHz. If the input frequency is dropped to 1 MHz, what will be the new gain?

12. If the op-amp of problem 11 had a typical open-loop gain of 100,000, what would be the bandwidth under open-loop conditions?

13. If the input of an op-amp changes such that the output of the op-amp rises from 0 V to 5 V, how long will it take to reach the new output level if the op-amp has a slew rate of 70 V/μs?

14. An op-amp has a slew rate of 0.5 V/μs. How much of a change can there be after 18 μs?

15. An op-amp is to be used with a 3-V_p output signal. The slew rate of the op-amp is 1 V/μs. Determine the power bandwidth of the amplifier.

16. A source drives the input of Figure 11.20. If this source has an internal impedance of 600 Ω, how much output offset voltage will be produced when the input bias current is 100 nA? Assume that A_{ol} is 1, and $I_{B1} = I_{B2}$.

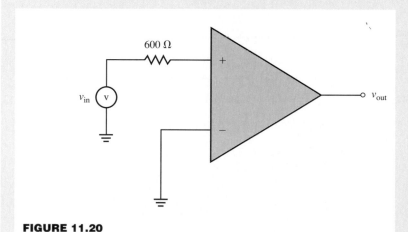

FIGURE 11.20

17. Assuming that I_{B1} and I_{B2} are balanced for the op-amp in Figure 11.20, what can be done to eliminate the output offset voltage, other than nulling?

18. An op-amp has an input bias current of 1000 nA and an input offset current of 250 nA. Determine the maximum and the minimum value of I_{B1}.

19. Suppose the op-amp of Figure 11.21 has an I_{B1} of 200 nA and an I_{B2} of 175 nA. The open-loop gain of the op-amp is 50,000. Determine the output offset voltage.

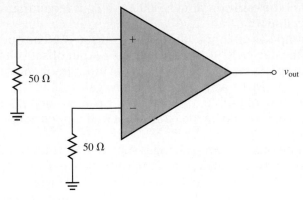

FIGURE 11.21

20. An op-amp has an $A_{OL:}$ of 50,000 and a common-mode gain, A_{cm}, of 2.5. Determine the common mode rejection ratio.
21. An op-amp has a common mode rejection ratio of 10,000. What is this ratio in decibels?

BASIC OPERATIONAL
AMPLIFIER CIRCUITS

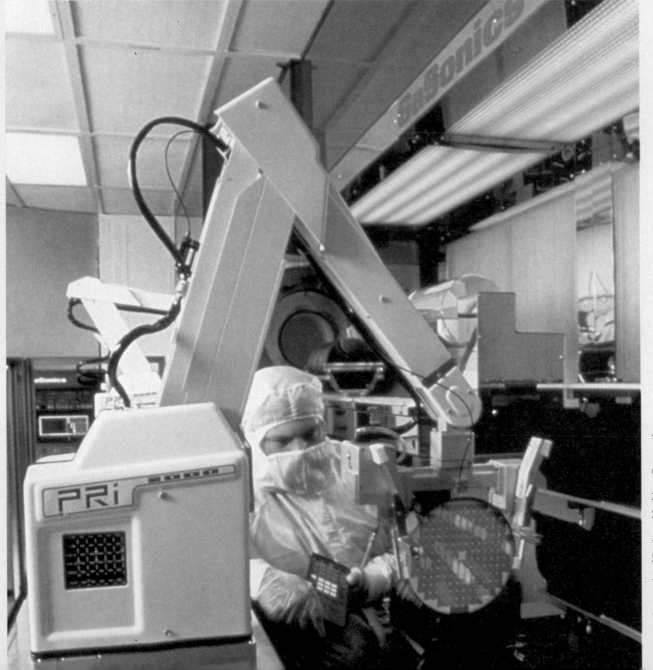

✓ **As you read this chapter, concentrate on learning how to :**

1. Identify an op-amp circuit using series-parallel negative feedback
2. Determine the gain of a series-parallel op-amp circuit
3. Determine the input and output impedances of an op-amp circuit using series-parallel feedback
4. Identify an op-amp circuit using parallel-parallel negative feedback
5. Determine the gain of an inverting voltage amplifier
6. Determine the output of a current-to-voltage transducer
7. Determine the input and output impedances of an op-amp circuit using parallel-parallel feedback
8. Identify an op-amp circuit using parallel-series negative feedback
9. Determine the current gain of an op-amp wired in a parallel-series configuration
10. Determine the input and output impedances of an op-amp circuit using parallel-series feedback
11. Identify an op-amp circuit using series-series negative feedback
12. Determine the output of a voltage-to-current transducer
13. Determine the input and output impedances of an op-amp circuit using series-series feedback
14. Figure out how sacrificing gain affects bandwidth in an op-amp circuit using negative feedback

INTRODUCTION

Figure 12.1 is the schematic of a transistor amplifier using collector feed-back bias. In Chapter 4, we learned that this type of bias arrangement re-turned a portion of the output signal to the front end of the amplifier so that

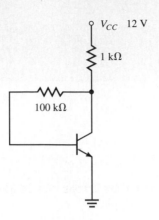

V_{CC} 12 V

1 kΩ

100 kΩ

FIGURE 12.1 Collector feedback circuit

the signal being returned to the front end of the circuit was 180° out of phase with the input signal. This arrangement is known as negative feedback.

The negative feedback in the collector feedback circuits studied in Chapter 4 was used to stabilize the circuit. In particular, the feedback compensated for changes in beta and helped to prevent thermal runaway.

In this chapter we will study the effects, both advantages and disadvantages, of using negative feedback arrangements in operational amplifier circuits.

12.1 FOUR METHODS OF FEEDBACK

Two parts of an amplifier use feedback. Figure 12.2a shows the block diagram of an amplifier circuit. In this part of the circuit the signal flows from left to right. The gain of this part of the circuit will be designated A.

The second part of an amplifier that uses feedback is the feedback circuit itself, shown in Figure 12.2b. The signal in this part flows from right to left. The gain of this part of the circuit will be designated B.

The feedback circuit that returns part of the output voltage to the input of the circuit could be another amplifier, a transformer, a coupling capacitor, or, as shown in Figure 12.2c, a resistive voltage divider.

The two parts to the circuit can be arranged in four ways to provide negative feedback to operational amplifiers: series-parallel, series-series, parallel-parallel, and parallel-series. In all of these arrangements, the circuit forms a closed loop. In a closed-loop circuit, part of the output is sampled and returned to the input of the circuit to modify circuit operation.

The overall gain of a closed-loop system is often designated AB and is known as the **loop gain.** In many op-amp circuits and texts, open-loop gain is designated A_{ol}. Closed-loop gain is designated A_{cl}. We will use the A_{ol} and A_{cl} designations.

THE SERIES-PARALLEL FEEDBACK ARRANGEMENT

In the block diagram in Figure 12.3a, an amplifier with a feedback circuit is connected in a series-parallel feedback arrangement. At the input side of the amplifier, the signal is applied across the amplifier and the feedback circuit,

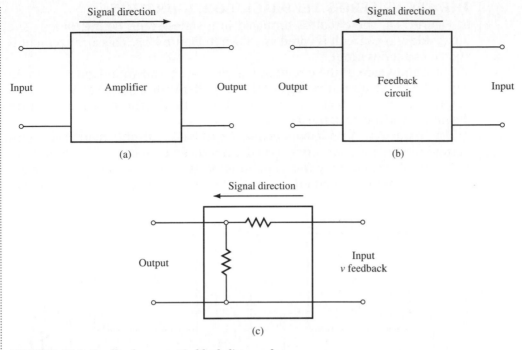

FIGURE 12.2 Feedback concept in block diagram form

shown in Figure 12.3b. In this situation, the output voltage developed across the load resistor is used to drive the inputs of the feedback network.

Note that the voltage at the input of the feedback network is in parallel with the voltage developed across the load, hence parallel feedback. This is why this type of feedback arrangement is called series-parallel feedback.

This type of arrangement is also known as voltage feedback. It is often used as a noninverting voltage amplifier. We will look at its characteristics in section 12.2.

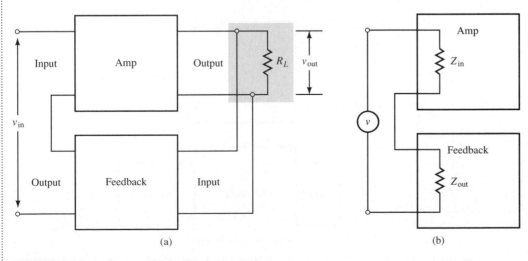

FIGURE 12.3 Series-parallel feedback connection

THE SERIES-SERIES FEEDBACK CONFIGURATION

In Figure 12.4, the circuit is arranged in a series-series configuration. The input side of the circuit is wired as shown in Figure 12.3. This arrangement is known as a series input.

The output side of the circuit in Figure 12.4a is shown in Figure 12.4b. In Figure 12.4b the output voltage of the amplifier is designated as $V_{TH(out)}$, the output Thevenin voltage. This voltage drives the output load in series with the input impedance of the feedback circuit.

This particular output connection feeds back current; therefore, series output connections are often called **current feedback** circuits. In section 12.5 we will discuss how this configuration is used as a voltage-to-current transducer, that is, a circuit that converts an input voltage into a current that flows through a load resistance.

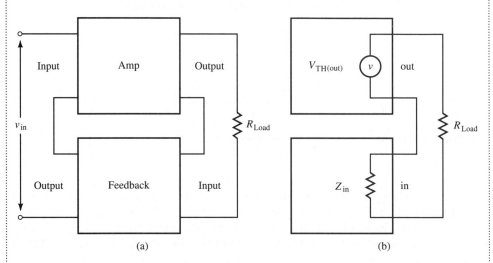

FIGURE 12.4 Series-series feedback connection

PARALLEL-PARALLEL FEEDBACK ARRANGEMENT

The third feedback arrangement is the parallel-parallel feedback configuration, Figure 12.5. As you look at the input side of the overall circuit, note that

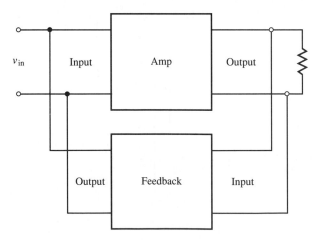

FIGURE 12.5 Parallel-parallel feedback connection

the input voltage is across the input to the amplifier as well as the output of the feedback network. This is a parallel configuration. The output of Figure 12.5 is similar to the parallel output connection in Figure 12.3.

The parallel-parallel feedback connection can be used for two different functions. One function is an inverting voltage amplifier. The second function is to act as a current-to-voltage transducer, that is, a circuit that converts an input current into an output voltage. These will be studied in section 12.3

PARALLEL-SERIES FEEDBACK CONFIGURATION

The last of the four feedback arrangements is the parallel-series feedback configuration, Figure 12.6. This type of circuit is often used as a current amplifier. Detailed analysis of this type of feedback circuit can be found in section 12.4.

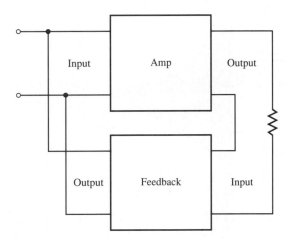

FIGURE 12.6 Parallel-series feedback connection

REVIEW SECTION 12.1

1. Collector feedback is a form of _____ feedback.
2. There are _____ basic ways to apply negative feedback to operational amplifiers.
3. List the types of negative feedback that can be applied to an op-amp circuit.
4. What are some circuit connections that can be used to apply negative feedback to the input of an op-amp circuit?
5. Overall loop gain is designated _____.
6. What does A_{ol} mean? A_{cl}?
7. The series-parallel negative feedback arrangement can be used as a

    _____ _____ _____.
8. Which two basic negative feedback arrangements feed back voltage? Which two feed back current?
9. What are two functions for which the parallel-parallel negative feedback op-amp circuit can be used?
10. Which of the four basic negative feedback arrangements can be used as a current amplifier?

12.2 NONINVERTING VOLTAGE AMPLIFIER

The series-parallel form of negative feedback can be used to implement a noninverting voltage amplifier. An op-amp circuit wired for series-parallel negative feedback is shown in Figure 12.7. In this circuit the entire output voltage is used to drive a voltage divider consisting of R_f (feedback) and R_b (bottom). The ratio of R_f to R_b determines the amount of output voltage that reaches the input of the op-amp at the inverting terminal.

The input to the op-amp is in series with the input impedance of the device. This signal can be either an AC signal or a DC signal. You should remember that the input of an operational amplifier is a differential amp, so it can be direct-coupled into the IC. This means that the low frequency cutoff is all the way down to DC. In section 12.6 we'll look at how to determine the high frequency cutoff of basic op-amp circuits.

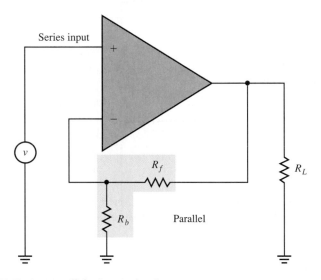

FIGURE 12.7 Series-parallel op-amp circuit

IDEAL ANALYSIS

One critical idea in understanding op-amp circuits (not the op-amp itself, but the circuit) is illustrated in Figure 12.8. In this circuit we see a series connection with two voltage sources connected across a resistance. The two batteries are equal and opposite; therefore there is no current through the resistor, and the voltage potentials at the test points marked + and − are equal.

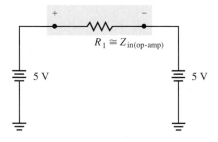

FIGURE 12.8 No current flowing through R_1 because of equal and opposite voltages

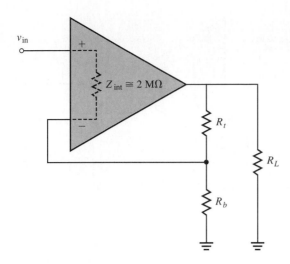

FIGURE 12.9 Another way to draw an SP op-amp circuit

Figure 12.9, shows an alternate representation of a series-parallel configuration. In this circuit, the internal input impedance of the op-amp is indicated by a dashed line. This input impedance is sometimes listed as being infinite. In practical terms the input impedance of an op-amp is in the millions of ohms, with 2 MΩ being a typical value. The series-parallel connection of Figure 12.9 forces the input impedance of the stage toward infinity. For now, we can treat the input impedance of the op-amp as infinite. Thus, there is no current flowing through the input impedance Z_{int} in Figure 12.9.

If there is no current flowing through the impedance, then we have a situation similar to Figure 12.8. If the left-hand battery represents the input voltage at the noninverting terminal, then the right-hand battery represents the feedback voltage at the inverting terminal.

When this situation occurs, the effect is often referred to as **bootstrapping.** This means that the op-amp and the feedback arrangement try to make the input to the two terminals equal. Remember that for our ideal analysis we consider the open-loop gain to be infinite and the voltage drop between the two inputs as 0 V. Only when we want an exact analysis will we assume differently.

Progress Check
You have now completed objective 1.

EXAMPLE 12.1

Using ideal analysis, determine the output voltage for the series-parallel op-amp circuit in Figure 12.10.

Solution
The input to the noninverting terminal is 10 mV. This voltage is bootstrapped to the inverting terminal. Therefore there is 10 mV across the 470-Ω resistor. The polarity of this voltage is marked with respect to ground.

Determine the current flowing through the 470-Ω resistor using Ohm's law.

$$I = \frac{10\,\text{mV}}{470\,\Omega}$$
$$I = 21.3\,\mu\text{A}$$

This current flows up through the 470-Ω resistor and into node 1 (N1).

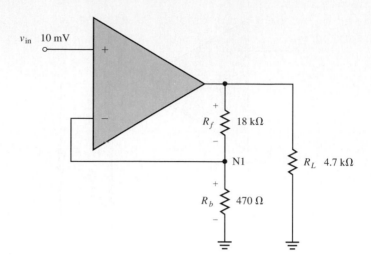

FIGURE 12.10   Series-parallel op-amp circuit

Applying Kirchhoff's current law, the sum of currents entering a node is equal to the sum of currents leaving a node.

Going left from node 1, we see that the top of the 470-Ω resistor is connected to the inverting terminal of the op-amp. As shown in Figure 12.9, this direction leads to an infinite impedance. In this case, typical DC bias currents are significantly smaller than the currents produced by the input signal. Therefore, we can ignore their effects and assume that no current flows left out of node 1. Consequently, all the current flowing through the 470-Ω resistor *must* flow up through the 18-kΩ resistor to satisfy Kirchhoff's current law.

We can now calculate the voltage dropped across the 18-kΩ resistor using Ohm's law:

$$V = 21.3 \, \mu A \times 18 \, k\Omega$$
$$V = 383 \, mV$$

Adding the voltages across the 470-Ω and the 18-kΩ resistor gives us the total voltage dropped across the voltage divider.

$$V = 10 \, mV + 383 \, mV$$
$$V = 393 \, mV$$

Notice in Figure 12.10 that the load resistor is in parallel with the voltage divider. Since we know the voltage across the divider and we know that voltage is the same across all branches in a parallel circuit, we know the output voltage across the load resistor.

$$V_{out} = V_{divider} = 393 \, mV$$

EXAMPLE 12.2

Determine the voltage gain for the circuit of Figure 12.10.

Solution

Voltage gain is a ratio of V_{out} to V_{in}. It is almost entirely independent of the operational amplifier. From this point forward, voltage gain of the series-parallel noninverting amplifier is identified as $Av_{(cl)}$. In this circuit we have

$$Av_{(cl)} = \frac{V_{out}}{V_{in}}$$
$$Av_{(cl)} = \frac{393\,mV}{10\,mV}$$
$$Av_{(cl)} = 39.3$$

This gain is a constant for the circuit of Figure 12.10 and is determined by the ratio

$$\frac{R_f + R_b}{R_b}$$

To understand this, remember that in Figure 12.10, the load resistor was in parallel with the voltage divider. Consequently:

$$V_{out} = V_{divider}$$

where

$$V_{divider} = I_{divider} \times (R_f + R_b)$$

Further, the input voltage is bootstrapped across R_b, so:

$$V_{in} = I_{divider} \times R_b$$

And, since gain is equal to V_{out}/V_{in},

$$Av_{(cl)} = \frac{V_{out}}{V_{in}} = \frac{I_{divider} \times (R_f + R_b)}{I_{divider} \times R_b}$$

Factor out $I_{divider}$ to obtain

$$Av_{(cl)} = \frac{R_f + R_b}{R_b}$$
$$Av_{(cl)} = \frac{R_f}{R_b} + \frac{R_b}{R_b}$$
$$Av_{(cl)} = \frac{R_f}{R_b} + 1$$

EXAMPLE 12.3

Change the input to the circuit of Figure 12.10 to 30 mV. Determine the output voltage.

Solution
Using the gain, $Av_{(cl)}$, determined in example 12.2, we can determine the output voltage:

$$V_{(out)} = Av_{(cl)} \times V_{in}$$
$$V_{out} = 39.3 \times 30\,mV$$
$$V_{out} = 1.18\,V$$

We can verify the answer by using the bootstrapping concept and ideal analysis.

EXAMPLE 12.4

Verify the answer of example 12.3.

The input voltage of 30 mV to the circuit of Figure 12.10 is bootstrapped to the inverting terminal and is placed across the 470-Ω resistor. The current through the 470-Ω resistor is

$$I = \frac{30\,\text{mV}}{470\,\Omega}$$
$$I = 63.8\,\mu A$$

As before, all this current flows through the 18-kΩ resistor. The voltage across the 18-kΩ resistor is

$$V = 63.8\,\mu A \times 18k\Omega$$
$$V = 1.15\,V$$

Add the voltage across R_f and R_b to find the total voltage across the voltage divider.

$$1.15\,V + 30\,\text{mV} = 1.18\,V$$

We know that this voltage is also across the load, therefore

$$V_{out} = 1.18\,V$$

PRACTICE PROBLEM 12.1

Determine the output voltage for the circuit in Figure 12.11.

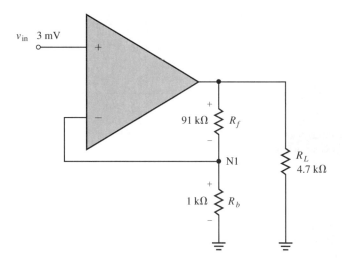

FIGURE 12.11   A series-parallel op-amp circuit with an open-loop gain of 100,000

PRACTICE PROBLEM 12.2

Determine the closed-loop voltage gain, $Av_{(cl)}$, for the circuit of Figure 12.11.

Determine the output voltage for the circuit of Figure 12.11, if the input voltage is changed to 20 mV.

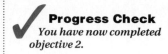

Progress Check
You have now completed objective 2.

FEEDBACK CONTROL LOOP

A "classic" feedback control loop is diagrammed in Figure 12.12. There are three main parts to the control loop shown in the figure.

The circle with the X is a differencing circuit. The upper rectangle is a non-inverting amplifier with an open-loop gain of A, and the lower rectangle is a feedback network. The amount of feedback reaching the differencing network is determined by the feedback ratio B.

If you think this looks like the feedback loops shown in Section 12.1, you are right. Figure 12.13a shows the control loop symbols, and Figure 12.13b shows the equivalent op-amp symbols. In the case of the op-amp the differencing circuit is the input stage of the op-amp itself, the differential amplifier.

We have mentioned the feedback control loop of Figure 12.12 to lay the foundation for what is to come next: a look at the difference between ideal op-amp analysis and a more exact op-amp analysis using equations from feedback theory.

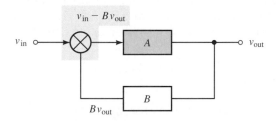

FIGURE 12.12 Classic feedback control loop

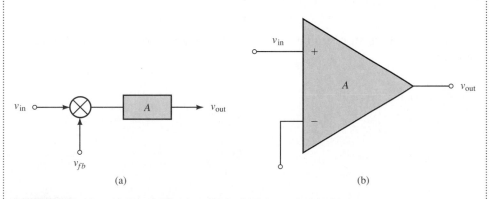

FIGURE 12.13 (a) Control loop symbols. (b) Op-amp equivalents.

ERROR VOLTAGE

EXAMPLE 12.5

Determine the output voltage and gain of the op-amp circuit of Figure 12.14 using ideal analysis.

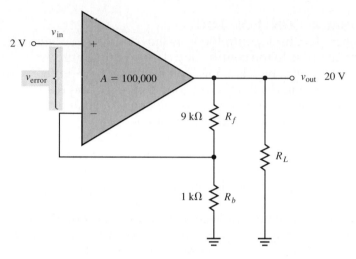

FIGURE 12.14 A series-parallel circuit showing v_error

Solution

The input voltage to the circuit is 2 V. This voltage is bootstrapped down to the inverting terminal so there is 2 V across R_b.

The current through R_b can be determined using Ohm's law.

$$I_{Rb} = \frac{2\,V}{1\,k\Omega}$$

$$I_{Rb} = 2\,mA$$

As before, all of this current must flow through R_f.

Find the voltage across R_f.

$$V_{Rf} = 2\,mA \times 9\,k\Omega$$
$$V_{Rf} = 18\,V$$

Adding the voltage across R_b and R_f, we obtain

$$V_{Rb} + V_{Rf} = 2\,V + 18\,V = 20\,V$$

Since the load resistor is in parallel with the voltage across $R_f + R_b$, the output voltage is

$$V_{out} = 20\,V$$

The gain of the circuit is

$$A = \frac{V_{out}}{V_{in}}$$

$$A = \frac{20\,V}{2\,V}$$

$$A = 10$$

The solution to example 12.5 assumes that the voltage between the noninverting terminal and the inverting terminal is 0 V. It turns out that for the op-amp to work properly there must be a small voltage difference between the two input terminals. We have designated this voltage as the **error voltage,** v_{error}.

If v_{error} were equal to 0 V the output voltage of the op-amp would be 0 V. Remember that the input stage of the op-amp is a differential amplifier with an output voltage equal to

$$v_{out} = A(v_1 - v_2)$$

The closed-loop gain of the classic feedback control loop is

$$Av_{(cl)} = \frac{A}{(1 + AB)}$$

where $Av_{(cl)}$ represents the closed-loop voltage gain, A is the open-loop gain of the noninverting amplifier, and B is the ratio of the feedback network.

The proof of the equation above can be found in Appendix A.

In Figure 12.12, we see a series-parallel circuit in which the voltage across the op-amp input terminals has been identified as v_{error}.

EXAMPLE 12.6

Determine the output voltage and the gain of the circuit in Figure 12.14 using equation 12.1.

Solution
Equation 12.1 says

$$Av_{(cl)} = \frac{A}{(1 + AB)}$$

The open-loop gain of the op-amp is given in the schematic of Figure 12.14.

$$A = 100,000$$

We next determine the feedback ratio using the voltage divider principle:

$$B = \frac{R_b}{(R_b + R_f)}$$
$$B = \frac{1\,k\Omega}{10\,k\Omega}$$
$$B = 0.1$$

Plugging these values into equation 12.1, we get

$$Av_{(cl)} = \frac{A}{1 + AB}$$
$$Av_{(cl)} = \frac{100,000}{1 + (100,000 \times 0.1)}$$
$$Av_{(cl)} = \frac{100,000}{1 + 10,000}$$
$$Av_{(cl)} = \frac{100,000}{10,001} = 9.999001$$

We can now determine the output voltage.

$$V_{out} = Av_{(cl)} \times V_{in}$$
$$V_{out} = 9.999 \times 2\,V$$
$$V_{out} = 19.998\,V$$

When using the ideal method, we calculated $V_{out} = 20\,V$.

Although the difference between the ideal answer and the answer obtained with equation 12.1 is very small, the error voltage must be considered.

EXAMPLE 12.7

Using the V_{out} answer obtained in example 12.6 determine the error voltage at the input of the op-amp in Figure 12.14.

Solution
We determined that the output voltage was
$$19.998\,V$$
We also discovered that the feedback ratio of Figure 12.14 was 1:10.
    From these two facts, we can determine the voltage at the input to the inverting terminal of the op-amp.
$$19.998\,V \times 0.1 = 1.9998\,V$$
The difference between the input voltage at the noninverting terminal and that at the inverting terminal is
$$2\,V - 1.9998\,V = 200\,\mu V$$
This voltage is the error voltage.

We can further verify our V_{out} answer by multiplying the open-loop gain by the error voltage.
$$100,000 \times 199.8\,\mu V = 19.98\,V$$
We now know that there is a small amount of voltage between the input terminals of the op-amp, the error voltage, which is very small and can be approximated as 0 V. When we approximate the error voltage as 0 V, we can use the ideal method of analyzing the series-feedback op-amp circuit.

PRACTICE PROBLEM 12.4

Determine the output voltage of the circuit of Figure 12.15. Use the ideal approach.

PRACTICE PROBLEM 12.5

Determine the closed-loop gain of the circuit of Figure 12.15. Use the ideal approach.

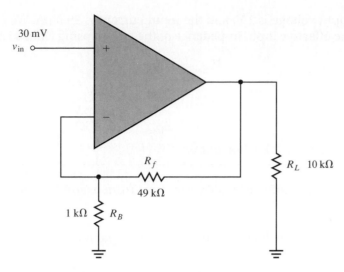

30 mV
v_{in} o——

R_f
49 kΩ

R_L 10 kΩ

1 kΩ R_B

FIGURE 12.15   A series-parallel op-amp circuit with a gain of 50,000

PRACTICE PROBLEM 12.6

Determine the gain and the output voltage of the circuit of Figure 12.15 using equation 12.1.

PRACTICE PROBLEM 12.7

Determine the error voltage between the input terminals of Figure 12.15 using equation 12.1.

INPUT IMPEDANCE

So far we have assumed that the input impedance of the op-amp circuit using series-parallel negative feedback is infinite. We will now determine the effects of series-parallel negative feedback on circuit operation using the concept of error voltage.

EXAMPLE 12.8

Consulting a specification sheet for a μ741 op-amp we find that the typical input impedance of the device is 2 MΩ. This is the input of the op-amp before it is hooked into the circuit. We use this information and the error voltage calculated in example 12.7 to determine the current flowing through the Z_{in} of the op-amp.

$$I_{in} = \frac{V_{error}}{Z_{int}}$$

$$I_{in} = \frac{200\,\mu V}{2\,M\Omega}$$

$$I_{in} = 100\,pA$$

The input voltage is 2 V, and the input current is 99.9 pA. We can now calculate the effective input impedance of the circuit using Ohm's law.

$$Z_{in(eff)} \text{ circuit} = \frac{V_{in}}{i_{in}}$$

$$Z_{in(eff)} = \frac{2 \text{ V}}{100 \text{ pA}}$$

$$Z_{in(eff)} = 20 \text{ G}\Omega$$

Yes, that's right, 20 billion ohms!

There are two conclusions we can draw from example 12.8. First, the series-parallel negative feedback arrangement drives the effective input impedance of the circuit up significantly. Second, using infinity as the ideal approximation of the input impedance of the series-parallel op-amp configuration is reasonable.

PRACTICE PROBLEM 12.8

If the op-amp in the circuit of Figure 12.15 has an input impedance of 1 MΩ, determine the circuit's input impedance.

OUTPUT IMPEDANCE

The output of a series-parallel op-amp circuit using negative feedback produces an output voltage across a load, as shown in Figure 12.16. In this situation, a variable-load resistor is placed across the output.

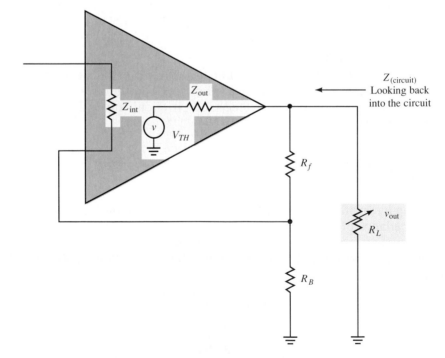

FIGURE 12.16   Series-parallel op-amp output impedance

Looking back into the op-amp output terminal the output voltage is represented as a combination of the Thevenin voltage V_{TH} seen by the load resistor and the Thevenin resistance R_{TH} of the op-amp's output terminal.

In this situation, Z_{out} represents both the R_{TH} seen by the load looking back into the op-amp and the internal resistance of the Thevenin voltage driving the load. Remember that an ideal voltage source should have no internal impedance.

If we refer to a specifications sheet for a $\mu741$ op-amp, the output impedance open loop is typically 75 Ω. The 709 op-amp has an typical output impedance of 150 Ω. Most op-amps already have a fairly low output impedance.

It is already close to zero because of the design of the internal circuitry. The series-parallel feedback configuration makes it effectively zero.

Suppose we lowered the resistance of the load in Figure 12.16. At some point the lower resistance of the potentiometer will cause the output voltage to drop. This drop is due to voltage losses across the internal resistance of the voltage source, in this case, Z_{out} of the amplifier.

As this voltage drops, the amount of voltage feeding the voltage divider made up of R_f and R_b also drops. Consequently the voltage being returned to the inverting terminal of the op-amp goes down. As a result of the voltage drop at the inverting terminal, the error voltage rises. This increase in error voltage is amplified by the op-amp and returned to the output as an increase in output voltage.

Notice that we started out assuming that the output voltage was dropping, then claimed that the output voltage was rising! Actually, this isn't the contradiction it seems. The series-parallel negative feedback arrangement works like this. As in the collector feedback circuit, negative feedback serves to stabilize circuit operation, trying to keep it constant. What appears to be a contradiction is really the circuit reacting to a change and trying to compensate for that change by doing just the opposite.

In fact, the series-parallel negative feedback arrangement is so good at compensating for drops in output voltage caused by loading conditions that the effective output impedance of this type of circuit arrangement is often near 0 Ω.

VOLTAGE GAIN

Figure 12.17 shows a series-parallel op-amp circuit using negative feedback. The voltage divider is made up of R_f and R_b. These two resistors form a feedback network. The ratio of R_b to the total resistance of the voltage divider is proportional to the amount of voltage fed back to the inverting terminal of the op-amp.

$$B = \frac{R_b}{R_b + R_f} = \frac{1\,k\Omega}{1\,k\Omega + 99\,k\Omega}$$
$$B = 0.01$$

Using the bootstrapping concept discussed earlier, we can determine the voltage gain of the op-amp. For convenience we will assume an input of 1 mV. Consequently 1 mV will be across R_b, and the current through this resistor is

$$I = \frac{1\,mV}{1\,k\Omega}$$
$$I = 1\,\mu A$$

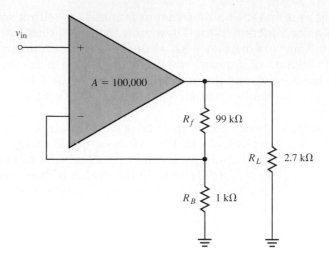

FIGURE 12.17   A series-parallel op-amp circuit with a 2.7-kΩ load

As we know, all of this current must flow through R_f. Thus the voltage across R_f would be

$$V_{Rf} = 1 \text{ μA} \times 99 \text{ kΩ}$$
$$V_{Rf} = 99 \text{ mV}$$

The total voltage across the voltage divider is

$$1 \text{ mV} + 99 \text{ mV} = 100 \text{ mV}$$

Completing our calculations, the voltage gain of the circuit is

$$A = \frac{V_{out}}{V_{in}}$$
$$A = \frac{100 \text{ mV}}{1 \text{ mV}}$$
$$A = 100$$

Therefore $Av_{(cl)} = 100$. In this case, $Av_{(cl)} = 1/B$, where B is the feedback ratio of the voltage divider.

PRACTICE PROBLEM 12.9

Assume that the input to the op-amp circuit of Figure 12.17 is 100 μV, then 2 mV, then 5 mV. Use the bootstrap method to determine the corresponding output and, from this, the voltage gain of the circuit for each of these inputs.

You could continue to calculate the voltage gain of the op-amp circuit of Figure 12.17 for an infinite number of input voltages, but if you are doing the mathematics correctly, you will find that the $Av_{(cl)}$ gain of the circuit continues to be 100.

We now state this conclusion: The voltage gain of the op-amp circuit is independent of the op-amp in use and is set by the inverse of the feedback ratio.

Thus:

$$Av_{(cl)} = \frac{1}{B}$$

This relationship is valid in almost all situations in which series-parallel feedback is used. As long as the open-loop gain times the feedback ratio is larger than 10, you will have accurate answers using the equation above. Use this as your guideline:

$$\text{If } AB > 10$$
$$\text{then } Av_{(cl)} = \frac{1}{B}$$

If AB is not greater than 10, you can use the equation for Av_{cl} on page 549 to determine the voltage gain.

We see that the open-loop gain in Figure 12.17 is given as 100,000. We already know the feedback ratio to be 0.01, so

$$AB = 100,000 \times 0.01$$
$$AB = 1000$$

THE SACRIFICE FACTOR

Next we turn to the sacrifice factor. The **sacrifice factor** (S) is the product of the open-loop gain and the feedback factor. In symbols it is

$$S = 1 + AB$$

where A is the open-loop gain and B is the feedback factor. Remember too that

$$Av_{(cl)} = \frac{1}{B}$$

EXAMPLE 12.9

Determine the sacrifice factor for the circuit of Figure 12.17.

Solution

$$B = \frac{1\,k\Omega}{(99\,k\Omega + 1\,k\Omega)} = 0.01$$
$$S = 1 + AB$$
$$S = 1 + (100,000)(0.01)$$
$$S = 1001$$

This means that the closed-loop gain is 1001 times smaller than the open-loop gain.

PRACTICE PROBLEM 12.10

Determine the sacrifice factors for these situations. (HINT: Convert $Av_{(cl)}$ to B.)
   (a) A = 50,000; $Av_{(cl)}$ = 200
   (b) A = 20,000; $Av_{(cl)}$ = 10
   (c) A = 100,000; $Av_{(cl)}$ = 500
   (d) A = 50,000; $Av_{(cl)}$ = 25

The answers to practice problem 12.10 are (a) 251, (b) 2001, (c) 201, and (d) 2001. If you got those answers you know how to calculate a sacrifice factor. What insights to series-parallel negative feedback op-amp circuits does the sacrifice factor provide?

EXAMPLE 12.10

To see how the sacrifice factor can be useful, let's examine the circuit in Figure 12.17 using the equation on page 549.

Solution
The equation, says

$$Av_{(cl)} = \frac{A}{(1 + AB)}$$

where A is the open-loop gain of the op-amp.
   We determine the gain of the circuit to be

$$Av_{(cl)} = \frac{100,000}{1 + (0.01 \times 100,000)}$$
$$Av_{(cl)} = 99.9000999$$

whereas before we obtained 100.
   This means that if the input to the circuit is 1 mV and the output is 99.9 mV, the voltage fed back to the inverting terminal of the op-amp would be

$$0.01 \times 99.9 \text{ mV} = 999 \ \mu V$$

This in turn produces an error voltage equal to

$$v_{error} = 1 \text{ mV} - 999 \ \mu V$$
$$v_{error} = 1 \ \mu V$$

If the input impedance of the op-amp is 1 MΩ, the current draw from the source is

$$I_{in} = \frac{1 \ \mu V}{1 \ M\Omega}$$
$$I_{in} = 1 \text{ pA}$$

Since the source is providing 1 mV, the effective impedance seen by the source is

$$Z_{in(eff)} = \frac{1 \text{ mV}}{1 \text{ pA}}$$
$$Z_{in(eff)} = 1 \ G\Omega$$

As we saw before, the series-parallel input configuration drives the input impedance upward. The sacrifice factor can tell us by how much.

$$Z_{in(eff)} = S \times Z_{in}$$

In the circuit of Figure 12.17, the input impedance of the op-amp is 1 MΩ. The sacrifice factor is

$$S = 1 + AB$$
$$S = 100,000 \times 0.01$$
$$S = 1001$$

The AB factor swamps out the 1, so we ignore it in the following.

Using the equation for $Z_{in(eff)}$ on page 556, we get

$$Z_{in(eff)} = S \times Z_{int}$$
$$Z_{in(eff)} = 1000 \times 1 \text{ M}\Omega$$
$$Z_{in(eff)} = 1 \text{ G}\Omega$$

This matches what we calculated earlier incorporating the error voltage.

PRACTICE PROBLEM 12.11

An op-amp circuit using series-parallel negative feedback has a sacrifice factor of 2000. If the input impedance of the op-amp is 2 MΩ, determine the input impedance of the circuit.

Earlier we discussed how the output impedance of the series-parallel negative feedback configuration is close to zero because the feedback circuitry tries to offset drops in output voltage caused by loading effects by increasing the output voltage. We now look at how we can determine the output impedance of a series-parallel op-amp circuit using the sacrifice factor.

$$Z_{out(circuit)} = \frac{Z_{out}}{S}$$

EXAMPLE 12.11

Determine the output impedance of the circuit in Figure 12.17 using the equation for $Z_{out(circuit)}$ above. The output impedance of the op-amp is 75 Ω. The sacrifice factor for the circuit of Figure 12.17 is 1000.

Solution
We have

$$Z_{out(circuit)} = \frac{75\ \Omega}{1000}$$
$$Z_{out(circuit)} = 75 \text{ m}\Omega$$

PRACTICE PROBLEM 12.12

Change R_f in Figure 12.17 to a 49-kΩ resistor and recalculate the circuit output impedance.

Note that the series-parallel negative feedback configuration sacrifices voltage gain. As a consequence, the input impedance is driven upward toward infinity and the output impedance is driven downward toward 0 Ω. Figure 12.18 diagrams the ideal voltage amplifier, one in which the input impedance is infinite and the output impedance is zero. The noninverting op-amp circuit using negative feedback comes very close to this ideal.

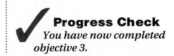

Progress Check
You have now completed objective 3.

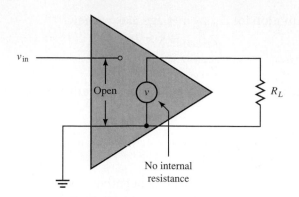

FIGURE 12.18   The perfect voltage amplifier

REVIEW SECTION 12.2

1. The front end of an op-amp is a _____. Consequently the op-amp can amplify all the way down to DC.
2. Under ideal analysis the input impedance of an op-amp is considered to be _____.
3. What does the term *bootstrapping* mean?
4. Where is v_{error} measured?
5. If the error voltage were truly zero, what would be the output voltage of the op-amp?
6. The sacrifice factor is _____.
7. What guideline should you use when calculating closed-loop gain using the relationship $Av_{(cl)} = 1/B$?
8. What are the characteristics of an ideal voltage amplifier?

12.3 OP-AMP CIRCUITS USING PARALLEL–PARALLEL NEGATIVE FEEDBACK

Figure 12.19 shows an op-amp circuit with negative feedback, but this time the circuit configuration is parallel-parallel.

In Figure 12.19, the input voltage is applied to the op-amp through an input resistance labeled R_{in}. The current through R_{in} reaches node 1 (N1), where it "sees" two possible pathways for current flow. One pathway is into the op-amp itself. This pathway is a high-impedance pathway, as the input impedance of the op-amp is fairly high. If we continue to use ideal analysis and treat the input impedance of the op-amp as infinite, we can conclude that all the input current that flows through R_{in} must also flow through the feedback resistor, R_f. This is a key point.

A second major point in the analysis of the parallel-parallel feedback configuration shown in Figure 12.19 is the presence of a virtual ground at N1. A point in a circuit that is at ground potential, but is not connected to ground directly, is known as a **virtual ground.**

Note that the noninverting terminal is connected to ground through an input bias compensating resistor, R_L. The voltage dropped across this compensating resistor is minimal due to the very small bias currents needed to operate the differential input stage. (See Chapters 10 and 11.)

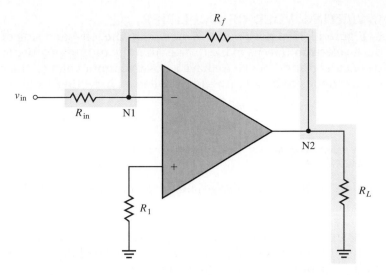

FIGURE 12.19 A parallel-parallel op-amp circuit, used as an inverting voltage amplifier

In the ideal analysis of an op-amp, we stipulate that the voltage drop across R_1 is 0 V. This implies that the noninverting terminal is at ground potential.

This ground potential is then bootstrapped over to the inverting terminal, placing the inverting terminal at virtual ground.

There is one other key point that you should know before we analyze parallel-parallel feedback circuits. R_f and R_L are essentially in parallel. It may not look like it, but the two resistors are connected together at the output node, N2. The other side of R_L is connected to ground, but the opposite side of R_f is also at ground potential, connected to the virtual ground point at N1. Thus, whatever voltage is developed across R_f is also developed across R_L.

✔ **Progress Check**
You have now completed objective 4.

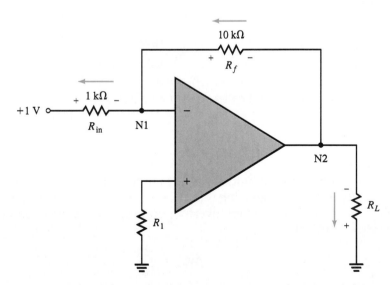

FIGURE 12.20 A parallel-parallel op-amp circuit used as an inverting voltage amplifier. Arrows denote electron flow.

INVERTING VOLTAGE AMPLIFIER

In Figure 12.20, a positive 1 V is applied to the left-hand side of R_{in}. This produces electron flow as marked. Note that the output polarity is negative with respect to ground. Since we have a positive input voltage, the input and output are opposite in polarity. This is called an **inverting amplifier.**

EXAMPLE 12.12

Determine the output voltage and the gain of the circuit in Figure 12.20.

Solution

First, find the current through R_{in}. Because the bootstrapping effect has placed ground potential on the right-hand side of R_{in}, there is 1 V across the input resistor and we can apply Ohm's law.

$$I_{R(in)} = \frac{1\,V}{1\,k\Omega}$$
$$I_{R(in)} = 1\,mA$$

Treating the op-amp terminals as open circuits, none of this current comes from the inverting terminal, all of it must come from R_f. Applying Kirchhoff's current law, we obtain the current flowing through R_f: The current leaving N1 is 1 mA, therefore the current entering node N1 is 1 mA. Since none of the current comes from the inverting terminal, the current through R_f is 1 mA.

Second, apply Ohm's law to find the voltage across R_f:

$$V_{Rf} = 1\,mA \times 10\,k\Omega$$
$$V_{Rf} = 10\,V$$

Because of the virtual ground at N1, R_f and R_L are essentially in parallel, so the voltage across R_L is also 10 V, with the polarity as marked on the circuit.

The gain of the circuit can now be computed:

$$A = \frac{V_{out}}{V_{in}}$$
$$A = \frac{-10\,V}{1\,V}$$
$$A = -10$$

where the minus sign indicates an opposite polarity.

It is *not a coincidence* that the ratio of R_f to R_{in} is also 10 for the circuit of Figure 12.20. In fact, recognizing this relationship gives us a simple way to calculate the voltage gain of an inverting op-amp circuit using parallel-parallel negative feedback.

$$Av_{(cl)} = \frac{-R_f}{R_{in}}$$

where the minus sign indicates the phase inversion.

Change R_{in} to a 2.2-kΩ resistor in Figure 12.20. Compute the gain using the technique shown in example 12.12. Then verify this answer with the ratio R_f/R_{in}.

EXAMPLE 12.13

Determine the output voltage and the gain for the circuit of Figure 12.21.

Solution

In this circuit −200 mV is applied to the input. R_{in} = 100 Ω, and the feedback resistor equals 2 kΩ.

Because this circuit is using parallel-parallel negative feedback we can use the idea of a virtual ground.

The left side of R_{in} is at −200 mV and the right side is at ground. First, use Ohm's law to determine the current flowing through R_{in}.

$$I = \frac{200\,mV}{100\,\Omega}$$

$$I = 2\,mA$$

All of this current must flow through R_f with the polarity as marked on the schematic.

Second, use Ohm's law to determine the voltage dropped across R_f:

$$V_{Rf} = 2\,mA \times 2\,k\Omega$$
$$V_{Rf} = 4\,V$$

The voltage across R_f is also across the output load because R_f and R_L can be considered to be in parallel. Therefore the output voltage is 4 V and is positive with respect to ground.

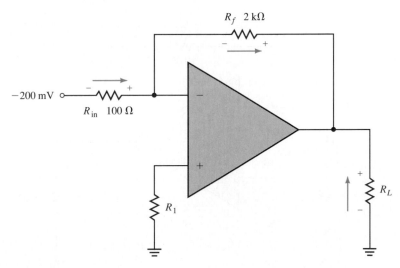

FIGURE 12.21 Direction of electron flow around a PP op-amp circuit

The gain of the circuit is

$$A = \frac{V_{out}}{V_{in}}$$

$$A = \frac{4\,V}{-200\,mV}$$

$$A = -20$$

where the negative sign indicates a polarity inversion.

PRACTICE PROBLEM 12.14

Verify the gain answer obtained in example 12.13 for the circuit of Figure 12.21 using the ratio of R_f to R_{in}. That is, R_f/R_{in}.

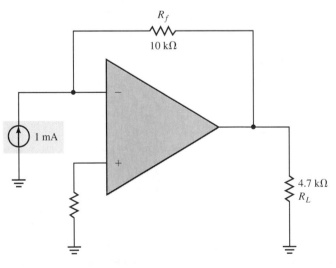

Progress Check
You have now completed objective 5.

CURRENT-TO-VOLTAGE TRANSDUCERS

The circuit of Figure 12.22 shows an op-amp wired for parallel-parallel negative feedback, but in this case the input circuit is driven by a current source. When you see this type of arrangement, the op-amp is being used as a current-to-voltage transducer.

In general, **transducers** are circuits or devices that take one type of quantity and convert it into something else. For example, a thermistor is a temperature-sensitive resistor. It converts changes in temperature into changes in resistance. A JFET is a transducer. It converts changes of input voltage to output current, hence the importance of the transconductance curve in analyzing JFET circuitry.

The circuit of Figure 12.22 is also a transducer. It is converting input current into output voltage. We know that the current flowing into the circuit is forced to flow through R_f. Since R_f is a 10-kΩ resistor, we can determine the voltage across it.

$$V_{Rf} = 1\,mA \times 10\,k\Omega$$

$$V_{Rf} = 10\,V$$

Since R_f can be considered to be in parallel with R_L, the output voltage is 10 V.

FIGURE 12.22   A current-to-voltage transducer

Determine the output voltage for the circuit of Figure 12.22 if the input current is changed to 150 μA.

✓ **Progress Check**
You have now completed objective 6.

Consider Figure 12.23. Did the circuit designer intend to have this circuit act as an inverting voltage amplifier or as a current-to-voltage transducer?

The highlighted section in the circuit indicates that the 50-kΩ resistance and the 5-V_p signal are internal to some device. That is, the 50 kΩ is acting like the internal impedance of the circuit or device that is generating the 5-V_P signal.

If we apply Ohm's law to the 5-V_P signal and the 50-kΩ impedance we can calculate the current that flows into the virtual ground point.

$$I = \frac{5\,V_P}{50\,k\Omega}$$
$$I = 100\,\mu A$$

As before, all of this current flows through R_f, producing

$$100\,\mu A \times 4.7\,k\Omega = 470\,mV_P$$

If the circuit designer was trying to build an inverting voltage amplifier, then a mistake has been made. The input was at 5 V_P and the output has been attenuated, down to 470 mV_P.

A key to the designer's intentions and whether the circuit is an amplifier is the ratio of R_f to R_{in}. If R_{in} is larger than R_f, there will be a loss of voltage. Therefore assume that the circuit is supposed to be acting as a current-to-voltage transducer. In addition, while not infinite, a 50-kΩ internal impedance is a far cry from the near 0 Ω of internal impedance that a voltage source is supposed to have. Thus we can treat the 5-V_P and 50-kΩ combination in Figure 12.23 as a current source.

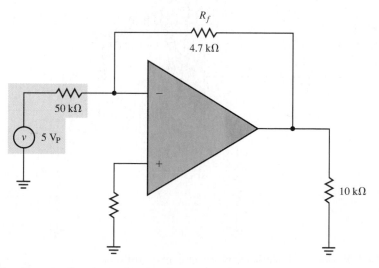

FIGURE 12.23   How to tell an inverting amp from a transducer

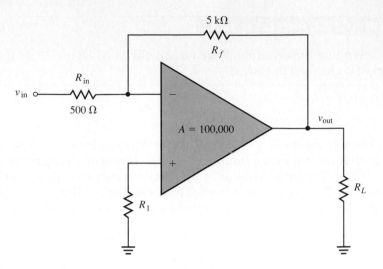

FIGURE 12.24   An inverting voltage amplifier with a gain of 100,000

INPUT AND OUTPUT IMPEDANCE

Figure 12.24 shows a parallel-parallel negative feedback amplifier. Since R_{in} is much smaller than R_f, it is an inverting voltage amplifier. The input impedance of the circuit can be determined by applying Miller's theorem (see Chapter 10) to the feedback resistor.

EXAMPLE 12.14

Determine the input impedance of the circuit in Figure 12.24 by millerizing the feedback resistor.

Solution
The result of millerizing the feedback resistor can be seen in Figure 12.25.

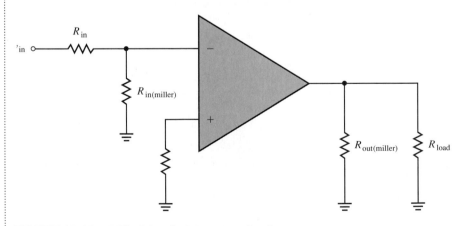

FIGURE 12.25   Millerizing the PP op-amp circuit

$R_{in(miller)}$ is across the op-amp's input impedance. With an open-loop gain, $R_{in(miller)}$ is

$$R_{in(miller)} = \frac{R_{in}}{A_{OL} + 1}$$
$$R_{in(miller)} = \frac{5000\,\Omega}{(100,000 + 1)}$$
$$R_{in(miller)} = 50\,m\Omega$$

which is almost zero. So we have additional proof of the virtual ground concept.

This 49 mΩ is in parallel with the input impedance of the op-amp. Therefore the input impedance to the op-amp can be visualized as drawn in Figure 12.26.

$$Z_{in(eff)} = R_{in(miller)} \parallel Z_{int}$$
$$Z_{in(eff)} = 50\,m\Omega \parallel 2\,M\Omega$$
$$Z_{in(eff)} = 50\,m\Omega$$

Hence the actual input impedance the source sees is

$$R_{in} + Z_{in(eff)} = 500\,\Omega + 50\,m\Omega = 500.05\,\Omega$$

Because the op-amp's input impedance is so small, it is not significant. So the input impedance to the circuit is equal to R_{in}.

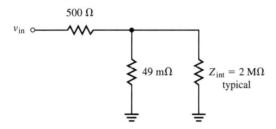

FIGURE 12.26 Equivalent input impedance of a parallel-parallel op-amp circuit

You will find that the input impedance to most of the parallel-parallel negative feedback op-amp circuits is equal to R_{in}. When this configuration is used as a current-to-voltage transducer as in Figure 12.22, the input impedance to the circuit is effectively 0 Ω.

The output impedance of the circuit in Figure 12.25 includes the $R_{out(miller)}$ in parallel with the output impedance of the op-amp.

In op-amp circuits you can rely on this guideline: When the open-loop gain of the op-amp is very large, the $R_{out(miller)}$ equals R_f because the quantity $A/(1 + A)$ is very, very close to unity.

Based on this discussion the output impedance of the circuit in Figure 12.25 is similar to the output impedance in Figure 12.27. The load "sees" back into the circuit, $R_{out(miller)}$, which equals R_f in parallel with the Z_{out} of the op-amp. Since in most cases

$$Z_{out} \ll R_f$$

the output impedance of the amplifier is equal to Z_{out}; however, the feedback arrangement drives this impedance toward zero.

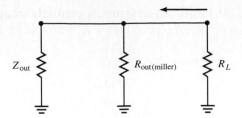

FIGURE 12.27 Equivalent output impedance of a parallel-parallel op-amp circuit

The sacrifice ratio of the circuit in Figure 12.24 is

$$S = 1 + AB$$

B in this case is $R_{in}/(R_{in} + R_f) = 0.0909$, so

$$S = 1 + 100{,}000 \times 0.0909 = 9091$$

Thus

$$Z_{out(circuit)} = \frac{Z_{out}}{9091}$$

EXAMPLE 12.15

Determine the output impedance of the circuit of Figure 12.24 if $Z_{out} = 50\ \Omega$.

Solution
From our previous discussion:

$$Z_{out(circuit)} = \frac{Z_{out}}{9092}$$

$$Z_{out(circuit)} = \frac{50\ \Omega}{9091}$$

$$Z_{out(circuit)} = 5.5\ m\Omega$$

✓ Progress Check
You have now completed objective 7.

In conclusion, the parallel-parallel negative feedback circuit provides very low input impedance and very low output impedance. It can be used as an inverting voltage amplifier when R_f is larger than R_{in} and as a current-to-voltage transducer when the input to the circuit is a current source.

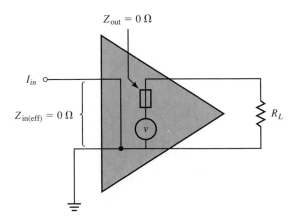

FIGURE 12.28 The ideal current-to-voltage transducer

Figure 12.28 shows a diagram of the ideal current-to-voltage transducer. As you now know, the parallel-parallel negative feedback configuration approaches these characteristics.

REVIEW SECTION 12.3

1. All of the current that flows through R_{in} must flow through _____.
2. Which resistor in the parallel-parallel feedback arrangement is essentially in parallel with the load resistor?
3. What is meant by the term *virtual ground?*
4. Where is the virtual ground point in a parallel-parallel feedback op-amp circuit?
5. R_f is 10 kΩ, R_{in} is 500 Ω. If the op-amp circuit is wired as an inverting voltage amplifier, what is the gain?
6. A _____ is a circuit or device that converts one quantity into another quantity.
7. How can you decide if a circuit designer meant the parallel-parallel feedback configuration to be used as an inverting voltage amplifier or as a current-to-voltage transducer?
8. What is the input impedance of an inverting voltage amplifier?
9. What are the characteristics of an ideal current-to-voltage transducer?

12.4   THE INVERTING CURRENT AMPLIFIER

Figure 12.29 shows the schematic of an **inverting current** amplifier. The op-amp is connected in a parallel-series negative feedback configuration. The input is applied to the inverting terminal, which is bootstrapped to ground.

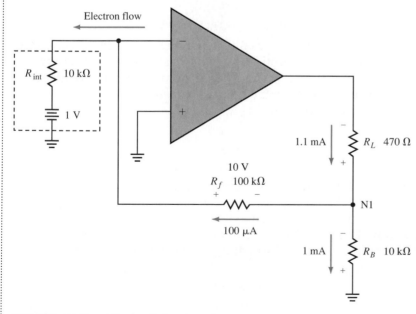

FIGURE 12.29   The parallel-series op-amp circuit

The output side of the op-amp circuit is a series feedback connection. The load resistor is no longer in parallel with the feedback network but is part of the feedback network. In addition, the load resistor is "floating"—it does not have a direct connection to ground. This is one easy way to distinguish between series feedback and parallel feedback. If the load is across the feedback network and has a ground connection, then the feedback arrangement is parallel; if not, the feedback arrangement is series.

CURRENT GAIN

EXAMPLE 12.16

Determine the current gain of the circuit in Figure 12.29.

Solution
We begin our analysis using two basic concepts. First, assume that the input impedance of the op-amp is infinite. Therefore, no current flows into or out of the inverting terminal. Second, the inverting terminal is at virtual ground because of the bootstrapping effect.

First, determine the current through the 10-kΩ resistor, R_{int}, using Ohm's law.

The bottom side of R_{int} is connected to +1 V with respect to ground. The top side of R_{int} is connected to ground potential via the virtual ground. Consequently the voltage across R_{int} is 1 V. Using Ohm's law we have:

$$I_{in} = \frac{1\,V}{10\,k\Omega}$$
$$I_{in} = 100\,\mu A$$

None of this current is supplied by the inverting terminal of the op-amp, therefore it must come from R_f. This means that the current flowing through R_f is equal to the current flowing through R_{int} (Kirchhoff's current law).

Second, determine the voltage drop across R_f.

$$V_{Rf} = 100\,\mu A \times 100\,k\Omega = 10\,V$$

We conclude that R_f and R_b are essentially in parallel, since they are connected together at node 1 (N1). The other sides of these resistors are at ground potential. So whatever voltage is across one resistor must be across the other. Therefore, we conclude that the voltage across R_b is 10 V.

Third, determine the current flowing through R_b:

$$I_{Rb} = \frac{10\,V}{10\,k\Omega}$$
$$I_{Rb} = 1\,mA$$

The polarities and the direction of current flow are marked on the circuit diagram of Figure 12.29. As you can see, the current flowing through R_b is leaving N1. The current flowing through R_f is also leaving N1.

Since the sum of currents leaving a node must equal the sum of currents entering it, the current flowing through R_L is entering N1 and is the sum of the other two currents.

$$I_{RL} = I_{Rb} + I_{Rf}$$
$$I_{RL} = 1.1\,mA$$

We can now determine the current gain of the circuit, A_I.

$$A_I = \frac{I_{out}}{I_{in}}$$

I_{out} is the current through the load, and I_{in} is the current through R_{int}

$$A_I = \frac{I_{out}}{I_{in}}$$

$$A_I = \frac{1.1 \, mA}{100 \, \mu A}$$

$$A_I = 11$$

PRACTICE PROBLEM 12.16

Change R_b in Figure 12.29 to a 4.7-kΩ resistance. Recompute the current gain of the circuit.

Refer to Figure 12.29 to calculate the ratio of $R_b/(R_b + R_f)$.

$$\frac{R_b}{R_b + R_f} = \frac{10 \, k\Omega}{110 \, k\Omega}$$

$$\frac{R_b}{R_b + R_f} = 0.90909$$

If we invert this fraction, we get

$$\frac{1}{0.90909} = 11$$

As shown in example 12.16 this is the current gain of the amplifier. Therefore

$$A_I = \frac{1}{B_I}$$

where A_I is the current gain of an inverting current amplifier and B_I is the ratio of the feedback circuit.

EXAMPLE 12.17

Determine the gain of the inverting current amplifier in Figure 12.30.

Solution
In Figure 12.30, the combination of 5 V and the R_{int} of 1 kΩ provides the input current:

$$I_{in} = \frac{5 \, V}{1 \, k\Omega}$$

$$I_{in} = 5 \, mA$$

In Figure 12.30 the polarity of the voltage driving R_{int} is different from the polarity of the input voltage in Figure 12.29. Therefore the direction of electron flow is different.

All of the current flowing into the circuit will flow through R_f because the input impedance of the op-amp itself is infinite; therefore the current flowing through R_f is equal to I_{in}. Apply Ohm's law to determine the voltage drop across R_f.

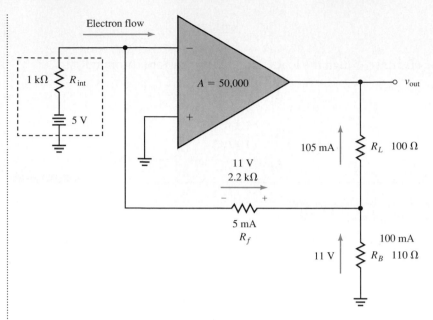

FIGURE 12.30   A parallel-series op-amp circuit used as a current amplifier

$$V_{Rf} = 5\text{ mA} \times 2.2\text{ k}\Omega$$
$$V_{Rf} = 11\text{ V}$$

Since R_b is in parallel with R_f, the voltage across R_b is also 11 V. The current flowing through R_b:

$$I_{Rb} = \frac{11\text{ V}}{110\text{ }\Omega}$$
$$I_{Rb} = 100\text{ mA}$$

The current flowing through R_L is the sum of the currents flowing through R_b and R_f.

$$I_{RL} = I_{Rb} + I_{Rf}$$
$$I_{RL} = 5\text{ mA} + 100\text{ mA}$$
$$I_{RL} = 105\text{ mA}$$

The current gain of the circuit is equal to I_{out}/I_{in}, so:

$$A_I = \frac{I_{out}}{I_{in}} = \frac{105\text{ mA}}{5\text{ mA}} = 21$$

PRACTICE PROBLEM 12.17

Verify the current gain answer of example 12.17 by using the current feedback ratio method of determining current gain.

EXAMPLE 12.18

In Figure 12.31, the inverting current amplifier is used directly with a current source driving the input. In this case, the input current is 200 µA. As before, all of this current flows through R_f, producing a voltage drop of 11 V.

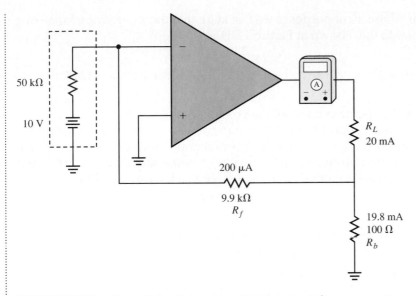

FIGURE 12.31   **A parallel-series op-amp circuit connected to an ammeter**

$$V_{Rf} = 200\ \mu A \times 9.9\ k\Omega$$
$$V_{Rf} = 1.98\ V$$

This voltage is across R_b and produces a current of

$$I_{Rb} = \frac{1.98\ V}{100\ \Omega}$$
$$I_{Rb} = 19.8\ mA$$

The total current sensed by the ammeter is

$$I_{out} = 19.8\ mA + 200\ \mu A$$
$$I_{out} = 20\ mA$$

The current gain of the circuit is

$$A_I = \frac{I_{out}}{I_{in}}$$
$$A_I = \frac{20\ mA}{200\ \mu A}$$
$$A_I = 100$$

which can be verified by finding the inverse of the feedback ratio.

$$B_I = \frac{100}{(9.9\ k\Omega + 100\ \Omega)}$$
$$B_I = 0.01$$

The inverse of B_I is

$$\frac{1}{0.01} = 100$$

Progress Check
*You have now completed
objective 9.*

INPUT AND OUTPUT IMPEDANCE

The input side of the inverting current amplifier is a parallel connection with an input node connected to a virtual ground point. We have already seen that this arrangement drives the input impedance of the circuit down toward ground. For all practical purposes we can state that the input impedance of a circuit similar to that shown in Figure 12.31 is

$$Z_{in(eff)} = 0 \, \Omega$$

The input impedance of a circuit similar to that shown in Figure 12.30 is

$$Z_{in} = R_{int}$$

as seen from the "point of view" of the voltage source.

Earlier, we stated that R_b acts like a current source, providing current for the load resistance. We know that the internal impedance of a current source is supposed to be infinite. Therefore without doing any analysis, we could state that the output impedance of a series feedback arrangement is

$$Z_{out} = \infty = \text{infinite}$$

To obtain a more accurate answer for the output impedance of a circuit using parallel-series feedback, use this relationship:

$$Av_{OL} \times B \times (R_f \parallel R_b)$$

Since in most cases $R_b \ll R_f$, we have:

$$Z_{out} = Av_{OL} \times B \times R_b$$

where Av_{OL} is the open-loop gain of the op-amp and B is the voltage feedback factor:

$$B = \frac{R_b}{R_b + R_f}$$

A proof for the equation above can be found in Appendix C.

EXAMPLE 12.19

Determine the output impedance of the circuit of Figure 12.30 using the equation above.

Solution
The open-loop gain of the op-amp is given as 50,000. R_b is a 110-Ω resistor.
First, determine the value of B:

$$B = \frac{R_b}{R_b + R_f}$$

$$B = \frac{110 \, \Omega}{110 \, \Omega + 2.2 \, k\Omega}$$

$$B = 0.0476$$

Therefore the output impedance of the circuit is

$$Z_{out(circuit)} = Av_{(ol)} \times B \times R_b$$
$$Z_{out(circuit)} = 50,000 \times 0.0476 \times 110 \, \Omega$$
$$Z_{out(circuit)} = 262 \, k\Omega$$

PRACTICE PROBLEM 12.18

The open-loop gain of the circuit of Figure 12.31 is 100,000. Determine the output impedance of the circuit.

Note that the circuit is driven by an ideal current source so $R_{in} = \infty$, which makes B = 1. To see this, suppose $R_{in} = 1$ MΩ, then B = 1 MΩ/(1 MΩ + 9.9 kΩ) = 0.99, and as R_{in} approaches ∞, B approaches 1.

✓ **Progress Check**
You have now completed objective 10.

An ideal current amplifier is diagrammed in Figure 12.32. In this case, the circuit provides current gain, the input impedance is zero, and the output impedance is infinite because a current source is driving the output load. The parallel-series feedback configuration that turns an op-amp into an inverting current amplifier comes close to this ideal.

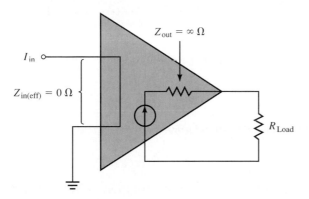

FIGURE 12.32   The ideal current amplifier

REVIEW SECTION 12.4

1. What two load resistor connections can you look for when you are deciding if the feedback method is series?
2. The input impedance of a parallel-series feedback circuit is ideally _____.
3. What is the circuit relationship between R_f and R_b in a parallel-series feedback circuit?
4. Which resistor in a parallel-series op-amp feedback acts like a current source?
5. The ideal output impedance of a parallel-series op-amp feedback circuit is _____.
6. What ratio can be used to determine the current gain of an inverting current amplifier?
7. What are the characteristics of an ideal current amplifier?

12.5   A VOLTAGE-TO-CURRENT TRANSDUCER

Figure 12.33 shows an op-amp connected in a series-series negative feedback configuration. Notice that the output load is part of the feedback network and that it is floating. Consequently we can identify the output part of the circuit as a series configuration.

The input signal is applied to the noninverting terminal. This is similar to the series-parallel input that we saw in section 12.2. It is then bootstrapped to the top of R_f.

✓ **Progress Check**
You have now completed objective 11.

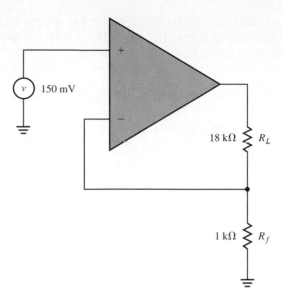

FIGURE 12.33 The series-series feedback op-amp circuit

OUTPUT CURRENT

The combination of R_f and the bootstrapped voltage provides a current for the load resistance.

EXAMPLE 12.20

Determine the output current for the circuit shown in Figure 12.33.

Solution

There is a 150-mV signal applied to the noninverting terminal of the op-amp, which is bootstrapped to the top of R_f. Calculate the current flowing through R_f:

$$I_{Rf} = \frac{150\,\text{mV}}{1\,\text{k}\Omega}$$
$$I_{Rf} = 150\,\mu\text{A}$$

Because of the infinite impedance looking into the op-amp, all the current flowing through R_f must flow through R_L. Thus the output current equals the current through R_f.

$$I_{out} = 150\,\mu\text{A}$$

PRACTICE PROBLEM 12.19

Determine the current flowing through the load resistance in Figure 12.34.

INPUT AND OUTPUT IMPEDANCE

The input side of the circuit of Figure 12.35 is a series configuration. In this case, the input impedance is driven upward toward infinity. As we learned

Progress Check
You have now completed objective 12.

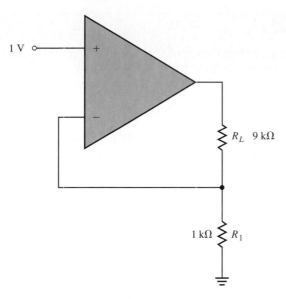

FIGURE 12.34   A series-series op-amp circuit used as a voltage-to-current transducer

when studying the series-parallel configuration in section 12.2, the input impedance of a series input is calculated by:

$$Z_{in(eff)} = S \times Z_{int}$$

where Z_{int} is the input impedance of the amplifier and S is the sacrifice ratio expressed by the relationship:

$$S = 1 + (Av_{(ol)} \times B)$$

to determine the effects on the input impedance. $Av_{(ol)}$ continues to represent the open-loop gain of the amplifier, and in this case B is the feedback ratio:

$$B = \frac{R_f}{(R_f + R_L)}$$

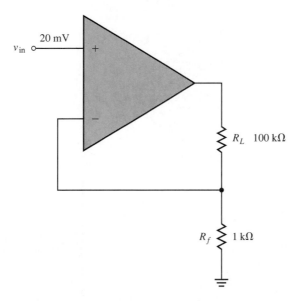

FIGURE 12.35   Another series-series op-amp circuit used as a voltage-to-current transducer

EXAMPLE 12.21

Determine the input impedance of the circuit in Figure 12.35. Let $Z_{int} = 1\,M\Omega$ and the open-loop gain = 50,000.

Solution
Using this relationship:

$$Z_{in(eff)} = (1 + AB) \times Z_{int}$$

The feedback ratio is

$$B = \frac{1\,k\Omega}{101\,k\Omega}$$
$$B = 0.0099$$

So we have

$$Z_{in(eff)} = [1 + (50,000 \times 0.0099)] \times 1\,M\Omega$$
$$Z_{in(eff)} = 496\,M\Omega$$

PRACTICE PROBLEM 12.20

Change the values in the circuit in Figure 12.35 so that the input impedance is now 2 $M\Omega$ and the open-loop gain is 20,000. Determine the new input impedance to the circuit.

The output side of the amplifier shown in Figure 12.36 is a series feedback circuit. It is similar to the output side of the parallel-series circuit already discussed in section 12.4. In addition, R_f is acting like a current source for the load, so we expect the output impedance to be driven upward toward infinity.

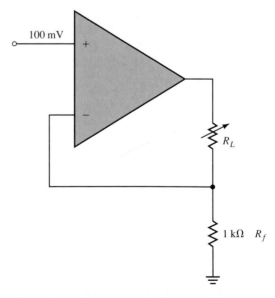

FIGURE 12.36   The series-series op-amp circuit. R_f acts like a constant current source.

The output impedance of the series-series circuit can be approximated by the relationship

$$Z_{out(circuit)} = A \times R_f$$

where A is the open-loop gain of the amplifier.

EXAMPLE 12.22

Determine the output impedance of the circuit in Figure 12.36.

Solution

$$Z_{out(circuit)} = A \times R_f$$
$$Z_{out(circuit)} = 50{,}000 \times 1 \text{ k}\Omega$$
$$Z_{out(circuit)} = 50 \text{ M}\Omega$$

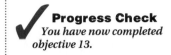

Progress Check
You have now completed objective 13.

> **REVIEW SECTION 12.5**
> 1. The series-series negative feedback op-amp circuit acts like a
>   _____ -to-_____ transducer.
> 2. Which component in the circuits studied in this section acts like a constant current source?
> 3. If 50 mV is placed at the noninverting input of an op-amp wired in a series-series feedback arrangement, how much voltage will be dropped across R_f?
> 4. How can the expression $1 + AB$ be used to determine the input impedance of an op-amp wired with series-series negative feedback?
> 5. An ideal approximation for the input impedance of an op-amp circuit wired in series-series negative feedback is
>   _____.
> 6. An ideal approximation of the output impedance for an op-amp circuit wired in series-series negative feedback is
>   _____.

12.6 OP-AMP APPLICATIONS

In this section we will look at a few op-amp applications. This should give you some idea of the many possible uses of op-amps.

THE VOLTAGE FOLLOWER

Figure 12.37 shows an op-amp circuit wired in a series-parallel configuration. The input signal is wired to the noninverting terminal and is bootstrapped to the inverting terminal. Then it is connected directly to the output. This type of configuration is known as a **voltage follower,** or a **unity-gain amplifier.**

Since the inverting terminal voltage is equal to the input voltage, the output voltage equals the input voltage. This means that the circuit has unity gain—the input and output signal amplitudes are the same. There is no volt-

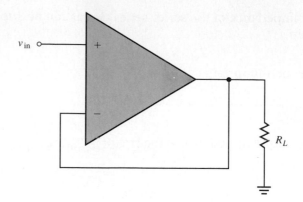

FIGURE 12.37   The voltage follower configuration

age gain or phase inversion. Since we have no voltage gain, we have sacrificed the entire open-loop gain of the op-amp. To understand the purpose, look at example 12.23.

EXAMPLE 12.23

Determine the sacrifice factor of the op-amp circuit of Figure 12.37. Use this information to determine the input impedance of the circuit. The input impedance of the op-amp is 1 MΩ.

Solution
Since the circuit is a voltage follower, it has unity gain.

$$S = 1 + AB$$
$$S = 100,000 + 1$$
$$S = 100,000$$

Sacrificing gain drives the AC circuit impedance of a series input up toward infinity:

$$Z_{in(circuit)} = S \times Z_{int}$$
$$Z_{in(circuit)} = 100,000 \times 1 \text{ M}\Omega$$
$$Z_{in(circuit)} = 100 \text{ G}\Omega$$

Yes, one hundred billion ohms.

The voltage follower with its *extremely* high input impedance is almost a perfect impedance buffer. In addition to the very high input impedance, the output voltage looks like an ideal voltage source:

$$Z_{out\,(circuit)} = \frac{Z_{out}}{S}$$

If we have a typical output impedance, Z_{out} is low to begin with, for example, 75 Ω. $Z_{out(circuit)}$ is

$$Z_{out\,(circuit)} = \frac{75\,\Omega}{S}$$
$$Z_{out\,(circuit)} = \frac{75\,\Omega}{100,000}$$
$$Z_{out\,(circuit)} = 750\,\mu\Omega$$

We can treat 750 μΩ as zero. This circuit configuration allows us to sample signals without loading them down. If the circuit needs to be amplified, other stages can be connected to the output of Figure 12.37.

NONINVERTING AC VOLTAGE AMPLIFIER

Most of the op-amp circuits studied so far in this chapter had DC input voltages. In most cases, DC sources were used to simplify circuit analysis. In addition, the repeated use of DC inputs should have made the point that op-amps can work all the way down to DC. The lower cutoff frequency is 0 Hz, the upper cutoff frequency can be extended up toward f_{unity} by sacrificing gain.

Figure 12.38 shows a noninverting voltage amplifier with an AC signal applied.

We can use the feedback ratio to determine the gain of the amplifier:

$$B = \frac{R_b}{R_b + R_f}$$

$$B = \frac{1\,k\Omega}{1\,k\Omega + 99\,k\Omega}$$

$$B = \frac{1}{100}$$

$$B = 0.01$$

and, since $Av_{(cl)} = 1/B$, the closed-loop gain of the amplifier is

$$Av_{(cl)} = \frac{1}{B}$$

$$Av_{(cl)} = \frac{1}{0.01}$$

$$Av_{(cl)} = 100$$

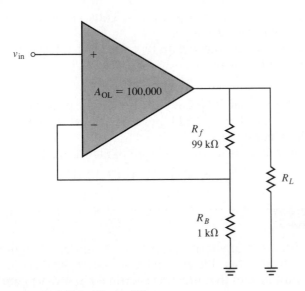

FIGURE 12.38 A noninverting AC amplifier

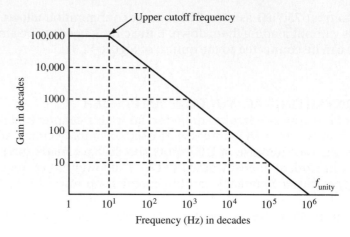

FIGURE 12.39 Gain-bandwidth graph

Using the open-loop gain of the amplifier given as 100,000, we can now determine the sacrifice ratio:

$$S = \frac{A_{(ol)}}{A_{cl}}$$

$$S = \frac{100,000}{100}$$

$$S = 1000$$

To see how this affects the AC operation of the circuit, look at Figure 12.39, a plot of gain versus frequency. There are several points of note in this diagram. First, the upper cutoff frequency of the op-amp is listed as 10 Hz. The open-loop gain of the circuit below this frequency is typically 100,000.

Second, f_{unity}, the frequency at which gain drops to 1, is listed as 1 MHz. This frequency is also the gain-bandwidth product for the op-amp.

$$A_{ol} \times f_{c(ol)} = 100,000 \times 10 \text{ Hz} = 1 \text{ MHz}$$

As stated in Chapter 11, the gain-bandwidth product is a constant. If bandwidth goes up, then gain must go down. Thus, in order to obtain a larger bandwidth, we must sacrifice gain.

In Figure 12.38, we sacrificed gain by a factor of 1000. Thus the bandwidth goes up by the same factor.

$$f_{c(cl)} = S \times f_{c(ol)}$$

where $f_{c(cl)}$ is the critical frequency of the circuit under closed-loop operations, S is the sacrifice factor, and $f_{c(ol)}$ is the critical frequency of the op-amp under open-loop conditions.

Applying this relationship to the circuit of Figure 12.38, we obtain

$$f_{c(cl)} = 1000 \times 10 \text{ Hz} = 10 \text{ kHz}$$

EXAMPLE 12.24

An op-amp with the characteristics given in the graph of Figure 12.39 is to be used as an audio amplifier. We want a bandwidth from 0 Hz to 20 kHz. Determine the amount of gain that must be sacrificed in order to obtain this bandwidth.

Solution

$$f_{c(cl)} = S \times f_{c(ol)}$$

$$S = \frac{f_{c(cl)}}{f_{c(ol)}}$$

$$S = \frac{20\,\text{kHz}}{10\,\text{Hz}}$$

$$S = 2000$$

This means that the open-loop gain of the op-amp must be reduced by a factor of 2000.

PRACTICE PROBLEM 12.21

An op-amp has an $f_{c(ol)}$ = 15 Hz and an open-loop gain of 50,000. Determine f_{unity}.

PRACTICE PROBLEM 12.22

The op-amp mentioned in practice problem 12.21 has circuit gain reduced by a factor of 1000. Determine the closed-loop upper cutoff frequency.

Progress Check
You have now completed objective 14.

SUMMING AMPLIFIERS

The circuit of Figure 12.40 shows an inverting voltage amplifier. The circuit differs from what has been studied before in that there are two input resistances connected to the virtual ground point. The inputs have been marked L(eft) and R(ight). They represent the left and right signals in which it is desired to broadcast a stereo signal. Without going into details about FM stereo broadcasting, we can state that two signals are broadcast; one is the L + R signal and the other is the L − R signal. If you take a communications class,

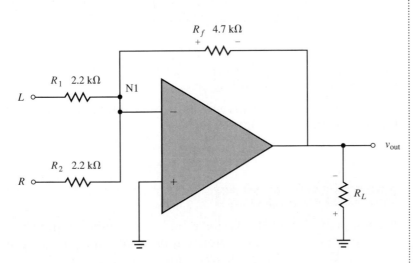

FIGURE 12.40 A summing amplifier

you will learn how your receiver responds to these signals. For now, it is enough to know that an L + R signal is needed and that the amplifier shown in Figure 12.40 can provide the summing action.

When signals are applied to the left side of the input resistances, currents flow. The two currents combine together at the virtual ground node. According to Kirchhoff's current law, all this current flows through R_f.

EXAMPLE 12.25

Determine the output of the amplifier in Figure 12.40, when the L input is +1 V to ground and the R input is +2 V to ground.

Solution
The L input produces

$$I_L = \frac{1\,V}{2.2\,k\Omega}$$
$$I_L = 455\,\mu A$$

The R input produces

$$I_R = \frac{2\,V}{2.2\,k\Omega}$$
$$I_R = 909\,\mu A$$

The sum of these two currents flows through R_f.

$$I_{Rf} = 455\,\mu A + 909\,\mu A$$
$$I_{Rf} = 1.36\,mA$$

The voltage drop across R_f is

$$V_{Rf} = 1.36\,mA \times 4.7\,k\Omega$$
$$V_{Rf} = 6.4\,V$$

Because the current flows from the op-amp output to virtual ground, the voltage developed across R_f will be negative. Since R_f is in parallel with R_L, the output voltage is

$$V_{out} = -6.4\,V$$

PRACTICE PROBLEM 12.23

The R input to Figure 12.40 remains at +2 V, but the L input changes to +3 V. Determine the new output voltage.

PRACTICE PROBLEM 12.24

The R input to Figure 12.40 remains at +2 V, but the L input changes to −1 V. (Note that this changes the direction of current flow through the left channel—R_{in}.) Determine the voltage output.

Another common use of a summing amplifier is illustrated in Figure 12.41. In this circuit the input voltages are binary, either 5 V (logic 1) or 0 V (logic 0).

The R_{in} resistors are chosen so that the currents produced when the inputs are high can be divided by powers of two.

Example 12.26 illustrates how this circuit works.

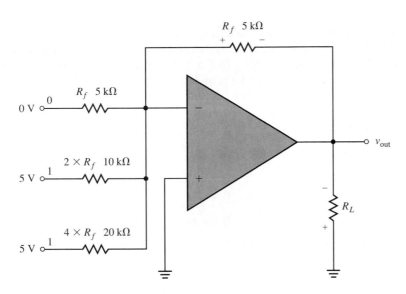

FIGURE 12.41 A summing amplifier with a binary ladder

EXAMPLE 12.26

Determine the output voltage for the circuit of Figure 12.41, when the binary input is 011.

Solution
The top resistor produces no input current. The middle resistor produces

$$\frac{5\,V}{10\,k\Omega} = 0.5\,mA$$

The bottom resistor produces

$$\frac{5\,V}{20\,k\Omega} = 0.25\,mA$$

The total current flowing through the feedback resistor is

$$0.5\,mA + 0.25\,mA = 0.75\,mA$$

The output voltage is

$$V_{out} = 0.75\,mA \times 5\,k\Omega$$
$$V_{out} = 3.75\,V$$

Because the input is on the inverting terminal, we have phase inversion. Thus, V_{out} is -3.75 V, which is the analog equivalent of the digital input. Thus the circuit of Figure 12.41 is the basis of a simple **digital-to-analog converter (DAC)**.

Determine the analog output of the circuit in Figure 12.41 for a binary input of 101.

INSTRUMENTATION APPLICATION

Figure 12.42, shows an op-amp wired in a parallel-series feedback configuration. The op-amp is acting like a current amplifier. This application shows how an op-amp is used to boost the current input to a level sufficient to make the 1-mA meter movement go full scale when the input to the circuit is at the maximum listed value of 1 μA.

The feedback ratio is

$$B = \frac{R_b}{R_f + R_b}$$

$$B = \frac{100\,\Omega}{(99.9\,k\Omega + 110\,\Omega)}$$

$$B = 0.001$$

The closed-loop current gain is

$$A_I = \frac{1}{B}$$

$$A_I = 1000$$

Thus, when the input is at the maximum listed value, 1 μA, the output current is 1 mA, and the meter deflects full scale. The meter face is then relabeled to reflect the microamp range. The two trimmer potentiometers are in the circuit so that the gain of the circuit can be set precisely to 1000.

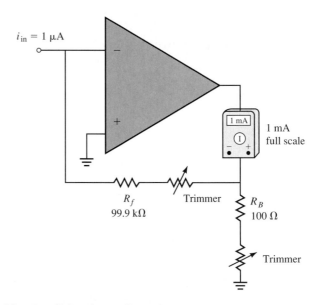

FIGURE 12.42 Parallel-series configuration

TROUBLESHOOTING

Figure 12.43 shows the schematic of an op-amp circuit using parallel-parallel feedback. The photograph of this circuit, shown in Figure 12.44, shows just how few components there are in a typical op-amp circuit.

Because there are so few components, op-amp circuits are usually easy to troubleshoot. Perhaps the greatest difficulty is in isolating a particular stage. Signal-tracing techniques are used to verify that the correct signal has reached the input of the circuit. If the signal is not present at the output, then we turn our attention to the components of the circuit.

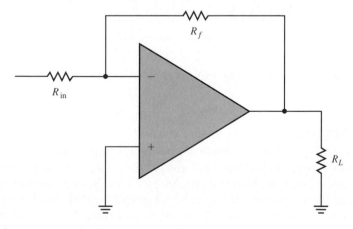

FIGURE 12.43

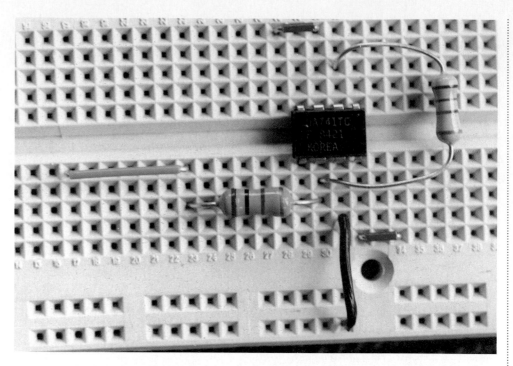

FIGURE 12.44 The operational amplifier reduces external components, making troubleshooting fairly easy. Here is a PP feedback circuit; power supplies connect to the red leads, black is connected to ground, and the input signal is applied via the blue wire. The output is taken from pin 6.

IF R_f OPENS

If R_f opens, the feedback loop is opened. The circuit operates as an open-loop. The input is amplified by A_{ol} and drives the output into positive and negative saturation; see Figure 12.45.

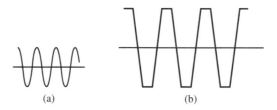

(a)　　　　　(b)

FIGURE 12.45 (a) Normal output. (b) Output with R_f open.

IF R_{in} OPENS

Obviously, if R_{in} opens, the input is no longer applied to the circuit. The output should drop to 0, but this isn't always the case. What is not so obvious is that the circuit can start to oscillate. An oscillator is a circuit that converts the DC energy of the power supply into an output signal. Such circuits will be discussed in detail in Chapter 13. For now, you should know that if R_{in} opens, the output of the circuit will be 0 or the output will oscillate. If the circuit does go onto oscillations, the output will look very much like Figure 12.45b. To determine if the problem is R_{in} or R_f, disconnect the input signal going into the circuit. If the output continues to appear as shown in Figure 12.45b, the problem is R_{in}.

IF THE OP-AMP IS DEFECTIVE

If the IC is defective, there is no way to fix it. It must be replaced. In addition, there is no way to probe inside the device, so we must check the inputs to and the outputs from the chip to decide if the IC is defective. It is best to verify that the external components are working properly. This includes checking to see if the power supply voltages are present and correct. If the power supplies and external components are working properly and the circuit still doesn't work, replace the op-amp.

CHECKING THE FEEDBACK NETWORK

Figure 12.46 shows a noninverting voltage amplifier using series-parallel feedback. In this circuit, if R_f opens, the circuit operates as an open-loop. The output will go into positive and negative saturation. If R_b opens, the circuit operation is changed to that of a voltage follower. The gain of the circuit will drop, and the output signal will be the same size as the input signal.

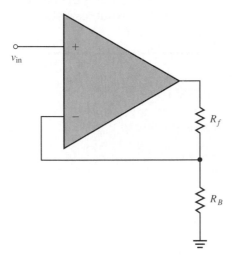

FIGURE 12.46

PRACTICE PROBLEM 12.26

Figure 12.47 shows an op-amp circuit using series-series feedback. Predict the output of the circuit should R_f open.

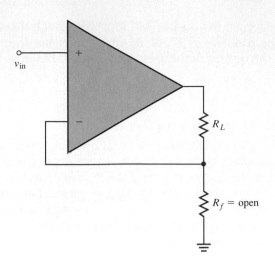

FIGURE 12.47 R_f open in a series-series feedback circuit

TROUBLESHOOTING QUESTIONS
1. If R_{in} opens in a parallel-parallel feedback arrangement, the output will always go to 0 V. True or false?
2. If R_f opens in a parallel-parallel feedback circuit, the circuit will operate _____ loop.
3. When R_b opens in an op-amp circuit wired with SP feedback, the gain of the circuit increases. True or false?
4. Checking the inside of an op-amp for a fault is of course impossible. If you suspect the op-amp is defective, what tests should be performed before replacing the device?

WHAT'S WRONG WITH THIS CIRCUIT?

For the circuit below, the expected value of $V_{out} = 11 \, V_p$, the actual value of V_{out} is

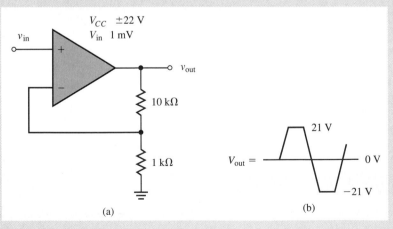

Why does this discrepancy occur?

This chapter looked at the four basic negative feedback configurations that can be used with op-amps.

The series-parallel feedback connection can be used as a noninverting voltage amplifier. It has very high input impedance, very low output impedance, and a voltage gain that is determined by the feedback ratio. It approaches the characteristics of an ideal voltage amplifier.

The parallel-parallel negative feedback connection can be used as an inverting voltage amplifier or a current-to-voltage transducer. When used as an inverting voltage amplifier, the input impedance of the circuit is equal to R_{in}, and the voltage gain is equal to R_f/R_{in}. When used as a current-to-voltage transducer, all the input current flows through R_f and produces an output voltage. Finally, this circuit has relatively low input impedance and very low output impedance.

The parallel-series negative feedback configuration is often used as a current amplifier. The input impedance of the circuit is very low, and the output impedance is very high. The current gain of the circuit is determined by the current feedback ratio, where $A_I = 1/B_I$. This configuration approaches the characteristics of a perfect current amplifier.

The series-series negative feedback configuration is most often used as a voltage-to-current transducer. When wired in this configuration, the op-amp circuit has very high input impedance and very high output impedance. To determine the output current, divide V_{in} by R_f, that is, $I_{out} = V_{in}/R_f$.

PROBLEMS

Refer to Figure 12.48 for problems 1–9.

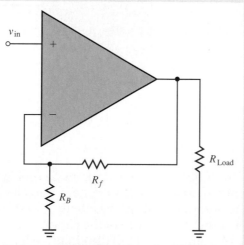

FIGURE 12.48 Series-parallel configuration. A = 50,000.

1. The input to the op-amp is 2 mV. R_f is 4.7 kΩ, R_b is 1 kΩ. Determine the output voltage.
2. If R_f is 100 kΩ and R_b is 2.2 kΩ, determine the gain of the op-amp circuit.
3. R_f is 39 kΩ and R_b is 3.9 kΩ. What is the feedback ratio?
4. If R_f is 47 kΩ and R_b is 3.3 kΩ, what is the sacrifice factor?
5. The sacrifice factor for the circuit is 1000. If the op-amp has an input impedance of 1 MΩ, determine the input impedance of the circuit.
6. The op-amp has a sacrifice factor of 2000. The op-amp's output impedance is 75 Ω. What is the output impedance of the circuit?
7. R_f is 47 kΩ and R_b is 2.7 kΩ. Determine the value of $(R_f/R_b) + 1$. Determine the feedback ratio, B. Divide B into 1 to get circuit gain. Does $(R_f/R_b) + 1$ equal $Av_{(cl)}$? Verify this for problems 2, 3, and 4.
8. The op-amp of Figure 12.48 has a v_{error} of 50 μV across an input of 1 MΩ. What does an input voltage of 2 V see as the input impedance to the circuit?
9. The op-amp of Figure 12.48 has an input voltage of 50 mV. R_f is 56 kΩ and R_b is 2.2 kΩ. Determine the exact voltage out using the relationship

$$Av_{(cl)} = \frac{A}{1 + AB}$$

Use the answer to determine the error voltage. The input impedance of the op-amp is 1 MΩ.

Refer to Figure 12.49 for problems 10–17.

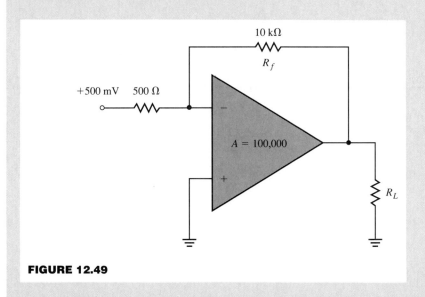

FIGURE 12.49

10. Determine if this PP feedback configuration is being used as a voltage amplifier or as a current-to-voltage transducer.
11. Determine the output voltage of the circuit.
12. Determine the input impedance of the circuit.
13. Locate the virtual ground point of the circuit. What does this mean?

14. How much of the input current flows through R_f?
15. What is the relationship between R_f and R_L?
16. What is $R_{in(miller)}$ for this circuit?
17. What is $R_{out(miller)}$ for this circuit?

Refer to Figure 12.50 for problems 18–22.

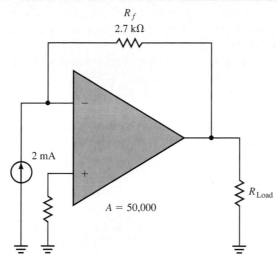

R_f
2.7 kΩ

2 mA

$A = 50,000$

R_{Load}

FIGURE 12.50

18. What feedback arrangement does this circuit use?
19. What is the function of this circuit?
20. Determine the output voltage.
21. What is the ideal input impedance?
22. What is the actual input impedance?

Refer to Figure 12.51 for problems 23–33.

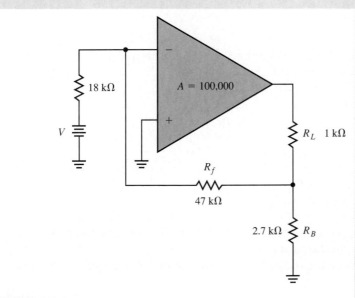

18 kΩ

$A = 100,000$

V

R_L 1 kΩ

R_f

47 kΩ

2.7 kΩ R_B

FIGURE 12.51

23. What feedback arrangement is this circuit using?
24. Determine the input current.
25. Ideally, what is the input impedance to the circuit?
26. What is meant by a floating load?
27. What is the output current?
28. What is the output load voltage?
29. What is the output voltage of the op-amp to ground?
30. What is the output impedance of the circuit?
31. What is the output current of the circuit if the load resistor is reduced to 500 Ω?
32. What is the output current of the circuit when R_b is changed to 3.9 kΩ and R_L is kept at 1 kΩ?
33. As diagrammed, what is the gain of the circuit?

Refer to Figure 12.52 for problems 34–40.

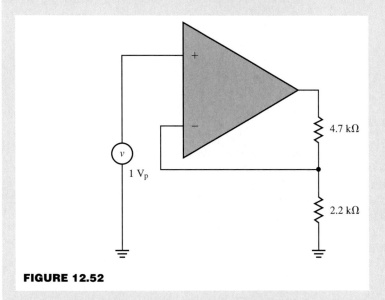

FIGURE 12.52

34. What feedback arrangement is this circuit using?
35. What is the function of this circuit?
36. Determine the output current.
37. Determine the output voltage across the load.
38. Determine the output voltage of the op-amp to ground.
39. Ideally, what is the input impedance of this circuit?
40. What is the output impedance of this circuit?
41. Figure 12.53 shows a frequency-versus-gain plot for an op-amp. Determine the f_{unity} of the op-amp.
42. An op-amp has an f_{2ol} (upper cutoff frequency) of 10 Hz. The open-loop gain of the op-amp is 20,000, and the closed-loop gain of the circuit in which it is used is 100. Determine the closed-loop upper cutoff frequency.
43. Determine the output voltage for the circuit in Figure 12.54.

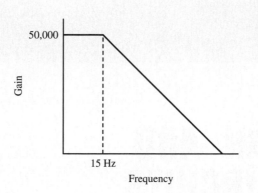

FIGURE 12.53

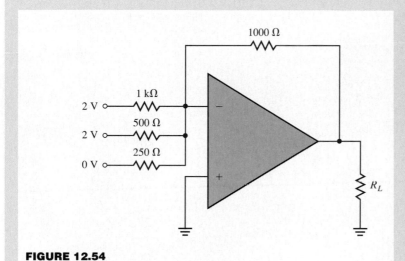

FIGURE 12.54

WAVE GENERATION AND WAVESHAPING CIRCUITS

✓ **As you read this chapter, concentrate on learning how to:**

1. Describe how oscillators use circuit noise to help create a signal
2. Determine the critical frequency of an RC network
3. Determine the amount of phase shift in a circuit
4. Determine the resonant frequency of an LC circuit
5. Determine the output frequency of a Wien-bridge oscillator
6. Determine the effects of loading on an LC oscillator
7. Determine the resonant frequency of an LC oscillator
8. Describe how differentiators work
9. Describe how integrators work
10. Describe how differentiators and integrators can be used to change the shape of an input waveform

INTRODUCTION

In this chapter we will study two types of circuits: circuits that generate signals, or oscillators, and circuits that distort the input waveform, or waveshaping circuits.

Oscillator circuits use a form of feedback to produce output signals from DC sources. In Chapter 12 we studied circuits using negative feedback. With negative feedback, we were able to change circuit parameters such as gain, bandwidth, input impedance, and output impedance. In this chapter we will study *positive* feedback. Positive feedback means that we take a portion of the output and return it to the input of the circuit, but, in this case, we cause the signal to return to the input of the circuit in phase with the original signal. Positive feedback is often undesirable because it causes the circuit to be unstable, that is, to change drastically from one level to another level and then back again. Sometimes, however, an unstable circuit is exactly what we want. Using such a circuit, we can select the frequency at which the voltage level changes and we can select a particular waveform shape.

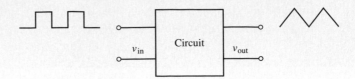

FIGURE 13.1 A waveshaping circuit

Circuits of the other type distort the input waveform. Whereas in oscillator circuits we take advantage of instability, in waveshaping circuits we take advantage of the distortion. For example, we may input a square wave into a circuit and deliberately distort that wave so that the output forms a triangular wave, as shown in Figure 13.1. Such a circuit is known as a waveshaping circuit.

13.1 GENERAL OSCILLATOR THEORY

An oscillator can be thought of as a transducer, converting constant DC voltages into a changing AC voltage.

To build an oscillator, we need a circuit with positive feedback. If the feedback signal is large enough and has the correct phase there will be an output signal, even if the only voltage driving the circuit is DC.

A basic oscillator consists of two functional blocks. In the block diagram in Figure 13.2, A represents the amplifier gain, which is used to increase the signal strength to a usable level. B represents the feedback part of the circuit, but, unlike the feedback circuits studied in Chapter 12, this feedback circuit will present a signal at the input that is in phase with the original input, thus positive feedback.

Oscillators use electrical noise to help convert the DC voltage into some form of AC. Electronic circuits, particularly solid state devices, have some circuit noise, electrical signals that are usually not desired. In most cases the signal strength is so large that it drowns out the circuit noise and we are not

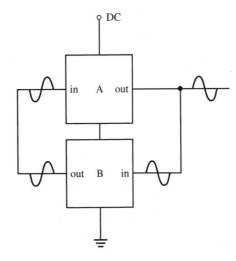

FIGURE 13.2 A sinusoidal oscillator

aware of these unwanted signals. In situations where the noise is a problem, it is necessary to incorporate a noise reduction system into the circuit.

We use noise, however, to start up an oscillator. For example, a carbon composition resistor with current flowing through it produces wideband noise. The noise voltage is produced by random motion of electrons within the device. Frequencies range from very low to upwards of a billion hertz. We normally don't hear this noise because its amplitude is so small.

The oscillator circuit selects one particular frequency out of the many available. This is done in the feedback circuit. Then the amplifier part of the circuit amplifies this particular frequency. A portion of this signal is then fed back through the feedback and frequency selection circuitry so that it is in phase with the original selection.

Because the two signals are in phase, they are additive. The second time through the amplifier, the output will get a little larger. This process is repeated until the desired amplitude is reached.

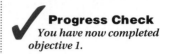

Progress Check
You have now completed objective 1.

START-UP CONDITIONS

Even an oscillator will not oscillate unless conditions are proper. The closed loop in Figure 13.2 has an overall loop gain equal to A × B. In order for oscillations to start, the closed-loop gain must be greater than 1, and the output of the feedback circuit must select one frequency to feed back to the input of the circuit, which has a 0° (or 360°) phase shift. If these conditions exist, the oscillations will build up in amplitude.

MAINTAINING OSCILLATIONS

Once oscillations have built up to a suitable output amplitude, we need to maintain the output. The closed-loop gain must decrease to unity to produce a stable output amplitude. Thus either A or B or both need to be adjusted so that the product of A × B is equal to 1.

REVIEW SECTION 13.1

1. In order to have positive feedback, the portion of the signal returned to the input of the amplifier must be _____ _____ _____ the original input.

2. Unwanted electrical signals are referred to as _____ _____.

3. What part of an oscillator circuit selects the frequency to be amplified?

4. Closed-loop gain for an oscillator can be expressed as A × B. At start-up, this product must be greater than _____ for oscillations to begin.

5. In order to maintain oscillations, the circuit must adjust the A × B loop gain to _____.

6. If an amplifier has an A of 10 and a B of 0.01, will oscillations start?

LEAD AND LAG NETWORKS

Figure 13.3 shows a simple frequency selection circuit. This circuit is a high-pass filter that is labeled as a lead network.

At very high input frequencies, the capacitor appears as a short. This is because as frequency rises, the reactance of the capacitor decreases toward zero.

At a very low frequency, the reactance of the capacitor rises. Therefore a large portion of the input signal is dropped across the plates of the capacitor, never reaching the output terminal.

At one particular frequency, a special condition arises. This frequency, known as the **critical frequency,** is the point at which the output amplitude has dropped to 70.7% of the input voltage. It is also the frequency at which the reactance of the capacitor is equal to the resistance of the resistor and in which the phase shift caused by the circuit is 45°.

Determining how the circuit of Figure 13.3 works cannot be done with simple algebra, because of the nature of the interaction of the capacitor with the resistance of the circuit. As you recall, the voltage across the plates of a capacitor is 90° out of phase with the current flowing through the capacitor. In addition, the voltage across a resistor is in phase with the current flowing through it. Therefore, the voltage across the capacitor peaks 90° after the voltage across the resistor.

It is usually possible to solve a circuit like Figure 13.3 by plotting the different voltages on the polar plane. As shown in the polar plane graph in Figure 13.4 the voltage across the circuit resistance is plotted along the X axis at 0°, indicating no phase shift.

The voltage across the capacitor is plotted down along the Y axis at 270°, indicating a phase shift of $-90°$. Although we are not discussing inductors at the moment, you can see from the polar graph that if we were dealing with inductors, the voltage across the component would be plotted up at $+90°$.

In Figure 13.5 we see a plot of V_R and V_C on the polar plane as well as a phasor, V_A, representing the applied voltage (input voltage). A **phasor** is

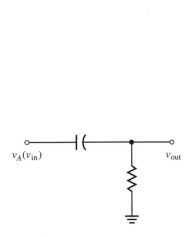

FIGURE 13.3 A lead network   **FIGURE 13.4** The polar plane

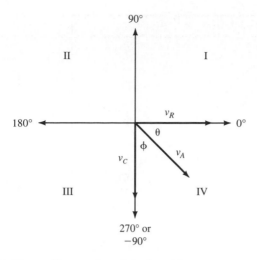

FIGURE 13.5 Phasor diagram in standard position

some force displaced over time. Voltages shifted in time are phasors. Two angles are given designations. The angle between V_R and V_A is theta (θ), and the angle between V_A and V_C is phi (ϕ). (Using Greek letters to represent these angles is standard practice.)

Figure 13.6 shows the three phasors rotated around the polar plane toward the positive 90° arm of the Y axis. V_R precedes, or leads, V_A, while V_C lags behind the other two voltages. Thus if we take the output voltage across R, as in Figure 13.3, and compare it with the input voltage, we see that V_R leads V_A.

This is not an arbitrary graphical representation. In Figure 13.7 we see the faceplate of an oscilloscope showing two different waveforms. Channel 1 is across the input to the circuit, V_A. The oscilloscope controls are set so that the input waveform is at 0° of rotation at the start of the trace. Channel 2 is across the output resistor of Figure 13.3. The output voltage is already at some positive angle, while the input voltage is at 0°. Thus V_R leads V_A in that V_R peaks earlier than V_A. In this instance, the angle between V_R and V_A is illustrated as 45°. As already mentioned, a 45° phase shift indicates that the circuit is at the critical frequency.

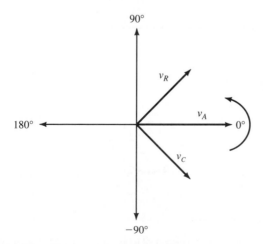

FIGURE 13.6 Rotating V_A to the reference position

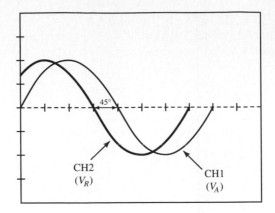

FIGURE 13.7 Oscilloscope phase measurement of a lead angle. Note that the CH2 V_R scale has been adjusted so that the two images are the same size.

Figure 13.8 shows a lag network. In this circuit arrangement the output voltage is developed across the capacitor. As we saw in Figure 13.6, this voltage lags behind the input voltage; hence the name *lag network*.

Figure 13.9 depicts the faceplate of an oscilloscope. Channel 1 is connected to the input voltage of Figure 13.8, while channel 2 is connected across the output voltage. The channel 1 trace is set up so that its 0° point is at the start of the image on the left of the screen. Channel 2 is at a negative value, indicating that it is lagging behind channel 1. You can see that the voltage sampled by channel 2 peaks at a later time than that of channel 1, producing the lagging angle.

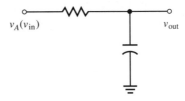

FIGURE 13.8 A lag network

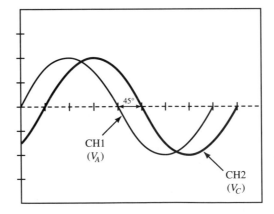

FIGURE 13.9 Oscilloscope phase measurement of a lag angle. The CH2 vertical sensitivity has been adjusted so that the two images are the same size.

DETERMINING THE CRITICAL FREQUENCY

By definition the critical frequency of an RC circuit is the frequency at which $X_C = R$. From AC electronics you may recall that $X_C = 1/(2\pi f C)$. At the critical frequency of an RC circuit, $R = X_C$, therefore R is also equal to $1/(2\pi f C)$. Rearranging this relationship, we can solve for f:

$$f = \frac{1}{2\pi RC}$$

EXAMPLE 13.1

Determine the critical frequency of the circuit in Figure 13.8 if $R = 10$ kΩ and $C = 0.1$ μF.

Solution
Using the equation above, we have:

$$f = \frac{1}{2\pi RC}$$

$$f = \frac{1}{2\pi \times 10 \text{ k}\Omega \times 0.1 \text{ } \mu\text{F}}$$

$$f = 159 \text{ Hz}$$

EXAMPLE 13.2

Verify that the capacitive reactance of the capacitor mentioned in example 13.1 is in fact 10 kΩ.

Solution

$$X_C = \frac{1}{2\pi f C}$$

At 159 Hz, we have an X_C of

$$X_C = \frac{1}{2\pi \times 159 \text{ Hz} \times 0.1 \text{ } \mu\text{F}}$$

$$X_C = 10,014 \text{ } \Omega$$

$$X_C = 10 \text{ k}\Omega$$

The discrepancy of 14 Ω is due to rounding off.

Figure 13.10 shows a frequency versus gain plot for a lag network. Above the critical frequency of the network, the amplitude drops off. Thus we are able to select a particular range of frequencies for use. This is also the frequency versus gain plot for a low-pass filter.

Figure 13.11 is the frequency versus gain plot for a lead network, which is a high-pass circuit. Frequencies above the critical frequency of the network

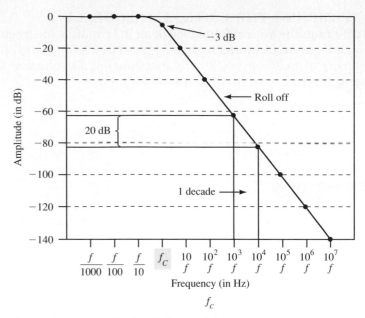

FIGURE 13.10 Frequency versus gain plot of a lag network

pass through to the output without loss of amplitude. Again, we have a way in which we can select frequencies.

We are interested not only in the critical frequencies, but also in the lead or lag angles.

If we use lead and lag circuits as part of a feedback network, we must not only select a particular frequency, we must also return the portion of the output selected to the input of the circuit with an overall phase shift of 0°; therefore we *must* know the phase angles of the circuits being used!

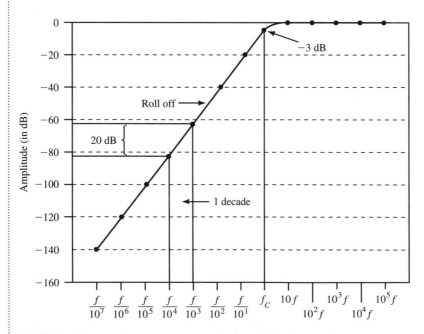

FIGURE 13.11 Frequency versus gain plot of a lead network

Determine the critical frequency for the following combinations of R and C:
(a) R = 100 kΩ; C = 1 μF
(b) R = 56 kΩ; C = 100 pF
(c) R = 4.7 kΩ; C = 3.3 μF
(d) R = 22 kΩ; C = 470 μF

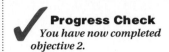

Progress Check
You have now completed objective 2.

PHASE SHIFT

Figure 13.12 shows a lag network in which the reactance of the capacitor is 3 kΩ and the resistance of the resistor is 5 kΩ. Figure 13.13 shows a graphical solution to the circuit with X_C plotted down, and R plotted out along the 0° axis.

The total opposition of the circuit, Z, is plotted in the fourth quadrant of the polar plane; thus we have a negative phase angle.

While it is possible to graph solutions to circuits such as that shown in Figure 13.12, then measure Z and theta, the scientific calculator can quickly give us more-exact answers. We will look at two ways in which this can be done.

A note about calculator keystroke sequences: Not all scientific calculators have data entered in the same way. If your model doesn't correspond to the sequence given here, refer to your calculator's instruction manual or ask your instructor for help.

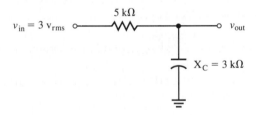

FIGURE 13.12   Lag network

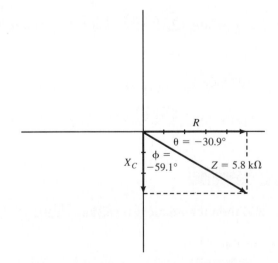

FIGURE 13.13   Graphical solution to a lag network

EXAMPLE 13.3

Determine the impedance of the circuit shown in Figure 13.12 using the trigonometric functions of a scientific calculator.

Solution

The plot of the circuit shown in Figure 13.13 can serve as a visual aid to help you remember which functions you need to use.

In addition, here is an explanation of the keystroke sequence.

First, solve for the phase angle of the RC circuit and then solve for the magnitude of the impedance.

$$\theta = \arctan \frac{X_C}{R}$$

θ represents the phase shift between R, the component across which we will take the output, and Z. We can use the trigonometric function of Tangent to find that phase angle. Observe that X_C is opposite θ and that R is adjacent, and we know their values.

Calculator keystrokes are

$$\boxed{3}\,\boxed{0}\,\boxed{0}\,\boxed{0}\,\boxed{+/-}\,\boxed{/}\,\boxed{5}\,\boxed{0}\,\boxed{0}\,\boxed{0}\,\boxed{=}$$

The display should read $\boxed{-0.6}$.

Now press the INV key and then the TAN key. Some calculators have a key called "2nd shift" rather than the INV key.

The display should now read $\boxed{-30.9}$, which is the phase angle between R and Z.

To find the magnitude of the impedance use X_C and R again. This time, however, we will apply the Pythagorean theorem.

$$Z = \sqrt{X_C{}^2 + R^2}$$

To enter the Pythagorean theorem into your calculator use the following keystrokes. First enter X_C by pressing $\boxed{3}\,\boxed{0}\,\boxed{0}\,\boxed{0}$ and then square it by pressing $\boxed{x^2}$.

The display should read $\boxed{90000000}$. Save the display in your calculator's memory. Next enter R by pressing $\boxed{5}\,\boxed{0}\,\boxed{0}\,\boxed{0}$ and then square it by pressing $\boxed{x^2}$.

The display should read $\boxed{25000000}$.

Next press the $\boxed{+}$ key and then the memory recall key $\boxed{MR}$.

The next key to press is the equal key.

The display should read $\boxed{34000000}$.

The last key to press is the square root key $\boxed{\sqrt{}}$.

The display should read $\boxed{5830.95}$.

We round off the answer to 5.8 kΩ.

EXAMPLE 13.4

Rework the solution to Figure 13.12 using the rectangular to polar function(s) of a scientific calculator.

Solution
Enter the 5 k, which represents the resistance of the resistor.

Press the *a* button (some calculators use *x*).

The display should still read $\boxed{5\,k}$

Next enter −3 k. This represents X_C.

Press the *b* button (some calculators use *y*).

The display should still read

$\boxed{-3\,k}$

Find the R→P function on your calculator. It is usually a secondary function. (Look at the print above the *a* or *b* button—a very common location.)

Activate the R→P function.

On some calculators the key sequence is 2nd, then *b*.

The display should now read

$\boxed{-5.83\,k}$

This represents the impedance of the circuit.

Press *b* a second time (do not press 2nd again).

The display now reads

$\boxed{-30.9}$

which is the phase angle of the circuit.

In the calculator sequence used in example 13.4, we entered a set of numbers known as an ordered pair. In mathematics this pair is usually given as *(x,y)* [(R, X_C) in example 13.4]. This pair determines the location of the end point of the phasor representing Z. The starting point is at the origin. The R→P function converts this information into a magnitude and an angle.

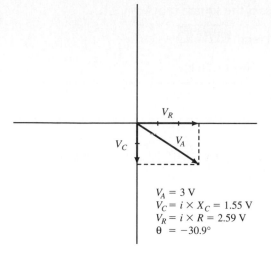

$$V_A = 3 \text{ V}$$
$$V_C = i \times X_C = 1.55 \text{ V}$$
$$V_R = i \times R = 2.59 \text{ V}$$
$$\theta = -30.9°$$

FIGURE 13.14   Equivalence of a voltage triangle and an impedance triangle

We now know the phase angle for the circuit and the impedance of the circuit. Figure 13.14 shows that there is a direct correlation between the impedance phasor diagram and the voltage phasor diagram (when analyzing series circuits).

Ohm's law is used to determine the current flowing through the circuit; see Figure 13.14. When the applied voltage is divided by Z, we get:

$$i = \frac{v}{z} = \frac{3 \text{ V}}{5.8 \text{ k}\Omega} = 517 \text{ µA}$$

This current flows through both X_C and R, producing voltage drops, V_C and V_R. These voltage drops are calculated using Ohm's law:

$$V_R = i \times R$$
$$V_R = 517 \text{ µA} \times 5 \text{ k}\Omega$$
$$V_R = 2.59 \text{ V}$$

$$V_C = i \times X_C$$
$$V_C = 517 \text{ µA} \times 3 \text{ k}\Omega$$
$$V_C = 1.55 \text{ V}$$

When these voltages are plotted in the polar plane, we get a voltage phasor diagram with the same phase angle as the impedance phasor diagram. Remember that Ohm's law tells us that voltage is proportional to resistance. As a result, we know that the larger the resistance or reactance, the larger the voltage drop. The voltage drop will also have the same phase angle as the resistance or reactance that develops it. Hence, the phasor diagrams for impedance and voltage appear to have the same shape. The size of the impedance phasors is increased by a factor of i, where i is the current of the circuit.

Determine the phase angle and the impedance of the circuit shown in Figure 13.15 using the trigonometric functions on a scientific calculator.

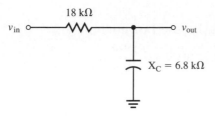

FIGURE 13.15 Lag network

PRACTICE PROBLEM 13.3

Verify the answers you obtained in practice problem 13.2 using the R→P function of a scientific calculator.

PRACTICE PROBLEM 13.4

Determine V_R and V_C in Figure 13.15 if the applied voltage is 15 V.

Figure 13.16 shows a lead network. In this circuit the capacitive reactance is 35 kΩ and the resistance is 10 kΩ. A plot of the impedance phasor is shown in Figure 13.17.

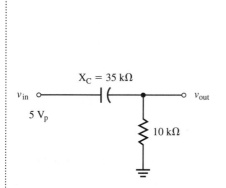

FIGURE 13.16 Lead network

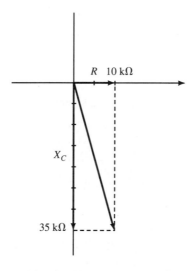

FIGURE 13.17 Graphical solution of a lead network. Each unit = 5 kΩ.

EXAMPLE 13.5

Determine the phase angle and the impedance of the circuit shown in Figure 13.16 using the trigonometric functions on a scientific calculator.

Solution
First, find the phase angle using the Tangent function.

$$\text{TAN } \theta = \frac{\text{OPP}}{\text{ADJ}}$$
$$\text{TAN } \theta = \frac{X_C}{R}$$
$$\text{TAN } \theta = \frac{-35 \text{ k}\Omega}{10 \text{ k}\Omega}$$
$$\text{TAN } \theta = -3.5$$

Enter X_C into your calculator. The display reads

$-35\,\text{k}$

Divide this number by R, 10 kΩ. Your display should read

-3.5

Press (INV), (TAN), to obtain

$-74.1°$

This is the phase angle between R and Z, with R at the 0° axis.
　　Now we will find Z using trigonometry rather than the Pythagorean theorem.

Press the (SIN) button. The display now reads

-0.96

Invert this by using the $(1/x)$ button to obtain

-1.04

and multiply by X_C, 35 kΩ, to get

$36.4\,\text{k}$

That is, Z = 36.4 kΩ.
　　Since R leads Z, V_R leads V_A and the phase angle that would be measured would be

$-74.1°$

EXAMPLE 13.6

Verify the answers obtained in example 13.5 using the R→P function of a scientific calculator.

Solution

Enter R:

$\boxed{10\,\text{k}}$

Press *a*. The display should still read:

$\boxed{10\,\text{k}}$

Enter X_C:

$\boxed{-35\,\text{k}}$

Press *b*. The display should still read:

$\boxed{-35\,\text{k}}$

Activate the R→P function (2nd, *b*).

The display should now read:

$\boxed{36.4\,\text{k}}$

which is Z.

Press *b* again:

$\boxed{-74.1°}$

which is the phase angle when R is plotted along the 0° part of the *x* axis. On an oscilloscope, V_A represents the position of Z and V_R represents the position of R. Since the output is taken across R, the phase angle is leading and we measure:

$\boxed{-74}$

PRACTICE PROBLEM 13.5

Determine the impedance of the lead network shown in Figure 13.18. Determine the phase angle. Use the trigonometric functions on your calculator.

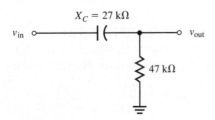

$X_C = 27\,\text{k}\Omega$

v_{in} v_{out}

$47\,\text{k}\Omega$

FIGURE 13.18   **Lead network**

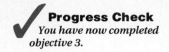

PRACTICE PROBLEM 13.6

Verify the answers of practice problem 13.5 using the R→P function on your calculator.

LEAD-LAG COMBINATIONS

Lead and lag circuits can be connected together as shown in Figure 13.19. In this case, the series string of R_1 and C_1 forms a lead network, while the parallel string of R_2 and C_2 forms a lag network. This circuit is a complex circuit consisting of Z_1 and Z_2. The boldface type indicates that $\mathbf{Z_1}$ and $\mathbf{Z_2}$ are phasors. These quantities cannot be added algebraically, but must be added using j operators (complex numbers). See Appendix A.

If $\mathbf{Z_1}$ produces a lead angle at node 1, this angle can be minimized or eliminated altogether by the lag angle produced by $\mathbf{Z_2}$. This is important in oscillator operation because we use lead and lag angles to select a particular critical frequency to feed back to the input of the amplifier and eliminate any phase shift produced by these networks. This keeps the phase at N1 at 0°, which is necessary to obtain positive feedback.

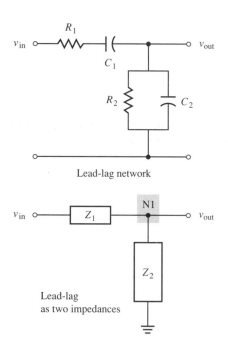

FIGURE 13.19   **Lead-lag networks**

LC FILTER CIRCUITS: PARALLEL RESONANCE

Figure 13.20 shows a parallel LC network in shunt to the signal path. This LC circuit has a resonant frequency (f_r) that can be determined by

$$f_r = \frac{1}{2\pi\sqrt{LC}}$$

In addition, $X_L = X_C$ at the resonant frequency. At other frequencies, say very low frequencies, the inductor appears as a low impedance to signals applied to the circuit, and the voltage is shunted to ground through the inductor.

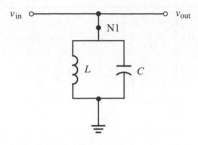

FIGURE 13.20   Resonant parallel LC circuit

At very high frequencies the capacitor appears as a short, shunting the high-frequency voltages to ground. The resonant frequency "sees" a very high impedance to ground. With ideal components, this impedance to ground is infinite.

In Figure 13.21 $X_L = X_C$ at resonance and the LC network is a parallel configuration. The two branches have the same amount of current flowing through them; however, current through a capacitor leads the voltage across its plates by a phase angle of 90°. Also, the current through an inductor lags the voltage across it by a −90° angle. If the currents I_c and I_l are equal and opposite, there is no current flow from N1 *through* the LC circuit to ground. Thus, to the input voltage the parallel LC circuit at resonance is an open loop.

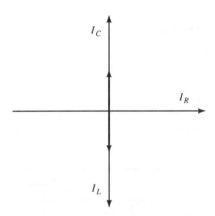

FIGURE 13.21   Effects of parallel resonance on circuit current

Consequently, any voltage at the resonant frequency reaches the output with no attenuation because no path leads to ground except through the load; see Figure 13.22. Other frequencies lose signal strength, which means we have a frequency selection circuit that can be used as part of a feedback network in an oscillator.

This circuit is sometimes referred to as a *bandpass filter*. Only the resonant frequency and those frequencies near resonance reach the output with significant amplitude. Very low frequencies and very high frequencies are shunted to ground as already explained.

In Figure 13.23, the LC circuit is placed in series with the input signal. As we know, the resonant frequency causes a parallel LC circuit to appear as if it

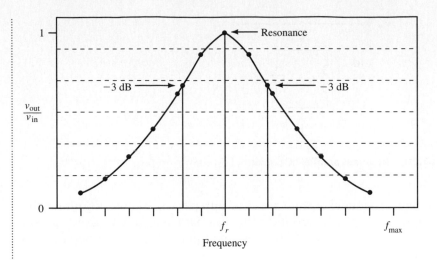

FIGURE 13.22   The relationship between V_{out}/V_{in} and frequency

were an open loop. Low frequencies reach the output by passing through the inductor. Very high frequencies reach the output through the capacitor. Frequencies near resonance are blocked. This type of arrangement is called a **bandstop filter,** or sometimes a **notch filter.** A plot of the relationship between $\dfrac{V_{out}}{V_{in}}$ and frequency for a bandstop filter circuit is shown in Figure 13.24.

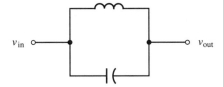

FIGURE 13.23   Bandstop filter circuit

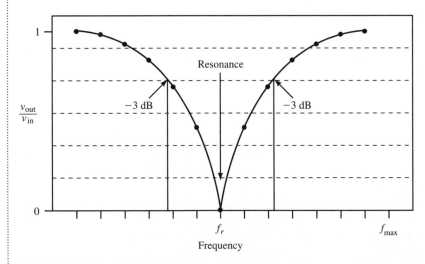

FIGURE 13.24   The relationship between V_{out}/V_{in} and frequency for a bandstop filter circuit

EXAMPLE 13.7

Let L = 1000 μH and C = 1000 pF in Figure 13.20. Determine the resonant frequency of the circuit.

Solution

$$f_r = \frac{1}{2\pi\sqrt{LC}}$$

First, $L \times C = (1000 \times 10^{-6}) \times (1000 \times 10^{-12}) = 1 \times 10^{-12}$
The square root of 1×10^{-12} is

$$\sqrt{1 \times 10^{-12}} = 1 \times 10^{-6}$$

Multiply by 2π, to get

$$2\pi (1 \times 10^{-6}) = 6.28 \times 10^{-6}$$

Invert this to get

$$f_r = 159.2 \text{ kHz}$$

which we would round to 159 kHz.

Determine the resonant frequency of a circuit in which L = 100 mH and C = 1 μF.

✓ Progress Check
You have now completed objective 4.

REVIEW SECTION 13.2
1. What is the formula used to determine the critical frequency of an RC circuit?
2. Explain what is meant by the phrase "a high-pass filter."
3. A Bode plot graphs _____ versus output _____ .
4. List three conditions that arise when an RC circuit's critical frequency is applied to it.
5. What is meant by the phrase "lag network"?
6. Why would a circuit use a lead network in combination with a lag network?
7. A bandpass circuit can be built by placing a parallel LC circuit in _____ with the signal path.
8. What is the theoretical impedance of a parallel LC circuit at resonance?
9. Notch filter is another term used for a _____ filter.

13.3 WIEN-BRIDGE OSCILLATOR

The first type of oscillator circuit we will study is the Wien-bridge oscillator. The Wien-bridge oscillator circuit is a popular choice for oscillators that produce frequencies lower than 1 MHz. The amplifier section is built using an operational amplifier. There are two feedback networks in Figure 13.25. First is the negative feedback network, consisting of the resistive voltage divider, R, and 2R, which sets the gain (A) of the amplifier. Second is the circuitry providing the positive feedback, consisting of a lead-lag network. The series string consisting of R_1 and C_1 forms a lead network. The parallel combination of R_2 and C_2 forms a lag network. In addition, $R_1 = R_2$ and $C_1 = C_2$. This means that the critical frequency of the lead network will match the critical frequency of the lag network.

We have already discussed the phase angle of 45° associated with critical frequency. The combination of R_1 and C_1 will produce a leading 45° phase shift, while the combination of R_2 and C_2 produces a lagging 45° phase shift. The net result of these two phase shifts is 0°.

The node (N1) between the lead and lag networks in Figure 13.25 is connected to the noninverting input of the op-amp. Thus, the critical frequency of the lead and lag networks is applied to the noninverting input of the op-amp with no phase shift. The result is positive feedback.

A mathematical analysis of the Wien-bridge oscillator is in Appendix A. The following needs to be remembered about Wien-bridge oscillators:

1. The critical frequency of a lead-lag network in which $R_1 = R_2$ and $C_1 = C_2$ is

$$f_c = \frac{1}{2\pi RC}$$

In the circuit in Figure 13.25

$$f_c = \frac{1}{2\pi \times 4.7 \text{ k}\Omega \times 0.01 \text{ μF}}$$
$$f_c = 3.39 \text{ kHz}$$

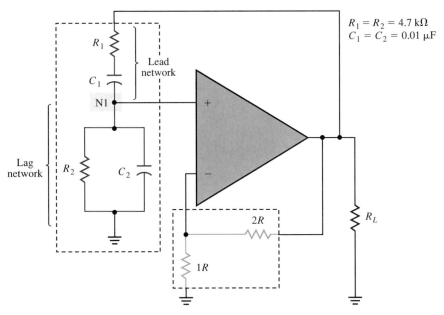

$R_1 = R_2 = 4.7 \text{ k}\Omega$
$C_1 = C_2 = 0.01 \text{ μF}$

Wien-bridge has two parts:
in blue, voltage divider; in red, lead-lag network

FIGURE 13.25   **Wien-bridge oscillator**

2. This type of lead-lag network will reduce the feedback voltage by 1/3 at the critical frequency. If V_{out} is 30 V_{p-p}, then the N1 voltage will be 10 V_{p-p}.

3. The phase shift at N1, which is between the lead and lag networks, is 0°.

4. In order to sustain oscillations, the negative feedback must provide a gain of 3. The amplifier is configured as a noninverting amplifier. Also R_f is twice the resistance of R_i. The gain formula is:

$$A = \frac{R_f}{R_i} + 1$$
$$A = \frac{2R}{1R} + 1$$
$$A = 3$$

Thus, the closed-loop gain, AB = 1/3 × 3 = 1.

Remember that in order for the circuit to produce an output signal at the critical frequency, the closed-loop gain, AB, must be greater than unity at start-up. This is why it is important to know that B = ⅓, which is the feedback ratio of the lead-lag network. To achieve an AB > 1, A must be greater than 3 at start-up. Here are a few ways in which this can be done.

1. Use a tungston lamp in place of R in the negative feedback loop. At start-up, the filament is cold and has very low resistance. This means that the A gain of the op-amp will be much larger than 3, causing oscillations to start. As the filament of the lamp heats, its resistance will rise until it reaches R. At that time, the A gain of the op-amp will be 3. Now AB will be 1, and oscillations will be sustained.

2. Replace R in the negative feedback loop with a thermistor (a temperature-sensitive resistor). The idea is the same as that given for the bulb. The resistance of R starts out low while the device is cold. The resistance of the thermistor will rise as it self-heats, therefore we want a thermistor with a positive temperature coefficient. That is, as heat increases, resistance in-

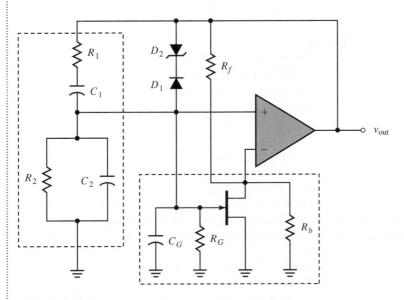

FIGURE 13.26 Wien-bridge oscillator with JFET control

creases. Eventually the thermistor will reach half of the 2R value and the A gain of the op-amp will settle at 3.

3. Use a JFET as an electrically variable resistance. In Figure 13.26 the JFET is self-biased at start-up and provides a low-impedance path to ground for R_f. This ensures that $A > 3$; thus oscillations will start. As the signal builds, it eventually reaches a level where the zener diode breaks down and conducts. This will cause C_G to charge with a negative voltage at the upper plate with respect to ground. This moves the gate voltage closer to $V_{GS(off)}$ and forces more current through R_b, causing A to settle at 3. This is enough to sustain oscillations. Diode D_1 is in the circuit to prevent positive voltages from reaching the gate of the JFET, thus protecting the JFET from damage.

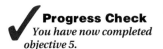
Progress Check
You have now completed objective 5.

PRACTICE PROBLEM 13.8

Determine the critical frequency of the circuit in Figure 13.26 if $R_1 = R_2 = 10$ kΩ. $C_1 = C_2 = 100$ pF.

REVIEW SECTION 13.3
1. What is a generally accepted upper frequency limit of an op-amp-driven Wien-bridge oscillator?
2. What is the attenuation of the lead-lag network in a Wien-bridge oscillator operating at the critical frequency?
3. The values of R in the lead network and R in the lag network of a Wien-bridge oscillator will be the same. True or false?
4. List three ways in which the closed-loop gain of a Wien-bridge oscillator can be reduced to unity after start-up.

13.4 TWIN-T OSCILLATOR

Figure 13.27 shows another type of lead-lag network, the twin-T filter. This particular filter arrangement is a bandstop, or notch, filter. Low frequencies reach the output through path 1 as indicated by the arrows. These low frequencies are not shunted to ground by the 2C capacitor because of its relatively high reactance to low frequencies.

High frequencies reach the output through path 2, again indicated by an arrow in Figure 13.27a. The capacitors in path 2 have very low reactances for high frequencies, so R/2 has a relatively high impedance to ground. Thus high frequencies are not shunted to ground, instead reaching the output.

Only at frequencies near the critical frequency of the lead-lag combination is there substantial opposition to current flow between the input and output terminals.

The critical frequency of the twin-T combination is approximately

$$f_c = \frac{1}{2\pi RC}$$

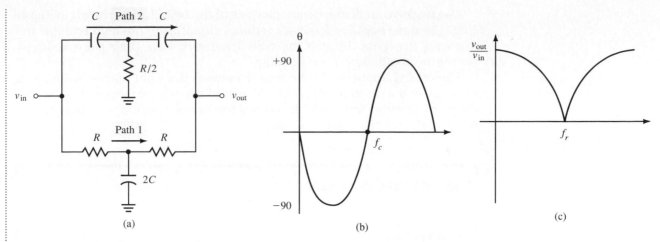

FIGURE 13.27   Twin-T notch filter

when the capacitor in the bottom T is twice the value chosen for the capacitors in the top T, and the resistor in the top T is half the value chosen for the resistors in the bottom T. Even with these qualifications, the R/2 component often needs to be a potentiometer in order to adjust the circuit to the theoretical critical frequency.

The phase angle response of a twin-T network is shown in Figure 13.27b. At the critical frequency, the phase shift reduces to 0°. At very high frequencies, the phase shift of the circuit approaches +90°. At very low frequencies the phase angle begins to lag, approaching a −90° phase angle.

The output-to-input ratio is plotted in Figure 13.27c. Notice that at either very high frequencies or very low frequencies, the output is equal to the input. As the input frequency approaches the critical frequency of the filter, the amplitude of the output drops off until the output is essentially zero at the critical frequency of the circuit.

Figure 13.28 shows how a twin-T network can be used in an oscillator circuit. In this circuit the positive feedback is applied to the noninverting input through a resistive voltage divider.

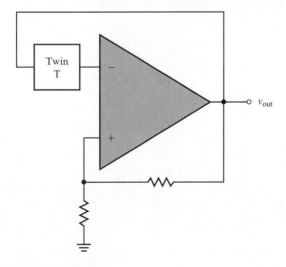

FIGURE 13.28   Twin-T oscillator

The negative feedback circuit consists of the twin-T filter shown in Figure 13.27. Because negative feedback reduces circuit gain, the high and low frequencies that pass through the twin-T network have their gains reduced, preventing oscillations from building.

The critical frequency of the twin-T network does not reach the inverting terminal of the op-amp. Because the gain of the critical frequency is not reduced by negative feedback, the critical frequency builds signal strength as a result of the positive feedback path.

REVIEW SECTION 13.4
1. The twin-T filter is an example of a _____ filter.
2. The output of a twin-T filter is essentially _____ at the critical frequency.
3. In a twin-T oscillator, the twin-T filter is part of the positive feedback network. True or false?
4. At very high frequencies (well above the critical frequency of the filter), the output of a twin-T filter will be equal to _____ _____.

13.5 LC OSCILLATORS

In this section we will look at four types of LC oscillators: the Colpitts oscillator, the Clapp oscillator, the Hartley oscillator, and the Armstrong oscillator. We will discuss the Colpitts oscillator in detail, then briefly discuss the other three, pointing out differences in circuit operation.

THE COLPITTS OSCILLATOR
Figure 13.29 shows a very common form of oscillator circuit known as the Colpitts oscillator. This circuit can be recognized by the center-tapped capacitive voltage divider in the parallel LC circuit.

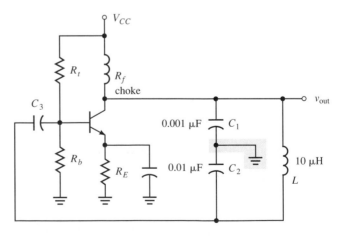

FIGURE 13.29   Colpitts oscillator

When you see an oscillator with a discrete active component, the circuits have been designed to oscillate at very high frequencies. This eliminates op-amps, which typically have an f_{unity} of approximately 1 MHz. A circuit like that shown in Figure 13.29 might be designed to oscillate in the hundreds of millions of hertz.

The voltage divider provides the bias needed to operate the transistor. This bias sets up an emitter current, which in turn establishes r'_e. The voltage gain, A_v, of the transistor is set high enough to ensure that oscillations will start.

Frequency of Oscillations

The feedback network is through the frequency selection circuit, the parallel LC oscillator. This LC network has a resonant frequency equal to

$$f_r = \frac{1}{2\pi\sqrt{LC}}$$

where $C = (C_1 \times C_2)/(C_1 + C_2)$.

Note that this is the product over the sum: C_1 and C_2 are in series.

EXAMPLE 13.8

Determine the frequency of oscillations for the circuit in Figure 13.29. Use the values for C_1, C_2, and L given in the schematic.

Solution

First, determine the overall capacitance of the parallel LC circuit using the product-over-the-sum method:

$$C_T = \frac{C_1 \times C_2}{C_1 + C_2}$$
$$C_T = 909 \text{ pF}$$

Now determine the resonant frequency.

$$f_r = \frac{1}{2\pi\sqrt{LC}}$$
$$f_r = 1.67 \text{ MHz}$$

PRACTICE PROBLEM 13.9

Change the value of C_1 in Figure 13.29 to 0.005 μF. Recalculate the resonant frequency.

Positive Feedback

The output voltage is developed across C_1. The feedback voltage is developed across C_2. The ground tap between C_1 and C_2 forces the AC voltage across C_2 to be 180° out of phase with the output voltage.

Since the common-emitter configuration used in the amplifier has already produced a 180° phase shift, the overall phase shift of the circuit is 0° (180° + 180° = 360° = 0°). This produces positive feedback at the input of the circuit.

Feedback Ratio

The feedback ratio (B) can be determined by taking the value of C_1 and dividing it by C_2:

$$B = \frac{C_1}{C_2}$$

Here's how we arrived at that relationship:

$$B = \frac{V_f}{V_{out}}$$

$$B = \frac{i \times X_{C_2}}{i \times X_{C_1}}$$

Because the two capacitors are in series, the current cancels out, leaving

$$B = \frac{X_{C_2}}{X_{C_1}}$$

$$B = \frac{1/2\pi f C_2}{1/2\pi f C_1}$$

Invert and multiply:

$$B = \frac{1}{2\pi f C_2} \times \frac{2\pi f C_1}{1}$$

The $2\pi f$ terms cancel out, leaving

$$B = \frac{C_1}{C_2}$$

EXAMPLE 13.9

Determine the feedback ratio for the oscillator circuit of Figure 13.29. Use the values given in the schematic.

$$B = \frac{C_1}{C_2}$$

$$B = \frac{0.001\,\mu F}{0.01\,\mu F}$$

$$B = 0.1$$

Because B is 0.1, A, which is amplifier gain, must be greater than 10 for oscillations to begin.

PRACTICE PROBLEM 13.10

Let $C_1 = 0.002\ \mu F$ and $C_2 = 0.033\ \mu F$ in a Colpitts oscillator. Determine the feedback ratio. How large must A be to start oscillations?

Forward Gain

Now that we know the feedback ratio for the oscillator in Figure 13.29, we can determine whether the oscillator's closed-loop gain, AB, is greater than unity at start-up if we know the gain of the amplifier, A.

In Chapter 6, in the section on class C amplifiers, we studied how to determine the effective AC resistance of a parallel resonant tank circuit. Start with this assumption: The RF choke connected between the DC power supply and the collector prevents high frequencies from reaching the DC power supply. Thus, we treat it as an AC open loop.

The effective AC resistance seen by the collector becomes the effective AC resistance of the tank circuit. This can be determined if we know the Q of the coil.

EXAMPLE 13.10

The Q of the coil in Figure 13.29 is 25. Determine the effective AC resistance of the tank circuit.

Solution
From example 13.8, we know that the resonant frequency is 1.67 MHz. From this, we can determine X_L:

$$X_L = 2\pi f L$$
$$X_L = 104 \ \Omega$$

The effective AC resistance of the entire tank circuit is found by multiplying the Q of the coil by the inductive reactance:

$$r_{AC} = Q \times X_L$$
$$r_{AC} = 25 \times 104 \ \Omega$$
$$r_{AC} = 2.6 \ k\Omega$$

We can now determine if oscillations will start. Remember that $A \times B$ must be greater than 1.

Let $r'_e = 25 \ \Omega$.

$$A = \frac{r_{AC}}{r'_e}$$
$$A = \frac{2.6 \ k\Omega}{25 \ \Omega}$$
$$A = 104$$

In example 13.9, we found that $B = 0.1$.

$$A \times B > 1$$
$$104 \times 0.1 = 10.4$$

Oscillations will start.

Loading Effects
As drawn in Figure 13.29, the only load connected to the oscillator is the tank circuit. (Remember we're treating the RF choke as an open.) If a load resistance is placed between the output terminal and ground, it is in parallel with the tank circuit. Several things would change.

Because the load resistance is in parallel with the tank, it lowers the AC resistance seen by the collector; see Figure 13.30.

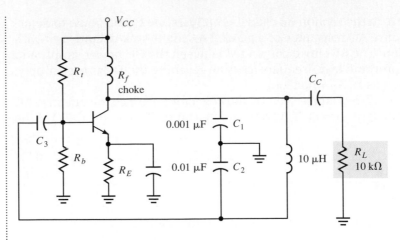

FIGURE 13.30 Loaded Colpitts oscillator

EXAMPLE 13.11

Determine the AC resistance seen by the collector of the transistor in Figure 13.30.

Solution

From example 13.10 we know that the effective AC resistance of the tank circuit is 2.6 kΩ. So

$$r_{AC(loaded)} = 2.6 \text{ k}\Omega \parallel 10 \text{ k}\Omega = 2.1 \text{ k}\Omega$$

Now A becomes

$$A = \frac{r_{AC(loaded)}}{r'_e}$$

$$A = \frac{2.1 \text{ k}\Omega}{25 \text{ }\Omega}$$

$$A = 84$$

The product of $A \times B = 84 \times 0.1 = 8.4$.
Therefore oscillations will still start.

The Q of the circuit also changes with a load resistance in parallel with the tank circuit.

$$Q_{coil} = \frac{X_L}{R_w}$$

where R_w is the DC winding resistance of the coil. In the analysis of the circuits in Figures 13.29 and 13.30, we used a $Q_{coil} = 25$. With no resistive load, the Q of the coil is equal to the Q of the circuit.

In the case of Figure 13.30, the resistive load reduced the effective AC resistance to 2.1 kΩ. This reduces the Q of the circuit:

$$Q_{cir} = \frac{r_{AC(loaded)}}{X_L}$$

$$Q_{cir} = \frac{2.1 \text{ k}\Omega}{104 \text{ }\Omega}$$

$$Q_{cir} = 20$$

What are the implications of reducing Q? If Q_{cir} is less than 10, the resonant frequency we have predicted will be higher than the actual resonant frequency.

The reduction in the actual resonant frequency is determined by the expression

$$\sqrt{Q^2/(Q^2+1)}$$

Example 13.12 shows how to use this relationship.

EXAMPLE 13.12

In example 13.8, we determined that the resonant frequency of the circuit in Figure 13.29, in which we had no resistive load, was 1.67 MHz. In our discussion of the circuit in Figure 13.30, we determined that placing an AC load across the tank of Figure 13.29 reduces the circuit Q from 25 to 20. Apply the above expression to see how this lower Q affects the resonant frequency.

Solution

$$\sqrt{Q^2/(Q^2+1)} = \sqrt{20^2/(20^2+1)}$$
$$\sqrt{400/401} = \sqrt{0.9975} = 0.9987$$

We multiply this result by f_r:
$$f_r = 0.9987 \times 1.67 \text{ MHz} = 1.667 \text{ MHz}$$

As you can see, lowering Q_{cir} from 25 to 20 had little effect on the resonant frequency of the circuit.

EXAMPLE 13.13

Determine the effect on the resonant frequency of the circuit in Figure 13.30 if the resistive load decreases the circuit Q to 5.

Solution
We start with the above expresssion and substitute values:

$$\sqrt{Q^2/(Q^2+1)} = \sqrt{25/(25+1)}$$
$$\sqrt{25/26} = \sqrt{0.961538} = 0.980$$

Now multiply by f_r to determine the new resonant frequency.
$$0.980 \times 1.67 \text{ MHz} = 1.63 \text{ MHz}$$

This is 32.4 kHz lower than ideal (with no resistor across the tank circuit). In many applications, a frequency shift of 32 kHz would cause our equipment to malfunction.

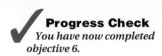
The 10-kΩ load in Figure 13.30 has been replaced with a 2.2-kΩ load; r'_e remains 25 Ω. First, determine the new Q of the circuit. Next, determine if oscillations will start. Finally, use the expression $\sqrt{Q^2/(Q^2 + 1)}$, to see how the 2.2-kΩ load changes the resonant frequency of the circuit.

THE CLAPP OSCILLATOR

The Clapp oscillator is a variation of the Colpitts. Figure 13.31 is a schematic diagram of a Clapp oscillator. Notice that there is an additional capacitor in the tank circuit. This extra capacitor in the tank circuit minimizes the effects stray circuit capacitance and capacitances inside the transistor have on C_1 and C_2. In most cases this third capacitor (C_3) is much smaller than either C_1 or C_2. The total capacitance of the tank (C_T) is now

$$\frac{1}{C_T} = \frac{1}{C_1} + \frac{1}{C_2} + \frac{1}{C_3}$$

The resonant frequency of the Clapp oscillator is determined by

$$f_r = \frac{1}{2\pi\sqrt{LC_T}}$$

In terms of the feedback ratio, the gain of the amplifier, and loading effects, the Clapp works like the Colpitts oscillator.

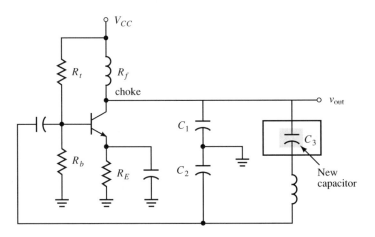

FIGURE 13.31   **Clapp oscillator**

THE HARTLEY OSCILLATOR

Figure 13.32 is the schematic of a Hartley oscillator. This oscillator uses a center-tapped inductor pair to set up the feedback circuit. As with the other LC oscillators discussed, the tank circuit selects the frequency of oscillations. In a Hartley oscillator, the total inductance (L_T) of the tank is

$$L_T = L_1 + L_2$$

The resonant frequency of the oscillator is determined by

$$f_r = \frac{1}{2\pi\sqrt{L_T C}}$$

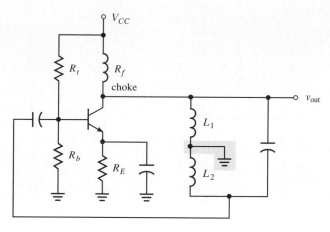

FIGURE 13.32   Hartley oscillator

The output voltage is developed across L_1, and the feedback voltage is developed across L_2. Thus the feedback ratio is

$$B = \frac{v_f}{v_{out}} = \frac{i \times X_{L_2}}{i \times X_{L_1}}$$

Since the inductors are in series with each other, i cancels out, leaving:

$$B = \frac{X_{L_2}}{X_{L2}}$$

$$B = \frac{2\pi f L_2}{2\pi f L_1}$$

The $2\pi f$ cancels out, leaving:

$$B = \frac{L_2}{L_1}$$

The gain of the amplifier and the effects on loading are determined by the same method used for the Colpitts oscillator.

THE ARMSTRONG OSCILLATOR

Figure 13.33 is a schematic of an Armstrong oscillator. The resonant frequency of the oscillator is determined by C_1 and L_1, using

$$f_r = \frac{1}{2\pi\sqrt{L_1 C_1}}$$

The positive feedback is developed across the secondary of the transformer, T_1. The transformer is wound so that the voltage across L_2 is 180° out of phase with the voltage across L_1. Since the voltage across L_1 is already 180° out of phase with the input, the overall phase shift is 0°, producing positive feedback.

Finally, the feedback ratio for an Armstrong oscillator is approximately

$$B = \frac{M}{L}$$

where M is the mutual inductance of L_1 and L_2, and L is the value of L_1.

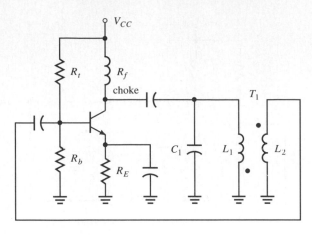

FIGURE 13.33   Armstrong oscillator

REVIEW SECTION 13.5

1. The Colpitts oscillator uses a center-tapped _____ pair to provide feedback.
2. The feedback ratio of a Colpitt's oscillator is equal to C_1/C_2. True or false?
3. Give the formula that can be used to determine the frequency of oscillations in a Colpitts oscillator.
4. Explain how positive feedback is developed in a Colpitts oscillator.
5. Loading the tank circuit in a Colpitts oscillator will never affect the resonant frequency of oscillations. True or false?
6. The three capacitors inside the tank circuit of a Clapp oscillator are in (series/parallel) with each other.
7. The Hartley oscillator uses a center-tapped _____ pair to provide feedback.
8. The feedback ratio of a Hartley oscillator is equal to L_1/L_2. True or false?
9. The Armstrong oscillator uses a _____ to provide feedback.
10. What is the advantage of using a crystal-controlled oscillator over an LC or RC oscillator?

Progress Check
You have now completed objective 7.

13.6   RELAXATION OSCILLATORS

Figure 13.34 is the schematic of a relaxation oscillator. In this type of circuit the frequency of oscillations depends on the rate at which the capacitor charges and discharges.

Positive feedback is applied to the noninverting terminal of the operational amplifier through a resistive voltage divider, $R_f + R_b$. When first powered up, the V_{out} of the op-amp usually goes to $+V_{sat}$ or $-V_{sat}$.

Let's arbitrarily stipulate that V_{out} of the circuit in Figure 13.34 is $+14$ V. (We could have just as easily chosen -14 V. The analysis that follows would be the same.)

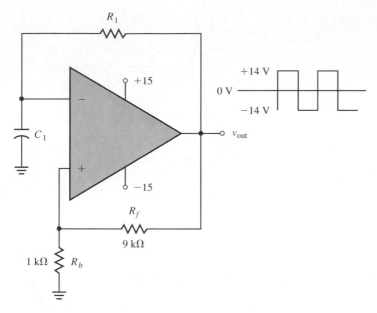

FIGURE 13.34 Relaxation oscillator

The resistive voltage divider, $R_f + R_b$, provides positive feedback from V_{out} and divides it down:

$$B = \frac{R_b}{R_b + R_f}$$

$$B = \frac{1\ k\Omega}{1\ k\Omega + 9\ k\Omega}$$

$$B = 0.1$$

This means that the feedback voltage applied to the positive terminal of the op-amp is

$$v_f = B \times V_{out}$$

$$v_f = 1.4\ V$$

We refer to this 1.4 V at the positive input of the op-amp as the trip point. The **trip point** is the voltage that the capacitor must exceed before the output of the op-amp can change levels.

This is what is happening along the positive feedback loop. In the negative feedback loop, there are two components, R_1 and C_1. These two components are also driven by the V_{out}, which we have stipulated is +14 V. This voltage causes current to flow through R_1, charging the capacitor C_1.

C_1 will continue to charge toward V_{out} until it exceeds the trip point, which is +1.4 V. When the inverting input is more positive than the noninverting input, the op-amp will drive V_{out} from $+V_{sat}$ to $-V_{sat}$. The reason is that the op-amp is operating at its open-loop gain of several thousand. At that very high gain, only a few millivolts of difference at the input will cause V_{out} to drive to $-V_{sat}$.

When V_{out} swings to -14 V, the trip point voltage changes from +1.4 V to -1.4 V.

Because capacitor C_1 still has a positive voltage on it and the noninverting input is more negative than the inverting input, V_{out} stays at -14 V.

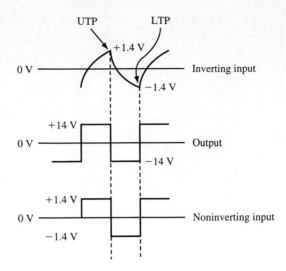

FIGURE 13.35   Relaxation oscillator waveforms

The circuit will not stabilize in this new condition. Capacitor C_1 will start to charge toward the negative polarity because R_1 is now connected to a -14 V. This happens by first discharging the positive $+1.4$ V, then charging to the new trip point voltage of -1.4 V.

So we have a circuit that oscillates back and forth, with the output swinging between $+14$ V and -14 V, while the capacitor charges and discharges between $+1.4$ V and -1.4 V.

The waveforms generated by a relaxation oscillator are drawn in Figure 13.35. The capacitor swings between $+1.4$ V, called the upper trip point (UTP) and -1.4 V, called the lower trip point (LTP). This change in voltage is exponential. Because of this exponential change in voltage across the plates of the capacitor, we use the equation below to figure out the frequency of oscillations.

$$T = 2RC \ln \frac{1+B}{1-B}$$

where T = the period of the output signal
   R = R_1
   C = C_1
   B = the feedback ratio

EXAMPLE 13.14

Determine the frequency of oscillations for the relaxation oscillator in Figure 13.34 if $R_1 = 2.2 \text{ k}\Omega$ and $C_1 = 1.0 \text{ }\mu\text{F}$.

Solution
From the discussion of trip points, we know that the feedback ratio is 0.1.

$$\frac{1+B}{1-B} = \frac{1.1}{0.9} = 1.222$$

With the 1.222 showing in the calculator's display, activate the ln (base e log) function to obtain:

0.2006

We must multiply this answer by 2RC.

$$2RC = 2 \times 2.2 \text{ k}\Omega \times 1.0 \text{ }\mu\text{F}$$
$$2RC = 4.4 \text{ ms}$$

Finally,

$$T = 0.2006 \times 4.4 \text{ ms}$$
$$T = 883 \text{ }\mu\text{s}$$

Invert time to get frequency:

$$f = \frac{1}{883 \text{ }\mu\text{s}}$$
$$f = 1.1 \text{ kHz}$$

PRACTICE PROBLEM 13.12

In a relaxation oscillator similar to that of Figure 13.34, the feedback ratio is 0.25. $R_1 = 2.7$ kΩ, and $C_1 = 0.1$ μF. Determine the frequency of oscillations.

CRYSTAL-CONTROLLED OSCILLATORS

Some electronic systems, especially in communications, require a very stable critical frequency. The critical frequency of LC and RC oscillators can be changed by one of the following:

1. Loading. Can change the Q of the tank.

2. Replacing the active device. An active device such as a 2N3904 has a range of parameters that is acceptable to the manufacturers but may not be acceptable to the circuit needs.

3. Changing any component in the feedback path. May alter the critical frequency.

Where a very stable frequency is needed, crystal control is used. A crystal is a quartz material that when placed in an electric field will vibrate. Its vibrations are dependent upon the shape of the crystal. That means that we can cut a crystal to a particular frequency. The Q of a crystal can be as high as 500,000.

EXAMPLE 13.15

Find the bandwidth of a crystal with a Q of 10,000 and a critical frequency of 1 MHz.

Solution

We can find bandwidth (BW) by using:

$$BW = \frac{f_c}{Q}$$

$$BW = \frac{1\,\text{MHz}}{100,000}$$

$$BW = 100\,\text{Hz}$$

Figure 13.36 is a crystal-controlled Colpitts oscillator. The oscillator works the same with or without the crystal. The difference is that the crystal will allow only a very narrow band of frequencies to pass. Although parts may be replaced in the circuit, the crystal will keep the oscillator on the needed critical frequency.

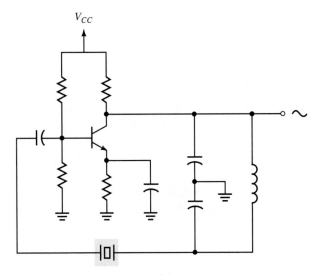

FIGURE 13.36   A crystal-controlled Colpitts oscillator

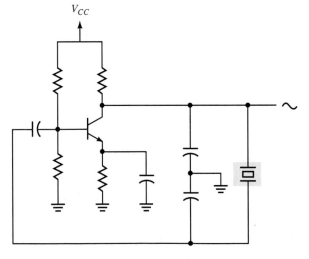

FIGURE 13.37   A crystal-controlled Pierce oscillator

Figure 13.37 is a Pierce oscillator. It looks like the Colpitts oscillator except that the tank inductor has been replaced by a crystal. To change critical frequencies, the technician would replace the crystal with another crystal that vibrated at the new critical frequency.

REVIEW SECTION 13.6

1. Relaxation oscillators depend on the rate at which a _____ charges and discharges to establish output frequency.
2. In a relaxation oscillator, what is meant by a trip point? How many trip points are there?
3. Write the equation used to determine the T(time) of a period of oscillations of the relaxation oscillator discussed in this section.

13.7 WAVESHAPING CIRCUITS

In this section we will look at circuits that are designed to distort the shape of the input signal. In most of these cases, the input signal is available somewhere else in the system or circuit, but it is not suited for a particular task. In these cases it is the function of the waveshaping circuit to change the input into a different shape at the output, a shape that is suitable for the application at hand.

DIFFERENTIATORS

The first circuit that we will discuss is drawn in Figure 13.38. At first glance it might appear that the circuit is a high-pass filter or a lead network, but the input to the RC differentiator is either a square wave or a triangular wave. The corresponding outputs are shown for each of these input waves.

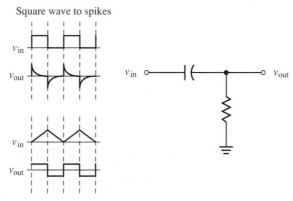

FIGURE 13.38   RC differentiator

The circuit is named from the mathematical function known as the derivative. This means that the output is proportional to the rate of change of the input. The faster the input voltage changes levels, the greater the value of the voltage dropped across the resistor. As technicians, we are not overly concerned with the mathematical derivative, but rather with how the circuit works and how we can predict the output shape given the input shape.

The first thing on which we need to focus our attention is the relationship between the pulse width of the square wave input and the RC time constant of the circuit. If you want narrow output spikes from a differentiator while applying a square wave to the input of the circuit, the pulse width (PW) of the input waveform should be 10 times the RC time constant.

$$PW \geq 10RC$$

When this relationship is true, the capacitor can fully charge and discharge during each half of the input cycle.

At start-up, the capacitor is discharged. The rising edge of the square-wave input appears to the capacitor as a very high frequency. Therefore, this edge passes through the capacitor's low reactance and appears across the output.

While the pulse amplitude is held constant, the capacitor begins to charge toward the amplitude of the pulse. Keeping the pulse amplitude constant for 10 RC time constants ensures that the capacitor fully charges. As the capacitor charges, the voltage across the resistor begins to decay. When the capacitor reaches full charge, the voltage across the resistor drops to zero.

In Figure 13.39, the graphs of voltage in, voltage out, and the capacitive voltage are shown for a ±2-V square wave. At start-up you can see that the voltage across the resistor reaches a peak value of 2 V. Thereafter, each time the polarity of the input changes, the output across the resistor is 4 V, twice the original input. Why? At turnaround time, the input voltage and the capacitor are temporarily series-aiding, thus the 2 V from the source and the 2 V from the capacitor push current the same way through the resistor, producing output spikes twice the original input amplitude.

In Figure 13.40, an op-amp is connected to a differentiator. The basic concept is still the same; the input pulse width needs to be 10 times larger than

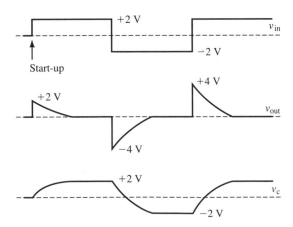

FIGURE 13.39   Sample RC differentiator waveforms. Vertical scale: 1 div = 1 V; horizontal scale: 1 div = 1 time constant.

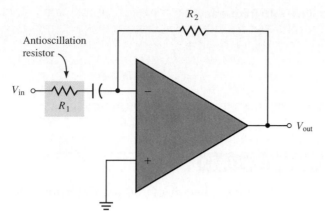

FIGURE 13.40 Op-amp differentiator

the RC time constant. The advantage of adding the op-amp into the circuit is this: The op-amp provides a low impedance out of the circuit, making it easier to couple the output spikes into a load without reducing the amplitude of the spikes.

You may see an "antioscillation" resistor in series with the capacitor in Figure 13.40. This resistance, typically 1/100 to 1/10 of R_2, helps to prevent unwanted oscillations from developing.

EXAMPLE 13.16

An RC circuit has an RC time constant of 100 μs. Determine the minimum pulse width that will make this circuit work as a differentiator. Determine the corresponding input frequency for a 50% duty cycle waveform.

Solution
The relationship PW ≥ 10RC is an accepted relationship for a differentiator. Accordingly:

$$PW = 10 \times 100 \ \mu s$$
$$PW = 1000 \ \mu s$$
$$PW = 1 \ ms$$

Duty cycle (DC) is found by

$$DC = \frac{PW}{PRT} \times 100\%$$

where PRT is pulse repetition time.

If the input has a 50% duty cycle, the input frequency would be found by rearranging and solving for PRT.

$$PRT = \frac{PW}{DC} \times 100\%$$
$$PRT = \frac{1 \ ms}{50\%} \times 100\%$$
$$PRT = 2 \ ms$$

To convert PRT into frequency:

$$f = \frac{1}{T}$$

$$f = \frac{1}{2 \text{ ms}} = 500 \text{ Hz}$$

Frequencies higher than this would not be acceptable.

PRACTICE PROBLEM 13.13

A circuit has an RC time constant of 150 μs. Determine the minimum PW that will make this circuit a differentiator. Determine the lowest input frequency, when the input is a 50% duty cycle.

We have just discussed one of the common uses of a differentiator, that is, to take an input square wave and produce a spiked waveform out. These spikes have a variety of uses, for example, to trigger industrial devices such as silicon-controlled rectifiers (SCRs) and Triacs. In broadcast applications the spikes are used to modulate and demodulate signals produced in various pulse modulation systems, where pulses of information, rather than continuous signals, are broadcast.

Differentiators can also be used to convert triangular waveforms into square waves. In this application, the amplitude of the square wave is proportional to the slope of the ramp of the triangle. (Slope is a measure of a change in voltage divided by a change in time.) The more quickly a voltage changes over a given time, the higher the slope and the higher the output amplitude of the square wave created by the differentiator.

Progress Check
You have now completed objective 8.

INTEGRATORS

An integrator circuit takes its name from the calculus function of integration. We do not need to know exactly what integration is to make use of and repair these types of circuits. We do need to understand what the circuit does with a square-wave input.

Figure 13.41 shows an RC circuit with the output taken across the capacitor. That is the most obvious difference between an integrator and a differentiator. The other important difference is the relationship between the RC time constant of the circuit and the input pulse width. In the case of an integrator, this relationship is

$$PW \leq \frac{RC}{10} \text{ time constant}$$

or, alternatively,

$$10PW \leq RC$$

When this relationship is true, the output capacitor charges only a small amount. The amount of charge across the output capacitor is less than 10% of the input. Under these conditions, the voltage across the output capacitor changes in a near linear fashion. In Figure 13.41, the output voltage is drawn as a triangular voltage. This is the primary purpose of an integrator: to take

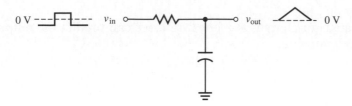

FIGURE 13.41   RC integrator

an input square wave and convert it to an output with positive- and negative-going ramps. (A ramp is a positive- or negative-going voltage that increases or decreases at a linear rate.)

In Figure 13.42, we see an op-amp integrator. When a square wave is input, resistor R_1 acts like a constant current source. This is due to the constant amplitude of the input waveform as it settles at its pulse amplitude and to the virtual ground at the input to the op-amp. So if the voltage across a fixed resistor remains constant, then the current it produces will remain constant.

All of this current flows through C_1, which charges at a linear rate, producing an output ramp. R_2 is present to prevent any DC voltages from saturating (fully charging) C_1. Without R_2 in the circuit, any DC voltages would create constant currents, which would quickly fully charge C_1. Once saturated, the circuit could not function as an integrator. R_2 provides an alternative path for low-frequency currents and with R_1 makes the circuit a low-frequency inverting amplifier.

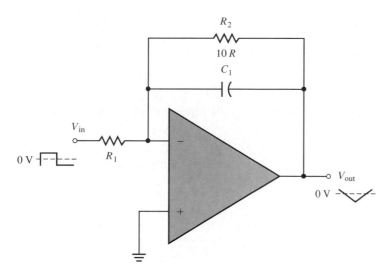

FIGURE 13.42   Op-amp integrator

EXAMPLE 13.17

Determine the output of the circuit in Figure 13.42 when the input is a ± 1-V_p square wave, 50% duty cycle, with a frequency of 10 kHz. R_1 is a 1-kΩ resistor, and the feedback capacitor is 0.01 μF.

Solution

When a capacitor charges at a linear rate, we can predict the output voltage with this relationship:

$$V = \frac{I \times t}{C}$$

where I is the charging current

  t is the time the charging current is applied

  C is the value of the capacitor

Start by assuming that the input is positive with respect to ground. This forces electron current to flow from the op-amp through C_1 and then through R_1, producing a negative-going ramp. (The circuit configuration is inverting.)

The amplitude of this ramp can be determined by:

$$I = \frac{+V_{in}}{R_1}$$

$$I = \frac{1\,V}{1\,k\Omega}$$

$$I = 1\,mA$$

We can find the pulse width by

$$PW = \frac{DC + PRT}{100\%}$$

Substitute $\dfrac{1}{freq}$ for PRT.

$$PW = \frac{DC}{100\% \times freq}$$

$$PW = \frac{50\%}{100\% \times 10\,kHz}$$

$$PW = \frac{0.5}{10\,kHz}$$

$$PW = 50\,\mu s$$

We plug these values into the equation for V above to obtain:

$$V = \frac{I \times t}{C}$$

$$V = \frac{1\,mA \times 50\,\mu s}{0.01\,\mu F}$$

$$V = 5\,V$$

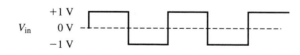

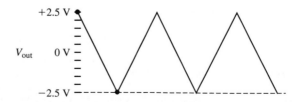

FIGURE 13.43   Output waveforms from an op-amp integrator. Vertical scale: 1 div = 0.5 V; horizontal scale: 1 div = 10 ms.

When the input goes positive, the output ramp goes negative. Since the duty cycle is 50%, the capacitor will discharge from −5 V back to zero.

The waveforms for this circuit are shown in Figure 13.43.

Progress Check
You have now completed objective 9.

PRACTICE PROBLEM 13.14

An integrator circuit like that shown in Figure 13.42 has an input voltage of ±10 V. The capacitor is 47 μF; the input frequency is 250 Hz; and R_1 = 6.8 kΩ. Determine the peak amplitude of the output ramp waveform.

TRIANGULAR WAVEFORM GENERATOR

In Figure 13.44, the first stage of the circuit is a relaxation oscillator. This first stage produces a square wave with an output frequency equal to the rate at which the capacitor charges and discharges between the trip points.

The output of the relaxation oscillator feeds the input of the second stage of the circuit, which is an op-amp integrator. R_1 and C_1 are chosen to produce the desired ramp amplitude, while the output triangular waveform has a frequency equal to the frequency out of the relaxation oscillator. The overall effect of the circuit is to convert input energy in the form of DC to an output signal in the form of a triangular waveform.

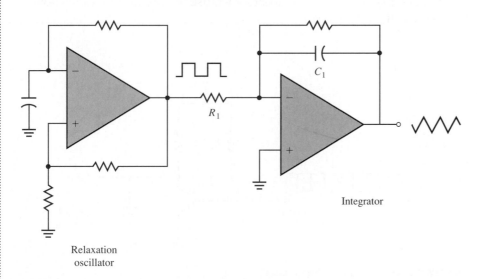

FIGURE 13.44 Triangular waveform generator

SAWTOOTH GENERATORS

A sawtooth waveform consists of a ramp (either positive or negative) and a step voltage change back to the starting value. You can see graphically in Figure 13.45. Triangular waves and sawtooth waves are both types of ramp waveforms, so you may hear someone calling the circuits studied in the section on integrators and in this section ramp generators. Whatever they are called, the circuits will work as described.

(a)   (b)

FIGURE 13.45   (a) Sawtooth waveforms with positive-going ramp. (b) Sawtooth waveform with negative-going ramp.

RC Ramp or Sawtooth Generator

In Figure 13.46, we see a simple circuit often referred to as an RC ramp, or sawtooth, generator. In this circuit, the transistor is biased on through R_b. The resistors R_b and R_c are chosen in conjunction with the β of the transistor to ensure that the transistor reaches saturation. Thus the normal output is approximately $V_{CE(sat)}$, which is typically between 0.2 and 0.3 V.

When a negative input pulse is applied to the circuit, the transistor shuts off and the capacitor starts to charge up to V_{CC} through the collector resistor. This produces the exponential ramplike output shown in Figure 13.46. When the input pulse is removed, the transistor turns back on again, providing a low-impedance path through which the output capacitor can discharge. This rapidly returns the output voltage to the saturation level. This circuit is not widely used because of the nonlinearity of the output ramp. We have discussed it because we can use it as a foundation to study the next circuit.

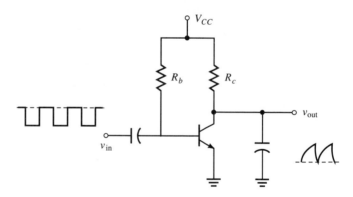

FIGURE 13.46   RC sawtooth generator

Constant-Current Sawtooth Generator

In the circuit in Figure 13.47, transistor Q_2's collector provides a constant current source feeding the output capacitor. This capacitor is normally at 0.2 V because transistor Q_1 is normally on. When a negative input pulse is applied to the base of Q_1, C_2 charges through Q_2. The charging current is determined by analyzing the biasing for Q_2.

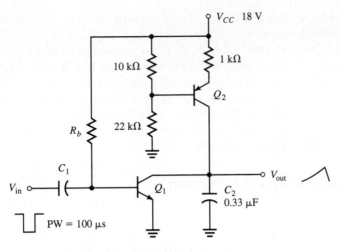

FIGURE 13.47 Constant-current sawtooth generator

Determine the output amplitude of the ramp in Figure 13.47.

Solution
The voltage divider places 12.4 V on the base of Q_2 with respect to ground.

$$\frac{22\ \text{k}\Omega}{22\ \text{k}\Omega + 10\ \text{k}\Omega} \times 18\ \text{V} = 12.4\ \text{V}$$

Since Q_2 is a PNP transistor, the emitter is at 13.1 V to ground.
Therefore, there is 4.9 V across the emitter resistor connected to Q_2:

$$V_{RE} = V_{cc} - V_E$$
$$V_{RE} = 18\ \text{V} - 13.1\ \text{V}$$
$$V_{RE} = 4.9\ \text{V}$$

This in turn implies $I_E = 4.9$ mA:

$$I_E = \frac{4.9\ \text{V}}{1\ \text{k}\Omega}$$
$$I_E = 4.9\ \text{mA}$$

This is the charging current applied to C_2.
This current is applied to C_2 while Q_1 is off. Now determine the output voltage:

$$V = \frac{(I \times t)}{C}$$
$$V = \frac{4.9\ \text{mA} \times 100\ \mu\text{s}}{0.33\ \mu\text{F}}$$
$$V = 1.5\ \text{V}$$

where the 1.5 V is the amplitude of the output ramp.

In order for the circuit in Figure 13.47 to work properly, Q_1 must stay off during the entire time the input pulse is low. This means that C_1 should not charge so fast that it blocks the input pulse. If C_1 does charge too quickly, the base of Q_1 will return to 0.7 V, and Q_1 will turn on. In that case, the output ramp would not reach its full amplitude.

In addition, the collector of Q_2 must remain a constant current source. To ensure that it does, most circuit designers keep at least 1 V from collector to emitter (V_{CE}) to prevent Q_2 from saturating. In the circuit in Figure 13.47, this means that the collector of Q_2 rises no higher than 12.1 V with respect to ground. This restriction limits the output amplitude.

Finally, Q_1 must be given enough time to fully discharge C_2 before the next negative input pulse is applied to the base.

Bootstrap Sawtooth Generator

Figure 13.48 shows circuit that can be used to produce a sawtooth waveform. The first stage of the circuit is an RC sawtooth generator. The transistor is biased on, keeping the voltage across C_2 at $V_{CE(sat)}$. When the transistor is turned off by a negative-going pulse, C_2 begins to charge.

The voltage across C_2 is applied to the output through the second stage. In this case, the second stage is an operational amplifier wired as a voltage follower. So whatever voltage is developed across C_2 also appears at the output.

The key to this circuit is C_3, which is known as the bootstrap capacitor. C_3 is chosen so that it is very large in comparison to C_2.

Because the voltage at N3 is the same as the voltage at N2, C_3 is effectively in parallel with R_1. They are connected together at N1, while N2 and N3 have the same voltage on them because of the voltage follower.

The current through R_1 is determined by:

$$I_{R1} = \frac{V_{C3}}{R_1}$$

It is this current that charges C_2.

We return to this point: C_3 is deliberately made large so that it does not *significantly discharge*. If it doesn't, then V_{C3} is almost constant and thus I_{R1}, which charges C_2, is almost constant. This forces C_2 to charge at a linear rate, which provides the output ramp desired.

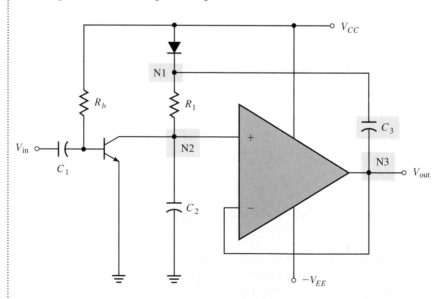

FIGURE 13.48   Bootstrap sawtooth generator. C_3 is the bootstrap capacitor.

Finally, in a bootstrap sawtooth generator the top plate of C_3 rises to a voltage level above the power supply. The diode prevents current from C_3 from flowing through the power supply.

✓ **Progress Check**
You have now completed objective 10.

REVIEW SECTION 13.7

1. Waveshaping circuits are designed to _____ the shape of the input waveform.
2. The ratio of the pulse width (PW) to the RC time constant for a differentiator is PW ≤ 10RC. True or false?
3. In an RC differentiator, the amplitude of the output spikes will be _____ the amplitude of the input pulse.
4. An op-amp is connected to a differentiator so that the output can be coupled into a load without significantly reducing the amplitude of the output spikes. True or false?
5. In an integrator, the output voltage is taken across the _____ .
6. What equation is used to predict the output voltage across a capacitor when it is charged at a linear rate?
7. A triangular waveform generator is a two-stage circuit. The first stage is a _____ _____ and the second stage is an _____ .
8. A constant-current sawtooth generator produces an output ramp when the input transistor is turned _____ .
9. List three conditions that must exist for the constant-current sawtooth generator.
10. Explain how the bootstrap capacitor in Figure 13.48 can be effectively in parallel with R_1, even when they are not connected directly across each other.

 TROUBLESHOOTING

Oscillators are among the most difficult circuits to troubleshoot because every component is required to produce the alternating output. If one component in the circuit fails, the output does not alternate. However, standard troubleshooting procedures are helpful. First, look and see. Sometimes problems are obvious. A burned resistor or a leaky electrolytic capacitor may be observed. Second, check for power. An oscillator is a device that converts DC energy to AC energy. If the supply voltage is too low, the oscillator will not oscillate. Third, with the power off, perform an ohmic check of the resistors and capacitors. When testing resistors in a circuit, there may be other components in parallel. The meter reading will be the parallel resistance of all the components. There are also several good in-circuit component testers that would help troubleshoot components. Fourth, check the bias of the active devices. DC bias voltages should follow the rules we learned earlier. In op-amp circuits check the amplifier gain resistors. Fifth, if all else looks good, re-

(Photograph courtesy of DeVry Institutes)
Students troubleshoot oscillators in the lab.

move the active device and replace it with a known good device. Even if the removed active device tests good, do not reinstall it into the circuit. It has been heated too many times.

WHAT'S WRONG WITH THESE CIRCUITS?

For the circuit below, the expected output is a square wave with a frequency of 19.4 kHz. The actual output is closer to 10 kHz. Why does this discrepancy occur?

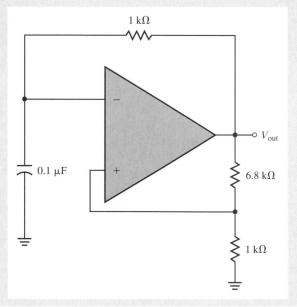

SUMMARY

This chapter concentrated on two broad categories of circuits, oscillators and waveshaping circuits. Oscillators develop AC signals from DC sources, and waveshaping circuits are designed to distort the shape of the input waveform, turning it into some other useful shape.

Oscillators rely on positive feedback to help them generate output signals. In addition, the positive feedback circuits in oscillators serve as frequency selection circuits.

The closed-loop gain of an oscillator, $A \times B$, must be greater than 1 if oscillations are to start. Once oscillations start, the closed-loop gain of an oscillator, $A \times B$, should be reduced to 1 in order to maintain oscillations.

The critical frequency of an RC network can be determined by $f_c = 1/(2\pi RC)$. At this critical frequency, the output has dropped to 70.7% of the input, and the phase angle is $\pm 45°$. It is also the frequency at which $X_C = R$.

In a lead RC network, the output is taken across the resistor. The output voltage leads the input voltage by the phase angle theta (θ). In a lag network, the output voltage is taken across the capacitor. The output voltage lags the input voltage by phi (ϕ). In either a lead or a lag circuit a Bode plot graphs the relationship of V_{out}/V_{in} versus frequency.

Lead-lag networks can be built so that the positive phase shift of the lead network is canceled out by the lagging angle of the lag network. This produces phase shifts of 0°.

The resonant frequency of a parallel tank circuit can be found by the relationship $f_r = 1/\left(2\pi\sqrt{LC}\right)$. At resonance $X_L = X_C$.

Parallel LC circuits can be used to form bandpass or bandstop filters. In a bandpass filter, the frequencies near resonance reach the output terminal of the circuit. In a bandstop filter, those frequencies near resonance are prevented from reaching the output terminal.

Remember this: The theoretical impedance of a parallel LC circuit at resonance is infinite.

The lead-lag network in a Wien-bridge circuit has an output equal to 1/3 the input voltage at the critical frequency of the circuit. The critical frequency of a lead-lag network in which $R_1 = R_2$ and $C_1 = C_2$ is equal to $1/(2\pi CR)$.

Twin-T filters are notch, or bandstop, filters. In a twin-T oscillator, the twin-T filter prevents signals at the critical frequency of the filter from reaching the inverting input of the op-amp. Therefore this frequency is not canceled out by the differential input of the op-amp, thus allowing a signal to build up at this frequency.

The Colpitts oscillator is usually used at frequencies well above 1 MHz. It can be identified by the center-tapped capacitive pair in the LC tank circuit. The frequency of oscillations in a Colpitts oscillator is equal to the resonant frequency of the tank circuit. The feedback ratio of a Colpitts oscillator is $B = C_1/C_2$.

The Clapp oscillator works in essentially the same way as the Colpitts oscillator. A third capacitor is inserted in series with the coil inside the tank. The overall capacitance of the tank becomes $1/C_T = 1/C_1 + 1/C_2 + 1/C_3$.

A Hartley oscillator uses a center-tapped inductor pair to provide feedback. The feedback ratio of a Hartley oscillator is $B = L_2/L_1$.

The Armstrong oscillator uses a transformer wound in such a way that the secondary voltage is 180° out of phase with the output voltage to provide feedback. Since the output is already 180° out of phase with the input, the overall phase shift in the circuit is 0°.

Crystals can be added to oscillator circuits. This is done to increase the accuracy of the output frequency. In a Colpitts crystal oscillator, the crystal acts as an inductor that resonates with the capacitance of the tank. In addition, crystals can have very high Qs.

A relaxation oscillator relies on the rate of charge and discharge in a capacitor to determine the output frequency.

Differentiators turn input square waves into output spikes or input triangular waves into output square waves. PW ≥ 10RC is the ratio of the pulse width to the RC time constant for a differentiator.

Integrators having input square waves produce output triangular waves. PW ≤ RC/10 time constant is the PW-to-RC time constant ratio of an integrator. Op-amp integrators provide a constant current to a feedback capacitor to develop output ramps.

A relaxation oscillator can be used to drive the input of an op-amp integrator. The overall circuit will generate triangular waveforms.

A sawtooth waveform is a waveform with either a positive-going or a negative-going ramp and a step change back to the initial level.

PROBLEMS

Refer to Figure 13.49 for problems 1–3.

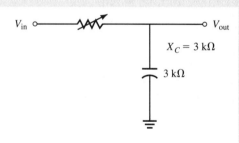

FIGURE 13.49

1. Is the circuit a lead or a lag network?
2. What is the phase angle theta when the potentiometer is set to 2 kΩ?
3. What is the phase angle theta when the potentiometer is set to 8 kΩ?

Refer to Figure 13.50 for problems 4–8.

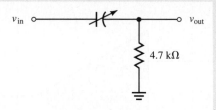

FIGURE 13.50

4. Is the circuit a lead or a lag network?
5. What is the critical frequency of the circuit when C is set to 0.1 μF?
6. What is the critical frequency of the circuit when C is set to 0.033 μF?
7. Graph the frequency versus amplitude of the circuit. Be sure to label the critical frequency.
8. Is the circuit a high-pass or a low-pass filter?

Refer to Figure 13.51 for problems 9–11.

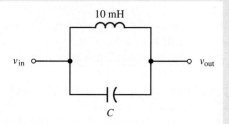

FIGURE 13.51

9. Determine the resonant frequency of the circuit when C = 0.1 μF.
10. Determine the resonant frequency of the circuit when C = 0.002 μF.
11. Is the circuit a bandpass or a bandstop filter?
12. An oscillator circuit must have an overall phase shift of _____ ° if oscillations are to occur.
13. The closed-loop gain, A × B, of an oscillator must be greater than _____ for oscillations to begin.

Refer to Figure 13.52 for problems 14–17.

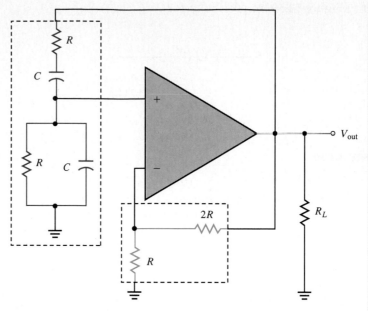

Wien bridge has two parts:
in blue, voltage divider; in red, lead-lag network

FIGURE 13.52

14. Determine the frequency of oscillations for the Wien-bridge oscillator when R = 100 kΩ and C = 0.01 μF.
15. Determine the frequency of oscillations when R = 56 kΩ and C = 100 pF.
16. A practical upper limit for most Wien-bridge oscillators built with op-amps is _____ Hz.
17. The feedback ratio at the noninverting terminal of a Wien-bridge oscillator (for the critical frequency) is _____ .
18. The twin-T RC filter arrangement is an example of a _____ filter.
19. A Colpitts oscillator can be identified by looking at the tank circuit, which has one inductor and center-tapped _____ .
20. Determine the resonant frequency for the circuit in Figure 13.53

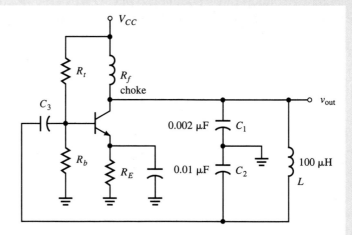

FIGURE 13.53

21. Determine the feedback ratio for the circuit in Figure 13.53.
22. A Clapp oscillator differs from a Colpitts oscillator in that an additional _____ has been added to the tank circuit.
23. Hartley oscillators use center-tapped _____ in the tank circuit.
24. Refer to Figure 13.54. If the crystal has a critical frequency of 4 mHz and and a Q of 10,000, what is the oscillator's bandwith?.

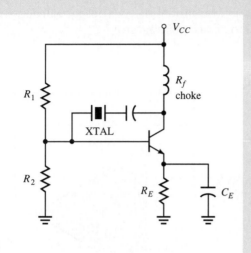

FIGURE 13.54

Refer to Figure 13.55 for problems 25–27.

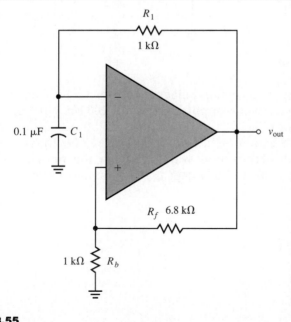

FIGURE 13.55

25. Determine the frequency of oscillations for the relaxation oscillator.

26. R_f is changed to a 5.6-kΩ resistor. Determine the new frequency of oscillations.
27. The output of the relaxation oscillator in Figure 13.55 is a:
    (a) Sine wave
    (b) Ramp wave
    (c) Triangular wave
    (d) Square wave
28. List two reasons why an op-amp would be used in a differentiator circuit.
29. Determine the minimum pulse width that will allow the circuit in Figure 13.56 to act as an integrator.

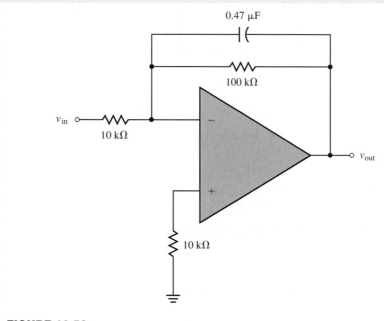

FIGURE 13.56

30. The input to the circuit in Figure 13.56 is a ±1 full peak-to-peak square wave whose input frequency is 200 Hz. What does the output look like?
31. Refer to the op-amp integrator of Figure 13.56. Determine the amplitude of the output waveform when the input frequency is 250 Hz and the input voltage is ±10 V_p square wave.

14

ACTIVE DIODE CIRCUITS
AND COMPARATORS

(Photograph of a prelaunch check of a LANDSAT satellite, courtesy of NASA)

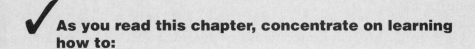

✓ **As you read this chapter, concentrate on learning how to:**

1. Tell how an op-amp changes the effective diode barrier voltage
2. Predict the output of an active half-wave rectifier
3. Predict the output of an active peak detector
4. Predict the output of an active limiter
5. Predict the output of an active clamper
6. Use hysteresis to minimize the effects of circuit noise on comparator circuits
7. Determine the upper and lower trip points for an inverting Schmitt trigger
8. Determine the upper and lower trip points for a noninverting Schmitt trigger
9. Determine the center voltage of a hysteresis loop shifted off of a 0-V center
10. Determine the upper trip point and lower trip point of a window comparator

INTRODUCTION

In this chapter we will look at two more ways in which op-amps can be used. The first part of this chapter discusses active diode circuits. Many of the basic ideas behind these circuits were discussed in Chapter 3. Clippers, clampers, and peak detectors will be discussed again; however, the focus this time will be on how amps improve the performance of active diode circuits.

The second part of this chapter discusses crossing detectors (circuits designed to detect when the input voltage crosses a particular value or values) and comparators.

The third section of the chapter deals with window comparators. A regular comparator detects when an input voltage exceeds or crosses a reference voltage. A window comparator detects when the input voltage is between two limits.

Op-amps may also be used as crossing detectors. These circuits sense when a reference voltage has been crossed. A zero crossing detector senses when an input voltage crosses 0 V. This type of detector is used in high-power applications. Large AC voltage will sometimes damage components if instantly applied to them. The zero crossing detector will allow power to be applied to the components only when the AC power is at 0 V. An example would be a large AC motor. If power were applied when the voltage was near peak voltage, the motor's windings would develop a very large induced voltage, which could short the windings. A zero crossing detector would prevent start-up damage.

Crossing detectors can also be used to monitor transducer output voltages. Thermistors are temperature-sensitive transducers. If a temperature decreases, the transducer produces a voltage that reflects this change. The crossing detector can then be set to turn on a heat source when the voltage into the crossing detector drops below a certain voltage, referred to as a trip point. Light detectors, smoke detectors, and sound detectors can also be built with the appropriate transducer and a crossing detector circuit.

Applications for many of the circuits in Section 14.1 involve converting AC signals into DC voltages. Chapter 3 developed the idea behind how rectifiers can be used to implement and build power supplies. As before, clamper circuits can be used to shift the AC signal off of ground, either in a positive or negative direction. This is useful when a signal going both above ground and below ground can cause problems—for example, with a polarized capacitor or when the signal is meant to be the input to a JFET and we must ensure that the JFET's gate diode does not become forward-biased.

14.1 ACTIVE DIODE CIRCUITS

Placing an op-amp into a diode circuit makes it possible to change the way in which the diode circuit operates. First, the op-amp can be used as a buffer between a signal source and a diode circuit. This allows us to connect rectifiers, clippers, clampers, and peak detectors to signal sources without loading down the signal. As we learned in Chapter 11, the input impedance of the op-amp itself is quite high. And, as we learned in Chapter 12, with negative feedback arrangements the circuit input impedance is driven toward infinity by the sacrifice factor.

Second, the op-amp can prevent loading effects on the output side of the diode circuit. We learned, again in Chapter 12, that the series-parallel negative feedback connection makes the op-amp appear as an ideal voltage source, that is, a voltage source with no internal impedance. Placing diode circuits in this type of feedback arrangement makes it possible to connect low-impedance loads to the diode circuit without loading down the output signal.

Third, the diode barrier potential is ideally reduced to zero when placed in an op-amp circuit with negative feedback. Consider example 14.1.

EXAMPLE 14.1

In Figure 14.1a a diode has been placed in an op-amp circuit. The sacrifice factor of the op-amp circuit is 2000. Determine the effective diode barrier voltage on circuit operation.

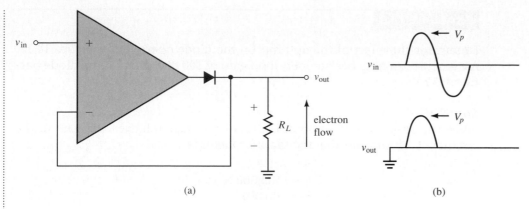

(a)   (b)

FIGURE 14.1 (a) An active half-wave rectifier circuit. (b) The input and output waveforms for the circuit.

Solution

The rest of the circuit "sees" what it thinks is a diode with a barrier potential equal to

$$V_{D(cir)} = \frac{V_D}{S}$$
$$V_{D(cir)} = \frac{0.7\,V}{2000}$$
$$V_{D(cir)} = 350\,\mu V$$

where $V_{D(cir)}$ is the barrier potential "seen" by the rest of the circuit, V_D is the barrier potential of the diode, and S is the sacrifice factor of the op-amp feedback arrangement.

There are two important points we can draw from example 14.1. First, signals smaller than the diode barrier can now be rectified. Second, for all practical purposes, the diode is a piece of wire when forward-biased. This means practically no voltage is lost at the output of a diode circuit due to the diode barrier voltage.

PRACTICE PROBLEM 14.1

An active diode circuit has an op-amp feedback arrangement that produces a sacrifice factor of 50,000. If the diode is germanium, the diode barrier $V_D = 0.3$ V. Determine $V_{D(cir)}$.

✓ Progress Check
You have now completed objective 1.

ACTIVE HALF-WAVE RECTIFIERS

Figure 14.1a is an active half-wave rectifier. The op-amp configuration is a voltage follower. The input and output waveforms are shown in Figure 14.1b.

In this case, when the input goes positive, the output goes positive and the diode conducts. Current flows through the load as indicated in Figure 14.1a. When the input goes negative, the diode stops conducting and the current through the load drops to zero. This circuit action produces the half-wave output.

EXAMPLE 14.2

Determine the effect of the op-amp on the diode operation in Figure 14.1a. Assume the op-amp has an open-loop gain of 100,000 and that the diode barrier is 0.7 V.

Solution
Since the op-amp configuration is that of a voltage follower, the gain of the circuit is 1. This means that the sacrifice factor is

$$S = AB + 1$$
$$S = (100,000 \times 1) + 1$$
$$S = 100,001$$

We can now determine what the rest of the circuit "sees" as the diode barrier:

$$V_{D(cir)} = \frac{V_D}{S}$$
$$V_{D(cir)} = \frac{0.7\,V}{100,001}$$
$$V_{D(cir)} = 7\ \mu V$$

In example 14.2 the peak output of the circuit is only 7 μV less than the peak input. For practical circuit analysis, we can treat the peak output as equal to the peak input:

$$V_{p(out)} = V_{p(in)}$$

HALF-WAVE RECTIFIERS WITH GAIN

Figure 14.2 shows a half-wave rectifier that has a gain greater than 1. The op-amp configuration is series-parallel with negative feedback.

When the input goes positive, the output goes positive and, in this case, the diode shuts off because of the reverse bias.

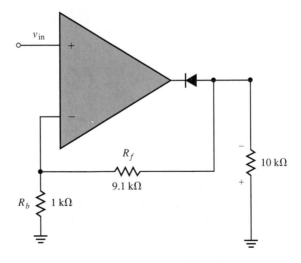

FIGURE 14.2 An active half-wave rectifier with gain. This circuit is also called a positive clipper.

When the input goes negative, the output goes negative and the diode conducts. Both the load and the feedback part of the circuit are now electrically connected to the output of the op-amp. The polarity of the voltage developed across the load is marked in Figure 14.2.

EXAMPLE 14.3

Determine the output of the circuit in Figure 14.2, when the input to the circuit is a 1-V_p sine wave.

Solution
Notice that the diode allows current to flow only downward through the load resistor. This half-wave rectifier circuit is also called a positive clipper because its output is negative half waves.

The feedback ratio of the feedback network is:

$$B = \frac{R_b}{R_b + R_f}$$

$$B = \frac{1\,k\Omega}{1\,k\Omega + 9.1\,k\Omega}$$

$$B = \frac{1}{10.1}$$

$$B = 0.099$$

Since $A_{cl} = 1/B$ the gain of the circuit is 10.1

Figure 14.3 illustrates the input and output waveforms of an active positive clipper.

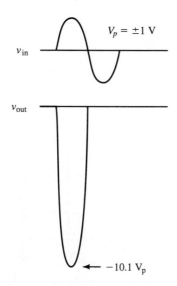

FIGURE 14.3 Input and output waveforms for the circuit in Figure 14.2

PRACTICE PROBLEM 14.2

Determine the output of the circuit in Figure 14.4.

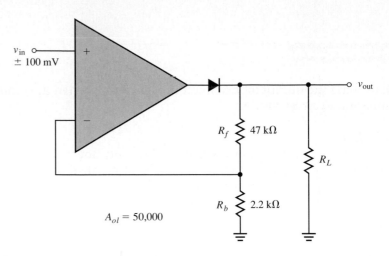

FIGURE 14.4

ACTIVE PEAK DETECTORS

The circuit in Figure 14.5 is an active peak detector. The basic op-amp configuration is a voltage follower. This means that the sacrifice factor is equal to

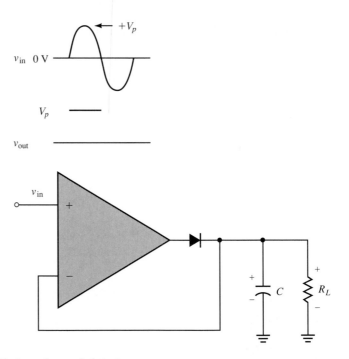

FIGURE 14.5 An active peak detector

the open-loop gain of the op-amp, and the diode barrier potential seen by the rest of the circuit is very small; $V_{D(cir)}$ is typically in the microvolt range.

When the circuit input goes positive, the diode conducts. The polarity of the capacitor will be as marked. After a few input cycles, the capacitor will be fully charged, and the voltage across the plates will be equal to $V_{p(in)}$.

In order to keep a constant voltage across the plates of the capacitor, the R_LC time constant should be at least 100 times larger than the discharge period of the lowest input frequency.

EXAMPLE 14.4

Determine the lowest input frequency into the circuit if the R_LC time constant of Figure 14.5 is set at 100 times the value of the discharge time. Assume that $C = 0.1\ \mu F$ and $R_L = 4.7\ k\Omega$.

Solution
First, determine the R_LC time constant:

$$R_LCT = 4.7\ k\Omega \times 0.1\ \mu F$$
$$R_LCT = 470\ \mu s$$

The discharge period of the waveform should be:

$$R_LCT = 100 \times \text{discharge period}$$

Because we know R_LCT, we can find discharge time by rearranging the formula:

$$\text{discharge period} = \frac{R_LCT}{100}$$
$$\text{discharge period} = \frac{470\ \mu s}{100}$$
$$\text{discharge period} = 4.7\ \mu s$$

The capacitor charges during the positive half cycle and partially discharges through R_L on the negative half cycle. It will discharge for only 4.7 μs. This means that the total time for an input cycle is:

Period = negative alternation time + positive alternation time
Period = 4.7 μs + 4.7 μs
Period = 9.4 μs

This can be converted into frequency by:

$$f = \frac{1}{T}$$
$$f = \frac{1}{9.4\ \mu s}$$
$$f = 106\ kHz$$

From example 14.4 we see that if the input frequency is kept above 106 kHz, there is almost no ripple out of the circuit, and the output is a constant DC voltage equal to V_p. The op-amp in Figure 14.5 acts like a buffer at the input side of the circuit and like a perfect voltage source at the output of the circuit. In addition, the peak level output is not reduced by the diode barrier voltage.

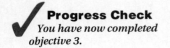
PRACTICE PROBLEM 14.3

Determine the lowest input frequency to the circuit of Figure 14.5 if $C = 0.33$ μF and $R_L = 22$ kΩ.

ACTIVE LIMITERS

Figure 14.6a shows an active diode limiter circuit with an adjustable potentiometer that is used to set the point where limiting (clipping) will occur. In this circuit, the diode acts like a shunt, preventing a portion of the input signal from reaching the output.

As long as the voltage applied to the inverting input is more negative than the voltage applied to the noninverting input, the op-amp output is a positive voltage. D_1 is reverse-biased and does not conduct. The current going through R_{in} also goes through R_L. The input signal is applied to the load.

However, once the inverting input is more positive than the noninverting input, the op-amp output is negative. D_1 is forward biased. Since the op-amp output impedance is near zero, the input signal is shunted to ground. Because the inverting input voltage follows the noninverting voltage, a voltage equal to the noninverting input voltage is applied to the load resistor.

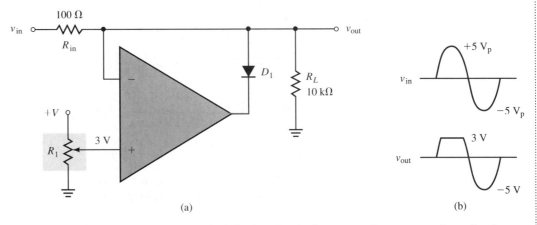

FIGURE 14.6 (a) An active diode limiter circuit. (b) Input and output waveforms for the circuit.

EXAMPLE 14.5

Describe the output waveform. In Figure 14.6a, the potentiometer, R_1, is set to +3 V. A waveform of ±5 V_P is applied to R_{in}.

Solution
As the input waveform goes in a positive direction from 0 V, the noninverting input is more positive than the inverting input. The op-amp output is positive $(+V_{sat})$, which reverse biases D_1.

The load voltage may be found by using the voltage divider method.

$$v_L = v_{in} \times \frac{R_L}{R_L + R_{in}}$$

$$v_L = v_{in} \times \frac{10\,k\Omega}{10\,k\Omega + 100\,\Omega}$$

$$v_L = v_{in} \times \frac{10\,k\Omega}{10.1\,k\Omega}$$

$$v_L = v_{in} \times 0.99$$

$$v_L = 5\,V \times 0.99$$

$$v_L = 4.95\,V$$

Because the load voltage is only 1% less than the applied voltage, we consider the load voltage equal to the supply voltage.

This is true until the inverting input is greater than $+3$ V. Because the noninverting input is more negative than the inverting input, the op-amp output goes negative, which forward biases D_1. As previously explained, D_1 acts like a piece of wire. The input signal sees an inverting op-amp circuit whose gain is

$$A = \frac{R_f}{R_{in}}$$

where R_f is a piece of wire

$$A = \frac{0\,\Omega}{100\,\Omega}$$

$$A = 0$$

The input signal produces an output voltage of

$$v_{out} = v_{in} \times A$$

$$v_{out} = v_{in} \times 0$$

$$v_{out} = 0\,V$$

However, the $+3$ V at the noninverting input sees a voltage follower circuit whose gain is 1.

$$V_{out} = V_{in} \times A$$

$$V_{out} = +3\,V \times 1$$

$$V_{out} = +3\,V$$

The two output voltages are summed at the load.

$$V_L = 0\,V + 3\,V$$

$$V_L = +3\,V$$

Figure 14.6b compares the input and output voltages. Notice that as long as the input voltage is below $+3$ V, $V_{in} = V_{out}$. However, when the input voltage is above $+3$ V, the output voltage equals the noninverting input voltage.

If the wiper arm of the potentiometer is moved closer to ground potential, more of the positive alternation is clipped. When the wiper arm is at 0 V, the entire positive alternation is clipped off.

Raising the voltage at the wiper arm allows more of the positive alternation to reach the output. In fact, if the noninverting input is greater than the peak input signal, the entire positive half cycle reaches the output.

Figure 14.7 is an op-amp limiter circuit with a parallel-parallel negative feedback op-amp configuration. The circuit is an inverting voltage amplifier

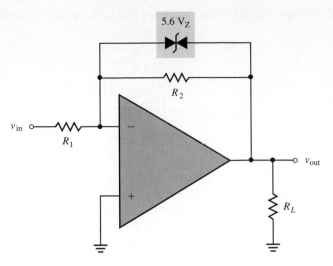

FIGURE 14.7 An active zener diode limiter circuit

and can provide gain if R_2 is larger than R_1. The zener diodes prevent the output signal from exceeding limits selected by the circuit designer. When zener diodes are used in this way, the output is *bounded*.

EXAMPLE 14.6

In the circuit in Figure 14.7, $R_2 = 10$ kΩ and $R_1 = 1$ kΩ. The zener diodes have the same zener knee; $V_z = 5.6$ V. Determine the output of the circuit when the input is a sine wave with a 1 V_p.

Solution
First, determine the gain of the circuit:

$$A_{cl} = \frac{R_f}{R_{in}}$$

$$A_{cl} = \frac{R_2}{R_1}$$

$$A_{cl} = \frac{10\,k\Omega}{1\,k\Omega}$$

$$A_{cl} = 10$$

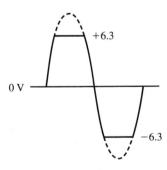

FIGURE 14.8 The effects of output bounding. The dashed portion of each of the waveforms is clipped off.

The output of the circuit is normally a 10-V_p signal, but the zener diodes limit the amplitude of the output swing.

When the voltage across R_2 reaches the zener knee plus 0.7 V, current flows through the zener that is in reverse breakover and through the zener that is forward-biased. The sum of these voltages is

$$5.6 \text{ V} + 0.7 \text{ V} = 6.3 \text{ V}$$

Therefore the output is bounded at ± 6.3 V (see Figure 14.8).

There are two main reasons why an active zener diode limiter circuit is used. First, it is used to protect the output load from too large a voltage. Second, the input can be made large enough to ensure that the signal is large enough to force the zener diodes into conduction. In this case, a sinewave input is distorted into an "approximate" square wave output.

PRACTICE PROBLEM 14.4

Change the zener diodes in Figure 14.7 so that both have a zener knee of 7.8 V. R_2 is changed to 22 kΩ, while R_1 is left at 1 kΩ. The input signal is reduced to 0.5 V. Graph the expected output.

ACTIVE CLAMPERS

Figure 14.9a depicts an active positive clamper. In this circuit when the input goes negative, the output goes positive and the diode conducts. Electron current flows around the feedback loop as indicated. The capacitor charges quickly during this time, building up a voltage with polarity as indicated in Figure 14.9a. The capacitor charges up to V_p because the op-amp reduces the effective diode barrier to near 0 V. The blue arrow

Progress Check
You have now completed objective 4.

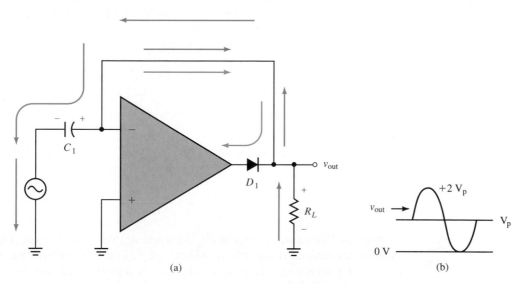

(a)

(b)

FIGURE 14.9 (a) An active positive clamper circuit. The green arrows indicate the closed loop after $V_C = V_p$. (b) The summed output waveform for the circuit.

indicates the C_1 charge path. Once the capacitor is fully charged, it electrically "removes" the diode from the circuit by applying a negative voltage to the anode of the diode, keeping the diode reverse-biased. The circuit loop then becomes the loop indicated by the green arrows. In this situation, the capacitor acts like a DC source and develops a DC voltage across the load. The AC signal is superimposed on the DC load voltage and is offset in the positive direction by the capacitor voltage, which is equal to V_P.

EXAMPLE 14.7

Find the output waveform for the circuit in Figure 14.9a when the input voltage is 1 V_P.

Solution
The input voltage goes negative and is applied to the inverting input. Because the noninverting input is more positive than the inverting input, the op-amp output is positive. D_1 is forward-biased.

Because the output impedance is near zero, the capacitor C_1 quickly charges to 1 V, the peak voltage.

With C_1 fully charged, the inverting input is now more positive than the noninverting input. D_1 is reverse-biased. The op-amp is effectively removed from the circuit.

The source and capacitor act as AC and DC sources, respectively. C_1 slowly discharges through the load resistor, causing a DC voltage to be developed at the output equal to $+V_P$. The AC source voltage is also developed at the output. Figure 14.9b shows the summed output waveform. The 1-V_P AC voltage rides (is referenced) at 1 VDC. The input voltage is clamped to a 1-V reference.

Whenever C_1 discharges to the point that the inverting input is more negative than the noninverting input, D_1 is forward-biased by the op-amp. The op-amp quickly recharges C_1 back to V_P.

PRACTICE PROBLEM 14.5

Draw the output waveform when the AC input waveform is 1.5 V_P.

Figure 14.10a shows a negatively clamped active diode circuit. Figure 14.10b shows the summed output waveform for this circuit. As in the active circuits discussed earlier in this chapter, the op-amp provides a way in which the effects of the diode barrier voltage are reduced to zero.

Progress Check
You have now completed objective 5.

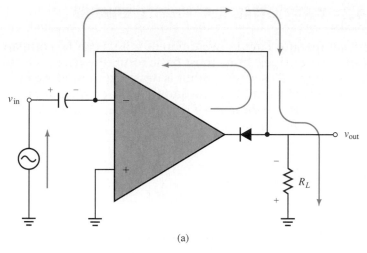

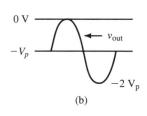

(a)

(b)

FIGURE 14.10   (a) An active negative clamper circuit. The green arrows indicate the closed loop after $V_C = V_p$. (b) The summed output waveform for the circuit.

REVIEW SECTION 14.1

1. An op-amp used in an active diode circuit can be used as a _____, helping to prevent loading of the input signal.
2. The output of an active diode circuit in which an op-amp is wired in a series-parallel negative feedback acts like a constant current source. True or false?
3. The effective diode barrier is decreased/increased by the _____ factor of the circuit.
4. In active diode circuits, the effective diode voltage can be ignored. True or false?
5. An active half-wave rectifier can rectify a signal of 200 mV$_p$. True or false?
6. It is possible to design an active half-wave rectifier with gain. True or false?
7. When setting up an active peak detector, the $R_L C$ time constant is at least _____ times the discharge time of the capacitor. This ratio produces an almost constant _____ voltage out with no ripple.
8. The output of an active peak detector is equal to ($V_p - V_D$), where V_D is the diode barrier potential. True or false?
9. What is meant by "output bounding"?
10. Active diode clampers use capacitors that are in series/parallel with the load.
11. Once a capacitor in a clamper circuit is fully charged, the diode in the circuit continues to conduct for half cycles. True or false?
12. In a clamper circuit, the capacitor acts like a temporary DC source. True or false?

14.2 COMPARATORS AND SCHMITT TRIGGERS

In a comparator circuit, two different inputs can be compared to see which of the two is larger. In its most basic form, the circuit has two inputs and a single output. This single output is designed to produce one of two different responses. In Figure 14.11, the output of the comparator circuit is set to saturate in the positive direction if A is larger than B. If B is larger than A, then the output saturates in the negative direction.

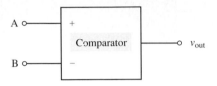

FIGURE 14.11 **Basic comparator operation. $V_{out} = +V_{sat}$ if A > B; $V_{out} = -V_{sat}$ if A < B.**

It is easy to see how an op-amp is used for such an application. The input side of the comparator in Figure 14.11 is a differential input. The output side is equivalent to the single-ended output of an op-amp. As we know from studying Chapter 11, an op-amp easily saturates when it is used in an open-loop condition.

When A is larger than B, the noninverting input "dominates" the circuit and forces the output toward $+V_{sat}$. When the input to terminal B is larger, then the inverting input dominates the circuit and forces the output toward $-V_{sat}$.

While it is possible to use op-amps in comparator applications, many circuit designers use a variation of the op-amp specifically designed for use in comparator applications. This is done because in many cases the slew rate of the op-amp limits the ability of the circuit to respond to quickly changing input signals. So a circuit designer, when building a circuit to compare two input voltages, either selects a very fast op-amp, that is, an op-amp with a very high slew rate, or builds the comparator circuit around a specially designed IC comparator.

Figure 14.12 shows a comparator circuit built around a 339 comparator. As you can see, the schematic symbol for a comparator is identical to that used

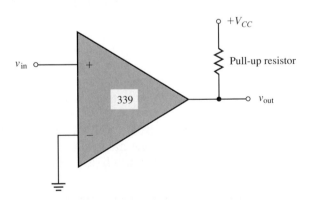

FIGURE 14.12 **LM339 comparator as a zero crossing detector**

for the op-amp. The only way that you could tell for sure is to look up the part number on the schematic in a data book.

The LM339, LM311, and NE529 comparators are different in two significant ways from a standard op-amp. First, the output is an open collector circuit. This is shown in Figure 14.13. In an open collector circuit, there is no path to power. An external resistor, known as a pull-up resistor, must be connected to the output pin to provide the path to power.

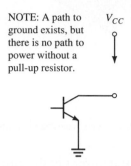

NOTE: A path to ground exists, but there is no path to power without a pull-up resistor.

V_{CC}

FIGURE 14.13 **An open collector transistor**

Second, the internal compensating capacitor found inside an op-amp is missing. This internal compensating capacitor has one major function. It helps to prevent unwanted oscillations from building up. It does this by decreasing open-loop voltage gain at high frequencies. You could think of this compensating capacitor as part of a lag network, designed to shunt high frequencies to ground.

In comparators, the output of the circuit is nonlinear. The circuit is designed to switch from one level to another quickly. A compensating capacitor helps to reduce oscillations and limits slew rate. Take the compensating capacitor out of the IC and you get much faster slew rates and therefore a circuit that responds much more quickly to changing inputs.

As a technician you would not place a comparator in a circuit in which the output is supposed to be linear. Also, if a comparator is required, you should be alert to the need for a pull-up resistor.

NOTE: *In the following comparator applications it is assumed that the circuit designer has made the correct choice, that is, a very-high-speed op-amp (fast slew rate) or a special-purpose IC comparator. Pull-up resistors are not shown. We will concentrate on how to predict circuit outputs. For our purposes it does not matter if the IC element is a comparator or an op-amp.*

CROSSING DETECTORS

Figure 14.14 shows a noninverting zero crossing detector. The input voltage is applied to the noninverting terminal, while the inverting terminal is held at 0 V by connecting it to ground. In this particular case, the output of the circuit is designed to saturate positively when the input rises above 0 V. If the input voltage goes below 0 V, the output of the circuit saturates negatively.

Figure 14.15, shows another type of crossing detector, the inverting detector. In this circuit, a reference voltage is applied to the noninverting terminal

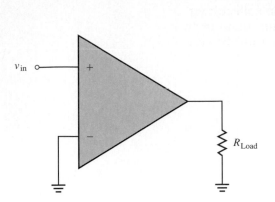

FIGURE 14.14   A noninverting zero crossing detector

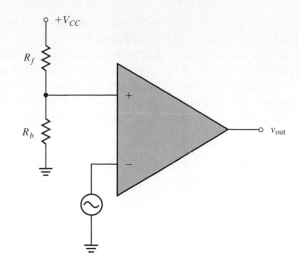

FIGURE 14.15   A nonzero crossing detector

of the comparator. With no input signal, the output of the circuit is positively saturated.

When the input signal rises above the reference voltage, the output is negatively saturated. When the input signal drops below the reference voltage, the output of the comparator returns to the positive saturation level. A graph of V_{in} versus V_{out} for this circuit is shown in Figure 14.16. This type of graph is often called a **transfer characteristics graph.**

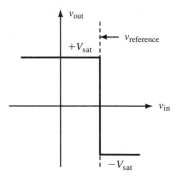

FIGURE 14.16   A transfer characteristics graph for the circuit in Figure 14.15

PRACTICE PROBLEM 14.6

Assume that the circuit of Figure 14.14 is designed to saturate at ± 12 V. Draw the transfer characteristic graph for the circuit.

CROSSING DETECTORS WITH HYSTERESIS: SCHMITT TRIGGERS

Figure 14.17 shows how noise on an input signal can affect a zero crossing detector like that shown in Figure 14.14. As you can tell from looking at the diagram, the noise component embedded in the signal causes the input to cross the zero level several more times than if the signal had been clean.

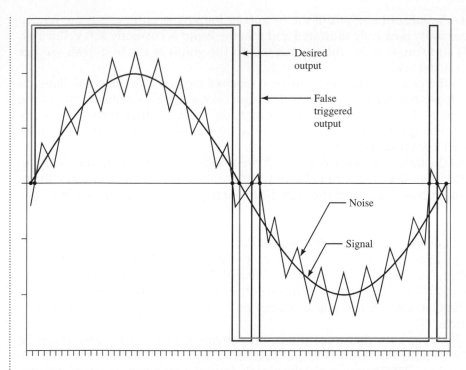

FIGURE 14.17 The effects of noise on comparator operation. The blue lines = desired operation.

These noise-produced crossings cause unwanted or false outputs. If the output of the comparator were connected to a relay driving a load, the load would be cycling on and off at the incorrect times. This, at best, is annoying and at worst can be damaging to the equipment being operated or dangerous to the equipment operator.

The effects of noise on comparator circuits can be minimized by crossing detectors built with hysteresis. These are called **Schmitt triggers.**

A hysteresis loop is pictured in Figure 14.18. This graph represents a situation in which a comparator has two different switching points. One point is known as the upper trip point (UTP); the other switching point is known as the lower trip point (LTP).

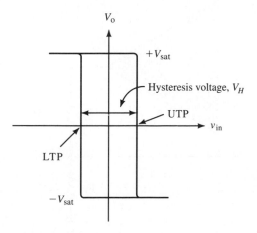

FIGURE 14.18 A hysteresis loop

Let's start our analysis of the hysteresis loop by assuming that the output is currently positively saturated and that the input is currently at 0 V. Further, let's stipulate that the voltage marked on the graph as the hysteresis voltage (V_H) is 200 mV.

As the input signal goes positive, it reaches the point on the graph listed as the upper trip point. As soon as the input exceeds the UTP, the output of the circuit switches to the negative saturation level. Further increases in the input voltage will have no effect on the output of the circuit. It will remain negatively saturated.

As the input voltage is reduced, it eventually passes through the UTP, but, in this case, the output remains negatively saturated. The input must continue to drop until it passes through the lower trip point. Once this happens, the output of the circuit returns to the positive saturation level. Further decreases in the input voltage will have no effect on the output of the circuit. It will continue to remain at positive saturation.

How does this improve the operation of a crossing detector? As mentioned earlier, the voltage (V_H) is equal to 200 mV. Suppose, with no signal present, we had 100 mV of noise at the input of the circuit. Since this noise is not large enough to cross a trip point, it does not cause the output of the circuit to switch.

Next, suppose that once the input voltage went more positive than the UTP, a voltage spike on the input pulled the input voltage below the UTP but not below LTP. The output of the circuit continues to remain negatively saturated because the input voltage and spike combination did not drop below the LTP.

In both situations, the hysteresis loop improves the noise immunity of the crossing detector. As we shall see in the remaining portions of this section, it is possible to control both the position of the UTP and LTP and the amount of hysteresis voltage.

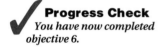

Progress Check
You have now completed objective 6.

INVERTING SCHMITT TRIGGERS

Figure 14.19 shows an inverting Schmitt trigger. As you can see, a voltage divider runs between the output and the noninverting input of the circuit. This is positive feedback.

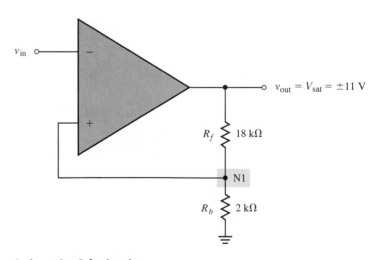

FIGURE 14.19   **An inverting Schmitt trigger**

EXAMPLE 14.8

Analyze the circuit of Figure 14.19 by assuming that the output is currently positively saturated.

Solution

If V_{out} is currently at $+11$ V, then the voltage divider theorem is used to predict the voltage at node 1 (N1).

$$V_{N1} = \frac{R_b}{R_b + R_f} \times V_{out}$$

$$V_{N1} = \frac{2\,k\Omega}{2\,k\Omega + 18\,k\Omega} \times V_{out}$$

$$V_{N1} = \frac{2\,k\Omega}{20\,k\Omega} \times +11\,V$$

$$V_{N1} = +1.1\,V$$

This voltage is fed back to the noninverting terminal of the comparator. In order to cause the output voltage to change to the negative saturation level, we must place a voltage greater than $+1.1$ V on the inverting terminal. Therefore, $+1.1$ V is the UTP of the circuit, and the circuit goes negative when the UTP is reached. This means that we have an inverting Schmitt trigger circuit.

EXAMPLE 14.9

Conclude the analysis of the circuit in Figure 14.19.

Solution

The circuit is now negatively saturated. This means that the voltage at node 1 is:

$$V_{N1} = \frac{R_b}{R_b + R_f} \times V_{out}$$

$$V_{N1} = \frac{2\,k\Omega}{20\,k\Omega} \times -11\,V$$

$$V_{N1} = -1.1\,V$$

If the input voltage goes in a negative direction until its potential exceeds -1.1 V, the output once again goes to $+V_{sat}$.

Therefore, the LTP of the circuit is -1.1 V. Crossing the LTP forced the output high, reinforcing our previous conclusion that the circuit is an inverting Schmitt trigger.

Figure 14.20 depicts the hysteresis loop for the circuit of Figure 14.19.

Change the value of R_f in Figure 14.19 to equal 27 kΩ. Recompute the UTP and LTP and draw the corresponding hysteresis loop.

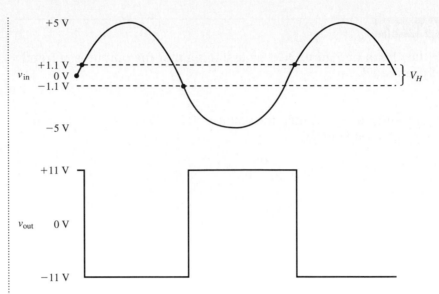

FIGURE 14.20

Adjusting the Trip Points

As you can see from the hysteresis loop in Figure 14.20, the UTP and LTP are symmetrical about zero. (Your practice problem should have had the same result.) It is possible to change the trip points so that they are not symmetrical about zero.

Figure 14.21 shows one way in which trip points can be made asymmetrical. In this circuit, the inverting Schmitt trigger of Figure 14.19 has been modified by the placement of a diode across R_b. When the output is saturated positively, the diode is reverse-biased and has no effect on circuit operation. The UTP remains 1.1 V.

When the output of the circuit is negatively saturated, the diode is forward-biased and shunts R_b, placing 0.7 V across it. Since this affects the feedback node N1, it also affects the LTP, which is now -0.7 V rather than the -1.1 V calculated earlier.

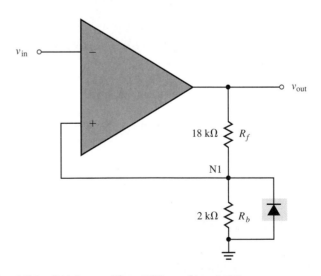

FIGURE 14.21   A Schmitt trigger with an LTP equal to -0.7 V

Draw the hysteresis loop for the circuit of Figure 14.21.

Figure 14.22 shows a Schmitt trigger with a diode connected between the output of the op-amp and the top of R_f. When the op-amp output is positive, the diode, D_1, is forward-biased. The ratio of $R_b/(R_f + R_b)$ determines the UTP. When the output of the op-amp is negatively saturated, the diode, D_1, is reverse-biased and ideally there is no current flow through the feedback network, placing the LTP close to ground potential. Example 14.10 looks at this type of circuit in more detail.

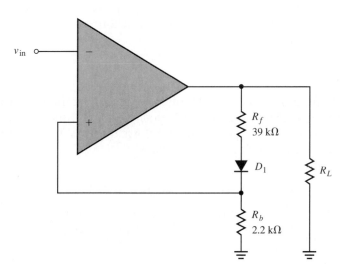

FIGURE 14.22 **A Schmitt trigger with an LTP close to ground potential**

EXAMPLE 14.10

Determine the UTP and LTP for the circuit of Figure 14.22. Assume that the power supply connections are at ± 22 V. In addition, assume that when the diode is reverse-biased there is a 500-nA leakage current.

Solution
Let's start with the output positively saturated. The diode is forward-biased. This means that the top of the voltage divider is at $+20$ V ($+22$ V $-$ 10%). Therefore, we have:

$$UTP = \frac{R_b}{R_b + R_f}(V_{out} - V_D)$$

$$UTP = \frac{2.2\,k\Omega}{2.2\,k\Omega + 39\,k\Omega}(20\,V - 0.7\,V)$$

$$UTP = +1.03\,V$$

When the input reaches or exceeds 1.03 V, the output switches to the negative saturation level. Thus, this circuit is an inverting Schmitt trigger.

Now, the output of the op-amp is negatively saturated. The only current flowing is the reverse leakage current through the diode. This was given as 500 nA.

$$V_{Rb} = 500 \text{ nA} \times 2.2 \text{ k}\Omega$$
$$LTP = -1.1 \text{ mV}$$

Since V_{Rb} produces the voltage at the node between R_f and R_b, this value, -1.1 mV, is the LTP.

$$V_{Rb} = LTP = 1.1 \text{ mV}$$

PRACTICE PROBLEM 14.9

Change the value of R_b in Figure 14.22 to equal 2.7 kΩ. Reverse the polarity of the diode. Determine the UTP and LTP when the power supply levels have been set to ±18 V.

Figure 14.23 shows a circuit in which we can adjust the UPT and LTP to some nonzero value. When the output is positively saturated, D_1 is forward-biased while D_2 is reverse-biased, so current flows through R_{f1}, then through R_b. This divides the output voltage by the ratio of

$$\frac{R_b}{R_{f1} + R_b}$$

When the output is negatively saturated, D_1 is reverse-biased and D_2 is forward-biased. R_{f2} is now electrically connected while R_{f1} is effectively taken out of the circuit. The ratio for feedback voltage is now:

$$\frac{R_b}{R_{f2} + R_b}$$

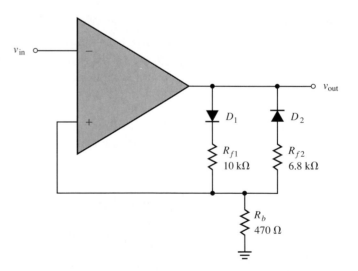

FIGURE 14.23   A Schmitt trigger with different UTP and LTP nonzero levels

EXAMPLE 14.11

Determine the UTP and LTP for Figure 14.23 with these values: $R_b = 470\ \Omega$, $R_{f1} = 10\ k\Omega$, $R_{f2} = 6.8\ k\Omega$, power supply voltages $= \pm18$ V. Ignore the effects of reverse leakage currents.

Solution
Starting with the output positively saturated, we have:

$$\text{UTP} = \frac{R_b}{R_b + R_{f1}}(V_{out} - V_{Dl})$$

$$\text{UTP} = \frac{470\ \Omega}{470\ \Omega + 10\ k\Omega}(16.2\ V - 0.7\ V)$$

$$\text{UTP} = 696\ mV$$

Now with the output negatively saturated, we have:

$$\text{LTP} = \frac{R_b}{R_b + R_{f2}}(V_{out} - V_{Dl})$$

$$\text{LTP} = \frac{470\ \Omega}{470\ \Omega + 6.8\ k\Omega}(-16.2\ V + 0.7\ V)$$

$$\text{LTP} = -1.0\ V$$

Determine the UTP and LTP for the circuit in Figure 14.24.

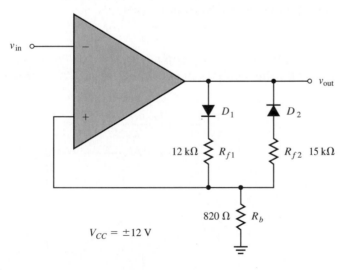

FIGURE 14.24

✓ **Progress Check**
You have now completed objective 7.

NONINVERTING SCHMITT TRIGGERS

Figure 14.25 shows a noninverting Schmitt trigger. In this circuit, when the UTP is crossed with the input going positive, the output switches to a posi-

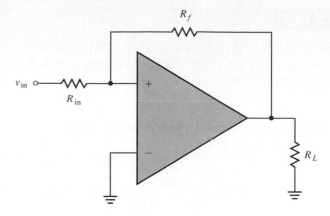

FIGURE 14.25   A noninverting Schmitt trigger

tive saturation level. When the LTP is crossed, with the input going negative, the output of the circuit will switch to the negative saturation level.

This is different from the inverting Schmitt triggers we have discussed up to this point. In the inverting Schmitt triggers, when the UTP was crossed with the input going positive, the output switched to negative saturation, while the LTP, with the input going negative, when crossed, forced the output to rise to positive saturation.

EXAMPLE 14.12

Determine the UTP and LTP for the noninverting Schmitt trigger shown in Figure 14.26. Power supply levels are ± 19 V.

Solution
In Figure 14.26, the output is positively saturated. This sets up current flow and polarities for the feedback network as shown by the arrows.

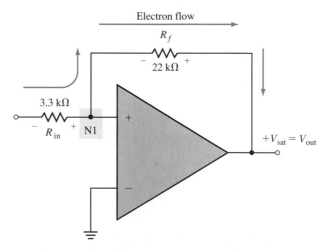

FIGURE 14.26   An analysis of noninverting Schmitt trigger with a positively saturated output

With no input to the circuit, we can determine the amount of voltage present at the noninverting input (N1) of the comparator by using the voltage divider principle:

$$V_{N1} = \frac{R_{in}}{R_{in} + R_f} \times V_{out}$$

$$V_{N1} = \frac{3.3\,k\Omega}{3.3\,k\Omega + 22\,k\Omega} \times +17\,V$$

$$V_{N1} = +2.2\,V$$

V_{N1} is applied to the noninverting terminal of the comparator, which keeps the output positively saturated, but it is not a trip point.

In order for the comparator to switch states, the two input terminals must be equal. Since the inverting terminal is tied to ground, the input voltage to the circuit must pull the noninverting terminal down to ground potential, thus lowering the voltage at the noninverting terminal from $+2.2$ V to 0 V.

Figure 14.27 shows the feedback network when the input to the noninverting terminal is 0 V. At this point, the comparator output drops to zero, then continues to slew in the negative direction until it is negatively saturated. We can analyze this moment by using Ohm's law.

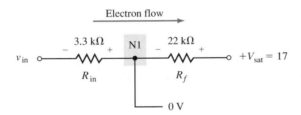

FIGURE 14.27   **The comparator for Figure 14.26**

As indicated in Figure 14.27, N_1 is at 0 V. This means that there is 17 V across R_f.

The current through R_f is

$$I_{Rf} = \frac{V_{Rf}}{R_f}$$

$$I_{Rf} = \frac{17\,V}{22\,k\Omega}$$

$$I_{Rf} = 773\,\mu A$$

All of this current must flow through R_{in}, producing a voltage drop:

$$V_{Rin} = 773\,\mu A \times 3.3\,k\Omega$$

$$V_{Rin} = 2.55\,V$$

As you can see, V_{in} must be -2.55 V in order to pull down the voltage at the noninverting terminal to 0 V. This means that the LTP of this circuit is

$$LTP = -2.55\,V$$

Figure 14.28 shows the circuit after it has switched over to the negative saturation level. With no input to the circuit, we can determine the voltage at the noninverting terminal using the voltage divider principle:

$$V_{N1} = \frac{R_{in}}{R_{in} + R_f} \times V_{out}$$

$$V_{N1} = \frac{3.3 \, k\Omega}{3.3 \, k\Omega + 22 \, k\Omega} \times (-17 \, V)$$

$$V_{N1} = -2.2 \, V$$

Once again, it must be emphasized that this is not the trip point voltage. It merely reinforces the present circuit condition; that is, a negative voltage at the noninverting terminal keeps the output of the circuit negatively saturated.

To determine the UTP, look at Figure 14.29. There is 17 V across R_f, so the current flowing through it is

$$I_{Rf} = \frac{V_{Rf}}{R_f}$$

$$I_{Rf} = \frac{17 \, V}{22 \, k\Omega}$$

$$I_{Rf} = 773 \, \mu A$$

This current is flowing in the opposite direction of that shown in Figure 14.27. The polarities across R_f and R_{in} are as marked.

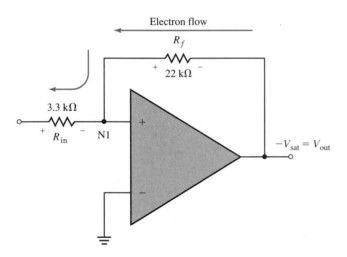

FIGURE 14.28 An analysis of a noninverting Schmitt trigger with a negatively saturated output

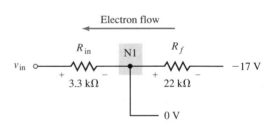

FIGURE 14.29 The comparator for Figure 14.28

Now remember that with no input to the circuit and the output negatively saturated, the noninverting terminal was at −2.2 V.

This means that if switching is to take place the voltage at the noninverting terminal must be pulled up by the input voltage until it reaches zero. We can determine that voltage by using Ohm's law:

$$UTP = I_{RF} \times R_{in}$$
$$UTP = 733 \, \mu A \times 3.3 \, k\Omega$$
$$UTP = 2.55 \, V$$

This is a positive voltage as indicated by the voltages and polarities marked in Figure 14.29. It is also the UTP.

Determine the UTP and LTP for the noninverting Schmitt trigger shown in Figure 14.30.

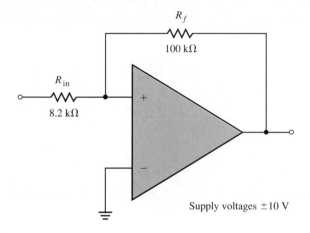

FIGURE 14.30

Adjusting the Trip Points

Figure 14.31 shows a noninverting Schmitt trigger circuit where the UTP and LTP are not symmetrical around 0 V. The difference between the previously

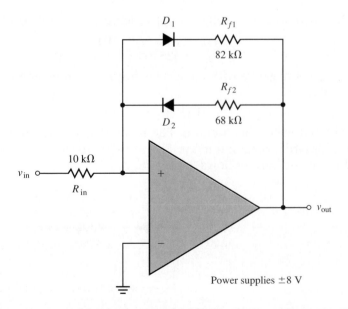

FIGURE 14.31   A noninverting Schmitt trigger with UTP and LTP at different levels

discussed Schmitt trigger circuits and Figure 14.31 is that Figure 14.31 has two feedback paths. However, D_1 and D_2 allow only one path to be used at a time. The diodes function as electronic switches.

EXAMPLE 14.13

Find the UTP and LTP of the Schmitt trigger in Figure 14.31.

Solution
When the output is positively saturated, the output is +7 V. Diode D_2 is forward-biased. Current flows through R_{f2} and R_{in} as shown in Figure 14.32.

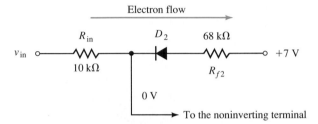

FIGURE 14.32   An analysis of Figure 14.31 with a positively saturated output

When the noninverting terminal is at 0 V, the voltage across R_{f2} is

$$V_{Rf2} = V_{sat} - 0.7\ V$$
$$V_{Rf2} = 6.3\ V$$

The current through R_{f2} is

$$I_{Rf2} = \frac{6.3\ V}{68\ k\Omega}$$
$$I_{Rf2} = 92.6\ \mu A$$

All of this current flows through R_{in}, producing a voltage drop of:

$$V_{Rin} = 92.6\ \mu A \times 10\ k\Omega$$
$$V_{Rin} = 926\ mV$$

This voltage is negative with respect to ground as indicated in Figure 14.32, so the LTP is:

$$LTP = -926\ mV$$

When the output of the circuit of Figure 14.31 is in negative saturation, we have the situation diagrammed in Figure 14.33. In this case, diode D_1 is forward-biased and current flows through R_{f1}.

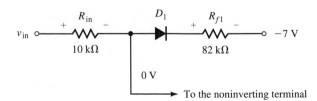

FIGURE 14.33   An analysis of Figure 14.31 with a positively saturated output

$$I_{Rf1} = \frac{6.3\,\text{V}}{82\,\text{k}\Omega}$$
$$I_{Rf1} = 76.8\,\mu\text{A}$$

All of this current flows through R_{in}, producing a voltage drop of

$$V_{Rin} = 76.8\,\mu\text{A} \times 10\,\text{k}\Omega$$
$$V_{Rin} = 768\,\text{mV}$$

This voltage is positive with respect to ground as shown in Figure 14.33, so this voltage is the UTP:

$$\text{UTP} = 768\,\text{mV}$$

PRACTICE PROBLEM 14.12

Change R_{in} in Figure 14.31 to a 27-kΩ resistor and recompute the UTP and LTP.

Progress Check
You have now completed objective 8.

SCHMITT TRIGGERS WITH NONZERO CENTERED HYSTERESIS LOOPS

We have now looked at Schmitt triggers that invert, that don't invert, that have trigger points that are symmetrical about 0 V, and that have trigger points that are not symmetrical about 0 V. In those circuits, the UTP was above ground potential, while the LTP was below ground potential.

Now we will look at circuits in which the entire hysteresis loop is moved either above or below ground. This means that both the UTP and the LTP will be above ground potential or both will be below ground potential. Figure 14.34a shows how the hysteresis loop would look when both trip points are positive with respect to ground, and Figure 14.34b shows it when both trip points are negative with respect to ground.

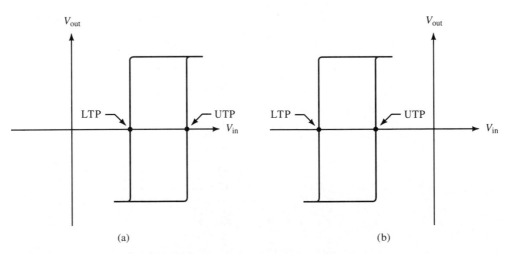

(a)   (b)

FIGURE 14.34 (a) Hysteresis loop shifted positive with respect to ground. (b) Hysteresis loop shifted negative with respect to ground.

EXAMPLE 14.14

Figure 14.35 shows an inverting Schmitt trigger in which a DC voltage is applied to the bottom of R_b. Determine the UTP and LTP for this circuit.

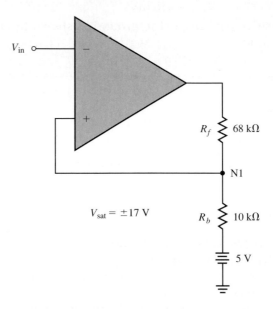

FIGURE 14.35 An inverting Schmitt trigger with a nonzero centered hysteresis loop

Solution

When the output is saturated positively with respect to ground, we have the situation diagrammed in Figure 14.36. In this figure, the saturation voltage is +17 V. At the bottom of the voltage divider, a DC voltage source is placed between R_b and ground. This means that the total potential difference across

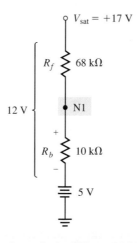

FIGURE 14.36 An analysis of Figure 14.35 with a positively saturated output

the voltage divider is 12 V. Calculate the current flowing through the voltage divider using Ohm's law:

$$I = \frac{12\,V}{78\,k\Omega}$$
$$I = 154\,\mu A$$

None of this current flows into the input of the comparator, so we can now determine the voltage drop across R_b:

$$V_{Rb} = 154\,\mu A \times 10\,k\Omega$$
$$V_{Rb} = 1.54\,V$$

The polarity across R_b when V_{sat} is positive is marked in Figure 14.36. The top of R_b is more positive with respect to ground. This means that the total voltage drop from ground to N1 is

$$V_{N1} = 5\,V + 1.54\,V$$
$$V_{N1} = 6.54\,V$$

Now analyze the circuit of Figure 14.35 when the output is at negative saturation; see Figure 14.37. The top of R_f is now at -17 V with respect to ground. This places a total of 22 V across the combination of R_f and R_b. The current flowing through these resistors is

$$I = \frac{22\,V}{78\,k\Omega}$$
$$I = 282\,\mu A$$

From this we can predict the voltage drop across R_b:

$$V_{Rb} = 282\,\mu A \times 10\,k\Omega$$
$$V_{Rb} = 2.82\,V$$

The polarity of voltage across R_b when the output of the comparator is negatively saturated is marked in Figure 14.37. The top of R_b is at a lower potential with respect to ground than the bottom of R_b. The voltage at N1 is

$$V_{N1} = 5\,V - 2.82\,V$$
$$V_{N1} = 2.18\,V$$

Both the calculated voltages, 6.54 V and 2.18 V, are positive with respect to ground. When an oscilloscope is connected to the circuit of Figure 14.35 in

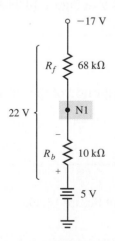

FIGURE 14.37   **An analysis of Figure 14.35 with a negatively saturated output**

the x-y mode of operation, a hysteresis loop similar to that shown in Figure 14.38 is displayed. Because of component tolerances, observed voltages may vary. However, you will see that the entire hysteresis loop is shifted above ground potential.

$$UTP = +6.5 \text{ V}$$
$$LTP = +2.2 \text{ V}$$

The center of the hysteresis loop can be found by:

$$V_{cen} = \frac{(UTP + LTP)}{2} = \frac{(6.5 \text{ V} + 2.2 \text{ V})}{2} = 4.3 \text{ V}$$

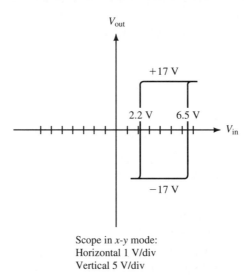

Scope in *x-y* mode:
Horizontal 1 V/div
Vertical 5 V/div

FIGURE 14.38   The hysteresis loop for the circuit in Figure 14.35

PRACTICE PROBLEM 14.13

Change the DC source voltage from 5 V to 3 V in Figure 14.35 and recompute the UTP and LTP for the circuit.

In Figure 14.39, the polarity with respect to ground of the DC voltage applied to the bottom of R_b is reversed.

EXAMPLE 14.15

Determine the UTP and LTP for the circuit of Figure 14.39.

Solution
Assume that the output is at positive saturation; see Figure 14.40.

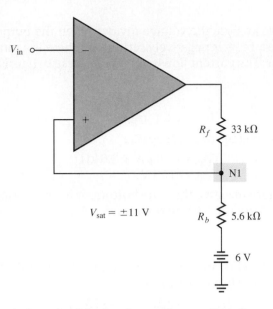

FIGURE 14.39   An inverting Schmitt trigger with a negative reference voltage

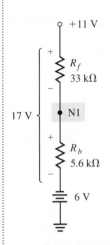

FIGURE 14.40   An analysis of Figure 14.39 with a positively saturated output

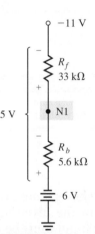

FIGURE 14.41   An analysis of Figure 14.39 with a negatively saturated output

The total voltage across the voltage divider is 17 V. From this we can determine the current through the resistors:

$$I = \frac{17\,V}{38.6\,k\Omega}$$
$$I = 440\,\mu A$$

This leads to the voltage drop across R_b:

$$V_{Rb} = 440\,\mu A \times 5.6\,k\Omega$$
$$V_{Rb} = 2.47\,V$$

This voltage is more positive at the top of R_b, with respect to ground, than at the bottom. Therefore we add it to the -6-VDC source to determine the voltage at N1:

$$V_{N1} = -6\,V + 2.47\,V$$
$$V_{N1} = -3.53\,V$$

In Figure 14.41, we analyze the voltage divider when the output of the comparator from Figure 14.39 is negatively saturated. There is a total of 5 V across the voltage divider. The current flowing through these two resistors is

$$I = \frac{5\,V}{38.6\,k\Omega}$$
$$I = 130\,\mu A$$

This leads us to the voltage drop across R_b:

$$V_{Rb} = 130\,\mu A \times 5.6\,k\Omega$$
$$V_{Rb} = 0.73\,V$$

The top of R_b is more negative than the bottom of R_b with respect to ground, so we subtract V_{Rb} from the -6 V to obtain:

$$V_{N1} = -6\,V - 0.73\,V$$
$$V_{N1} = -6.73\,V$$

We conclude that the UTP and LTP are equal to -3.53 V and -6.73 V, respectively. The center of the hysteresis loop is:

$$V_{cen} = \frac{-3.53\,V + (-6.73\,V)}{2}$$
$$V_{cen} = -5.13\,V$$

PRACTICE PROBLEM 14.14

Graph the hysteresis loop for the conclusions of example 14.15.

Figure 14.42 shows a noninverting Schmitt trigger with a DC reference voltage applied to the inverting input. This reference voltage means that the noninverting input must be changed to V_{ref} in order for the circuit to change states.

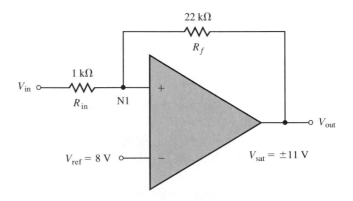

FIGURE 14.42   A noninverting Schmitt trigger with a nonzero centered hysteresis loop

EXAMPLE 14.16

Determine the UTP and LTP for the circuit of Figure 14.42.

Solution
As shown in Figure 14.43, the output saturated negatively. In order to force the comparator to switch states, we must make the voltage at the noninverting terminal equal to the reference voltage at the inverting terminal. When this is done, the voltage across R_f is 19 V (see Figure 14.43).

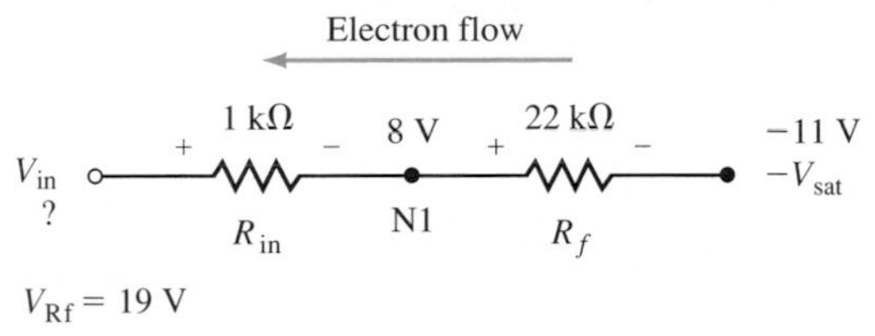

FIGURE 14.43   The feedback loop for Figure 14.42 with a negatively saturated output

From this we determine the current flowing through R_f:

$$I_{Rf} = \frac{19\,V}{22\,k\Omega}$$
$$I_{Rf} = 864\,\mu A$$

All of this current must flow through R_{in}. The voltage drop across R_{in} is:

$$V_{Rin} = 863.6\,\mu A \times 1\,k\Omega$$
$$V_{Rin} = 864\,mV$$

The left-hand side of R_{in} is more positive than the right-hand side, so this voltage adds to the reference voltage to produce:

$$V_{in} = V_{ref} + V_{Rin}$$
$$V_{in} = 8\,V + 0.864\,V$$
$$V_{in} = 8.86\,V$$

Therefore

$$UTP = 8.86\,V$$

In Figure 14.44, the output is positively saturated. The total voltage drop across R_f, when the noninverting terminal is brought to 8 V, is 3 V. From this we determine the current through R_f:

$$I_{Rf} = \frac{3\,V}{22\,k\Omega}$$
$$I_{Rf} = 136\,\mu A$$

All of this current flows through R_{in}, producing a voltage drop of:

$$V_{Rin} = I_{Rf} \times R_{in}$$
$$V_{Rin} = 136\,\mu A \times 1\,k\Omega$$
$$V_{Rin} = 136\,mV$$

The flow of electrons is from left to right when the output is saturated positively. This produces voltage drops as marked in Figure 14.44. The left-hand

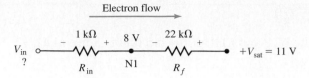

FIGURE 14.44 An analysis of Figure 14.42 with a positively saturated output

side of R_{in} is at a lower potential than the right-hand side. This means that when the noninverting terminal is brought to 8 V, V_{in} is

$$V_{in} = V_{ref} - V_{Rin}$$
$$V_{in} = 8\ V - 136\ mV$$
$$V_{in} = 7.86\ V$$

Therefore the LTP is

$$LTP = 7.86\ V$$

PRACTICE PROBLEM 14.15

Draw the hysteresis loop for the circuit of Figure 14.42.

EXAMPLE 14.17

Figure 14.45 shows a noninverting Schmitt trigger with a negative reference voltage. Determine the UTP and LTP for the circuit of Figure 14.45.

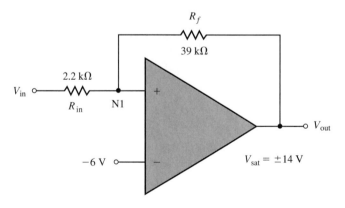

FIGURE 14.45 A noninverting Schmitt trigger with a negative reference voltage

Solution
As Figure 14.46 shows, the output is at positive saturation. We start our analysis by realizing that there must be 20 V across R_f when the noninverting terminal is brought to −6 V. From this we can determine the current through the feedback resistor:

Electron flow →

R_{in} −6 V R_f

V_{in} o—/\/\/\—•—/\/\/\—• +14 V
      − 2.2 kΩ + N1 − 39 kΩ +

FIGURE 14.46   An analysis of Figure 14.45 with a positively saturated output

$$I_{Rf} = \frac{20\ V}{39\ k\Omega}$$
$$I_{Rf} = 513\ \mu A$$

All of this current flows through R_{in}, producing a voltage drop:

$$V_{Rin} = 513\ \mu A \times 2.2\ k\Omega$$
$$V_{Rin} = 1.13\ V$$

Electron flow is from left to right when the output is positively saturated, so the left-hand side of R_{in} is at a lower potential than the right-hand side. We can now determine V_{in} when the voltage at the noninverting terminal is −6 V:

$$V_{in} = -6\ V - 1.13\ V$$
$$V_{in} = -7.13\ V$$

Therefore:

$$LTP = -7.13\ V$$

Figure 14.47 shows the output is saturated negatively. Note, electron flow is now from right to left.

← Electron flow

R_{in} −6 V R_f

v_{in} o—/\/\/\—•—/\/\/\—o − 14 V
      + 2.2 kΩ − N1 + 39 kΩ −

FIGURE 14.47   An analysis of Figure 14.45 with a negatively saturated output

The voltage across R_f is 8 V, so

$$I_{Rf} = \frac{8\ V}{39\ k\Omega}$$
$$I_{Rf} = 205\ \mu A$$

All of this current flows through R_{in}, producing a voltage drop:

$$V_{Rin} = 205\ \mu A \times 2.2\ k\Omega = 451\ mV$$

Because of the polarity of R_{in}, when electrons flow from right to left, add this voltage to the reference voltage to obtain:

$$V_{in} = V_{ref} + V_{Rin}$$
$$V_{in} = -6\ V + 451\ mV$$
$$V_{in} = -5.55\ V$$

Therefore:

$$UTP = -5.55\ V$$

The hysteresis loop for the circuit of Figure 14.45 is graphed in Figure 14.48.

Progress Check
*You have now completed
objective 9.*

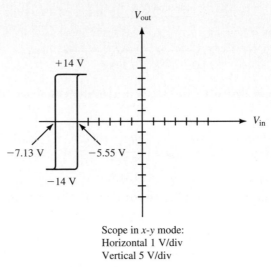

Scope in x-y mode:
Horizontal 1 V/div
Vertical 5 V/div

FIGURE 14.48   **Hysteresis loop for the circuit of Figure 14.45**

PRACTICE PROBLEM 14.16

Determine the UTP and LTP for the circuit of Figure 14.49.

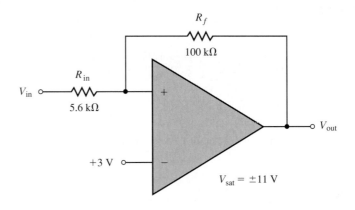

FIGURE 14.49

REVIEW SECTION 14.2

1. In its most basic form a comparator has three inputs and one output. True or false?
2. Op-amps can never be used for comparator applications. True or false?
3. _____ rate limits the speed at which an op-amp can respond as a comparator.
4. Many comparator ICs have _____ collector outputs.
5. Internal compensating capacitors are found inside _____ _____ but not _____ ICs.

6. A comparator designed with hysteresis has two trip points known as the _____ _____ _____ and the _____ _____ _____.

7. In a circuit designed with two trip points, these trip points are always equal and opposite. True or false?

8. Once an input signal has exceeded the UTP, further increases in the input signal cause the output to switch states. True or false?

9. An input signal crosses the LTP but is going positive. This forces the output of the comparator to change states. True or false?

10. Including hysteresis in a crossing detector improves its _____ immunity.

11. In an inverting Schmitt trigger, when the input rises above the UTP (going positive), the output of the comparator goes positive. True or false?

12. The center of the hysteresis loop for a noninverting Schmitt trigger is always equal to the reference voltage applied to the circuit. True or false?

13. The UTP of a Schmitt trigger must be greater than 0 V. True or false?

14. The hysteresis voltage of a circuit can be determined by subtracting the _____ from the _____.

14.3 WINDOW COMPARATORS

Figure 14.50 shows a window comparator in which the input voltage is applied to two different comparators at the same time. Assume that the UTP voltage applied to the upper comparator is 5 V. In addition, the LTP voltage applied to the lower comparator is equal to 2 V.

When the input voltage is at 0 V, the upper comparator has a negatively saturated output. This is because the voltage at the UTP input terminal is +5 V. This is the inverting terminal to the comparator, which means that diode D_1 is reverse-biased. No current flows through this diode, so no voltage is dropped across R_L due to currents from the upper comparator.

The lower comparator is at positive saturation. When the input voltage is at 0 V, the voltage at the LTP forces the output of the comparator positive, turning on D_2. Current flows up through R_L, through D_2, producing an output voltage across R_L.

The only time the output voltage is equal to zero is when both comparators are in negative saturation. For that to occur the input voltage must be greater than +2 V (the LTP) and less than +5 V (the UTP). This gap (in this case, a gap of 3 V) is known as a "window."

A voltage of +4 V with respect to ground is within the window, and the output voltage will be zero. Why? With an input of 4 V, the +5 V at the UTP input terminal causes the upper comparator to have an output at negative saturation. D_1 is reverse-biased, and no current flows through the load, R_L.

In addition, the same +4 V is larger than the +2 V at the LTP input of the lower comparator. This causes the output of the lower comparator to be negatively saturated, reverse biasing D_2, which means no current flows through R_L. Consequently there is 0 V on the output.

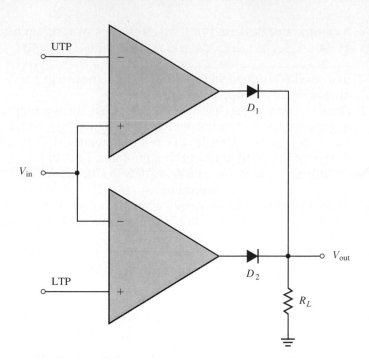

FIGURE 14.50   A window comparator

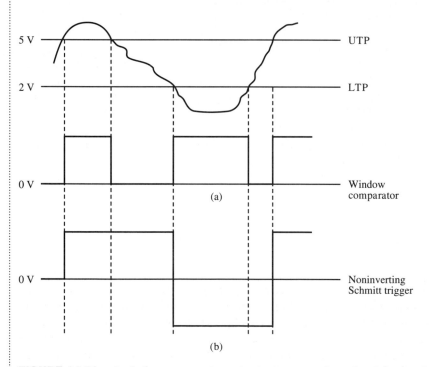

FIGURE 14.51   A window comparator output versus a noninverting Schmitt trigger output

The output response of a window comparator and the output of a noninverting Schmitt trigger with the same trip points are shown in Figure 14.51. As you can see, the two types of circuits do not produce the same output response.

EXAMPLE 14.18

Figure 14.52 shows a window comparator where resistive voltage dividers are used to establish trip points. Determine the UTP and LTP of this circuit.

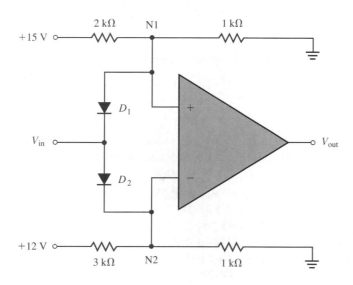

Power supplies ± 17 V

FIGURE 14.52   A single-comparator window comparator

Solution
The noninverting input is connected to N1 in a voltage divider powered by +15 V. This divider produces:

$$V_{N1} = \frac{1\,k\Omega}{2\,k\Omega + 1\,k\Omega} \times 15\,V$$

$$V_{N1} = 5\,V$$

The voltage divider connected to the inverting terminal produces

$$V_{N2} = \frac{1\,k\Omega}{1\,k\Omega + 3\,k\Omega} \times 12\,V$$

$$V_{N2} = 3\,V$$

When the input is grounded, the diode D_2 is reverse-biased while diode D_1 is forward-biased. This means that the voltage at the noninverting terminal is reduced to +0.7 V. Since diode D_2 is reverse-biased the voltage at the inverting terminal remains at +3 V. This causes the output of the comparator to be at negative saturation. Figure 14.53 shows the transfer characteristic graph for this window comparator.

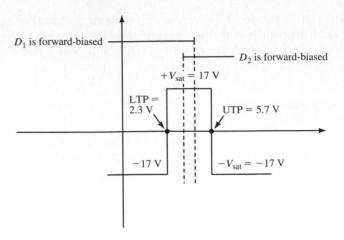

FIGURE 14.53   Transfer characteristic graph for the circuit of Figure 14.52

As the input voltage is raised, a point is reached where the input voltage plus diode drop D_1 combine to equal 3 V. Diode D_2 is still reverse-biased.

$$V_{in} = 2.3 \text{ V}$$
$$V_{N1} = V_{in} + V_{D1} = 3 \text{ V}$$

The comparator begins to switch states, with the output to be at positive saturation. This is also the LTP of the window.

Increasing the input voltage to $+3.7$ V forward biases D_2. At $+3.7$ V, D_1 continues to be forward-biased. The noninverting input voltage, V_{N1}, is

$$V_{N1} = V_{in} + V_{D1}$$
$$V_{N1} = +3.7 \text{ V} + 0.7 \text{ V}$$
$$V_{N1} = +4.4 \text{ V}$$

Because the noninverting input is greater than the inverting input, the output continues to be at $+V_{sat}$.

Increase the input voltage to $+4.3$ V. D_1 is now at its reverse bias point. However, V_{N1} is still larger than V_{N2}, causing the output to remain at $+V_{sat}$.

In order for the inverting input to be a greater voltage than the noninverting input, V_{in} must be slightly more than $+5.7$ V.

V_{N1} is at $+5$ V because D_1 is reverse-biased. N2's voltage may be calculated:

$$V_{N2} = V_{in} + V_{D2}$$
$$V_{N2} = +5.7 \text{ V} - 0.7 \text{ V} = +5 \text{ V}$$

Any further increases in input voltage cause the inverting input to have a greater voltage. The output switches to $-V_{sat}$.

$$UTP = +5.7 \text{ V}$$

Progress Check
You have now completed objective 10.

PRACTICE PROBLEM 14.17

Determine the UTP and LTP for the window comparator of Figure 14.54.

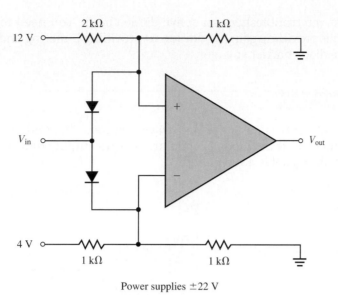

12 V ○───WWW───┬───WWW───┤
    2 kΩ   1 kΩ

Power supplies ±22 V

FIGURE 14.54

REVIEW SECTION 14.3

1. Explain what is meant by a window.
2. A window comparator whose output is $-V_{sat}$ when the input is within the window works exactly the same as a noninverting Schmitt trigger with the same trip points. True or false?
3. When the input to a window comparator crosses the UTP the output switches. It does not matter whether the input voltage is going positive or negative. True or false?
4. In the window comparator of Figure 14.52 it is not possible to have the two diodes forward-biased at the same time. True or false?

TROUBLESHOOTING

This troubleshooting section will discuss problems in active diode circuits and problems in comparators and Schmitt trigger circuits.

ACTIVE DIODE CIRCUITS

Active diode circuits can fail in ways similar to those found in passive diode circuits. You may wish to review material in Chapter 3. In addition, the types of failures found in op-amp circuits can also be found in active diode circuits. Troubleshooting suggestions involving the op-amp part of the circuit can be found in Chapter 12.

When you troubleshoot an active diode circuit, you need to decide if the op-amp is providing gain in addition to providing buffering and reducing the effective diode barrier voltage.

EXAMPLE 14.19

Figure 14.55 shows an active half-wave rectifier. The diode is in series with the signal path. If the diode is open, there is no output voltage. If the diode is shorted, the signal is not rectified.

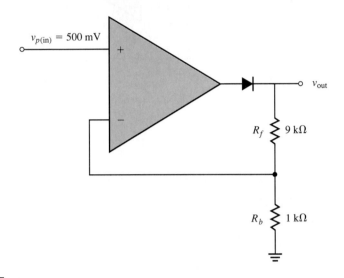

FIGURE 14.55

The op-amp circuit is set up to provide gain. The ratio of $R_f/(R_f + R_b)$ provides a gain of 10. If the output of the circuit is rectified, but not at 10 x $V_{p(in)}$, it is necessary to check the op-amp or the resistors as discussed in Chapter 12.

EXAMPLE 14.20

Figure 14.56 is an active positive clamper. If the diode is open, the feedback path is no longer functional. The input signal passes through the capacitor to the output with no shift in signal level.

If the diode is shorted, the input capacitor receives equal amounts of current on both alternations and does not charge. Thus, the output signal is again the same as the input signal.

Whether the diode is shorted or open, no clamping occurs. Turn off the power and with an ohmmeter test the diode in the circuit. In Figure 14.56 connect the meter's positive lead to the anode of the diode and the negative lead to the cathode. The meter selector should be set on its highest range. Because this forward biases the diode, you should read some resistance. Reverse the leads, and you should read an open, as the diode is now reverse-biased. CAUTION: In some circuits there may be current paths that are parallel to the diode. Be sure to take those parallel paths into consideration.

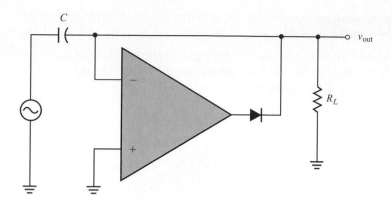

FIGURE 14.56

In the total absence of an output signal, you should check to see if the input capacitor is opened.

PRACTICE PROBLEM 14.18

First determine the function of the circuit in Figure 14.57. Determine the normal output. Then:

1. Determine if the diode is open or shorted if v_{out} is 0 V.
2. Determine if the diode is open or shorted if there is an AC signal present at the output, but the amplitude is much smaller than the input.

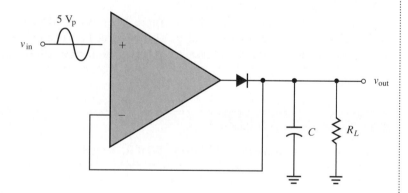

FIGURE 14.57

TROUBLESHOOTING COMPARATORS AND SCHMITT TRIGGERS

Troubleshooting a simple comparator is relatively easy since it is acting as a switch. If the circuit doesn't switch as expected, check the power supplies. If the power supply voltages are correct, replace the comparator. If the comparator switches at the wrong voltage, check to see if the input reference voltage is correct. If the comparator switches erratically, the input signal may be excessively noisy. If so, signal filtering or the addition of hysteresis into the circuit is necessary. This brings us to Schmitt triggers.

EXAMPLE 14.21

Figure 14.58 depicts an inverting Schmitt trigger, designed to have hysteresis. If the UTP or LTP is not correct, something is wrong with one of the resistors in the feedback network.

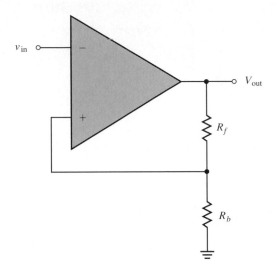

FIGURE 14.58

If R_b is too small or R_f too big, then the trip points are closer together than expected. On the other hand, if R_b is too large or R_f too small, the trip points are further apart than expected.

Next consider what happens if R_f is open. Then the feedback path is broken, and R_b acts like a pull-down resistor. This means that the circuit will now act like a comparator with no hysteresis and has a reference voltage at 0 V.

PRACTICE PROBLEM 14.19

Determine what happens in Figure 14.58 if R_f opens.

PRACTICE PROBLEM 14.20

Determine what happens in Figure 14.58 if R_b opens.

TROUBLESHOOTING QUESTIONS
1. List three functions of an op-amp in an active diode circuit.
2. Troubleshooting skills learned in _____ diode circuits can be applied to active diode circuits.

3. An active half-wave rectifier with gain amplifies the input signal but does not rectify it. Is the diode open or shorted?
4. An active diode clamper does not change the DC reference of the output to the input peak, but it does shift the signal off of ground. The diode is shorted. True or false?
5. A comparator circuit switches erratically. To improve circuit performance, hysteresis should be added to the circuit. True or false?

WHAT'S WRONG WITH THIS CIRCUIT?

In the following circuit, determine why the actual output differs from the expected output. (R_b is known to be good.)

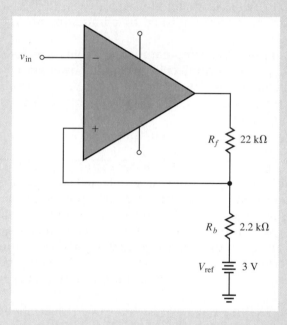

	Expected Output	**Actual Output**
UTP	4.36 V	10.5 V
LTP	1.09 V	−7.5 V

SUMMARY

Using an op-amp in an active diode circuit provides several important advantages. First, the op-amp can be used as a buffer between a signal source and a diode circuit. Second, the op-amp can prevent loading effects on the output side of the diode circuit. Third, the diode barrier potential is, ideally, reduced to zero when placed in an op-amp circuit with negative feedback. In addition to the foregoing, you should remember that active rectifiers can be built with gain, and active peak detectors can be built that have almost no ripple. Active clippers and limiters can be built so that the diode drop is not a consideration.

Finally, active clampers offset the AC signal by a level equal to V_p. Again, the diode barrier voltage does not affect the output.

A comparator circuit is one in which two different inputs can be compared to see which of the two is the larger. In its most basic form, the circuit has two inputs and a single output. This single output is designed to produce one of two different responses: $+V_{sat}$ if the noninverting input is larger than the inverting input and $-V_{sat}$ if the noninverting input is less than the inverting input.

While it is possible to use op-amps in comparator applications, many circuit designers use a variation of the op-amp specifically designed for use in comparator applications. Comparators have much higher slew rates than typical op-amps. This is accomplished by eliminating an internal compensating capacitor.

It is possible to purchase comparators with open collector outputs. If you have such a chip, you must connect its output to power via a pull-up resistor.

Crossing detectors determine when an input voltage is moving across a predefined reference voltage. In the case of a zero crossing detector, the reference voltage is ground.

A graph of V_{in} versus V_{out} for a crossing detector is known as a transfer characteristic graph.

To minimize the effects of noise on comparator circuits, crossing detectors are built with hysteresis. A crossing detector with hysteresis is called a Schmitt trigger. Schmitt triggers have two trip points known as the upper trip point (UTP) and the lower trip point (LTP). Simply crossing a trip point in a Schmitt trigger is not enough to get the output to change states. When crossing the UTP, the input must be going positive if the output is to change. When crossing the LTP, the input must be going negative for the output to change. Hysteresis improves the noise immunity of the crossing detector.

The UTP and LTP do not have to be symmetrical about zero. Circuit arrangements are possible in which the UTP and LTP do not have the same absolute magnitude.

There are inverting Schmitt triggers and noninverting Schmitt triggers. In a noninverting Schmitt trigger, when the UTP is crossed with the input going positive, the output moves to positive saturation. In an inverting Schmitt trigger, the same situation at the input causes the output to be at negative saturation.

By applying reference voltages to Schmitt triggers it is possible to move the entire hysteresis loop either above or below ground.

Window comparators produce "active" outputs when the input voltage is between two different voltages, the UTP and LTP. The trip points in window comparators do not act exactly the same as trip points in Schmitt triggers. Crossing a trip point causes the output of a window comparator to change states. It does not matter if the input is going negative or positive when it crosses a trip point in a window comparator.

1. Refer to Figure 14.59. Determine the output voltage. Sketch the output waveform. Label the peaks with respect to ground.

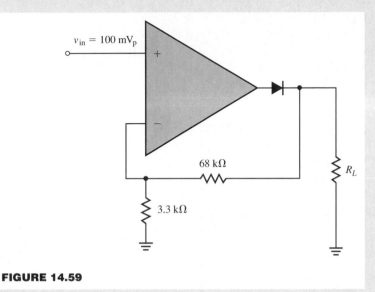

FIGURE 14.59

2. Refer to Figure 14.59. Determine the output voltage when the diode is turned around, that is, the cathode is connected to the output of the op-amp.

Refer to Figure 14.60 for problems 3–6.

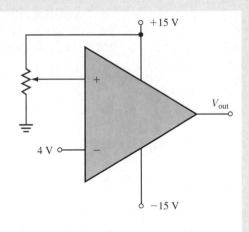

FIGURE 14.60

3. Determine the output voltage when the potentiometer delivers 0 V to the noninverting input of the op-amp.

4. Determine the output voltage when the potentiometer delivers 2 V to the noninverting input of the op-amp.
5. Determine the output voltage when the potentiometer delivers 4 V to the noninverting input of the op-amp.
6. Determine the output voltage when the potentiometer delivers 12 V to the noninverting input of the op-amp.

Refer to Figure 14.61 for problems 7–12.

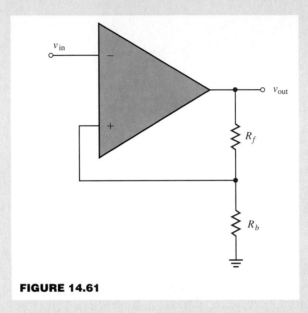

FIGURE 14.61

7. $R_f = 22$ kΩ, $R_b = 820$ Ω. The power supply saturates at ± 12 V. Determine the UTP and LTP.
8. Determine the hysteresis voltage for the conditions listed in problem 7.
9. $R_f = 22$ kΩ, $R_b = 820$ Ω, but the power supplies are changed so that saturation now occurs at ± 20 V. Determine the UTP and LTP.
10. $R_f = 22$ kΩ, but $R_b = 3.3$ kΩ. The power supplies saturate at ± 20 V. Determine the UTP and LTP.
11. $R_f = 39$ kΩ, $R_b = 3.3$ kΩ, and the power supplies continue to saturate at ± 20 V. Determine the UTP and the LTP.
12. Graph the hysteresis loop for the conditions given in problem 11.

Refer to Figure 14.62 for problems 13 and 14.

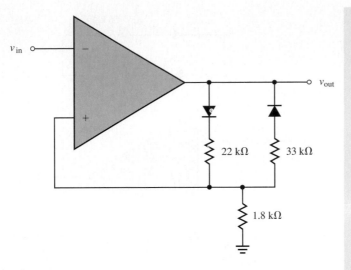

FIGURE 14.62

13. The power supplies saturate at ± 11 V. Determine the UTP and LTP.
14. Graph the hysteresis loop for the conditions given in problem 14.

Refer to Figure 14.63 for problems 15–17.

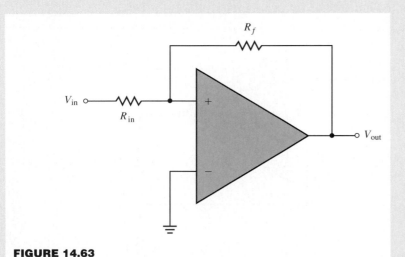

FIGURE 14.63

15. $R_f = 22$ kΩ, $R_{in} = 1.8$ kΩ. The power supplies saturate at ± 15 V. Determine the UTP and LTP.
16. Change R_{in} to 560 Ω. Determine the UTP and LTP.
17. Graph the hysteresis loop for the conditions given in problem 16.

Refer to Figure 14.64 for problems 18–20.

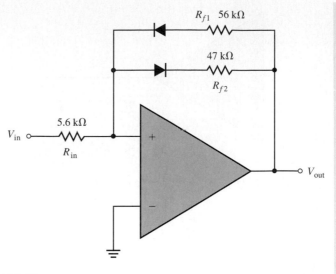

FIGURE 14.64

18. The power supplies saturate at ±21 V. Determine the UTP and the LTP.
19. Graph the hysteresis loop for the conditions given in problem 18.
20. Change R_{f1} to a 39-kΩ resistor. Recompute the UTP and LTP.

Refer to Figure 14.65 for problems 21–27.

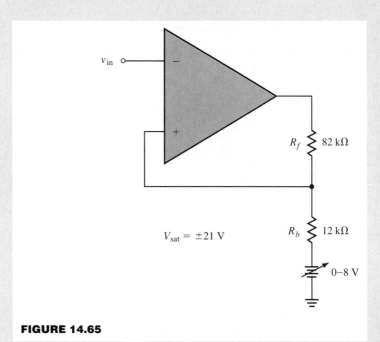

FIGURE 14.65

21. The variable battery is set to 0 V. Determine the UTP and LTP.
22. The variable battery is set to 2 V. Determine the UTP and LTP.
23. Graph the hysteresis loop for the conditions given in problem 22.
24. The variable battery is set to 5 V. Determine the UTP and LTP.

25. Graph the hysteresis loop for the conditions given in problem 24.
26. The variable battery is set to 8 V. Determine the UTP and LTP.
27. Graph the hysteresis loop for the conditions given in problem 26.

Refer to Figure 14.66 for problems 28–35.

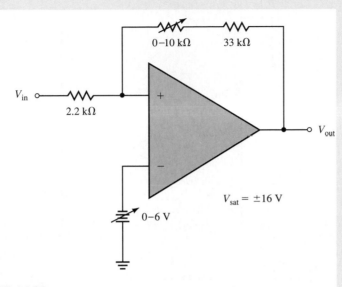

$0{-}10\ \text{k}\Omega$ $33\ \text{k}\Omega$

V_{in} $2.2\ \text{k}\Omega$

V_{out}

$V_{sat} = \pm 16\ \text{V}$

$0{-}6\ \text{V}$

FIGURE 14.66

28. The battery is set to 2 V. The potentiometer is set to 2 kΩ. Determine the UTP and LTP.
29. Graph the hysteresis loop for the conditions given in problem 28.
30. The battery is set to 4 V. The potentiometer is set to 10 kΩ. Determine the UTP and LTP.
31. Graph the hysteresis loop for the conditions given in problem 30.
32. The battery is set to 0 V. The potentiometer is set to 0 kΩ. Determine the UTP and LTP.
33. Graph the hysteresis loop for the conditions given in problem 32.
34. The battery is set to 6 V. The potentiometer is set to 7 kΩ. Determine the UTP and LTP.
35. Graph the hysteresis loop for the conditions given in problem 34.
36. Refer to Figure 14.67. Graph the transfer function of the window comparator.
37. Draw the output for the window comparator given the input shown in Figure 14.67.

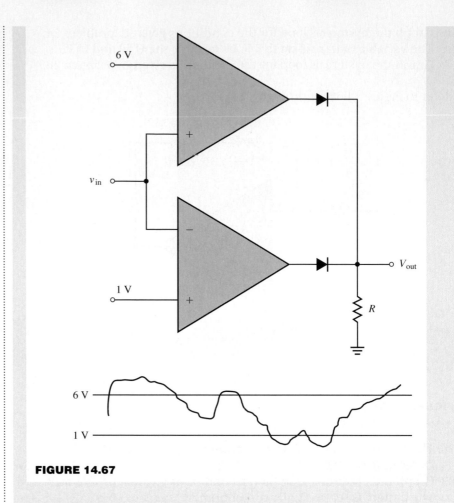

FIGURE 14.67

38. Explain the difference between a Schmitt trigger and a window comparator.

ACTIVE FILTERS

✓ **As you read this chapter, concentrate on learning how to:**

1. Define filter performance by configuration, frequency response, and time response
2. Recognize the differences between low-pass, high-pass, band-pass, and band-reject filters
3. Find the upper and lower critical frequencies of low- and high-pass filters
4. Recognize the differences between Butterworth, Chebyshev, and Bessel active filters
5. Recognize the differences between passive and active filters
6. Describe how operational amplifiers work as active filters
7. Determine component values for an active filter circuit
8. Describe VCVS, twin-notch T, and state-variable filter circuits

INTRODUCTION

The term *filter* can describe many different and commonly used items. We use filters in furnaces, automobiles, and even in home aquariums. Filters are also used in electronics. Whether a filter cleans particles from air, oil, or water or changes a signal, it removes some unwanted portion of a whole.

As we have moved through this book, we have seen how capacitors can work as filters. A filter capacitor in a power supply removes or attenuates the unwanted part of the pulsating DC signal. Indeed, we can define filtering as the process where the frequency response of a signal is modified, reshaped, or manipulated according to some desired specification.

In addition to the power supply filter, there are also filters that attenuate a range of frequencies, or that reject or isolate one part of a frequency range. Some filters remove noise or distortion introduced by transmission lines. Other filters separate previously mixed signals or limit the bandwidth of signals.

In this chapter, we will discuss how active filters function. Essentially, the chapter mixes a little of the old with a little of the new. Like the more basic passive filters that we touched on during previous chapters, active filters change the waveform of a circuit. As with passive filters, there are four basic types of active filters: low-pass, high-pass, band-pass, and band-reject. As we discuss signals and waveform analysis, you may find the discussion increasing your understanding of how resistors and capacitors affect frequencies.

15.1 DEFINING FILTER PERFORMANCE

The layout of the parts of a circuit in relation to one another defines its configuration. Figure 15.1 shows the schematic diagram for a pi filter. We can say that the filter has the pi configuration because the layout of its parts in the schematic diagram gives the appearance of the Greek letter pi (π).

Two possible filter configurations exist. In the lattice configuration shown in Figure 15.2, the components have a symmetrical arrangement. The sides of the lattice configuration mirror each other. Boxes labeled with a Z or as separate impedances in the figure represent individual resistors, capacitors, and inductors. With the lattice configuration, balanced loads may either drive or be driven by the filter.

The ladder configuration shown in Figure 15.3 is not symmetrical. Again, boxes labeled with a Z represent individual components. Because of the lack of symmetry, the ladder configuration can drive only unbalanced loads.

We can define differences between circuits by referencing circuit terminals to ground. For a filter, **balance** means with balance respect to ground. The lattice configuration has balance because we could ground any of the

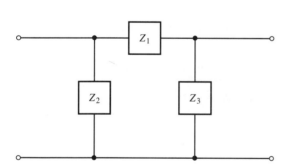

FIGURE 15.1 The configuration of a π filter

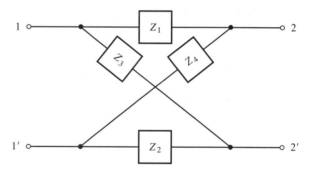

FIGURE 15.2 A lattice configuration of a filter

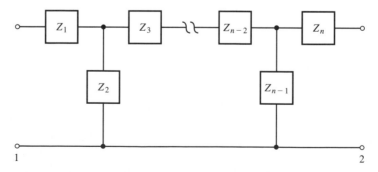

FIGURE 15.3 A ladder configuration of a pi filter

terminals without affecting performance. With the ladder configuration, we could ground only terminals 1′ and 2′ without affecting filter performance.

FREQUENCY AND FILTERS

One key characteristic of how filters work is the gain of the filter versus frequency. This relationship is graphically depicted in a frequency response curve. Look at Figure 15.4 as you read through the definitions for passband, cutoff frequency, ripples and the ripple band, transition region, and stopband.

The **passband** is the set of frequencies that the filter does not attenuate. Usually, the passband occurs between 0 dB and −3 dB. Amplitude within the passband may vary up to 3 dB.

Within the passband, variations of amplitude with frequency show up as **ripples.** As Figure 15.4 shows, the ripples sometimes define a small area called the **ripple band.**

On the horizontal scale the critical frequency, f_c designates the end of the passband frequencies. When the passband frequencies reach the critical frequency, the response begins to drop off. Attenuation increases as frequency increases past f_c.

At the critical frequency, the passband stops and the response of the filter drops. The area where the fall-off occurs is called the **transition region.** Starting at the critical frequency point, the transition region extends to the stopband. The **stopband** is the set of frequencies that have the most attenuation. Generally, the beginning of the stopband is defined by some minimum amount of attenuation such as 40 dB. For our purposes, the transition region is a part of the stopband.

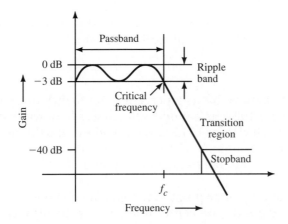

FIGURE 15.4 Frequency properties of a filter response curve

BANDWIDTH

Figure 15.5 shows a typical filter response curve. The points designated f_1 and f_2 are −3 dB or half-power points and indicate where the filter response is down 3 dB from the maximum point on the curve. In addition, those points represent a specified insertion loss on the curve.

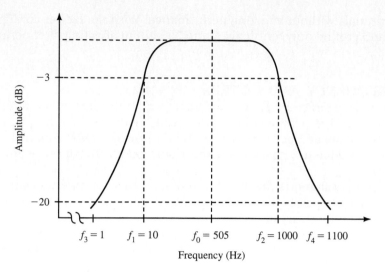

FIGURE 15.5 Illustration of bandwidth using a typical filter response curve

The **bandwidth** of a filter is the difference between the two points or the difference between the lower frequency and the upper frequency. Point f_1 represents the lower frequency at a point 3 dB down from the maximum, while point f_2 is the upper frequency at a point 3 dB down from the maximum. The equation for bandwidth is:

$$BW = f_2 - f_1$$

where f_1 and f_2 are at the −3-dB points.

PRACTICE PROBLEM 15.1

Two points on a response curve that are the −3-dB points are at 150 Hz and 2000 Hz for f_1 and f_2. What is the bandwidth of the filter?

We also can show bandwidth as a percent of frequency or percent bandwidth. As an example, a 100-kHz bandpass filter has a bandwidth of 10%. The bandwidth is 10% of 100 kHz, or 10 kHz. We would express this as ±5% or ±5 kHz with respect to the original 100 kHz. Figure 15.6 shows a technician checking the bandwidth and the frequency response of a filter used in digital communications.

In Figure 15.5, the frequency point that splits the response curve, or f_0, is the **center frequency point.** The center frequency is the algebraic center of the response curve, or the average of the lower and upper cutoff frequencies. Mathematically, we can express the center frequency as:

$$f_0(\text{Hz}) = \frac{f_1 + f_2}{2}$$

$$f_0 = \frac{1000\ \text{Hz} + 10\ \text{Hz}}{2} = 505\ \text{Hz}$$

for the algebraic center. In Figure 15.5, the center point fits halfway between the −3-dB points.

FIGURE 15.6 Technician checking bandwidth and frequency response. *(Courtesy of Hewlett-Packard)*

EXAMPLE 15.1

A filter has a lower cutoff frequency of 1000 kHz and an upper cutoff frequency of 1400 kHz. Calculate the center frequency (f_0) of the filter.

Solution
Substitute the lower and upper cutoff frequency values into the equation:

$$f_0 = \frac{f_1 + f_2}{2}$$

$$f_0 = \frac{1000 \text{ kHz} + 1400 \text{ kHz}}{2}$$

$$f_0 = \frac{2400 \text{ kHz}}{2} = 1200 \text{ kHz}$$

From one perspective, the quality factor, or Q, is a measure of the sharpness of the peak. So, Q equals the center frequency divided by the width at the −3-dB points. The center frequency of a tuned circuit is the resonant frequency. The value of Q equals the center frequency divided by the bandwidth value, or

$$Q = \frac{f_0}{\text{BW}}$$

By changing the equation slightly, we can use a known value of Q to find the bandwidth of a filter circuit. The bandwidth of a tuned circuit equals:

$$\text{BW} = \frac{f_0}{Q}$$

If the filter circuit used in example 15.1 has a center frequency of 10.7 MHz and a Q of 53.5, what is the bandwidth of the circuit?

A frequency value called the **critical frequency** divides the high-pass frequency region from the low-pass frequency region. The critical frequency, or f_c, occurs when the capacitive reactance equals the value of the resistance in a filter. Figure 15.7 illustrates the location of the critical frequency on a characteristic curve.

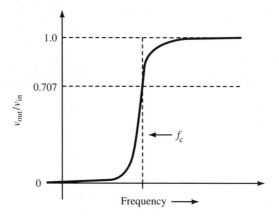

FIGURE 15.7 The critical frequency

POLES

Figure 15.8 shows the frequency response curve of a low-pass filter. To the right of the response curve knee, or beyond the critical frequency, the output amplitude proportionally drops at a rate of -6 dB per one octave. An octave is a doubling of frequency. Thus, a simple RC (resistor-capacitor) filter has a 6-dB-per-octave fall-off.

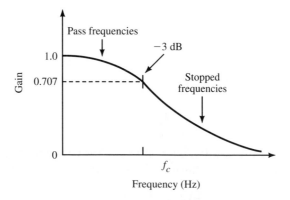

FIGURE 15.8 Frequency response curve of a low-pass filter

We can construct filters with different numbers of RC sections. A filter with two RC sections has a 12-dB-per-octave fall-off, while a filter with three RC sections has an 18-dB-per-octave fall-off. We call a filter that either has three RC sections or acts like a filter with three RC sections a "three-pole filter." Each RC section in the filter represents a pole and has a given amount of frequency fall-off.

FREQUENCY LOSS

In all filters, the range of frequencies that pass through the filter experience some loss. We define that loss as **insertion loss.** Removing the filter from a circuit and directly connecting the source and load to one another would increase the output signal by the amount of insertion loss. Some filter designs create more loss than others.

Parasitic elements cause unwanted attenuation of desired frequencies and can occur with any filter design. Along with unwanted attenuation, parasitic elements can change the shape of the frequency response curve. Losses in iron cores of chokes, distributed wiring capacitance, capacitance between windings of a choke, resistance between wiring connections, and leakage inductance are examples of parasitic elements.

SELECTIVITY

We can measure the performance of a filter through its selectivity. Selectivity is the ability of the filter to pass certain frequencies while rejecting others. A perfect filter has no insertion loss throughout the passband and infinite attenuation of all frequencies in the stopband. Unfortunately, perfect filters do not exist. A given band of frequencies causes the change from relatively little attenuation to greater attenuation.

Selectivity is measured by finding the shape factor of a filter. The shape factor is the ratio of -3-dB bandwidth to the bandwidth where the insertion loss equals 20 dB. For the response curve shown in Figure 15.5, the shape factor (S) is:

$$S = \frac{f_4 - f_3}{f_2 - f_1}$$

While points f_4 and f_3 are down 20 dB from the maximum, points f_1 and f_2 are down 3 dB from the maximum. The ideal shape factor is 1.

PRACTICE PROBLEM 15.3

Referring to Figure 15.5, use the values for f_1, f_2, f_3, and f_4 to find the shape factor of the filter.

PHASE SHIFT

As signals pass through a filter, a phase shift occurs between the input and output signals. Ideal filter action has phase shift increasing linearly with frequency. A filter with ideal phase characteristics is called a linear-phase filter.

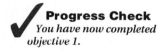
TIME AND FILTERS

We can use time to describe filter action. Using time, we refer to rise time, overshoot, ringing, and settling time. Figure 15.9 illustrates each variable. As the figure shows, timing errors can cause problems such as overshoot and ringing.

The time required for the filter response to reach 90% of the final value is called **rise time** t$_r$.

Settling time t$_s$ is the amount of time that the filter response needs to reach a specified amount of the final value and then rcmain stationary.

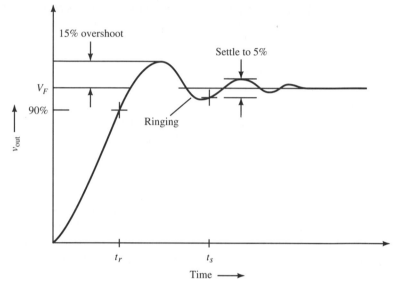

FIGURE 15.9 Time properties of a filter response curve

REVIEW SECTION 15.1

1. The two basic types of configurations seen with filters are the
   _____ and _____ arrangements.
2. Name a key characteristic of filter performance.
3. The area between the cutoff frequency and the passband frequency
   is called the _____ _____.
4. Describe phase shifting.
5. If the phase shift varies linearly with frequency, the filter produces a
   _____ .
6. Define bandwidth.
7. What is a pole?
8. Insertion loss and parasitic elements are two desirable features of a
   filter. True or false?
9. How does the Q factor affect bandwidth?
10. Define attenuation.
11. What is the transition region?

Many different ways exist for identifying filter types. We can identify filters through the shape of characteristic curves. For example, with a sharp cutoff filter, the characteristic curve shows a sharp distinction between the frequencies that pass through the filter and the frequencies that do not pass through the filter. The most popular scheme of classifying filters relies on the range of frequencies that a filter passes and rejects. With that scheme, only four basic types of filters exist: low-pass, high-pass, band-pass, and band-reject.

Sometimes, the layout of filter components classifies the filter as a particular type. A pi filter resembles the Greek letter while a ladder circuit looks like a ladder. The method of design sometimes identifies filters. For example, a Miller feedback filter is based on the use of capacitive feedback to produce high-frequency cutoff. Other filters—such as the Chebyshev, Butterworth, and Bessel filters—take their names from the name of the person who originated the design or method of calculation.

Before applying a specific filter to a given task, a technician must identify the desirable and undesirable frequencies for the circuit. He or she should know about requirements such as passband flatness and the amount of attenuation of a frequency outside the passband. Only then can a technician choose the correct filter configuration and calculate the values of the components used in the filter.

LOW-PASS, HIGH-PASS, BAND-PASS, AND BAND-REJECT PASSIVE FILTERS

Figure 15.10 shows a passive filter network. We know that a passive filter consists of a series of interconnected elements. The network features components such as inductors and capacitors connected in such a way that they respond to certain signals differently. A perfect bandstop filter network will produce the response shown in Figure 15.11.

Many stereo amplifier designs use low-pass filters to eliminate radio interference. Generally, filters used for smoothing power sources are low-pass filters similar to the one pictured in Figure 15.12. As you may recall, a low-pass filter allows all frequencies from DC to the upper cutoff frequency to pass. To find the value of the upper frequency, we use:

$$f_2 = \frac{1}{2\pi\,RC}$$

The equation tells us the frequency where the capacitive reactance equals the resistance.

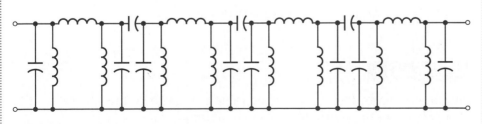

FIGURE 15.10 A complex passive filter

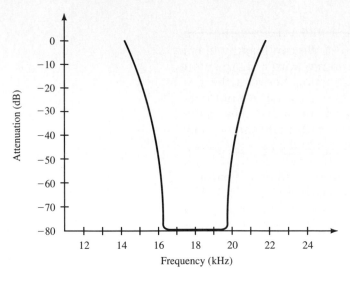

FIGURE 15.11   Ideal filter response curve

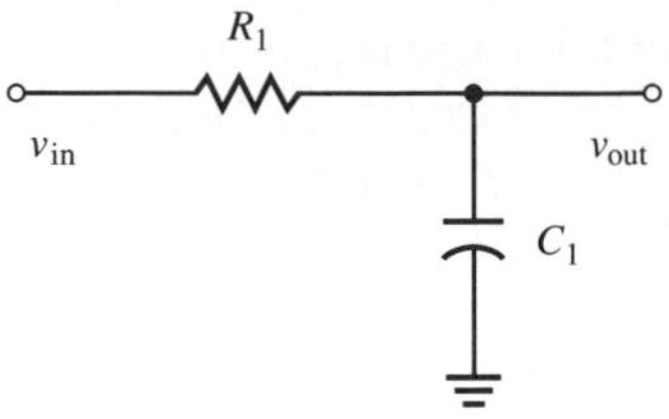

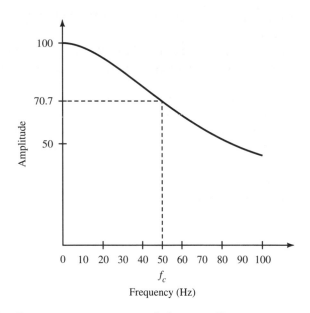

FIGURE 15.12   Low-pass passive filter

Figure 15.13 shows the frequency response of a low-pass filter. Any frequencies above the upper cutoff frequency do not pass. The "pass-through" zone lies between the upper and lower bandwidth limits, which are 50 Hz and 0 Hz respectively.

FIGURE 15.13   Frequency response curve of a low-pass filter

Referring to Figure 15.12, assume the value of R_1 is 10 kΩ and the value of C_1 is 0.15 μF. Find the value of the upper cutoff frequency.

Solution

We can substitute the circuit values for the resistor and capacitor into the equation. This gives us:

$$f_2 = \frac{1}{2\pi RC}$$

$$f_2 = \frac{1}{2\pi\left(10\text{ k}\Omega \times 0.15\,\mu\text{F}\right)}$$

$$f_2 = 106\text{ Hz}$$

PRACTICE PROBLEM 15.4

Using the circuit from Figure 15.12, find the upper cutoff frequency if R_1 equals 33 kΩ and C_1 equals 0.05 µF.

A high-pass filter like the one shown in Figure 15.14 does the opposite of the low-pass filter. Note that the resistor and capacitor have switched places. Frequencies above the lower cutoff frequency pass through, while the filter stops frequencies below the cutoff point. As with the low-pass filter, the frequency or set of frequencies that pass through the filter depend on the filter parameters. The equation for finding the lower cutoff frequency is:

$$f_1 = \frac{1}{2\pi RC}$$

Figure 15.15 shows the frequency response curve for a high-pass filter.

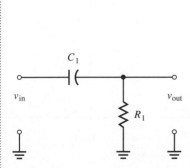

FIGURE 15.14 High-pass passive filter

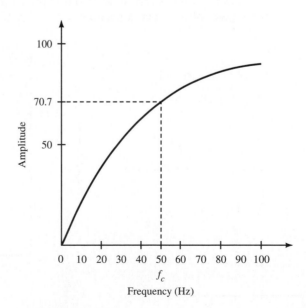

FIGURE 15.15 Frequency response curve of a high-pass filter

EXAMPLE 15.3

Find the lower cutoff frequency of a high-pass RC filter like the one shown in Figure 15.14. The resistor has a value of 15 kΩ, while the capacitor equals 0.012 μF.

Solution
We can substitute the circuit values for the resistor and capacitor into the equation. This gives us:

$$f_1 = \frac{1}{2\pi\left(15\ k\Omega \times 0.012\ \mu F\right)}$$
$$f_1 = 884\ Hz$$

Using the circuit from Figure 15.14, find the lower cutoff frequency if R_1 equals 18 kΩ and C_1 equals 0.02 μF.

While band-pass filters pass all frequencies within their bandwidths, band-reject filters eliminate all frequencies within their bandwidths. Both the band-pass and band-reject filters rely on the cascading of low- and high-pass filters. Figure 15.16 shows a passive band-pass filter. The frequency response curve shown in Figure 15.17 illustrates which frequencies are allowed to pass through the filter.

A band-pass filter also may result from the cascading of overlapping low-pass and high-pass filters with some circuits utilizing multistage band-pass filters. Because of some circuit requirements, though, cascading does not always work when making band-pass filters. Extremely sharp or high-Q band-pass filters have a greater sensitivity to component values than other band-

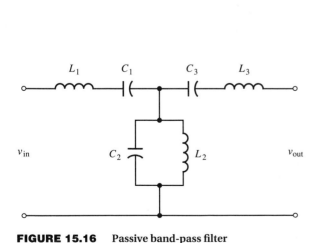

FIGURE 15.16   Passive band-pass filter

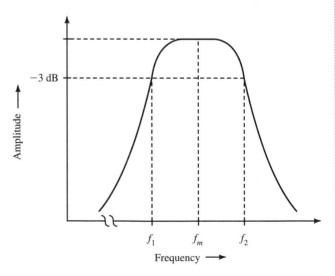

FIGURE 15.17   Frequency response curve of a band-pass filter

pass filter types. A high-Q, single-stage band-pass filter works better for that type of circuit.

Like the multistage amplifiers that you were introduced to in Chapter 5, band-pass filters can have several stages. While studying the simple band-pass filter and the response curve shown in Figures 15.16 and 15.17, think about the purpose of a band-pass filter. Since we want to pass a certain set of frequencies, we can use the combination of a low-pass and a high-pass filter to cut off other frequencies.

Figure 15.18 shows a two-stage band-pass filter using operational amplifiers. Note how stage affects the response of the filter. The first, or low-pass, stage of the filter eliminates all frequencies above its upper cutoff frequency. All frequencies below that point pass through the filter and move into the second stage. There, a high-pass filter passes all frequencies above its lower cutoff frequency. All frequencies below that point are stopped.

When we began this discussion, we noted that the cascaded filters overlapped. In other words, the passbands of the low- and high-pass filters have certain frequencies in common. Those frequencies pass through the band-pass configuration. To find the value of the frequencies that pass through, we simply subtract the value of the lower cutoff frequency from the value of the upper cutoff frequency. The bandwidth of a band-pass filter equals:

$$BW = f_2 - f_1$$

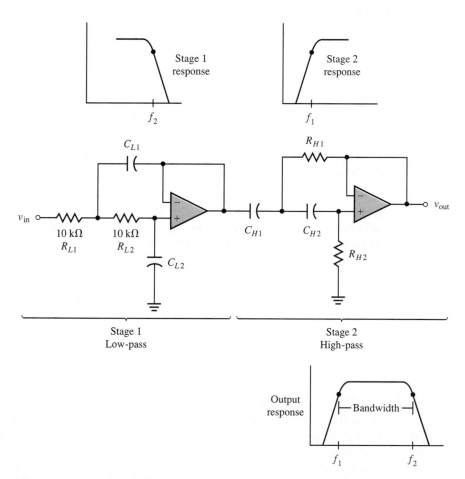

FIGURE 15.18   Two-stage band-pass filter using operational amplifiers

EXAMPLE 15.4

Determine the bandwidth, center frequency, and Q of a band-pass filter.

For this example, the low-pass filter section sets an upper frequency cutoff value of 2 kHz. The high-pass filter section sets a lower frequency cutoff value of 500 Hz.

Solution
The band-pass filter circuit passes all frequencies between 2 kHz and 500 Hz. Since we know the values of the upper and lower cutoff frequencies, we can find the bandwidth of the circuit. The bandwidth is:

$$BW = f_2 - f_1$$
$$BW = 2 \text{ kHz} - 500 \text{ Hz} = 1.5 \text{ kHz}$$

Going back to the definition of center frequency we can use that equation to find a center frequency value. The center frequency equals:

$$f_0 = \frac{(f_2 + f_1)}{2}$$
$$f_0 = \frac{(2000 \text{ Hz} + 500 \text{ Hz})}{2}$$
$$f_0 = 1250 \text{ Hz}$$

Now we can find a value for the circuit Q:

$$Q = \frac{f_0}{BW}$$
$$Q = \frac{1250 \text{ Hz}}{1500 \text{ Hz}}$$
$$Q = 0.83$$

PRACTICE PROBLEM 15.6

Draw a two-stage band-pass filter. For this problem, the values of R and C for the low-pass section of a band-pass filter are: R = 18 kΩ and C = 0.01 μF. The values of R and C for the high-pass section are: R = 10 kΩ and C = 0.025 μF. Calculate the upper and lower cutoff frequencies of the filter sections, the bandwidth of the filter circuit, the center frequency of the filter circuit, and the circuit Q.

Figure 15.19 shows a passive band-reject filter. Another name for a band-reject filter is notch filter. One look at the frequency response curve for a band reject filter shown in Figure 15.20 discloses where the term *notch* originated.

In a multistage notch filter, high-pass and low-pass filter sections work together to produce the set of rejected frequencies. With the high-pass filter setting the value for the lower cutoff frequency and the low-pass filter setting the value for the upper cutoff frequency, the area between the two frequencies is the bandwidth of the filter. As Figure 15.20 shows, all frequencies outside the filter bandwidth pass through the filter.

Progress Check
You have now completed objectives 2 and 3.

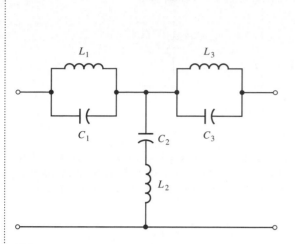

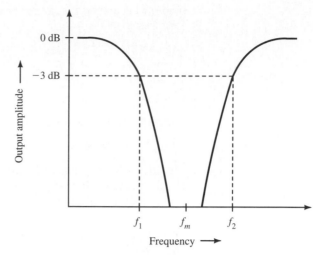

FIGURE 15.19 Passive band-reject filter

FIGURE 15.20 Frequency response curve of a band-reject filter

CHOOSING THE RIGHT FILTER

Filter selection usually begins with the technician choosing the filter best suited for the task. Several filter designs answer specific requirements for passband flatness, attenuation of a frequency outside the passband, and passing undistorted signals within the passband. Along with deciding on the number of poles necessary for the task, a technician must also look at the frequency response of the filter.

Three filter types, the Butterworth, the Chebyshev, and the Bessel, provide several options when considering passband response, sharp transition curves, and time response. Of the three, Butterworth filters have the flattest passband. Chebyshev filters have the steepest transition from passband to stopband. A Bessel filter gives the best response when working with a time delay. All the filter types have low-pass, high-pass, band-pass, and band-reject configurations.

Although each filter type has a different set of characteristics, their general appearances are similar. The different characteristics occur because of different passive and active component values and configurations. Consequently, you may see the same circuit with different values representing the Chebyshev, Butterworth, or Bessel characteristics.

Each filter type, although similar in appearance to the others, answers specific criteria regarding passband flatness, attenuation of some frequency outside the passband, steep transition from passband to stopband, or the ability of the filter to pass certain signals without distortion. Recognizing that the three filter types represent two extremes and a compromise, we can optimize gain and phase by selecting different component values. Since Butterworth filters have become more popular than the Chebyshev or Bessel filters, most examples cover Butterworth filters.

THE BUTTERWORTH FILTER

Butterworth filters have the same configuration as the filter shown in Figure 15.21 and have the flattest passband response when compared with other filter types. Another name for a Butterworth filter is the maximally

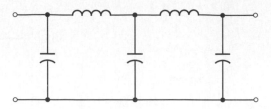

FIGURE 15.21   Passive Butterworth filter

flat filter. No or very little ripple is present in the passband. Any noticeable ripple occurs when filter components change from specified tolerance ratings. Because of its flat amplitude response, the Butterworth filter offers a constant impedance to signals with frequencies in its passband range. The constant impedance translates into constant gain across the passband of the circuit.

Butterworth filters have a good—but less than ideal—phase response. Phase response is a measure of how much phase shift occurs as the signal passes through the filter. If the phase shift in a filter varies linearly with frequency, the filter acts as a constant time delay for any signal within the passband. With good phase response, the passband waveform displays little or no distortion.

Figure 15.22 shows a two-pole, low-pass Butterworth filter circuit. An emitter-follower transistor and a network of resistors and capacitors make up the filter. Bias is supplied through the division of the 10-kΩ series input resistor into two 20-kΩ resistors. Figure 15.23 shows the characteristic curve for a Butterworth filter. The curve has a 3-dB loss at the cutoff frequency and shows a straight line in the cutoff region. The phase curves of a Butterworth filter are linear with frequency below cutoff.

Analyzing the Butterworth filter is a straightforward process. Since we also know that the filter is a low-pass filter, we can label the frequency as the

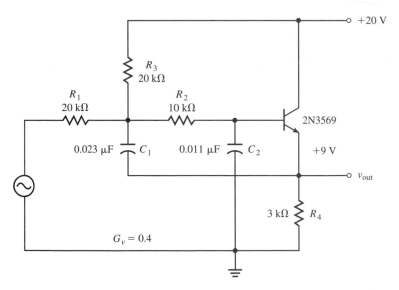

FIGURE 15.22   Two-pole Butterworth filter

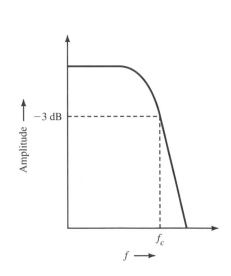

FIGURE 15.23   Frequency response curve of a Butterworth filter

upper cutoff frequency. Consequently, our equation for the single-pole filter should look like this:

$$f_2 = \frac{1}{2\pi\,RC}$$

As you solve the following example, note that the equation to find the upper cutoff frequency for a two-pole Butterworth filter changes to:

$$f_2 = \frac{1}{2\pi\sqrt{(R_1 \parallel R_3)R_2C_1C_2}}$$

EXAMPLE 15.5

Refer to the two-pole Butterworth filter shown in Figure 15.22. If resistors R_1 and R_3 equal 10 kΩ, R_2 equals 1500 Ω, capacitor C_1 equals 0.010 µF, and C_2 equals 0.25 µF, what is the value of the upper cutoff frequency?

Solution
The upper cutoff frequency for a two-pole Butterworth filter is

$$f_2 = \frac{1}{2\,\pi\sqrt{(R_1 \parallel R_3)R_2C_1C_2}}$$

Substituting component values into the equation, we find that:

$$f_2 = \frac{1}{2\pi\sqrt{5\ \text{k}\Omega \times 1.5\ \text{k}\Omega \times 0.01\ \text{µF} \times 0.25\ \text{µF}}}$$

$$f_2 = 8.49\ \text{MHz}$$

PRACTICE PROBLEM 15.7

Refer to Figure 15.22. If R_1, R_2 and R_3 equal 10 kΩ, C_1 equals 0.022 µF, and C_2 equals 0.047 µF, what is the value of the upper cutoff frequency?

Refer to Figure 15.23. Butterworth filters trade off other characteristics for the maximum flatness of amplitude response. Because of this tradeoff, the graph of a Butterworth filter response starts out flat at zero frequency and bends over near the cutoff frequency or near the -3-dB point.

The intentional flat response of the Butterworth filter sometimes hurts its performance. Some applications require filters that sharply cut off adjacent frequencies. Variation in passband response causes a gradual roll-off near the cutoff frequency point on the frequency response curve. To compensate for the variation in passband response and to give a sharper knee on the response curve, some designs for Butterworth filters add even more poles.

The flat response comes at the expense of steepness in the transition region. In the transition region, the response from passband to stopband flattens. Mathematically, the amplitude response of a Butterworth filter equals:

$$A_{V_{\text{Filter}}} = \frac{1}{\left(1 + (F/F_c)^2\right)^{n/2}}$$

where n equals the number of poles. Increasing the number of poles flattens the passband response and causes a steeper fall-off of the stopband.

Even considering the tradeoffs needed to achieve a flat response, Butterworth filters are the most commonly used active filters. As you will see when you look at the Chebyshev and Bessel filter designs, the Butterworth design offers a compromise. The compromise involves decent phase and roll-off characteristics combined with the desirable flat response.

THE CHEBYSHEV FILTER

As Figure 15.24 shows, Chebyshev filters have the same LC (inductor-capacitor) configuration as Butterworth filters. Like the Butterworth filter, the Chebyshev filter does not exhibit ideal phase characteristics. The important difference between Butterworth and Chebyshev filters is the flatness of the passband response. While Butterworth filters exhibit no ripple, Chebyshev filters have ripple in the passband response. Also, Chebyshev filters do not have the consistent gain and phase response seen with Butterworth filters.

Chebyshev filter specifications depend on the number of poles and the amount of passband ripple. All frequencies in the passband have equal importance with the Chebyshev design. Other filter designs make the zero frequency the most important frequency. With a low ripple, the phase and amplitude characteristics of the Chebyshev do not differ much from the characteristics of the Butterworth design.

Although Chebyshev filters have the disadvantage of allowing ripples throughout the passband, the filter has a sharper cutoff above the cutoff frequency. Figure 15.25 shows a two-pole Chebyshev filter using a common-emitter transistor and a Darlington amplifier. In the filter circuit, the sharp cutoff requires a large spread of capacitor values. The Darlington amplifier provides the close-tolerance amplifier characteristics needed to produce the Chebyshev response.

Figure 15.26 shows the frequency response of a Chebyshev filter. Even with ripples that have an amplitude of only 0.1 dB, the Chebyshev filter has a sharper knee than the Butterworth filter. Greater ripple amplitude produces steeper attenuation outside the passband edge. More passband ripple gives a sharper response knee.

The Chebyshev or equiripple filter improves the transition region because it spreads equally sized ripples throughout the passband. In effect, the Chebyshev balances the disadvantage of having ripples in the passband against the advantage of a steeper transition region.

Some circuit applications accept some ripple in the passband, with the only requirement being that the ripple stay below a given level. Chebyshev filters are useful for sinusoidal signals such as speech signals that accept the

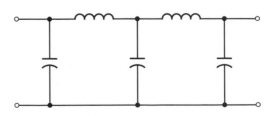

FIGURE 15.24   Configuration of a Butterworth or Chebyshev filter

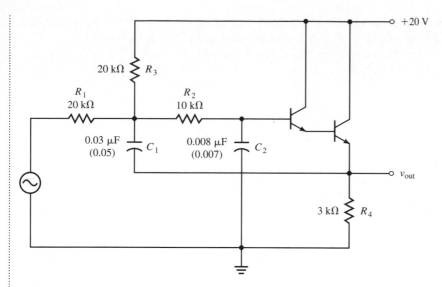

FIGURE 15.25   Two-pole Chebyshev filter

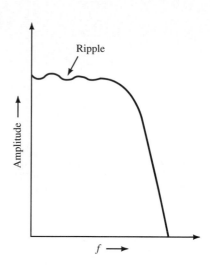

FIGURE 15.26   Frequency response of a Chebyshev filter

waveform distortion caused by the poor phase-frequency characteristics. Considering the acceptance of ripple and the sharper knee, the Chebyshev filter is a practical filter design.

THE BESSEL FILTER

Neither the Butterworth nor the Chebyshev filter has ideal phase characteristics. For some applications, the flat response of a filter or a sharp roll-off is not as important when compared with its phase characteristics. Large phase shifts distort the waveforms of signals in the passband. Situations that demand a waveform with a constant shape need a linear-phase or constant-time-delay filter. With the constant time delay, the passband waveform does not become distorted.

Bessel filters have a maximally flat time delay within the passband. Figure 15.27 shows a two-pole Bessel filter designed around a type 2N3569 transistor and a network of resistors and capacitors. The low-pass, time delay circuit provides the linear phase-frequency characteristics shown in Figure 15.28.

Look again at Figure 15.23. Butterworth filters have a maximally flat amplitude response within the passband. Figure 15.29 compares the time delay versus frequency characteristics of six-pole Butterworth and Bessel filters. When driven with pulse signals, the less-than-ideal phase response of the Butterworth filter shows as overshoot on the response curve. The response curve for the Bessel filter shows that the flat time delay characteristic of the filter cuts phase distortion.

However, a tradeoff between the two filter designs exists. Bessel filters have ideal phase characteristics but sacrifice amplitude response. Given the exceptional time constancy of the Bessel, its amplitude response between the passband frequencies and the stopband frequency is flatter than that of the Butterworth filter.

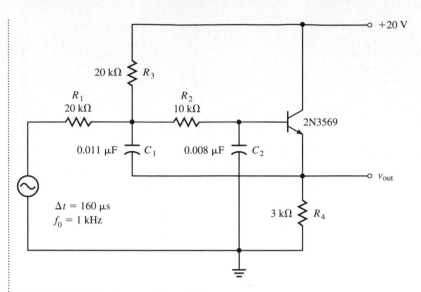

FIGURE 15.27   Two-pole Bessel filter

For the Bessel design, the cutoff frequency is used to design a network that supplies a time delay. In equation form, the cutoff frequency to time relationship appears as:

$$2\pi f_0 = \frac{1}{t_0}$$

Using that equation, a cutoff frequency equal to 1 kHz produces a time delay of 160 μs. Also, the phase shift of a Bessel filter remains linear up to the number of poles (n) multiplied by 45 degrees, or n(45°).

As an example, a 1.6-s time delay that uses a four-pole Bessel filter network has a phase shift that is approximately 180° at a frequency of 0.2 Hz. The phase shift is twice the cutoff or design frequency. When the cutoff frequency equals 1 kHz, the phase shift remains linear with frequency and is 180° at 2 kHz.

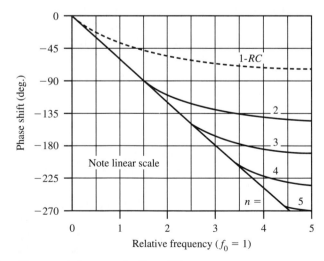

FIGURE 15.28   Characteristic curves for Bessel filters

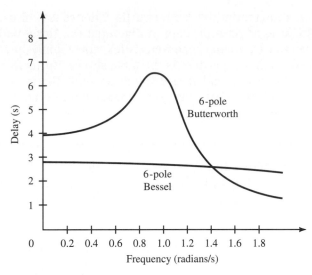

FIGURE 15.29 Comparison of time delay versus frequency characteristics of six-pole Butterworth and Bessel filters

PRACTICE PROBLEM 15.8

Determine the time delay of a Bessel filter that has a cutoff frequency equal to 3 kHz.

Since the phase characteristics needed to produce a time delay without changing the signal waveform control the frequency response of a Bessel filter, the design does not produce a flat frequency response. Also, the design does not have a specific 3-dB cutoff frequency. With the Bessel design, the 3-dB cutoff frequency equals approximately:

$$1.2 \, n(f_0)$$

Table 15.1								
Filter type	**f (3 dB/Hz)**	**No. of poles**	**Rise time (s)**	**Overshoot (%)**	**Settling time**		**Stopband (f = 2f_c)**	**Attenuation (f = 10f_c)**
					(to 1%)	**(to 0.1%)**		
Bessel	1.0	2	0.4	0.4	0.6	1.1	10	36
	1.0	4	0.5	0.8	0.7	1.2	13	66
	1.0	6	0.6	0.6	0.7	1.2	14	92
Butterworth	1.0	2	0.4	4	0.8	1.7	12	40
	1.0	4	0.6	11	1.0	2.8	24	80
	1.0	6	0.9	14	1.3	3.9	36	120
Chebyshev (0.5-dB ripple)	1.0	2	0.4	11	1.1	1.6	8	37
	1.0	4	0.7	18	3.0	5.4	31	89
	1.0	6	1.1	21	5.9	10.4	54	141

Table 15.1 shows how the Butterworth, Chebyshev, and Bessel designs fare when looking at time domain characteristics. Along with its desirable amplitude versus frequency characteristics, the Chebyshev filter has the poorest time domain properties. As the table shows, Butterworth filters fall in between the Chebyshev and Bessel filters on both the time and frequency scales. If the circuit requires only good phase characteristics, the Bessel is a desirable filter design.

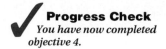

Progress Check
You have now completed objective 4.

REVIEW SECTION 15.2

1. We describe filters by their_____ , _____ and _____.

2. When choosing a filter, a technician should consider _____ _____, _____ _____, and _____ .

3. In one sense, only four types of filters exist. What are the four types of filters?

4. The purpose of a low-pass filter is _____ _____ _____.

5. A low-pass filter passes frequencies that are _____ _____ _____.

6. What is the equation for finding the upper frequency?

7. A high-pass filter passes frequencies that are _____ _____ _____.

8. What is the equation for finding the lower frequency?

9. Draw the response curves for low- and high-pass filters.

10. Band-pass and band-reject filters do the same job but have different names because of their configurations. True or false?

11. Describe the differences between Butterworth, Chebyshev, and Bessel filters. List at least five differences. Draw the response curve for each filter type.

15.3 ACTIVE FILTERS

We call a filter "active" because it employs active devices such as transistors, FETs, or operational amplifiers. Each of the two-pole filter circuits used to describe the Butterworth, Chebyshev, and Bessel designs is an example of an active filter. The active devices allow the filter to use amplifying elements to produce the same filter characteristics seen with passive filters.

However, active filters produce better characteristics at a lower cost. Along with size, weight, and cost savings, active filters do not have the signal loss seen with both RC and RL passive filters. In addition, component tolerances are not as critical as with passive filters. Using active filters also eliminates the signal loss generally seen with both RC and RL passive filters.

Within the frequency range of 0.01–100 Hz, the gain given by the active elements gives response characteristics not found with passive filters. Given adequate feedback, active filters have stable frequency characteristics between 100 Hz and 100 kHz. The resistors and capacitors that make up the feedback network control the response characteristics.

For some circuit applications, a filter design may require that the filter output signal work as the input for the active filter. Since op-amps exhibit considerable gain and have both positive and negative feedback inputs, they have gained widespread use in active filters.

Active filters may be configured as low-pass, high-pass, band-pass, or band-reject filters. Like passive filters, active filters consist of interconnected elements. The design of an active filter determines how the network responds to some sort of excitation. Depending on the degree of specialization, active filters may function in applications ranging from dedicated computers to communication systems.

As a whole, active filters should have a minimum number of parts and a small spread of parts values. In addition, active filters should be easily adjustable. From the perspective of the operational amplifier, an active filter should not have extreme requirements for slew rate, bandwidth, and output impedance. The filter characteristics should not be sensitive to component values and the op-amp gain-bandwidth product.

ACTIVE LOW-PASS FILTERS

Figures 15.30 shows operational amplifiers configured as low-pass active filters. Both filter designs provide the Butterworth response that we saw in Figure 15.23. Also, the filters are configured as noninverting amplifier. While Figure 15.30a shows a first-order low-pass section, Figure 15.30b shows a second-order low-pass section.

To analyze each filter, we need to find the voltage gain of the filter section, the cutoff frequency, and the relationship between the output voltage and the input voltage of the operational amplifier. Connected to the negative input of the operational amplifiers, resistors R_2 and R_3 control the voltage gain of the filter. The voltage gain of each filter section equals:

$$A_v = 1 + \frac{R_3}{R_2}$$

The relationship between resistance and capacitance in each section determines the half-power cutoff frequency. That frequency equals:

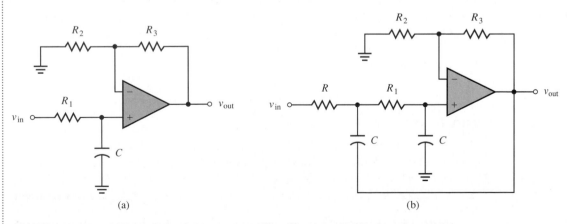

(a)   (b)

FIGURE 15.30 (a) First-order, low-pass active filter. (b) Second-order, low-pass active filter.

$$f_c = \frac{1}{2\pi RC}$$

where f_c is in hertz, R is resistance, and C is capacitance.

EXAMPLE 15.6

Use the equations for finding voltage gain, half-power cutoff frequency, and amplitude response for a Butterworth filter to analyze a first-order, low-pass active filter like the one shown in Figure 15.30a. Assume that $R_2 = 3.3$ kΩ; $R_3 = 6.2$ kΩ; $R_1 = 1$ kΩ; C = 0.01 μF, $f = 20$ kHz; and $V_{in} = 2$ V.

Solution
To find the voltage gain of the amplifier, substitute the values for R_2 and R_3 into the voltage gain equation. The voltage gain equals:

$$A_v = 1 + \left(\frac{6.2 \text{ k}\Omega}{3.3 \text{ k}\Omega} \right)$$
$$A_v = 1 + 1.8$$
$$A_v = 2.8$$

Next, find the value of the half-power cutoff frequency. That value equals:

$$f_c = \frac{1}{2\pi R_1 C}$$
$$f_c = 15.9 \text{ kHz}$$

Now, use the amplitude response equation to determine a value for the operational amplifier output voltage. The equation is:

$$A_{v(\text{filter})} = \frac{1}{[1 + (f/f_c)^{2n}]^{1/2}}$$

Substituting values into the equation gives:

$$A_{v(\text{filter})} = \frac{1}{\left[1 + \left(\dfrac{20 \text{ kHz}}{15.9 \text{ kHz}} \right)^{2 \times 1} \right]^{1/2}}$$

$$A_{v(\text{filter})} = \frac{1}{[1 + (1.25)^2]^{1/2}}$$

$$A_{v(\text{filter})} = \frac{1}{(2.58)^{1/2}}$$

$$A_{v(\text{filter})} = \frac{1}{1.61}$$

$$A_{v(\text{filter})} = 0.623$$

Last, find the overall ciruit output voltage

$$V_{out} = V_{in} \times A_v \times A_{v(\text{filter})}$$
$$V_{out} = 2 \text{ V} \times 2.88 \times 0.623$$
$$V_{out} = 3.59 \text{ V}$$

Figure 15.31 shows the frequency response of the low-pass Butterworth filter for different orders with n equaling the number of orders. By cascading

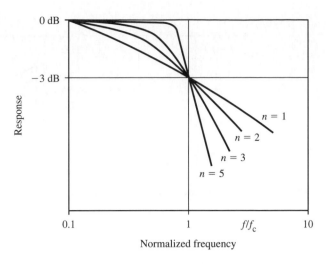

FIGURE 15.31   Frequency response curve of low-pass Butterworth filters for different orders

the basic sections and adding to the value of n, the shape factor of the characteristic curve improves. If n = 1, the first-order section from Figure 15.30a is used. When n = 2, the second-order section from Figure 15.30b is used. If n = 3, we would cascade the first- and second-order sections. A fourth-order filter, where n = 4, uses two cascaded second-order sections.

Table 15.2 lists the gain coefficients for a second-order filter that has different values of n. The coefficient is represented by the letter *a*. While a first-order filter may have some arbitrary value for its gain, second-order and other filters have defined values. The relationship between the filter section voltage gain and the coefficient value is:

$$A_v = 3 - a$$

Table 15.2 Gain coefficients for a second-order, low-pass Butterworth filter

n	a
2	1.414
3	1
4	0.765, 1.848
5	0.618, 1.618
6	0.518, 1.414, 1.932

EXAMPLE 15.7

Figure 15.32 shows a schematic drawing for a third-order, low-pass Butterworth filter. Determine the value of resistor R_3 and the output voltage for the low-pass active filter shown in Figure 15.32.

For this example, the filter has a cutoff frequency of 1 kHz. R_1 and R_2 have values of 1 kΩ and 5 kΩ. The first-order section of the filter has a gain of 10. The output frequency is 4 kHz. The input voltage equals 10 V peak to peak.

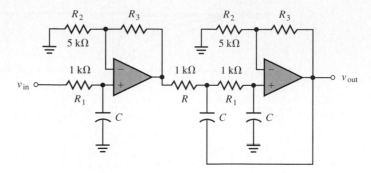

FIGURE 15.32 Active low-pass, third-order Butterworth filter

Solution

First, we find the value for the capacitance by manipulating the equation for cutoff frequency. The equation for capacitance appears as:

$$C = \frac{1}{2\pi f_c R_1}$$

Substituting values, we have:

$$C = \frac{1}{2\pi \times \left(1\text{ kHz} \times 1\text{ k}\Omega\right)}$$
$$C = 0.159\ \mu F$$

Using the equation for the voltage gain, we can find the value for R_3. We already have values for R_2 and the first-order section voltage gain. Rearranging the equation to solve for R_3 gives us:

$$A_v = \frac{R_3}{R_2} + 1$$

$$\frac{R_3}{R_2} = A_v - 1$$

$$R_3 = R_2(A_v - 1)$$
$$R_3 = 5\text{ k}\Omega\,(10 - 1)$$
$$R_3 = 45\text{ k}\Omega$$

Now, refer to Table 15.2 to find the gain coefficient for the second-order filter section. Since the example provided a third-order filter, use 0.618 for the first second-order filter section. Your equation should be:

$$A_v = 3 - a$$
$$A_v = 3 - 0.618$$
$$A_v = 2.38$$

Substitute the voltage gain value into the voltage gain equation to find a value for R_3 in the second-order section:

$$R_3 = R_2(A_v - 1)$$
$R_3 = 6.91\text{ k}\Omega$ in the first second-order filter section.

Substituting the values for frequency and the capacitance into the reactance equation, we find that the capacitive reactance equals:

$$X_c = \frac{1}{2\pi f C}$$

$$X_c = \frac{1}{2\pi \left(4 \text{ kHz} \times 0.159 \, \mu\text{F}\right)}$$

$$X_c = 250 \, \Omega$$

To calculate the output voltage, each stage gain and filter must be considered.

First, calculate the voltage applied to the first op-amp. R_1 and c form a single-pole, low-pass filter whose critical frequency is 1 kHz. The filter's gain is found by using

$$A_{v\text{filter 1}} = \frac{1}{[1+(f \, / \, f_c)^2]^{n/2}}$$

$$A_{v\text{filter 1}} = \frac{1}{[1+(4 \text{ kHz}/1 \text{ kHz})^2]^{1/2}}$$

$$A_{v\text{filter 1}} = \frac{1}{[1+(4)^2]^{1/2}}$$

$$A_{v\text{filter 1}} = \frac{1}{[1+16]^{1/2}}$$

$$A_{v\text{filter 1}} = \frac{1}{4.123} = 0.243$$

The first stage input voltage is:

$$V_{\text{in(op-amp 1)}} = V_{\text{in}} \times A_{\text{filter}}$$

$$V_{\text{in(op-amp 1)}} = 10 \, V_{\text{p-p}} \times 0.243 = 2.43 \, V_{\text{p-p}}$$

The output voltage of the first stage is:

$$V_{\text{out(op-amp 1)}} = V_{\text{in(op-amp 1)}} \times A_{v\text{(op-amp 1)}}$$

$$V_{\text{out(op-amp 1)}} = 2.43 \, V_{\text{p-p}} \times 10 = 24.3 \, V_{\text{p-p}}$$

Next calculate the effect of the second stage gain and filtering.
The second stage is a low pass, two pole filter. The filter's gain is found by:

$$A_{v\text{filter 2}} = \frac{1}{[1+(f \, / \, f_c)^2]^{n/2}}$$

$$A_{v\text{filter 2}} = \frac{1}{[1+(4 \text{ kHz}/1 \text{ kHz})^{2 \times 2}]^{1/2}}$$

$$A_{v\text{filter 2}} = \frac{1}{[1+(4)]^2}$$

$$A_{v\text{filter 2}} = \frac{1}{[1+16]^{1/2}}$$

$$A_{v\text{filter 2}} = \frac{1}{17} = 0.0588$$

The input voltage applied to the second op-amp is:

$$V_{\text{in(op-amp 2)}} = V_{\text{out(op-amp 1)}} \times A_{\text{filter 2}}$$

$$V_{\text{in(op-amp 2)}} = 24.3 \, V_{\text{p-p}} \times 0.0588 = 1.43 \, V_{\text{p-p}}$$

The output voltage of the second stage is:

$$V_{out(op\text{-}amp\ 2)} = V_{in(op\text{-}amp\ 2)} \times A_{v(op\text{-}amp\ 2)}$$

$$V_{out(op\text{-}amp\ 2)} = 1.43\ V_{p-p} \times 2.38 = 3.4\ V_{p-p}$$

PRACTICE PROBLEM 15.9

Using Figure 15.32 and Table 15.2 find values for resistor R_3 in each section and the filter output voltage when the operating frequency is 3 kHz. $R_1 = 100\ \Omega$ and $R_2 = 3\ k\Omega$ in each section. The first-order section of the filter has a gain of 7.4 and the input voltage is 1.5 V. The cutoff frequency is 2 kHz.

ACTIVE HIGH-PASS FILTERS

Figures 15.33 shows a basic first-order and second-order, high-pass Butterworth filter. The filters have almost the same appearance as the basic low-pass Butterworth filters shown in Figure 15.30. Because of the interchanging of the resistor and the capacitor, though, the circuits function as high-pass filters. Even with the change, the equations and the coefficient table used to analyze the low-pass filters still apply to the high-pass filters.

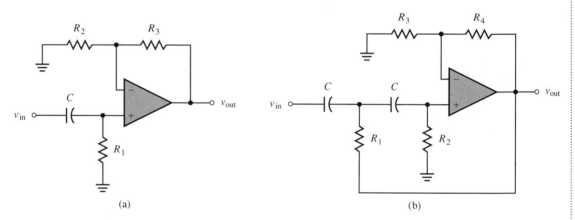

(a) (b)

FIGURE 15.33   (a) Active high-pass, first-order Butterworth filter. (b) Active high-pass, second-order Butterworth filter.

PRACTICE PROBLEM 15.10

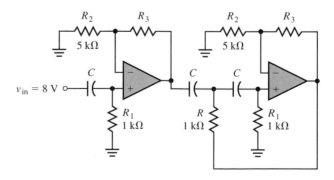

FIGURE 15.34   Active high-pass, thrid-order Butterworth filter

The filter shown in Figure 15.34 is configured as a third-order, high-pass Butterworth filter. The cutoff frequency is 10 kHz. Resistor R_1 equals 1 kΩ while resistor R_2 equals 5 kΩ for each section. The first-order section of the filter has a gain of 7.4. The input voltage is 8 V. Calculate the value of C, the values of R_3 for each section, and the filter output voltage when the operating frequency is 8 kHz.

ACTIVE BAND-PASS FILTERS

Figure 15.35 shows an operational amplifier configured as a wide band-pass filter while Figure 15.36 shows the response of the filter. The ratio of R_F to R_R determines the maximum gain of the circuit. On a response curve, the gain is shown as the flat portion of the curve. The ratio of R_F to R_1 determines the minimum circuit gain. A difference of approximately 10 dB exists between maximum and minimum gain. If the minimum gain is 10 dB, the maximum gain is 20 dB.

The relationship between capacitors and resistors sets up the shape of the passband. Ratios of the relationships between R_F, R_R, and R_N determine the relationships of the passband frequencies. R_N has a value that is 30% of the R_F value, while R_R has a value that is 30% of the R_1 value. With all other factors with the same values, an increase of the value of R_F with respect to the value of R_R will cause frequency f_1 to decrease while frequency f_2 remains unchanged. Thus, a greater spread exists between the two frequencies.

We can find the value for C_N with the equation:

$$C_N = \frac{1}{2\pi f_3 (R_N + R_F)}$$

and the value for C_R with the equation:

$$C_R = \frac{1}{2\pi f_2 R_R}$$

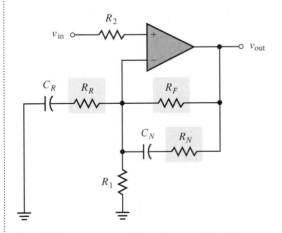

FIGURE 15.35 Wide band-pass active filter

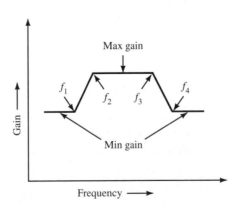

FIGURE 15.36 Frequency response curve of the wide band-pass active filter

If the value of capacitor C_N increases and all other values stay the same, the frequencies f_3 and f_4 decrease. Frequencies f_1 and f_2 remain unchanged. With the increase in capacitance, the passband narrows. An increase in the value of R_N creates the same conditions. In equation form, the frequency relationships appear as:

$$f_1 = \frac{10}{2\pi \times [C_R (R_F + 10 R_R)]}$$

$$f_2 = \frac{1}{2\pi \times C_R \times R_R}$$

$$f_3 = \frac{1}{2\pi \times [C_N (R_N + R_F)]}$$

$$f_4 = \frac{40}{2\pi \times C_N (40 R_N + R_F)}$$

EXAMPLE 15.8

Looking at Figure 15.35, determine the values for capacitors C_R and C_N. Also, calculate the values for the first and fourth frequencies of the passband.

The values of the minimum gain and the amount of gain at the passband are 100 dB and 110 dB, respectively. Frequency f_2 has a value of 200 kHz, while frequency f_3 is 400 kHz. Resistors R_1 and R_2 equal 1 kΩ. R_N has a value that is 30% of the R_F value, while R_R has a value that is 30% of the R_1 value.

With all other factors with the same values, an increase of the value of R_F with respect to the value of R_R will cause frequency f_1 to decrease while frequency f_2 remains unchanged. The ratios given by each equation provide a difference of approximately 10 dB between minimum and maximum gain. If the minimum gain equals 20 dB, the maximum gain equals 30 dB. The ratio of R_1 to R_F sets the minimum gain.

Solution
Since R_1 has a value of 1 kΩ, R_F equals 100 kΩ and R_R equals 300 Ω. Thus, R_N has a value of 30 kΩ. Given those resistance values, the minimum gain is 100 dB and the passband gain is 110 dB.

Substituting the values of R_R and f_2 into the equation for finding C_R gives us:

$$C_R = \frac{1}{2\pi \times 200 \text{ kHz} \times 300 \text{ } \Omega}$$

$$C_R = 0.00265 \text{ μF}$$

In real life, we will not find capacitors that have a value of 0.00265 μF. Instead, we would use a 0.0033-μF capacitor. Using the values for R_N, R_F, and f_3, we can find the value of C_N:

$$C_N = \frac{1}{2\pi \times 400 \text{ kHz} \times (30 \text{ k}\Omega + 100 \text{ k}\Omega)}$$

$$C_N = 3 \text{ pF}$$

In this case, the standard size for a capacitor is 3.3 pF.

Now that we have values for C_R and C_N, we can calculate the first and fourth frequencies of the passband. The first frequency equals:

$$f_1 = \frac{10}{2\pi C_R(R_F + 10R_R)}$$

$$f_1 = \frac{10}{2\pi \times 0.0033\ \mu F(100\ k\Omega + 10 \times 300\ \Omega)}$$

$$f_1 = 4.68\ kHz$$

The fourth frequency is:

$$f_4 = \frac{40}{2\pi C_N(R_F + 40R_N)}$$

$$f_1 = \frac{10}{2\pi 3.3\ pF(100\ k\Omega + 40 \times 30\ k\Omega)}$$

$$f_1 = 1.48\ MHz$$

PRACTICE PROBLEM 15.11

Determine the selectivity of the filter circuit used for example 15.8.

CASCADING FILTERS TO FORM AN ACTIVE BAND-PASS FILTER

If we cascade high- and low-pass active filters, the result can work as an active band-pass filter. The cutoff frequency of the low-pass filter section must be higher than the cutoff frequency of the high-pass filter section. If the cutoff frequency of the low-pass section is f_2 and the cutoff frequency of the high-pass section is f_1, the bandwidth of the band-pass filter equals $f_1 - f_2$.

Figure 15.37 shows a block diagram of the cascaded filter sections, while Figure 15.38 shows the schematic diagram for the band-pass filter. Figure

FIGURE 15.37 Cascading low- and high-pass filter sections result in an active band-pass filter

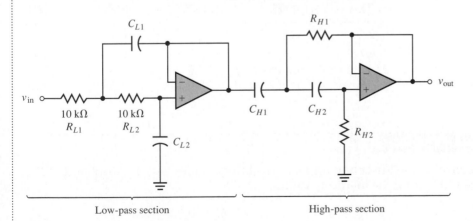

FIGURE 15.38 Schematic diagram for the band-pass filter in Figure 15.37

15.39 compares the characteristic curves of the low-pass and high-pass filter sections. Each curve represents how component values used in the filter affect its response; Table 15.3 lists the values. While any of the curves will work for the band-pass filter, curve 3 provides the sharpest knee at cutoff. Curves 1 and 2 have less slope at cutoff, while curve 4 displays peaking at the breakpoint.

The lower critical frequency (f_1) occurs when the reactance of C_L is equal to the resistance of R_L. The lower critical frequency (f_1) occurs when the reactance of C_L is equal to the resistance of R_L.

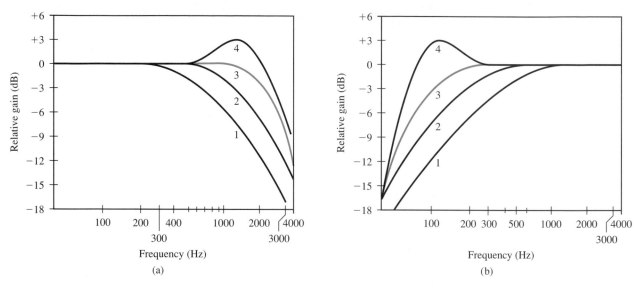

FIGURE 15.39   (a) Frequency response curve of the low-pass filter section. (b) Frequency response curve of the high-pass filter section.

Table 15.3					
Low-pass filter section			**High-pass filter section**		
Curve	**C_{L1}** **(μF)**	**C_{L2}** **(μF)**	**C_{H1}** **(μF)**	**C_{H2}** **(μF)**	**R_{H1}** **(Ω)**
1	0.003	0.015	0.3	0.1	15
2	0.007	0.010	0.3	0.3	5.1
3	0.015	0.005	0.5	0.5	1.5
4	0.05	0.002	1.0	1.0	0.51

EXAMPLE 15.9

Determine the lower and upper critical frequencies and the stopband of Figure 15.42 using the following values.

$$R_L = 1.2 \text{ k}\Omega \qquad\qquad C_L = 0.033 \text{ }\mu\text{F}$$
$$R_H = 1 \text{ k}\Omega \qquad\qquad C_H = 0.022 \text{ }\mu\text{F}$$

Solution

First, find the lower critical frequency.

$$f_L = \frac{1}{2\pi \times R_L \times C_L}$$

$$f_L = \frac{1}{2\pi \times 1.2 \text{ k}\Omega \times 0.033 \text{ }\mu\text{F}}$$

$$f_L = 4 \text{ kHz}$$

Next, find the upper critical frequency.

$$f_H = \frac{1}{2\pi \times R_H \times C_H}$$

$$f_H = \frac{1}{2\pi \times 1 \text{ k}\Omega \times 0.022 \text{ }\mu\text{F}}$$

$$f_H = 7.2 \text{ kHz}$$

Last, find the stop bandwidth.

$$BW = f_H - f_L$$

$$BW = 7.2 \text{ kHz} - 4 \text{ kHz} - 3.2 \text{ kHz}$$

PRACTICE PROBLEM 15.12

Find the upper and lower cutoff frequencies for the two-stage band-pass filter shown in Figure 15.40. In addition, calculate the center frequency and the bandwidth of the filter. What is the Q of the filter circuit?

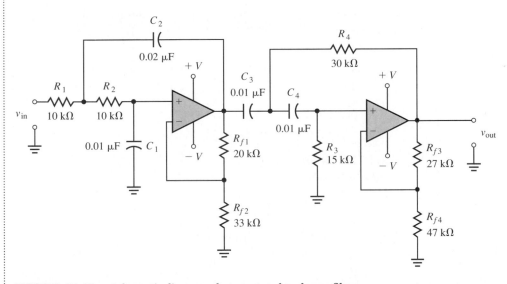

FIGURE 15.40 Schematic diagram of a two-stage band-pass filter

ACTIVE BAND-REJECT FILTERS

Figure 15.41 shows a block diagram for an active band-reject filter. In the diagram, the inputs and outputs of a low- and a high-pass filter connect to a summing amplifier. In this case, the cutoff frequency of the low-pass stage

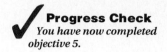
must be less than the cutoff frequency of the high-pass stage. As with the bandpass filter, the cutoff frequencies of the two filter stages make up the upper and lower frequencies of the filter. The stopband of the band-reject filter equals $f_2 - f_1$.

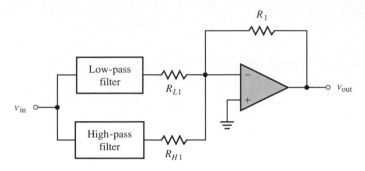

FIGURE 15.41   Block diagram of an active band-reject filter

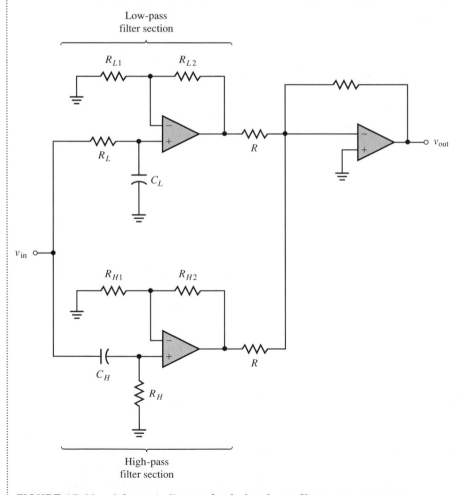

FIGURE 15.42   Schematic diagram for the band-pass filter in Figure 15.41

Determine the lower and upper critical frequencies and the stopband of Figure 15.42 using the following values.

$$R_L = 2.2 \text{ k}\Omega \qquad\qquad C_L = .047 \text{ }\mu\text{F}$$
$$R_H = \mathbf{3}.3 \text{ k}\Omega \qquad\qquad C_H = .01 \text{ }\mu\text{F}$$

REVIEW SECTION 15.3

1. Why are some filters called active while others are called passive?
2. List the advantages that active filters have over passive filters.
3. Active filters use inductors. True or false?
4. The filter characteristics should be sensitive to component values and the op-amp gain-bandwidth product. True or false?
5. Operational amplifiers are used in active filters because of

   _____ , _____ , _____ ,
   and _____ .
6. Cascading a low- and a high-pass filter section produces a

   _____ _____.
7. If the cutoff frequency of the low-pass section is f_L and the cutoff frequency of the high-pass section is f_U, the bandwidth of the bandpass filter equals _____ .
8. Figure 15.39 shows sets of characteristic curves from low- and high-pass active filters. As the text describes the cascading of the low- and high-pass filter sections, it advises using the third curve from each set. Why does the third curve show the best response?
9. In Figure 15.41, the inputs and outputs of the low- and high-pass sections connect to a _____ amplifier.
10. What is the equation for finding the voltage gain of an active low-pass filter?
11. Draw block diagrams for band-pass and band-reject active filters. Show the frequency response curve for both filters.

15.4   ACTIVE FILTERS USING MILLER-EFFECT FEEDBACK

Very simple active filters use Miller-effect feedback. The collector-to-base feedback capacitance, or C_{OB}, in the first stage of an amplifier often produces the high-frequency cutoff of an amplifier. Figure 15.43 shows the input current in the feedback capacitor. Since the AC voltage at the collector equals the product of voltage gain, or A_v, and the signal input voltage, the signal input current through the feedback capacitance increases. The increase occurs as if the stage input capacitance equals the product of the voltage gain and the feedback capacitance.

We can define the apparent increase of the input capacitance by the amount of voltage gain as the Miller input capacitance. Figure 15.44 shows the Miller input capacitance as the equivalent circuit of Figure 15.43. With the Miller effect, a small capacitance becomes more effective in reducing high-frequency response as a multiple of the gain value. In high-impedance, high-frequency stages, the value of the feedback capacitance should be held

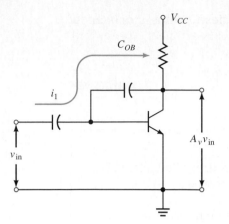

FIGURE 15.43 Illustration of the Miller effect

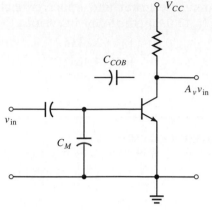

FIGURE 15.44 Equivalent circuit of the circuit shown in Figure 15.43

to a minimum. When the Miller effect limits the high-frequency cutoff of an amplifier, three factors—reducing the input impedance, reducing the feedback capacitance, and reducing the voltage gain—can increase the cutoff frequency.

For RF (radio frequency) transistors, the feedback capacitance usually equals 1 pF. However, low-power audio transistors have a feedback capacitance of around 10 pF. Consequently, a voltage gain of 30 can cause the input capacitance to exceed 300 pF. The increase of input capacitance causes the shunt impedance to fall to 25 kΩ at 25 kHz. With a lower shunt impedance and a 10-kΩ input impedance, the high-frequency cutoff extends to just above the audio frequency range. A higher input impedance further limits the frequency response of the amplifier.

Figure 15.45 shows a simple active filter circuit using Miller-effect feedback. Through the use of Miller-effect feedback, active filters utilize voltage gain and either a small feedback capacitor or the transistor feedback capacitance to produce a high-frequency cutoff at a low frequency. The upper cutoff frequency depends on the amplifier current and voltage gains. This de-

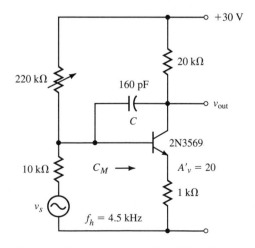

FIGURE 15.45 An active filter circuit using the Miller effect

pendency is a disadvantage in circuits that require a stable upper cutoff frequency since both temperature and the supply voltage affect current and gain.

If a stable upper cutoff frequency is not important, then the circuit has the advantage of having a small collector-to-base feedback capacitance replacing a large base-to-ground capacitance. Another limitation also exists with active filters. High-frequency loss cannot decrease indefinitely with frequency. A feedback capacitor with a very low reactance value can cause the stage gain to decrease with frequency. When that situation occurs, the amplifier gain decreases until the reactance equals the value of the load resistor divided by the voltage gain.

EXAMPLE 15.10

Find the critical frequency of the circuit shown in Figure 15.45.

As shown in Figure 15.45, the signal source connects to a transistor base through a 10-kΩ resistor. Also, a 160-pF capacitor ties from the collector to the base.

Solution
Because of the collector-emitter resistor ratio, the base-to-collector voltage gain equals 20. The Miller input capacitance is found by using:

$$C_{M(in)} = C_{OB}(A_v + 1)$$
$$C_{M(in)} = 160 \text{ pF}(20 + 1)$$
$$C_{M(in)} = 3360 \text{ pF}$$

The input signal is divided by the 10-kΩ resistor and the Miller capacitance. The critical frequency occurs when the input signal is reduced to its −3 dD point. That happens when the reactance equals the resistance or:

$$f_c = \frac{1}{2\pi R \, C_{M(in)}}$$
$$f_c = \frac{1}{2\pi \times 10 \text{ kΩ} \times 3360 \text{ pF}}$$
$$f_c = 4.7 \text{ kHz}$$

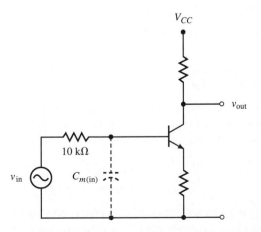

FIGURE 15.46 Equivalent circuit of the circuit shown in Figure 15.45

Figure 15.47 shows an active filter using a high-gain IC amplifier and Miller-effect feedback. The source "sees" an input impedance, represented by R_1, of approximately 10 kΩ. As the ratio shows, A_v in the low-frequency passband equals R_2 divided by R_1. The upper cutoff frequency occurs at the point when the reactance of the feedback capacitor equals the value of R_2 or:

$$-3 \text{ dB } f_h \text{ when } X_c = R_2$$

Instead of using the amplifier gain to find the upper cutoff frequency, f_h is found by balancing the source impedance with the reactance of the feedback capacitance and the open-loop gain. The simple active filter has unity gain at the upper cutoff frequency when the reactance of the feedback capacitor equals the value of the source impedance.

To ensure that the amplifier operates at unity gain, R_3 has a value equal to the parallel value of R_1 and R_2. R_3 minimizes the bias currents. The external feedback resistors and capacitors precisely determine the response characteristic of the operational amplifier. In turn, the high-frequency gain of the amplifier must maintain sufficient open-loop gain, high input impedance, and low output impedance. Most active filters have enough high-frequency gain.

Internally compensated operational amplifiers may not have adequate gain above 1 kHz. For an internally compensated op-amp, the gain and cutoff frequency should have values that make the response with external feedback 20 dB below the open-loop response of the op-amp. If the gain-frequency characteristic is too close to the open-loop characteristic, the external components cannot control the feedback response.

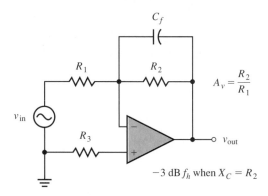

FIGURE 15.47   Active filter using a high-gain IC amplifier and Miller-effect feedback

---

PRACTICE PROBLEM 15.14

For this problem, R_1 has a value of 22 kΩ. The capacitive reactance, or X_C, equals 10 kΩ and the value of R_2 when the upper cutoff frequency is 15 kHz. If the value of R_3 is equal to the parallel value of R_1 and R_2, what is the value of R_3?

VOLTAGE-CONTROLLED VOLTAGE SOURCE FILTERS (VCVS)
Various circuit applications call for different filter designs. Depending on the need, technicians can utilize active filters that emulate the frequency and

time responses of Butterworth, Chebyshev, and Bessel filters. Voltage-controlled voltage source or controlled source filters offer simple designs based on a noninverting amplifier that has a gain higher than 1.

Along with a simple design, VCVS filters rely on fewer precision parts than other filter designs. When a filter circuit requires a larger number of high-precision components, it loses its ease of adjustability. In addition, VCVS filters have a low output impedance, a small spread of component values, easily adjusted noninverting gain, and the capability of operating at a high Q.

Yet, VCVS filters have a higher sensitivity to component value changes. As high-precision components age, they fall out of tolerance, causing the cutoff frequency to drift. Also, VCVS filter performance suffers with changes in amplifier gain. Thus, VCVS filters do not fit applications that require a tunable filter with stable characteristics.

Figure 15.48 shows examples of low-pass, high-pass, and band-pass VCVS filters. For each configuration, the resistors at the outputs of the operational amplifiers create a noninverting amplifier. The values of the other resistors

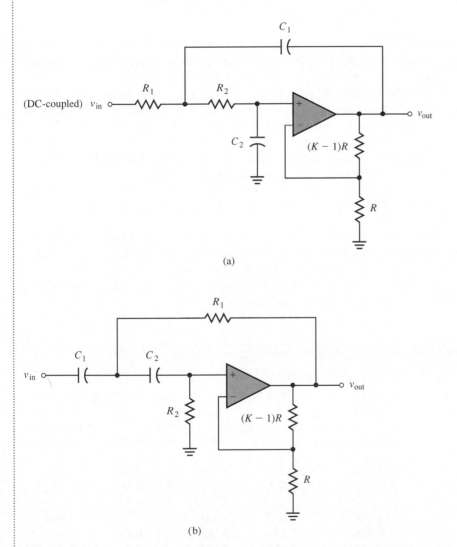

(a)

(b)

FIGURE 15.48   (a) VCVS active low-pass filter. (b) VCVS active high-pass filter. (c) VCVS active band-pass filter.   **(continued on next page)**

FIGURE 15.48 (continued)

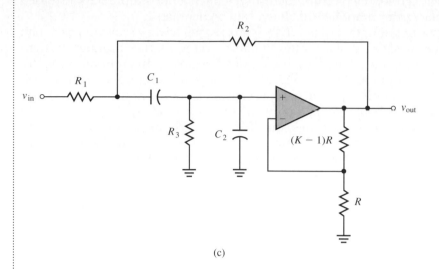

(c)

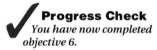

Progress Check
You have now completed objective 6.

and capacitors affect the circuit Q and, consequently, the frequency response of the filter.

Table 15.4 lists different parameters needed when setting up VCVS filters. In the table, K represents the value of gain, while f_n represents the normalizing factor. Each set of parameter values corresponds with a given number of poles. The parameters produce certain filter characteristics. A particular circuit may require a flat passband, a fast roll-off from passband to stopband, good phase characteristics, or a specified amount of gain.

To find the number of filter sections needed for a given application, we simply divide the number of poles by 2. Within each filter section, $R_1 = R_2 = R$ and $C_1 = C_2 = C$.

Since VCVS filters use op-amps, the value of R will range from 10 kΩ to 100 kΩ. At high frequencies, smaller resistances add to the rising open-loop output impedance of the op-amp and change circuit calculations. After finding the stage values, use the table to set the gain for each stage. Each n-pole filter will have n/2 entries, or one for each section.

Table 15.4 Low-pass VCVS filter values

No. of poles	Butterworth gain	Bessel gain		2.0-dB Chebyshev gain	
		f_n	K	f_n	K
2	1.586	1.274	1.268	0.907	2.114
4	1.152	1.432	1.084	0.471	1.924
	2.235	1.606	1.759	0.964	2.782
6	1.068	1.607	1.040	0.316	1.891
	1.586	1.692	1.364	0.730	2.648
	2.483	1.908	2.023	0.983	2.904

LOW-PASS BUTTERWORTH VCVS FILTERS

Within a Butterworth filter, all sections have the same values of resistors and capacitors. The reactance value of each section equals:

$$RC = \frac{1}{2\pi f_c}$$

with f_c equaling the desired -3-dB frequency of the entire filter. A six-pole, low-pass Butterworth filter would require the cascading of three low-pass sections. Consecutively, the filter sections would have gains of 1.068, 1.59, and 2.48.

EXAMPLE 15.11

Find the RC value of a Butterworth filter section. The cutoff frequency is 50 Hz.

Solution
Substituting the cutoff frequency value into the equation, we have:

$$RC = \frac{1}{2\pi(50 \text{ Hz})}$$

$$RC = \frac{1}{314}$$

$$RC = 3.18 \text{ m}\Omega\text{-F}$$

PRACTICE PROBLEM 15.15

Find the RC value of a Butterworth VCVS filter section. The cutoff frequency is 200 Hz.

LOW-PASS CHEBYSHEV AND BESSEL VCVS FILTERS

Defining VCVS Chebyshev and Bessel filters calls for a similar process. Again, R_1 equals R_2 and C_1 equals C_2 in each filter section and we cascade several sections with given gain values. But the products of the combined RC sections are different for Chebyshev and Bessel filters. So, we must apply the normalizing factor labeled as f_n on Table 15.4. In the case of the Chebyshev and Bessel filters, the RC product equals:

$$RC = \frac{1}{2\pi f_n f_c}$$

For Bessel filters, f_c is the -3-dB point and a fixed frequency. For Chebyshev filters, the value of f_c indicates the end of the passband. The Chebyshev filter cutoff frequency equals the frequency at which the amplitude response falls out of the ripple band on its way into the stopband. An example showing how to find the Chebyshev cutoff frequency may show what the value represents.

EXAMPLE 15.12

Using Table 15.4, determine the value of the resistance-capacitance combination used to construct each section of a two-pole Chebyshev filter. The cutoff frequency is 120 Hz.

The normalizing frequency for a two-pole Chebyshev filter is 0.907 Hz. To find the value for the resistance-capacitance, use the equation:

$$RC = \frac{1}{2\pi f_n \, f_c}$$

Solution
Substituting the frequency values, we have:

$$RC = \frac{1}{2\pi(0.907 \times 120 \text{ Hz})}$$

$$RC = \frac{1}{2\pi \times 109}$$

$$RC = \frac{1}{683}$$

$$RC = 1.46 \text{ m}\Omega\text{-F}$$

Using Table 15.4, determine the value of the resistance-capacitance combination used to construct a six-pole Chebyshev filter. The cutoff frequency is 5000 Hz.

Determine a value for the RC combination used in each section of a six-pole Bessel filter. The cutoff frequency equals 10,000 Hz.

HIGH-PASS VCVS FILTERS

The procedures for constructing high-pass Butterworth, Chebyshev, and Bessel filters do not differ significantly from the procedures shown in the last section. Butterworth high-pass filters differ only in the physical interchanging of the resistors and the capacitors. Values given for the resistors, capacitors, and gain in Table 15.4 remain the same.

While Chebyshev and Bessel high-pass filters require the interchanging of the resistors and capacitors and retain the gain values, another change is also required. Making either the Chebyshev or Bessel filter calls for inverting the normalizing factor value seen in the table. For each section of a Chebyshev or Bessel filter, the new normalizing value equals:

$$\frac{1}{f_n}$$

(Note that f_n is found in Table 15.4.)

 Progress Check
You have now completed objective 7.

BAND-REJECT TWIN-T FILTERS

Most RC band-reject filters have a set rate of attenuation. However, the parallel-T network pictured in Figure 15.49 has a theoretical infinite attenuation. This occurs through the adding of two exactly equal 180° out-of-phase signals at the cutoff frequency. For a narrow frequency range, the signal through one T network becomes canceled by the signal through the other T network.

Figure 15.50 shows the amplitude characteristic for the parallel-T circuit at a Q of 0.3. Despite any changes in the source and load impedances, the filter gives a deep null at the notch frequency, or f_0. Since the required response is a narrow rejection band and equal low- and high-frequency responses away from the band, the series resistor R must be 1.4 times the geometric mean value of the source and load impedances. Using an equation, we find that the value of R equals:

$$R = 2 \times R_S \times R_L$$

In its passive form, the twin-T notch filter has several disadvantages. The Q is always less than 0.5. When the filter has equal source and load impedances, the loss one decade away from the null frequency equals 14 dB. Because of the high passband loss, the load impedance usually is 100 times the value of the source impedance in practical applications. Passive RC filters have soft cutoff characteristics. The passive twin-T acts like other passive filters unless the filtered frequency nears the cutoff value. Then the response drops sharply.

Figure 15.51 shows an active band-reject twin-T filter. Combining the twin-T configuration with an emitter follower improves the rejection characteristics of the twin-T design. The variable resistors in the circuit allow the separate adjustment of the Q and the null. With the addition of the amplifier,

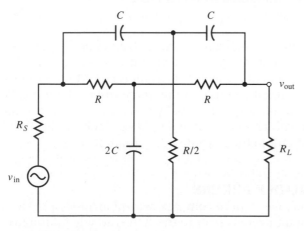

FIGURE 15.49 Parallel-T network

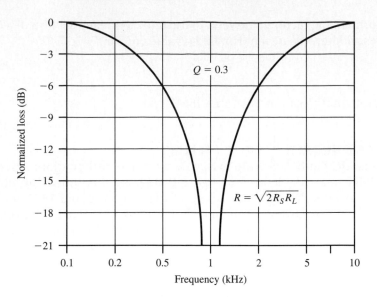

FIGURE 15.50   Amplitude characteristic of the parallel-T circuits

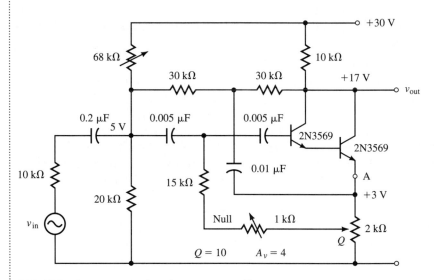

FIGURE 15.51   Active band-reject twin-T filter

the circuit produces high gain at the high and low frequencies. The use of the emitter follower cuts the gain to 3 and raises the Q to 5 with a 50-dB null.

Most twin-T filters are available as modules. Modular designs offer stable, matched components with good temperature coefficients. Within the module, an operational amplifier usually replaces the emitter-follower transistor. Voltage gain with an operational amplifier is achieved by connecting point A of the filter circuit to the output of the op-amp.

STATE-VARIABLE FILTERS

Figure 15.52 shows a more complex twin-pole filter called a state-variable filter. Many circuit designs feature the state-variable filter because it has better stability than the VCVS filters. In addition, the state-variable filter is easier to

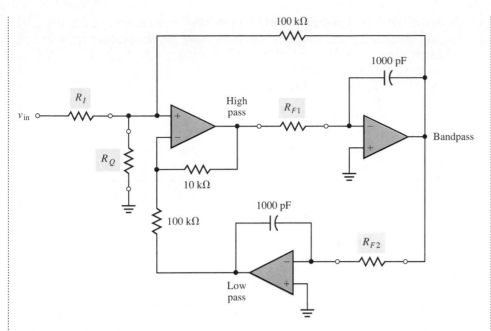

FIGURE 15.52 Schematic of a state-variable filter

adjust. Manufacturers package the state-variable or biquad filter as a module. Designers also prefer state-variable filters because of the integration of the capacitors. All components except the R_I, R_Q, R_{F1}, and R_{F2} arrive as part of the module.

State-variable filters can provide low-pass, high-pass, and band-pass outputs from the same circuit. Also, the filters are easily modified for a band-reject response by adding another operational amplifier. Some state-variable packages include the additional op-amp as part of the module. Filter configurations include the Butterworth, Chebyshev, and Bessel types. Along with excellent versatility, state-variable filters also can be tuned while maintaining either a constant Q or a constant bandwidth. Some filter designs will use cascaded state-variable filters when requiring higher-order filters.

Progress Check
You have now completed objective 8.

REVIEW SECTION 15.4
1. Describe the Miller effect.
2. Considering the Miller effect, name the three factors that can cause the cutoff frequency of an amplifier to increase.
3. VCVS filters provide the advantages of _____ and _____.
4. VCVS filters are sensitive to changes in component values. True or false?
5. Where is the cutoff frequency of a Chebyshev filter?
6. The cutoff frequency for a Bessel filter is a variable frequency. True or false?
7. Many circuits use twin-T style filters for _____.
8. As the filtered frequency approaches cutoff, the response of the twin-T filter increases. True or false?

9. An active twin-T filter compensates for the _____ and the _____ normally seen with a passive twin-T filter.
10. State-variable filters are:
    (a) used only for complex situations
    (b) not easily adjustable
    (c) easily adjustable and stable
    (d) not easily adjustable but do provide good stability.
11. A state-variable filter produces only one kind of frequency response. True or false?
12. Adding an operational amplifier to a state-variable filter can produce a _____.

 TROUBLESHOOTING

As with all electronic circuits, a few preliminary procedures aid the troubleshooting of active filters. Always verify the type of filter that you are troubleshooting. Even though low- and high-pass filters seem similar, they have different fault characteristics and different output waveforms. Also, check the input signal and the supply voltages for the filter. Finally, isolate the filter from its load to confirm that the filter is malfunctioning.

Figure 15.53 shows a technician troubleshooting an active filter circuit with an oscilloscope. Because there are few components in an active filter

FIGURE 15.53 Technician troubleshooting an active filter circuit with an oscilloscope. *(Courtesy of Hewlett-Parkard Company)*

circuit, troubleshooting can be reduced to a simple process. In a two-pole, low-pass active filter, the input resistors and capacitors control the frequency response of the filter. If either of the two resistors seen in the input stage opens, the filter will have no output. If either of the two capacitors opens, the cutoff frequency of the filter will decrease. Also, the roll-off rate of the circuit will decrease.

The other components in a two-stage, low-pass active filter control the gain of the filter. If either of the two resistors connected to the output of the operational amplifier opens, the gain will change. An open feedback resistor opens the feedback path and causes the output to drive into saturation. An open lower resistor reduces the gain to unity. Because of the similarities between low-pass and high-pass filters, high-pass filters will experience the same symptoms.

As electrolytic capacitors age, they begin to leak. A leaky cap allows some DC current to flow. The capacitor acts like a resistor and a capacitor in parallel. The result is that f_c has a lower frequency than it should. A very leaky capacitor has a low DC resistance, which shorts part of the input. If the chip fails, there is no corresponding output to a specific input.

For troubleshooting, we can break band-pass and band-reject filters down into their low- and high-pass filter sections. If either stage has a fault, a loss of gain for that stage will result. Checking the stages becomes a matter of checking the low- and high-frequency roll-off rates and the quiescent DC levels along the signal paths. After troubleshooting the passive components of any active filter and not finding any defects, replace the operational amplifier.

Troubleshooting filters combine both passive and active device troubleshooting techniques. Figure 15.54a is a low-pass filter. Its bandwidth response curve should look like Figure 15.54b. An open R_2 would make the filter act like a comparator because the device would be operating open loop.

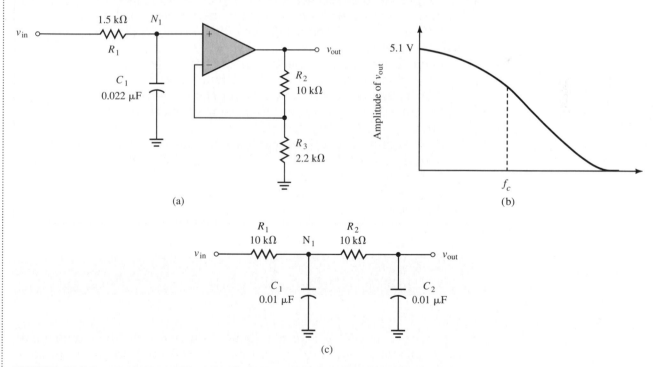

FIGURE 15.54   (a) A low-pass filter. (b) The frequency response curve. (c) The circuit.

The output voltage would be $\pm v_{sat}$ depending on the input voltage. If R_3 were open, the circuit gain would be reduced to 1. The circuit now would act like a voltage follower. The output voltage would have the proper shape but its amplitude during bandpass would be about 1 V.

WHAT'S WRONG WITH THIS CIRCUIT?

Refer to the circuit in Figure 15.54c and assume an input voltage of 1.5 VDC. The output voltage is 0 V. The following are voltage readings you have taken.

$$V_{R1} = 1.5 \text{ VDC}$$
$$V_{Rf} = 0 \text{ VDC}$$
$$V_{R(in)} = 0 \text{ VDC}$$
$$C_1 = 0 \text{ VDC}$$

Solution
If all the input voltage is being dropped over R_1, then there must be a current path to ground at N_1. A shorted (very leaky) C_1 would be the most likely culprit.

What is the problem with this circuit?

TROUBLESHOOTING QUESTIONS
1. List the preliminary procedures for troubleshooting filters.
2. Why should you isolate the filter from its load?
3. In a two-stage, low-pass filter, the input resistors and capacitors control the frequency response of a filter. True or false?
4. Because of their different configurations, you will need to use different troubleshooting methods with low- and high-pass filters. True or false?
5. Before troubleshooting band-pass and band-reject filters, (a) separate or (b) combine the individual sections.
6. A faulty filter stage will exhibit (a) more or (b) less gain.
7. Describe the simple process used when troubleshooting active filters.
8. If either of the two resistors connected to the (a) output or (b) input of the operational amplifier _____ , the gain changes.
9. If either of the two _____ _____ opens in a two-pole, low-pass active filter circuit, the cutoff frequency of the filter decreases.

SUMMARY

Filters remove unwanted parts of a signal. As they remove the signal, the resulting waveform may become modified or reshaped according to some need. While some filters reject or isolate one part of a frequency, others attenuate a range of frequencies. In electronics, filters work to remove unwanted noise from transmission lines or to separate previously mixed signals.

Both the time and frequency aspects are affected by filter action. Every filter circuit has frequency characteristics that show in its output waveform. In addition, cutoff frequencies mark the response of a filter. The effects of filtering action on time appear in properties such as rise time, overshoot, ringing, and settling time.

Whether working with passive or active filters, four basic types of filters exist. Low-pass filters eliminate frequencies above their upper bandwidth limit, while high-pass filters eliminate frequencies below their lower bandwidth limit. Band-pass filters allow all frequencies between the lower and upper cutoff frequencies to pass through the filter. Band-reject filters reject all frequencies between the lower and upper limits.

Active filters have many advantages over passive filter designs. Those advantages include size, cost, and reduced signal loss. Active filters do not have inductors. We call filters "active" when they contain devices such as transistors, FETs, or operational amplifiers. Usually, active filters have a minimum number of parts. Of the active devices, operational amplifiers provide the best performance when used as active filters.

With response dependent on circuit Q, different component values can provide different amplitude and time responses in active filters. Butterworth filters have a flat amplitude response. Chebyshev filters introduce ripple into their amplitude response but have a sharper response knee. Another type of filter, called the Bessel filter, has a flat time delay within its passband frequencies.

Voltage-controlled voltage source filters offer simple designs and use few parts. Low-pass, high-pass, band-pass, and band-reject filters can be designed to work as Butterworth, Chebyshev, and Bessel filters. While VCVS filters answer many needs, the filter design has a low tolerance for component value changes.

State-variable filters have better stability than VCVS filters. Also, state-variable filters are easier to adjust than VCVS filters. Normally, state-variable filters feature modular integration. Another modular design, the twin-T notch filter, has also gained widespread use. Notch filters remove interfering signals.

PROBLEMS

1. The roll-off of Figure 15.55 is _____ dB per octave.

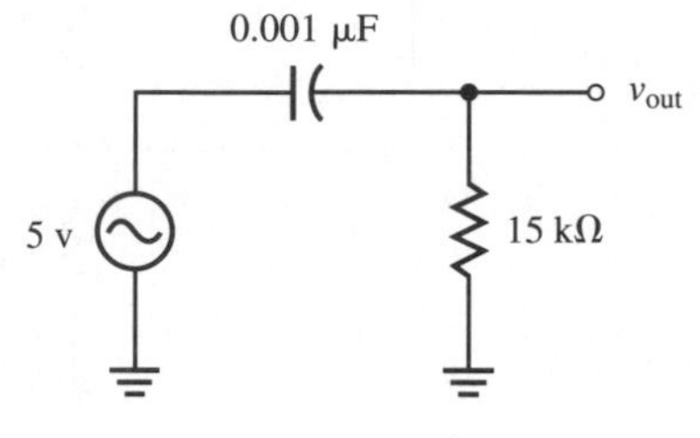

FIGURE 15.55

Refer to Figure 15.56 for problems 2–4.

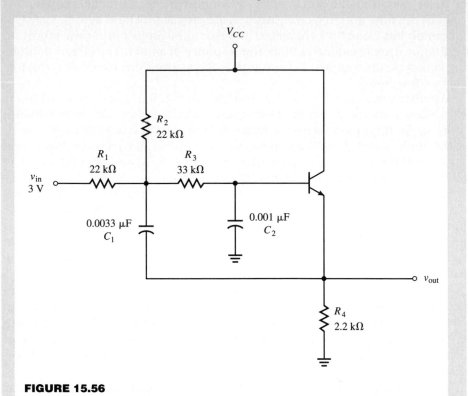

FIGURE 15.56

2. Solve for the critical frequency.
3. At critical frequency the output voltage is _____ .
4. The circuit is a:
   (a) low-pass filter
   (b) high-pass filter
   (c) band-pass filter
   (d) bandstop filter

Refer to Figure 15.57 for problems 5–7.

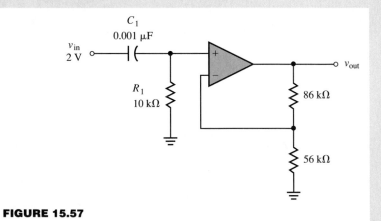

FIGURE 15.57

5. The critical frequency is _____ .
6. The circuit is a:
   (a) low-pass filter
   (b) high-pass filter
   (c) band-pass filter
   (d) bandstop filter
7. If C_1 opens, will the critical frequency increase or decrease?

Refer to Figure 15.58 for problems 8–12.

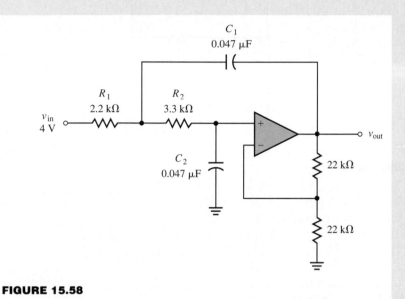

FIGURE 15.58

8. The circuit is a:
   (a) low-pass filter
   (b) high-pass filter
   (c) band-pass filter
   (d) bandstop filter
9. The critical frequency is _____ .
10. The roll-off is _____ dB per octave.
11. At critical frequency the output voltage is _____ .
12. The circuit has _____ pole(s).

Refer to Figure 15.59 for problems 13–20

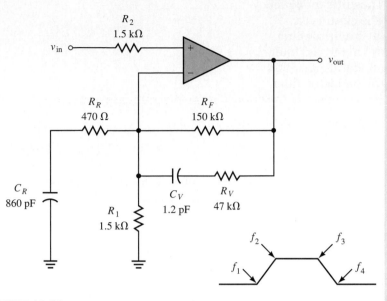

FIGURE 15.59

13. The critical frequency is_____ .
14. At critical frequency the output voltage is _____ .
15. The roll-off is _____ dB per octave.
16. The circuit is a:
    (a) low-pass filter
    (b) high-pass filter
    (c) band-pass filter
    (d) bandstop filter
17. The circuit is a:
    (a) first-order filter
    (b) second-order filter
    (c) third-order filter
18. Solve for the circuit gain.
19. The filter is called a _____ filter.
20. Solve for the frequencies at:
    $f_1 =$
    $f_2 =$
    $f_3 =$
    $f_4 =$

16

ANALOG AND DIGITAL SIGNALS

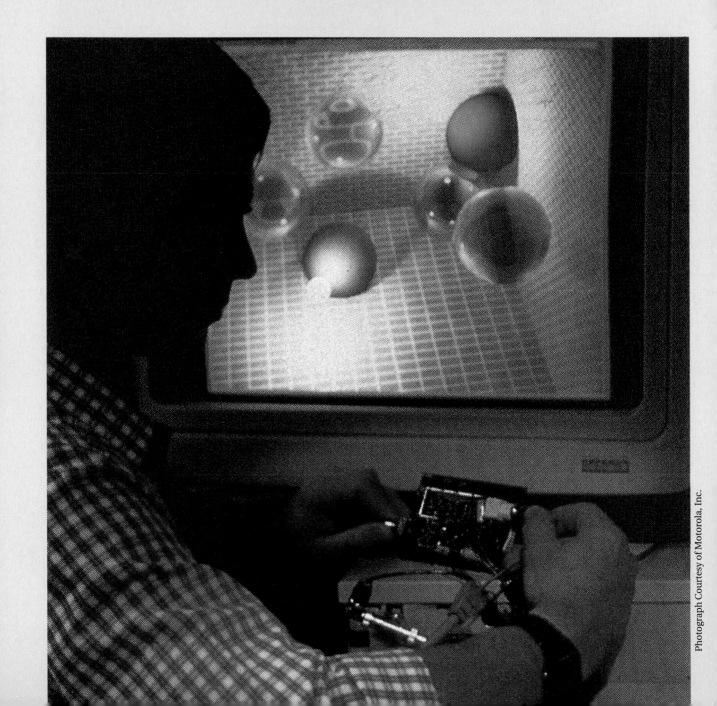

OBJECTIVES

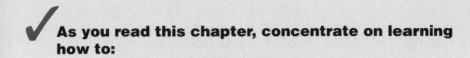

As you read this chapter, concentrate on learning how to:

1. List the types of transducers
2. Discuss the physical properties that affect transducers
3. Define the negative temperature coefficient of thermistors
4. Describe how a thermocouple works
5. Describe how a photoresistor works
6. Discuss the effect of light on a reversed PN junction
7. Discuss where optoisolators may be used
8. Explain how a strain gauge works
9. Define the purpose of instrumentation amplifiers
10. Describe different instrumentation amplifier configurations
11. Define how analog-to-digital and digital-to-analog converters function

INTRODUCTION

This chapter includes information about transducers and differential and instrumentation amplifiers. Even though you may not realize it, transducers surround us. While many types of transducers exist, this chapter concentrates on temperature-activated, light-activated, and movement-activated transducers.

The coverage of differential amplifiers includes a review of common-mode gain and differential gain, the common mode rejection ratio, and other operating characteristics. You will also learn about using differential amplifiers as instrumentation amplifiers to take advantage of their high gain and DC coupling features.

When we work with transducers and instrumentation amplifiers, we are working with devices that accomplish two basic tasks. Transducers convert some basic quantity, such as temperature, to a voltage or some other electrical quantity. Instrumentation amplifiers use the properties of differential amplifiers to distinguish signal from noise.

Analog-to-digital (A/D) converters change analog signals into digital data. Digital-to-analog (D/A) converters provide analog voltages that correspond to the input digital data. Microprocessors and support circuits analyze, control, and record the processes.

16.1 TRANSDUCERS

A transducer connects the physical world to the world of electronics. Transducers exist in forms that respond to temperature, light, strain, movement, and position. For the purposes of this text, we will limit our discussion to transducers that respond to temperature, light, and strain. Figure 16.1 shows how a transducer may fit into a practical electronics application.

Radiant energy has become a useful tool in semiconductor electronics. By using light, an engineer can design a circuit that senses the presence or absence of an opaque object or a circuit that provides electrical isolation. Many circuits use light-activated semiconductors as controllers for power devices.

Several key variables—wavelength, geometry, light filtering, distance, focus, and intensity—exist when considering light sources for light-sensitive semiconductors. In addition, designers look at the response of the sensor to a particular light source and the optical coupling between the source and the sensor.

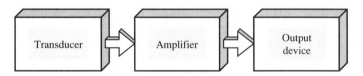

FIGURE 16.1 How transducers fit into electronic applications

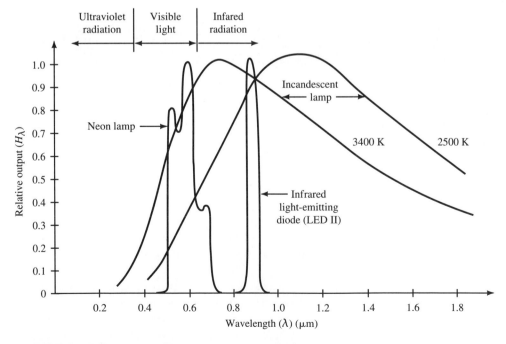

FIGURE 16.2 Outputs of light sources versus wavelength

The response of a sensor to a light source depends on the characteristics of the sensor and the wavelength of the light radiation. Light sources include incandescent lamps, neon bulbs, light-emitting diodes, and lasers. Each source has an output in a different part of the light spectrum. Figure 16.2 shows the output of most light sources versus wavelength.

Incandescent lamps provide an inexpensive source of light with a wide spectral output. Neon lamps have longer operational lives than incandescent lamps, require less input power, and have a short rise time. However, the output from a neon lamp covers only a limited part of the spectrum.

Light-emitting diodes and lasers use semiconductor PN junctions that emit photons. The emission of diodes made of gallium arsenide and gallium phosphide occurs as a result of hole-electron recombinations. Light-emitting semiconductors emit light in either a narrow bandwidth or a narrow line. Light-emitting diodes (LEDs) and lasers have extremely fast response times good distance range.

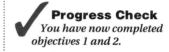

Progress Check
You have now completed objectives 1 and 2.

THERMISTORS

A **thermistor** is a semiconductor device whose resistance varies with a change in temperature. If the resistance increases when temperature increases, then the device is said to have a positive temperature coefficient. If the resistance decreases when temperature increases, then the device is said to have a negative temperature coefficient. Most thermistors have negative temperature coefficients. Table 16.1 shows how resistance decreases as the temperature of a typical thermistor increases.

All thermistor circuit designs must consider the self-heating of a thermistor. A typical thermistor may have a dissipation constant of 1 mW/°C.

Table 16.1						
Temperature (°C)	0	25	50	100	150	200
Resistance (Ω)	5700	2000	810	185	59	25

EXAMPLE 16.1

A thermistor has a resistance of 15,000 Ω at 25°C. Determine the resistance of a thermistor at 0°C, or R_0.

Solution
Use the values from Table 16.1 to set up a ratio of temperature versus resistance or:

$$\frac{R_0}{15 \text{ k}\Omega \text{ @ } 25°} = \frac{5.7 \text{ k}\Omega \text{ @ } 0°}{2 \text{ k}\Omega \text{ @ } 25°}$$

$$R_0 = \frac{\left(5.7 \text{ k}\Omega \text{ @ } 0°\right) \times \left(15 \text{ k}\Omega \text{ @ } 25°\right)}{2 \text{ k}\Omega \text{ @ } 25°}$$

$$R_0 = 42.8 \text{ k}\Omega \text{ @ } 0°$$

A thermistor has a resistance of 7.5 kΩ at 25°C. Determine the resistance, or R_0, of a thermistor at 50°C.

Figure 16.3 shows the schematic symbol for a thermistor, and Figure 16.4 shows how several different types of thermistors appear. Because of their thermal properties, thermistors work well for temperature measurement and control. Thermistors used for temperature measurement or compensation usually have a resistance of a few thousand ohms at room temperature and can work with temperatures within a range of −50°C to +300°C. Also, standard thermistors can detect 0.0005°C temperature changes.

Figure 16.5 expands the low temperature end of the thermistor range by making use of the exponential resistance change seen in thermistors and then coupling that property with an amplifier. If we looked at that exponential change as an equation, it would appear as:

$$R = R_0 e^{+V/kt}$$

FIGURE 16.3 Schematic symbol for a thermistor

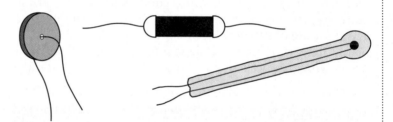

FIGURE 16.4 Different types of thermistors

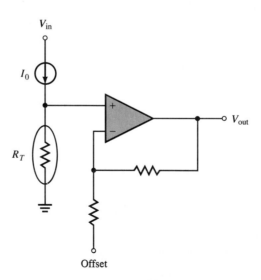

FIGURE 16.5 Circuit that expands the low temperature end of a thermistor

where R is the resistance of the thermistor, R_0 and E represent constants for the unit, and e represents exponential change. The Centigrade temperature (°C) is represented by t, and k represents Boltzmann's constant, or 1.38×10^{-23} joules/°C. If nothing else, the equation for exponential behavior of a thermistor tells us that resistance changes rapidly with temperature and that the thermistor is a sensitive device.

THERMOCOUPLES

When two dissimilar metals are joined, the junction, or **thermocouple,** generates a small voltage in the millivolt range. In addition, the junction has a low source impedance and a temperature coefficient of 50 μV/°C. We can use the thermocouple to measure temperatures over a large range. Different pairs of metal alloys allow the measurement of temperatures from −270°C to +2500°C.

Figure 16.6 shows a basic thermocouple circuit. In the circuit, iron and constantan are welded together to form a thermocouple. (Constantan is an alloy of copper and nickel.) The circuit provides a voltage that is dependent on the temperatures of both junctions. That voltage is proportional to the difference between the reference and sensing junction temperatures.

Since the reference junction provides a comparison voltage, the junction remains at a fixed temperature. Without the reference junction, any measurements would be erratic. However, the changing temperature at the sensing junction is the key variable for thermocouple measurements.

Figure 16.7 illustrates one method used to maintain a stable reference junction temperature. In the circuit, an AD590 temperature-sensing integrated circuit (IC) and other additional circuitry compensate for differences between the actual junction temperature and 0°C. The integrated circuit produces an output current equal to the temperature in kelvins (K).

The values of R_1, R_2, and R_3 control any corrections made between the desired 0°C and the 51.5 μV/°C thermoelectric coefficient of the two metals. The thermoelectric coefficient changes with the type of metals used in the thermocouple and is listed as a thermocouple specification. When the reference junction of the thermocouple is at 0°C, no corrections occur. When the reference junction temperature varies from 0°C, the offset of the AD590 changes, and the temperature corrects back to 0°C.

Progress Check
You have now completed objective 3.

Progress Check
You have now completed objective 4.

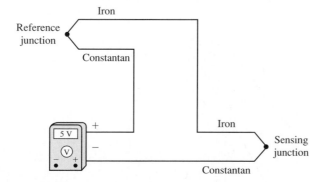

FIGURE 16.6 **A basic thermocouple circuit**

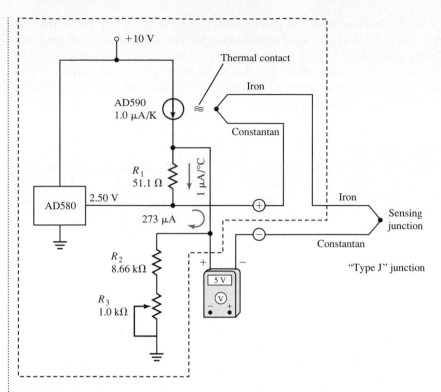

FIGURE 16.7 Thermocouple circuit with a stable reference junction temperature

PHOTORESISTORS

A **photoresistor** is a single piece of semiconductor material that changes its resistance with a change in light intensity. The semiconductor material may be either N- or P-type material. While photoresistors made from pure semiconductors have a high resistance, N- and P-type photoresistors have a lower resistance. Figure 16.8 shows how a typical photoresistor appears.

Applications for photoresistors include photographic light meters and automatic lens-adjusting cameras. Along with those applications, photoresistors have uses as movement detectors, heart rate monitors, and fluid level

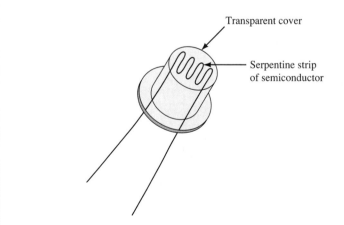

FIGURE 16.8 Typical photoresistor

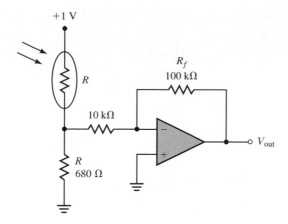

+1 V

R_f
100 kΩ

R

10 kΩ

R
680 Ω

V_{out}

FIGURE 16.9 Photoresistor exposed to high-intensity light

detectors. When used as movement detectors, a passing object blocks a light beam that illuminates a photoresistor. The light blockage causes a voltage variation across the photoresistor. Then a voltage divider connected to the photoresistor passes the voltage signal to a counter. Movements can range from passing automobiles to a slight finger tremor.

A light source and photoresistor connected to the opposite sides of an earlobe can work as part of a heart monitor. When the heart beats, blood pumps into the earlobe capillaries and makes the flesh seem more opaque. Consequently, the light reaching the photoresistor changes. Changes in resistance of the photoresistor are then monitored by another electronic circuit.

Figure 16.9 shows a photoresistor that can be exposed to high-intensity light. With the increased application of light, more electrons become freed in the semiconductor material. Lowered resistance in the material allows more conduction.

EXAMPLE 16.2

Find the output voltage of Figure 16.9 when $R_0 = 560 \, \Omega$.

Resistance of a photoresistor depends on the intensity of the light or:
$$R = R_0 I^{-K}$$

where R equals resistance, I equals intensity, and K is a constant having a range of 0.5 to 1.0. Instead of having a linear dependence on light intensity, the resistance follows an inverse power law. Photoresistors have resistances that vary from several MΩ when darkened to hundreds of kΩ when exposed to light.

Once we know the resistance of the photoresistor, we can solve for input voltage and then the output voltage.

Solution
To find the resistance of the photoresistor, we use the formula
$$R_1 = R_0 \times I^{-K}$$
$$R_1 = 560 \, \Omega \times 3^{-0.65}$$
$$R_1 = 274 \, \Omega$$

The voltage drop across R_2 may be found using the voltage divider method.

$$V_{R2} = \frac{V_s \times R_2}{R_1 + R_2}$$

$$V_{R2} = \frac{+1 \text{ V} \times 680 \text{ } \Omega}{274 \text{ } \Omega + 680 \text{ } \Omega}$$

$$V_{R2} = +0.71 \text{ V}$$

The output voltage is found by multiplying the amplifier gain by the input voltage

$$V_{out} = A \times V_{in}$$

$$V_{out} = \frac{-R_f}{R_{in}} \times V_{in}$$

$$V_{out} = \frac{-100 \text{ k}\Omega}{10 \text{ k}\Omega} \times \left(+0.71 \text{ V}\right)$$

$$V_{out} = -7.1 \text{ V}$$

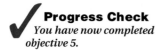

Progress Check
You have now completed objective 5.

PRACTICE PROBLEM 16.2

Find the resistance if the constant equals 0.76 and the intensity is 4. The value of R_0 is 330 Ω.

PHOTODIODES AND PHOTOTRANSISTORS

A **photodiode** is a PN junction diode with an opening in its case that allows light to focus on the junction. Figure 16.10 shows the schematic symbol for a photodiode. The light on the junction increases the number of unlike charges at the junction. As a result, the diode has greater conductivity.

In Figure 16.11, a photodiode operates with reverse bias and makes use of minority current carriers. Exposing the diode to light causes the diode's minority current to increase. Cutting off light from the junction decreases the flow of minority current carriers.

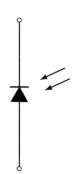

FIGURE 16.10   Schematic symbol for a photodiode

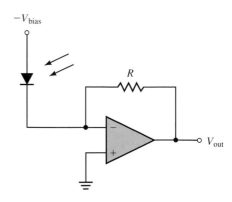

FIGURE 16.11   A photodiode operates with reverse bias

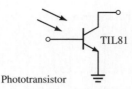

Phototransistor

FIGURE 16.12 Schematic symbol for a phototransistor

Phototransistors also respond to light and, in addition, amplify the current generated by the light. We can define a phototransistor as a bipolar transistor with light energy acting as the input to the base terminal. The schematic diagram of a phototransistor in Figure 16.12 shows light radiation working as the base current source. Phototransistors have a proportional response to light.

In the bipolar phototransistor, the light-induced current is base current. Changes in light intensity will not cause current changes in a forward-biased diode. The emitter junction diode is forward-biased. However, the collector junction diode will be reverse-biased. When light stimulates an increase in collector-base leakage current, the base current increases. Thus, a phototransistor is a high-gain, low-current semiconductor device. A phototransistor can work with the base open or with biasing resistors. Biasing will adjust the sensitivity of the transistor.

Phototransistors depend on the biasing arrangement and light frequency for proper operation. All phototransistors respond to the entire visible and infrared radiation ranges. Transistor action multiplies the light-generated photocurrent in the base-to-collector junction and creates a sensitive amplification device.

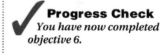

Progress Check
You have now completed objective 6.

OPTOISOLATORS

Figure 16.13 shows the schematic diagram for an optoisolator. **Optoisolators** provide isolation from high voltages in electronic equipment. Other common names for the device are optocoupler and photocoupler. A typical LED-phototransistor optoisolator uses a transparent dielectric channel to isolate and insulate a LED from the phototransistor. Through that insulation, the device achieves an isolation value of 25,000 to 75,000 V.

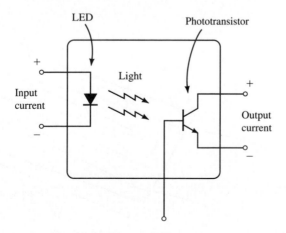

FIGURE 16.13 Schematic diagram for an optoisolator

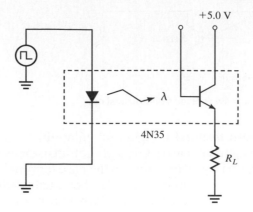

FIGURE 16.14 4N35 optocoupler-isolator

Figure 16.14 shows a 4N35 optocoupler-isolator consisting of an NPN phototransistor and a PN light-emitting diode. The LED portion of the optocoupler connects to the pulse generator, and the collector of the phototransistor attaches to the +5-V voltage supply. The emitter is connected to the load. The load will often be another amplifier circuit.

STRAIN GAUGE

Longer length and a smaller cross section cause the resistance of a metal or a semiconductor to increase if it is stretched. If the material is not stretched to its breaking point, the elasticity of the material allows it to regain its original shape and resistance after the stress becomes removed.

Fine wire used to measure the amount of strain or movement is called a **strain gauge.** Figure 16.15 shows one type of strain gauge. In the figure, fine wire is bound to an elastic insulating sheet. The entire insulating sheet is bound to the object to be studied. When the section of the object under the strain gauge stretches or compresses, the strain gauge stretches or compresses by an equal amount. The measure of strain of the object is the resistance change of the gauge.

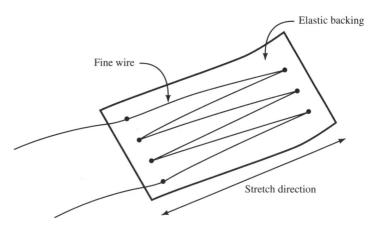

FIGURE 16.15 One type of strain gauge

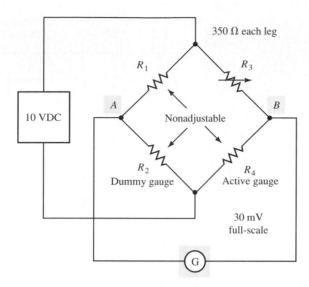

FIGURE 16.16   Strain gauges working as part of a Wheatstone bridge

Figure 16.16 shows another type of strain gauge. In the figure, two strain gauges, R_1 and R_4, work as part of a Wheatstone bridge. This arrangement gives the advantage of minimizing the effects of resistance change versus temperature. Resistance change with strain is small, while resistance change with temperature can be large.

With the Wheatstone arrangement, one strain gauge acts as a dummy and the other acts as the active gauge. Measurement involves the use of three known resistor values and one unknown resistor value and an excitation voltage. Adjusting R_3 causes the bridge to balance. G represents a galvanometer, which is a very sensitive ammeter. No current flows through the galvanometer because the voltage at point A equals the voltage at point B.

The circuit contains two voltage dividers. We can set up a ratio equation to find the value of the unknown resistance.

$$\frac{R_1}{R_2} = \frac{R_3}{R_4}$$

$$R_4 = \frac{R_3 R_2}{R_1}$$

In the balanced condition, any change in the excitation voltage will have no effect on the galvanometer. In Figure 16.16, R_2 is the dummy strain gauge and R_4 is the strain gauge actually being used for measurement. The two gauges are exposed to the same temperature, but only R_4 will measure strain. Because any resistance changes caused by temperature will affect R_2 and R_4 equally, the effect of temperature will be canceled out. When a Wheatstone bridge configuration is used for strain gauge measurements, the measurements will reflect on the strain, as temperature changes will be negated.

Progress Check
You have now completed objective 8.

DYNAMIC MICROPHONES AND MAGNETIC PHONOGRAPH CARTRIDGES

Dynamic microphones and magnetic phonograph cartridges use the effect of a changing magnetic field producing an electromotive force in a coil circuit.

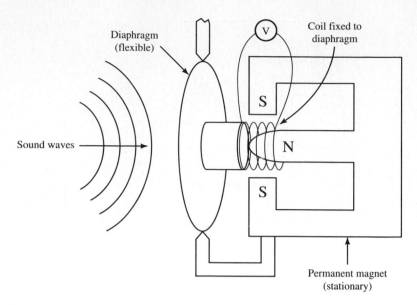

FIGURE 16.17   A dynamic microphone

Typically, a small permanent magnet close to a sensing coil produces the magnetic field. Any movement of the magnet relative to the coil produces a change in the magnetic flux. The combination produces a voltage only while magnetic flux changes.

Figure 16.17 shows a dynamic microphone. Here, a coil attaches to a flexible diaphragm that is fixed at the edges. When sound waves strike the diaphragm and cause it to flex and vibrate, the coil oscillates in the field of the fixed magnet. This oscillation causes the magnetic field through the coil to oscillate. An AC voltage proportional to the amplitude of oscillation and hav-

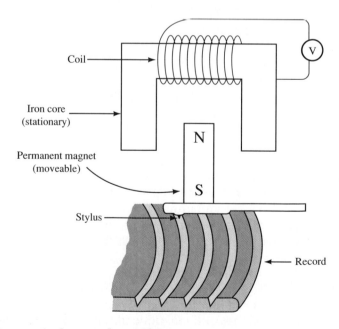

FIGURE 16.18   A magnetic phonograph cartridge

ing the same frequency is induced in the coil. The voltage will usually have a value in the range of tenths of volts.

In the magnetic phonograph cartridge shown in Figure 16.18, a stylus oscillates from side to side in a record groove. The small permanent magnet attached to the stylus oscillates between the iron core tabs that pass through a stationary coil. Since the magnetic field easily passes through the iron core, the core gathers the magnetic field through the coil from the air. An oscillating magnetic flux occurs in the coil and an AC voltage is generated in response to the audio recording.

REVIEW SECTION 16.1

1. A transducer connects the _____ world to the world of _____ .
2. Transducers respond to _____ , _____ , _____ , and _____ _____ .
3. Describe several sources of light energy.
4. A thermistor responds to _____ changes.
5. A photoresistor changes its _____ with a change in _____ _____ .
6. Photodiodes should be (forward/reverse) biased to operate normally.
7. Phototransistors amplify the _____ generated by _____ .
8. Optoisolators provide _____ .
9. What is the advantage of a Wheatstone strain gauge arrangement over a single strain gauge?
10. Dynamic microphones and magnetic phonograph cartridges use coils and magnets to change mechanical motion into an electrical current. True or false?

16.2 DIFFERENTIAL AND INSTRUMENTATION AMPLIFIERS

DIFFERENTIAL OUTPUT CONFIGURATIONS

In Chapter 10, you learned that differential amplifiers have become the workhorses of the linear integrated circuit family. Also, you concentrated on the most common configuration of the differential amplifier—the single-ended output seen in Figure 16.19.

Figure 16.20 shows a simple schematic of another differential amplifier configuration. The amplifier pictured in Figure 16.20 has a differential output. The output is taken between the collectors of the two transistors that make up the amplifier. In effect, the symmetrical differential amplifier acts like the Wheatstone bridge seen in Figure 16.21.

Usually, the circuit requires two power supplies for operation: $+V_{CC}$ and $-V_{EE}$. Since the differential amplifier is DC-coupled, it can amplify either DC or AC signals. Either input signal alone or both input signals are applied to the respective transistors. If the input signals equal one another, the output voltages also equal one another. From the perspective of a bridge, the bridge is balanced and V_{out} is equal to zero.

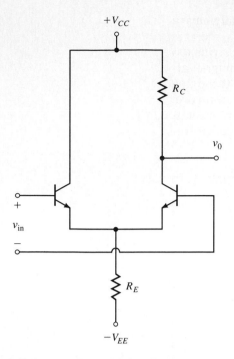

FIGURE 16.19   A basic differential amplifier configuration

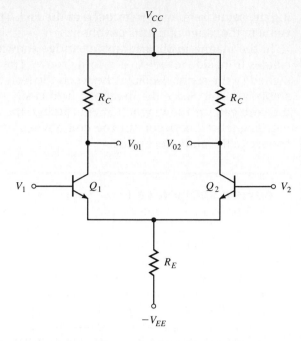

FIGURE 16.20   A differential amplifier with a differential output

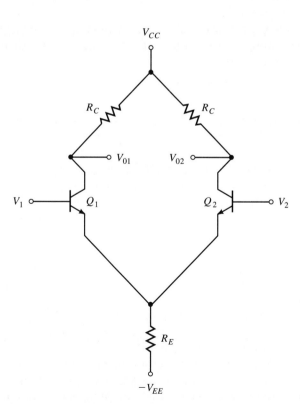

FIGURE 16.21   Differential amplifier with a Wheatstone bridge

If input voltage V_1 is greater than V_2, more collector current flows in transistor Q_1 and less in Q_2. So, the V_{01} output voltage is less than the V_{02} voltage. If V_2 is greater than V_1, more collector current flows in transistor Q_2 and V_{02} is less than V_{01}.

Practical signals contain two basic components: the common-mode portion of the signal that is common to both inputs and the difference portion that is not common to both inputs. Any voltage applied equally and simultaneously to two inputs is a common-mode voltage. Examples of common-mode signals are:

1. noise produced by drift,

2. power supply hum,

3. the effects of temperature changes, and

4. noise produced by differences in ground potential.

The symmetrical design of differential amplifiers reduces undesired noise that can occur on both input lines. Because of the symmetrical design, the collectors of a differential amplifier generate equal amplitude output signals. Effectively, the common-mode signals cancel each other. We call the attenuation of undesired voltages by differential amplifiers **common-mode rejection.**

A differential amplifier amplifies voltage differences. The difference-mode portion of the signal is the desired signals that the differential amplifier amplifies. In an ideal differential amplifier, the gain of the common-mode signals, or A_{vc}, would equal zero, while the gain of the difference-mode signals, or A_{vd}, would have a high value. If a common-mode voltage brings both inputs of the differential amplifier off ground, then the output voltage equals zero. If a difference occurs between the two input voltages, then the output voltage is a nonzero voltage.

The behavior of a differential amplifier responds to the equation

$$V_{out} = A_{vd}(V_{in+} - V_{in-}) = A_{vd}V_{id}$$

This equation shows the noninverting and inverting characteristics of the differential amplifier inputs. V_{in+} and V_{in-} are the voltages into the positive and negative inputs, while V_{out} is the output voltage of the differential amplifier. A_{vd} is the difference voltage gain, while V_{id} is the input difference voltage. V_{id} results from the difference of the input voltages.

In reality, the common-mode voltage, or A_{vc}, equals the ratio of the output voltage, or V_{oc}, to a common input voltage, or V_{ic}. In equation form, this appears as

$$A_{vc} = V_{oc} / V_{in}$$

Even with the same or common-mode voltage at the differential amplifier inputs, a slight nonzero voltage appears at the output.

Because we consider the effect of the common-mode voltage undesirable, we want the common-mode voltage gain ratio to remain small when compared to the desirable difference voltage gain of the amplifier. The difference voltage gain equals:

$$A_{vd} = V_{od} / V_{id}$$

or the ratio of the voltage difference at the input terminals to the output voltage.

EXAMPLE 16.3

Determine the common-mode and difference-mode signal gain values of the differential amplifier shown in Figure 16.20.

For this example, the output voltage, or V_{oc}, for a common-mode voltage input equals 0.005 V while the common input voltage, or V_{ic}, equals 1 V. The output voltage, or V_{od}, for a voltage difference equals 2 V, while the voltage difference between the input terminals, or V_{id}, equals 0.1 V.

Solution

To find the common-mode and difference-mode values, substitute the example values into the gain equations. The common-mode signal gain equals:

$$A_{vc} = \frac{V_{oc}}{V_{ic}}$$

$$A_{vc} = \frac{0.005 \text{ V}}{1 \text{ V}}$$

$$A_{vc} = 0.005$$

The difference-mode signal gain equals:

$$A_{vd} = \frac{V_{od}}{V_{id}}$$

$$A_{vd} = \frac{2 \text{ V}}{0.1 \text{ V}}$$

$$A_{vd} = 20$$

Calculate the common-mode and difference-mode signal gain values of the differential amplifier shown in Figure 16.20. For this problem, the output voltage, or V_{oc}, for a common-mode voltage input equals 0.006 V while the common voltage, or V_{ic}, equals 1.5 V. The output voltage, or V_{od}, for a voltage difference equals 3 V, while the voltage difference between the input terminals, or V_{id}, equals 0.2 V.

We measure the ability to amplify difference-mode signals and to attenuate common-mode signals through the **common mode rejection ratio** (CMRR), or the ratio of difference-mode gain to common-mode gain.

$$CMRR = \frac{A_{vd}}{A_{vc}}$$

The common mode rejection ratio is expressed in decibels. A greater value of CMRR indicates a better differential amplifier.

EXAMPLE 16.4

Determine the common mode rejection ratio for the circuit values given in example 16.3.

Solution

In example 16.3, we found that the common-mode signal gain equals 0.005 and the difference-mode signal gain equals 20. Using those values in the common mode rejection ratio gives:

$$\text{CMRR} = \frac{A_{vd}}{A_{vc}}$$

$$\text{CMRR} = \frac{20}{0.005}$$

$$\text{CMRR} = 4000$$

To express CMRR in decibels use

$$dB = 20 \log \text{CMRR}$$

$$\text{CMRR} = 72 \text{ dB}$$

PRACTICE PROBLEM 16.4

Determine the common mode rejection ratio for the circuit values given in practice problem 16.3

CASCADING DIFFERENTIAL AMPLIFIERS

Cascading an additional amplifier stage to a differential amplifier provides higher gain. Figure 16.22a shows a schematic of a cascaded pair of differential amplifiers, and Figure 16.22b shows the equivalent circuit. Two transistors—Q_1 and Q_2—and a current sink consisting of a transistor (Q_5 in Figure 16.22b) and two resistors make up the first stage. Symbolized by the circled arrow, the current sink simulates a higher value of resistance for R_E.

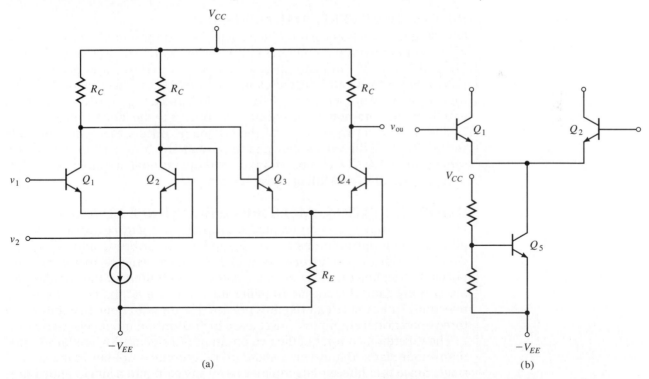

FIGURE 16.22 (a) Schematic diagram of a cascaded pair of differential amplifiers. (b) Equivalent circuit.

Transistors Q_3 and Q_4 in Figure 16.22a increase the gain and convert the double-ended output of the differential amplifier to a single-ended output. During operation, Q_4 inverts the output of Q_2 and compares the inverted signal with the output of Q_1. With that action, the full differential output of the differential amplifier is across Q_4. Transistor Q_4 amplifies the differential signal and supplies the single-ended output.

The circuit shown in Figure 16.22a is not sensitive to changes in the positive supply voltage, V_{CC}. If V_{CC} rises, the collector currents in Q_3 and Q_4 rise, while the collector voltages drop. In the diagram, the bases of the two transistors are fed from a common voltage point through the equal-value resistors R_C.

When the collector voltages fall, the common voltage point becomes less positive and reduces base current to Q_4. Since a reduced base current causes an increase in collector voltage, the circuit has a regulating mechanism.

INSTRUMENTATION AMPLIFIERS

Since differential amplifiers have a symmetrical design and do not use coupling capacitors, they have important uses as DC amplifiers. **Instrumentation amplifiers** are high-gain, DC-coupled circuits. Differential amplifiers with a single-ended output, a high input impedance, and a high common mode rejection ratio often work as instrumentation amplifiers.

Instrumentation amplifiers amplify small differential signals produced by transducers that may have a large common-mode signal level. Common mode rejection ratios of up to 120 dB are not uncommon for instrumentation amplifiers. While most instrumentation amplifiers have a common-mode input voltage range that does not go above the 5-V and 10-V supply voltages, others have common-mode input voltage ranges of ± 200 V.

Progress Check
You have now completed objective 9.

PRECISION DIFFERENTIAL AMPLIFIER

One common method used for increasing the common mode rejection ratio of an instrumentation amplifier is to use closely matched, precision resistors. Figure 16.23 shows an instrumentation amplifier that includes precision resistors. Since the CMRR decreases with frequency, the stability of the resistors becomes a key factor. With 0.01% resistors and a high-CMRR operational amplifier, the common mode rejection ratio stays in the 80-dB range.

The trimmer at the bottom of the schematic offsets the common-mode sensitivity. With the values shown in the circuit, the trimmer can trim out errors up to 0.05%. The combination of good operational amplifiers and trimming can result in a CMRR of above 100 dB.

INSTRUMENTATION AMPLIFIERS USING THREE OP-AMPS

Figure 16.24 shows an instrumentation amplifier using three operational amplifiers. The input op-amps are configured as noninverting amplifiers. The circuit provides the advantage of a high input impedance that greatly reduces any loading on the source. Since the input buffers to the circuit operate at unity gain, the output amplifier must provide all the common-mode rejection. Thus, the circuit requires precise resistor matching. In addition, all three operational amplifiers must have a high common mode rejection ratio.

The differential output of the two op-amps is a signal with a reduced common-mode signal that drives a standard differential amplifier in the second stage. Since the differential amplifier has unity gain and a single-ended output, it eliminates any remnants of the common-mode signal.

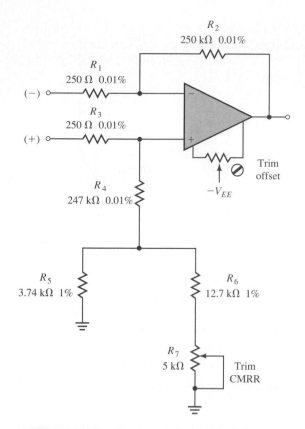

FIGURE 16.23 Precision differential amplifier

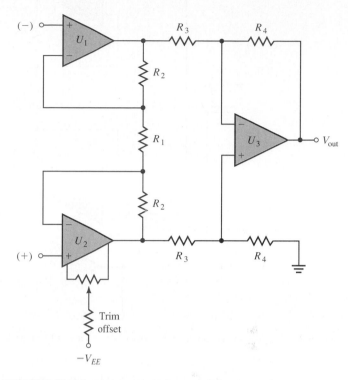

FIGURE 16.24 Instrumentation amplifier using three op-amps

Because of the configuration, the third operational amplifier does not require a high common mode rejection ratio or critical resistor matching. However, the operational amplifiers used in the input stage must have a high CMRR. In addition, the R_2 resistors are precision-matched resistors.

Instrumentation amplifiers using three operational amplifiers are available as modules with nearly all internal components. At the bottom of the circuit diagram, a trimmer tied to a negative supply voltage accomplishes offset trimming for the entire circuit. A single external resistor, R_1, sets the gain for the circuit.

Typical three-op-amp instrumentation amplifiers are the LH0036, AD552, and the 3630. Those amplifiers have a gain that ranges from 1 to 1000, a CMRR of 100 dB, and input impedances of 100 MΩ. An LH0036 will operate with supply voltages as low as ± 1 V. The 3630 provides gain linearity of 0.0002%, an initial offset voltage of 25 μV, and an offset drift of 0.25 μV/°C.

In the circuit shown in Figure 16.25, two operational amplifiers provide high differential gain and unity common-mode gain without close resistor matching. As the figure shows, the first-stage gain equals:

$$A_v = 1 + \left(\frac{2\,R_2}{R_1} \right)$$

Figure 16.25 depicts the complete instrumentation amplifier from Figure 16.24 with the addition of guard, sense, and reference terminals. At the output of U_4, the buffered common-mode voltage works as a "guard" voltage and reduces the effects of cable capacitance. As the schematic shows, the

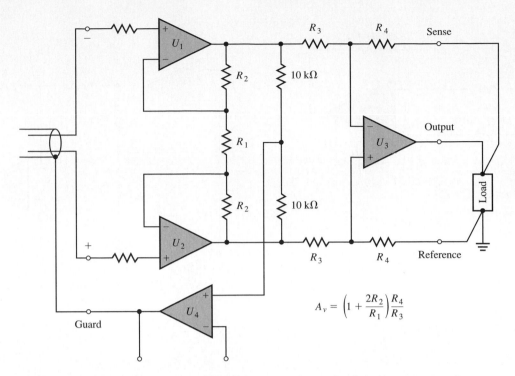

FIGURE 16.25   Three-op-amp instrumentation amplifier with guard, sense, and reference terminals

guard voltage connects to the shield of the input cables. Sense and reference terminals allow the offsetting of the output signal by a DC level. The impedance from the reference terminal to ground is small so that the CMRR is not degraded. Gain for the circuit equals:

$$A_V = \frac{R_4}{R_3}\left[\left(\frac{2\,R_2}{R_1}\right)+1\right]$$

EXAMPLE 16.5

Determine the first-stage gain and the circuit gain for the three-op-amp circuit instrumentation amplifier shown in Figure 16.25.

The R_2 precision resistors have a value of 100 kΩ. Resistor R_1 sets the gain for the circuit and has a value of 27 kΩ. R_4 has a value of 10 kΩ, and R_3 4.7 kΩ.

Solution
During our discussion of three-op-amp instrumentation amplifiers, we found that the first-stage gain equals:

$$A_V = 1 + \left(\frac{2\,R_2}{R_1}\right)$$

Substituting the R_1 and R_2 values into the equation gives us:

$$A_V = 1 + \left[\frac{2\,(100\ \text{k}\Omega)}{27\ \text{k}\Omega}\right]$$
$$A_V = 1 + 7.4$$
$$A_V = 8.4$$

To find the total circuit gain, we use the equation:

$$A_v = \frac{R_4}{R_3}\left[\left(\frac{2R_2}{R_1}\right) + 1\right]$$

and substitute the values for R_1, R_2, R_3, and R_4. That yields:

$$A_v = \frac{10 \text{ k}\Omega}{4.7 \text{ k}\Omega}\left[\left(\frac{200 \text{ k}\Omega}{27 \text{ k}\Omega}\right) + 1\right]$$

$$A_v = 2.13 \times 8.4$$

$$A_v = 17.9$$

The first-stage gain is 8.4, and the total circuit gain is 17.9.

PRACTICE PROBLEM 16.5

Find the first-stage and circuit gains of a three-op-amp instrumentation amplifier. R_1 has a value of 47 kΩ, and R_2 has a value of 500 kΩ. R_4 equals 100 kΩ, and R_3 equals 33 kΩ.

AN INSTRUMENTATION AMPLIFIER WITH A BOOTSTRAPPED POWER SUPPLY

Noise can be injected through power supplies. The **power supply rejection ratio** defines the quantity of change seen at the output of an operational amplifier when supply voltages change. If an operational amplifier has a power supply rejection ratio (PSRR) of 100 μV/V (maximum), the DC output voltage of the amplifier will change by no more than 100 μV when the power supply changes by 1 V. Thus, a 4-V change in the supply voltage would result in a change of no more than 400 μV, or

$$4 \times 100 \text{ }\mu\text{V} = 400 \text{ }\mu\text{V}$$

PRACTICE PROBLEM 16.6

Determine the amount of change in the output of an operational amplifier if the amplifier has a PSRR of 180 μV/V (maximum) and the supply voltage varies by 2.5 V.

The common mode rejection ratio of the input stage can limit the common-mode rejection of the instrumentation amplifier shown in Figure 16.25. Often, however, applications require a CMRR greater than 120 dB.

In Figure 16.26, the addition of an operational amplifier compensates for the common-mode signal from the input-stage operational amplifiers. Labeled U_4, the extra op-amp buffers the common-mode signal level and drives the common terminal of a floating split supply for U_1 and U_2.

Since the op-amp buffers the common-mode signal, op-amps U_1 and U_2 do not have the swing at their input terminals with respect to the power supply. The system power supply provides power for the buffer op-amp and the second-stage operational amplifier. Although the bootstrapped instrumentation amplifier works well for DC circuits that require a higher common mode rejection ratio, frequency still affects performance. Increasing frequencies present problems with matching impedance to the input capacitances.

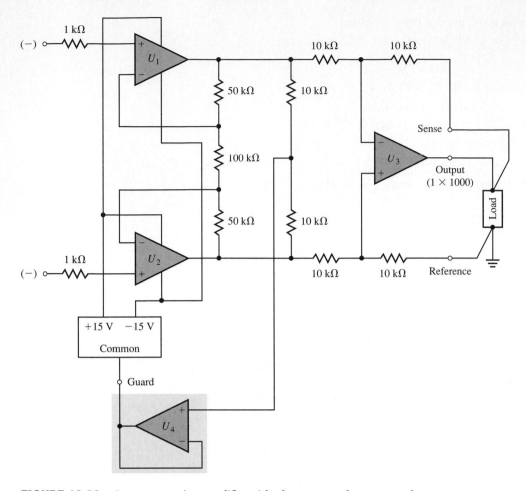

FIGURE 16.26 Instrumentation amplifier with a bootstrapped power supply

TRANSDUCERS COMBINED WITH DIFFERENTIAL AND INSTRUMENTATION AMPLIFIER CIRCUITS

Thermistor Circuits

Figure 16.27 shows two variations of thermistor circuits. With the configuration change shown in Figure 16.27a, the output voltage from the differential amplifier varies linearly with temperature. Figure 16.27b features a thermistor working as part of a Wheatstone bridge. The bridge balances when the ratio of resistance equals:

$$\frac{R_T}{R_2} = \frac{R_1}{R_3}$$

When coupled with a high-gain amplifier, the circuit can detect small changes in a reference temperature.

A Strain Gauge Circuit

One common type of transducer–instrumentation amplifier circuit is the strain gauge shown in Figure 16.28. The bridge arrangement of four metal, thin-film 350-Ω resistors converts the elongation or strain of attached material to resistance changes.

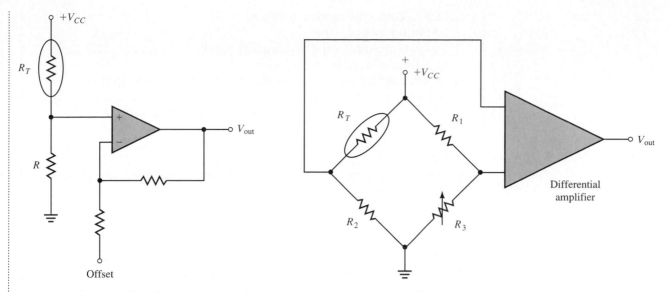

FIGURE 16.27 (a) Thermistor circuit with the output of the differential amplifier varying linearly with temperature. (b) Thermistor working as part of a Wheatstone bridge.

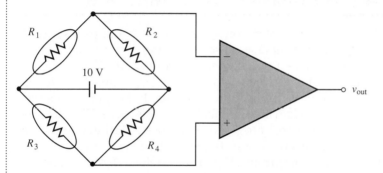

FIGURE 16.28 Transducer–instrumentation amplifier circuit

Even though the resistors are equal in value, they are subjected to different levels of strain. Applying a DC voltage across two of the terminals and then measuring the voltage between the other two terminals shows a voltage difference. The resistance changes result in a small change of differential output voltage driven by a fixed bias voltage.

With a full-scale sensitivity of 3 mV/V, the full-scale output of the bridge is 30 mV for the +10 VDC bias. The small differential output voltage that is proportional to strain rides on a +5-VDC level. Amplifying the millivolt voltages while rejecting variations in the common-mode 5-V signal requires a differential amplifier with a high common mode rejection ratio. As an example, a differential amplifier amplifying a 0.3-mV signal riding on 5000 mV would require a CMRR of 15,000 to 1, or approximately 90 dB.

Progress Check
You have now completed objective 10.

REVIEW SECTION 16.2

1. The elimination of undesired signals by differential amplifiers is called (common-mode/difference-mode) rejection.
2. Any voltage applied to one input of a differential amplifier is a common voltage. True or false?

3. Define common-mode rejection.
4. List four types of noise.
5. Instrumentation amplifiers amplify small _____ signals produced by transducers.
6. Differential amplifiers are used as instrumentation amplifiers because they have _____ and _____ .
7. Why should instrumentation amplifiers have high input impedances?
8. The power supply rejection ratio is the ratio of _____ to _____ .
9. Instrumentation amplifiers have common mode rejection ratios in the range of
   (a) 50 dB
   (b) 120 dB
   (c) 0 dB
   (d) 75 dB
10. CMRR (increases/decreases) with frequency.

16.3 CONVERTING THE DIFFERENTIAL OUTPUT SIGNAL TO DATA

Many applications require the conversion of an analog signal to a series of pulses called digital signals that represent the amplitude of the analog signal. This text is not designed to teach you digital electronics, but linear devices have many applications in digital circuits. As you learned earlier, transducers convert physical measurements into electrical signals. For example, a strain gauge converts a change in strain into a change in resistance. Thermocouples convert temperature changes into voltage changes. Manufacturers use computers to process data received from transducers. Because transducers produce analog voltages at their outputs, digital-to-analog (D/A) and analog-to-digital (A/D) conversion is required.

D/A converters accept input voltage pulses and output an analog voltage that is proportional to the numerical equivalent of the input pulses. The analog output may control a mechanical valve, a robot arm motor, or a heating and cooling system. In addition, D/A converters control the level of music signals and generate complex waveforms. For music signal level control, the music signal acts as a reference and a digital signal is sent to the D/A converter that controls the amount of signal reaching the output. Also, waveforms repeatedly generated by D/A converters are used in radar and sonar systems.

Several A/D techniques exist with each having advantages and limitations. All arrive as prepackaged modules or integrated circuits. However, every A/D circuit has specific characteristics. Since A/D converters incorporate amplifiers, they have impedance, drift, nonlinearity, and noise characteristics. The quality of the A/D converter affects the quality of its conversion accuracy.

DIGITAL-TO-ANALOG CONVERTERS

Weighted-Resistor Network

Figure 16.29 is a special kind of summing amplifier called a weighted resistor D/A converter. The term *weighted* means that even though the input voltage

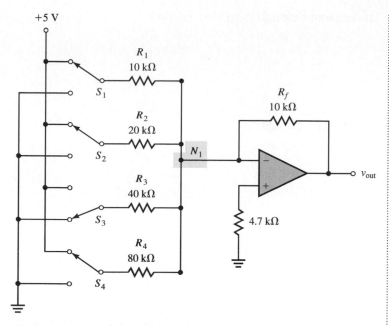

FIGURE 16.29 Weighted-resistor network

is the same for all inputs, each input affects the output differently. Starting from the top, observe that the resistors double in size as we go down. This means that the current flowing through each resistor will decrease by one-half. Because the current flowing through R_1 is twice the current flowing through R_2, the R_1 input affects the output twice as much as the R_2 input. All the currents sum at node N_1 and then flow through R_f. The feedback voltage is the product of the summed current times the feedback resistance. V_{out} will equal the feedback voltage.

$$V_{out} = V_{Rf} = I_f \times R_f$$

EXAMPLE 16.6

Using Figure 16.29, solve for the output voltage given the pictured switch settings.

The switches may be connected to either +5 V or 0 V. If all the switches were connected to 0 V, no current would flow through the feedback resistor and the output would be zero. If all the switches were connected to +5 V, then the maximum current would flow through the feedback resistor.

Solution

In Figure 16.29, switches S_1, S_2, and S_4 are connected to +5 V, and switch S_3 is connected to 0 V. First, we need to find the current flowing through each input resistor. The current flowing through R_1 is

$$I_1 = \frac{5\ V}{R_1}$$

$$I_1 = \frac{5\ V}{10\ k\Omega}$$

$$I_1 = 0.5\ mA$$

The current through R_2 is

$$I_2 = \frac{5\ V}{R_1}$$
$$I_2 = \frac{5\ V}{20\ k\Omega}$$
$$I_2 = 0.25\ mA$$

The current through R_3 is

$$I_3 = \frac{0\ V}{R_1}$$
$$I_3 = \frac{0\ V}{40\ k\Omega}$$
$$I_3 = 0.0\ mA$$

The current through R_4 is

$$I_4 = \frac{5\ V}{R_1}$$
$$I_4 = \frac{5\ V}{80\ k\Omega}$$
$$I_4 = 0.0625\ mA$$

Second, find the current flowing through N_1.

$$I_{N1} = I_1 + I_2 + I_3 + I_4$$
$$I_{N1} = 0.5\ mA + 0.25\ mA + 0.0\ mA + 0.0625\ mA$$
$$I_{N1} = 0.8125\ mA$$

Last, find the feedback voltage, which will equal the output voltage.

$$V_f = I_{N1} \times R_f$$
$$V_f = 0.8125\ mA \times 10\ k\Omega$$
$$V_f = 8.125\ V$$

Because the current flows from the op-amp output, the output voltage is negative.

$$V_{out} = -8.125\ V$$

PRACTICE PROBLEM 16.7

Determine the output voltage of Figure 16.29 when S_1, S_3, and S_4 are connected to 5 V and S_2 is connected to 0 V.

R-2R Ladder Network

There are two major problems with the weighted resistor D/A converter. First, standard resistors are not available in doubling units. Hence, a weighted resistor D/A converter would require a custom-made resistor, which is very expensive. Second, a D/A converter may have 16 inputs. If the smallest resistor was 1 kΩ, the largest resistor would have to be 32.768 MΩ. The range of precision resistors is impractical. In Figure 16.30 the 100-kΩ resistors would be standard resistors. The easiest way to get 50 kΩ is to connect two 100-kΩ resistors in parallel.

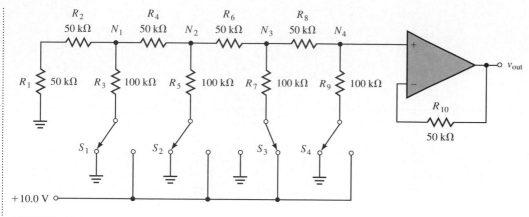

FIGURE 16.30   R-2R ladder network

To see how the R-2R ladder works, we will thevenize the resistor network by opening the load at each node. We will set all but one of the switches to 0 V.

EXAMPLE 16.7

Determine the voltage at N_4 when S_4 is set to 5 V and S_1, S_2, and S_3 are set to 0 V.

With S_4 set to 5 V, there is only one voltage source. However, there are three grounds. First, open the circuit at N_1. Figure 16.31a shows that R_1 and R_2 are in series and may be added together. The equivalent circuit is called R_{TH1}. R_{TH1} equals 2R. Figure 16.31b shows that R_{TH1} is in parallel with R_3, whose value is also 2R. Two equal resistors in parallel equal one-half either's value, which is R, and is called R_{TH2}. Figure 16.31c shows that R_{TH2} is in series

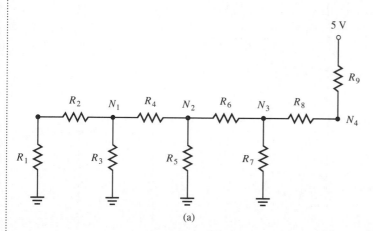

(a)

FIGURE 16.31   (a–h) Equivalent circuits of network shown in Figure 16.30 with S_4 at V_s

(Continued on next page)

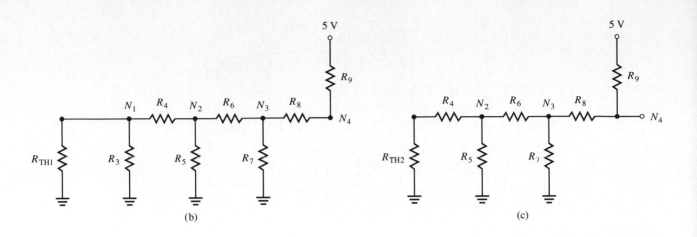

(b)

(c)

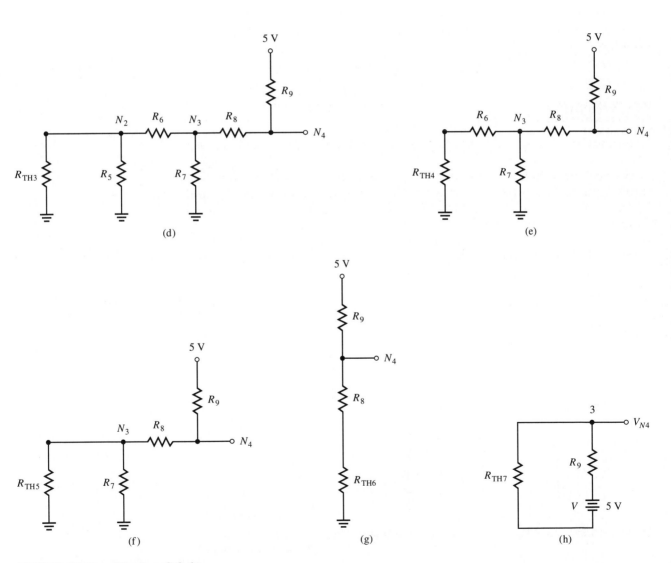

(d)

(e)

(f)

(g)

(h)

FIGURE 16.31 *(Continued)* **(b–h)**

with R_4 when we open the load at N_2. R_4 and R_{TH2} will equal $2R$, which is called R_{TH3}. Figure 16.31d shows that R_{TH3} is in parallel with R_5. Their equivalent resistance, R_{TH4}, has a value of R. The thevenizing process is illustrated in Figure 16.31e through 16.31g. Figure 16.31h shows the resultant thevenized circuit, which is a series circuit. R_1 through R_8 have been reduced to one equivalent Thevenin resistance R_{TH7}. The voltage at N_4 may now be solved using the voltage divider method.

Solution
Beginning with the left side of the resistor network, add R_1 to R_2.

$$R_{TH1} = R_1 + R_2$$
$$R_{TH1} = 50\ k\Omega + 50\ k\Omega$$
$$R_{TH1} = 100\ k\Omega$$

R_{TH1} is in parallel with R_3.

$$R_{TH2} = R_{TH1} \parallel R_3$$
$$R_{TH2} = 100\ k\Omega \parallel 100\ k\Omega$$
$$R_{TH2} = 50\ k\Omega$$

R_{TH2} is in series with R_4.

$$R_{TH3} = R_{TH2} + R_4$$
$$R_{TH3} = 50\ k\Omega + 50\ k\Omega$$
$$R_{TH3} = 100\ k\Omega$$

R_{TH3} is in parallel with R_5.

$$R_{TI4} = R_{TH3} \parallel R_5$$
$$R_{TH4} = 100\ k\Omega \parallel 100\ k\Omega$$
$$R_{TH4} = 50\ k\Omega$$

R_{TH4} is in series with R_6.

$$R_{TH5} = R_{TH4} + R_6$$
$$R_{TH5} = 50\ k\Omega + 50\ k\Omega$$
$$R_{TH5} = 100\ k\Omega$$

R_{TH5} is in parallel with R_7.

$$R_{TH6} = R_{TH5} \parallel R_7$$
$$R_{TH6} = 100\ k\Omega \parallel 100\ k\Omega$$
$$R_{TH6} = 50\ k\Omega$$

R_{TH6} is in series with R_7.

$$R_{TH7} = R_{TH6} + R_8$$
$$R_{TH7} = 50\ k\Omega + 50\ k\Omega$$
$$R_{TH7} = 100\ k\Omega$$

Figure 16.32 illustrates that R_{TH7} is in series with R_9 and both have a value of $100\ k\Omega$. The voltage at N_4 is

$$V_{N4} = \frac{V_S \times R_{TH7}}{R_{TH7} + R_9}$$
$$V_{N4} = \frac{5V \times 100\ k\Omega}{100\ k\Omega + 100\ k\Omega}$$
$$V_{N4} = 2.5\ V$$

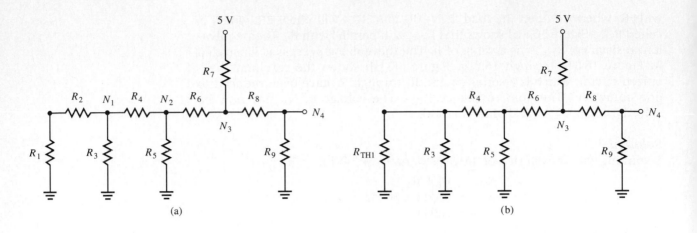

(a)

(b)

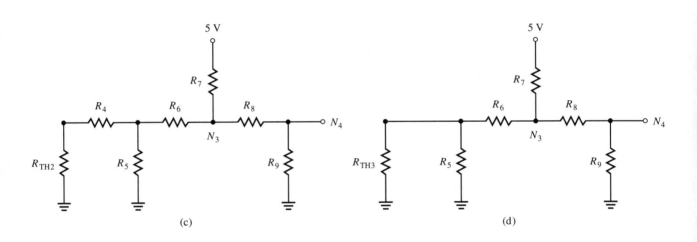

(c)

(d)

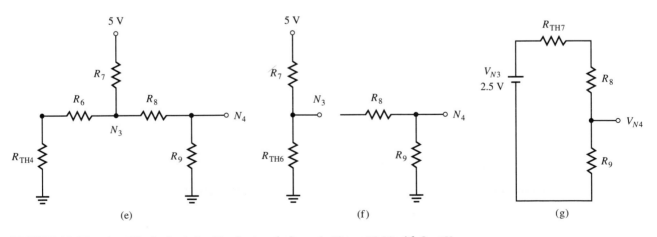

(e)

(f)

(g)

FIGURE 16.32 (a–g) Equivalent circuits of network shown in Figure 16.30 with S_3 at V_s

EXAMPLE 16.8

Determine the voltage at N_4 when S_3 is set to 5 V and S_1, S_2, and S_4 are set to 0 V.

The circuit is solved the same way except the voltage at N_3 must be determined before solving for the voltage at N_4.

Solution

Beginning with the left side of the resistor network, add R_1 to R_2.

$$R_{TH1} = R_1 + R_2$$
$$R_{TH1} = 50\ k\Omega + 50\ k\Omega$$
$$R_{TH1} = 100\ k\Omega$$

R_{TH1} is in parallel with R_3.

$$R_{TH2} = R_{TH1} \| R_3$$
$$R_{TH2} = 100\ k\Omega \| 100\ k\Omega$$
$$R_{TH2} = 50\ k\Omega$$

R_{TH2} is in series with R_4.

$$R_{TH3} = R_{TH2} + R_4$$
$$R_{TH3} = 50\ k\Omega + 50\ k\Omega$$
$$R_{TH3} = 100\ k\Omega$$

R_{TH3} is in parallel with R_5.

$$R_{TH4} = R_{TH3} \| R_5$$
$$R_{TH4} = 100\ k\Omega \| 100\ k\Omega$$
$$R_{TH4} = 50\ k\Omega$$

R_{RH4} is in series with R_6.

$$R_{TH5} = R_{TH4} + R_6$$
$$R_{TH5} = 50\ k\Omega + 50\ k\Omega$$
$$R_{TH5} = 100\ k\Omega$$

Solve for V_{N3}.

$$V_{N3} = \frac{V_S \times R_{TH5}}{R_7 + R_{TH5}}$$
$$V_{N3} = \frac{5\ V \times 100\ k\Omega}{100\ k\Omega + 100\ k\Omega}$$
$$V_{N3} = 2.5\ V$$

R_{TH5} is in parallel with R_7.

$$R_{TH6} = R_{TH5} \| R_7$$
$$R_{TH6} = 100\ k\Omega \| 100\ k\Omega$$
$$R_{TH6} = 50\ k\Omega$$

Solve for V_{N4}.

$$V_{N4} = \frac{V_{N3} \times R_9}{(R_{TH6} + R_8) + R_9}$$
$$V_{N4} = \frac{2.5\ V \times 100\ k\Omega}{(50\ k\Omega + 50\ k\Omega) + 100\ k\Omega}$$
$$V_{N4} = 1.25\ V$$

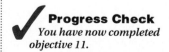

Progress Check
You have now completed objective 11.

16.4   APPLICATIONS

PHOTODIODE AUTOMOBILE LAMP DIMMER

A photodiode lamp dimmer such as the one shown in Figure 16.33 automatically changes automobile headlights from high beam to low beam. Pressing the high-beam switch (S_1) reverse-biases the photodiode with the battery voltage. A reverse-biased photodiode will not allow current to flow through the relay coil. Reverse biasing causes the diode to have high device resistance that prevents current from flowing through the relay coil. Relay contacts A and C complete the high-beam circuit.

Oncoming headlights start the dimming process with light activating the diode. Light focused on the photodiode PN junction increases conductivity, which increases minority current. The current flows through the photodiode and the relay coil. When enough current is flowing, the relay is energized, completing the low-beam headlight circuit.

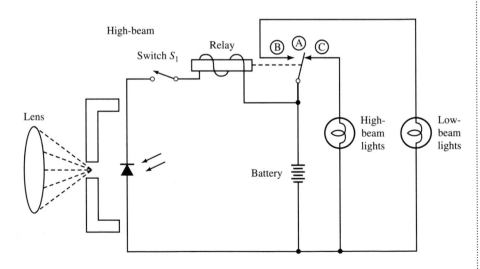

FIGURE 16.33   Photodiode automobile lamp dimmer

After the oncoming car passes, the light source becomes removed from the diode. With no light source for the PN junction, the minority current decreases, causing the relay to de-energize, which turns on the high-beam headlights.

OPTOISOLATOR IN A HIGH-SPEED LINE RECEIVER

Figure 16.34 uses an infrared-emitting diode and a very high speed phototransistor to provide line isolation in a high-speed receiver. The phototransistor has an output of +5 V. In addition to providing isolation, the optoisolator package provides high-speed coupling for the receiver stages. The entire circuit is a part of a larger high-speed data transmission line.

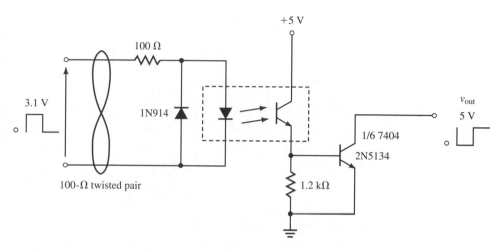

FIGURE 16.34 Optoisolator in a high-speed line receiver

BALANCED THERMOCOUPLE AMPLIFIER

Figure 16.35 shows a balanced thermocouple amplifier. In the circuit, an AD517K differential amplifier provides a high voltage gain and works as an instrumentation amplifier for a thermocouple. The T connection in the amplifier feedback path keeps the input impedance large so that loading by the source impedance does not cause errors.

Two 0.1-μF input bypass capacitors at the sensing junction reduce common-mode interference at 60 Hz and at radio frequencies. Since thermocouples sometimes utilize long cables, some protection against radio frequency interference is necessary. In addition, the two 0.01-μF capacitors placed across the feedback resistors provide protection against radio frequency problems by limiting the circuit bandwidth.

The reference junction compensation circuit at the bottom of the figure acts at the output instead of compensating the voltage from the thermocouple at the input. Because of this configuration, the input remains differential. Thus, the circuit takes advantage of the good common-mode rejection of the differential amplifier. Since the AD590 amplifier is a precision low-offset amplifier with a drift less than 1 μV/°C, any measurement error stays below 50 μV; 50 μV corresponds to a 1°C error. With an amplifier voltage gain of 200, the compensation circuit adds 200 × 51.5 μV/°C, or 10.3 mV/°C at the output.

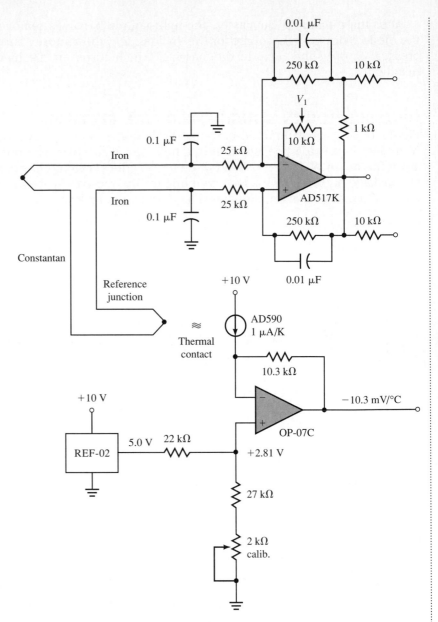

FIGURE 16.35   Balanced thermocouple amplifier

REVIEW SECTION 16.4

1. Reverse biasing a photodiode causes a high device (resistance/conductance) that (prevents/causes) current flow.
2. The automobile lamp-dimming circuit contains a phototransistor. True or false?
3. The optoisolator shown in Figure 16.34 provides _____ and _____.
4. Why does the thermocouple amplifier have input bypass capacitors?

5. The differential amplifier used in the balanced thermocouple amplifier circuit has ＿＿＿＿＿＿ ＿＿＿＿＿＿ and a high ＿＿＿＿＿＿ .

6. The balanced thermocouple circuit utilizes the good ＿＿＿＿＿＿ ＿＿＿＿＿＿ ＿＿＿＿＿＿ of the differential amplifier.

TROUBLESHOOTING

Troubleshooting instrumentation amplifiers is similar to troubleshooting any op-amp circuit. The difference is found in the application environment. Instrumentation amplifiers are often used in noisy situations. If the output is noisy, the cause may be a defective filter. Look for capacitors that are leaking. A leaking capacitor does not provide DC isolation; rather it provides a high resistance path. In addition, sensors tend to fail over time. Thermistors may be heated and their change in resistance noted. (CAUTION: Do not use an open flame, as it will damage the thermistor.) Photoresistors, photodiodes, and phototransistors may be tested by alternately exposing the device to light and shielding it from light. The device's conductance should change.

TROUBLESHOOTING QUESTIONS
1. One difference between troubleshooting instrumentation amplifiers and op-amp circuits is the application environment. True or false?
2. A defective ＿＿＿＿＿＿ may cause a noisy output in an instrumentation amplifier circuit.
3. A leaky capacitor provides DC isolation. True or false?
4. When working with photoconductors, the alternating absence and presence of light will cause the ＿＿＿＿＿＿ to change.
5. Sensors improve with age. True or false?

WHAT'S WRONG WITH THIS CIRCUIT?

In the circuit in Figure 16.36, the voltage at N_1 is 0 V. The output voltage is −5 V. Which of the components has failed?

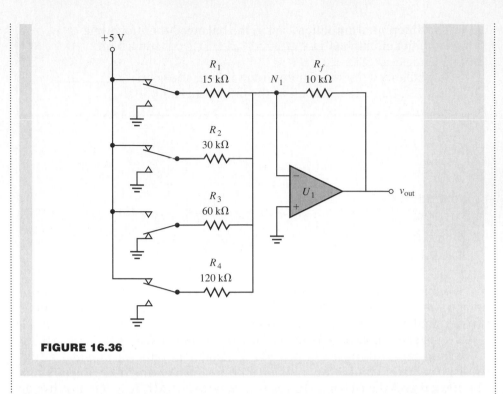

FIGURE 16.36

SUMMARY

Transducers connect the physical world to the world of electronics and respond to temperature, light, strain, movement, and position. Thermistors and thermocouples respond to changes in temperature. Other devices—such as photoresistors, photodiodes, phototransistors, and optoisolators—respond to changes of light intensity.

Instrumentation amplifiers are high-gain, DC-coupled differential amplifiers that exhibit good common mode rejection ratios. Several different configurations of instrumentation amplifiers match CMRR characteristics to specific needs. Instrumentation amplifiers may use two or three operational amplifiers or may have bootstrapped power supplies.

Converters change digital signals to analog signals and analog signals to digital signals. The output signals from digital-to-analog converters may control mechanical systems, heating and cooling systems, or music signals.

PROBLEMS

1. Use Figure 16.37 and Table 16.1. The Wheatstone bridge is balanced at 25°C. At 50°C, what is the output of the op-amp?
2. Calculate the output voltage of Figure 16.38 if the photoresistor has a resistance of 540 Ω.

3. At a given amount of light the diode current is 200 μA. Use Figure 16.39 to calculate the output voltage.
4. State the advantage of using a bridge strain gauge configuration over a single strain gauge configuration.

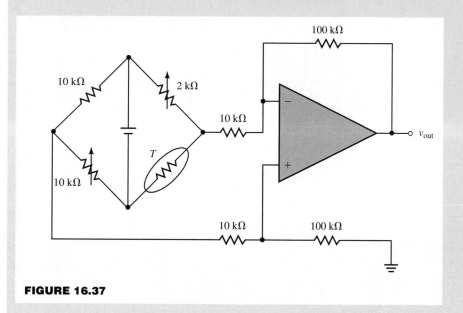

FIGURE 16.37

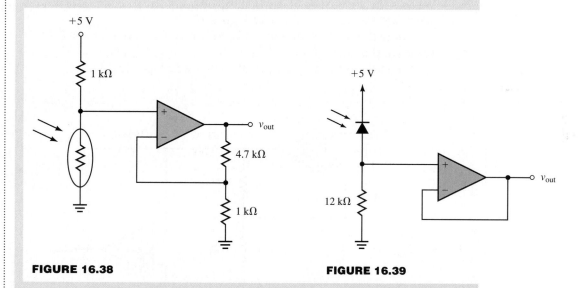

FIGURE 16.38   **FIGURE 16.39**

Refer to Figure 16.40 for problems 5 through 8.

5. Find the common-mode gain.
6. Find the differential-mode gain.
7. If 1 V is applied to both inputs, what is the output?
8. If V_2 is grounded and V_1 has +100 mV applied, what is the output voltage?

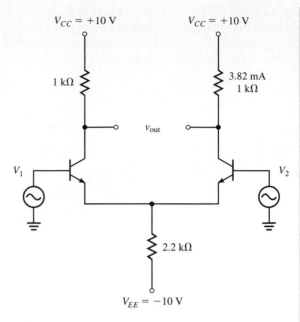

FIGURE 16.40

Refer to Figure 16.41 for problems 9 through 11.

9. Determine the gain of U_1.
10. Determine the gain of U_3.
11. What is the advantage of using three op-amps over using only one?
12. Calculate the output voltage of Figure 16.42.
13. Calculate the output voltage of Figure 16.43 if S_1 and S_2 are at 10 V and S_3 and S_4 are at 0 V.
14. Calculate the output voltage if all switches in Figure 16.43 are at 10 V.

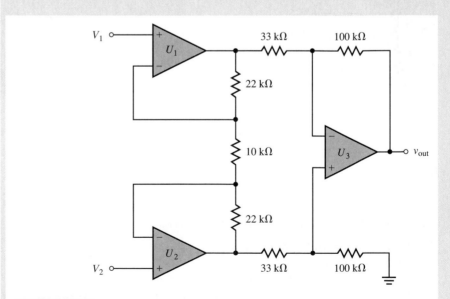

FIGURE 16.41

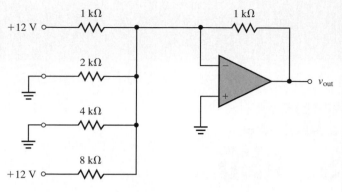

FIGURE 16.42

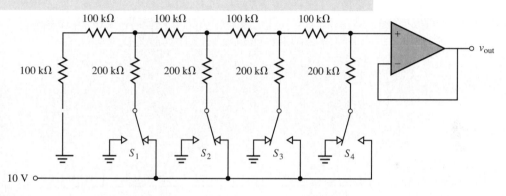

FIGURE 16.43

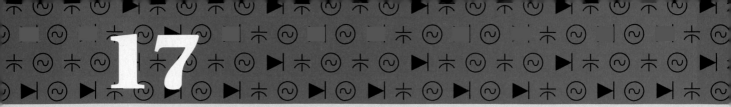

17

MULTIVIBRATORS, OSCILLATORS, AND TIMERS

As you read this chapter, concentrate on learning how to:

1. Discuss the different purposes of bistable, astable, and monostable multivibrators
2. Analyze a bistable multivibrator circuit
3. Analyze an astable multivibrator circuit
4. List the characteristics of monostable multivibrators
5. Analyze a monostable multivibrator circuit
6. Discuss the functions of a 555 timer
7. Describe various circuit applications for multivibrators and timers

INTRODUCTION

This chapter introduces an interesting facet of electronic circuitry called multivibrators and timers. Along with introducing a different type of electronic circuitry, the chapter also reinforces knowledge that we have about RC circuits. As you will see, multivibrators and timers rely on simple RC circuits for time constants.

Multivibrators generate rectangular voltage waveforms that electronic equipment utilizes for timing. A rectangular waveform like that shown in Figure 17.1 has four basic, desirable characteristics:

1. a stable low level,

2. a stable high level,

3. a sharp rise from the low to the high level, and

4. a sharp fall from the high to the low level.

Multivibrator circuits control the period of time that a rectangular waveform stays at the low and high levels. Also, multivibrators can supply high-speed, symmetrical rectangular waveforms. In addition, all multivibrators exhibit some amplification and use positive feedback.

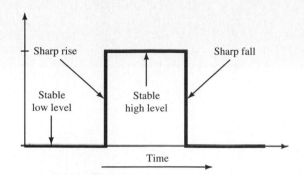

FIGURE 17.1 A rectangular waveform

Three types of multivibrators exist. Bistable multivibrators have two stable states, astable multivibrators have no stable states, and monostable, or one-shot, multivibrators have one stable state. Astable, or free-running, multivibrators oscillate without any triggering by external signals. Bistable and monostable multivibrators require an external triggering signal before they can generate a waveform.

At the end of the chapter, you will see how multivibrators play a vital role in the makeup of waveform generators, alarm circuits, and electronic test equipment. Also, you will know how combinations of different multivibrator types can shape waveforms for specific needs.

Progress Check
You have now completed objective 1.

17.1 ASTABLE MULTIVIBRATORS

Astable, or free-running, **multivibrators** have no stable states and no input signals. Because of that characteristic, the output oscillates between the two semistable states of the multivibrator. The output of an astable multivibrator is a square wave. Since the astable multivibrator runs by itself, it is an oscillator.

External components in the astable circuit determine the amount of time given to each semistable state. Primarily, the astable multivibrator is used as a nonsinusoidal, or relaxation, oscillator or a square wave generator. Digital computers often utilize astable multivibrators as clocks to ensure that operations, such as addition, synchronize with other operations in the computer system.

Figure 17.2 shows a basic circuit of a collector-coupled astable multivibrator using NPN transistors. In the circuit, the coupling capacitors, represented by C, couple the collector of each transistor to the base of the other transistor. Figure 17.3 shows how the waveforms at the collectors of each transistor would appear. In that Figure, Q represents the output voltage across the Q_1 collector and ground, and $\overline{Q}$ represents the output voltage across the Q_2 collector and ground. The line above Q means that its output is opposite the other output. When Q is at a high level $\overline{Q}$ is at a low level.

Figure 17.3 illustrates that each transistor always generates a rectangular-type waveform. The length of the waveform (T) equals:

$$T = 0.7RC$$

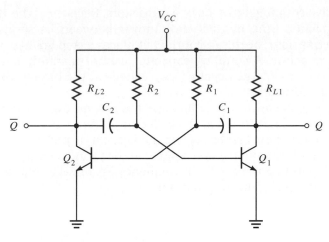

FIGURE 17.2 Basic astable multivibrator circuit

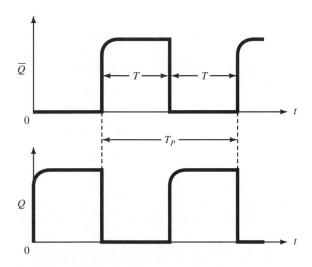

FIGURE 17.3 Waveforms of an astable multivibrator circuit

Thus, the total period, or T_P, equals

$$T_P = 2T = 1.4RC$$

The frequency *(f)* of the generated square wave equals 1 divided by the total period. The frequency of the generated square wave equals:

$$f = \frac{1}{1.4RC}$$

ANALYZING A SIMPLE ASTABLE MULTIVIBRATOR CIRCUIT

Figure 17.2 shows a simple astable multivibrator circuit. Even though the circuit seems symmetrical in its design, differences in the component values cause one transistor to turn on more quickly than the other as power is applied. As the collector voltage of the conducting transistor falls, the second transistor goes toward cutoff.

Capacitive coupling links the two transistors. Because of the storage ability of the capacitors, the multivibrator moves between its semistable states. When one transistor begins to conduct, it places a step voltage input into the coupling capacitor. A coupling capacitor cannot instantly change voltage. Thus, it produces a large negative step voltage at the base of the other transistor in the multivibrator.

With a large negative bias at its base, the second transistor goes into cutoff. The rising voltage found at the collector of the second transistor couples back to the base of the first transistor and keeps it turned on.

Complementary square waves appear at the Q and Q outputs as shown in Figure 17.3. Finding the oscillator frequency becomes a matter of substituting values into the following equation:

$$f = \frac{1}{0.7\left[\left(R_1 C_1\right) + \left(R_2 C_2\right)\right]}$$

EXAMPLE 17.1

Determine the oscillator frequency for the astable multivibrator shown in Figure 17.2.

The components have the following values: $R_1 = 470\ \Omega$; $C_1 = 0.27\ \mu F$; $R_2 = 330\ \Omega$; $C_2 = 0.1\ \mu F$.

Solution
The equation for finding the oscillator frequency of an astable multivibrator is

$$f = \frac{1}{0.7\left[\left(R_1 C_1\right) + \left(R_2 C_2\right)\right]}$$

If we substitute the component values into the equation, we have:

$$f = \frac{1}{0.7\left[\left(470\ \Omega \times 0.27\ \mu F\right) + \left(330\ \Omega \times 0.1\ \mu F\right)\right]}$$
$$f = 8.9\ \text{kHz}$$

PRACTICE PROBLEM 17.1

Using Figure 17.2, match the circuit conditions with the waveforms shown in Figure 17.3.

(a) Q_1 conducting and Q_2 cut off.
(b) Changing states.
(c) Q_2 conducting and Q_1 cut off.

Figures 17.4 and 17.5 illustrate what happens when Q_1 reaches saturation. When Q_1 reaches saturation, its collector voltage (V_{C1}) decreases from its cutoff voltage of 5 V to its saturation voltage of 1 V. The battery, $V_{C1\ eqv}$, is the equivalent of Q_1's collector saturation voltage. The rapid 4-V decrease is coupled across C_1, driving Q_2 into hard cutoff. Q_2's base voltage at that instant will be -3.3 V ($+0.7$ V $- 4$ V). The rapid decrease in collector voltage has charged C_1. R_1, the source voltage, and Q_1 saturated form the discharge loop for C_1. Kirchhoff's voltage law says that the algebraic sum of all the voltages in a closed loop is zero. As C_1 discharges the voltage at Q_2's base will become

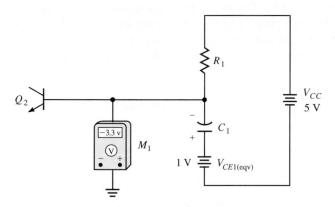

FIGURE 17.4 When Q1 reaches saturation

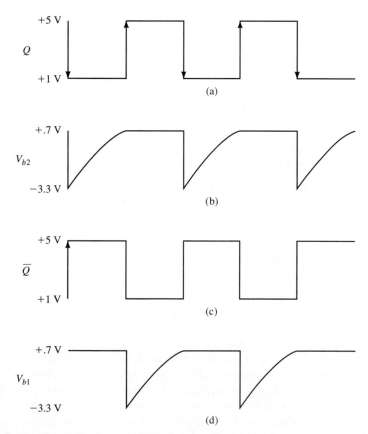

FIGURE 17.5 Waveforms for the circuits in Figures 17.4 and 17.6

more positive. Figure 17.5b shows the discharging voltage that is present on Q_2's base. The rate of discharge is determined by the values of R_1 and C_1. Once Q_2's base voltage is at 0.7 V, Q_2 starts to become saturated.

Figure 17.6 illustrates what happens when Q_2 reaches saturation. When Q_2 becomes saturated, its collector voltage (V_{CE2}) decreases from its cutoff voltage of 5 V to its saturation voltage of 1 V. The battery, $V_{CE2\ eqv}$ is the equivalent of Q_2's collector saturation voltage. The rapid 4-V decrease is coupled across C_2, driving Q_1 into hard cutoff. Q_1's base voltage at that instant will be -3.3 V $(+0.7$ V $- 4$ V). The rapid decrease in collector voltage has charged C_2. R_2, the source voltage, and Q_2 saturated form the discharge loop for C_2. Kirchhoff's voltage law says that the algebraic sum of all the voltages in a closed loop is zero. As C_2 discharges, the voltage at Q_1's base will become more positive. Figure 17.5d shows the discharging voltage that is present on Q_1's base. The rate of discharge is determined by the values of R_2 and C_2. Once Q_1's base voltage is at 0.7 V, Q_1 starts to become saturated and the next cycle begins.

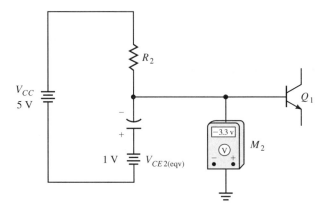

FIGURE 17.6 When Q_2 reaches saturation

Progress Check
You have now completed objective 2.

PRACTICE PROBLEM 17.2

Determine the output waveform at Q of Figure 17.4 with the following component values: $R_1 = 2.2$ kΩ; $C_1 = 0.033$ µF; $R_2 = 3.3$ kΩ; $C_2 = 0.047$ µF.

IMPROVING RISE TIME IN AN ASTABLE CIRCUIT

The square waves generated by the simple astable circuit will have the leading edge rounded at both outputs. However, many applications require a sharper square wave. Adding charging resistors and diodes to each side of the circuit shown in Figure 17.2 improves the rise time of the output waveforms. Figure 17.7 shows the circuit with the additional components.

If we make the parallel combination of R_L and R_C equal to the value of the original resistor, the astable multivibrator will have a faster rise time. Also, the R_L must be smaller to compensate for any stray circuit capacitance. With the addition of the diodes and resistors, the negative charge for each cycle discharges through a transistor, diode, and resistor. The diodes in the circuit allow the signals at the outputs to increase without charging a capacitor.

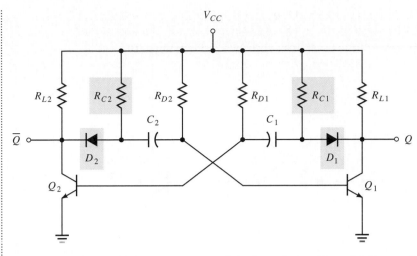

FIGURE 17.7 Adding charge resistors and diodes to the astable multivibrator circuit

REVIEW SECTION 17.1

1. Astable multivibrators have two stable states and no input signals. True or false?
2. Astable multivibrators work as
   (a) oscillators
   (b) regulators
   (c) flip-flops
   (d) multivibrators
3. Adding resistors and diodes to an astable multivibrator circuit improves the circuit delay time. True or false?
4. An astable multivibrator has (no/two) input terminals and (no/two) output terminals.
5. The two transistors that make up an astable multivibrator are linked by _____ coupling.
6. Another name for an astable multivibrator is _____ multivibrator.
7. Does an astable multivibrator require triggering from an external source?

17.2 BISTABLE MULTIVIBRATORS

Figure 17.8 shows a bistable multivibrator circuit. A **bistable multivibrator** has two resting states. An identifying circuit characteristic is that the bistable multivibrator uses only resistive feedback paths. No capacitors are in the feedback path. As the figure shows, two symmetrical feedback paths exist in the circuit. Those feedback paths provide quick transitions and stability for both states.

Assume that during power-up transistor Q_1 turns on and Q_2 remains off. With the transistors in those states, the collector-to-emitter voltage for Q_1 equals zero, and the collector-to-emitter voltage for Q_2 almost equals the col-

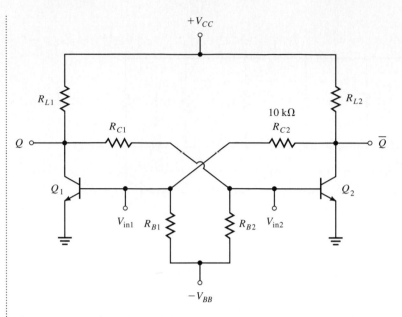

FIGURE 17.8 A bistable multivibrator circuit

lector supply voltage. In addition, the base of Q_1 is positive, while the base of Q_2 has a negative voltage.

Both the collector and base supply voltages going through R_{L2}, R_{C2}, and R_{B1} keep the Q_1 base positive. Since only the negative base supply voltage connects to the base of Q_2 through R_{C1} and R_{B2}, the base of Q_2 cannot go positive. With the collector voltage of Q_1 at zero, the positive collector supply voltage does not affect Q_2. Both transistors remain in the initial states until a triggering pulse is applied.

A **triggering pulse** is a positive or negative input signal used to push a multivibrator stage into a desired state. The signal is above or below the normal operating level of the stage. Triggering pulses can exist as the leading edge or trailing edge of waveforms. Or they can take the form of pulses that have carefully controlled amplitudes and shapes.

If a negative input signal arrives at the base of Q_1 or a positive input signal arrives at the base of Q_2, the transistors switch states. A positive signal at the base of Q_2 causes the transistor to conduct and reduces its collector voltage to zero. The falling collector voltage of Q_2 changes the bias on Q_1 from positive to negative. With that bias change, Q_1 turns off and its collector-to-emitter voltage rises. Now, Q_1 remains off and Q_2 remains on until another input pulse arrives and causes a change.

EXAMPLE 17.2

Using the circuit shown in Figure 17.8 find the collector-to-emitter voltage for Q_1 after one positive pulse has been applied to the base of Q_2.

The voltage at V_{CC} equals 7.6 V, and the voltage at V_{BB} equals -25 V. At startup, Q1 is on and Q2 is off.

Solution

If a positive pulse arrives at the Q_2 base, switching begins. The collector-to-emitter voltage for Q_2 is slightly greater than zero. The collector-to-emitter voltage for Q_1 equals the V_{CC+} supply voltage, or

$$V_{CE1} = V_{CC+} = 7.6 \text{ V}$$

PRACTICE PROBLEM 17.3

Using the circuit conditions from example 17.2, determine what needs to occur for the two transistors to switch states again.

In Figure 17.9, two transistors alternately switch on and off when triggered and form a flip-flop. Since positive feedback introduces instability into the circuit, each Q point moves until one transistor goes completely off and the other goes completely on. In the stable state of the circuit, one transistor is saturated while the other is cut off. Since the transistors always have opposite states, the outputs always have opposite logic levels. In the diagram, the Q_2 output is labeled as "NOT Q," or $\overline{Q}$. (The overbar symbol means "not." $\overline{Q}$ is stated as "Q not.")

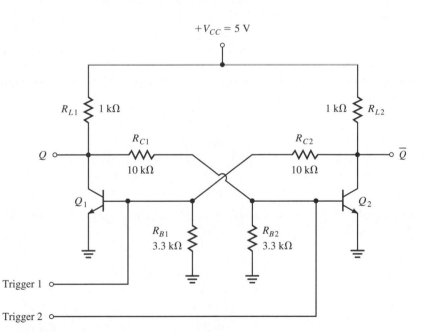

FIGURE 17.9 Basic discrete flip-flop circuit

EXAMPLE 17.3

Analyze the circuit shown in Figure 17.9. Determine the saturation point of transistor Q_2.

If Q2 is saturated, the potential at the $\overline{Q}$ output is 0.2 V. The 10 kΩ resistor between the collector of Q2 and the base of Q1 and the 3.3 kΩ resistor from the base of Q1 to ground form a voltage divider.

Solution

Since the voltage divider determines the potential on the base of Q_1, that value equals

$$V_{B1} = \frac{V_{C2} \times R_{B1}}{R_{B1} + R_{C2}}$$

$$V_{B1} = \frac{0.2\,V \times 3.3\,k\Omega}{10\,k\Omega + 3.3\,k\Omega}$$

$$V_{B1} = 50\,mV$$

Because 50 mV will not forward bias Q_1, it remains cut off.

With Q_1 cut off, the only significant amount of current through the Q_1 load resistor flows through the voltage divider, $R_{L1} + R_{C2}$.

That current may be calculated by using Ohm's law.

$$I_{L1} = \frac{V_{CC} - V_{B2}}{R_{L1} + R_{C1}}$$

$$I_{L1} = \frac{5\,V - 0.7\,V}{1\,k\Omega + 10\,k\Omega}$$

$$I_{L1} = \frac{4.3\,V}{11\,k\Omega}$$

$$I_{L1} = 391\,\mu A$$

The collector voltage of Q_1 may be found also using Ohm's law.

$$V_{C1} = V_{CC} - (I_{L1} \times R_{L1})$$
$$V_{C1} = 5\,V - (391\,\mu A \times 1\,k\Omega)$$
$$V_{C1} = 5V - 0.391\,V$$
$$V_{C1} = 4.6\,V$$

To be saturated Q_2 must have a base current equal to or greater than I_{C2}/h_{FE}. The collector current is

$$I_{C2} = \frac{V_{CC} - V_{CE}}{R_{L2}}$$

$$I_{C2} = \frac{5\,V - 0.2\,V}{1\,k\Omega}$$

$$I_{C2} = 4.8\,mA$$

With Q_1 cut off, its collector current comes from R_{B2} and Q_2's base. The current source by R_{B2} is

$$I_{RB2} = \frac{V_B}{R_{B2}}$$

$$I_{RB2} = \frac{0.7\,V}{3.3\,k\Omega}$$

$$I_{RB2} = 212\,\mu A$$

Q_2's base current may be found by rearranging Kirchhoff's current law.

$$I_{L1} = I_{RB2} + I_{Q2B}$$
$$I_{Q2B} = I_{L1} - I_{RB2}$$
$$I_{Q2B} = 391\,\mu A - 212\mu A$$
$$I_{Q2B} = 179\,\mu A$$

Any transistor with an h_{FE} of 26 or more will saturate with this circuit.

✓ **Progress Check**
You have now completed objective 3.

17.3   MONOSTABLE, OR ONE-SHOT, MULTIVIBRATORS

Monostable, or one-shot, **multivibrators** have one DC stable state and generate rectangular pulses. When a monostable multivibrator receives a pulse as an input, it immediately generates an output pulse of a given duration that is independent of the length of the input triggering pulse. Monostable multivibrators such as the one shown in Figure 17.10 generate only one output pulse for every input triggering pulse. Figure 17.11 shows the input and output waveforms for a one-shot multivibrator.

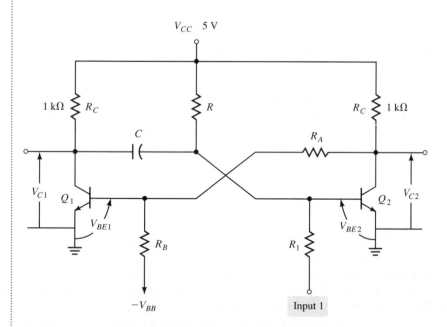

FIGURE 17.10   A monostable, or one-shot, multivibrator

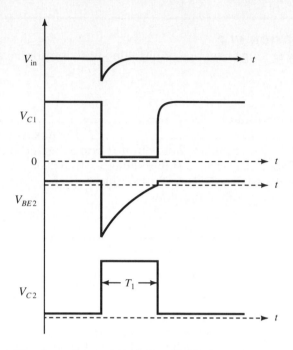

FIGURE 17.11 Waveforms from a one-shot multivibrator

A monostable multivibrator can be triggered into its "semistable" state for a given period of time and then spontaneously return to the initial stable state. Component values and circuit parameters—not the triggering pulse—determine the length of time that a monostable multivibrator remains in the semistable state.

One-shot multivibrators may be called triggered, adjustable-width pulse generators. They also function as pulse shapers that provide uniform-width pulses from a variable-width pulse train. Along with those two uses, monostable multivibrators also work as delay elements. Working as a delay, a monostable provides a pulse output transition at a fixed time after the triggering pulse.

Compared with the bistable multivibrator, the monostable has asymmetrical outputs. Another difference between bistable and monostable multivibrators is the coupling method used in the circuit. While bistable multivibrators use resistive coupling, monostable multivibrators rely on both capacitive coupling and resistive coupling.

TRIGGERING

Typically, monostable multivibrators use negative-going trigger pulses. The output pulse goes high at the time the negative slope of the trigger cuts off Q_2. As you saw earlier, the RC time constant of the monostable circuit controls the output pulse duration.

To allow different input signals to act as triggering pulses, one-shot multivibrators may have several inputs. The rising or falling edge of an input pulse at a specific input triggers a one-shot multivibrator to change states.

While triggering pulses for monostable multivibrators must have a minimum width that usually varies between 25 and 100 ns, they can be shorter or longer than the output pulse.

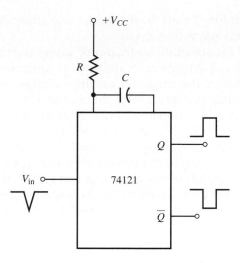

FIGURE 17.12   Block diagram of a 74121 one-shot multivibrator

Although we have shown you discrete one-shots, there are also several types of one-shots made of integrated circuits. The pulse width will be determined by the size of external components attached to the IC. Figure 17.12 illustrates a 74121, a one-shot multivibrator.

EXAMPLE 17.4

Find the pulse width of the triggered one-shot multivibrator in Figure 17.10. $C = 0.0015\ \mu F$; $R = 10\ k\Omega$.

Q2 is normally saturated. Its collector voltage is about 0.2 V. Q1 is normally cut off. Its collector voltage is near VCC. A negative-going trigger at input 1 will cut off Q2. Its collector voltage will rise rapidly, causing Q1 to become saturated. The Q1 collector voltage will decrease, charging C1. C1 will discharge through R, keeping Q2 cut off. Once C1 has discharged, Q2 will become saturated again, causing its collector voltage to decrease to about 0.2 V. Q1 will again cut off. The one-shot is now ready for the next trigger pulse.

Solution
The time of the pulse width (PW) is found by multiplying the RC time constant by 0.7.

$$PW = C \times R \times 0.7$$
$$PW = 0.0015\ \mu F \times 10\ k\Omega \times 0.7$$
$$PW = 10.5\ \mu s$$

The output of V_{C2} will be 0.2 V until triggered. Then the output will be 5 V for 10.5 μs. The output of V_{C1} will be 5 V until the one-shot is triggered. Then the output will be about 0.2 V for 10.5 μs.

MAXIMUM DUTY CYCLE

Since monostable multivibrators work with time, we can apply a time measurement called **duty cycle** to the circuits. The duty cycle is the fraction of time that the multivibrator spends in its semistable state. All monostable multivibrators have a maximum duty cycle.

Figure 17.13 shows a monostable multivibrator where the rate of charging capacitor C when the circuit returns to its stable state limits the duty cycle. If a trigger pulse arrives before the capacitor can fully charge, the time period becomes shorter. The capacitor has a reduced charge. One triggering pulse arriving directly after a previous triggering pulse does not allow the capacitor to charge. C charges through R_L. The charge on the capacitor is too small either to turn Q_1 off or to keep it off. No output pulse can occur.

If there are not enough charging time constants after the previous pulse, some error in the pulse duration will occur. As a rule, the triggering pulse should not occur until after three charging time constants after the previous pulse. Therefore, the minimum stable-state time of a monostable multivibrator is

$$3R_LC$$

The pulse width of a monostable multivibrator is

$$0.7RC$$

Dividing the semistable state time by the minimum time of one complete cycle gives the maximum duty cycle (DC_{max}), or:

$$DC_{max} = \frac{0.7RC}{0.7\,RC + 3R_LC}$$

$$DC_{max} = \frac{R}{R + 4.3R_L}$$

Duty cycle becomes important when working with switching circuits. Rating sheets for multivibrators always specify a maximum duty cycle. If the duty cycle goes beyond the rating, the multivibrator will not operate properly.

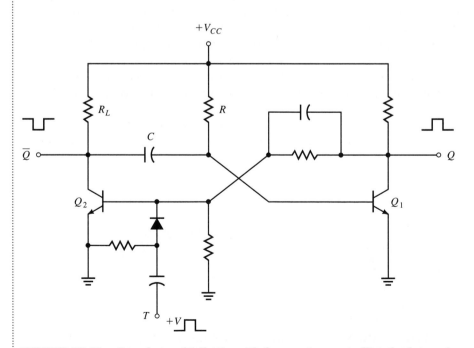

FIGURE 17.13   One-shot multivibrator with the capacitor controlling the duty cycle

EXAMPLE 17.5

Find the maximum duty cycle for the circuit shown in Figure 17.13.

For this example, the following values exist in the circuit: R = 500 W; C = 0.1 mF; R_L = 330 W.

Solution
The equation for maximum duty cycle is:

$$DC_{max} = \frac{R}{R + 4.3R_L}$$

If we substitute the component values into the final equation, we have:

$$DC_{max} = \frac{500\ \Omega}{500\ \Omega + (4.3 \times 330\ \Omega)}$$

Calculating, we find that the maximum duty cycle equals

$$DC_{max} = \frac{500\ \Omega}{1919\ \Omega}$$
$$DC_{max} = 0.26$$

TIMING

Monostable multivibrators contain a combination of linear and digital methods. Since transistors have problems with the base-to-emitter voltage and beta varying with temperature, monostable multivibrators may exhibit temperature and supply voltage sensitivity through the output pulse width. Typically, a monostable multivibrator, such as a 9602, will show pulse width variations over a 0–50°C temperature range and over a ±5-V supply voltage range. Other examples of monostable multivibrators, such as the 74121 and 74221, exhibit variations of a few tenths of a percent over the temperature and voltage ranges.

PULSE WIDTH

In the introductory comments about monostable multivibrators, you learned that monostable multivibrators sometimes work as triggered, adjustable-width pulse generators. The decay curve of the base-emitter voltage controls the amount of pulse width for the output signal. Standard monostable multivibrators can provide pulse widths that range from 40 ns to several seconds. Figure 17.14 shows the output waveform for a monostable multivibrator.

If a monostable multivibrator generates long pulses, the capacitor in the RC circuit must have a value of several microfarads. Some circuit designs use electrolytic capacitors because of the need for increased capacitance. With larger electrolytic capacitors, leakage becomes a problem. The problem with leakage becomes greater because a monostable multivibrator applies both polarities of voltage across the capacitor.

Adding a diode or transistor to the circuit can cut the leakage problem to a minimum. However, using a diode or transistor in the circuit may introduce problems with temperature and voltage stability. Plus, the transistor or diode can add problems with predicting the pulse width. While some monostable

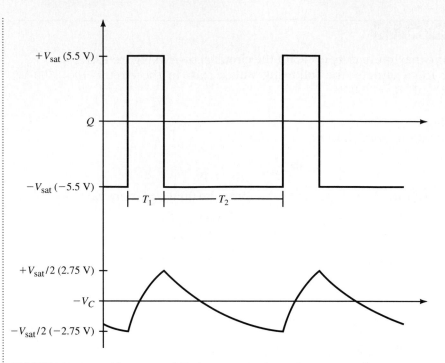

FIGURE 17.14 Charging and discharging waveforms for the circuit shown in Figure 17.13

circuits have a ±10% prediction accuracy, most of the popular one-shot circuits have an accuracy within 1%.

PRACTICE PROBLEM 17.4

Using Figure 17.13, match the waveforms from Figure 17.14 with the circuit conditions listed below:

(a) Q_2 conducting and Q_1 cut off.

(b) Applications of trigger pulses.

(c) Q_1 conducting and Q_2 cut off.

(d) Steady-state.

NOISE IMMUNITY

Monostable multivibrators have low noise immunity; that is, small noise spikes will not trigger the one-shot. However, the linear circuits within the monostable lessen the noise immunity, Figure 17.15. Also, the capacitive coupling present around the RC circuit adds the possibility of noise spikes. Some monostable designs have problems with voltage spikes on the supply voltage lines. Those spikes can greatly affect the pulse width.

Progress Check
You have now completed objective 4.

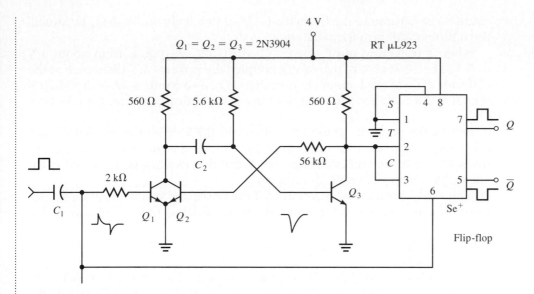

FIGURE 17.15 Adding a flip-flop to ensure noise immunity

ANALYSIS OF A SIMPLE MONOSTABLE MULTIVIBRATOR CIRCUIT

Figure 17.16 shows a basic collector-coupled monostable multivibrator using NPN transistors. In the figure, a 10-kΩ resistor couples the collector of Q_1 to the base of transistor Q_2. However, capacitor C couples the collector of Q_2 to the base of Q_1. Because a positive voltage biases its base and only a capacitor

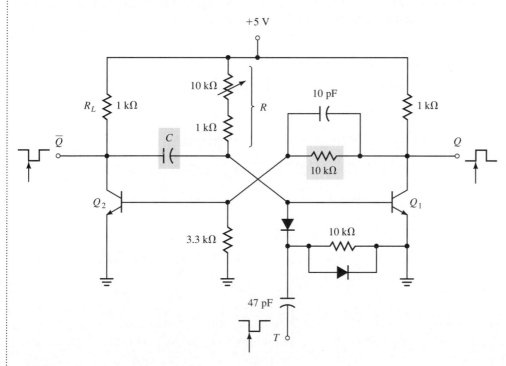

FIGURE 17.16 A simple monostable multivibrator

couples the Q_1 base to the Q_2 collector, Q_1 is normally on. With Q_1 saturated, the multivibrator is in its stable state.

When a triggering pulse arrives, Q_1 turns off. V_{CE1} goes from about 1 V to 5 V. The 10-kΩ coupling resistor couples the positive 4-V difference at the collector of Q_1 to the base of Q_2, turning on Q_2. As a result, a 4.8-V drop in potential at the collector of Q_2 couples through coupling capacitor C to the base of Q_1.

Before, the potential at the base of Q_1 was 0.7 V. With the applied voltage drop, the potential at the base drops to -4.1 V. The negative voltage keeps transistor Q_1 turned off until capacitor C discharges through R. Once the capacitor discharges enough, Q_1 turns on.

Figure 17.17 shows the waveforms at the bases and collectors of Q_1 and Q_2 for the circuit shown in Figure 17.16. As we discuss the charging and discharging of the coupling capacitor, carefully study each waveform. Earlier, we saw that Q_2 is cut off and Q_1 is saturated when the monostable remains in its stable state.

Only leakage current passes through resistor R_L. Therefore, the collector of Q_2 has a potential of $+5$ V. Also, the potential at the base of Q_1 equals the voltage across the base-emitter junction, or 0.7 V. The difference between the Q_2 collector potential and the Q_1 base potential is 4.3 V and the voltage seen across the coupling capacitor.

When the monostable enters its semistable state and Q_2 turns on, Q_1 turns off. Consequently, the voltage applied to the Q_2 collector drops 4.8 V to approximately 0.2 V. The voltage applied across a capacitor cannot change instantly. As a result, the potential at the base of Q_1 also drops 4.8 V to -4.1 V.

Since a 9.1-V difference exists across the resistance R, C begins to discharge through Q_2 and R. Figure 17.18 shows the discharge path. The resistance of Q_2 is much less than the resistance value of R. Because of that factor, the discharge time constant, or T_d, equals RC.

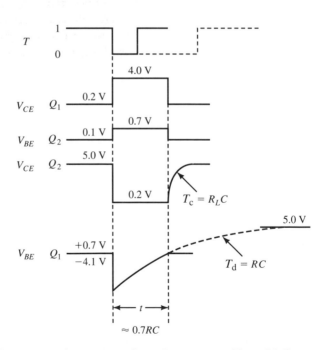

FIGURE 17.17 Input and output waveforms for a monostable multivibrator

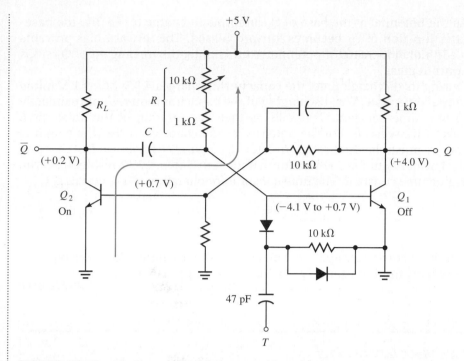

FIGURE 17.18 Discharge path of the monostable multivibrator

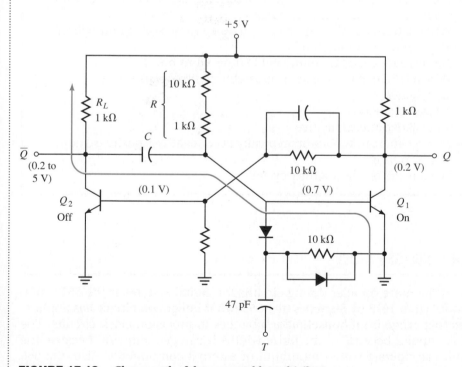

FIGURE 17.19 Charge path of the monostable multivibrator

As the potential at the base of Q_1 attempts to charge to +5.0 V, the base-emitter junction of Q_1 becomes forward-biased. The forward bias prevents any additional increase of potential. At this time, the turning on of Q_1 stops the output pulse.

During its discharge time, the capacitor discharges 4.8 V of a 9.1-V initial potential difference. The discharging of the capacitor requires approximately 7/10 of a time constant. This tells us that the duration of the pulse, or t, equals 0.7RC seconds. For the potential at the collector to rise, the coupling capacitor must charge through resistor R_L. Figure 17.19 shows the charging path. The curvature in the waveform shown in Figure 17.19 results from the charging time constant. The time constant for charging, or T_C, equals $R_L C$.

PRACTICE PROBLEM 17.5

Given the following component values, find the discharge time constant and the charging time constant. $R_L = 1.5$ kΩ; $C = 0.1$ μF; $R = 12$ kΩ.

REVIEW SECTION 17.3

1. How many stable states do monostable multivibrators have?
   (a) none
   (b) one
   (c) two
2. Another name for the monostable multivibrator is
   _____ .
3. Monostable multivibrators generate _____ output pulse for every _____ pulse.
4. What determines the length of time that a monostable multivibrator stays in a semistable state?
5. Name several uses for monostable multivibrators.
6. Which kind of coupling do monostable multivibrators use?
   (a) resistive
   (b) capacitive
   (c) resistive and capacitive
7. Monostable multivibrators typically use (negative-/positive-) going trigger pulses.
8. What is the maximum duty cycle?

17.4 THE 555 TIMER

One of the more popular timing circuits for digital systems is the 555 timer. Introduced in 1973 by Signetics, the 555 timer integrated circuit has applications that range from household appliances to precision clock circuits. The 555 is popular because of the pulse widths it can generate and because the circuit can operate with a minimum of external components. Also, the 555 works as either an astable or a monostable multivibrator.

555 timers can operate from supply voltages that range from 4.5 V to 18 V. The height of the output pulse corresponds to the supply voltage used. 555s have temperature-stable pulse widths that usually vary only 0.005% with every Celsius degree change. Also, the 555 is stable with regard to power supply variations with a 0.1% change in pulse length per volt change in the power supply. Since the 555 can supply up to 200 mA of current through a 20-Ω impedance, it can be used to operate LEDs and relays.

Pulse widths for standard 555 timers may vary from 10 μs up to 100 s. The value of external resistances + capacetances determine the duration of the pulse width. Typically, R has a maximum value of 10 MΩ, and C has a maximum value of 10 μF.

Figure 17.20 shows the internal structure of the timer. A voltage divider made of three equal series resistors divides a supply voltage into three equal parts. One-third of the supply voltage goes across each resistor. Two high-gain comparators control a set-reset flip-flop by comparing portions of the supply voltage with the trigger and threshold voltages. When a flip-flop is "set" its Q output is a low voltage and Q out is a high voltage. "Reset" causes the opposite condition.

While the bottom comparator compares one-third of the supply voltage with the trigger voltage, the top comparator compares two-thirds of the voltage against the threshold voltage. If the trigger voltage drops below the reference supply voltage, the output from the bottom comparator sets the flip-flop. The top comparator resets the flip-flop if the threshold voltage exceeds the reference supply voltage. Connected to the R_d input of the flip-flop, Q_2 buffers the reset input from the flip-flop, and Q_1, the discharge switch.

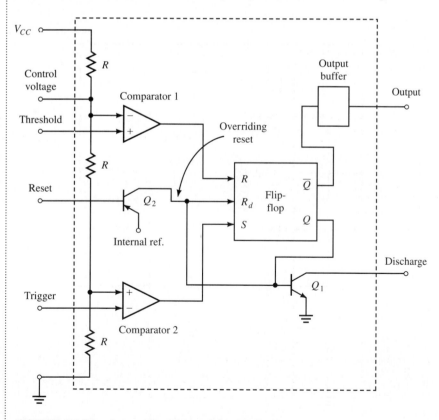

FIGURE 17.20   Internal structure of the 555 timer

At the bottom of the diagram, a transistor switch completes the integrated circuit package. A high voltage at the Q output saturates Q_1, while a low voltage cuts off the transistor. Q_1 may be used to discharge a capacitor that is part of the external RC timing component.

OPERATING THE 555 AS AN ASTABLE MULTIVIBRATOR

In Figure 17.21, we see a 555 timer connected as an astable multivibrator. In the diagram, the upper threshold and lower trigger comparator leads connect to the top of the capacitor C_1. A discharge switch connects to the junction of resistors R_a and R_b. The resistors are in series with the capacitor.

With the discharge switch open, the external capacitor charges through R_a and R_b toward the V+ supply voltage. Charging continues until the voltage across the capacitor exceeds the two-thirds reference supply voltage, which closes the discharge switch.

Closing the discharge switch places a ground between R_a and R_b. Then the discharge transistor allows the capacitor to discharge through R_b. Discharging continues until the voltage across the capacitor goes below the one-third reference supply voltage. When the voltage across the capacitor reaches the one-third point, the lower comparator sets the flip-flop. Setting the flip-flop causes the output to go low and opens the discharge switch.

Thus, capacitor C charges through R_a and R_b toward V_{CC} with the time constant equal to $(R_a + R_b)C$. When the charging voltage reaches 2/3 V_{CC}, the

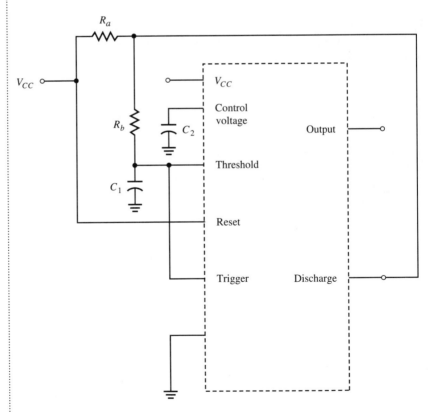

FIGURE 17.21   A 555 timer connected as an astable multivibrator

upper comparator resets the flip-flop so that Q goes high, which causes the discharge switch to close. Then, the cycle repeats.

Since the external capacitor charges through the combination of R_a and R_b but discharges through R_b, the time constants for high and low outputs are different. This difference produces an asymmetrical square wave. The time for the high output equals

$$t_H = 0.7 (R_a + R_b)C$$

The time for the low output equals

$$t_L = 0.7R_bC$$

In equation form, the period of the square wave equals

$$T = t_H + t_L$$
$$T = 0.7R_aC + 0.7R_bC + 0.7R_bC$$
$$T = (1.4R_b + 0.7R_a)C$$
$$T = \frac{1}{f}$$

C_2 is used to decouple any circuit noise that may falsely trigger the 555.

EXAMPLE 17.6

Find the period of a square wave. Using Figure 17. 21, capacitor C has a value of 0.1 µF. Resistor R_a equals 10 kΩ and Resistor R_b equals 15 kΩ.

Remember that when the charging voltage reaches 2/3 V_1, the upper comparator resets the flip-flop to low. Thus, the output goes low and the discharge switch closes. Then the cycle repeats. The external capacitor charges through the combination of R_a and R_b but discharges through R_b. Therefore, the time constants for high and low outputs are different. This produces an asymmetric square wave.

Solution
The time for the high output equals

$$t_H = 0.7(R_a + R_b)C.$$
$$t_H = 0.7(10 \text{ k}\Omega + 15 \text{ k}\Omega)0.1 \text{ µF}$$
$$t_H = 0.7 \times 25 \text{ k}\Omega \times 0.1 \text{ µF}$$
$$t_H = 1.75 \text{ ms}$$

The time for the low output equals

$$t_L = 0.7R_bC.$$
$$t_L = 0.7 \times 15\text{k} \times 0.1 \text{ µF}$$
$$t_L = 1.05 \text{ ms}$$

In equation form, the period of the square wave equals

$$T = t_H + t_L$$
$$T = 0.00175 + 0.00105$$
$$T = 2.80 \text{ ms}$$

PRACTICE PROBLEM 17.6

Determine the total charge/discharge period for the circuit shown in Figure 17.21. R_a = 270 Ω; R_b = 330 Ω; C = 0.33 µF.

If we connect an external voltage to the control voltage input of the timer, we can vary the frequency of the timer. Changing the control voltage also changes the threshold and trigger voltages. From the description of the 555 circuit, we know that the threshold and trigger voltages affect charge and discharge times of the capacitor. Since the output frequency depends on the time required to charge and discharge the capacitor, the output frequency also varies.

OPERATING THE 555 AS A MONOSTABLE MULTIVIBRATOR

Figure 17.22 shows a 555 timer set up as a monostable multivibrator. Here, a capacitor C_3 and resistor R_3 connect between ground and the supply voltage. The threshold of the upper comparator—pin 6—and the discharge—pin 7—connect to the junction between the resistor and the capacitor.

In its rest state, the monostable 555 has a low output voltage. By holding the trigger voltage at the supply voltage level, the internal flip-flop of the 555 resets and gives a logic zero output. The C_3 shorts to ground through the discharge switch.

When an input pulse on the trigger input, or pin 2, goes below one-third of the supply voltage, the lower comparator goes high. This sets the flip-flop and opens the switch. At pin 3, or the 555 output, the level goes high.

At this point, a pulse begins. The discharge transistor becomes cut off and the external capacitor, C_3, begins to charge. Current flows through the external resistor while the voltage across the capacitor charges toward the supply voltage. When the charging voltage reaches two-thirds of the supply voltage, the voltage at the threshold lead of the upper comparator rises above the threshold level of the comparator.

The comparator resets the flip-flop to a low output. Consequently, the switch closes and discharges the capacitor. Then, the 555 timer output

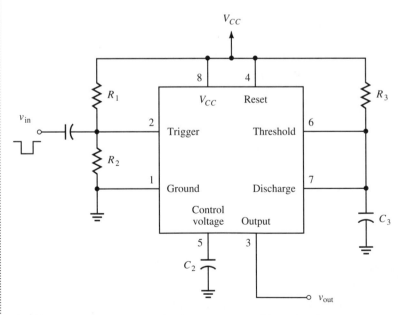

FIGURE 17.22   Operating the 555 as a monostable multivibrator

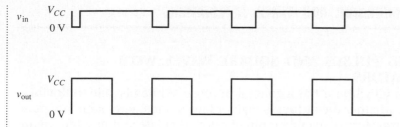

FIGURE 17.23   How the output waveform of the 555 monostable compares with its input waveform

goes low and the pulse ends. The monostable 555 timer is back in its rest state.

Figure 17.23 shows how the output waveform compares with the input signal. Pulse duration equals the amount of time required for the capacitor to charge through the resistor to two-thirds of the supply voltage.

We can calculate the duration of the output pulse with the equation

$$T_P = 1.1 \times RC$$

Progress Check
You have now completed objective 6.

PRACTICE PROBLEM 17.7

Find the output pulse width of the circuit shown in Figure 17.24. R = 560 Ω and C = 0.15 µF.

REVIEW SECTION 17.4
1. Why are 555 timers popular?
2. 555 timers can work as
   (a) bistable multivibrators
   (b) astable multivibrators
   (c) monostable multivibrators
3. 555 timers can operate from supply voltages that range from
   _____ to _____ V. The _____ of the output
   pulse corresponds to the supply voltage used.
4. 555 timers are unstable with regard to temperature and power
   supply variations. True or false?
5. Describe the internal structure of a 555 timer.
6. In Figure 17.23, a 555 timer is configured as a(n) _____
   multivibrator. The _____ _____ and _____
   _____ _____ leads connect to the top of the capacitor C.
7. In Figure 17.24, a 555 timer is configured as a(n) _____
   multivibrator. The _____ _____
   connects to the junction between the resistor and capacitor.

GENERATING PULSES AND SQUARE WAVES WITH MULTIVIBRATORS

Each type of multivibrator has a general purpose. While bistable multivibrators are stable whether the output is high or low, monostables generate pulses and shape waveforms. Astable multivibrators oscillate and provide square waves. In electronics, combinations of the multivibrator outputs are sometimes needed to complete a given task.

If an application requires a series of pulses at a constant repetition rate, the combination of astable and monostable multivibrators may be the best. Shown in Figure 17.24, an astable multivibrator triggers a monostable multivibrator. The arrangement gives two advantages:

1. The circuit can provide a small pulse width without affecting frequency.

2. The repetition rate does not depend on the pulse width.

Because of those advantages, the combination works better than a single asymmetrical astable multivibrator.

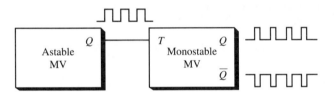

FIGURE 17.24   Block diagram of an astable multivibrator triggering a monostable multivibrator

USING 555 TIMERS IN AN AUTOMOBILE BURGLAR ALARM CIRCUIT

Figure 17.25 depicts two 555 timers working within an automobile alarm circuit. Both timers use the automobile battery as a voltage source. Timer A provides a delay that allows the driver to enable the alarm system and exit the car. In addition, the first timer also provides the same amount of time delay when the driver enters the car and disables the alarm circuit. A switch hidden under the dash disables the alarm.

At the bottom of the circuit, a silicon-controlled rectifier (SCR) keeps the first timer from triggering the second timer. Timer B activates the alarm. If an intruder causes the parallel sensor switches to close, timer B generates a signal for the alarm.

USING MONOSTABLE MULTIVIBRATORS TO DESIGN AN OSCILLOSCOPE-TRIGGERED TIME BASE

Oscilloscopes such as the one pictured in Figure 17.26 use a triggered time base for accurately measuring elapsed time. By using the triggered time base, a technician can view the duration of any part of a displayed waveform. Trig-

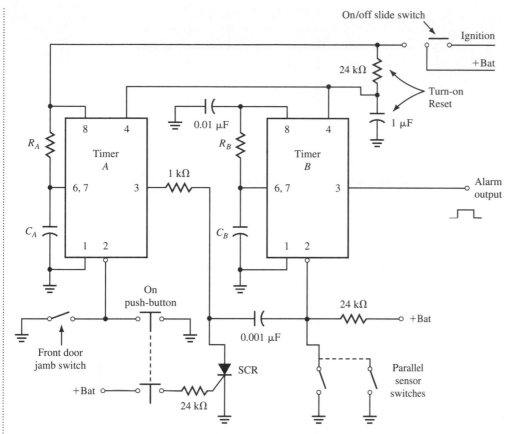

Timer: Signetics NE555

FIGURE 17.25 Two 555 timers working in an automobile alarm system

FIGURE 17.26 Students using advanced oscilloscopes Courtesy of De Vry Inc.

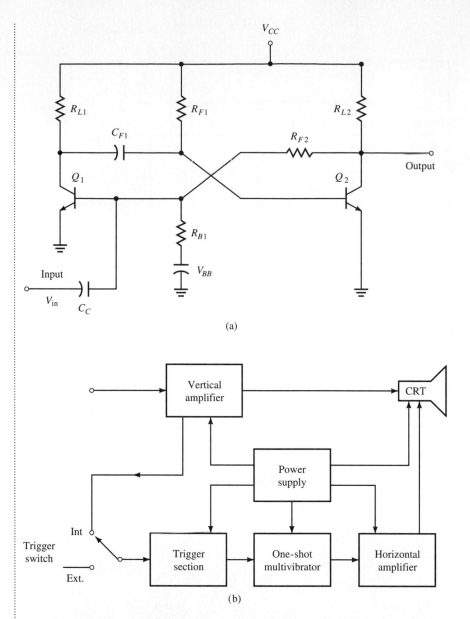

FIGURE 17.27 (a) Basic multivibrator used in a triggered time base. (b) How a monostable multivibrator fits into a triggered time base.

gered time bases use a sawtooth oscillator biased beyond cutoff. When a trigger pulse becomes applied to the oscillator, the oscillator cycles through one operation and generates a single sawtooth waveform. This allows the oscilloscope to display a small section of a waveform.

The design of a triggered time base builds around a monostable multivibrator such as the one shown in Figure 17.27a. With transistor Q_1 cut off, the multivibrator has no output in its stable state. As a result, the oscilloscope CRT beam does not deflect. Figure 17.27b shows how the multivibrator action fits into the oscilloscope action.

Oscilloscopes may be either externally or internally triggered. The trigger section provides the one-shot multivibrator with a negative-going trigger. When triggered, the one-shot multivibrator produces an output pulse of pre-

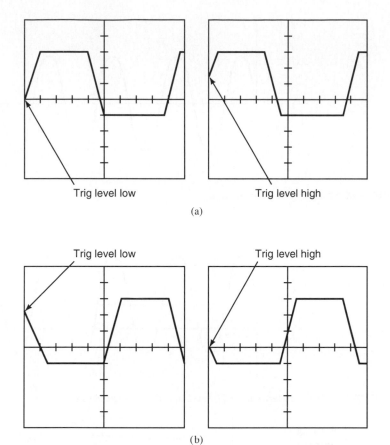

(a)

(b)

FIGURE 17.28   (a) Waveforms triggered at the low and high points of the positive slope.
(b) Waveforms triggered at the low and high points of the negative slope.

determined width to the horizontal amplifier. The horizontal amplifier outputs a sawtooth that causes the horizontal portion of the oscilloscope scan.

Figure 17.27a, which is the one-shot multivibrator section, requires a negative-going trigger to fire. The duration of the output pulse is controlled by
C_{FI} and R_{FI}

Figure 17.28 shows waveforms triggered at the low and high points of the
positive slope and the low and high points of the negative slope. For triggering to occur, the sync signal must reach a specified amplitude. In the oscilloscope circuitry, the addition of a DC voltage to the AC sync signal raises and
lowers the sync signal to the needed level. Figure 17.30 shows the insertion of
the DC signal into the AC sync signal shown in Figure 17.29.

A complete block diagram of the triggered-sweep system is shown in Figure 17.31. Controls at the sync amplifier section allow triggering by either the
negative or positive portion of the sync waveform. Also, the controls set the
point of triggering on the waveform leading edge. When the leading edge
reaches this point, the trigger shaper generates a sharp trigger pulse.

At that time, the pulse generator sends an unblanking pulse to the oscilloscope CRT and turns the sawtooth generator on. Also, the unblanking pulse
turns on the CRT electron beam only for the duration of the sawtooth. The
sawtooth generator drives the horizontal amplifier and initializes the sweep
hold off circuit.

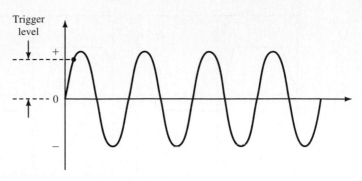

FIGURE 17.29   The AC sync signal

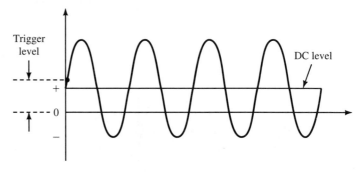

FIGURE 17.30   Inserting the DC level into the AC sync signal

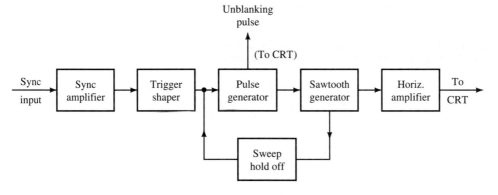

FIGURE 17.31   Block diagram of the triggered-sweep system

Progress Check
You have now completed objective 7.

If the sawtooth output reaches the level set by the sweep-level control, the sweep hold off circuit generates a turn-off pulse for the sawtooth generator. The sweep hold off circuit blocks any output from the trigger shaper until a turn-off pulse arrives at the sawtooth generator. In short, the sweep hold off circuit prevents any false triggering of the sawtooth generator.

REVIEW SECTION 17.5
1. Discuss the general purposes of each type of multivibrator.
2. What are the advantages of having an astable multivibrator trigger a monostable multivibrator?

3. In the automobile alarm circuit shown in Figure 17.25, timer A provides a _____ that allows the driver to enter the automobile.
4. What is the purpose of a triggered time base in an oscilloscope?
5. The design of the triggered time base for the oscilloscope in Figure 17.31 builds around which type of multivibrator?
   (a) astable
   (b) monostable
   (c) bistable
6. In the oscilloscope circuitry, the addition of a DC voltage to the AC sync signal _____ and _____ the sync signal level to the needed level.

 # TROUBLESHOOTING

Troubleshooting 555 timers is a matter of deductions. First visually inspect the circuit for obvious defects. Then make sure the supplied voltages are correct. Next, where possible, check each external component for its correct value. If you have determined they are within tolerances, then the 555 timer should be replaced. As with any troubleshooting procedure, involve as many senses as you can. Look for damage to external components. Excessive heat may cause a resistor or IC to bulge or crack. A defective electrolytic capacitor may leak some of its dielectric material. Like resistors, capacitors may also show external signs of damage. Also, components that are too hot may emit a burnt odor.

A capacitor whose dielectric is letting some DC current pass through is called a leaky capacitor. That is, it is leaking DC current. The equivalent circuit for a leaky capacitor is a capacitor and a resistor in parallel. Because the resistor provides a parallel path for current flow, the capacitor does not charge as much, decreasing the output frequency. After visually examining the circuit, take voltage readings at the 555's pins. Compare the readings with the values that should be present on each pin. Solve any differences by theoretically making each component attached to that pin defective. Change the component most likely to cause the problem.

TROUBLESHOOTING QUESTIONS
1. Troubleshooting a 555 timer circuit is a matter of _____.
2. Checking the supply voltages in a 555 timer circuit is something that you can overlook when troubleshooting the circuit. True or false?
3. What is the equivalent circuit for a leaky capacitor?

4. A leaky capacitor (allows/prevents) current (to/from) pass(ing) through its dielectric.
5. A leaky capacitor causes a(n) (decrease/increase) in the output frequency.
6. When checking the 555 pin voltages, you would compare those voltages with _____ .

WHAT'S WRONG WITH THIS CIRCUIT?

What is wrong with the circuit in Figure 17.32? An oscilloscope shows that the output frequency is 181 Hz and the duty cycle is 0.24. The output waveform is a square wave. How would you troubleshoot this circuit? Explain what component is defective and why you think it is faulty.

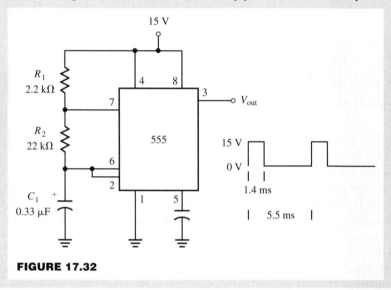

FIGURE 17.32

SUMMARY

Multivibrators generate rectangular waveforms for electronic equipment while using positive feedback. Rectangular waveforms should have stable low and high levels, a sharp rise from the low to the high level, and a sharp drop from the high level to the low level. Well-designed multivibrator circuits have the advantage of supplying high-speed, symmetrical rectangular waveforms.

Astable multivibrators do not have stable states and do not require an external triggering signal for operation. An astable multivibrator circuit will work as an oscillator or square wave generator. Modifications, such as the addition of charging resistors and diodes, to the basic astable design can improve square wave rise time.

Bistable multivibrators have two stable states. A bistable multivibrator relies on an external triggering pulse that arrives via an RC circuit. A flip-flop is an example of a bistable multivibrator. Flip-flops are two transistors that alternately turn on and off.

Monostable multivibrators have one stable state. An external triggering pulse causes the monostable multivibrator to change from the stable state to its semistable state. Also, monostable multivibrators have an asymmetrical path and use either capacitive or resistive coupling. Characteristics of monostable multivibrator circuits are pulse width, duty cycle, and noise immunity.

The 555 timer works in diverse items such as household appliances and precision clocks. 555 timers are popular circuits that have longer pulse widths and a minimum of external components. A 555 timer can be configured as either an astable or a monostable multivibrator.

PROBLEMS

1. How many inputs does an astable multivibrator have?
2. The output of an astable multivibrator is a
   (a) sine wave
   (b) square wave
   (c) triangle wave
   (d) sawtooth wave

Refer to Figure 17.33 for problems 3–7.
3. For how long is V_{out1} low?
4. For how long is V_{out1} high?
5. Calculate the output frequency.
6. If both capacitors were 0.1 µF, what would be the output frequency?
7. What is the difference between V_{out1} and V_{out2}?
8. Astable multivibrators are also called _____.

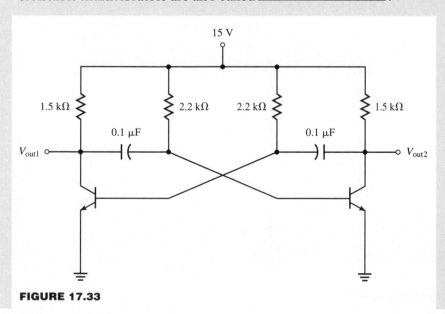

FIGURE 17.33

Refer to Figure 17.34 for problems 9–13.

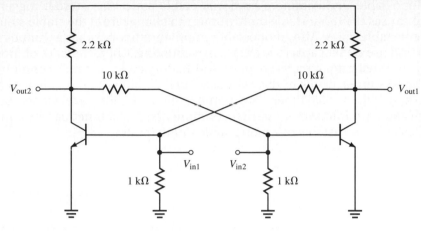

FIGURE 17.34

9. Bistable multivibrators may oscillate without input signals. True or false?
10. Another name for a bistable multivibrator is _____.
11. _____ coupling connects the two transistors to each other.
    (a) Resistive
    (b) Capacitive
    (c) Inductive
12. Bistable multivibrators use (positive/negative) feedback to switch states.
13. What controls the output voltage levels of the bistable multivibrator?

Refer to Figure 17.35 for problems 14–19.

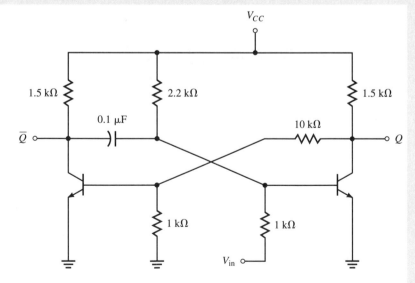

FIGURE 17.35

14. Why is the circuit called a monostable multivibrator?
15. Another name for the circuit is _____ .
16. The output Q is at a (high/low) voltage when in the rest state.
17. Draw a diagram of the input and output waveforms.
18. What is the maximum duty cycle?
19. The input trigger is (positive-/negative-) going.

Refer to Figure 17.36 for problems 20–23.

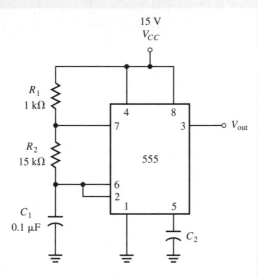

FIGURE 17.36

20. The 555 is configured as which type of multivibrator?
    (a) bistable
    (b) monostable
    (c) astable
21. What is the output frequency?
22. What is the purpose of C_2?
23. Describe the effect on output frequency if R_1 was increased in value.

Refer to Figure 17.37 for problems 24–26.
24. The 555 is configured as which type of multivibrator?
    (a) bistable
    (b) monostable
    (c) astable
25. What is the output pulse width?
26. What are the trigger magnitude and direction required to trigger this circuit?

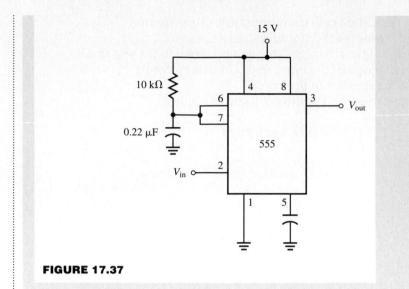

FIGURE 17.37

18

THYRISTORS AND
THYRISTOR DEVICES

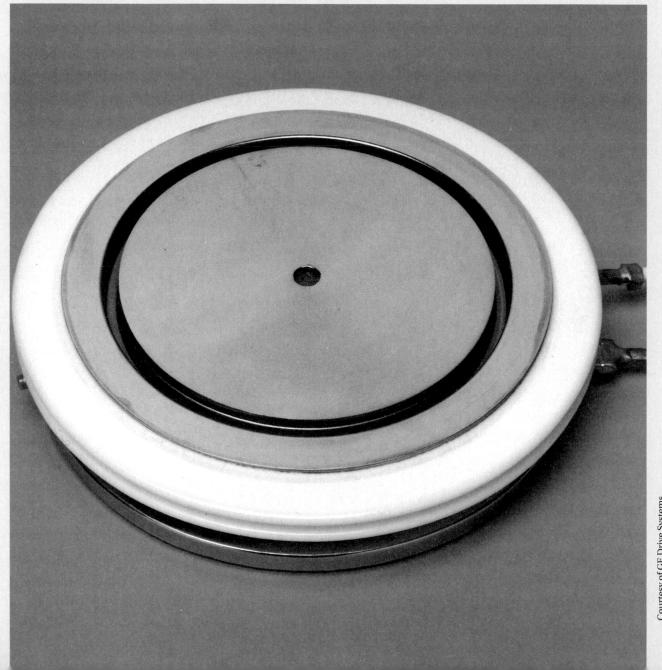

OBJECTIVES

✓ **As you read this chapter, concentrate on learning how to:**

1. Describe the arrangement of the P-type and N-type materials of a Shockley diode
2. Fire and open a Shockley diode
3. Describe how an SCR is fired and opened
4. Show how SCRs and triacs differ
5. Discuss triggering methods for triacs
6. Describe how a gate turn-off switch is fired and opened
7. State the function of silicon bidirectional switches
8. Describe why the diac does not have a holding current
9. Show how UJT and PUT circuits can provide triggering for SCRs
10. Describe how a crowbar circuit works
11. Discuss how precision time delay circuits are able to delay firing an SCR

INTRODUCTION

This chapter introduces a family of four-layer semiconductor devices called thyristors. This family is used as electronic switches. They have only two states: open and closed. There is no active or linear operating region. Thyristors are either saturated or cut off.

You will find that thyristors have two or more leads. As a family, they are constructed to handle current from a few milliamps to hundreds of amps. Their principal use is in industrial control circuits and power protection circuits. Some of the names for these devices are Shockley diode, SCR, diac, and triac.

Another part of the thyristor family is used in oscillator and timing circuits. These are called unijunction transistors (UJTs) and programmable unijunction transistors (PUTs). Like the other members of the thyristor family, these devices also have four layers and have two states of operation.

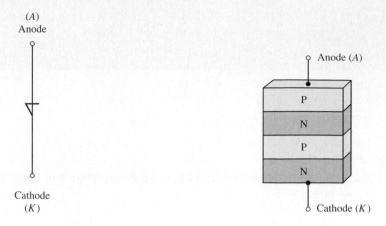

FIGURE 18.1 All thyristors have the same basic characteristics.

FIGURE 18.2 Block diagram of a four-layer PNPN semiconductor device

18.1 THE SHOCKLEY DIODE

The first thyristor we will examine is the Shockley diode. It has two leads and four layers. Figure 18.1 shows the schematic symbol for the Shockley diode, and Figure 18.2 shows the block diagram.

Arranging the two P-type and two N-type semiconductor materials in a series array consisting of alternate N-type and P-type layers produces a device that behaves in two different ways. In the reverse direction, the Shockley diode behaves as a conventional rectifier. In the forward direction, it behaves as a series combination of an electronic switch and a rectifier.

As a switch, the Shockley diode can be switched from a high-impedance off state to a conducting on state and from the on state back to the off state.

In Figure 18.2 the PNPN device may be viewed as two transistors. If we start at the anode, the top three layers appear as a PNP transistor. If we look from the cathode, the bottom three layers appear as an NPN transistor.

Figure 18.3 shows the block diagram of the two-transistor device. The diagram depicts a complementary NPN-PNP transistor feedback pair. Each transistor uses a positive feedback loop created when its collector direct-couples to the base of the other transistor.

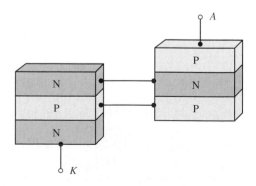

FIGURE 18.3 Block diagram of a two-transistor device

On the NPN side, a small increase in base current causes a larger change in its collector current. The collector current is also the base current for the PNP transistor because of the direct coupling. On the PNP side, any increase in its collector current also affects the base current of the NPN transistor. Essentially, positive feedback causes an initial small change in the base current of the NPN transistor to become a larger change.

Figure 18.4 shows a schematic diagram of the complementary PNPN transistor pair. If forward-biased, the pair of transistors works as a regenerative, bistable switching device. As Figure 18.4 shows, the terminals of the complementary pair are labeled as the anode (A) and cathode (K). The anode is positive with respect to the cathode when the device has forward bias.

To close the Shockley diode requires that forward biasing voltage in excess of the breakover voltage be applied. **Breakover voltage** is the minimum anode-to-cathode voltage required to cause the thyristor to conduct. The term *fired* is commonly used regarding thyristors. If a thyristor is fired, then it has been made to conduct.

Figure 18.5 shows the complete electrical path for the complementary transistor pair shown in Figures 18.3 and 18.4. The emitter of the PNP transistor returns to the positive side of the DC power supply through a limiting resistor (R). The emitter of the NPN transistor is connected to the negative side of the supply voltage.

Shockley diodes are manufactured to fire at breakover voltage. Before crossing breakover voltage, the thyristor is open and no current flows through the device. As voltage is increased toward the breakover point, a very small current begins to flow. When the breakover point is crossed, there is enough current to start regeneration. That current is called the latching current. **Latching current (I_L)** is the minimum current required to initiate regeneration. When the thyristor fires, it saturates and a maximum current flows in the circuit. The thyristor remains closed until some external device reduces the thyristor current to a point where regeneration cannot be sustained. That current is called the holding current. **Holding current (I_{HX})** is the minimum current before a thyristor opens. Typically, thyristor current is

✓ **Progress Check**
You have now completed objective 1.

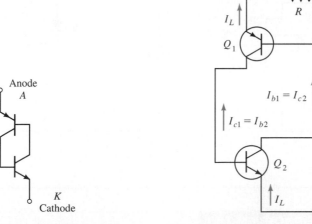

FIGURE 18.4 Schematic diagram of the complementary PNPN transistor pair

FIGURE 18.5 Electrical path for the complementary transistor pair dropped below the holding current point by opening the current flow with a manual switch or, in the case of a power protection circuit, a fuse. A second way to open a thyristor is to reverse the current flow, as with an AC signal.

With this quick sketch of a normally operating thyristor, we also have an outline of the three basic features of thyristors:

1. A Shockley diode requires a voltage in excess of breakover voltage before the regenerative process can begin.

2. Initiating regeneration requires a minimum anode current called the latching current. Thyristor specification sheets list the latching current as I_L.

3. If the latching current falls below a certain level, called the holding current, the device turns off. Usually, the holding current is a little more than zero. On thyristor specification sheets, the holding current is listed as I_{HX}. Holding current is normally shown at room temperature. The required amount of holding current decreases as temperature increases.

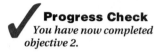

Progress Check
You have now completed objective 2.

EXAMPLE 18.1

Using Figure 18.6a, discuss how the circuit operates.

The figure is a simplified heater control circuit. The source provides power for the heater. The relay is normally open and must be closed before current will flow through the heater coil. The Shockley diode prevents current flow through the relay coil during the entire negative alternation. It also prevents current flow during the positive alternation up to its breakover voltage. At breakover voltage the Shockley diode saturates, and current will flow through the relay coil, closing its contacts. Current will now flow through the heater coil. As the source voltage crosses 0 V, the Shockley diode opens, stopping current from flowing through the relay coils. The relay contacts open, and current stops flowing through the heater coil. D_1 provides a rapid discharge path for the relay coil.

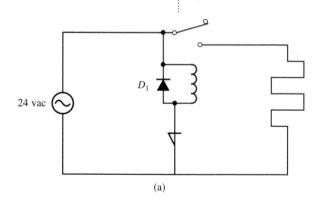

(a)

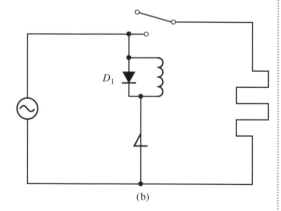

(b)

FIGURE 18.6

PRACTICE PROBLEM 18.1

Using Figure 18.6b, discuss how the circuit operates.

REVIEW SECTION 18.1

1. Thyristors are two-state devices. True or false?
2. Draw the schematic diagram for a Shockley diode.
3. Thyristors are _____ -layer semiconductors.
4. When reverse-biased, a Shockley diode behaves like a _____ diode.
5. When a Shockley diode is fired, does it open or close?
6. Shockley diodes are closed by _____ voltage.

18.2   SILICON-CONTROLLED RECTIFIERS (SCRs)

The silicon-controlled rectifier (SCR) is a three-terminal, four-layer thyristor. In addition to an anode and a cathode, the SCR has a gate terminal. The gate terminal is used to fire, but not open, an SCR. SCRs are opened by reducing current flow below holding current levels.

SCRs function as current-controlled devices, rectifiers, and latching switches. Because SCRs conduct in only one direction, they make excellent rectifiers for DC load applications.

Because of its latching characteristics, an SCR can work as an on-off switch. A short, 1-μs pulse of control current applied to the gate will switch the SCR into conduction. Dropping cathode-to-anode current below the holding current turns the SCR off. Acknowledging the use of an SCR as a switch, specification sheets rate SCRs on the maximum amount of carried current, operating voltage, and peak voltages.

BASIC SCR OPERATION

Figure 18.7 shows that the gate terminal is connected to the P-type material nearest the cathode. That P-type material is also the NPN transistor's base. Applying a positive voltage at the gate will forward bias the gate-to-cathode PN junction.

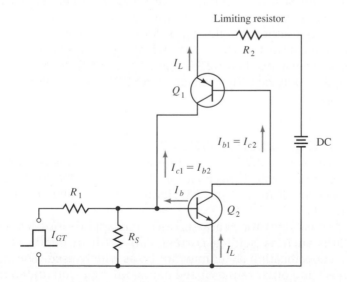

FIGURE 18.7 Resistive termination at the base of the NPN transistor

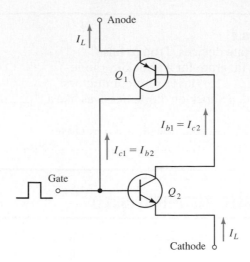

FIGURE 18.8 An SCR equivalent circuit

Figure 18.8 shows a simple schematic for the two-transistor concept of an SCR. A gate trigger current applied to the base of the NPN transistor causes the transistor Q_2 to conduct and initiates regeneration. The gate trigger current takes the form of a positive pulse. Conduction forces the collector voltage of the NPN transistor (Q_2) and the base voltage of the PNP transistor (Q_1) lower. Current then flows along the path marked I_L.

Figure 18.7 shows how resistive termination at the base of the NPN transistor affects the latching and holding currents. Collector current flowing through the PNP transistor must increase since it supplies the base current for the NPN transistor and the shunt current through the terminating resistor. Since the increase of PNP collector current depends on an increase in the principal current flow, the latching and holding current requirements also increase. Terminating the NPN part of the SCR gives the device immunity from false, or nongated, turn-on.

Switching the SCR on without the presence of a gate signal is undesirable. However, a SCR may switch on when a source voltage is applied to the thyristor through either the closing of an AC line switch or transient voltages. Energizing transformers, load switching, and solenoid closings can cause voltage transients. During AC operations, the false, or nongated, turn-on is limited to only half the cycle since a SCR switches off at the zero current crossing.

DEFINING THE SPECIFICATIONS

When you work with thyristors, a different set of terms begins to surface. Table 18.1 shows a typical specification sheet for a 2N885 SCR. Most specification sheets for SCRs will provide information similar to that shown in Table 18.1.

As we discuss thyristor characteristics and applications, we cover some specifications such as holding current and dv/dt in depth. Other listings shown on a specification sheet may not be as widely used. The short definitions of those and other ratings listed below will help you understand thyristor specifications.

Table 18.1 Maximum Allowable Ratings and Characteristics for a 2N885 SCR

RMS forward current, on-state I_F	0.5 A
Average forward current, on-state I_F (av)	Depends on conduction angle
Peak one-cycle surge forward current (nonrepetitive), I_{FM} (surge)	7 A
Peak reverse gate voltage, V_{GRM}	6 V
Operating temperature T_J	$-65°C$ to $+150°C$
Forward and reverse blocking current, I_{FX}, I_{RX}	Typ. 10/max. 20 μADC
Holding current, I_{HX}	Typ. 1.1/max. 3 mADC
Turn-off time, t_{off}	Typ. 15 μs
dv/dt	Typ. 40 V/μs

RMS Forward Current (I_F) and Average Forward Current (I_F(av))

The RMS forward current (on-state) is the maximum continuous RMS current that can flow in the forward direction from cathode to anode. The average forward current (on-state) is the maximum continuous DC current that can flow in the forward direction.

Peak One-Cycle Surge Forward Current (I_{FM} (surge))

The peak one-cycle surge forward current is the maximum allowable nonrecurring peak current during a single forward cycle. This specification lists the limits of anode current the device can withstand without damage. The I_{FM} rating should not be exceeded.

Forward Blocking Current (I_{FX}) and Reverse Blocking Current (I_{RX})

The forward blocking current value (I_{FX}) is the value of instantaneous anode current when the thyristor is in its forward blocking state. The reverse blocking current value (I_{RX}) is the value of instantaneous anode current for a negative anode voltage.

Peak Forward Blocking Voltage (V_{FOM})

The peak forward blocking voltage with the gate open (V_{FOM}) is the maximum instantaneous value of forward blocking voltage with the anode terminal positive.

Peak Forward Voltage (PFV)

The peak forward voltage (PFV) is the maximum instantaneous forward voltage permitted that will cause a reverse blocking thyristor to switch to the on state.

Forward Breakover Voltage ($V_{(BR)F}$)

The forward breakover voltage ($V_{(BR)F}$) is the voltage value that causes the thyristor to switch into its conductive state. It is the maximum positive voltage on the anode terminal with respect to the cathode terminal when the small-signal resistance is zero.

Peak Reverse Gate Voltage (V_{GRM})

The peak reverse gate voltage (V_{GRM}) is the maximum allowable peak reverse voltage between the gate terminal and the cathode terminal.

Repetitive Peak Reverse Voltage with Gate Open (V_{DROM})

In its reverse direction or with its anode negative with respect to its cathode, a thyristor acts like a conventional rectifier diode. The repetitive peak reverse voltage with the gate open (V_{DROM}) is the maximum allowable instantaneous value of negative voltage that may be applied to the reverse blocking thyristor anode terminal with the gate terminal open. Although the value of V_{DROM} does not represent a breakdown voltage, it should be exceeded only by the transient rating of the device.

Junction Temperature (T_J)

The junction temperature range (T_J) of a thyristor varies with individual thyristor types. A low temperature rating limits stress in the silicon crystal to safe values. The rated maximum operating temperature is used to determine steady-state and recurrent overload capability for operating temperatures and heat sinks.

Delay Time (t_d)

The delay time (t_d) is the time interval between the time the gate current pulse reaches 10% of its final value and the time when the resulting forward current reaches 10% of its maximum value. This interval occurs during switching from the off state to the on state into a resistive load.

Rise Time (t_r)

The rise time (t_r) is the time interval between the time the gate current pulse reaches 10% of its maximum value and the time when the resulting forward current reaches 90% of its maximum value. This interval occurs during switching from the off state to the on state into a resistive load.

Turn-on Time (t_{on})

The turn-on time (t_{on}) is the sum of the delay time and the rise time values.

Turn-off Time (t_{off})

The turn-off time (t_{off}) is the time interval between zero current and the time of reapplication of the forward blocking voltage. The interval occurs with the device remaining in the off state after leaving the on state.

Reverse Recovery Time (t_{rr})

The reverse recovery time (t_{rr}) is the time interval between zero current and the time when the reverse current through the thyristor reaches 10% of the peak reverse recovery current.

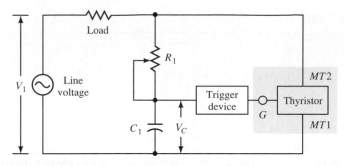

FIGURE 18.9 Block diagram of a thyristor triggering circuit

TRIGGERING A THYRISTOR

Figure 18.9 shows a block diagram of a thyristor triggering circuit. The amount of gate voltage and current prior to switching a thyristor to its on state at maximum-rated forward anode current are the gate voltage and current needed to trigger the device. Increasing the MT1 current to a high enough level switches the PNPN device on. The MT1 current may rise because of the voltage, the rate of voltage change, temperature, or radiant energy.

Increasing the collector-to-emitter voltage of a transistor increases the leakage current. Because of the increased current, avalanche occurs and the collector current sharply increases. The avalanche current in a PNPN semiconductor causes the switching to take place.

The amount of gate current and voltage needed to trigger vary inversely with temperature. When the junction temperature increases, the level of the gate signal required to trigger the thyristor is smaller. At high operating temperatures, the gate voltage level required to trigger a thyristor approaches a minimum value. Consequently, undesirable noise signals can trigger the thyristor. The maximum nontriggering gate voltage at the maximum operating junction temperature of the thyristor is a measure of the noise-rejection of the device.

The gate current value on specification sheets for thyristors is the DC gate trigger current needed to switch the thyristor into its on state.

The gate current pulse usually exceeds the DC value required to trigger the thyristor. Using large trigger currents cuts variations in the turn-on time and reduces the effect of temperature variations on the triggering characteristics of the thyristor. In addition, the larger trigger currents cut switching time. When a thyristor is triggered by a gate signal just large enough to turn on the device, the entire junction area of the thyristor does not conduct instantaneously. Thus, the critical rate of rise of on-state current or "dv/dt effect" comes into play.

THE DV/DT EFFECT

Any PN junction exhibits capacitance. In Figure 18.10, closing S_1 causes a step voltage to be applied across the anode-to-cathode terminals and produces a charging current that flows from cathode to anode. The current charges the thyristor capacitance and switching may occur. For thyristor devices, this property is called the critical rate of rise of on-state current or the "dv/dt effect."

For DC applications, the rate of rise of forward voltage must be governed so that the thyristor will not switch on until desired. The thyristor could switch to its on state and not switch off until some external circuit action occurs.

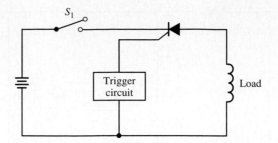

FIGURE 18.10 The step function of the voltage across the anode-to-cathode terminals produces a charging current that flows from anode to cathode.

Manufacturers' specification sheets give a value to the rate of rise of forward voltage. The dv/dt is always listed at the maximum junction temperature with the gate open. Higher junction temperatures make the thyristors even more susceptible to dv/dt problems because the device requires less current to turn on.

EXAMPLE 18.2

A circuit uses an SCR whose specification sheet lists dv/dt as 20 V/μs. If the waveform in Figure 18.11 is applied to the circuit, will the SCR false trigger?

Solution
The linear rising edge of the waveform increases 22 V in 500 ns. Convert the waveform dv/dt to the same time base as the SCR is using:

$$\frac{22\ V}{500\ ns} = \frac{dv}{1\ \mu s}$$

Rearranging the equation results in:

$$dv = \frac{22\ V \times 1\ \mu s}{500\ ns}$$
$$dv = 44\ V$$

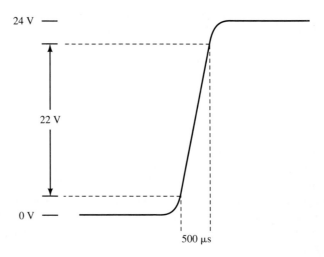

FIGURE 18.11

The rising edge of the waveform has a dv/dt of 44 V/μs, which is faster than the maximum dv/dt of the SCR. The SCR will false trigger.

PRACTICE PROBLEM 18.2

A circuit uses an SCR whose specification sheet lists dv/dt as 30 V/μs. The applied waveform has a linear rise time of 100 ns. The voltage rises 500 mV. Will the SCR false trigger?

One solution to dv/dt causing an SCR to trigger falsely is the snubber circuit. Figure 18.12 shows two possible snubber circuits. First, look at the series coil and resistor, which are in series with the SCR. Coils oppose a change in current flow. The faster the current rise the more the coil opposes the change. That opposition will keep the SCR from firing. Often an additional circuit coil will not be required because the inductance of a motor's windings may be enough to prevent dv/dt firings. The snubber is an LR circuit with a time constant equal to L/R. Charging the coil reduces the steepness of the applied step voltage to a level below the dv/dt of the SCR.

The second snubber circuit is the series resistor and capacitor connected in parallel with the SCR. The resistor and capacitor form an RC circuit with a time constant equal to $R_3 \times C_1$. The capacitor will reduce the steepness of the applied step voltage to a level below the dv/dt of the SCR.

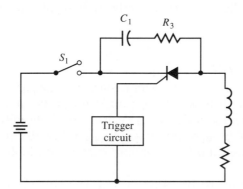

FIGURE 18.12   An RC or snubber circuit connected between the main terminals of an SCR

EXAMPLE 18.3

Solve for the RC time constant (RCT) required to prevent false triggering of the circuit in Figure 18.12.

The SCR has a dv/dt of 20 V/μs. The RC time constant must keep the linear rising rate of the waveform below 20 V/μs. The maximum nonrepetitive forward blocking voltage (V_{DSM}) for the SCR is 200 V.

Solution
Use the equation

$$RCT = \frac{0.63 \times V_{DSM}}{dv/dt \text{ of the SCR}}$$

$$RCT = \frac{0.63 \times 200 \text{ V}}{20 \text{ V/}\mu s}$$

$$RCT = 6.3 \text{ }\mu s$$

To ensure that the RC circuit will prevent false triggering a safety factor of 10 should be used.

$$RCT = 6.3 \text{ }\mu s \times 10 = 63 \text{ }\mu s$$

PRACTICE PROBLEM 18.3

Solve for the RC time constant required to prevent false triggering if the SCR has a dv/dt of 40 V/μs. The maximum nonrepetitive forward blocking voltage (V_{DSM}) for the SCR is 100 V.

Temperature can falsely trigger an SCR. Every 8°C increase in temperature doubles the leakage current in a reverse-biased PN junction. An increase in temperature in a PNPN device can cause the loop gain to rise. If current exceeds the latching level, the thyristor fires.

Some thyristors use light or radiant energy as a triggering device. Light penetrating the silicon lattice releases hole-electron pairs and causes leakage current to increase. The increasing leakage current switches the thyristor on. Light-activated SCRs (LASCRs) have a window that allows light to reach the silicon structure. Either light or an externally applied gate current can trigger a LASCR.

TURNING THE SCR ON

Figure 18.13 illustrates the voltage-current characteristics of an SCR. A small amount of forward current flows through an SCR during the forward-blocking state. Increasing the forward bias (I_A) causes the SCR voltage to reach a point called the forward breakover voltage, $V_{(BR)F}$. When the SCR reaches the forward breakover voltage point, the forward current suddenly increases and the SCR turns on.

After reaching the forward breakover point, the voltage drop across the SCR goes to a low value. We call this value the forward on-state voltage. The feedback within the SCR equivalent circuit is large enough that it causes each transistor to drive the other into saturation. At this point the SCR is in its on state. Since the SCR is conducting as hard as possible, any further increases in forward voltage provide only small increases in forward current. For practical SCR applications, the injection of a small pulse of current, or trigger, into the "gate" region causes the SCR to conduct. In the equivalent circuit shown in Figure 18.8, the base regions of either one of the complementary transistors is a gate region for the SCR. If an SCR uses the P-type base as the injection point, it is a conventional SCR. If an SCR uses the N-type base as its injection point, it is a complementary SCR.

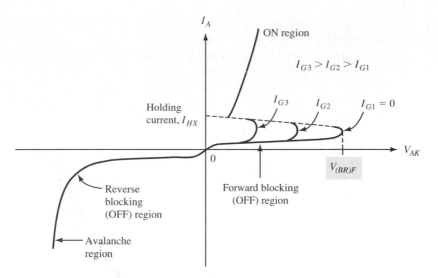

FIGURE 18.13 Characteristic curve for an SCR

TURNING THE SCR OFF

Depending on whether the SCR is operating under DC or AC conditions, the removal or a decrease of the supply voltage turns the SCR off. DC circuits require the removal of the anode-to-cathode supply voltage before the SCR turns off. AC circuits require a sufficient decrease in the anode-to-cathode voltage before the SCR turns off. As you know, the latching current must exceed the holding current level to keep the SCR conducting.

When an SCR conducts, it is saturated. Practical applications require the turning off of an SCR in a minimum amount of time. Applying a reverse voltage across the SCR causes the electrons and holes in the end junctions of the SCR structure to diffuse. As a result, a reverse current occurs in the external circuit and keeps the voltage across the SCR at approximately 0.7 V. Once the holes and electrons have been completely removed from the region of the junctions, the reverse current stops. Then the junctions are in the blocking, or off, state.

The applied supply voltage now determines the reverse voltage across the SCR. Before the SCR can completely recover, any charge carriers must be removed from the center junction through recombination. Once the process of recombination removes the excess charge carriers, a forward voltage can be applied to the SCR without turning the device on. Usually, the SCR requires about 10 μs after the forward current flow stops before any forward voltage can be safely applied. The 10-μs delay time is the turn-off time for the SCR.

✓ **Progress Check**
You have now completed objective 3.

REVERSE-BIAS CONDITIONS FOR THE SCR

Take another look at the characteristic curve for an SCR shown in Figure 18.13. Under reverse-bias conditions, or with its anode negative with respect to its cathode, an SCR works much like any reverse-biased silicon rectifier. The device has an internal high impedance with only a small amount of reverse blocking current flowing. When the reverse voltage grows larger than the reverse breakdown voltage, the reverse current increases quickly.

Most SCR specification sheets will show a rating for the peak reverse voltage with the gate open, called the V_{DROM} rating. If an external force causes the reverse voltage to go above its specified rating, the SCR may become damaged.

At times, manufacturers will build rectifier diodes into SCR circuits to control excessive transient reverse voltages. The rectifier diode works in series with the SCR. In addition, manufacturers supply heat sinks for SCRs such as those shown in Figure 18.14. For reverse voltage stability, the junction temperature must stay below a minimum value.

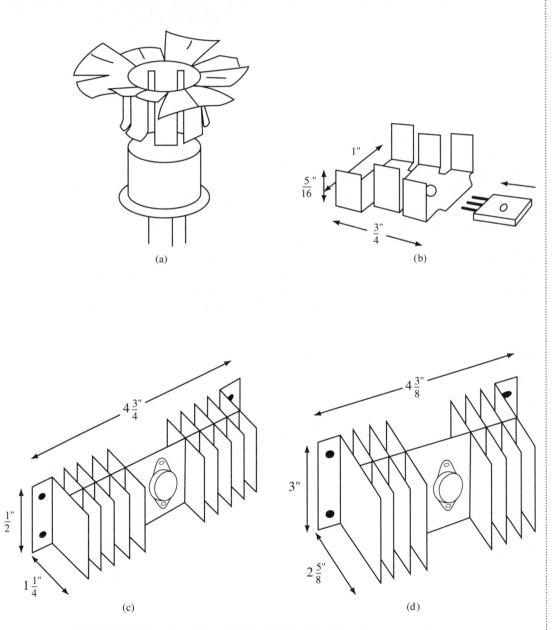

FIGURE 18.14   **Examples of SCR heat sinks** *(Continued on next page)*

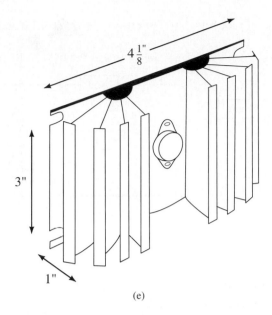

(e)

FIGURE 18.14 *(Continued)*

PHASE CONTROL WITH THYRISTORS

Thyristors often work as phase control devices. Phase control is rapid on-off switching that connects an AC power supply to a load for a controlled fraction of each cycle. The rapid switching results in a highly efficient method of controlling the average power to loads such as lamps, heaters, motors, or DC supplies. Phase control occurs through the control of the phase angle of the AC wave that triggers the thyristor. The thyristor conducts for the remaining half-cycle of the AC wave.

Figure 18.15 illustrates several different kinds of phase control. Figure 18.15a shows one SCR controlling a half-wave. With current flowing in only one direction, the circuit works with loads that require power control from zero to one-half of the maximum full wave. Figure 18.15b adds one rectifier to the circuit shown in Figure 18.15a. Adding a rectifier allows a fixed half-cycle of power that shifts the power control range to half-power minimum and full-wave maximum.

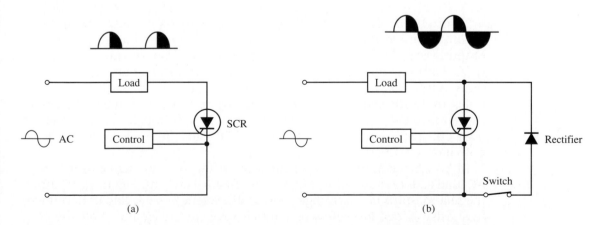

FIGURE 18.15 **(a) One SCR controlling a half-wave. (b) Adding a rectifier to the circuit shown in Figure 18.15a.** *(Continued on next page)*

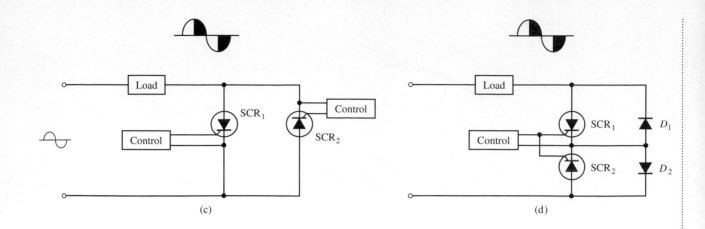

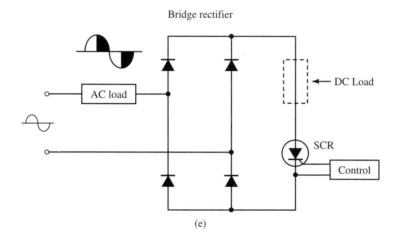

FIGURE 18.15 *(Continued)* **(c) Two SCRs controlling zero to full power. (d) Two SCRs with common cathode and gate connections configured for full-wave control. (e) One SCR working inside a bridge rectifier for AC full-wave control or full-wave rectified DC control.**

Figure 18.15c shows two SCRs to control zero to full power. Since the SCRs in the circuit have equal triggering angles, the circuit produces asymmetrical output waves. The circuit shown in Figure 18.15d also uses two SCRs for full-wave control. However, the SCRs have common cathode and gate connections to prevent any reverse voltage from appearing across the SCRs.

Figure 18.15e shows one SCR working inside a bridge rectifier. The circuit can control either AC or full-wave rectified DC. Although the circuit is the least efficient, it utilizes almost all the SCR capacity by using one SCR on both halves of the AC wave.

Although rectifiers and SCRs function in terms of average current, most AC loads are rated in terms of RMS, or effective current. Figure 18.16 shows the relationships of phase angle, power, and average, peak, and RMS voltages in a resistive load for half-wave and full-wave control circuits. Note the non-linearity of the curves. The first and last 30° of each half-cycle contribute only 6% of the total power to each cycle. Thus, a triggering range from 30° through

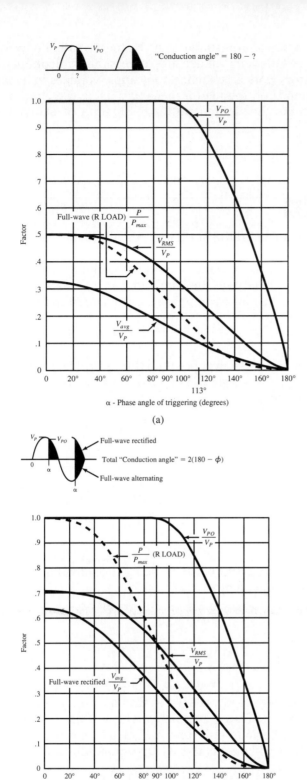

"Conduction angle" = 180 − ?

(a)

Total "Conduction angle" = 2(180 − ϕ)

(b)

FIGURE 18.16 (a) Relationships of phase angle, power, and average, peak, and RMS voltages in a resistive load for half-wave control circuits. (b) Relationships of phase angle, power, and average, peak, and RMS voltages in a resistive load for full-wave control circuits.

150° produces a power-control range of 3% to 97% of full power. The power-control range excludes any voltage drop across the semiconductors.

Since thyristors work as switches, they apply voltages to a load. The value of current depends on the load impedance. Using the chart as a reference, we can determine the desired power dissipation of a thyristor switching circuit.

EXAMPLE 18.4

Assume that we are operating a 1200-W resistive load, rated at 120 V, from a 240-V power supply. Connecting the load directly to the power supply will result in 4800 W. Thus, we want to operate the circuit at 1/4 of its maximum power capability. Determine the true and average amounts of power dissipated if we connect a half-wave phase control circuit to the load. Use Figure 18.16a as a reference.

Solution
In Figure 18.16a, the 1/4-power point on the vertical factor scale corresponds with a triggering phase angle of 90°. First, find the load resistance by using the load's rated values.

$$R_L = \frac{V^2}{P} = \frac{(120 \text{ V})^2}{1200 \text{ W}} = 12 \text{ } \Omega$$

Now find the peak input voltage.

$$V_p = \frac{V_{RMS}}{0.707 \text{ V}}$$

$$V_p = \frac{240 \text{ V}}{0.707 \text{ V}}$$

$$V_p = 340 \text{ V}$$

Because only 90° of the waveform is being developed across the load, load RMS voltage is calculated by

$$V_{RMS(load)} = \frac{V_{RMS}}{2}$$

$$V_{RMS(load)} = \frac{240 \text{ V}}{2}$$

$$V_{RMS(load)} = 120 \text{ V}$$

Next, find the average load voltage. To find average voltage developed by a load supplied by a half-wave rectifier, we have used $V_{avg} = V_p \times 0.318$. Because our circuit permits only one-half of the half-wave to be developed across the load, we will modify the average voltage formula

$$V_{avg(load)} = \frac{0.318 \times V_p}{2}$$

$$V_{avg(load)} = \frac{0.318 \times 340 \text{ V}}{2}$$

$$V_{avg(load)} = 54 \text{ V}$$

Now that voltage and resistance are known, current may be calculated using Ohm's law.

$$I_{RMS\,(load)} = \frac{V_{RMS\,(load)}}{R_L}$$

$$I_{RMS\,(load)} = \frac{120\ V}{12\ \Omega}$$

$$I_{RMS\,(load)} = 10\ A$$

$$I_{avg\,(load)} = \frac{V_{avg\,(load)}}{R_L}$$

$$I_{avg\,(load)} = \frac{54\ V}{12\ \Omega}$$

$$I_{avg\,(load)} = 4.5\ A$$

Lastly, calculate the power dissipated by the load.

$$P_{(load)} = V_{avg(load)} \times I_{avg(load)}$$
$$P_{(load)} = 54\ V \times 4.5\ A$$
$$P_{(load)} = 243\ W$$

PRACTICE PROBLEM 18.4

Determine the amount of power dissipated if we connect a full-wave phase control circuit to the same load conditions seen in example 18.4. Use Figure 18.16b as a reference for finding the phase angle of triggering.

REVIEW SECTION 18.2

1. SCRs have _____ terminals.
2. Name the SCR terminals.
3. SCRs allows current to flow in how many directions?
4. SCRs are fired by a _____ to the gate.
5. Firing an SCR by dv/dt effect is desirable. True or false?
6. What is a snubber?
7. LASCRs are _____ -activated.
8. Phase control is an efficient way to control _____ power to a load.
9. How is an SCR turned off?
10. What is the dv/dt effect?

18.3 TRIACS

Unlike the unidirectional SCR, triacs conduct in both directions and can switch either polarity of an applied voltage. Along with acting as a bidirectional device, triacs also differ from SCRs in one other important way. A triac is an NPNPN semiconductor device with three terminals. The three termi-

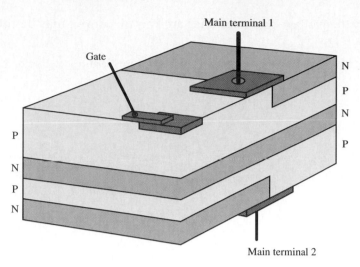

FIGURE 18.17   Structure of a typical triac

nals are labeled as main terminal 1, main terminal 2, and the gate. Figure 18.17 shows the structure of a basic triac.

General Electric developed the triac in response to a need for improving control over AC power. Comparatively, triacs provide smaller, cheaper, simpler answers for solid-state switching and phase control. Triacs have uses as lamp controls and dimmers, motor controls, and temperature controls. Almost all solid-state switching utilizes triacs.

Figure 18.18 shows the schematic symbol for a triac, and Figure 18.19 shows the equivalent circuit for a triac. Look at the equivalent circuit. The triac works like two SCRs in parallel, with the anode of one connected to the cathode of the other. A triac exhibits the same forward blocking characteristics as an SCR when main terminal 1 is positive with respect to main terminal 2. The same characteristics are true when main terminal 2 is positive with respect to main terminal 1. Figure 18.20 shows the characteristic curve of a triac.

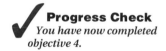

Progress Check
You have now completed objective 4.

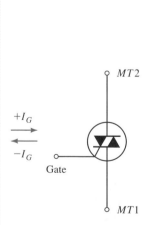

FIGURE 18.18   Schematic diagram for a triac

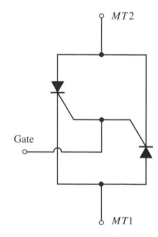

FIGURE 18.19   Equivalent circuit for a triac

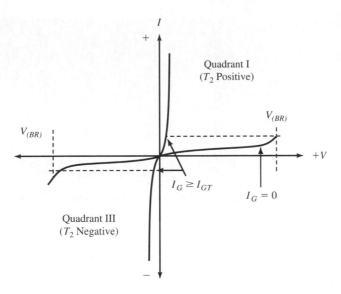

FIGURE 18.20 Characteristic curve of a triac

TRIAC PROPERTIES

The three electrodes of the NPNPN triac are labeled as main terminal 1 (MTI), main terminal 2 (MT2), and the gate (G). With the design of the triac gate, either a positive or negative gate voltage can trigger the triac for either polarity of the voltage across the main terminals. Once the triac turns on, however, the gate has no further control. Until the current through the main terminals is reduced below holding current, the triac remains on. For continuous conduction, the triac requires retriggering on each half-cycle.

Reversing the polarity of the voltage across the main terminals will not turn off the triac. Because of the internal construction—two back-to-back SCRs—reversing the main terminal voltages only causes current to flow in the opposite direction.

If we look at the semiconductor structure of the triac shown in Figure 18.17, another characteristic becomes apparent. A PNPN thyristor is in parallel with an NPNP thyristor, both of which are between the two main terminals.

TRIGGERING METHODS FOR TRIACS

A number of triggering methods for triacs also exist. Triggering may take the form of DC, rectified AC, AC, or pulses. Triacs use either positive gate voltage or negative gate voltage for triggering.

Progress Check
You have now completed objective 5.

COMMUTATION

The turn-off characteristics of thyristors do not apply to triacs. The physical design of the triac—two back-to-back SCRs—does not allow the application of a reverse voltage for turn-off. Instead, triacs commutate a fixed value of current. This ability is called the critical rate of rise of commutation voltage, or the commutating dv/dt capability of the triac.

For AC power-control applications, triacs must switch from the conducting state to the blocking state at each zero-crossing point, or twice each

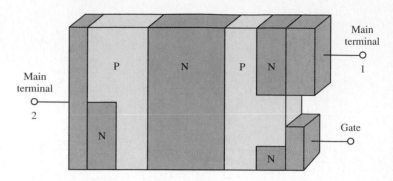

FIGURE 18.21   Considering the triac as two halves to understand commutation

cycle. This switching action is called **commutation.** Without commutating ability, the triac cannot control the load power.

The commutating dv/dt is a major operating characteristic of a triac. Figure 18.21 may help you to understand commutation. If we consider the triac as two halves, one half conducts in one direction while the other half conducts in the opposite direction. Blocking junctions and an N-type base region are common to both halves.

Charge becomes stored in the base region when the triac conducts current in either direction. The amount of charge stored in the base region at the end of each half-cycle depends on the rate of decrease of load current. Any junction capacitance within the triac is a function of the remaining charge. With a greater decrease in load current, the triac has a greater charge and a greater junction capacitance.

If the voltage changes direction, the remaining charge diffuses into the opposite half of the triac structure. The commutating dv/dt, or the rate of rise of voltage in the opposite half of the triac, results in a current flow. If the current flow is large enough, the triac goes back to its conducting state with no gate signal.

Commutating dv/dt is specified in volts per second and depends on the following conditions:

1. the maximum rate on-state current, or $I_{T(RMS)}$,

2. the maximum case temperature for the rated value of on-state current,

3. the maximum rate off-state voltage, or V_{VDROM}, and

4. the maximum commutating rate of decrease of load current.

Commutation for resistive circuits does not present any special problems because voltage and current remain in phase. Circuits that feature inductive loading offer different conditions. Since the current lags the voltage in an inductive circuit, an applied voltage opposite to the current and equal to the peak of the AC line voltage occurs across the triac.

Figure 18.22 shows a simple inductive circuit, and Figure 18.23 shows the waveforms for the circuit. Applying a gate signal that allows continuous conduction to the triac causes the load current (I_{Load}) to lag the line voltage (V_{Line}) by 90° for the inductive load. When the triac is in its on state, the voltage across the triac (V_T) stays in phase with the line current. Typically, the voltage across the triac will equal ± 1.5 V.

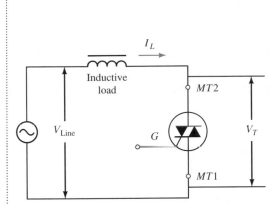

FIGURE 18.22 Simple inductive circuit

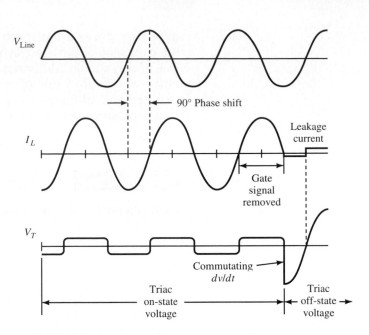

FIGURE 18.23 Waveforms for the circuit shown in Figure 18.22

Removing the gate signal causes the triac to commutate off near the end of the half-cycle. This occurs when the load current drops below the holding current of the triac. When the triac commutates off, the voltage across the triac simultaneously reverses direction. That voltage then increases to the peak of the line voltage. The rate of rise, or commutating dv/dt, and the overshoot of the voltage depend on the circuit components.

Once the triac successfully commutates off, the voltage across the triac is in phase with the line voltage. If the commutating dv/dt of the circuit is greater than the commutating dv/dt of the triac, the triac does not turn off. Instead, it reverts to the on state and, with no gate signal applied, tries to turn off the next half-cycle. If the triac turns off, it remains off. If it does not turn off, the triac stays on until an interruption of circuit power.

Figure 18.24 shows the addition of an RC snubber network to the simple inductive circuit. Placed across the main terminals of the triac, the snubber

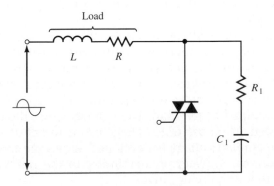

FIGURE 18.24 Adding components to the triac circuit to limit the dv/dt rise

network reduces the commutating dv/dt of the circuit so that it fits within the capability of the triac. Typically, an additional 0.1-μF capacitor limits the rate of rise, and an additional 100-Ω resistor damps the ringing of the capacitance. The combination of the capacitance and the load inductance can lead to overshoot. Adding the resistor also minimizes overvoltage ringing and limits any possible surge from the capacitor while the triac fires.

EXAMPLE 18.5

Using Figure 18.25, develop the waveforms for the circuit.

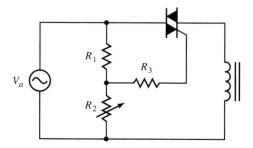

FIGURE 18.25

Figure 18.25 is a triac control circuit for an inductive load such as a motor. When the triac is open there is no current flowing to the load. R_1 and R_2 form a voltage divider network that will fire the triac. During the positive alternation, the triac will block current until the positive voltage developed across R_2, which is connected to the gate, is high enough to fire the triac. The triac saturates, and current is supplied to the load. When circuit current drops below holding current, the triac switches to its cutoff condition and once again blocks current flow. During the negative alternation the triac continues to block current flow until the negative voltage developed across R_2 is high enough to fire the triac. The triac saturates, supplying current to the load but in the opposite direction. Current is supplied throughout the remaining negative alternation until the current drops below holding current level. Then the triac opens blocking current to the load.

Figure 18.26 shows the resulting waveforms when R_2 is set to fire the triac at 30°. The applied voltage is sinusoidal. During the positive alternation, all the applied voltage is dropped across the triac while the triac is open. The voltage across R_2 follows the phase of the applied voltage and at a lower amplitude. There is no voltage across the load until the triac fires; then the applied voltage is developed by the load. The voltage across the triac increases until the triac is fired. Then the triac voltage drops to about 1 V. When the triac opens, all the applied voltage is developed across the triac. During the negative alternation, the waveforms are similar to the positive alternation waveforms except that they are negative.

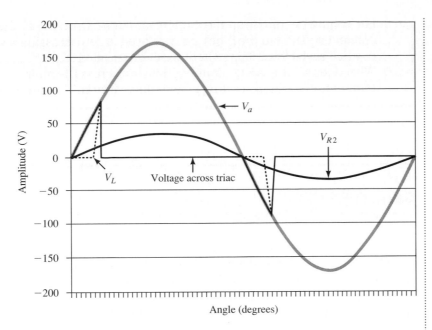

FIGURE 18.26

PRACTICE PROBLEM 18.5

Use Figures 18.25 and 18.26. If the resistance of R₂ is increased, what happens to the current supplied to the load?

REVIEW SECTION 18.3

1. Triacs conduct in only one direction and can switch either polarity of an applied voltage. True or false?
2. Triacs are NPNPN semiconductor devices with three terminals. True or false?
3. Triacs often work as _____ power controllers.
4. A triac works like two:
   (a) parallel diodes
   (b) parallel BJTs
   (c) parallel SCRs
5. The terminals of a triac are labeled as _____ , _____ , and _____ .
6. Either a positive or a negative gate voltage can trigger the triac for either polarity of the voltage across the main terminals. True or false?
7. Triggering of a triac may take the form of _____ , _____ _____ , _____ , or _____ .
8. The physical design of the triac allows the application of a reverse voltage for turn-off. True or false?

9. For AC power-control applications, triacs must switch from the conducting state to the blocking state at each zero-crossing point, or twice each cycle. This switching action is called _____ .
10. What does placing an RC snubber network across the main terminals of a triac operating in an inductive circuit accomplish?

18.4 OTHER THYRISTORS

GATE TURN-OFF

The gate turn-off switch shown in Figure 18.27 is similar to the SCR. Gate turn-off switches have special gate construction characteristics that allow a small reverse gate current to turn off the device. Within the gate turn-off switch, the base region's N-type and P-type regions are much narrower than the regions seen within the SCR. Even with the special characteristics of its gate, the gate turn-off switch can turn off like an SCR. For DC circuit applications, gate turn-off switches require less circuitry than SCRs.

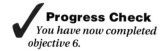

Progress Check
You have now completed objective 6.

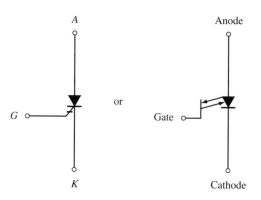

FIGURE 18.27   **Schematic symbol for a gate turn-off switch**

LIGHT-ACTIVATED SCRS

Light-activated SCRs (LASCRs) work much like phototransistors with one exception. While phototransistors have a proportional response to light, LASCRs operate as switches. Only light completely triggers a LASCR on. Figure 18.28 shows the general construction of a LASCR, and Figure 18.29 shows the schematic for a LASCR. The clear window on top of the container allows light to strike a light-sensitive silicon pellet.

LASCRs can handle up to 1.6 A of current and can block up to 200 V. Since the PNPN device has a high gain, a small amount of power controlled by light can switch large amounts of power directly.

SILICON BILATERAL SWITCH (SBS)

A silicon bilateral switch (SBS) is actually an integrated circuit that combines two identical silicon unilateral switches into one package and adds a gate lead. In addition, the arrangement of the SBS gate electrodes permits syn-

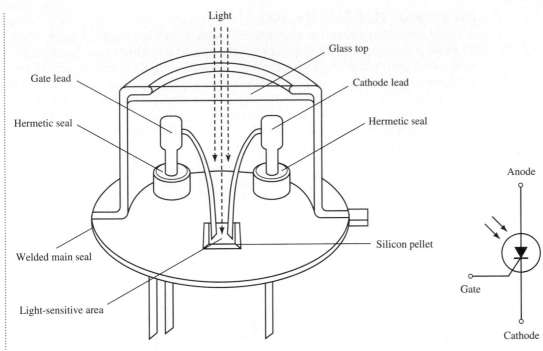

Light
Glass top
Gate lead
Cathode lead
Hermetic seal
Hermetic seal

Welded main seal
Silicon pellet

Light-sensitive area

FIGURE 18.28   General construction of a LASCR

Anode

Gate

Cathode

FIGURE 18.29   Schematic symbol
for a LASCR

Gate

FIGURE 18.30   A silicon bilateral switch

FIGURE 18.31   Characteristic curve for a silicon
bilateral switch

chronization. SBSs can switch both polarities of an applied voltage. Figure
18.30 shows the schematic symbol for the silicon bilateral switch.

Because of that property, they work as full-wave triggering devices for triacs. An AC voltage supply allows the SBS to supply the positive and negative
triggering pulses for a triac. Other uses for silicon bilateral switches include
SCR half-wave trigger applications and voltage level detection. Figure 18.31
shows the characteristic curve for a silicon bilateral switch.

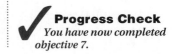

Progress Check
*You have now completed
objective 7.*

BILATERAL TRIGGER DIODES

A more-common name for the bilateral trigger diode is diac. Figure 18.32 shows the schematic symbol for the PNPN device and illustrates the close relationship that a diac has with a typical transistor. As the characteristic curve in Figure 18.33 shows, the negative resistance region of a diac extends over the entire operating range of currents above the breakdown point. Even though the diac is a thyristor, it does not have a holding current because of the large negative resistance region. Like the silicon bilateral switch, diacs work primarily as switching devices for triacs.

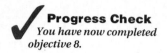

Progress Check
You have now completed objective 8.

FIGURE 18.32 Schematic symbol for a diac

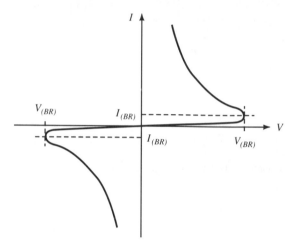

FIGURE 18.33 Characteristic curve for a diac

REVIEW SECTION 18.4

1. Gate turn-off switches have characteristics that allow a small _____ _____ to turn off the device.
2. Can gate turn-off switches turn off like SCRs?
3. A LASCR is a _____ device that operates as a switch.
4. What is a fundamental difference between phototransistors and LASCRs?
5. _____ triggers a LASCR on.
6. A silicon bilateral switch combines two identical silicon unilateral switches into one package. True or false?
7. SBSs work as full-wave triggering devices for _____ .
8. Another name for the bilateral trigger diode is a
   (a) SCS
   (b) diac
   (c) gate trigger
9. The negative resistance region of a diac extends over the entire operating range of currents above the breakdown point. True or false?

18.5 UNIJUNCTION TRANSISTORS

During our discussion about thyristors, we talked about different triggering methods. Along with using resistors, capacitors, diodes, and transistors as triggering devices, many circuits rely on unijunction transistors (UJTs). UJTs also work in oscillator and timing circuits. Compared with other semiconductors, UJTs offer stable characteristics that make them especially useful as triggering devices.

The UJT is a three-terminal device with only one PN junction. Figure 18.34 represents a UJT as a single bar of silicon with one PN junction. Two terminals—labeled base 1 (B_1) and base 2 (B_2)—extend from the N-type material that makes up most of the bar. Another terminal, the emitter, attaches to the center of the bar or into the P-type material. Figure 18.35 shows the schematic symbol for a UJT.

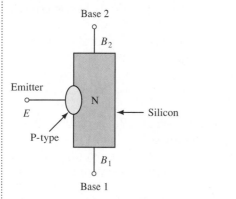

FIGURE 18.34   A UJT is a single bar of silicon with one PN junction.

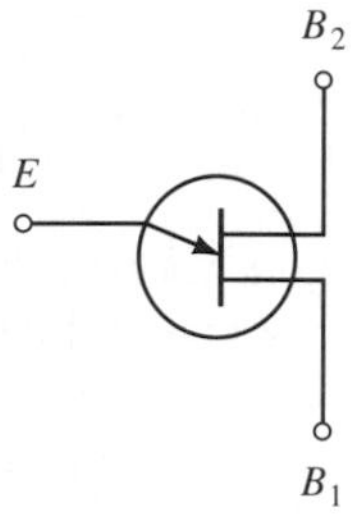

FIGURE 18.35   Schematic symbol for a UJT

BIASING THE UNIJUNCTION TRANSISTOR

A UJT switches from a blocking state to a conducting state when the applied voltage reaches a specific value. A second bias voltage can change the switching voltage. If we placed a UJT into a simple circuit, it would appear similar to Figure 18.36 and would work much like the equivalent circuit shown in

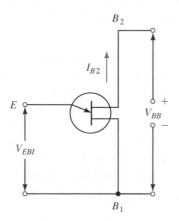

FIGURE 18.36   A simple UJT circuit

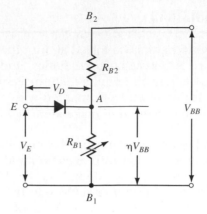

FIGURE 18.37   Equivalent circuit for the circuit shown in Figure 18.36

Figure 18.37. One voltage (V_{BB}) applied to the base terminals keeps base 2 more positive than base 1.

With that polarity, current flows between B_1 and B_2 and is limited only by an interbase resistance. That resistance equals approximately 7 kΩ or:

$$R_{BB} = \frac{V_{BB}}{I_{BB}}$$

When the base circuits are broken down into smaller units, each base has a resistance that contributes to the interbase resistance.

Essentially, the base sections set up a series-connected voltage divider identified as R_{B1} and R_{B2}. The interbase resistance equals the sum of the individual base resistances.

If the emitter voltage is less than the B_1 voltage, the PN junction is reverse-biased. The reverse-bias voltage allows a very small amount of leakage current to flow through the emitter lead.

If the voltage applied between the emitter and base 1 increases, the voltage will eventually equal the sum of the voltage across R_{B1} and the forward voltage drop across the junction, or

$$V_P = V_D + \left(\frac{R_{B1}}{R_{BB} \times V_{BB}} \right)$$

where

$$V_P = \text{peak point voltage}$$

and

$$V_D = \text{PN junction forward voltage}$$

The ratio caused by the relationship of the emitter-base 1 resistance to the interbase resistance is called the standoff intrinsic ratio and is designated by the Greek letter eta or η.

While the standoff intrinsic ratio always has a value between 0.51 and 0.82, the emitter diode voltage (V_D) is approximately 0.7 V. V_P decreases with increases in operating temperature. Placing a resistor in series with B_2 causes the voltage V_{BB} to compensate for changes in the peak-point voltage. The B_2 series resistor will have a value equal to:

$$R_{B2} = \frac{10,000}{\eta V_1}$$

If the voltage applied between the emitter and base is greater than the peak-point voltage, the PN junction has a forward bias. The P-type emitter injects holes into the N-type silicon bar and an emitter current flows. Because B_1 is more negative than the emitter, the holes move toward the B_1 terminal.

While the emitter injects holes into the silicon bar, B_1 injects an equal number of electrons. All this causes an increase of current flow through the silicon bar and a decrease of the R_{B1} resistance. Consequently, the small voltage between point A of the equivalent circuit in Figure 18.37 and B_1 decreases. In turn, the emitter current rises and the R_{B1} resistance decreases even more.

Figure 18.38 shows a set of out-of-scale characteristic curves for a unijunction transistor. When the voltage across R_{B1} decreases and the emitter current increases, a negative resistance region forms. On the curves, the negative resistance region lies between the peak-point voltage and the valley-point voltage, or V_V.

The valley point voltage and higher voltages have a higher density of charge carriers. Thus, the carriers have a shorter life. This cancels out the generation of new carriers and causes the emitter voltage to rise slowly at points above the valley current value. The area to the right of the dotted line represents the saturation, or positive resistance, region for the UJT.

Figure 18.39 depicts a scaled characteristic curve for a UJT. In the actual characteristic curve, only a small part of the cutoff region shows. Table 18.2 lists commonly used parameters for UJTs, parameter symbols, and definitions. As you look at the characteristic curve, try to associate the parameters with different locations on the curve.

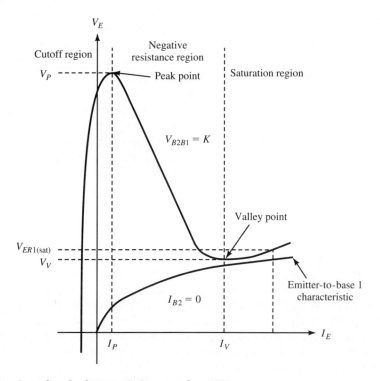

FIGURE 18.38 Out-of-scale characteristic curves for a UJT

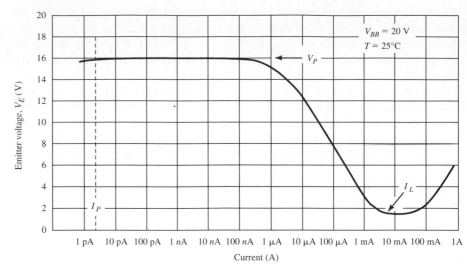

FIGURE 18.39   Scaled characteristic curve for a UJT

Table 18.2 Parameters for UJTs		
Parameter	**Symbol**	**Definition**
Peak emitter voltage	V_P	Maximum emitter voltage before the UJT enters the negative resistance region
Peak emitter current	I_P	Maximum emitter current before the UJT enters the negative resistance region; also thought of as the minimum emitter current required to turn on a UJT
Valley emitter voltage	V_V	Emitter voltage at the valley point
Valley emitter current	I_V	Emitter current at the valley point
Interbase voltage	V_{BB}	Voltage between base 1 and base 2
Emitter saturation voltage	$V_{BE1}(\text{sat})$	Voltage across the emitter and base 1 at a specified emitter current and interbase voltage
Interbase resistance	R_{BB}	DC resistance between base 1 and base 2 with the emitter open circuited
Intrinsic standoff ratio	η	Ratio of emitter-base 1 resistance interbase resistance, or R_{B1}/R_{BB}

At 10 pA, the peak-point voltage equals 16 V. The peak-point voltage sets the triggering point for most switching applications. As Figure 18.39 shows, the voltage holds at 16 V until the current reaches approximately 100 μA. Then the voltage decreases. At 8 mA, which is also the valley current level, the emitter voltage is 1.6 V and is at the valley voltage level.

Unijunction transistors have the disadvantage of being unilateral to current flow. In other words, a unijunction transistor requires a DC voltage to operate. An AC circuit that includes UJTs also must include diodes so that no reverse voltage is applied across the UJT. In addition, a UJT produces positive-going pulses. While those pulses will direct trigger SCRs, a UJT triggering a triac must use capacitive coupling.

EXAMPLE 18.6

Solve for the peak emitter voltage if the supply voltage is 20 V and the intrinsic standoff ratio is 0.57.

The peak emitter voltage is found by adding the barrier potential of the PN junction to the interbase voltage. Use the approximation value for the barrier potential of the PN junction, 0.7 V. The interbase voltage is found by multiplying the intrinsic standoff ratio and the supply voltage.

Solution
The interbase voltage is:

$$V_{BB} = \eta \times V_s$$
$$V_{BB} = 0.57 \times 20 \text{ V}$$
$$V_{BB} = 11.4 \text{ V}$$

The peak emitter voltage is:

$$V_p = V_{BB} + V_{BE}$$
$$V_p = 11.4 \text{ V} + 0.7 \text{ V}$$
$$V_p = 12.1 \text{ V}$$

PRACTICE PROBLEM 18.6

Solve for the peak emitter voltage if the supply voltage is 15 V and the intrinsic standoff ratio is 0.63.

A UJT TRIGGERING CIRCUIT
Figure 18.40 shows a basic UJT triggering circuit that works as a relaxation oscillator. During circuit operation, capacitor C_1 charges through R_1 until the emitter voltage of the UJT reaches the peak-point voltage. At that voltage, the

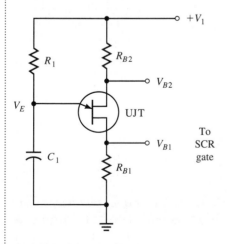

FIGURE 18.40 UJT triggering circuit

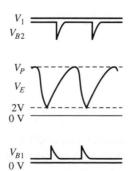

FIGURE 18.41 UJT waveforms

UJT saturates, providing a discharge path for C_1 through R_{B1}. When the capacitor is discharged to 2 V, the emitter stops conducting. Without the emitter conducting, the UJT turns off.

Figure 18.41 illustrates the resulting waveforms at the emitter and two bases. Base 1 has positive-going spikes and base 2 has negative-going spikes. The waveform developed at the emitter is called a sawtooth. Because R_{B1} is a low-value resistor, C_1 quickly discharges. Because emitter current is too low to sustain regeneration, the UJT opens. C_1 once again begins charging, which starts the next cycle of the oscillator. Example 18.7 provides circuit analysis of a UJT relaxation oscillator.

EXAMPLE 18.7

Use the following values in Figure 18.40: $V_1 = 12$ V; $R_{B1} = 100\ \Omega$; $R_{B2} = 1$ kΩ; $R_1 = 10$ kΩ; $C_1 = 0.1\ \mu$F; $\eta = 0.66$. Solve for the following peak values: V_E, I_E, V_{B1}, and V_{B2}.

Solution
First, solve for V_E. The peak emitter voltage is the voltage between B_1 and B_2 (V_{BB}) plus the PN junction barrier potential (V_{BE}). The peak emitter voltage will also equal the capacitor voltage. To find V_{BB} use the intrinsic standoff ratio.

$$V_{BB} = \eta \times V_1$$
$$V_{BB} = 0.66 \times 12\text{ V}$$
$$V_{BB} = 7.92\text{ V}$$

$$V_E = V_{BB} + V_{BE}$$
$$V_E = 7.92\text{ V} + 0.7\text{ V}$$
$$V_E = 8.62\text{ V}$$

Now to find the value of peak B_1 voltage, use the emitter Kirchhoff loop. When C_1 starts to discharge it becomes the source for the loop of C_1, R_{B1}, and the emitter junction. If C_1 has 8.62 V and the emitter junction drops 0.7 V, then

$$V_{B1} = V_{C1} - V_{BE}$$
$$V_{B1} = 8.62\text{ V} - 0.7\text{ V}$$
$$V_{B1} = 7.92\text{ V}$$

Next, solve for the peak emitter current. The peak emitter current is almost equal to the current through R_{B1}. To find I_{RB1}, use Ohm's law.

$$I_{RB1} = \frac{V_{B1}}{R_{B1}}$$

$$I_{RB1} = \frac{7.92\text{ V}}{100\ \Omega}$$

$$I_{RB1} = 79.2\text{ mA}$$

Last, solve for the peak voltage at B_2. When the UJT fires, the bases saturate, allowing a maximum current to flow through R_{B2}. The base current is found by:

$$I_B = \frac{V_1}{R_{B1} + R_{B2}}$$

$$I_B = \frac{12 \text{ V}}{100 \text{ } \Omega + 1 \text{ k}\Omega}$$

$$I_B = 10.9 \text{ mA}$$

V_{B2} is the supply voltage less the voltage drop across R_{B2}.

$$V_{B2} = V_1 - (I_B \times R_{B2})$$
$$V_{B2} = 12 \text{ V} - (10.9 \text{ mA} \times 1 \text{ k}\Omega)$$
$$V_{B2} = 1.1 \text{ V}$$

Use the following values in Figure 18.40: $V_1 = 15$ V; $R_{B1} = 68$ V; $R_{B2} = 1$ kΩ; $R_1 = 15$ kΩ; $C_1 = 0.22$ μF; $\eta = 0.66$. Solve for the following peak values: V_E, I_E, V_{B1}, and V_{B2}.

SYNCHRONIZING A UJT

Reducing either the interbase voltage or the supply voltage of the UJT in Figure 18.42 causes the triggering of the UJT. The decrease in either voltage causes the peak-point voltage to decrease. A UJT triggers if the peak-point voltage falls below the value of the emitter voltage. Because of this characteristic, we can use B_2 to synchronize the trigger circuit shown in Figure 18.42. Figure 18.42 shows a negative synchronizing pulse applied to B_2.

Moving to Figure 18.43, we can synchronize the UJT with the AC line voltage. A full-wave bridge supplies both power and a synchronizing signal to the trigger circuit. The zener diode clips the peaks of the AC signal and provides regulation. As each half-cycle ends, the B_2 voltage drops to zero and the UJT triggers.

The capacitor discharges at the beginning of each half-cycle. Triggering is now synchronized with the AC line. At the output at the end of the cycle, a

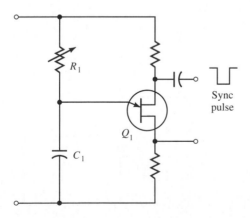

FIGURE 18.42 Negative synchronizing pulse applied to base 2

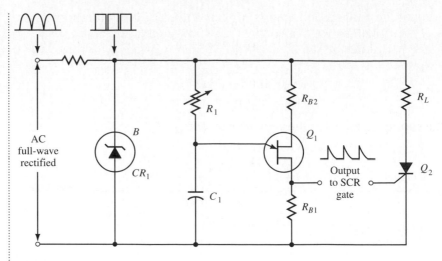

FIGURE 18.43   Synchronizing a UJT trigger circuit

pulse causes the SCR to fire. As a result, current flows through the load until load current decreases below holding current.

PROGRAMMABLE UNIJUNCTION TRANSISTORS

After looking at the schematic diagram for a programmable unijunction transistor (PUT) shown in Figure 18.44, you can see that a similarity exists between the PUT and the SCR. During operation, however, the gate potential of the PUT is brought closer to the anode potential. An SCR has its gate brought closer to its cathode. A programmable unijunction transistor blocks anode current until the voltage applied to the anode and cathode terminals reaches the same level as the voltage applied to the gate and cathode terminals.

After the voltages have equal levels, the PUT triggers and the anode current begins to quickly increase. After the PUT turns on, it will not turn off until the value of the anode current drops below the value of the peak-point current. Then, the PUT becomes cut off.

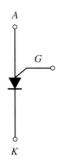

FIGURE 18.44   Schematic diagram for a programmable unijunction transistor

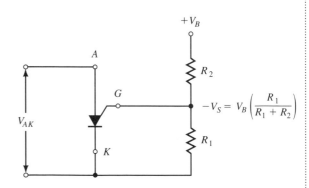

$$-V_S = V_B\left(\frac{R_1}{R_1 + R_2}\right)$$

FIGURE 18.45   Addition of program resistors to the PUT

Programmable unijunction transistors have uses in long-duration timer circuits. Figure 18.45 shows the connection of program resistors to the gate and cathode terminals of a PUT. The values of resistors can alter the operating parameters, such as η, R_{BB}, I_P, and I_V to meet the needs of a circuit design and give the device the programmable part of its name.

Table 18.3 shows the maximum ratings for programmable unijunction transistors. Before we can calculate the gate voltage (V_S) and find a value for V_{BB}, we need to refer to the maximum rating for the PUT. To find those values, we use the equation:

$$-V_S = \frac{R_3 \times V_B}{R_2 + R_3}$$

Table 18.3 Maximum Ratings for PUTs		
Rating	**Symbol**	**Value**
DC forward anode current mA	I_T	150
DC gate current	I_G	50 mA
Gate-to-cathode forward voltage	V_{GKF}	40 V
Gate-to-cathode reverse voltage	V_{GKR}	−5 V
Gate-to-anode reverse voltage	V_{GAR}	40 V

Progress Check
You have now completed objective 9.

EXAMPLE 18.8

Using Figure 18.46, calculate the time delay for the circuit when R_3 is adjusted for 2 kΩ. Use the following values: $R_1 = 470\ \Omega$; $R_2 = 2\ k\Omega$.

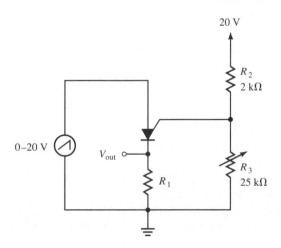

FIGURE 18.46

First, calculate $-V_S$.

$$-V_s = \frac{R_3 \times V_B}{R_2 + R_3}$$

$$-V_s = \frac{2 \text{ k}\Omega \times 20 \text{ V}}{2 \text{ k}\Omega + 2 \text{ k}\Omega}$$

$$-V_s = 10 \text{ V}$$

Next, calculate when the applied ramp voltage is at 10 V. The ramp voltage rises at a linear rate. That rate is found by dividing the peak amplitude by the time required to arrive at the peak amplitude. The time required to arrive at peak amplitude is 100 ms.

$$\text{Linear rate} = \frac{V_p}{\text{Time}}$$

$$\text{Linear rate} = \frac{20 \text{ V}}{100 \text{ ms}}$$

$$\text{Linear rate} = 0.2 \text{ V/ms}$$

To find the circuit time delay in milliseconds, divide 10 V by the linear rate.

$$\text{Time delay} = \frac{10 \text{ V}}{0.2 \text{ V/ms}} = 50 \text{ ms}$$

PRACTICE PROBLEM 18.8

Using Figure 18.46, calculate the time delay for the circuit when R_3 is adjusted for 20 kΩ. Use the following values: $R_1 = 680 \ \Omega$; $R_2 = 22.2 \text{ k}\Omega$. The time required to arrive at the peak amplitude is 100 ms.

REVIEW SECTION 18.5

1. Unijunction transistors work in _____ circuits, _____ circuits, and _____ circuits.
2. Why are UJTs useful as triggering devices?
3. A unijunction transistor has:
   (a) two PN junctions
   (b) no PN junctions
   (c) three PN junctions
   (d) one PN junction
4. List the terminals for a unijunction transistor. Draw the schematic symbol for a UJT.
5. What is interbase resistance? What is the standoff intrinsic ratio? What is the peak-point voltage?
6. A unijunction transistor switches from a _____ state to a _____ state when the applied voltage reaches a specific level.
7. What is a disadvantage of using UJTs?

8. A UJT triggers if the _____ _____ falls below the value of the emitter voltage.
9. A programmable unijunction transistor is similar to a(n):
    (a) diode
    (b) BJT
    (c) JFET
    (d) op-amp
    (e) SCR
10. The connection of _____ _____ to the PUT terminals makes the device programmable.

18.6 THYRISTOR APPLICATIONS

BASIC TRIAC CONTROL CIRCUIT

Figure 18.47 shows a basic triac control circuit. Closing S_1 applies power to the circuit. R_1 and C_1 form a voltage divider circuit. As long as the capacitor voltage is less than the gate voltage required to fire the triac, the triac remains off, blocking the line voltage. The load experiences no voltage drop. All the supplied voltage is dropped across the triac.

If a positive alternation is applied to the circuit, the capacitor will charge until its voltage exceeds the positive gate voltage required to fire the triac; the triac will close and the line voltage will appear across the load. R_2 acts as a current limiter to prevent excessive current from damaging the triac or the load. The triac remains closed until line current is less than holding current. The triac opens and once again blocks line voltage.

When the alternation goes negative, then a negative voltage will fire the triac and current will be supplied to the load during the negative alternation.

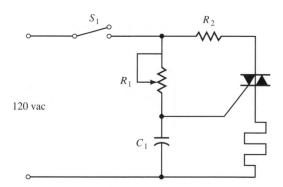

FIGURE 18.47   Basic triac control circuit

DC STATIC SWITCH

Figure 18.48 shows the use of SCRs as a switch in a DC circuit. A low power signal applied to the gate of the SCR_1 causes it to trigger and place a voltage across the load. When this happens, capacitor C_1 charges positively with respect to resistor R_1. When SCR_2 triggers on, capacitor C_1 connects across

SCR$_1$. This causes SCR$_1$ to become momentarily reverse-biased between anode and cathode. The reverse voltage turns SCR$_1$ off and interrupts the load current if the gate signal is not applied to both gates simultaneously.

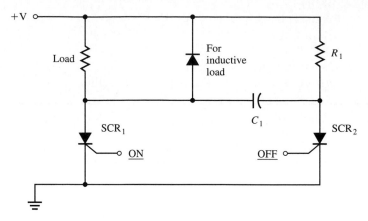

FIGURE 18.48   DC static switch

FULL-WAVE PHASE CONTROL CIRCUIT

Figure 18.49 shows a full-wave phase control circuit designed to control the brightness of a lamp. Notice that the circuit resembles Figure 18.47 except for the addition of a diac. A diac is a bidirectional thyristor that fires at a certain voltage. The triac is not fired until the capacitor voltage exceeds the sum of the diac firing voltage and the triac gate voltage. The lamp's brightness is controlled by R$_1$.

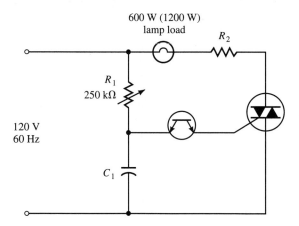

FIGURE 18.49   Full-wave phase control circuit

HIGH-SPEED SWITCH OR ELECTRONIC CROWBAR

In many DC circuits, SCRs protect DC circuits against input line voltage transients and short circuits within the load. Figure 18.50 shows two UJTs and an SCR working as a high-speed switch or "electronic crowbar." The circuit protects DC circuits from overvoltage or overcurrent conditions.

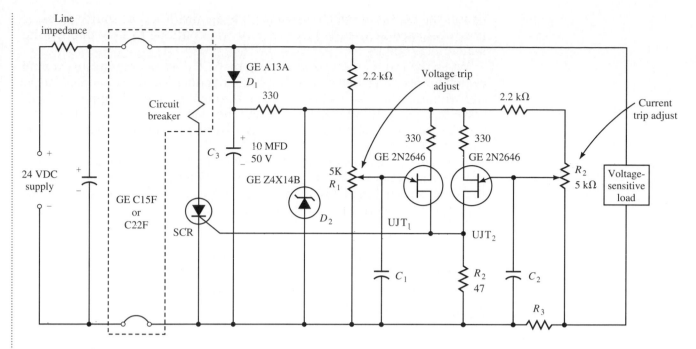

FIGURE 18.50 SCR used as a high-speed switch

To protect a DC circuit from excessive voltage, the SCR closes in a few microseconds after triggering, providing a current path through the coil of the circuit breaker, tripping the circuit breaker. A potentiometer, labeled R_1, sets the voltage trip for the DC supply. If the DC supply voltage exceeds that set amount, the voltage at the emitter of UJT_1 goes past its peak-point voltage and UJT_1 fires.

When UJT_1 fires, a trigger pulse is generated across R_2. The trigger fires the SCR. The SCR provides a current path through the circuit breaker's coil, popping the breaker. Aside from the action of the UJT and SCR speeding the circuit breaker response, it also shunts excessive current, preventing the voltage across the load from rising until the breaker opens.

UJT_2 and SCR action also protects the load and power supply against short-circuit conditions by monitoring the load current flowing through R_3. When the voltage across resistor R_3 goes past the desired maximum value set by potentiometer R_2, the voltage at the emitter of UJT_2 exceeds the peak-point voltage. This causes the UJT to fire, which fires the SCR.

False triggering of the circuit is prevented by the combined filtering action of rectifier D_1 and capacitor C_3. The filtering action excludes negative voltage transients. C_1 provides filtering for UJT_1 and C_2 provides filtering for UJT_2.

Progress Check
You have now completed objective 10.

LASCR TRIGGERING CIRCUIT

Some applications use a LASCR as a triggering device for a larger SCR. In those applications, the combination of the LASCR and SCR makes the equivalent of an electronic relay. When used with a DC power supply and with the LASCR off, the load appears as an open circuit. Exposing the LASCR to light causes the LASCR to switch on, which triggers the SCR. The source power is now dissipated across the load. Current will continue to flow through the load until S_1 opens.

When used with an AC power supply, the SCR does not require an external switch for resetting. Since an SCR will not conduct when the anode is negative with respect to the cathode, current through the load will appear as half-wave rectified. Current will flow through the load only when light strikes the LASCR and the AC alternation is positive. During the negative half of the AC cycle, the LASCR opens.

As Figure 18.51 shows, either DC power or a combination of AC power and pulsed light may work as the voltage source. When the DC source voltage is used, the optional capacitor allows larger values for resistors R_1 and R_2 since it carries the triggering pulse for the SCR.

If the circuit utilizes AC power and has a steady light applied to the LASCR, a charge will not build up on the capacitor. With no charge, the capacitor cannot aid triggering. In that instance, resistor R_1 must have a smaller value so that the triggering current can flow to the SCR. However, the resistor must have a large enough value so that the triggering current does not go past the gate current rating of the SCR.

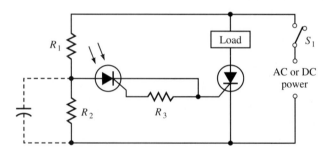

FIGURE 18.51   **LASCR triggering circuit**

PRECISION TIME-DELAY CIRCUIT

One of the most common UJT applications is shown in Figure 18.52. In the figure, a UJT provides a precise, time-delayed trigger pulse that turns on an SCR. The SCR works as part of a power-control circuit. Voltage across a capacitor supplies the emitter-base 1 voltage for the UJT.

When the circuit begins to operate, switch S_1 shorts the capacitor C_1. After the switch opens, the capacitor charges through the variable resistor R_4. The zener diode provides a regulated 18-V voltage supply. The zener voltage supply and the RC time constant of R_4 and C_1 set the amount of voltage across the capacitor.

Once the capacitor charges to the UJT peak-point voltage, the UJT fires and a triggering pulse is developed across resistor R_2. Then the SCR turns on and applies most of the supply voltage across the load. The amount of delay is a function of R_4 and C_1. Removing the DC supply voltage with an open switch S_2 resets the timing cycle and opens the SCR.

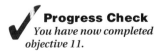

Progress Check
You have now completed objective 11.

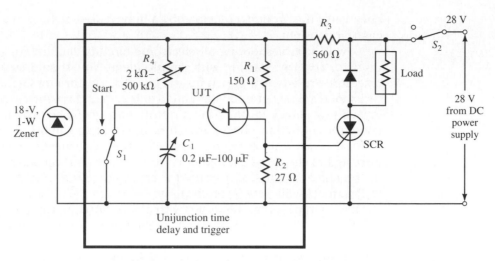

FIGURE 18.52 Precision time-delay circuit

REVIEW SECTION 18.6

1. In Figure 18.47 which component sets the triac trigger level?
2. In Figure 18.47 what is the purpose of R_2?
3. In Figure 18.48 describe how SCR_1 is turned off.
4. In Figure 18.49, what is the purpose of the diac?
5. What is the purpose of a crowbar circuit?
6. In Figure 18.50 does UJT_1 provide overvoltage or current protection?
7. In Figure 18.50 does UJT_2 provide overvoltage or current protection?
8. The circuit in Figure 18.51 is the equivalent of an _____ .
9. In Figure 18.52, time delay is provided by the _____ circuit.
10. Adjusting the amount of time delay is provided by _____ .

 TROUBLESHOOTING

The operation of a thyristor circuit depends on its ability to survive overcurrent conditions. Often, overcurrent protective systems in the circuit limit the duration of overloads and the frequency of the overloads. Also, protective circuits cut the duration and size of short circuits.

Although thyristors are durable devices, a lengthy overload will cause a breakdown in the gate-to-cathode region. To find the cause of SCR failure, look for voltage transients that may appear in the low-voltage rectifier circuit. Transients occur because of the energizing of step down transformers and because of the dropping of a load by a filter. The switching of a load also can cause a transient.

As with troubleshooting all electronic circuits, looking for problems in a thyristor requires starting with simple ideas. You should look for burnt or discolored resistors or blown fuses. Even if a resistor looks good, a change in resistance value can alter the equalization in a thyristor circuit. Check the low voltage supply for the proper output voltage. Since thyristors depend on good filtering action, check capacitors and inductors in the thyristor circuit.

If a thyristor fails to trigger, the first section to suspect is the triggering section. Again, check the components in the section for open circuits. A change in the characteristics of a transistor can change the triggering pulse for an SCR so that it will not trigger properly.

Diacs and triacs can be tested with an oscilloscope while remaining in the circuit. With an oscilloscope, you can check waveforms at the circuit input, at the device, and across the load. Comparing those waveforms should allow you to conclude if the diac or triac is switching. You also may use a multimeter to check diacs and triacs. If either device shorts, it will exhibit a low value of resistance. If either device opens, it will not conduct.

As with diacs and triacs, you can perform in-circuit checks on SCRs with an oscilloscope. Again, check the waveforms across the device and the load. At the SCR, check the gate, cathode, and anode voltages. An open SCR will not conduct even when triggered. A shorted SCR will conduct with or without changes in the polarity of the anode-cathode voltage.

WHAT'S WRONG WITH THESE CIRCUITS?

The heater is on continuously. Q_1 has been replaced and is a known-good component. Which other components could be defective, and what is wrong with them?

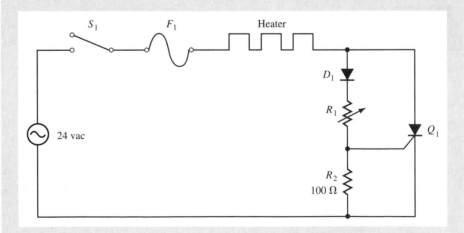

SUMMARY

Thyristors are four-layer PNPN semiconductor devices. In theory, thyristors are similar to a double transistor. Different types of thyristors are Shockley diodes, silicon-controlled rectifiers (SCRs), silicon bilateral switches (SBSs), bilateral trigger diodes (diacs), and triacs. Thyristors work because of regeneration, a latching current, and a holding current. Because of their applications, thyristors often require heat sinks and overcurrent protection.

SCRs are the most popular member of the thyristor family and function as current-control devices, latching switches, and on-off switches. SCRs have a gate, anode, and a cathode terminal.

Different types of SCR triggering circuits exist. Those configurations include resistance triggering, resistance-capacitance triggering, and semiconductor triggering circuits. When circuits require more switching power, designers will often connect SCRs in series. Series-connected SCRs require shunt resistors and capacitors.

While SCRs conduct in only one direction, triacs conduct in both directions. A triac can switch either polarity of an applied voltage. The terminals of a triac are labeled as main terminal 1 (MT1), main terminal 2 (MT2), and the gate (G). Triacs work as lamp controls and dimmers, motor controls, and temperature controls.

Along with resistors, diodes, capacitors, and transistors, UJTs act as thyristor triggering devices. UJTs provide a stable set of triggering characteristics. A UJT is a three-terminal semiconductor that has only one junction. The terminals are called the emitter, base 1, and base 2 terminals. UJTs switch from a blocking state to a conducting state because of different biasing arrangements. The intrinsic standoff ratio and peak-point voltage are two important characteristics of a UJT.

Although the programmable unijunction transistor (PUT) somewhat resembles an SCR, it is used as a UJT replacement. When used as a relaxation oscillator circuit, it can provide triggering to such devices as SCRs. By adjusting the anode resistors, the PUT oscillator frequency can be changed.

PROBLEMS

1. Identify the component in Figure 18.53 and label its leads.

FIGURE 18.53

2. A triac allows current to flow in both directions. True or false?

3. State two ways a thyristor may be caused to conduct.
4. What are the two possible states of a thyristor?
5. The minimum current required to cause a thyristor to conduct is called the _____ current.
6. If the thyristor current drops below the _____ current level, the thyristor opens.
7. List three possible functions of an SCR.
8. When an SCR is not conducting, its impedance is (high/low).
9. What is the purpose of the snubber current?
10. List three triggering methods for SCRs.
11. Which of the triggering methods allows for firing angles greater than 90°?
12. What is the purpose of the resistors in Figure 18.54?

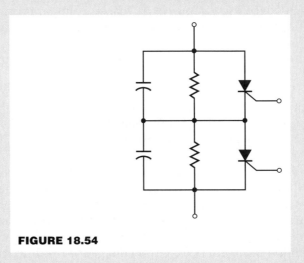

FIGURE 18.54

13. What is the purpose of the capacitors in Figure 18.54?
14. What is the advantage of connecting SCRs in series?
15. Triacs were developed for improved control over DC power. True or false?
16. Triacs are triggered by:
    (a) gate current
    (b) breakover voltage
    (c) either a or b
17. A(n) _____ _____ _____ has an anode gate.
18. A(n) _____ _____ _____ may be triggered by either a positive or negative gate signal.
19. A diac is triggered by _____ voltage.

REGULATORS

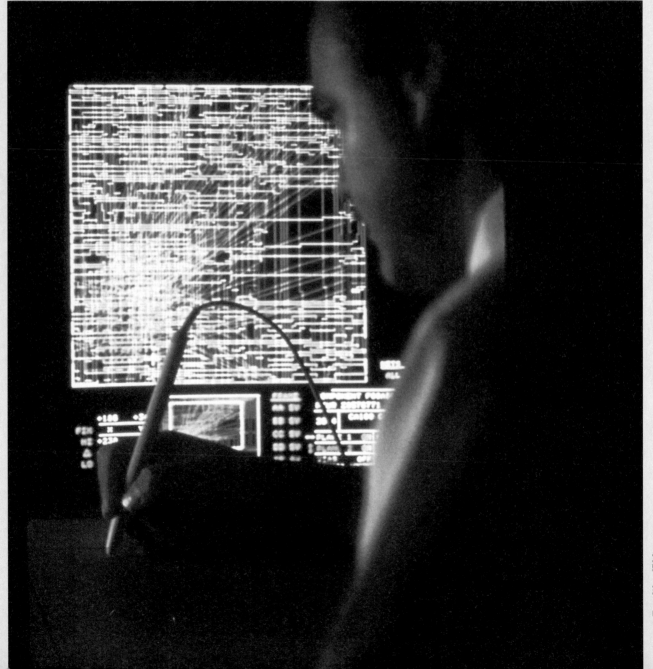

✓ **As you read this chapter, concentrate on learning how to:**

1. Define load regulation
2. Define line regulation
3. Determine factors that affect regulator performance
4. Identify the difference between series pass and shunt regulators
5. Identify the difference between linear and switching regulators
6. Define foldback current limiting
7. Understand current-regulating supplies
8. Determine uses for three-terminal regulators
9. Use regulator applications

INTRODUCTION

Chapter 19 builds on knowledge that you accumulated while studying the earliest chapters of this textbook. Even though you have gained some understanding of regulators, the theory behind regulator circuits and their applications go much further than anything we've studied so far. In this chapter, you will discover how modern power supplies utilize the regulator stage. Regulator circuits provide the constant DC voltages required by electronic circuitry in everything from computers to televisions.

Chapter 19 covers nearly everything that you have learned in this textbook—rectifiers, filters, transistors, differential amplifiers, op-amps, and SCRs—and shows how the components work together in regulated power supplies. In addition, you will study different types of regulators and configurations. As you look at regulator circuits, you will begin to understand why manufacturers choose specific components for different tasks.

LOAD REGULATION

All power supplies provide some quantity of load voltage regulation. If we think of **load regulation** as a ratio, it appears as the no-load voltage measurement minus the full-load voltage measurement divided by the no-load voltage value multiplied by 100, or:

$$\text{load voltage regulation} = \left(\frac{V_{nl} - V_{fl}}{V_{nl}} \right) \times 100$$

where

$$V_{nl} = \text{no-load voltage}$$
$$V_{fl} = \text{full-load voltage}$$

The no-load voltage is the output voltage with an open load, and the full-load voltage is the output voltage with the load demanding the maximum amount of current.

The lower the load voltage regulation value, the better the DC power supply stability. With the following example, we can see how the ratio works.

EXAMPLE 19.1

Determine the load voltage regulation of a regulated power supply.

The no-load output voltage of the power supply is 10 V, and the full-load output voltage is 9.7 V.

Solution

We can find the load regulation of the circuit by substituting the values into the equation. The load regulation is:

$$\text{load regulation} = \left(\frac{V_{nl} - V_{fl}}{V_{nl}} \right) \times 100$$

$$\text{load regulation} = \left(\frac{10\ V - 9.7\ V}{10\ V} \right) \times 100$$

$$\text{load regulation} = \frac{0.3\ V}{10\ V} \times 100$$

$$\text{load regulation} = 3\%,$$

This is the total percent of change occurring in the output voltage.

PRACTICE PROBLEM 19.1

A voltage regulator has a no-load output voltage equaling 15 V and a full-load voltage of 14.5 V. Find the load regulation percentage.

Under realistic operating conditions, load voltage regulation encounters different load conditions. Inductor and transformer resistance, flux leakage, eddy current losses, hysteresis loss, and the forward conduction resistance of rectifier diodes can all affect load voltage regulation, as can the amount of maximum load current that a power supply can provide.

Increasing the maximum load current value increases the full-load voltage regulation percentage. By going through the next example, you may be able to see how an increasing load current can affect regulation.

EXAMPLE 19.2

Compare the regulation percentage of a circuit that has 10 mA of load current with the regulation percentage of a circuit that has 30 mA of load current.

Using Figure 19.1, first consider the condition of minimum load current for the circuit equaling a no-load condition. Let V_S equal the no-load output voltage, and let all the loss factors that can affect load regulation equal Z_{int}. Using those factors, you can determine the load current. For this example, $V_S = 15$ V, $Z_{int} = 100$ Ω, and $R_L = 1.4$ kΩ.

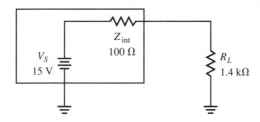

FIGURE 19.1

Solution
The load current equals:

$$I_L = \frac{V_S}{Z_{int} + R_L}$$

$$I_L = \frac{15 \text{ V}}{100 \text{ }\Omega + 1.4 \text{ k}\Omega}$$

$$I_L = \frac{15 \text{ V}}{1.5 \text{ k}\Omega}$$

$$I_L = 10 \text{ mA}$$

With 10 mA of load current, the voltage lost across Z_{int} equals 1 V and the output voltage equals 14 V. The regulation equals:

$$\text{regulation} = \left(\frac{V_{nl} - V_{fl}}{V_{nl}} \right) \times 100$$

$$\text{regulation} = \left(\frac{15 \text{ V} - 14 \text{ V}}{15 \text{ V}} \right) \times 100$$

$$\text{regulation} = 6.67\%$$

If we connect the same power supply to a smaller load resistance of 400 Ω, the load current equals:

$$I_L = \frac{V_S}{Z_{int} + R_L}$$

$$I_L = \frac{15\ V}{100\ \Omega + 400\ \Omega}$$

$$I_L = \frac{15\ V}{0.5\ k\Omega}$$

$$I_L = 30\ mA$$

With 30 mA of load current, the voltage lost across Z_{int} equals 3 V, and the output voltage equals 12 V. The regulation equals:

$$regulation = \left(\frac{V_{nl} - V_{fl}}{V_{nl}}\right) \times 100$$

$$regulation = \left(\frac{15\ V - 12\ V}{15\ V}\right) \times 100$$

$$regulation = 20\%$$

For this power supply, when providing a maximum load current of 10 mA, the regulation equals 6.67%. When providing a maximum load current of 30 mA, the regulation equals 20%.

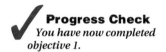
To improve the voltage regulation of a power supply, the power supply circuit will have an impedance value that is as low as practical. Low impedance values are possible because of high-quality power transformers and low-loss filters and rectifiers.

LINE REGULATION

Line regulation shows the change in the output voltage for a 10% change in the AC input voltage. In Figure 19.2, we can look at line regulation as the relationship between the output voltage and the input voltage. An ideal power supply has a line regulation rating of zero. Since we know that ideal power supplies do not exist, we can say that a power supply with a low line regulation rating has good line regulation quality.

To find the percentage of line regulation given by a voltage regulator divide the change in the output voltage by the change in the input voltage and multiply by 100, or:

$$\%\ line\ regulation = \left(\frac{\Delta V_{out}}{\Delta V_{in}}\right) \times 100$$

Normally, changes in the output voltage stay in the microvolt range. Changes in the input voltage occur in volts.

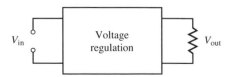

FIGURE 19.2 Line regulation is the relationship between V_{in} and V_{out}.

EXAMPLE 19.3

Determine the line regulation of a regulated power supply. The change in output voltage equals 124 μV, and the change in input voltage equals 5 V.

Solution
To find the line regulation for the circuit, substitute the listed values for change of output and input voltages into the equation. The line regulation for the example power supply equals:

$$\text{line regulation} = \left(\frac{\Delta V_{out}}{\Delta V_{in}} \right) \times 100$$

$$\text{line regulation} = \frac{124\,\mu V}{5\,V} \times 100$$

$$\text{line regulation} = 0.00248\%$$

Since the value given by the equation is very close to zero, the power supply circuit has good line regulation.

PRACTICE PROBLEM 19.2

Determine the line regulation for a power supply. The change in output voltage equals 250 μV, and the change in input voltage equals 2 V.

Progress Check
You have now completed objective 2.

VOLTAGE REGULATORS
In the earlier chapters of this textbook, we studied simple power supplies consisting of transformers, bridge rectifiers, and capacitors. That type of power supply—although still popular—cannot supply the exact DC voltages necessary for many applications. Output voltages from the simple supplies vary with load current and line voltage because of ripple and component factors.

Figure 19.3 shows how regulators fit into the power supply scheme. We can consider voltage regulators as electronic filters. After all, a voltage regulator maintains a constant output voltage despite any changes in the input voltage or the load resistance. Regulation occurs through a feedback circuit that senses a change in the DC output. Because of that effect, voltage regulators rely on a DC reference voltage. Without some type of reference, the percentage of load voltage regulation could change dramatically as the load resistance decreases or increases. The feedback circuit develops a control signal that eliminates the change.

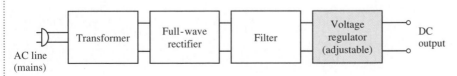

FIGURE 19.3 How regulators fit into the power supply scheme

Circuit configurations determine the amount of control that the feedback network has over regulation. There are two linear voltage regulator configurations—series and shunt. In addition to circuit configurations, the pass element also affects how a regulator operates. We can consider the **pass element** as a device whose resistance is variable. The load current of a voltage regulator "passes" through a pass element.

Both transistors and SCRs (silicon-controlled rectifiers) are used as pass elements. When a pass transistor operates anywhere between saturation and cutoff, a voltage regulator using a pass transistor is a linear voltage regulator. The conduction of the transistor varies linearly as it controls the output voltage. If the pass transistor works only in the cutoff and saturation regions, it is part of a switching regulator. SCRs are also used as switching regulators.

REGULATOR PARAMETERS

Overvoltage Protection for Voltage Regulators

Figure 19.4 shows an operational amplifier working as a high-current regulator. Most important, the figure illustrates an overvoltage protection scheme commonly used in voltage regulator designs and discussed in Chapter 18. At the lower right corner of the schematic, zener diode D_1, SCR Q_2, and a 33-Ω resistor make up an overvoltage crowbar protection circuit.

If the output voltage of the regulator goes past a predetermined limit, the SCR closes, shorting the output. In the circuit, an open resistor in the divider

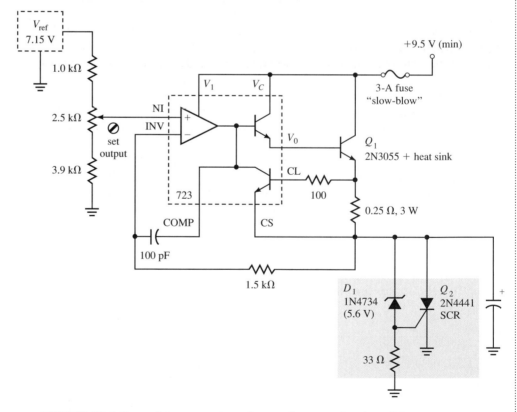

FIGURE 19.4 Overvoltage protection scheme using operational amplifier

or a failure within the voltage regulator integrated circuit could cause the output voltage to rise from $+5$ V to approximately 6.2 V. The increase in output voltage would forward bias the gate-cathode junction of the SCR and allow it to conduct. Q_2 would conduct until the anode current was removed.

As the output voltage goes higher, the zener voltage of D_1 will increase, increasing the gate current flow. This increase forces the regulator into a current-limiting mode, while the SCR holds the output near ground. One possible cause is a collector-to-emitter short in Q_1, which places the supply voltage at the load. With that condition, the SCR fires, sinking a large amount of current, and the 3-A fuse at the top of the circuit opens.

Power Dissipation in Voltage Regulator Circuits
Power transistors, SCRs, and power rectifiers dissipate a large amount of power. To prevent damage from the power handling, all power devices have external cases that tie the device to an external heat sink. Heat sinking keeps the junction of the device below its maximum specified operating temperature. Heat sink design takes the thermal resistance and the maximum specified operating temperature of the device into consideration.

Recovery Time
Another power supply parameter is called recovery time. **Recovery time** shows the amount of time required for either the regulated voltage or current to return to specified limits. The waveforms shown in Figure 19.5 illustrate the recovery time characteristics for regulated DC power supplies.

Recovery time is a function of the frequency response of the power supply feedback network. Voltage-regulated power supplies have a frequency "rolloff" because of high-frequency loop gain of the feedback networks. At high frequencies, the 20 dB/decade rolloff increases the output impedance of the regulator circuit. The impedance becomes inductive and introduces a voltage spike at the output of the regulator.

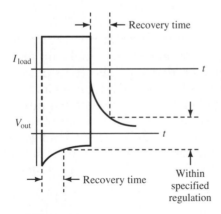

FIGURE 19.5 Waveforms illustrate recovery time characteristics for regulated DC power supplies.

Ripple Rejection Ratio
The quality of voltage regulation often depends on its ripple rejection capability. Voltage regulator specifications usually will list the **ripple re-**

jection ratio. This is a ratio of ripple output voltage to the ripple input voltage.

$$\text{ripple rejection ratio} = \frac{\text{ripple output voltage}}{\text{ripple input voltage}}$$

Drift

Another voltage specification is drift. Over a period of time, the output voltage in a regulated power supply changes from its initial value. **Drift** measures the amount of change of the output voltage against the initial value for a specified time. Since temperature affects drift, we measure the factor after allowing the power supply to warm up. In addition, drift is measured with a constant input voltage and an applied load.

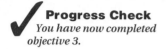

Progress Check
You have now completed objective 3.

REVIEW SECTION 19.1
1. Name some of the problems encountered with output voltages from simple power supplies.
2. What types of configurations exist for linear voltage regulators?
3. How do many voltage regulator designs protect for overvoltage?
4. Describe how heat sinks protect power devices.
5. In equation form, how is the load voltage regulation found?
6. The lower the voltage regulation, the better the DC power supply stability. True or false?
7. How does DC reference voltage affect circuit performance?
8. What is the line regulation a ratio of?
9. Recovery time is the time needed for current to shut off. True or false?

19.2 LINEAR REGULATORS

Two basic types of regulators—series and shunt—make up the group of linear regulators. The type of regulator used depends on the position of the regulating element with respect to the load. In each type of circuit, the regulator acts as a pass element. Located between the unregulated power supply and the output terminal, the regulator must pass the load current. The shunt regulator provides a bypass path for excess current that would go to the load. In both the series and shunt voltage regulator circuits, the output voltage equals the difference between the unregulated voltage and the voltage drop across the pass element.

SERIES PASS REGULATORS

In a series pass regulator, the pass element is connected in series with the load. Figure 19.6 shows a block diagram of a series pass regulator. The unregulated input voltage (V_{UN}) divides between the pass element and the load. As a result, the output voltage (V_O) equals the difference between the unregulated input voltage and the voltage drop across the regulating element.

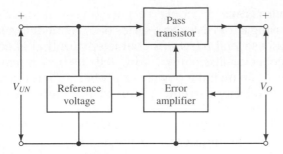

FIGURE 19.6 A series pass regulator circuit

In series pass regulators, the load current passes through the regulating element. Any voltage across the element is proportional to the load current. If the load current increases, the resistance of the pass element must decrease so that the voltage across the element remains constant. If the load current decreases, the resistance of the pass element must increase so that the voltage across the element remains constant.

If the unregulated voltage at the input increases, the regulating element must increase its resistance just enough to hold the current constant. With an increase in the input voltage and a constant current, the voltage drop across the pass element increases by the same amount as the increase of input voltage. Thus, the output voltage does not change.

Figure 19.7 shows a basic series regulator circuit. In the circuit, transistor Q_1 works as a pass transistor. A series pass transistor uses its internal resistance to maintain a constant voltage at the regulator output. If the load requires more current, the increase in voltage drop across the transistor causes the voltage output to drop.

Q_2 functions as an error amplifier. An error amplifier senses that a voltage drop has happened because of a change in the load current. Also, the error amplifier adjusts its output if there is a change in line voltage at the input of the power supply.

The decrease in output voltage causes Q_2 to conduct harder. As a result, more current flows through Q_1. Because of Ohm's law, we know that the increase in current flow means that the internal resistance of the transistor has decreased. Less internal resistance in the pass transistor allows the output voltage to increase back to its normal level. Zener diode ZD_1 provides a reference voltage for transistor Q_2.

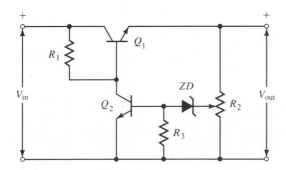

FIGURE 19.7 Series voltage regulator

Transistor Characteristics

When we consider using a transistor as a series pass element in a series regulator, we should study several transistor characteristics that affect regulator performance. The collector-dissipation rating tells us how much power the transistor can safely dissipate if the power supply has a short circuit. Another factor, the maximum collector current, or I_C(max), may limit the total current handled by the regulator.

To maintain circuit stability, the transistor must have low leakage current. In addition, the collector-to-emitter breakdown voltage, or V_{CEO}, limits the maximum output voltage of the power supply. Both the AC and DC current gains set the amount of drive current required at different collector current levels. In addition, the AC current gain sometimes determines the output impedance of the power supply. The total loop gain of the feedback circuit, as well as the current gain of the pass transistor, determines the output impedance. Usually, the loop gain is the dominant factor in the impedance equation.

Current Limiting for Series-Pass Transistors

Current overloads and short circuits can cause the series pass transistor to break down because of excessive power dissipation. If the output terminals of the regulator short together, the total input voltage and all available current become applied to the transistor. Although a fuse or large resistor placed in series with the transistor offers some protection, the best method of limiting current through the series pass transistor is with a current-limiting circuit.

As Figure 19.8 shows, a differential amplifier can work as an error amplifier or current-sensing device while a transistor functions as a current-limiting device. Transistor Q_L is the current limiter. The differential amplifier keeps the output voltage equal to the voltage at the top of potentiometer R_{ADJ}.

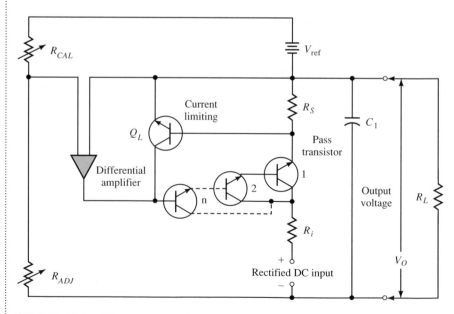

FIGURE 19.8 Differential amplifier working as a current-limiting device

Since differential amplifiers have high input impedances, only a small amount of current flows into the differential amplifier.

Because of that small amount of current, all the current from the reference voltage supply flows through the voltage divider consisting of R_{CAL} and R_{ADJ}. With no current-limiting action, the ratio of reference voltage to the output voltage equals the ratio of the calibration resistance to the adjustment resistance, or:

$$\frac{V_{REF}}{V_O} = \frac{R_{CAL}}{R_{ADJ}}$$

If the reference voltage stays constant, a constant amount of current will flow through the calibration resistor. An equal amount of current will flow through the adjustment resistor. Thus, the output voltage equals:

$$V_O = I_F \times R_{ADJ}$$

If 0.01 A of current flows through R_{ADJ}, then the output voltage is adjusted at a rate of 100 Ω per volt.

Placing a large capacitor in parallel with the output voltage of the power supply adds stability to the entire circuit.

EXAMPLE 19.4

Find the ratio of reference voltage to the output voltage for the circuit shown in Figure 19.8.

The value for R_{ADJ} is 100 kΩ, and the value for R_{CAL} is 47 kΩ.

Solution
The ratio of reference voltage to output voltage for the current-limiting circuit equals the ratio of the calibration resistance value to the adjustment resistance value. If we use the resistance values to find a ratio, we have also found the other ratio. The ratio is:

$$\frac{V_{REF}}{V_O} = \frac{R_{CAL}}{R_{ADJ}} = \frac{47\ k\Omega}{100\ k\Omega} = 0.47$$

PRACTICE PROBLEM 19.3

Determine the output voltage for the circuit if R_{ADJ} equals 100 kΩ, and I_F equals 0.03 A.

Figure 19.9 shows a series pass voltage regulator with a current-limiting circuit. Along with R_1, R_4, D_1, Q_1, Q_2, and Q_4 limit the amount of current applied to the transistor by acting as a large series resistance during overload conditions. Under normal operating conditions, the components appear as a small resistance.

The voltage across resistor R_4 is proportional to the circuit output current. If a current overload condition occurs, the increasing voltage across R_4 and the base-to-emitter voltages of Q_1 and Q_2 will cause diode DR_1 and transistor Q_4 to conduct. When the diode and transistor begin to conduct, Q_4 will shunt

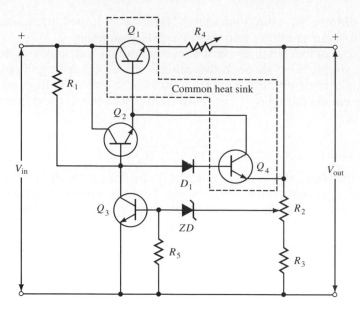

FIGURE 19.9 Series pass voltage regulator with a current-limiting circuit

part of the bias available to the series pass transistor. Consequently, the series resistance of Q_1 increases. Adjusting the value of R_4 changes the limiting value of the circuit output current.

SHUNT REGULATORS

In shunt regulators such as the one depicted in Figure 19.10, the pass transistor parallels or shunts the load circuit. We could look at the pass element as a variable resistance. The regulator has the function of varying its resistance to compensate for any changes in either load conditions or the input voltage so that the output voltage remains constant.

Figure 19.11 shows a simple shunt regulator circuit. Even though shunt regulators do not have the efficiency of series pass regulators, they have a

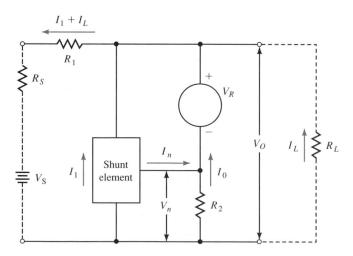

FIGURE 19.10 Shunt regulator circuit

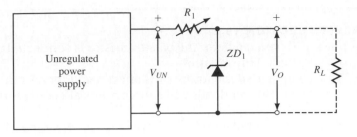

FIGURE 19.11   A simple shunt regulator circuit

much simpler design that includes the shunt and reference voltage elements. V_O represents the output voltage and the difference between the unregulated input voltage V_{UN} and the voltage drop across the pass element. R_1 represents the variable resistance of the voltage regulator.

In Figure 19.11, a voltage-dropping resistor R_1 is connected in series with the parallel network. For load regulation to occur, the pass element must maintain a constant current through R_1 so that voltage across R_1 remains constant. When the load current changes, current flowing through the pass transistor increases or decreases as it attempts to maintain a constant current through R_1.

For line regulation in the circuit, the voltage dropped across R_1 must be able to vary enough so that the output voltage may remain constant. If the unregulated input voltage increases, the pass element must increase the current through R_1 so that the voltage across R_1 increases the same amount as the input voltage increased. Any increase of the unregulated power supply voltage is dropped by the voltage-dropping resistor R_1 and does not appear at the output of the circuit.

Inside the Shunt Regulator
In Figure 19.12 the two common-emitter transistors function as the shunt pass element and are in parallel with the load. Transistors selected for the pass element must operate within the maximum voltage, current, and dissipated power limits set by the reference resistor and within the output voltage and load current specifications of the circuit.

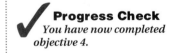

Progress Check
You have now completed objective 4.

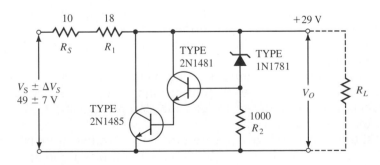

FIGURE 19.12   A shunt regulator

1. In a series pass regulator the pass transistor is connected in parallel with the load. True or false?
2. What characteristics of transistors affect regulator performance?
3. How do current overloads and short circuits affect the series pass transistor?
4. When a differential amp is used as a current-sensing device, what is the ratio of the reference voltage to the output voltage?
5. When a differential amp is used as a current-sensing device, more stability occurs when a large capacitor is added in parallel with the output voltage. True or false?
6. Shunt regulators have more efficiency than series pass regulators. True or false?
7. List the factors that input requirements, load conditions, and output voltage requirements depend on in a shunt regulator.

19.3 SWITCHING REGULATORS

Series and shunt regulators are called linear regulators because the pass elements conduct continuously over their linear operating ranges. Therefore, the pass element in a linear regulator dissipates power continuously. Because of the continuous power dissipation, the regulator circuit has a decreased efficiency. To make up for the larger power dissipation requirements, many linear regulator circuits have a number of pass elements working in parallel.

Because of power dissipation and efficiency requirements, conventional series pass regulators usually are not used in high-current power supplies. Compensating for the power dissipation and efficiency factors drives the cost of the circuit up. In those cases, another type of regulator called the switching regulator performs better.

With a switching regulator, a pass transistor switches rapidly from saturation to cutoff and transfers energy from the unregulated power supply to the load circuit. The switching speed ranges from 5 to 50 kHz. Switching action is controlled by the amplitude of the output voltage and a switching circuit. Often, an oscillator will function as a switching driver.

Figure 19.13 shows a basic switching regulator configuration with the pass transistor connected in series with the load. An output switching circuit controls on-off switching of the pass transistor through feedback. Feedback from the output sensing circuit determines the width of each pulse of the oscillator switching circuit. The oscillator switching circuit sets the on time of the switching transistor.

Comparison of the output voltage with a reference voltage occurs as the feedback circuit samples the output signal. Any difference between the two voltages makes up the error signal that controls the duration of the pass transistor on-off duty cycle.

If the output voltage decreases, the output voltage from the sensing circuit causes the switching circuit to increase the width of the driving pulse at the switching transistor base. Because of the longer pulse, the pass transistor is

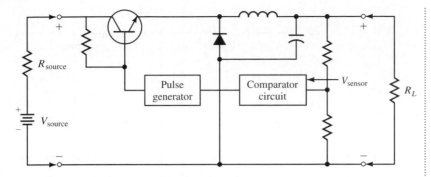

FIGURE 19.13   Switching regulator configuration

saturated for a longer period of time and the output voltage increases back to the reference level. When the output voltage rises above the reference voltage level, the on-time pulse length decreases. Then the pass transistor is saturated for a shorter time period, and the output voltage decreases back to the reference level.

A low-pass LC filter and a diode smooth the square wave output signal. High-frequency switching is easier to filter than low-frequency switching. As Figure 19.14 shows, on time increases under the valley points of the unregulated supply and decreases under the peaks. This removes 60-Hz ripple and leaves only the easily filtered high-frequency ripple. With the DC output signal smoothed, the regulator compares the output signal with a reference voltage. Unlike conventional voltage regulators that modify the base drive current, switching regulators modify the duty cycle of the switching waveform.

Since the pass transistor either saturates or is cut off and the inductors in the circuit dissipate only small amounts of power, switching regulators have a lower power dissipation and much better efficiency than linear regulators. Depending on the conduction time, switching regulators may have efficiencies of 90% or better. Linear voltage regulators typically have efficiencies in the 25 to 50% range. Thus, the switching regulator design yields a high-current, regulated power supply with high efficiency.

Switching regulators have several properties that make them valuable for computer applications. A switching regulator that has an output voltage lower than the unregulated input voltage delivers more current to the load than is seen at the unregulated input. Effectively, the unregulated input

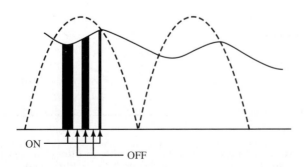

FIGURE 19.14   **High-frequency switching filtering**

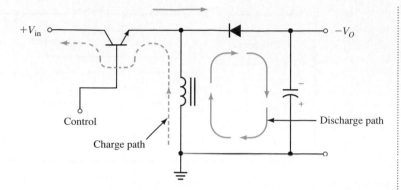

FIGURE 19.15   Simple switching regulator working as an inverter

power supply voltage can be much larger than the regulated output voltage. Computer designs use one regulated power supply to provide all the regulated voltages.

Many digital circuits require a stable DC negative voltage for proper operation. Switching regulators can provide output voltages that have a polarity opposite to that of the unregulated input voltage. Figure 19.15 shows a simple switching regulator working as an inverter. Current flows from the more negative output and through the diode.

When the transistor is switched on, the inductor charges, causing a positive voltage to be applied to the cathode of the diode. The diode reverse-biases, separating the inductor from the output circuit. Inductors try to keep current flowing in the same direction. When the transistor is switched off, the inductor causes current to flow in the same direction by reversing polarity, which forward-biases the diode. The capacitor is recharged by the inductor. Once the inductor's voltage is more positive than the capacitor's voltage, the diode reverse-biases once again.

SWITCHING REGULATOR CONFIGURATIONS

Generally, a switching regulator transistor will have two possible configurations. Figure 19.16a shows a configuration that works well under most conditions. Q_1 has the filter elements and load impedance in its collector circuit.

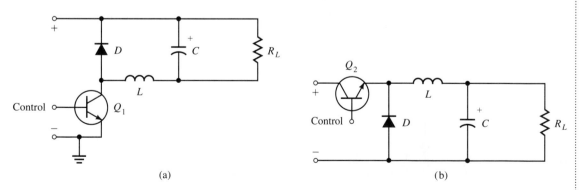

          (a)   (b)

FIGURE 19.16   **(a) Switching regulator transistor with the filter elements and load impedance in the collector circuit. (b) Switching regulator transistor with filter elements and impedance in the emitter circuit.**

With the collector of the driver stage connected to the positive side of the supply, the collector-to-emitter voltage of Q_1 can go lower than its base-to-emitter voltage. However, the current in the driver flows through the load, and any power associated with the driver current becomes lost.

Although the circuit shown in Figure 19.16b has the disadvantage of not having a V_{CE} that can go lower than its V_{BE}, it works better for circuits that can utilize the higher collector-to-emitter voltage. Since the filter elements and load impedance are in the emitter circuit, the base of Q_2 cannot connect to a point more positive than the positive voltage of the power supply. Both circuits work well with isolated power supplies.

A PULSE-WIDTH MODULATED STEP-DOWN CONVERTER

Some types of switching regulator circuits switch the pass element at the AC line frequency. The conduction angle of the element is varied to obtain the desired pulse width. Generally, this type of regulator works with SCRs because the thyristors have simple turn-on schemes. The SCR turns off automatically with the reversal of the line voltage.

Figure 19.17 shows the schematic diagram of a step-down switching DC-regulated power supply that uses an SCR as the pass element. The circuit provides approximately 125 V regulated at ±3% for both load and line regulation. Also, the circuit has a ripple of less than 0.5% RMS.

With a power supply that functions as a half-wave phase-controlled rectifier, the 5-μF capacitor between the cathode and gate of the SCR charges up during half of each cycle. The firing of the SCR discharges the capacitor. Charging current flowing into the capacitor increases or decreases the SCR firing angle.

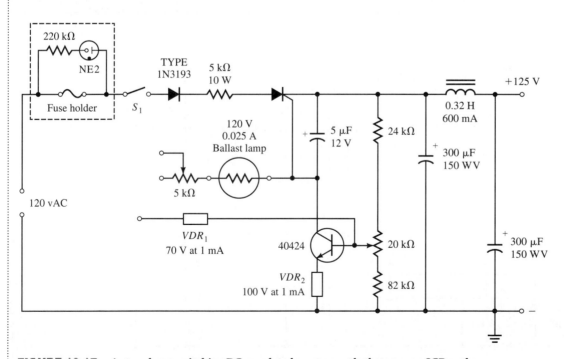

FIGURE 19.17   A step-down switching DC-regulated power supply that uses an SCR as the pass element

The 40424 control transistor shunts part of the current that normally charges the capacitor. As the current flowing through the control transistor increases, current shunts around the capacitor through the ballast lamp. This increases the charging time of the capacitor, decreases the firing angle of the SCR, and results in a lower output voltage.

Deriving the controlling voltage on the control transistor from both the DC output and the line voltage provides load and line regulation. A voltage-dependent resistor VDR$_2$ in the base circuit of the transistor decreases resistance for an increase in line voltage. The decrease in resistance increases the base and collector currents of the control transistor.

Changes in the DC output voltage that result from variations in the load current are fed back to the base of the control transistor. A voltage divider at the input to the filter circuit provides the feedback path. In addition, the voltage divider adjusts the collector current in a direction that compensates for changes in the DC output voltage.

Figure 19.18 shows another type of pulse-width modulated step-down converter. A capacitor and resistor network connected to pins 5 and 6 of the regulator provide the RC time constant for the sawtooth oscillator. The oscillator determines a frequency rate that interrupts the DC load current.

We can find the oscillation frequency by dividing 1.1 by the product of the resistor and capacitor values, or:

$$f_{osc} = \frac{1.1}{R_T \times C_T}$$

Thus the oscillator frequency of the circuit equals:

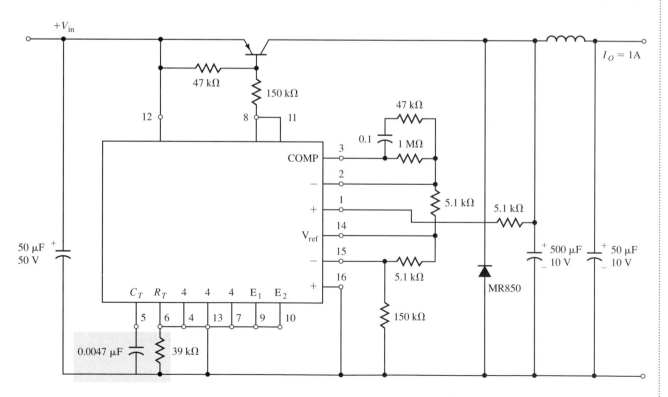

FIGURE 19.18   **A pulse-width modulated step-down converter**

$$f_{osc} = \frac{1.1}{47\ k\Omega \times 0.001\ \mu F} = 23.4\ kHz$$

With the DC load current chopped at a rate of 23.4 kHz, the duty cycle of the switching regulator is controlled by a changing duty cycle. At the top of the schematic, an inductor and filter capacitor average the DC level and help to remove the switching frequency and any harmonics.

EXAMPLE 19.5

Find the oscillation frequency of the circuit shown in Figure 19.18.

Use resistance and capacitance values of 39 kΩ and 0.0047 μF for this example.

Solution
To find the oscillation frequency, substitute the resistance and capacitance values into the following equation:

$$f_{osc} = \frac{1.1}{R_T \times C_T}$$

$$f_{osc} = \frac{1.1}{39\ k\Omega \times 0.0047\ \mu F}$$

$$f_{osc} = 6\ kHz$$

PRACTICE PROBLEM 19.4

Find the oscillation frequency if R_T has a value of 100 Ω and C_T has a value of 0.01 μF.

Progress Check
You have now completed objective 5.

REVIEW SECTION 19.3
1. Why don't conventional series pass regulators work for high-current power supplies?
2. What controls the duration of the pass transistor on-off duty cycle?
3. When does a pass transistor's conduction time increase? When does it decrease?
4. Switching regulators are different from conventional regulators in that they modify the duty cycle. True or false?
5. What characteristic do switching regulators have that makes them valuable for computer applications?
6. How can you choose the correct filter for a switching regulator?
7. Give the equation to find the energy stored by an inductor.
8. Name the transistor parameters influencing a switching regulator.
9. The fall and rise times of a transistor have no effect on its performance as a switch. True or false?
10. When does a switching regulator configured as a pulse-width modulated step-down converter switch?
11. How is the oscillation frequency found in a pulse-width modulated step-down converter?

If the output of a regulator shorts to ground, the excessive current flowing through the device can destroy it almost instantly. Foldback current limiting is a form of protection against excessive current that shuts the regulator off if an overload or short circuit occurs.

When the load impedance drops to a value that causes the pass element to draw more than its designed amount of maximum current, foldback current limiting reduces the output voltage and current. A further decrease of the load impedance causes a further decrease of output voltage and current. The voltage-current characteristics of a foldback current-limiting circuit are shown in Figure 19.19.

Foldback current limiting also works in the other direction. If the load impedance value increases during the limiting mode of the regulator circuit, the output voltage and current increase. Once the current reaches the switching threshold level for the regulator, the regulator turns on, and the power supply goes back to its normal operation.

Figure 19.20 shows a foldback current-limiting circuit used with a series voltage regulator. Since resistor R_5 has a value that results in zero bias for Q_5

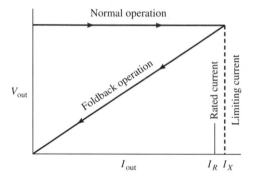

FIGURE 19.19   **Voltage-current characteristics of a foldback current-limiting circuit**

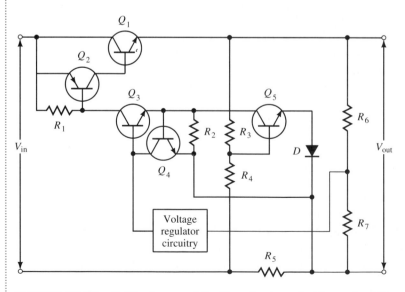

FIGURE 19.20   Foldback current-limiting circuit used with a series voltage regulator

when the output current reaches its rated value, or I_R, low output current sends transistor Q_5 into cutoff. Q_5 conducts when the load current or I_{out} reaches its limiting value designated as I_X.

With Q_5 conducting, current flows through R_2 and Q_4 turns on. The conduction of Q_4 reduces the base-to-emitter voltages of Q_2 and Q_3 along with the power supply output voltage. Since the output voltage decreases, the output current also decreases. Q_5 continues to conduct at the same emitter current. Reducing the load impedance even more causes Q_5 to conduct harder. As a result, the output voltage and current decrease again.

Analysis of the Foldback Current-Limiting Circuit

We can use a loop equation for the Q_5 base-emitter circuit to analyze the foldback current-limiting circuit shown in Figure 19.20. The equation is:

$$I_{out}R_5 = V_D + V_{BE} + V_{R4}$$

V_D equals the voltage drop across the diode, and V_{BE} equals the base-to-emitter voltage of Q_5. At the rated current, or I_R, V_{BE} should equal zero. If the rated current exists, we can rewrite the equation as:

$$I_R R_5 = V_D + V_{R4}$$

Just before foldback starts, the limiting current exists in the circuit. At the limiting current, the equation becomes:

$$I_X R_5 = V_D + V_{BE} + V_{R4}$$

If a short circuit occurs, the output voltage, or V_{out}, and the voltage across R_4, or V_{R4}, equal zero. The short circuit current, or I_{SC}, flows through the circuit. Now the equation appears as:

$$I_{SC} R_5 = V_D + V_{BE}$$

Using the equations, we can find the ratio of limiting current to the rated current, or:

$$\frac{I_X}{I_R} = \frac{V_D + V_{BE} + V_{R4}}{V_D + V_{R4}}$$

and the ratio of short circuit current to the rated current, or:

$$\frac{I_{SC}}{I_R} = \frac{V_D + V_{BE}}{V_D + V_{R4}}$$

For the limiting current to have a value close to the value of the rated current the following relationships must be true:

$$V_D + V_{BE} + V_{R4} = V_D + V_{R4}$$
$$(V_D + V_{R4}) > V_{BE}$$

EXAMPLE 19.6

Analyze the foldback current-limiting circuit shown in Figure 19.20. Find the limiting current and the shorted current.

R_5 has a value of 1 kΩ. The voltage drop across the diode, or V_D, is 0.7 V. The base-to-emitter voltage of the pass element transistor is also 0.7 V. The voltage dropped across R_4 is 1.3 V.

Solution

Substituting the listed values into the equations will give us the limiting current and short circuit current values. The limiting current has a value of:

$$I_X R_5 = V_D + V_{BE} + V_{R4}$$
$$I_X \times 1\text{ k}\Omega = 0.7\text{ V} + 0.7\text{ V} + 1.3\text{ V}$$
$$I_X \times 1\text{ k}\Omega = 2.7\text{ V}$$

$$I_X = \frac{2.7\text{ V}}{1\text{ k}\Omega}$$
$$I_X = 2.7\text{ mA}$$

The short circuit current value is:

$$I_{SC} R_5 = V_D + V_{BE}$$
$$I_{SC} \times 1\text{ k}\Omega = 0.7\text{ V} + 0.7\text{ V}$$
$$I_{SC} \times 1\text{ k}\Omega = 1.4\text{ V}$$

$$I_{SC} = \frac{1.4\text{ V}}{1\text{ k}\Omega}$$
$$I_{SC} = 1.4\text{ mA}$$

PRACTICE PROBLEM 19.5

Find the ratio of limiting current to the rated current and the ratio of the short circuit current to the rated current for the values presented in the last example.

If an op-amp were substituted for Q_5, circuit performance would improve because op-amps have greater sensitivity. Figure 19.21 shows an op-amp as part of the foldback current-limiting circuitry. The advantage of using an op-amp is that both the limiting current to rated current ratio and the shorted current to rated current ratio would be greatly improved. The limiting cur-

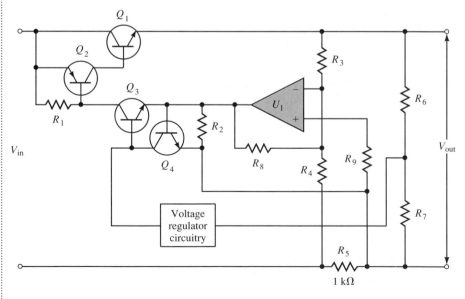

FIGURE 19.21   Using a differential amplifier in a foldback current-limiting circuit

rent to rated current ratio would be very close to 1, while the shorted current to rated current ratio would almost be zero. The actual values would depend on design.

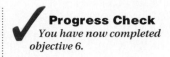

REVIEW SECTION 19.4
1. What happens in the transistor if the output of a regulator shorts to ground?
2. Foldback current limiting is a better answer to power dissipation than large heat sinks and high power rated transistors. True or false?
3. Describe the effects of foldback current limiting.
4. Foldback current limiting is a form of _____ against excessive current that _____ the regulator off if an overload or short circuit occurs.
5. What advantage does a differential amplifier offer a foldback current-limiting circuit?
6. If the load impedance value increases during the limiting mode of the regulator circuit, the output voltage and current _____ .

19.5 REGULATING CURRENT

CURRENT-REGULATING POWER SUPPLIES

Some applications, such as personal computers, require that the load current remain constant as the load resistance changes; that is, they need high-current regulator circuits. When you study the specifications for a voltage regulator, you also may see a specification for current regulation.

Current regulators, like the simple circuit shown in Figure 19.22, control load current by monitoring the voltage drop across a resistor in series with the load. We can find the amount of load current by dividing the unregulated voltage by the sum of the two resistances or:

$$I_L = \frac{V_{UN}}{R_1 + R_L}$$

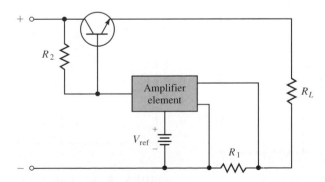

FIGURE 19.22 **Simple current regulator circuit**

While a voltage regulator controls current to maintain a constant output voltage, a current regulator automatically adjusts its output voltage so that the output current remains constant. Consequently, the supply delivers a varying output voltage and a constant flow of current. Also, current-regulating power supplies exhibit nearly an infinite output impedance.

Find the value of the load current for the circuit shown in Figure 19.22.

The unregulated input voltage is 10 V. Both resistor R_1 and the load resistor R_L have values of 50 Ω.

Solution

The load current for the circuit equals:

$$I_L = \frac{V_{UN}}{R_1 + R_L}$$

$$I_L = \frac{10 \text{ V}}{50 \ \Omega + 50 \ \Omega}$$

$$I_L = \frac{10 \text{ V}}{100 \ \Omega}$$

$$I_L = 0.1 \text{ A, or } 100 \text{ mA}$$

Using the circuit and the values from example 19.7, change the value of the load resistance from 50 to 60 Ω while maintaining the same constant load current. Does the resistance of R_1 need to decrease or increase to maintain 100 mA of current at the output of the circuit?

Figure 19.23 shows a series regulator modified to produce current regulation, and Figure 19.24 shows the output waveform from a current regulator. Since the voltage across the resistor in Figure 19.23 changes in direct propor-

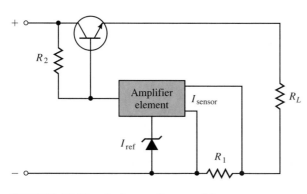

FIGURE 19.23 Series regulator working as a current regulator

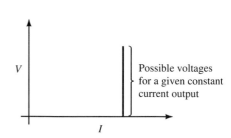

FIGURE 19.24 Output response characteristic of a constant current regulator

tion to the changes in the load current, an error signal can be used to cancel any detected change in load current from a set value. If a circuit design requires a constant, controlled 2-A current flow, any increase in load current would be sensed by R_1 as an increase in voltage. The amplifier element would modify the bias on the pass transistor so that less current would flow to the load resistor. A decrease in load current would be sensed by R_1 as a decrease in voltage, which would be used by the amplifier element to cause the pass transistor to pass additional current to the load. The effect is a constant current being supplied to the load.

VOLTAGE-REGULATING, CURRENT-REGULATING POWER SUPPLIES

Generally, power supplies do not feature only current regulators. One type of regulator combines the characteristics of current and voltage regulators. Figure 19.25 is a block diagram of a voltage-regulating, current-regulating power supply. A voltage-regulating, current-regulating power supply provides either voltage regulation or current regulation at its output. The type of regulation depends on load conditions, with the supply automatically switching from current regulation to voltage regulation at a given operating point. Controls at the output set the operating point.

After rectification of the input AC voltage, the supply filters the rectified voltage and applies it to an optional regulator stage called the pre-regulator. The pre-regulator stage utilizes switching regulators and adds high efficiency to the circuit. In the diagram, the output of the pre-regulator transfers to the series pass regulator element. The series pass element provides a fast response time for the entire circuit.

After comparing a sample of the output voltage with a reference voltage, the series pass element generates an error signal proportional to the difference between the two voltages. The error signal is amplified and applied to the base of the pass transistor to correct changes in the output voltage. Because of the role that the reference voltage plays in the control circuit, the output voltage of the circuit depends on the accuracy of the reference voltage. Usually, a temperature-compensated zener diode in series with a constant current source provides a steady reference voltage.

The regulator has a sensitivity that is the inverse function of the gain of the drive amplifier. If the comparator senses a small variation, the amplifier has

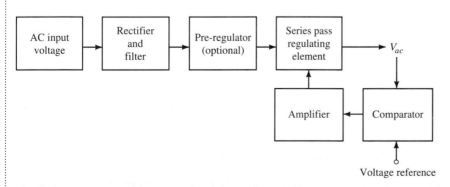

FIGURE 19.25   **A voltage-regulating, current-regulating power supply**

a higher required gain. However, the higher gain of the amplifier reduces the stability of the circuit.

Figure 19.26 shows a functional diagram of the pass transistor, driver amplifier, and the voltage and current error amplifiers. Two error amplifiers use their output signals to control a selection circuit that controls the driver amplifier. In turn, the driver amplifier controls the pass element. With the selection circuit, only the error amplifier with the largest output produces the output signals for controlling the pass transistor. Thus, the pass transistor regulates only one mode at a time.

Adjusting the reference inputs of the error amplifiers sets the desired operating levels for current and voltage control. Also, this sets the maximum output voltage, or V_O, and the load current, or I_L, that the circuit can provide. In addition, the settings establish a crossover point for the circuit to switch between the constant voltage and constant current modes of regulation. The driver amplifier steps up the output voltage to a level high enough to drive the pass transistor.

The voltage-current response of the circuit is shown in Figure 19.27. In the figure, the reference current, or I_L, equals 0.5 A, and the reference voltage, or V_O, equals 50 V. With the controls set, the regulator depends on the load resistance connected across its output terminals.

If we use Ohm's law, we can determine the crossover resistance. Dividing the reference voltage by the reference current gives us 100 Ω as a crossover resistance value. When the load resistance is greater than the crossover resistance, the regulator operates in the voltage regulation mode. This occurs because the sensed value of the load current is below the reference current level. Thus, the output of the current error amplifier remains small and blocked by the selection circuit. Since the voltage error amplifier has a larger value, it controls the pass transistor.

If the load resistance is less than the crossover resistance, the current error amplifier has the larger output. With that condition, the current error ampli-

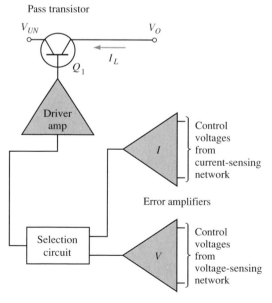

FIGURE 19.26 Functional diagram of the pass transistor and sensing circuits

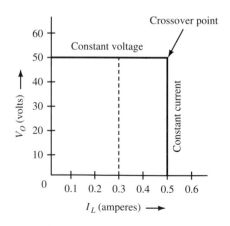

FIGURE 19.27 Characteristic curve of a voltage-regulating, current-regulating power supply

fier takes control of the regulator. Therefore, the current-regulating action holds the load current constant.

PRACTICE PROBLEM 19.7

The crossover point depends on the reference control settings. Using Figure 19.27, find the crossover resistance if the current level is 0.3 A and the voltage level is 50 V. When does the supply regulate voltage? When does the supply regulate current?

Progress Check
You have now completed objective 7.

> **REVIEW SECTION 19.5**
> 1. Draw the waveform produced from a current regulator.
> 2. How do current regulators control the load current?
> 3. A current regulator adjusts the output voltage to maintain a constant output current. True or false?
> 4. Describe the function of a current-voltage regulator.
> 5. Reference voltage plays a very small part in circuit operation of a current-voltage regulator. True or false?
> 6. The _____ gain of the amplifier reduces the stability of the voltage-regulating, current-regulating power supply.
> 7. Some applications, such as personal computers, require that the load _____ remain constant as the load _____ changes.
> 8. Figure 19.24 shows a shunt regulator modified to produce current regulation. True or false?
> 9. What does the driver amplifier contribute to a voltage-regulating, current-regulating supply?
> 10. Describe the crossover operation of a voltage-regulating, current-regulating regulator.

19.6 INTEGRATED CIRCUIT VOLTAGE REGULATORS

VOLTAGE REGULATORS FOR PRECISION POWER SUPPLIES
Figure 19.28 shows a basic integrated circuit voltage regulator. The regulator features temperature compensation for the voltage reference, a differential amplifier, a series pass transistor, and an overcurrent protective circuit.

AN INTEGRATED CIRCUIT SERIES VOLTAGE REGULATOR
Figure 19.29 shows a series voltage regulator built around an integrated circuit regulator. The RCA hybrid integrated circuit works as the voltage regulator. The CA3085A integrated circuit is rated for a maximum unregulated voltage of 40 V (terminal 3) and a feedback voltage to the inverting input

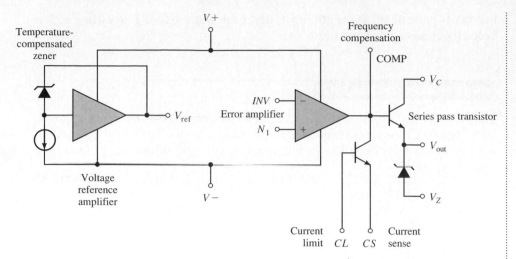

FIGURE 19.28 An integrated circuit voltage regulator

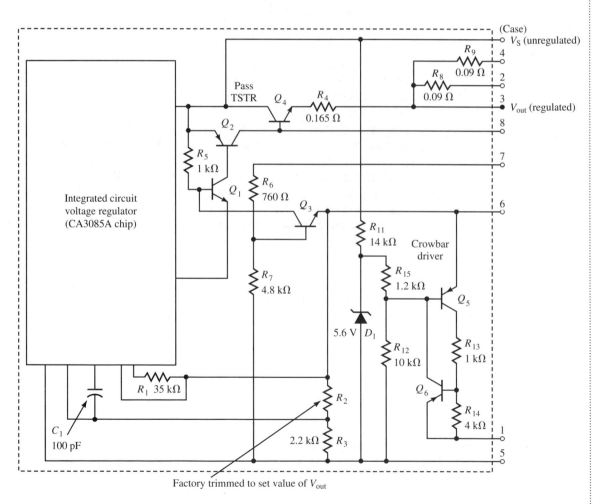

FIGURE 19.29 Series voltage regulator integrated circuit

(terminal 6) of 1.8 V. The regulator has a range of 2 to 32 V. As with many standard-design regulators, the CA3085A has an output voltage regulated to within $\pm 1\%$ for typical line voltage, load current, and temperature conditions.

An external booster circuit consisting of the PNP Q_2 and NPN Q_4 works as the pass element for the regulator circuit. The two-stage output circuit increases the load current capability of the regulator circuit to 4 A.

Voltage divider resistors R_2 and R_3 set the level of the output voltage by having their junction directly coupled to the inverting input of the voltage regulator IC. For the rated output, the voltage divider divides the output voltage so that 1.6 V is applied to the inverting input. Any change in the output voltage causes the voltage at the inverting input to change. With that change at its inverting input, the voltage regulator circuit develops an output that cancels the output voltage change.

As an example of regulator operation, an increase of load resistance causes the output voltage to increase. The increasing output voltage is applied to the inverting input voltage of the voltage regulator terminal 6. This causes terminal 2 to increase its output voltage, which is applied to the emitter of Q_1. Q_1 acts like a common-base amplifier. The increased voltage applied to Q_1 causes the collector voltage to increase. That increased voltage is applied to the base of Q_2. Because Q_2 is acting as a common-emitter amplifier, its collector output voltage decreases. This reduces the load current and causes the output voltage to return to its original value.

THREE-TERMINAL REGULATORS

Many electronic applications do not require extremely close-tolerance voltage regulation circuits. For those applications, a simple three-terminal voltage regulator such as the one pictured in Figure 19.30 may fit the bill. The popularity of the three-terminal regulator stems from its simplicity, low cost,

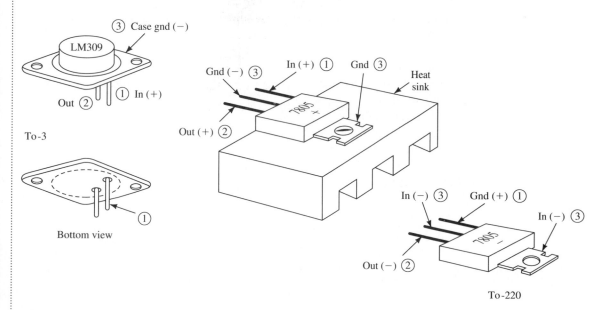

FIGURE 19.30   A three-terminal voltage regulator

packaging, high reliability, and performance. Most three-terminal regulators arrive in the same TO-3 or TO-220 style packages used by power transistors.

All three-terminal regulators stay within a general set of characteristics that allow the regulator to work with a variety of applications. Table 19.1 lists the characteristics and common values of an LM117 three-terminal adjustable regulator. Most three-terminal regulators will have a similar set of characteristics.

In Figure 19.31, the 7805 three-terminal regulator has only three connections—input, output, and ground. Its output is a fixed 5 V. The 7800 series of three-terminal regulators can provide up to 1 A of load current.

The figure also shows the ease of circuit construction when using a three-terminal regulator. A capacitor connected across the output keeps the circuit impedance low at high frequencies by removing high-frequency oscillations.

Other circuitry within the device keeps the regulator from operating outside the safe operating limits of the regulator. Instead of the regulator becoming damaged during overheating or overloading, the circuitry shuts the regulator down. That internal circuitry reduces output current when it senses a large difference between the input and output voltages.

Figure 19.32 provides a functional diagram of three-terminal voltage regulator operation. A voltage at the input of the regulator is slightly higher than its specified DC output voltage. The regulator reduces the input voltage to a

Table 19.1 Characteristics and Common Values of an LM117 Three-Terminal Adjustable Regulator

Characteristic	Common value
Package	TO-3
Rated power dissipation	20 W
Design load current	1.5 A
Input-output voltage differential	40 V
Operating junction temperature range	$-55°C$ to $+150°C$
Reference voltage	1.20(min) 1.25(typical) 1.30(max)
$3 <= (V_{in} - V_{out}) <= 40$ V	
Minimum input current	3.5 A(typical) 5 A(max)
$V_{in} - V_{out} = 40$ V	
Ripple rejection ratio	65%
$V_{out} = 10$ V, $f = 120$ Hz, $C_{adj} = 10$ μF	
Load regulation	0.3%(typical) 0.8%(max)
Over maximum load change	
Temperature stability	1%

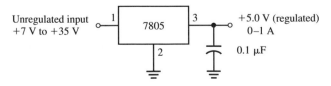

FIGURE 19.31 A three-terminal regulator

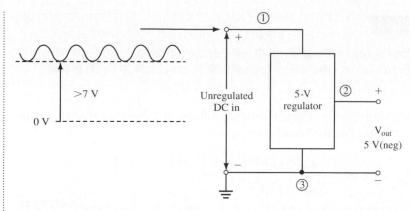

FIGURE 19.32   Theory behind three-terminal voltage regulator operation

steady DC output voltage. Depending on the manufacturer's specifications for the regulator, the input voltage has a set upper limit.

ADJUSTABLE THREE-TERMINAL REGULATORS

To improve three-terminal regulator flexibility, manufacturers have replaced the ground terminal of the fixed three-terminal regulator with an adjustment terminal. Figure 19.33 shows an LM117 adjustable three-terminal regulator. The LM117 can supply approximately 1.5 A of current over a 1.2- to 37-V output range. Since the LM117 requires only two external resistors to set the output, it is easy to use.

In Figure 19.33, the adjustment terminal connects to an external potentiometer and allows the adjustment of the output voltage. With the adjustment, a constant difference is maintained between the output terminal and the adjustment terminal.

During circuit operation, the regulator applies the 1.25-V reference voltage across resistor R_1. This allows a constant 5.2 mA of current to flow through the output set resistor R_2. With the adjustment terminal drawing a minimum amount of current, the output voltage equals:

$$V_{out} = 1.25 \text{ V} \left(\frac{R_2}{R_1} + 1 \right)$$

Using the values shown in the circuit, we can determine that the output voltage has an adjustable range of 1.25 to 27 V. If the circuit design requires a fixed output voltage, R_2 adjusts over only a narrow range.

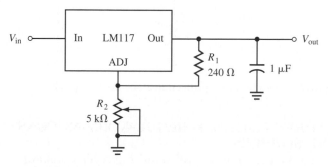

FIGURE 19.33   An adjustable three-terminal regulator

EXAMPLE 19.8

Determine the output voltage for a three-terminal regulator.

For this example, use the following values: R_2 is set at 470 Ω; R_1 is 1 kΩ.

Solution
Using the equation for finding the output voltage of a three-terminal regulator and substituting known values, we have:

$$V_{out} = 1.25 \text{ V}\left[\left(\frac{470\,\Omega}{1\,\text{k}\Omega}\right)+1\right]$$
$$V_{out} = 1.84 \text{ V}$$

As shown in Figure 19.33, a 1-μF tantalum capacitor may be used with the adjustable three-terminal regulator as an output bypass capacitor. Using an output bypass capacitor decreases problems that the regulator may have with the amplification of ripple. Also, the adjustment terminal can be bypassed to ground to improve ripple rejection.

Although the LM117 has good stability without output capacitors, certain values of external capacitances can cause ringing in the feedback circuit. A 1-μF solid tantalum capacitor on the output swamps ringing. By swamping the ringing, the output capacitor ensures good circuit stability.

In Figure 19.34 there are also two protection diodes. As long as input and output voltage are within tolerance range, the diodes are reversed-biased. If, however, the load shorted, D_2 would be forward-biased. C_2 would discharge through D_2 rather than through the regulator. If the input shorted, then D_1 would be forward-biased, providing a discharge path for C_1 rather than allowing C_1 to discharge through the regulator.

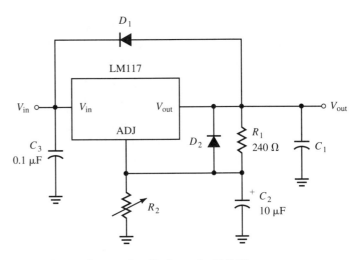

FIGURE 19.34   The addition of protection diodes to the LM117

USING THREE-TERMINAL REGULATORS AS CONSTANT CURRENT SOURCES

Often, electronic instruments will require flexible constant current supplies. Figure 19.35 shows an LM317 three-terminal regulator configured as a constant current source. In the figure, a resistor connects from the output of the regulator to the common terminal. The load connects from the common terminal to ground.

During operation, the three-terminal regulator maintains a fixed voltage and a constant current across the resistor, the constant current through the load. At low output currents, the circuit pictured in Figure 19.35 allows a higher-than-allowable margin of error into the output current. The few milliamps of operating current from the regulator increases the amount of current flowing through the load. At high output currents, the voltage drop across the resistor results in higher power dissipation.

To compensate for those problems, the circuit shown in Figure 19.36 adds an operational amplifier follower to the circuit. The follower connects to the adjust input and the output terminals of the three-terminal regulator. With that addition, the amount of output current of the regulator becomes limited to the 1.25-V reference voltage divided by the value of the resistor.

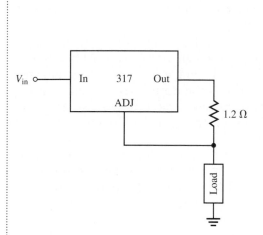

FIGURE 19.35 LM317 three-terminal regulator configured as a constant current source

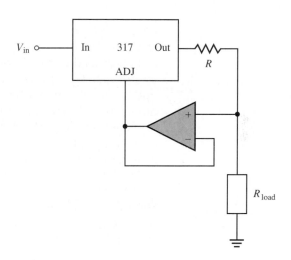

FIGURE 19.36 Adding an operational amplifier to the three-terminal constant current source

USING THREE-TERMINAL REGULATORS AS HIGH-CURRENT SOURCES

Adding an external pass transistor to a three-terminal regulator allows the regulator to output higher-value currents. Figure 19.37 shows a three-termi-

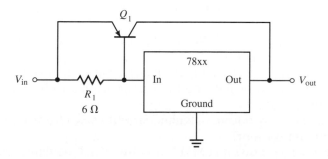

FIGURE 19.37 A three-terminal regulator configured with an external pass transistor

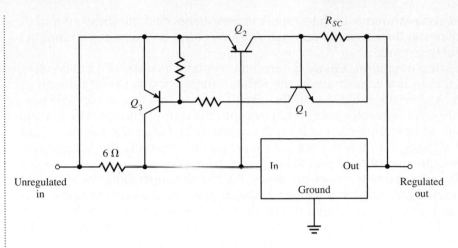

FIGURE 19.38   Another three-terminal current limiter

nal regulator configured with an external pass transistor. For load currents of less than 100 mA, the regulator operates conventionally.

Larger load currents, however, develop a voltage across R_1 that causes Q_1 to conduct. Q_1 limits the current through the regulator to approximately 100 mA. Although the output stays at the correct voltage—since the regulator reduces input current and the current to Q_1 if the output voltage increases—the load can draw more than 100 mA of current. The input voltage must exceed the output voltage by the sum of the dropout voltage and the base-to-emitter drop of the transistor.

Figure 19.38 shows another three-terminal current-limiting configuration. In the circuit, Q_2 functions as a high-current pass transistor and turns on at 100 mA of load current. Since the base of Q_1 connects through the collector of Q_2 to a voltage divider, the circuit features foldback current limiting. This lowers the regulator dropout voltage and cuts power dissipation.

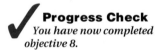

Progress Check
You have now completed objective 8.

REVIEW SECTION 19.6

1. What is in an RCA hybrid integrated circuit that allows it to regulate voltage so well?
2. In an RCA hybrid IC, what happens if an overvoltage of 105 to 125% occurs?
3. List the general characteristics of all three-terminal regulators. .
4. What functions does the capacitor connected across the output serve?
5. A three-terminal regulator could be thought of as a constant-value resistor. True or false?
6. How does a three-terminal regulator act as a variable resistor?
7. Give the equation of the output voltage from an adjustable three-terminal regulator.
8. What allows an output of higher current values from a three-terminal regulator?
9. Figure 19.35 shows an LM317 three-terminal regulator configured as a constant _____ source.

10. Why should some integrated circuit voltage regulators have protection diodes added to the circuit?
11. Why are three-terminal regulators so popular?
12. What advantage does an adjustable three-terminal regulator have when compared with a fixed three-terminal regulator?

19.7 VOLTAGE REGULATOR APPLICATIONS

AUTOMATIC SHUTDOWN FOR A REGULATOR

Figure 19.39 shows a circuit designed to protect a voltage regulator from a short circuit. The regulator supplies a regulated output voltage of +10 V. If the regulator short circuits, transistor Q_2 turns off. This causes the voltage at pin 2 of the regulator to rise and shuts down the regulator. After the short circuit has been diagnosed and removed, the circuit will restart with the push of a switch.

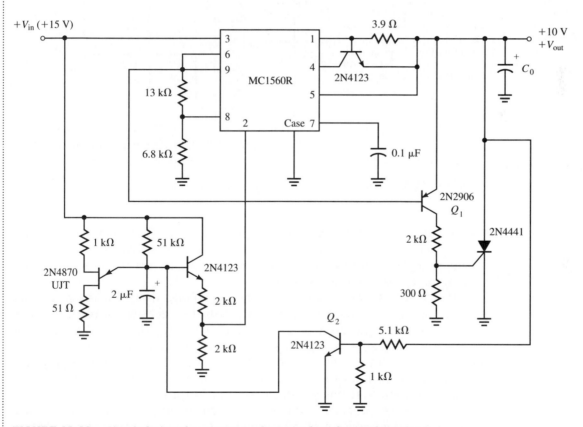

FIGURE 19.39   Circuit designed to protect a voltage regulator from a short circuit

SWITCHING POWER SUPPLY FOR A PERSONAL COMPUTER

Figure 19.40 shows a block diagram for a personal computer switching power supply. The power supply contains a full-wave AC-to-DC rectifier, an oscillator circuit, an AC-to-AC transformer, another AC-to-DC rectifier, filters, and

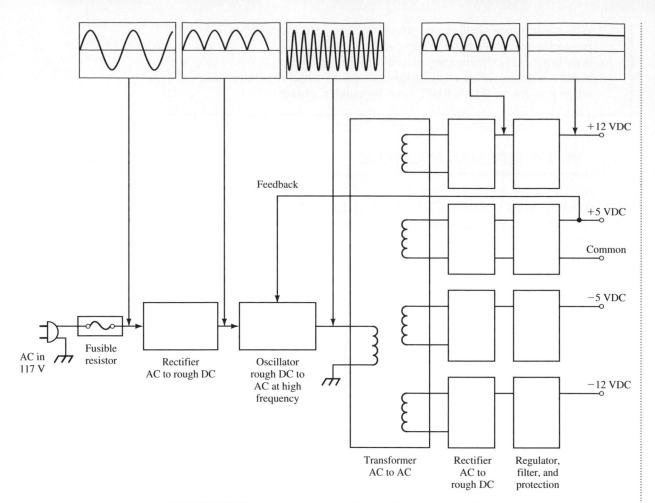

FIGURE 19.40 **A personal computer switching power supply**

a regulator. An unfiltered DC output from the first rectifier becomes "chopped" and converted back to high-frequency AC by the oscillator circuit. Chopping the DC output at a high frequency makes the power supply more resistant to 120-VAC electrical noise.

The AC-to-AC transformer steps the AC voltage down for the secondary taps. As Figure 19.40 shows, the second AC-to-DC rectifier converts the lower AC voltages into pulsating positive and negative DC voltages. After rectification and filtering, a voltage regulator smoothes the DC voltages into the clean, precise levels needed for computer circuitry. The +5-V supply also provides feedback for the oscillator circuit. If the +5-V supply goes beyond its specified limits, the oscillator corrects the output.

Since the computer circuitry requires equal positive and negative voltages, a dual-tracking regulator provides the power supply output voltages. Figure 19.41 shows a dual-tracking regulator. In the figure, Q_1 functions as a pass transistor for a positive regulated power supply. The positive output becomes a reference voltage for the negative regulated power supply.

Q_2, an error amplifier, controls the negative output voltage by comparing the two output voltages to ground. In this case, the average of the two voltages becomes the ± 15-V outputs. The rectifier diodes connected in the re-

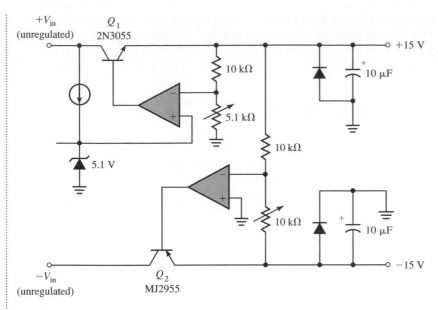

FIGURE 19.41   Dual-tracking regulator

verse direction at the output terminals protect the circuit against accidental voltage reversals.

Outputs from the regulator tie to points throughout the computer. In most computers, the ±12-V and ground connections connect to floppy and hard disk drives or CD-ROM readers. Along with the drives, most integrated circuits in the computer require a ±5-V supply and a ground connection.

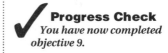

Progress Check
You have now completed objective 9.

TROUBLESHOOTING

Troubleshooting the power supply regulator is difficult because of the feedback control. The feedback control must remain active when you are troubleshooting the circuit, otherwise, there is no regulation. When troubleshooting power supply regulators, use basic troubleshooting techniques. First, make a visual inspection. Often, damage will be observable because of the power available in a power supply. Also, fuses and crowbars are designed to shut down equipment when a power surge is present. Sometimes replacing a blown fuse will fix the problem. If not, then disconnect the load and observe the power supply's operation. WARNING! Some power supplies are not supposed to be operated without a load. Check the manufacturer's literature before powering up a power supply without a load. Now, take voltage readings, looking for open or shorted active devices. If you change a component thinking that that is the problem but the problem remains unchanged, do not keep the component. Once a component is unsoldered from the circuit, heat damage may make the component unreliable.

The best advance in regulator troubleshooting has been the introduction of the IC regulator. If the output voltage is not regulated, replace the regulator.

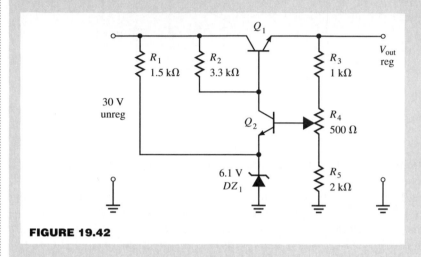
S U M M A R Y

All power supplies provide some quantity of load voltage regulation. The equation for finding the percentage of load voltage regulation is:

$$\% \text{ load voltage regulation} = \left(\frac{V_{nl} - V_{fl}}{V_{nl}} \right) \times 100$$

A low load voltage regulation value results in better DC power supply stability.

Line regulation is the relationship between the output voltage and the input voltage. An ideal power supply has a line regulation rating of zero. The equation for finding the percentage of line regulation given by a voltage regulator is:

$$\% \text{ line regulation} = \frac{\Delta V_{out}}{\Delta V_{in}} \times 100$$

A voltage regulator maintains a constant output voltage despite any changes in the input voltage or the load resistance. Regulation occurs through a feedback circuit that senses a change in the DC output. Circuit configurations and the pass element determine how a regulator operates.

Both transistors and SCRs work as pass elements. When a pass transistor operates anywhere between saturation and cutoff, a voltage regulator is a linear voltage regulator. The conduction of the transistor varies either linearly or as a switch as it controls the output voltage. If the pass transistor works only in the cutoff and saturation regions, it is part of a switching regulator. SCRs are switching regulators.

Two basic types of regulators—series and shunt—make up the group of linear regulators. The type of regulator depends on the position of the regulating element with respect to the load. In each type of circuit, the regulator acts as a pass element. Located between the unregulated power supply and the output terminal, the regulator must pass the load current.

Because of power dissipation and efficiency requirements, conventional series pass regulators usually are not used in high-current power supplies. Power dissipation and efficiency factors drive the cost of the circuit up. In those cases, the switching regulators perform better. With a switching regulator, a pass transistor switches rapidly from saturation to cutoff and transfers energy from the unregulated power supply to the load circuit. Switching action is controlled by the amplitude of the output voltage and a switching circuit.

Foldback current limiting is a form of protection against excessive current that shuts the regulator off if an overload or short circuit occurs.

Some circuit requirements require that the load current remain constant as the load resistance changes. Current regulators control load current by monitoring the voltage drop across a resistor in series with the load. Voltage-regulating, current-regulating power supplies combine the characteristics of current and voltage regulators.

Many electronic applications do not require extremely close-tolerance voltage regulation circuits. For those applications, a simple three-terminal voltage regulator works well. The popularity of the three-terminal regulator stems from its simplicity, low cost, packaging, high reliability, and performance. All three-terminal regulators stay within a general set of characteristics that allow the regulator to work with a variety of applications.

PROBLEMS

1. If the no-load output voltage is 20 V, and the full-load output voltage is 19 V, what is the load regulation?
2. The following input voltages and output voltages were recorded. What is the line regulation?

	V_{in}	V_{out}
record 1	20 V	15 V
record 2	18 V	14.9 V

3. In the regulator circuit in Figure 19.43, the SCR fires when the gate voltage is at 2.1 V. What is the maximum regulated output voltage?

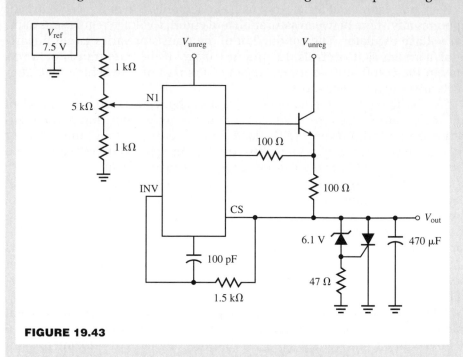

FIGURE 19.43

4. Calculate the ripple rejection ratio of Figure 19.44.

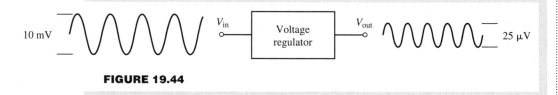

FIGURE 19.44

5. Refer to Figure 19.8. Find the output voltage if the reference voltage is 6.1 V, R_{CAL} is 1 kΩ, and R_{ADJ} is 1.96 kΩ.
6. If the unregulated voltage in Figure 19.45 rose to 16 V, how much current would the zener diode have to provide to maintain load voltage at 12 V?

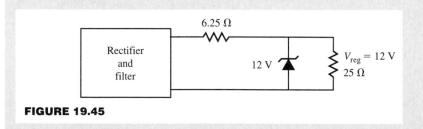

FIGURE 19.45

7. The 7812 in Figure 19.46 has an output voltage of 12 V. If its output tolerance is ±1%, what are the maximum and minimum acceptable output voltages?

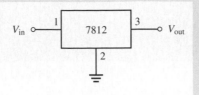

FIGURE 19.46

8. What is the percent of load regulation of the 7812 of Figure 19.46 if the change in output voltage is 24 mV going from no load to full load?
9. Calculate the ripple rejection ratio of the 7812 if the input ripple is 2 V and the output ripple is 200 μV.
10. Calculate the V_{out} of Figure 19.47 when $R_1 = 240 \ \Omega$ and $R_2 = 1 \ k\Omega$.

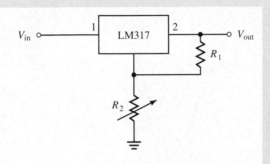

FIGURE 19.47

20

COMMUNICATIONS CONCEPTS

 As you read this chapter, concentrate on learning how to:

1. Determine the differences between amplitude and frequency modulation
2. Discuss frequency oscillator operation
3. List the differences between Hartley, Colpitts, crystal, and phase-shift oscillators
4. Show the differences between AM and FM detection
5. List the components of a phase-locked loop (PLL)
6. Describe how a PLL works
7. Describe how a tunnel diode works
8. Describe how a varactor works
9. List diodes used in microwave generation
10. Describe the steps in systems troubleshooting

INTRODUCTION

Semiconductors give us the ability to manipulate current flows. That manipulation can have meaning to us. In this chapter we will look at how semiconductors are used in communications. Some semiconductor circuits, which are called oscillators, are used to convert DC power into high-frequency AC power. Other circuits are designed to place an audio signal, such as your favorite radio station, onto a high-frequency carrier in such a way that it can be sent around the world. Other semiconductors help us tune into our favorite radio station, take the audio off the high-frequency carrier, and amplify the signal to a level that will drive our speakers.

One of the newer semiconductor devices in communications is the phase-locked loop. We will examine its operation and see how it can help us hear the world.

FREQUENCY SPECTRUM

Frequency can mean many different things. A technician working with audio systems may think of audio frequencies as falling between 15 Hz and 20,000 Hz. Another technician working with radio-frequency equipment may see frequency as something falling between 10 MHz and 30,000 MHz. In reality, frequencies are divided into many different subgroups. Table 20.1 lists the designations for the many different groups of frequency ranges that make up the frequency spectrum.

Table 20.1 Designations of Frequency Ranges

Frequency Range	Designation
30 to 300 Hz	ELF, extremely low frequency
300 to 3000 Hz	VF, voice frequency
3 to 30 kHz	VLF, very low frequency
30 to 300 kHz	LF, low frequency
300 to 3000 kHz	MF, medium frequency
3 to 30 MHz	HF, high frequency
30 to 300 MHz	VHF, very high frequency
300 to 3000 MHz	UHF, ultra high frequency
3 to 30 GHz	SHF, super high frequency
30 to 300 GHz	EHF, extremely high frequency

CONCEPT OF MODULATION

When we transmit frequency signals, we send those signals on a channel such as a cable or fiber-optic link. As we send the signals, we usually modulate the desired signal to a radio-frequency "carrier" instead of directly transmitting the frequency signal. We can define **modulation** as the process of adding information to a carrier. An example of a carrier is a radio-frequency carrier that allows us to send information from one place to another without wires.

Several modulation methods exist. Of those methods, we will discuss amplitude and frequency modulation. With all modulation methods, the modu-

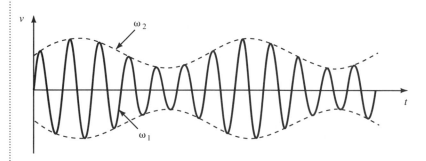

FIGURE 20.1 Modulated waveform

lated signal occupies a bandwidth least comparable to the bandwidth of the modulating signal. For example, a high-fidelity audio transmission will occupy the 20-Hz to 20-kHz portion of the spectrum regardless of the carrier frequency. A perfect unmodulated carrier has zero bandwidth and contains no information. Figure 20.1 shows a modulated waveform. In the figure, the high-frequency carrier ω_1 is modulated at the low-frequency ω_2.

AMPLITUDE MODULATION (AM)

Amplitude modulation (AM) is the simplest form of modulation. An example is the AM broadcast band. AM radio stations operate between 535 and 1605 kHz. With amplitude modulation, a simple carrier wave has its instantaneous amplitude changed by a modulating signal of a much lower frequency. The amplitude of the carrier waveform varies while its frequency remains constant. Figure 20.2a shows a high-frequency carrier waveform, and Figure 20.2b shows the modulating audio signal.

A brief study of the simple circuits shown in Figure 20.3 may increase your understanding of amplitude modulation. If we applied the carrier and modulating signals to the resistor network shown in Figure 20.3a, the output would resemble the waveform shown in Figure 20.4. With that output, the carrier is not amplitude-modulated. Instead, the carrier and modulating signals have simply been added together. The high-frequency waveform rides on the low-frequency varying amplitude.

Changing the amplitude of the carrier requires the addition of a nonlinear element into the circuit. Figure 20.3b illustrates this change. By adding a diode to the resistor circuit, we can produce an amplitude-modulated wave-

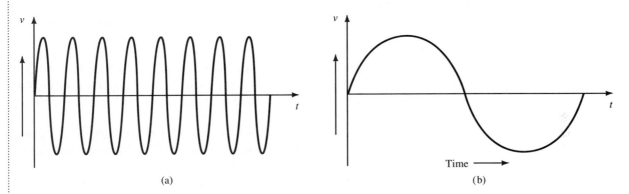

FIGURE 20.2 (a) Carrier waveform. (b) Modulating signal.

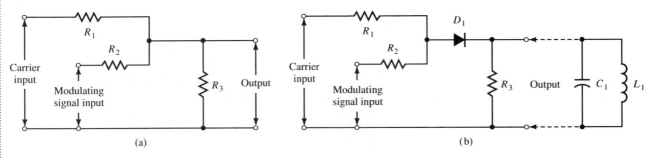

FIGURE 20.3 (a) Resistor network. (b) Adding a diode to the resistor network.

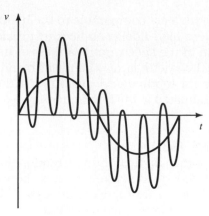

FIGURE 20.4 Unmodulated carrier waveform

form. The nonlinear properties of the diode produce an output that is not always exactly proportional to the input.

Voltage applied to diode D_1 is a combination of the carrier and modulating signals similar to the output shown in Figure 20.4. Since the diode conducts current in one direction only, current flows through D_1 and R_3 only when the anode of D_1 is more positive than its cathode.

Figure 20.5 shows the resulting current after applying the combination of the carrier and modulating signals to D_1. When the applied voltage is negative, the current will not have the same waveshape as the applied voltage. The amplitude of the current waveshape follows the modulating signal waveshape. Since the current flows through resistor R_3, the output voltage develops a shape similar to the current wave.

Figure 20.3b also features the addition of a coil and capacitor. The combination of L_1 and C_1 is resonant to the carrier frequency and modifies the shape of the output voltage wave. With the oscillating action of the LC network, the circuit produces an upside-down copy of the positive half of the cycle for the negative half of the cycle. If the inductor and capacitor have the proper values, the negative half will have the same amplitude as the positive half.

While Figure 20.4 shows an unmodulated waveform, Figure 20.6 shows the resulting output voltage and a waveshape normally associated with amplitude modulation. The amplitude of the carrier follows the shape of the modulating signal. Figure 20.7 shows the original carrier (a), modulating signal (b), and the result of amplitude modulation (c). Note the dashed lines

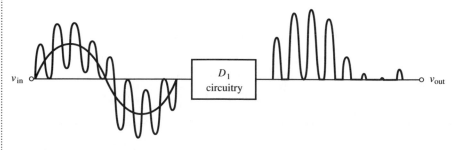

FIGURE 20.5 Result of applying carrier and modulating signals to D_1

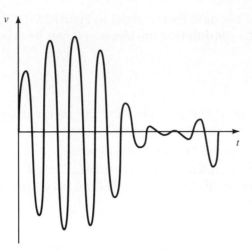

FIGURE 20.6 Amplitude-modulated waveform

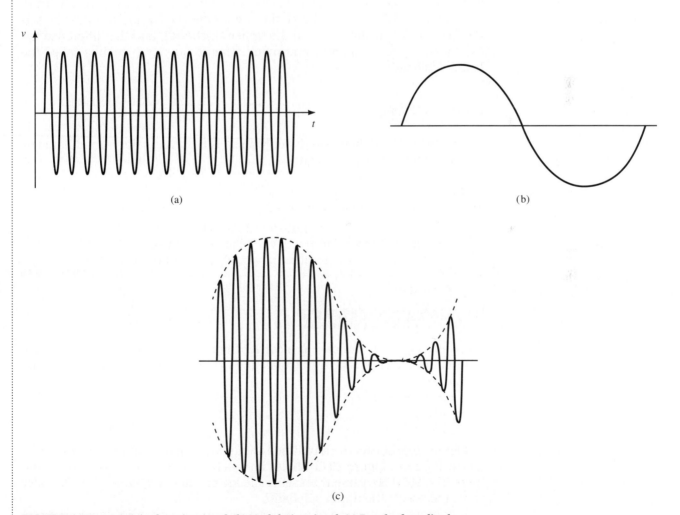

(a)

(b)

(c)

FIGURE 20.7 (a) Original carrier signal. (b) Modulating signal. (c) Result of amplitude modulation.

that follow the peaks of the modulated carrier in Figure 20.7c. The pattern of dotted lines is called the **modulation envelope** and can be seen with an oscilloscope.

AM SIDEBANDS

If we represent the carrier signal as $\cos \omega_c t$ and the modulating signal as $\cos \omega_m t$, the amplitude-modulated signal appears as:

$$\text{signal} = (1 + m \cos \omega_m t)\cos \omega_c t$$

where m is a modulation index that is less than or equal to 1. Amplitude modulation requires the biasing of the modulating waveform so that it will not have a negative value.

If we expand the product, the equation looks like this:

$$\text{signal} = \cos \omega_c t + \left[\frac{1}{2} m \cos \left(\omega_c + \omega_m\right)t\right] + \left[\frac{1}{2} \cos \left(\omega_c - \omega_m\right)t\right]$$

This lengthy equation tells us that the modulated carrier has power at the carrier frequency represented by ω_c and at frequencies on either side ω_m away. It also shows that the waveshape shown in Figure 20.7c results from the addition of the carrier signal, the sum of the carrier and modulating signal frequencies (which is called the **upper sideband**), and the difference between the carrier frequency and modulating frequency (which is called the **lower sideband**).

EXAMPLE 20.1

Calculate the frequencies produced when a 1000-kHz carrier is modulated by a 1-kHz audio modulating signal.

Solution
The resulting frequencies would be:

Carrier frequency = 1000 kHz
Upper sideband frequency = 1000 kHz + 1 kHz = 1001 kHz
Lower sideband frequency = 1000 kHz − 1 kHz = 999 kHz

PRACTICE PROBLEM 20.1

A certain AM radio station has a carrier frequency of 790 kHz. Its maximum modulating signal is 5 kHz. Calculate the maximum upper and lower carrier frequencies.

Figure 20.8a shows the amplitude-modulated signal that would be seen on an oscilloscope. Figure 20.8b shows its spectrum as seen on a spectrum analyzer. The 999-kHz value represents the lower sideband, while the 1001-kHz value represents the upper sideband.

Amplitude modulation of a carrier frequency by an audio signal produces four frequencies: (1) the carrier frequency, (2) the upper sideband frequency,

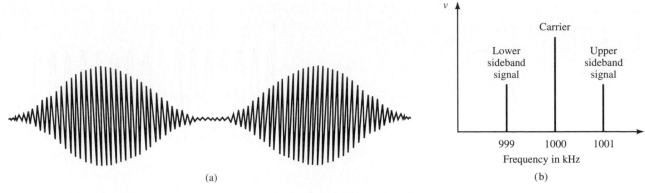

FIGURE 20.8 (a) Amplitude-modulated signal. (b) Amplitude-modulated signal spectrum.

(3) the lower sideband frequency, and (4) the modulating audio frequency. At the output, the modulating frequency is an undesired frequency. Therefore, the resonant LC circuit connected across R_3 in Figure 20.3b is tuned to pass the carrier and sidebands while attenuating the amount of audio signal.

AM GENERATION

Any technique that allows the control of the signal amplitude with a voltage in a linear manner will generate an amplitude-modulated radio-frequency (R-F) signal. Common methods include varying the R-F amplifier supply voltage while performing modulation at the output stage or using a multiplier circuit. Modulation performed at a low-level stage requires linear amplifier stages after the modulation stage.

FREQUENCY MODULATION (FM)

Frequency modulation involves modulating the frequency, rather than the amplitude, of a carrier. The FM waveform has a constant amplitude while the frequency varies. Frequencies in the FM broadcast band range from 88 through 108 MHz. Frequency modulation has the advantage of being able to reject noise better than AM and can carry more information because of its wider bandwidth. This provides a better quality sound than AM receivers.

Figure 20.9 illustrates frequency modulation. Figure 20.9a shows a 1000-Hz modulating signal, and Figure 20.9b shows a 1-MHz radio-frequency carrier. Figure 20.9c shows the result of frequency modulation in the modulated R-F carrier.

As the modulating signal increases to its maximum positive value, the carrier frequency increases. When the modulating signal drops to zero, the carrier frequency returns to its original value. As the modulating signal increases to its maximum negative value, the carrier frequency decreases.

As the modulating signal swings positive and negative, the carrier frequency deviates above and below its original value. We call the amount of carrier frequency change the **frequency** or **carrier deviation.** The rate of carrier deviation depends on the frequency of the modulated signal. An unmodulated carrier frequency is called the **center frequency.**

Carrier deviation is proportional to the amplitude of the modulating signal, with the maximum carrier deviation occurring at the peaks of the modu-

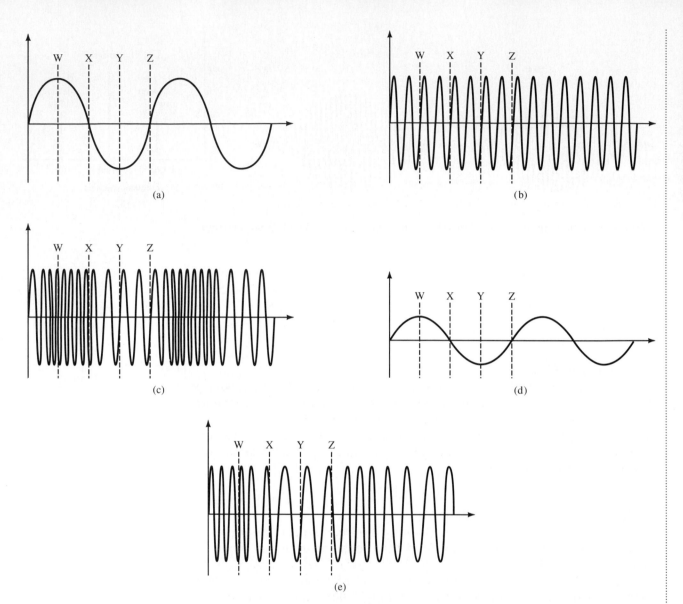

FIGURE 20.9 (a) 1000-Hz modulating signal. (b) 1-MHz R-F carrier. (c) Result of frequency modulation in the modulated R-F carrier. (d) Modulating signal with half the amplitude of signal seen in Figure 20.9a. (e) Frequency-modulated carrier that results from the signal seen in Figure 20.9d.

lating signal. Figure 20.9d shows a modulating signal with half the amplitude seen in Figure 20.9a. Figure 20.9e shows the frequency-modulated R-F carrier that results from the modulating signal of Figure 20.9d.

EXAMPLE 20.2

Using Figure 20.9a, find the carrier frequencies for points W and Y of the frequency-modulated R-F carrier shown in Figure 20.9e.

For this example, the carrier deviation in Figure 20.9a is ±0.1 MHz. With a center frequency of 1 MHz, the carrier frequency is 1.1 MHz at point W of Figure 20.9a. For the same figure, the carrier frequency at point Y is 0.9 MHz.

Solution

Since the modulating signal in Figure 20.9d is reduced by half, the carrier deviation seen in Figure 20.9e also reduces by one-half. One-half of 0.1 MHz is 0.05 MHz. Therefore, the carrier frequency at point W of Figure 20.9e is 1.05 MHz, and the carrier frequency at point Y is 0.95 MHz.

PRACTICE PROBLEM 20.2

Using Figure 20.9, calculate the carrier frequency at W and Y points, when the center frequency is 10.7 MHz and the deviation is ± 75 kHz.

FM SIDEBANDS

Frequency modulation of a carrier produces a large number of sideband signals. A carrier frequency-modulated by a sine wave has a spectrum similar to the one shown in Figure 20.10. As the figure shows, numerous sidebands are spaced at multiples of the modulating frequency from the carrier. Sideband signals far from the carrier frequency have smaller amplitudes than sideband signals close to the carrier frequency.

Progress Check
You have now completed objective 1.

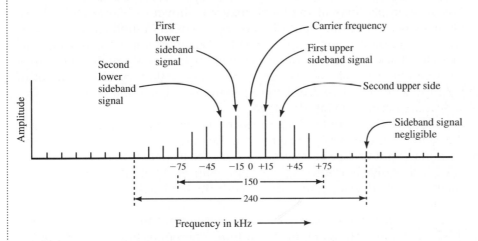

FIGURE 20.10   **FM spectrum**

REVIEW SECTION 20.1

1. List the frequency ranges and designations for the frequency spectrum.
2. Define modulation.
3. Describe amplitude modulation.
4. List the four frequencies produced by amplitude modulation.
5. Describe frequency modulation.
6. Define carrier deviation.
7. What is the spacing of FM sideband frequencies?

20.2 OSCILLATORS, MULTIPLIERS, AND MODULATORS

Throughout this text, we have briefly touched on oscillators. Any circuit that converts direct current to alternating current can be called an oscillator. Indeed, the astable multivibrators that we saw in Chapter 17 functioned as oscillators. Since astable multivibrators generate square waves, we can label them as nonsinusoidal oscillators. Any oscillator that generates a wave other than a sine wave, such as a square or a sawtooth wave, is a nonsinusoidal oscillator.

If the oscillator generates a sine wave, we define it as a sinusoidal oscillator. Sinusoidal oscillators work in circuits that test the performance of other circuits and generate radio frequencies in both receivers and transmitters.

Among the group of sinusoidal oscillators are crystal oscillators, Hartley oscillators, Colpitts oscillators, phase-shift oscillators, and voltage-controlled oscillators. The following section covers all but the voltage-controlled oscillators. When we move to the study of phase-locked loops, we will look at that type of sinusoidal oscillator.

The Hartley, Colpitts, and phase-shift oscillators are examples of feedback oscillators. Each type relies on having a portion of the output of an amplifier fed back to its input. When the feedback loop achieves unity gain, the feedback produces oscillating, or alternating, voltages or currents. Those oscillations occur at a specific frequency controlled by the feedback circuit.

For an example, look at the block diagram shown in Figure 20.11. If we apply a signal, V_{in}, to the input of a common-emitter amplifier, a reversed-polarity, larger output voltage, V_{out}, results. When we apply the output voltage to a feedback circuit, the components of the feedback circuit may change the amplitude of the signal. We can safely assume that the arrangement of the components in the feedback loop generates a signal that is equal in amplitude and phase to the input voltage.

Since the feedback voltage V_F is the same as the input voltage in amplitude and polarity, it is the opposite polarity of the output voltage. If we take away the original input voltage and replace it with the feedback voltage, the amplified feedback voltage produces the output voltage. Again, the output voltage feeds into the feedback circuit, which feeds the feedback voltage back to the input circuit.

Essentially, the circuit uses part of its own output for an input and then amplifies the input to maintain the output. Therefore, the circuit produces

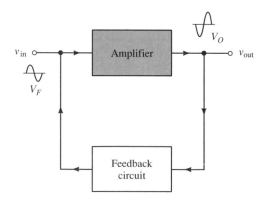

FIGURE 20.11   An oscillator

an alternating, or oscillating, output without the benefit of an external signal. Feedback oscillators have two basic requirements, which are called the **Barkenhausen criteria:**

1. The feedback circuit must provide a signal of proper phase to the input.

2. The feedback circuit must provide a signal of proper amplitude to the input.

OSCILLATOR CIRCUITS USING FEEDBACK

We will discuss three types of oscillators that use feedback: the Hartley, Colpitts, and phase-shift oscillators.

Hartley Oscillator

Figure 20.12 shows a tuned or resonant Hartley oscillator built around a bipolar junction transistor. Here, the tap on coil L_1 allows the coil to act as two inductors. Inductor L_1 and capacitor C_1 set the frequency of oscillation. The tapped winding distinguishes the Hartley oscillator from all other oscillator types. Mathematically, the frequency of oscillation for a Hartley oscillator appears as:

$$ f = \frac{1}{2\pi\sqrt{C_1 L_1}} $$

Resistors R_1 and R_2 supply forward bias to the base-emitter junction of transistor Q_1 to start the oscillator. With DC power applied to the circuit, the initial surge of current through the lower part of L_1 causes an oscillating current to set up in the $L_1 C_1$ tank circuit. Also, the collector of the transistor amplifier couples to the input side of the circuit by capacitor C_1. The additional pulses from the collector maintain the oscillating current in the tank circuit and an alternating voltage across coil L_1 and capacitor C_1.

Voltage developed across the lower part of L_1 is applied to the input of the circuit or the base of Q_1 as an alternating voltage. This alternating voltage

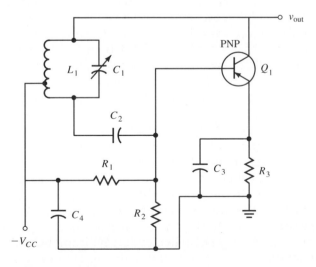

FIGURE 20.12 BJT Hartley oscillator

adds to and subtracts from the bias on the transistor base and causes a continuously pulsating collector current. As before, the pulsating collector current supplies energy to the tank and allows the circuit to continue oscillating. Capacitor C_2 provides DC isolation between the tank circuit and the base of Q_1. Thus, the tank and Q_1 can operate at different DC potentials while remaining at the same AC potential.

EXAMPLE 20.3

Determine the frequency of oscillation for a Hartley oscillator.
Coil L_1 has a value of 3.5 μH, and the capacitor has a value of 0.01 μF.

Solution
Using the equation for finding the oscillation frequency of a Hartley oscillator, we have:

$$f = \frac{1}{2\pi\sqrt{LC}}$$

$$f = \frac{1}{2\pi\sqrt{3.5\ \mu H \times 0.01\ \mu F}}$$

$$f = \frac{1}{1.18\ \mu s}$$

$$f = 851\ kHz$$

PRACTICE PROBLEM 20.3

Determine the frequency of oscillation for a Hartley oscillator that has a 0.0022-μF capacitor and a 10-μH inductor tank.

Colpitts Oscillator
The Colpitts oscillator shown in Figure 20.13 accomplishes phase shifting through the coupling of capacitors C_{1A} and C_{1B} and inductor L_1. Like the Hartley oscillator, the Colpitts oscillator uses a capacitor to couple the output of the transistor to the input side of the circuit. However, the Colpitts oscillator provides feedback by dividing the capacitive branch of the tank circuit.

The Colpitts oscillator is configured as a common-base amplifier. Capacitor C_2 places the base at AC ground potential. Since the output signal is in phase with the input in a grounded-base amplifier, the feedback components do not provide phase reversal. The output of the amplifier already has the proper phase needed to sustain oscillation.

Voltage across capacitor C_{1B} is the feedback voltage. Although the voltage across C_{1B} has a lower amplitude than the tank voltage, it has the same polarity as the total tank voltage. Feedback voltage applies to the emitter or the input of the transistor in the grounded-base configuration. In addition, feedback applies across the unbypassed emitter resistor R_3. A bypass capacitor across the resistor would short circuit capacitor C_{1B} at the oscillator frequency. Capacitor C_3 blocks any DC collector current from the tank circuit.

During operation, oscillations cause the ungrounded end of the tank circuit to swing positive. At this time, the feedback voltage across C_{1B}

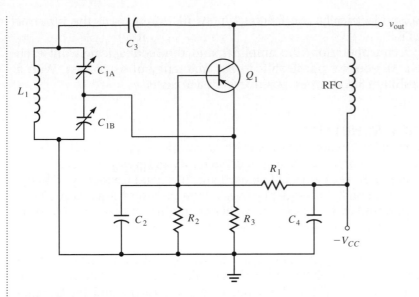

FIGURE 20.13 Colpitts oscillator

also swings positive. With the feedback voltage applied to the Q_1 emitter, the voltage change swings the base negative. Collector current increases as the voltage across the radio-frequency choke increases and causes the collector voltage to become less negative.

Capacitor C_3 couples the positive swing of the collector voltage to the upper end of the tank circuit. While the positive voltage pulse becomes coupled to the tank, the alternating tank voltage simultaneously swings positive. The applied pulse aids the tank voltage swing and keeps the feedback in the proper phase for maintaining oscillation.

Phase-Shift Oscillator

Another example of an oscillator using phase shift to satisfy the Barkenhausen requirements is the phase-shift oscillator pictured in Figure 20.14. In this circuit, a FET biased for class A operation has its output connected to three cascaded RC sections. Each of the three RC sections works as a feed-

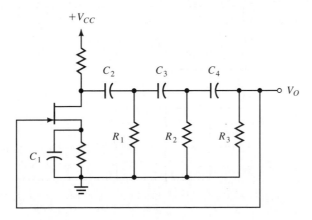

FIGURE 20.14 Phase-shift oscillator

back network. Electrically, the feedback network connects to the input or gate of the FET.

During circuit operation, the amplifier and the feedback network each contribute a 180-degree phase shift to the frequency of oscillation. With a total phase shift of 360 degrees, positive feedback exists.

FREQUENCY STABILITY

The frequency of oscillation varies or drifts as components age, as changes in amplification, junction resistance, and capacitance in transistors occur, or as the circuit temperature changes. If an oscillator has good frequency stability, it will maintain a constant frequency over a long period of time. Several circuit configurations work to keep an oscillator frequency constant. One circuit employs a capacitor with a negative temperature coefficient. The capacitor cancels the positive temperature coefficient of an inductor.

The frequency stability of an oscillator depends on the loaded Q, or Q_L, of the tank circuit. With a high loaded Q, the circuit has greater stability. We can find the value of Q_L by dividing the impedance Z_L across the tank by the reactance X of either the coil or capacitance in the tank. In equation form, this appears as:

$$Q_L = \frac{Z_L}{X}$$

Since X_L equals $2\pi f L$, a small inductance coil results in a small inductive reactance. Since X_C equals $1/2\pi f C$, a large capacitor in the tank circuit results in a small value of capacitive reactance. If we substitute a small value of inductance and a large value of capacitance into the loaded Q equation, the result is a large ratio of Z_L to X and a high Q_L. Thus, an oscillator has good frequency stability when the tank circuit of the oscillator has a low inductance-to-capacitance ratio.

EXAMPLE 20.4

In Figure 20.15, at the center frequency L_1 has a tank impedance of 800 Ω and the coil's reactance is 14.7 Ω. Find the loaded Q for the tank.

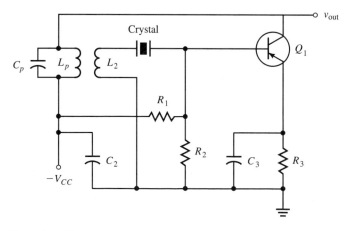

FIGURE 20.15 **Crystal oscillator**

Solution

The Q of the resonant tank is equal to the ratio of the tank impedance divided by the coil's reactance.

$$Q = \frac{Z_1}{X} = \frac{800 \ \Omega}{14.7 \ \Omega} = 54.4$$

PRACTICE PROBLEM 20.4

Find the loaded Q of an oscillator. The impedance across the tank equals 724 Ω, and the reactance of the coil in the tank equals 16 Ω.

CRYSTAL OSCILLATORS

Another method uses a very-high-Q tuned circuit. As Figure 20.15 shows, a quartz crystal works for this method. Quartz crystal becomes mechanically deformed when a voltage is impressed on its faces. Also, the piezoelectric effect or the effect of placing mechanical strain on a quartz crystal produces voltage. Because of those properties, the quartz crystal shown as part of the circuit in Figure 20.15 functions like a parallel LC tuned circuit. With a Q as high as 10,000, the crystal has excellent frequency stability.

The circuit shown in Figure 20.15 is a bipolar transistor crystal oscillator. With the crystal connected to the input circuit of the amplifier, the $L_P C_P$ tuned to the frequency of the crystal is the output circuit. Since the specifications of the crystal determine the frequency of oscillation, a 2.5-MHz crystal sets the oscillator frequency to 2.5-MHz.

MULTIPLIERS

Frequency multipliers increase an oscillator frequency to the desired carrier frequency. A **frequency multiplier** is an amplifier circuit that has its output circuit tuned to a harmonic of the input circuit. Consequently, the output frequency is a harmonic or multiple of the input frequency. A frequency doubler has its output circuit tuned to the second harmonic of its input, while a frequency tripler has its output circuit tuned to the third harmonic of the circuit.

Use of a frequency multiplier allows the oscillator to operate at a low, stable frequency. Furthermore, frequency multipliers multiply the amount of oscillator frequency deviation. If an oscillator has a frequency deviation of ± 2 kHz, a frequency quadrupler connected to a frequency doubler provides a carrier frequency multiplication of 4 $\times$ 2, or 8, and a carrier frequency deviation of 8 $\times$ 2 kHz, or 16 kHz.

MODULATORS

Amplitude modulation occurs as either high-level modulation or low-level modulation. High-level modulation takes place in the collector circuit of the final R-F stage that feeds an antenna. Low-level modulation takes place when the carrier is modulated at some point preceding the final R-F or power amplifier stage.

Frequency modulation also occurs as a product of two basic systems. Figure 20.16 shows a block diagram of a direct FM system. With the direct FM system, the modulating audio signal is amplified and applied directly to the next stage for modulation. Output from that stage causes the deviation of the oscillator frequency. Frequency-modulated output from the oscillator becomes amplified and applied to an antenna system.

In the indirect FM system shown in Figure 20.17, the oscillator operates at a constant frequency with frequency modulation occurring in a different stage. A fixed-frequency, highly stable oscillator operates in the indirect FM circuit. Frequency modulation occurs when the modulating signal shifts the phase space of the radio-frequency signal developed by the oscillator.

The stage where modulation occurs is called the **modulated stage,** or amplifier, and the stage that supplies the modulated signal is called the **modulator.** A modulator circuit changes the frequency of a signal for transmission over a communication channel. Usually, modulators utilize a diode or a nonlinear amplifier that multiplies one signal by the instantaneous amplitude of another. Then, the high-frequency modulated signal is transmitted. A diode detector or synchronous detector demodulates the signal at the receiving end.

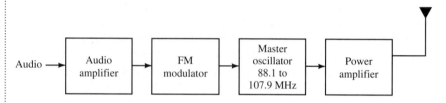

FIGURE 20.16   Direct FM system

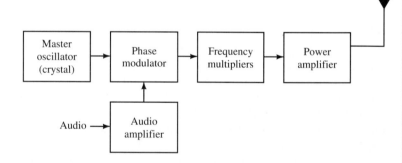

FIGURE 20.17   Indirect FM system

EMITTER AND COLLECTOR MODULATION

Figure 20.18a shows a transistor circuit supplying emitter modulation, and Figure 20.18b shows another circuit providing collector modulation. In Figure 20.18a, an audio-frequency modulating signal becomes applied to the emitter circuit. At the same time, the R-F carrier is applied to the base circuit of transistor Q_1. The audio-frequency signal both aids and opposes the voltage on the emitter of the transistor. This results in changes in the base-to-emitter bias voltage, which cause corresponding changes in the current through the collector circuit. As a result, the carrier is amplitude-modulated at the output.

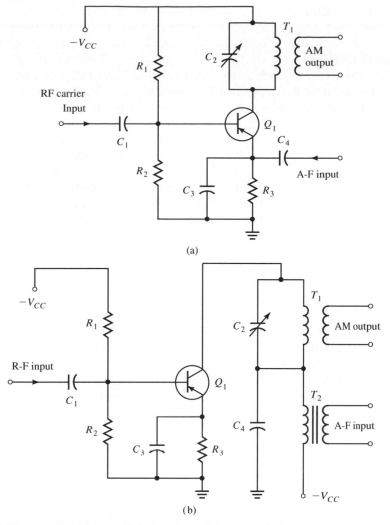

FIGURE 20.18   (a) Emitter modulation. (b) Collector modulation.

In Figure 20.18b, the audio-frequency signal feeds to the collector circuit through a modulation transformer, T_2. Here, the audio-frequency signal aids and opposes the transistor collector voltage. Any variations in the collector voltage cause changes in the collector current and result in an amplitude-modulated carrier output.

✓ **Progress Check**
You have now completed objectives 2 and 3.

REVIEW SECTION 20.2

1. Define the term oscillator.
2. List five sinusoidal oscillators.
3. What are the distinguishing characteristics of a Hartley oscillator?
4. What are the distinguishing characteristics of a Colpitts oscillator?
5. What are the distinguishing characteristics of an RC phase-shift oscillator?
6. Describe how a frequency doubler operates.
7. The two types of AM modulation are _____ and _____.
8. The two types of FM modulation are _____ and _____.

AM DETECTION

A simple AM receiver, such as the one shown in Figure 20.19a, contains several stages of tuned R-F amplification followed by a diode detector. The amplifier stages provide selectivity against near-frequency signals and amplify microvolt-level input signals for the detector. The detector rectifies the R-F waveform and recovers the smooth envelope with low-pass filtering. With the use of the low-pass filter, R-F signals become attenuated while audio frequencies pass.

Figure 20.19b shows the schematic of a simple AM detector circuit. In the circuit, a diode recovers the modulated signal from an amplitude-modulated carrier that is applied to the input terminals. Since the diode conducts in only one direction, current exists in one direction, and only the positive alternations of the input signal develop across resistor R_1.

To complete the recovery of the modulated signal, capacitor C_1 connects across R_1, and together they form the low-pass filter. The capacitor smooths out any high-frequency pulses while producing an output proportional to the changes in amplitude of the pulses. Voltage developed by the diode detector varies at the same rate as the original modulating signal. Obtaining the output shown in Figure 20.19b means that the detector circuit functions as both a rectifier and a filter.

Changes in amplitude of the incoming AM carrier determine the waveform of the varying DC. As Figure 20.19a shows, the incoming carrier passes through several amplifier stages before arriving at the detector. A double-

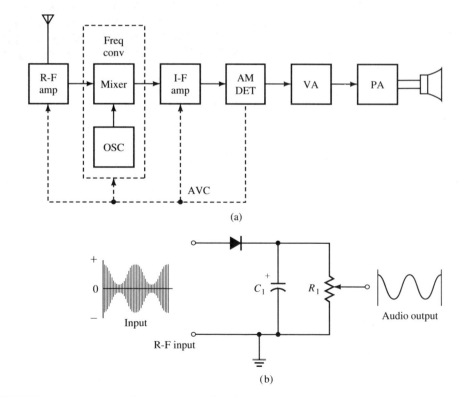

FIGURE 20.19 (a) Simple AM receiver. (b) Schematic of a simple AM detector circuit.

tuned transformer couples the modulated AM carrier from the amplifier stages to the detector.

EXAMPLE 20.5

Find the critical frequency for the low-pass filter in Figure 20.19b. Use the following values.

$$R_1 = 1 \text{ k}\Omega \qquad C_1 = 0.027 \text{ }\mu\text{F}$$

Solution
The formula for the critical frequency of a low-pass filter is

$$f_c = \frac{1}{2\pi RC} = \frac{1}{2\pi \times 1 \text{ k}\Omega \times 0.027 \text{ }\mu\text{F}} = 5.89 \text{ kHz}$$

PRACTICE PROBLEM 20.5

If, in Figure 20.19b, $C_1 = 0.033 \text{ }\mu\text{F}$ and $R_1 = 1.5 \text{ k}\Omega$, find the critical frequency for the low-pass filter.

FM DETECTION

Because of the nature of frequency modulation, FM detectors differ from AM detectors. An FM detector must respond to changes in the frequency of the modulated signal. Also, FM detectors operate with higher frequencies than AM detectors.

When we compare AM and FM signals, the most important difference, though, may be in the superior signal-to-noise ratio of FM signals. The signal-to-noise ratio improvement occurs because FM signals pass through an amplitude-limiting stage before detection. With limiting, the FM system becomes relatively insensitive to interfering signals and noise. AM systems can reject interference only with an increase in power.

FOSTER-SEELY DETECTOR

Figure 20.20 shows a common FM detector called the Foster-Seely detector. As the figure shows, the detector has two separate diode circuits. One cir-

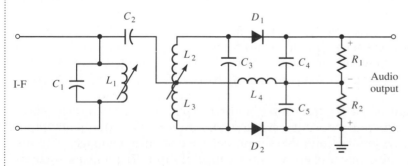

FIGURE 20.20   Foster-Seely detector

cuit consists of L_2, L_4, C_4, R_1, and D_1, and the other consists of L_3, L_4, C_5, R_2, and D_2. As you look at both circuits, note that inductor L_4 is common to both circuits. Capacitor C_3 tunes the transformer secondary, which is L_2, and L_3 to the center frequency of the 10.7-MHz intermediate carrier frequency signal.

The primary, or L_1, and the secondary windings, or L_2 and L_3 of the detector transformer, are tuned to resonance at the center frequency of the intermediate frequency. In addition, the primary voltage couples to the center tap of the secondary through capacitor C_2 and appears across the radio-frequency choke L_4.

The connection of the diodes allows current to flow through resistor R_1 in a direction opposite to the direction of current flow through R_2. As a result, the polarity of the voltage through R_1 is opposite the polarity of the voltage across R_2. The output voltage of the circuit equals the difference of the two voltages across the two resistors.

Breaking the Foster-Seely into two diode circuits shows that the voltage across L_2 and L_4 acts as a source voltage for the D_1 circuit. The voltages across L_3 and L_4 are the source voltage for the D_2 circuit. At resonance, the two voltages have the same phase relationship. Therefore, equal voltages become applied to both diode circuits. Equal, opposite-phase voltages develop across R_1 and R_2 and cancel each other, resulting in a zero voltage output.

With respect to the center tap, the voltages at the upper and lower ends of the transformer secondary or across windings L_2 and L_3 are 180 degrees out of phase. Off-resonant frequencies cause the L_2, L_3, and C_3 tank circuit to operate as either a capacitive or inductive circuit. As a result, the phase angle between the transformer secondary voltage and the voltage across L_4 varies with frequency.

Above resonance, the phase angle between the L_3 and L_4 voltages is less than the phase angle between the L_2 and L_4 voltages. Inductive reactance in the tank circuit increases while capacitive reactance decreases. Thus, R_2 has a greater voltage than R_1 and the output swings negative. Below resonance, the phase relationship reverses and the output voltage swings positive. Inductive reactance decreases while capacitive reactance increases in the tank circuit.

With its variations making an exact reproduction of the original modulating signal, the output voltage varies above and below the center frequency. Figure 20.21 shows the overall response characteristics of the Foster-Seely detector. The S curve shows the relationship between the variations of input frequency and the resulting variations of output voltage. With the knees of the curve marking frequencies equally spaced above and below the intermediate carrier frequency, or f_r, each knee is approximately 100 kHz from the carrier frequency.

The solid curve below the S curve represents the swing of the frequency-modulated I-F carrier frequency from frequency f_2 to frequency f_6. During each cycle of modulation, the I-F carrier frequency swings from f_r down through f_1 to f_2 and back up through f_1 to f_r and f_5 to f_6 and back down through f_5 and f_r.

To the right of the S curve, another curve plots the output voltage waveform. Each complete sinusoidal swing of the intermediate-frequency input frequency provides one complete sinusoidal cycle of output voltage. With the S curve remaining linear over the entire range of input frequencies, the amplitude of the output voltage is proportional to changes in frequency. Indi-

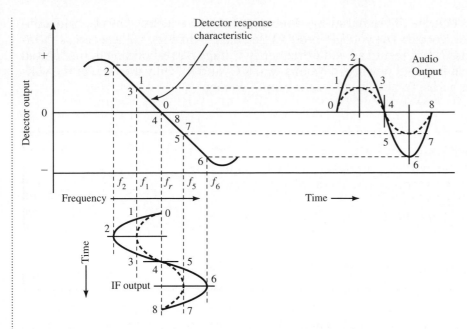

FIGURE 20.21 Response characteristic of a Foster-Seely detector

cated by the dashed-line input and output curves, if the frequency swing decreases, the output voltage amplitude decreases.

RATIO DETECTOR

Figure 20.22 shows an FM detector that has built-in limiting through capacitor C_6. Like the Foster-Seely detector, the ratio detector features two diodes. However, the diodes in the ratio detector are connected so that they series-aid instead of series-oppose voltages across the transformer secondary.

The series-aiding connection of the two diodes results in an output voltage that equals the ratio of the two diode voltages. When the circuit receives a carrier at an intermediate frequency, equal voltages apply across C_4 and C_5. As the FM signal swings below the center frequency, diode D_1 conducts harder than D_2. The increased conduction of D_1 causes the voltage across C_4 to increase while the voltage across C_5 decreases.

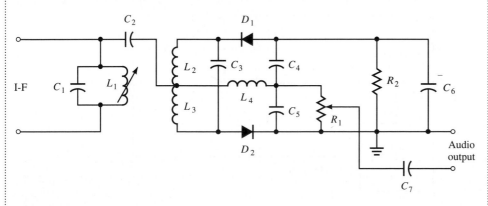

FIGURE 20.22 Ratio detector

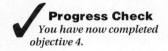
When the FM signal swings above the center frequency, the opposite conditions occur. The voltage across C_4 decreases while the voltage across C_5 increases. As a result, an audio-frequency voltage corresponding to modulated signal frequency swings develops across C_5 and R_1. Only the ratio of the voltages across C_4 and C_5 affects the output voltage.

REVIEW SECTION 20.3
1. What is the purpose of an AM detector?
2. An AM detector consists of a diode and a _____ _____ filter.
3. Name the two types of FM detectors.
4. In Figure 20.20, what component is common to both of the diode loops?
5. In Figure 20.22, what is the purpose of C_2?

20.4 PHASE-LOCKED LOOPS

A phase-locked loop (PLL) blends digital and analog technologies together into one integrated circuit package. PLL applications include tone decoding, demodulation of AM and FM signals, frequency multiplication, pulse synchronization, and regeneration of clean signals. Looking at Figure 20.23, you can see that phase-locked loops contain four basic components—a phase comparator, a low-pass filter, an amplifier, and a voltage-controlled oscillator (VCO).

Phase-locked loops have three states called free-running, capture, and phase lock. When the incoming signal has a frequency different from the VCO signal, the PLL enters its free-running state. Within the PLL, the phase comparator has a DC output error voltage that becomes applied to the VCO. The voltage causes the VCO frequency to change.

Capture begins when the VCO changes frequency to reduce the difference between its frequency and the incoming signal. As soon as the input frequency equals the VCO frequency, the PLL is phase locked. The phase of the VCO remains slightly different from the phase of the incoming signal. This differ-

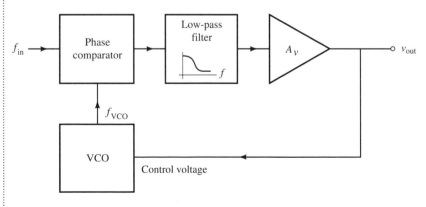

FIGURE 20.23   Block diagram of a phase-locked loop

ence causes the comparator to continue producing an error voltage needed to maintain the correct VCO frequency.

PLL COMPONENTS

Phase Comparator

A phase comparator compares an input signal and a reference signal and then generates an output that represents the phase difference of the two inputs. As an example, the two inputs may have different frequencies. The output is a periodic output signal at the difference frequency. We can express the phase relationship in terms of radians and use an equation to show the relationship between radians and degrees. The equation is:

$$\theta_{rad} = \left(\frac{\pi}{180°}\right)\theta$$

where θ_{rad} equals the angle in radians, and θ is the angle in degrees.

EXAMPLE 20.6

Determine the phase relationship of two signals 45° out of phase with each other.

For this example, both signals are sine waves.

Solution

If we use the radian/degree conversion equation, we have:

$$\theta_{rad} = \left(\frac{\pi}{180°}\right)\theta$$

After substituting the number of degrees into the equation and letting π equal 3.14, we have:

$$\theta_{rad} = \left(\frac{3.14}{180°}\right) \times 45°$$

$$0.0174 \times 45 = 0.78 \text{ rad}$$

PRACTICE PROBLEM 20.6

Draw the two 45° out-of-phase sine waves.

Voltage-Controlled Oscillator

Because the phase comparator output controls the frequency of an oscillator, the VCO is an essential part of the phase-locked loop. Filtered output from the phase comparator to the input of the VCO is a measure of the input frequency. The VCO input must not load down or change the characteristics of the low-pass filter that makes up the remainder of the loop. So the buffer in a phase-locked loop has an almost infinite input resistance. This allows the use of a wide range of filter components.

Output from the VCO consists of a locally generated frequency. That frequency equals the input frequency to the phase comparator, f_{in}, and provides a clean replica of that frequency.

We can measure a VCO in terms of frequency output per amount of control voltage input, or:

$$V_f = \frac{f_2 - f_1}{V_2 - V_1}$$

In the equation, V_f equals the voltage-to-frequency conversion of VCO in hertz per volt; $f_2 - f_1$ equals the change in VCO output frequency; and $V_2 - V_1$ equals the change in VCO control voltage.

EXAMPLE 20.7

Determine the voltage-to-frequency conversion of a VCO.

The VCO has an output frequency of 10 kHz at an input voltage of 2 V. The output frequency is 15 kHz at an input voltage of 3.2 V.

Solution
To find the voltage-to-frequency conversion of a VCO, we use the equation:

$$V_f = \frac{f_2 - f_1}{V_2 - V_1}$$

Substituting our known values into the equation gives:

$$V_f = \frac{15 \text{ kHz} - 10 \text{ kHz}}{3.2 \text{ V} - 2 \text{ V}}$$

Reducing the equation, we find that:

$$V_f = \frac{5 \text{ kHz}}{1.2 \text{ V}}$$
$$V_f = 4.17 \text{ kHz/V}$$

The VCO voltage-to-frequency conversion equals 4.17 kHz per volt.

PRACTICE PROBLEM 20.7

Determine the voltage-to-frequency conversion of a VCO that has an output frequency of 8 kHz at an input voltage of 1.5 V and an output frequency of 12 kHz with an input voltage of 3 V.

Amplifier and Filter
The phase difference between the input frequency and the VCO frequency generates a DC offset voltage. An amplifier within the PLL amplifies the DC offset voltage and increases the PLL sensitivity. The voltage from the operational amplifier controls the VCO frequency.

A low-pass filter determines the dynamic characteristics of the PLL. As you will see, the filter sets the range of frequencies for capturing the phase lock signal. Also, the filter determines the response speed of the PLL as it responds to input frequency variations.

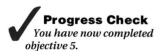
Progress Check
You have now completed objective 5.

PLL CAPTURE AND LOCK

A PLL will stay locked as long as the input frequency does not stray outside the feedback signal. Locking the PLL onto a signal can become a complicated issue. Instead of having a clean DC error signal, the initial frequency error from the comparator is a periodic difference frequency. Even after filtering, the signal continues to exist as a series of small-amplitude wiggles.

Figure 20.24 shows the waveform from the capture routine for a phase-locked loop. At the beginning of the capture routine, the phase error signal brings the VCO frequency closer to the reference frequency. As the frequencies get closer, the error signal changes more slowly over the part of the cycle that has the two frequencies the closest. The DC component of the error signal brings the PLL into lock.

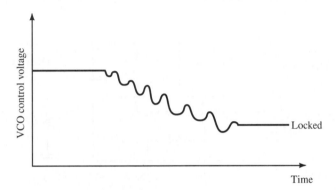

FIGURE 20.24 Capture routine waveform

PLL FM DEMODULATOR

In frequency modulation, information becomes encoded onto a carrier signal through varying the carrier signal frequency proportional to the information waveform. Recovering the modulated information requires a phase-locked loop. Figure 20.25 shows a block diagram of one demodulation method.

In the figure, a PLL locks onto the incoming signal. The voltage determining the VCO frequency has a value proportional to the input or modulated frequency. Response time of the PLL—set by the filter bandwidth—must be

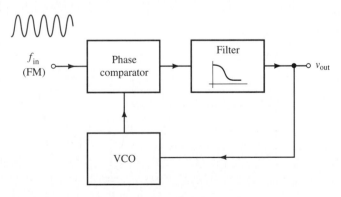

FIGURE 20.25 Block diagram of PLL FM demodulation method

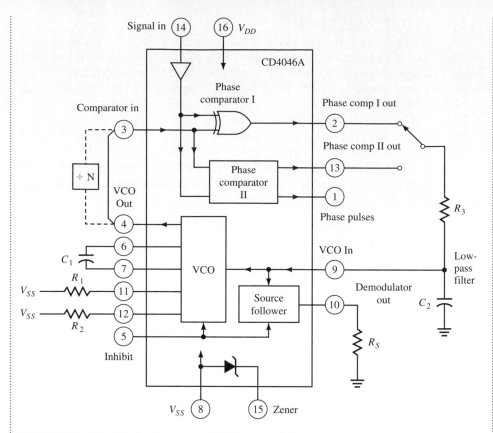

FIGURE 20.26 Block diagram of a PLL integrated circuit

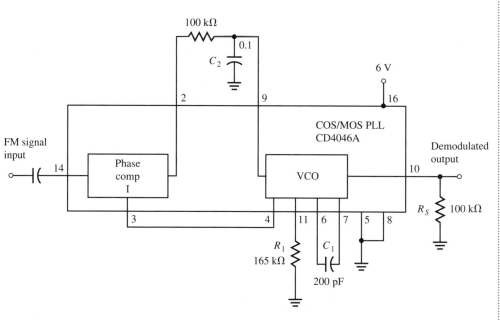

FIGURE 20.27 PLL configured as an FM demodulator

short compared with the time scale of the recovered signal variations. The VCO must have a high degree of linearity so that distortion in the audio signal stays low.

Figure 20.26 shows the block diagram of a PLL integrated circuit, and Figure 20.27 shows the phase-locked loop configured as an FM demodulator. When the PLL locks onto an FM signal, its voltage-controlled oscillator tracks the frequency of the signal. The filtered error voltage from the phase comparator becomes the input signal for the VCO. A demodulated signal at the output of the PLL corresponds to that VCO input signal.

During normal operation, an FM signal made up of a 10-kHz carrier frequency and modulated by a 400-Hz audio signal is AC coupled to the signal input, or terminal 14 of the PLL. The amplitude for the total FM signal is approximately 500 mV. While the center frequency of the VCO is set to equal the carrier frequency of 10 kHz, the capture range is about ±0.4 kHz. An FM signal amplitude of 150 mV causes a dcmodulated output of about 30 mV.

Progress Check
You have now completed objective 6.

REVIEW SECTION 20.4
1. Name the four basic components of a PLL.
2. List three PLL applications.
3. List the three states of PLL operation.
4. Describe how a VCO can be made to change frequencies.
5. When a PLL is used as an FM demodulator, from which basic component is the output taken?

20.5   HIGH-FREQUENCY DIODES

TUNNELING

When studying the zener diode, we found that at a certain reverse voltage, a large reverse current occurred. The current was called the avalanche current. If avalanche current occurs at near 0 V reverse voltage, it is called **tunneling.** Diodes that exhibit the characteristic of near-zero reverse-biased avalanche current are called **tunnel diodes.** If a tunnel diode were forward-biased, it would act like any other diode. Tunneling occurs only in heavily doped, reverse-biased diodes. Figure 20.28 is the schematic symbol for a tunnel diode.

FIGURE 20.28   Schematic symbol for a tunnel diode

TUNNEL DIODES

As a reverse voltage is applied to a tunnel diode, as expected, the reverse current increases. As shown by Figure 20.29, at a certain point current decreases while reverse voltage is increasing. This condition is called the **negative resistance region.** The negative resistance region is the area between V_p and

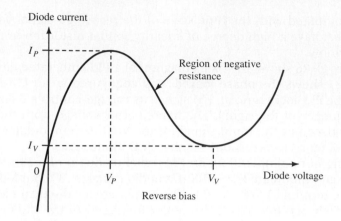

FIGURE 20.29   **Voltage-current characteristics of a tunnel diode**

V_v. V_p is the peak (maximum) voltage that produces a corresponding increasing current. The current produced by V_p is called peak current, I_p. V_v is called the **valley voltage.** It is the voltage at which the diode current changes from a decreasing to an increasing current. I_v is the minimum current produced in the negative resistance region. Voltages in excess of V_v cause the tunnel diode to act like any other diode.

The negative resistance makes tunnel diodes excellent high-frequency devices. They are used as R-F switches and oscillators. The disadvantage of tunnel diodes is that they are low-power devices.

Progress Check
You have now completed objective 7.

VARACTOR DIODES

A varactor diode, or variable reactor, uses the voltage-variable capacitance of a reverse-biased PN junction. Figure 20.30 shows the schematic diagram for a varactor diode. The device is a two-terminal semiconductor often used to control the operation of VHF and UHF oscillators. Many television receivers feature varactor diodes as part of the tuning assemblies. Other uses for varactor diodes include harmonic generation, microwave frequency multiplication, and active filters.

Varactors change capacitance as the reverse bias is changed. Increasing reverse bias lowers capacitance and decreasing reverse bias raises capacitance. Figure 20.31 shows how the capacitance decreases with an increased reverse bias, and Figure 20.32 shows how the capacitance increases with a

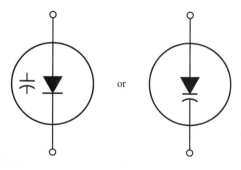

or

FIGURE 20.30   **Schematic symbols for a varactor diode**

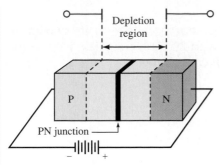

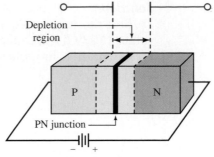

High reverse-bias voltage,
low capacitance

Low reverse-bias voltage,
high capacitance

FIGURE 20.31 Capacitance decreases in a varactor diode with an increase of reverse bias

FIGURE 20.32 Capacitance increases in a varactor diode with a decrease of reverse bias

decreased reverse bias. The depletion region between the P- and N-type materials determines the capacitance value of the varactor and works much like the dielectric of a capacitor. As reverse voltage is increased, the depletion region is increased. This is like pushing the plates of a capacitor farther apart. Because capacitance is inversely proportional to plate distance, capacitance decreases. When the reverse bias is decreased, the depletion region decreases, causing capacitance to be increased.

EXAMPLE 20.8

Calculate the upper and lower frequency limits of Figure 20.33.

Figure 20.33 is a parallel resonant circuit using a varactor in series with a capacitor. As the voltage source is increased, varactor capacitance will decrease. A 1N139 has 10 pF of capacitance at 1 V reverse bias and 3.5 pF at 20 V. To find the resonant frequency we have used

$$f_r = \frac{1}{2\pi\sqrt{LC_T}}$$

C_1 and D_1 are in series and their total capacitance is found by

$$C_T = \frac{C_1 \times C_{D1}}{C_1 + C_{D1}}$$

We will use the above formulas, solving first for the upper frequency limit and then for the lower frequency limit.

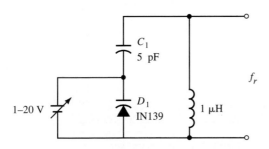

FIGURE 20.33 Varactor diode functioning in a UHF oscillator

Solution

To find the upper frequency limit, set the source voltage to its maximum value. 1N139 has a capacitance of 3.5 pF with 20 V reverse bias. We use the series capacitor formula to find total circuit capacitance

$$C_T = \frac{C_1 \times C_{D1}}{C_1 + C_{D1}}$$

$$C_T = \frac{5 \text{ pF} \times 3.5 \text{ pF}}{5 \text{ pF} + 3.5 \text{ pF}}$$

$$C_T = 2.06 \text{ pF}$$

The next step is to solve for the upper resonant frequency.

$$f_r = \frac{1}{2\pi\sqrt{LC_T}}$$

$$f_r = \frac{1}{2\pi\sqrt{1\ \mu H \times 2.06 \text{ pF}}}$$

$$f_r = 110.9 \text{ MHz}$$

To solve for the lower frequency limit, set the reverse bias at 1 V. At 1 V the 1N139 has 10 pF of capacitance. Use the above process again. You should get 87.2 MHz.

PRACTICE PROBLEM 20.8

Using Figure 20.33 solve for the lower and upper frequency limits if $C_1 = 10$ pF and L = 750 μH.

✓ Progress Check
You have now completed objective 8.

MICROWAVE POWER DIODES

Microwave frequencies are frequencies above the 1-GHz range. Along with tunnel and varactor diodes, IMPATT and Gunn diodes function as DC-to-microwave power converters. IMPATT and Gunn diodes make up a class of diodes called **microwave power diodes.** That class of diodes usually operates in resonant circuits such as high-Q resonant microwave cavities. Figure 20.34 shows a commonly used microwave cavity.

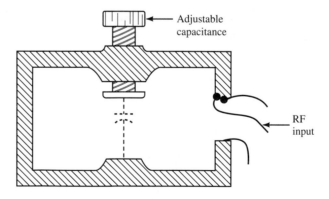

FIGURE 20.34   **A high-Q resonant microwave cavity**

Microwave power diodes utilize a semiconductor instability called negative conductance. With tunnel diodes, the current/voltage characteristic has a defined negative resistance region. However, amplification and oscillation can occur when the AC current is out of phase with the AC electrical field within the diode.

Negative conductance devices use a combination of avalanche and drift characteristics. A high DC reverse bias divides the N material of the PN diode into avalanche and drift regions as shown in Figure 20.35. Along with having avalanche breakdown, the diodes have an AC bias superimposed on the DC bias. The high electrical field within the avalanche region of the diode generates electron-hole pairs that become swept through the drift region to the device terminals. Flow of carriers is proportional to the voltage in the drift region.

The AC component of the current can be 180 degrees out of phase with the applied voltage under the proper bias conditions and device configurations. This feature allows the device to have negative conduction and oscillation in a resonant circuit. At the oscillating frequency, the avalanche and drift regions are 90 degrees out of phase, and the diode has a negative resistance. Connecting the diode to a tuned circuit allows the negative resistance to cancel out the positive resistance of the tuned circuit. As a result, the combination oscillates.

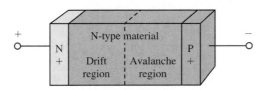

FIGURE 20.35   **Block diagram of a microwave power diode. A high DC reverse bias voltage divides the N material into avalanche and drift regions.**

IMPATT Diodes

The IMPATT (*imp*act ionization *a*valanche *t*ransit *t*ime) diode operates in the negative conductance mode when used as an R-F oscillator. Its design is similar to the varactor. However, the IMPATT is designed to operate with a reverse voltage in excess of breakdown voltage. Usually the reverse voltage is between 70 and 100 V. A large depletion area forms in the N-type material. An avalanche region forms near the P-type material.

Gunn Diodes

Although Gunn diodes are called diodes and have two terminals, the devices do not have PN junctions. Instead, Gunn effect devices use bulk instabilities that do not require junctions and work on the principle of transferred electrons. With the transferred electron mechanism, the conduction of electrons of some semiconductors shifts from a state of high mobility to a state of low mobility through the influence of a strong electrical field.

Progress Check
You have now completed objective 9.

TROUBLESHOOTING

As you have seen throughout this text, the basic troubleshooting procedure consists of preliminary, routine checks followed by more-specific tests. More than likely, your troubleshooting method will follow this outline:

1. Determine the symptom.

2. Make initial checks.

3. Locate the faulty stage.

4. Locate the faulty circuit.

5. Find the defect.

The nature of your troubleshooting methods depends on the type of defect seen and the type of equipment being tested. In addition, the method depends on the type of available test equipment.

Regardless of the circumstances, two basic methods should affect the way you test a circuit. Using the dynamic test method, you observe how the circuits of a faulty unit affect the signal. Two types of dynamic test methods are signal injection and signal tracing. With signal injection, we want to find how far ahead of the speaker the circuits are in good condition. With signal tracing, we want to find how far the received signal can travel through the receiver from the antenna before it reaches a defective circuit.

Dynamic testing offers the advantage of isolating the fault to a section or single stage. Any further testing can be limited to the defective stage. With static tests, you check the circuit conditions with no signal applied to the unit. Static tests are voltage, current, and resistance measurements.

SIGNAL INJECTION

A generator called a noise generator or signal injector is used to make signal injection tests. Usually, a signal injector consists of an RC oscillator or multivibrator that generates a fundamental frequency in the audio range. The waveform of the output contains harmonics that extend throughout the audio-frequency, intermediate-frequency, and radio-frequency ranges. Because of those harmonics, the output can serve as a test signal for many different kinds of receiver circuits.

Signal injection has the advantage of speed. You can inject test signals at any point in an audio receiver without selecting test frequencies. Unfortunately, the high impedance output of some signal injectors will not match well with transistor equipment.

Figure 20.36 shows a block diagram of a broadcast band radio receiver. R_1 represents the volume control, with the ungrounded terminal of R_1 used as a signal injection test point. Injecting the signal at the ungrounded terminal of the volume control allows us to divide the receiver into an audio-frequency section and a radio-frequency section.

Most modern signal generators have a blocking capacitor in series with the output terminal. However, for this test, we are connecting a 0.1-μF capacitor in series with the probe leads of the generator. In either case, the capacitor prevents the generator from changing the DC voltages of the circuit under test. Connect the generator ground lead to a ground point in the receiver.

To begin the test, set the volume control to its maximum setting and set the signal generator for a low-level output. Touching the generator probe to the test point should produce a normal audio tone. The audio tone indicates that the audio-frequency section of the receiver is in good condition and that the fault is before the volume control. An abnormal sound or no sound from the speaker indicates problems within the audio-frequency section.

We can isolate the faulty stage by injecting test signals at points between the stages of the suspected section. As an example, if you suspect problems with the audio-frequency section, touch the generator probe between the audio-frequency amplifier and the audio-frequency output.

If injecting a signal at the test point produces a normal output in a receiver that has no output when tuned to a station, suspect problems in the intermediate-frequency and radio-frequency sections of the receiver.

First, check the intermediate-frequency section of the receiver. Using a radio-frequency generator, set the generator to the intermediate frequency of the receiver and use a modulated signal. Turning R_1 to its maximum level, inject a low-level signal at the output or input of each stage in the intermediate-frequency section. Work from the detector toward the front end of the receiver. If the receiver produces a normal output, all stages between the test point and speaker are working.

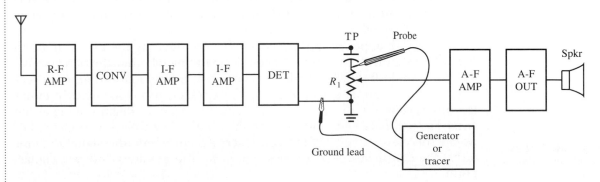

FIGURE 20.36 Block diagram of a receiver

A working intermediate-frequency section leaves the radio-frequency section. To test the radio-frequency section, use a modulated test signal, setting the receiver and the generator to the same frequency in the broadcast band. Choose a frequency between stations so that the signal does not provide false indications. Inject the signal at the radio-frequency amplifier; this checks the converter output. Then, inject a signal at the antenna terminals.

SIGNAL TRACING

We can use signal tracing as a troubleshooting method. With a radio receiver, we can use the signal produced by a station or a signal generator. With sound systems, we can use the signal received by the tuner, the signal from a tape playback head, the signal from a CD player, or the signal from a generator. Signal tracing consists of setting the unit to receive or accept one of these signals, connecting the signal source to the unit, and touching the probe of an indicating test instrument to various test points in the circuit. The test instrument may be an oscilloscope, a digital voltmeter, or a signal tracer.

Signal tracers can be divided into two types. One is an audio amplifier with a detector probe, and the other is a tuned radio-frequency radio receiver. Both types use speakers as indicators. Most modern signal injectors include some type of signal tracer.

Referring again to Figure 20.36, we can connect the signal tracer to the same tests points that we used for signal injection. Again, set the receiver volume control to its maximum point and connect the ground lead of the signal tracer to a ground point. Then, tune the receiver to a strong station to see if any signals can be received. If the receiver cannot receive any stations, touch the tracer probe to a test point in the front end of the receiver and tune the receiver until the tracer indicates a received signal.

After tuning to a station, touch the tracer probe to the volume control test point. Since the volume control section of the receiver follows the detector, audio-frequency signals should be present at the test point. If touching the probe to the volume control test point produces a normal signal at the tracer speaker, the radio-frequency and intermediate-frequency sections are functional.

Thus, you can limit your troubleshooting to the audio-frequency section of the receiver. Move the test probe to a point between the first audio-frequency amplifier and the next signal. A normal signal at the tracer speaker tells us that the stages following the audio-frequency amplifier have problems. No signal tells us that the first audio-frequency amplifier is defective.

In a preceding paragraph, we noted that by touching the tracer probe to the volume control test point, we could determine whether the defect existed in the audio-frequency, radio-frequency, or intermediate-frequency section. With no normal signal at the tracer speaker, problems exist in either the intermediate-frequency or radio-frequency section. Begin tracing at the antenna terminals and work toward the volume control as you check between successive stages. When you reach a point where the signal deteriorates, the defect lies between this point and the previous "normal signal" point.

Progress Check
You have now completed objective 10.

WHAT'S WRONG WITH THIS CIRCUIT?

The following circuit is an AM detector. The input signal has the correct amplitude and frequencies. The output is a straight line at 0 V. What are the possible causes?

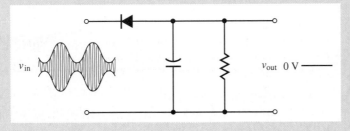

SUMMARY

We can define modulation as the process of adding information to a carrier. Usually the carrier is a radio-frequency carrier because that type of carrier will transmit without wires.

Oscillators that handle frequencies are often sinusoidal oscillators. If the oscillator generates a sine wave it is a sinusoidal oscillator. The Hartley, Colpitts, and phase-shift oscillators are examples of feedback oscillators. Each type relies on having the output of an amplifier fed back to its input.

An AM detector rectifies the R-F waveform and recovers the smooth envelope with low-pass filtering. Because of the nature of frequency modulation, FM detectors differ from AM detectors. An FM detector must respond to changes in the frequency of the modulated signal. Also, FM detectors operate with higher frequencies than AM detectors.

Phase-locked loops blend digital and analog technologies into a single integrated circuit. PLLs work as tone decoders, AM and FM demodulators, frequency multipliers, and synchronizers. A comparator, voltage-controlled oscillator, low-pass filter, and amplifier make up a PLL. Phase-locked loops compare incoming frequencies with self-generated frequencies and then lock onto a specified frequency range.

In order to utilize the upper portions of the frequency spectrum, we have high-frequency semiconductors such as tunnel diodes, IMPATT diodes, and Gunn diodes. Tunnel diodes provide us with high-speed switching. IMPATT and Gunn diodes are used as microwave oscillators.

These semiconductor devices are some of the devices that are used to provide us telephone service, radio and TV programs, and elaborate computer hookups. All this equipment requires technicians who understand semiconductor devices.

System troubleshooting is similar to troubleshooting a single circuit. First, determine the symptoms; second, make initial checks; third, locate the faulty stage; fourth, locate the faulty circuit; and, fifth, find the defect. We also learned that we can create our own signal with a signal injector or we can trace the system's generated signal with signal-tracing test equipment like oscilloscopes and DMMs.

PROBLEMS

1. What is the VHF frequency range?
2. An AM radio station broadcasts at the carrier frequency of 1060 kHz with a modulating signal of 5 kHz. Solve for the carrier and sideband frequencies the station would broadcast.
3. An FM radio station broadcasts at the carrier frequency of 99.9 MHz with a modulating signal of 15 kHz. Solve for the carrier and first and second sideband frequencies the station would broadcast.
4. A Hartley oscillator has a tank consisting of a 4.7-μH coil and a 4700-pF capacitor. What is the oscillator's output frequency?
5. List the Barkenhausen criteria.
6. The Hartley oscillator in problem 4 has a Z_L of 1.44 kΩ. What is the tank's Q?
7. If 10 MHz is input to a frequency tripler, what is the output frequency?
8. The input to Figure 20.37 is 450 kHz, 455 kHz, and 460 kHz. What is the output frequency?

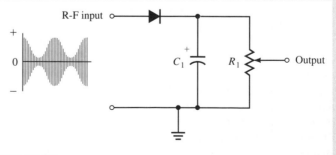

FIGURE 20.37

9. In Figure 20.38, is the output voltage positive or negative if the input frequency is 10.8 MHz? Carrier frequency is 10.7 MHz.

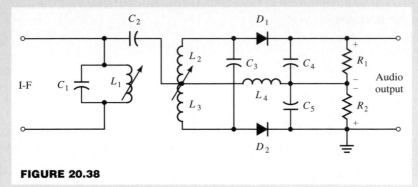

FIGURE 20.38

10. In Figure 20.39, what is the purpose of C_6?

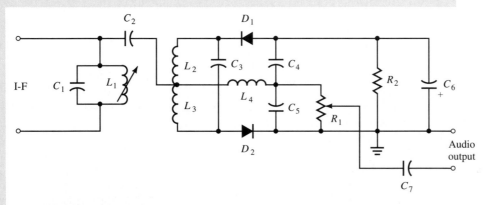

FIGURE 20.39

11. Convert 30 degrees into radians.
12. If a VCO has a voltage-to-frequency conversion of 5 kHz per volt, what will the frequency change be if the control voltage is increased 250 mV?
13. List the three steps the PLL goes through to lock onto a signal.
14. The input to Figure 20.40 has a carrier frequency of 10.700 MHz and first sideband frequencies of 10.715 MHz and 10.685 MHz. What is the output frequency?

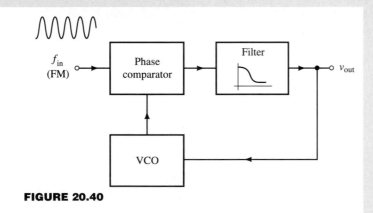

FIGURE 20.40

15. The minimum negative resistance current is called
_____.

16. If the voltage source in Figure 20.41 is increased, does f_r increase or decrease?

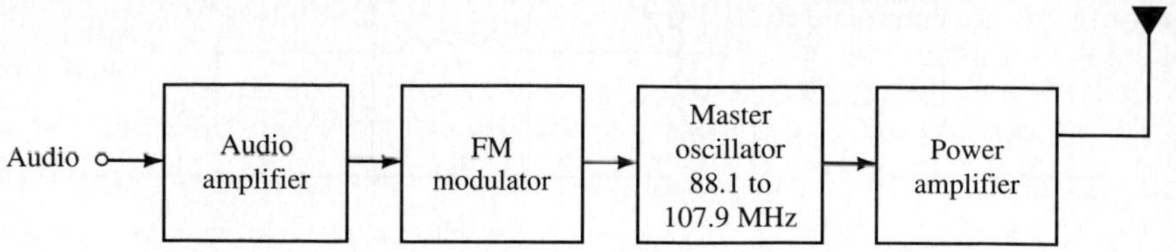

FIGURE 20.41

π (pi) Filter A circuit designed to remove ripple from DC levels, named for its schematic appearance to the π symbol.

AC Emitter Resistance (r'_e) The change in base-emitter voltage divided by the change in emitter current: $r'_e = \Delta V_{BE}/\Delta IE$, approximated by the formula $r'_e = 25$ mV/I_E.

Acceptor Impurity An element such as boron, aluminum, or gallium which is added to silicon to create an excess of holes.

Alpha The ratio of collector current to emitter current, I_c/I_e. For most small signal transistors, this value is greater than 0.99.

Ammeter A meter used for measuring current. To use this type of meter, the circuit must be broken and the meter inserted in series with the current path to be measured.

Amplifier A circuit consisting of one or more sections (stages) which has the feature that the output is a positive multiple of the input. This multiple is termed the gain of the circuit.

Amplifier Stage A section of an amplifier which usually contains one active device and its associated components.

Astable Multivibrator A multivibrator which constantly switches between two states and is used to produce a square wave output.

Atom The smallest particle of an element which still has the characteristics of the element. This particle consists of electrons, protons, and neutrons.

Avalanche Breakdown The phenomenon in which a reverse voltage of sufficient amplitude is applied to a PN junction to cause free electrons to accelerate enough to dislodge valance electrons. The valance electrons then dislodge other electrons causing the effect to continually increase.

Balance In reference to filters, circuits that are electrically alike and symmetrical with respect to ground.

Band-pass Filter A circuit which allows a narrow range of frequencies to pass through from input to output while rejecting all other frequencies.

Bandwidth The range of frequencies a circuit will pass from input to output without significant attenuation.

Barrier Potential The forward potential across a PN junction due to the difference in potential between ions on either side of the depletion region. This voltage is approximately 0.7 volts for silicon transistors and 0.3 volts for germanium.

Base-line Stabilizer A circuit that adds a DC voltage to an AC voltage, effectively anchoring the AC at a level other than zero.

Beta The current gain of a transistor; the ratio of collector current to base current (I_C/I_B).

Beta Box Model A method for simplifying transistors so that they can be solved using Ohm's law.

Biasing The application of DC voltages to a transistor's elements to produce the desired mode of operating conditions.

Bipolar Junction Transistor (BJT) An electronic device consisting of a semiconductor material sandwiched between two other dissimilar semiconductor materials, forming a transistor. These materials are called the emitter, base, and collector.

Bistable Multivibrator A multivibrator having two stable output states. Also known as a flip-flop circuit.

Bode Plot A graph which relates the gain of an amplifier to changes in frequency. The vertical axis represents the gain in dB and the horizontal axis represents the range of input frequencies.

Bootstrapping A circuit configuration in which a portion of the output is fed back to the input to force the circuit into a particular mode of operation.

Breakdown Diode A special type of diode, such as a Zener diode, designed to be operated with reverse voltage applied such as a Zener Diode.

Breakdown Region The range of reverse voltages applied to a semiconductor device which causes avalanche conduction to occur.

Breakdown Voltage (See Zener Voltage)

Breakover Voltage The minimum forward bias voltage (anode-to-cathode) required for a thyristor to begin conducting.

Bridge Rectifier A rectifier circuit consisting of four diodes arranged so that two at a time conduct to produce full-wave rectified DC. This circuit produces the highest DC output for a given AC input.

Capacitance A measure of the ability to store an electrical charge.

Capacitor Coupling The connection between circuits whereby AC signals are passed from one circuit to another while DC signals are prevented from passing through.

Capacitor Input Filter A passive filter in which a capacitor is placed across the circuit load to remove unwanted voltage variations.

Carrier Deviation (See Frequency Deviation)

Cascade To connect the output of one circuit to the input of another.

Cascaded Connected in series.

Center Frequency Point The geometric average of the upper and lower cutoff frequencies.

Center-tapped Full-wave Rectifier A circuit which requires a center-tapped transformer and two diodes to provide a full-wave rectified DC output.

Channel A semiconductor material connecting the source and drain in a Field-effect Transistor.

Chemically Stable Refers to a condition in which the outer orbit of an atom contains all the electrons it can hold, preventing the atom from combining with any other atoms.

Clamped Positive Sine Wave An AC signal which has its reference point shifted in the positive direction (i.e., a sine-wave riding on a positive DC level).

Clamper A circuit designed to produce a DC level from an AC signal and add this DC to the AC signal.

Clamping (DC Restoration) Shifting the DC reference level of an AC signal.

Class A Amplifier An amplifier biased so that the active device conducts for the entire 360° of the input cycle.

Class AB Amplifier An amplifier biased so that the active device conducts for more than 180° but less than 360° of the input cycle.

Class B Amplifier An amplifier biased so that the active device conducts for 180° of the input cycle.

Class C Amplifier An amplifier biased so that the active device conducts for less than 180° of the input cycle.

CMOS Circuit A circuit using N-MOS and P-MOS devices in a complementary arrangement.

Collector Feedback Bias A very stable BJT biasing method using a resistor connected between the collector and base so that IB is determined by V_C.

Collector Junction (See Collector-Base Junction)

Collector-Base Junction The union of the collector and base semiconductor materials in a BJT.

Common Gate Amplifier A configuration of a FET circuit in which the Gate terminal is common to both the input and the output signals.

Common Source JFET Amplifier A JFET amplifier in which the Source terminal is common to both the input and output signals.

Common Mode Rejection The ability of a differential amplifier to ignore signals which are applied to both input terminals simultaneously.

Common Mode Rejection Ratio (CMRR) A measure of a differential amplifier's ability not to respond to a signal present on both input terminals, usually measured in decibels.

Common Mode Signal A component of the input signal that is present equally on both the positive and negative input lines.

Commutating Capacitor (See Speed-up Capacitor)

Commutation The act of changing from a conducting state of operation to a non-conducting state in synchronization with a sine wave signal's zero crossing points.

Comparator A circuit which produces an output signal indicating the relative polarity of two input signals.

Compound Semiconductor (See Intermetallic Semiconductor)

Conductance A measure of the ease with which a circuit conducts, expressed in the units of Siemens (S).

Constant Current Source A circuit which maintains an unvarying current flow over a wide range of voltage changes.

Controlled Avalanche Rectifier A high-voltage diode designed to be safely operated in the breakdown region.

Covalent Bond A means by which atoms adhere together by sharing electrons.

Critical Frequency The frequency at which the circuit output has dropped 3 dB from the maximum output.

Crossover Distortion A form of distortion occurring in Class B push-pull amplifiers because both output transistors are biased off for a short period of time.

Crystal A quartz device which operates on piezoelectric effect and has stable resonant frequency properties.

Current (or power) Amplifier A circuit designed to provide a gain in current from input to output.

Current Divider A circuit designed to separate current into smaller amounts; the sum of these smaller amounts will equal the original current.

Current Feedback An amplifier circuit configuration in which a portion of the output current is returned to the input to modify the gain and/or frequency response of the circuit.

Current Gain The ratio of output current to input current represented by the symbol A_i: $A_i = I_{OUT}/I_{IN}$.

Current Mirror A circuit which acts as a current source of a value equal to the current through a resistor and diode.

Cutoff A biasing condition in a BJT which causes emitter current to drop to zero.

D MOSFET A metal oxide field effect transistor constructed by isolating the gate from the channel with a thin layer of silicon dioxide.

Darlington Pair Configuration A pair of similar transistors connected such that the emitter of the first supplies the base current for the second, resulting in a current gain equal to the product of the βs of the individual transistors.

DC Biasing A circuit connection used to provide the desired quiescent voltages and currents to an active device such as a BJT.

DC Drift An undesirable change in operating conditions caused by age or operating temperature of an active device.

DC Filters (See Base-line Stabilizers)

DC Inserters (See Base-line Stabilizers)

DC Restoration (See Clamping)

Decay Time The amount of time for a transistor to come out of saturation and for the current to drop to 10% of the saturated current value.

Decibel A unit used to measure gain based on the common logarithm.

Delay Time The amount of time required for a transistor in cutoff to begin conducting after application of base current.

Depletion Bias A form of MOSFET bias in which the voltage on the gate is used to provide reverse bias.

Depletion Mode A mode of operating JFETS in which the amount of current flowing is controlled by depleting the channel of free electrons using a reverse gate bias.

Depletion MOSFET An insulated gate FET that uses the action of the depletion layer to control drain current.

Dielectric An insulator placed between two conductors.

Differential Amplifier An amplifier which produces an output voltage proportional to the difference between two input voltages.

Differentiator A circuit which produces an output voltage approximating the rate of change of the input voltage.

Diffusion The process of conduction band electrons crossing a PN junction and recombining with valance band holes.

Diode Bias The use of diodes in place of resistors in a class B amplifier to reduce operating changes due to temperature by matching the diode forward voltage drop to V_{BE} of the output transistors.

Distortion An unwanted change in the shape of an AC signal.

Donor A material such as Arsenic, Antimony or Phosphorus which, when added to silicon, donates free electrons.

Doping A process of adding impurities to pure silicon to create an excess of either free electrons or holes.

Double L or Inverted Double L Filter A circuit used to remove unwanted variations or ripple from a DC signal, named for its shape as an upside-down letter "L".

Drain One of the three terminals of a FET, equivalent to a BJT collector.

Drain Feedback Bias A bias method in which a resistor is connected between the drain and the gate of a MOSFET such that it causes V_D to control V_G.

Dual-Gate MOSFET A MOSFET that comes with two gates to reduce gate input capacitance.

Dual Polarity Power Supply A supply with both positive and negative outputs, usually at the same voltage.

Duty Cycle The ratio of on time to waveform period for step voltage waveform, usually expressed in percent.

Dynamic Resistance A resistance whose value changes as a factor of circuit conditions.

E MOSFET A MOSFET designed and constructed to operate only in the enhancement mode.

Effective (or Root-Mean-Square) Value A measure of the heating effect of a sine wave; calculated by multiplying peak voltage times 0.707.

Efficiency The ratio of AC load power to the DC power supplied to the circuit, multiplied by 100 to be expressed as a percent.

Electrically Identical Refers to a circuit which, if replaced by another simpler circuit, exhibits exactly the same electrical characteristics.

Electricity The flow of electrons from a negative potential to a positive potential.

Electron The basic atomic particle in orbit around the nucleus which has a negative charge.

Electron Current Flow A convention where current is assumed to flow from the negative terminal of a source, through the circuit, and back to the positive terminal.

Element A fundamental substance containing only one kind of atom.

Elemental Semiconductor A type of semiconductor material which contains only one type of atom, that of the base material.

Emitter Bias A form of BJT bias using dual polarity supplies and a grounded base resistor.

Emitter Junction (See Emitter-base Junction)

Emitter-base Junction The depletion region and the junction between the emitter and base of a BJT.

Emitter-follower Circuit A circuit in which the input is applied to the base and the output taken from the emitter, also called a common-collector circuit.

Enhancement MOSFET (See E MOSFET)

Equipotential Points Nodes of a circuit having identical voltages with respect to a common.

Error Voltage The difference of potential between the inverting and the noninverting input terminals of a differential amplifier.

Extrinsic Semiconductor An intrinsic semiconductor material which has had impurities added.

Feedback Node A junction of two or more components in a circuit output used to provide a current path back to the circuit input.

Field Effect Transistor (FET) A three-terminal device that depends on an electric field to control current flow.

Firm Voltage Divider A divider in which the load is at least 10 times the value of the resistor with which it is parallel.

Flyback A large voltage potential generated by the collapse of a magnetic field around an inductive component.

Forward Biasing The bias applied to a PN junction which causes the depletion region to decrease, allowing current to flow.

Forward Current The current flowing through a PN junction when it is forward biased.

Forward Transfer Admittance (yfs) A measure of how the gate-to-source voltage affects the source current, which includes the effect of input reactance.

Forward Transfer Admittance Graph A plot of Forward Transfer Admittance versus Drain Current, used to determine Forward Transconductance.

Free Electron An electron that is not bound to a specific atom. Also called a conduction-band electron.

Frequency Deviation, or Carrier Deviation A measure of the amount a carrier signal changes frequency due to the modulating signal in frequency modulated systems.

Frequency Multiplier A circuit such as a Class C amplifier that produces an output signal higher in frequency than the input signal.

Frequency Response A measure of how a circuit responds to a range of frequencies.

Full-wave Rectifier A circuit which converts an AC sine-wave input into a pulsating DC voltage with two pulses occurring for each input cycle.

Gain The ratio of output to input of an amplifier. This may be voltage, current, or power gain.

Gain-bandwidth The product of the gain times the bandwidth of an amplifier which is always a constant for a particular amplifier.

Galvanometer A very sensitive electromagnetic current meter.

Gate The terminal of a FET which is equivalent to the base of a BJT.

General Voltage-divider Equation A method of finding the voltage drop across a resistor in a voltage divider by multiplying the ratio of the resistor in question to the total resistance by the applied voltage.

Half-power Point The frequency at which the output power of a circuit drops to one-half of the maximum power output.

Half-wave Rectifier Circuit A circuit that converts an AC sine-wave voltage input into a pulsating DC output voltage with one pulse occurring for each input cycle.

Half-wave Signal A pulse that appears for 180° of a symmetrical waveform.

Harmonic A sinewave that is an integer multiple of a base or fundamental frequency.

High-pass Filter A circuit which allows high frequency signals to reach the output while blocking low frequency signals.

Highpass Circuit A circuit designed to act as a high-pass filter.

Holding Current The minimum current level required to sustain current flow once a thyristor is in the regenerative mode of operation. This is less than the current required for regeneration to begin.

Hole A positive charge due to the absence of an electron in the valance band of an atom.

Hybrid (or H) Parameters A mathematical method of representing transistors under full-load and no-load conditions.

Hysteresis Voltage The voltage difference between the upper trigger point and the lower trigger point of a Schmitt trigger circuit.

Induced Channel A channel formed between the source and drain of a MOSFET by the action of a potential applied to the gate.

Input Bias Current The base current of the input transistors in an operational amplifier.

Input Offset Current The difference between the two input currents of a differential amplifier due a mismatch in Beta of these two transistors.

Insertion Loss An undesirable attenuation of a signal due to the circuit characteristics when connected, compared with the signal that would be present if the circuit were not there.

Instrumentation Amplifier A high gain differential DC amplifier used to increase the output of a transducer to a usable level.

Insulated Gate Field Effect Transistor (IGFET) Another name for a MOSFET transistor.

Integrator A circuit whose output is proportional to the area of the input waveform

Interbase Resistance The resistance between the bases of a unijunction transistor, measured when there is no emitter current.

Intermetallic (or Compound) Semiconductor A semiconductor material, such as gallium arsenide, containing more than one element.

Intrinsic Semiconductor A pure semiconductor crystalline material with no defects in the lattice structure.

Intrinsic Standoff Ratio The ratio of emitter-base 1 resistance to the interbase resistance in a unijunction transistor.

Inversion Layer A channel formed between the source and drain of a MOSFET by the action of a potential applied to the gate, also called an induced channel.

Inverted Double L Filter See Double L Filter.

Inverter A switching circuit in which the output voltage is 180_ out of phase with the input voltage.

Inverting Amplifier An amplifier in which the output is exactly 180_ out of phase with the input.

Junction Field Effect Transistor (JFET) A transistor created by diffusing a gate region into a channel with the current flow through the channel controlled by reverse voltage on the gate.

K A MOSFET constant determined during the manufacturing process which can be calculated by the ratio of I_D to $V_G{}^2$.

Kirchhoff's Current Law A statement that the total current entering a junction equals the total current leaving the junction.

Kirchhoff's Voltage Law A statement that the sum of the voltage drops around a closed loop equals zero.

Lagging Network A circuit in which the phase of the output lags behind the input signal.

Latching Current The minimum current level required for a thyristor to continue conducting once the breakover voltage is exceeded.

Lattice An organized arrangement of atoms in which atoms link with other like atoms by sharing valance electrons.

Law of Electrical Conduction The fundamental property that like charges repel and unlike charges attract.

Leading Network A circuit in which the phase output of the output voltage leads the phase of the input voltage.

Line Regulation A measure of the ability of a power supply to maintain a constant output voltage over a range of line voltage changes.

Linear Applications Circuit uses in which the input signal is reproduced exactly at the output with a difference of amplitude only.

Load Regulation A measure of the ability of a power supply to maintain a constant output voltage over the range of output current from no-load to full-load.

Low-pass Filter A circuit which allows low frequency signals to reach the output while blocking high frequency signals.

Lower Sideband Harmonic frequencies lower than the carrier frequency, resulting from the modulation of a carrier wave with a lower frequency signal.

Lower Threshold (or Lower Trip Point (LTP)) The lower input switching level of a Schmitt trigger. Crossing the LTP while going in a negative direction will cause the circuit output to switch.

Lower Trip Point (see Lower Threshold)

Lowpass Circuit A circuit which allows low frequency signals to reach the output while blocking high frequency signals.

Majority Carriers The primary charge carriers in doped semiconductor materials. In N-type material, the majority carriers are electrons; in P-type material the majority carriers are holes.

Majority Current Carriers (See Majority Carriers)

Matter Anything that has mass and takes up space.

Metal Oxide Semiconductor Field Effect Transistor (MOSFET) A transistor constructed by forming an insulating layer between the gate and channel.

Microwave Power Diodes Semiconductor devices such as the IMPATT or Gunn diodes which have been designed to operate at frequencies above one GHz.

Miller's Theorem A method of circuit analysis in which a feedback capacitor in an amplifier can be represented by equivalent capacitors across the input and output.

Milliammeter An instrument for measuring current scaled to read in thousandths of one ampere.

Minority Carriers The secondary charge carriers in doped semiconductor materials. In P-type material, the minority carriers are electrons; in N-type material the minority carriers are holes.

Minority Current Current flow of the secondary carriers. In N-type material, the hole flow is the minority current. In P-type material, the electron flow is the minority current.

Minority Current Carriers (See Minority Carriers)

Modulated Stage An amplifier stage which processes a modulated signal.

Modulation Envelope A curve connecting the peaks of a carrier signal which indicates the signal used to amplitude modulate the carrier wave.

Modulator The stage of a transmitter which combines the carrier signal with the input signal.

Molecules The smallest particle of a substance that still retains the properties of that substance.

Multistage Amplifier An amplifier consisting of two or more cascaded sections of amplification.

Multivibrator A switching circuit used to generate a repeatable squarewave output.

N-channel JFET A transistor constructed by forming a belt of P-type material around an N-type substrate. A voltage applied to the P-type material constricts current flow by effectively reducing the channel.

N-Type Semiconductor A semiconductor material that has been doped to provide an excess of free electrons.

Negative Alteration The portion of a sine wave which is of lower voltage than the reference voltage, usually the negative half of the sine wave.

Negative Feedback A circuit arrangement in which a portion of the output is returned to the input out of phase with the input.

Negative Ion An atom which has an excess of electrons and is electrically negative.

Negative Peak Clipper A circuit that removes the negative peaks of an input signal at a predetermined voltage.

Negative Resistance Region A characteristic of a Unijunction Transistor or tunnel diode by which a decrease in voltage causes an increase in current, or an increase in voltage causes a decrease in current.

Neutron A particle of matter found within the nucleus of an atom which has no electrical charge.

Node A connection point in a circuit of two or more components; a point where two or more current paths come together or diverge.

Nonlinear Distortion A type of distortion caused by operating the base-emitter junction of a BJT in its nonlinear range, resulting in a rounding of the peak of the output signal.

NPN Transistor A bipolar transistor in which the emitter and collector are made of N-type material and the base is made of P-type material.

Nulling The act of adjusting an operational amplifier for zero output volts when the input is zero.

Nulling Circuits Components used in conjunction with an operational amplifier to zero out the effects of the input offset voltage.

Ohm's Law A statement that current is directly proportional to voltage and inversely proportional to resistance, or I = V/R.

Ohmic Region A area of operation of a JFET approximating a variable resistor.

Ohmmeter An instrument for measuring resistance.

Open Circuit Voltage The voltage present at a pair of terminals when no current is flowing.

Optoisolator A device consisting of an LED and a phototransistor which is capable of transferring an electrical signal from input to output while providing extremely high isolation values.

Oscilloscope An instrument which displays the amplitude and waveshape of voltage versus time on a cathode ray tube.

Output Offset Current The amount of current flowing at the output of an amplifier with the inputs at zero.

Output Offset Voltage Any deviation from zero volts at the output of a differential amplifier when measured with both inputs connected to ground.

P-Channel JFET A transistor constructed by forming a belt of N-type material around a P-type substrate. A voltage applied to the N-type material constricts current flow by effectively reducing the channel.

Parallel Branch Portions of a circuit connected so that the voltage is the same across each branch.

Peak Inverse Voltage (See Peak Reverse Voltage)

Peak Negative Voltage The value of instantaneous voltage at 270°.

Peak Positive Voltage The value of instantaneous voltage at 90°.

Peak Reverse Voltage Rating The maximum amount of voltage that can be applied to a diode in the reverse direction, often called the Peak Inverse Voltage (PIV).

Peak Voltage The maximum positive or negative voltage of an alternating waveform.

Peak-to-Peak Voltage The total voltage change from the maximum positive voltage to the minimum negative voltage of an alternating waveform.

Pentavalent Atoms An atom, such as arsenic, antimony, or phosphorous, which has five valence electrons.

Period The amount of time it takes to complete one cycle of a periodic wave.

Phasor A graphical representation of the angular displacement of a waveform over time.

Pinch-off Voltage The amount of reverse bias voltage applied to the gate of a JFET that causes the channel current to become constant.

PN Junction The connection point between P-type material and N-type material.

PN Junction Diode An electronic component which uses the PN junction to restrict current flow to one direction: from the cathode to the anode.

PNP Transistor A bipolar junction transistor in which the emitter and collector are made of P-type material and the base is made of N-type material.

Positive Alteration The portion of a sine wave which is of higher voltage than the reference voltage, usually the positive half of the sine wave.

Positive Ion An atom having a positive charge due to a deficiency of electrons.

Positive Peak Clipper A circuit that removes the positive peaks of an input signal at a predetermined voltage.

Power Amplifiers A amplifier designed to increase the power of an input signal, usually by having a large current gain.

Power Gain The product of current gain times the voltage gain.

Product Over the Sum A method of finding the equivalent resistance of two resistors in parallel, $R_T = (R_1 \times R_2)/(R_1 + R_2)$.

Proton A positively charged particle of matter found within the nucleus of an atom.

Pulse Repetition Frequency (or Rate) The number of pulses generated per second.

Pulse Repetition Rate (see Pulse Repetition Frequency.)

Pulse Train A series of rectangular pulses appearing at regular intervals.

Push-Pull A circuit configuration in which two transistors, usually an NPN and a PNP, each conduct for one-half of the input cycle. The NPN delivers current to the load during the positive alteration while the PNP delivers current to the load during the negative alteration.

Q Point The point on the AC or DC load line that marks the current flowing and the voltage dropped by a transistor when no AC signal is present.

Quality Factor The ratio of inductive reactance (X_L) to the winding resistance (R_W) of a coil. In symbols: $Q = X_L/R_W$.

Quiescent Refers to a circuit condition in which there is no AC signal applied.

Ramp Generator A circuit which has a linear increase or decrease in voltage or current.

Ramp Signal A signal which has a linear increase or decrease in voltage or current.

RC Time Constant A time interval equal to the product of resistance times capacitance which determines circuit response.

Reciprocal of the Sum of the Reciprocals A mathematical method for determining the equivalent resistance of resistors in parallel.

Recovery (or Settling) Time The time required for a circuit to return to its normal quiescent state after a disturbing input is removed.

Regulation Maintaining a constant voltage or current over a range of load variations.

Resistor-capacitor (RC) Coupling A passive circuit in which a resistor and a capacitor are connected to provide a method of transferring an AC signal from one circuit to another.

Reverse Biasing Applying a reverse voltage to a PN junction, which widens the depletion region and inhibits current flow.

Reverse Current A small leakage current that flows when a PN junction is reversed biased.

Reverse Current Diodes (see Breakdown Diodes).

Ripples Variations of amplitude on a signal which cause the signal to deviate from being a true sine wave.

Ripple Band The region of a signal containing ripples.

Ripple Rejection Ratio The amount of variation in the output as compared with the variation on the input, calculated as: Ripple Rejection Ratio= (ripple output)/(ripple input).

Ripple Voltage Undesirable fluctuations in the output of a DC power supply.

Rise Time The time it takes a pulse to increase from 10% of its maximum value to 90% of its maximum value.

Roll Off A measure of how rapidly the gain of an amplifier decreases with a change in frequency, measured in dB per decade or dB per octave.

Root Mean Square Value (RMS) (see Effective Value.)

Root-mean-square The square root of the average of a group of numbers, each squared.

Sacrifice Factor The amount of gain sacrificed to increase the bandwidth of an amplifier, calculated as: $S = 1 + $ (gain times feedback).

Saturation A condition of a BJT in which an increase in base current

will not cause an increase in collector current.

Schmitt Trigger A circuit which detects when the input voltage exceeds a predetermined level or when the input voltage is less than another predetermined level; a comparator with hysteresis.

Self-biased Circuit The bias developed across the source resistor when using a JFET.

Self-heating A condition in BJT transistors in which the temperature of the transistor causes an increase in Beta, which causes an increase in collector current, which causes an increase in temperature, or thermal runaway.

Semiconductor A material that has a conductance value between that of an insulator and that of a conductor.

Series Aiding The connection of two or more sources in series in a manner such that their voltages add.

Series Opposing The connection or two or more sources in series in a manner such that their voltages subtract.

Series String A portion of a larger circuit in which components are connected end to end (in series).

Settling Time (see Recovery Time)

Shunt Clipper A circuit designed to remove either the positive or the negative alteration of an input waveform by placing a diode in parallel with the load.

Signal Clipper A circuit designed to remove part of a waveform and to pass only the part that is above or below a predetermined voltage level.

Signal Clamping Circuit (see Clamper)

Single Stage Amplifier An amplifier circuit that has only one active device.

Slew Rate The speed at which the output of an operational amplifier changes in response to a step change at the input, measured in volts per micro-second.

Slew Rate Distortion An undesirable change in the output signal of an operational amplifier due to the input signal changing at a rate which exceeds the switching rate of the amplifier.

Small-signal Amplifiers An amplifier intended for small input levels such that peak-to-peak AC collector current is less than 10% of the DC emitter current and power dissipation is less than 0.5 watt.

Solid State Device An electrical component that is constructed of semiconductor materials.

Source The terminal of a field effect transistor that is equivalent to the emitter of a BJT.

Source Follower A FET circuit in which the AC signal is applied to the gate and the output taken from the source, also known as a common drain amplifier.

Speed-up (or Commutating) Capacitor A capacitor placed in the base circuitry of a transistor switch to improve the switching speeds of the circuit.

Stage Efficiency The maximum power supplied to the load, divided by the DC power supplied to the stage, and multiplied by 100 to convert to percent.

Step Voltage Change Signals which exhibit a rapid change from one voltage level to another.

Step-down Transformer A transformer manufactured with fewer turns in the secondary than in the primary to cause the secondary voltage to be less than the primary voltage.

Stiff Voltage Divider A divider in which the load is at least 100 times the resistor with which it is in parallel.

Storage Time The amount of time for the collector current of a saturated transistor to drop from 100% to 90% after the base current has been removed.

Strain Gauge A device made of fine wire which will change resistance in proportion to the amount of tension or compression applied.

Superposition Theorem A method for analyzing circuits with two or more voltage sources by considering the effects of each independently and adding the effects algebraically.

Swamping Resistor A small resistor in the emitter circuit of an amplifier which is not bypassed and is used to stabilize gain and increase input impedance.

Symmetrical Limiter A clipper circuit using back-to-back zener diodes to limit the positive and negative peaks to a pre-determined voltage level.

Tail Current The current through the tail resistor in a differential amplifier.

Tail Resistor The common emitter resistor shared by the input transistors in a differential amplifier.

Thermal Runaway A condition in BJT transistors in which the temperature of the transistor causes an increase in Beta, which causes an increase in collector current, which causes an increase in temperature and continues until the transistor destroys itself.

Thermocouple The junction of two dissimilar metals which produces a voltage proportional to the temperature of the junction.

Thevenin's Equivalent Circuit A circuit analysis model in which any circuit is reduced to a single equivalent voltage source and a single equivalent resistance.

Thevenin's Theorem A circuit analysis method which reduces any circuit to an equivalent single source and single resistance.

Thevenize To reduce a circuit to an equivalent simpler circuit by applying Thevenin's theorem.

Threshold Voltage The amount of gate voltage in a MOSFET required to cause conduction to begin from source to drain, or in a JFET, the amount of gate to source voltage required to start drain current.

Transconductance A measure of how changes in gate voltage affect drain current in JFETS.

Transfer Characteristics Graph A graphical representation of the output voltage versus the input voltage of a circuit in which the output switches between positive and negative saturation.

Transformer Coupling A method of transferring AC signals from one circuit to another while blocking DC by using a transformer.

Transducers A device which converts one form of energy change to another, such as sound to voltage in a microphone.

Transient Refers to undesirable non-repetitive voltages or current spikes, usually at very high frequency.

Transistor-transistor Logic An integrated circuit construction method for digital circuits in which one transistor provides a path to ground while a second transistor provides a path to voltage.

Transition Region The range of frequencies from the point where the gain of an amplifier begins to decrease to the point where the output is considered zero.

Triangle Signal A waveform consisting of a linear positive going ramp and linear negative going ramp with each having the same absolute rate of change of voltage.

Tuned Coupling A method of transferring AC signals of a particular frequency from one circuit to another by using capacitors across the transformer windings resonant at a selected frequency.

Tunnel Diode A solid-state device which has been designed to exhibit tunneling, useful for high frequency due to its negative resistance characteristics.

Tunneling The process of a diode's avalanche current occurring near zero reverse voltage.

Turns Ratio A measure of transformer characteristics which uses the ratio of the number of turns of wire on the primary winding to the number of turns on the secondary.

UJT Relaxation Oscillator A circuit based on the unijunction transistor in which the output frequency is determined by the charging and discharging of a capacitor connected across the emitter-lower base terminals.

Unbalanced Bridge Circuit A series-parallel resistor circuit in which the ratio of the resistors in the series strings is different for each string.

Unbalanced Load An undesirable condition in which the amount of current required by a load is different for the positive alteration of a sine wave than for the negative alteration.

Unijunction Transistor A three terminal semiconductor device usually used as a relaxation oscillator whose trigger voltage is dependent on the applied bias voltages.

Unity-gain Amplifier (see Voltage Follower).

Upper Sideband Harmonic frequencies higher than the carrier frequency, resulting from the modulation of a carrier wave with a lower frequency signal.

Upper Threshold (or upper trip point (UTP)) The upper input switching voltage of a Schmitt trigger.

Upper Trip Point (see Upper Threshold)

Valance Crystals Crystal structures formed by covalent bonding of atoms.

Valance Shell The outermost electron orbit for an atom.

Valley Voltage The point at which the diode current changes from increasing to decreasing current for increases in reverse voltage, signifying the start of negative resistance characteristics.

Varactor A special diode which has a PN junction optimized for variable capacitance in a manner such that different levels of bias will change the capacitance value.

Virtual Ground A node in a circuit which is at zero volts with respect to ground but is not connected physically to ground.

Voltage Amplifiers Amplifiers in which the AC voltage increases from input to output.

Voltage Divider A series circuit in which different voltages are produced at the nodes between components with respect to ground.

Voltage Doubler A circuit which produces a DC voltage output which is twice the peak input voltage.

Voltage Follower, or Unity-gain Amplifier An amplifier in which the output is in phase with and exactly the same amplitude as the input signal.

Voltage Gain The ratio of output voltage to input voltage usually represented by the symbol AV.

Voltage Multiplier A circuit which produces a DC output voltage which is a multiple of the peak input voltage.

Voltmeter An instrument for measuring voltage.

Zener Diode A diode designed to operate in the reverse breakdown region and drop a constant voltage.

Zener Impedance The opposition to a change in current through Zener diode.

Zener Knee The point on a Zener diode curve at which the Zener begins to conduct in the reverse direction.

Zener Voltage The voltage across a Zener diode when it is operated in the reverse breakdown region of operation.

Zero Adjust The front panel adjustment of a measuring instrument which permits setting the indication to zero when there is a zero input.

Zero Crossing Points The points at which the AC signal is at zero volts in amplitude (for a sine wave, these points occur at $0°$, $180°$, and $360°$).

EXAMPLE 3.3

The output from a half-wave rectifier is depicted in Figure A.1, below:

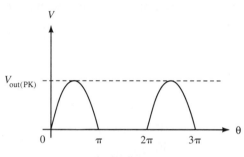

FIGURE A.1

During the first half cycle, the equation for V_{AVG} is

$$V_{AVG} = V_{pk}\sin\theta$$

for $0 < \theta < \pi$.

During the second half cycle, the equation for V_{AVG} is

$$V_{AVG} = 0$$

for $\pi < \theta < 2\pi$.

Therefore, integrating over the first half cycle, the average voltage from a half-wave rectifier becomes:

$$V_{AVG} = \frac{V_{out(pk)}}{2\pi} \int_0^\pi \sin\theta \, d\theta$$

Therefore

$$V_{AVG} = -\frac{V_{out(pk)}}{2\pi}[\cos\theta]_0^\pi = -\frac{V_{out(pk)}}{2\pi}(-2)$$

Thus the average voltage for a half-wave rectifier becomes:

$$V_{AVG} = \frac{V_{our(pk)}}{\pi} = 0.318 V_{out(pk)}$$

EXAMPLE 3.4

The output from the full-wave rectifier is depicted in Figure A.2. As you can see, the interval for a full-wave rectifier covers one complete cycle. Thus, the period of a full-wave rectifier equals π.

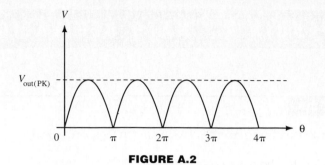

FIGURE A.2

Therefore, integrating over one complete cycle, the average voltage from a full-wave rectifier becomes:

$$V_{AVG} = \frac{1}{\pi} \int_0^\pi V_{pk} \sin\theta \, d\theta$$

Now, rearranging the equation gives

$$V_{AVG} = \frac{V_{pk}}{\pi} \int_0^\pi \sin\theta \, d\theta$$

Next, taking the integration of our equation gives

$$V_{AVG} = \frac{V_{pk}}{\pi} [\cos\theta]_0^\pi = -\frac{V_{pk}}{\pi} (-2)$$

Finally, the equation for the average voltage of a full-wave rectifier becomes

$$V_{AVG} = \frac{2V_{pk}}{\pi} = 0.636 V_{pk}$$

EXAMPLE 5.4

Using Shockley's equation for the total current through the base-emitter pn junction, which is

$$I_T = I_R(e^{\frac{VQ}{kT}} - 1)$$

Where

    I_T = Total forward current through the pn junction
    I_R = Reverse Saturation current through the pn junction
    V = Voltage across depletion layer
    Q = Charge of an electron, approximately 1.602×10^{-19}
    k = Boltzmann's constant, a number approximately equal to 1.381×10^{-23} J/°K
    T = Absolute temperature of the device

Solving the Q/kT at ambient room temperature 21°C; approximately 70°F yields:

$$\frac{Q}{kT} = \frac{1.602 X 10^{-19}}{\left(1.381 X 10^{-21}\right)(294)} = 40$$

Substituting above results into original equation gives us

$$I_T = I_R\left(e^{40V} - 1\right)$$

Expanding and rearranging the above equation gives us

$$I_T + I_R = I_R e^{40V}$$

Since the sum of the current must flow through the emitter, the emitter current I_E becomes

$$I_E = I_T + I_R$$

Therefore

$$I_E = I_R e^{40\,V}$$

Now, differentiating the original equation yields

$$\frac{dI}{dV} = 40\, I_R e^{40\,V}$$

Substituting into the above result gives us

$$\frac{dI}{dV} = 40\, I_E$$

Taking the reciprocal of the above differentiation will give the AC resistance of the base-emitter junction, therefore

$$r_e = \frac{dV}{dI} = \frac{1\,V}{40\,I_E} = \frac{25\,mV}{I_E}$$

EXAMPLE 9.6

This proof is done as a reminder of basic AC theory; the discharge of a capacitor is a decaying exponential function described by:

$$v_c\!\left(t\right) = v e^{-\frac{t}{\tau}}$$

Where

    v = total voltage charge on the plates of the capacitor

    t = number of time constants being evaluated

    τ = one time constant equal to resistance times capacitance

The circuit described in Example 9.6 has a base resistance of 47kΩ and a speed-up capacitor of 18pF. Therefore, one time constant would give us

$$\tau = R_B C_S = (47\ k\Omega)\,(18\ pF) = 846\ nS$$

Since we are interested in the voltage after three time constants, therefore

$$t = 3\tau = 3\ (846\ nS) = 2.5\ uS$$

Now, substituting these times into our original equation for the discharge of the capacitor gives us

$$v_c\!\left(t\right) = 2V(e^{-\frac{2.54\,uS}{846\,nS}}) - 99.34\,mV$$

If we take the ratio of initial charge to remaining charge after three time constants we get

$$\frac{99.34\,mV}{2.0\,V}\,X100 = 5\%$$

Since a capacitor has 5% of its initial charge left after three time constants, it must be 95% discharged.

EXAMPLE 10.4

Figure A.3 will aid us in developing the proof of Miller's Theorem:

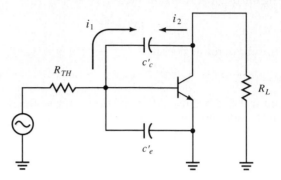

FIGURE A.3

The effect of $C'_{c'}$ as a feedback capacitor, on the input circuit is

$$i_1 = \frac{V_1 - V_2}{X_{c_c}}$$

Since the feedback reactance will attenuate gain

$$-A_V = \frac{V_2}{V_1}$$

Therefore

$$V_2 = -A_V V_1$$

Now, substituting and factoring gives us

$$i_1 = \frac{V_1 - \left(-A_V V_1\right)}{X_{C_C}} = \frac{V_1(1 + A_V)}{X_{C_C}} = \frac{V_1}{\left(\dfrac{X_{C_C}}{(1 + A_V)}\right)}$$

The effects of any reactance, as seen by the input circuit, would be

$$X_{C_{in(Miller)}} = \frac{X_{C_C}}{1 + A_V}$$

Since

$$C_{in(Miller)} = \frac{1}{2\pi f X_{C_{in(Miller)}}}$$

And

$$C_{OB} = \frac{1}{2\pi f X_{C_C}}$$

Substituting and rearranging gives us

$$\frac{1}{2\pi f C_{in(Miller)}} = \frac{1}{2\pi f C_{OB}(1 + A_V)}$$

Now, inverting and multiplying by 2πf gives us

$$C_{in(Miller)} = (1 + A_V)C_{OB}$$

Evaluating the effects of $C'_{c'}$ on the output circuit gives us

$$i_2 = \frac{V_2 - V_1}{X_{C_C}}$$

The effect of $C'_{c'}$ as a feedback capacitor, on the output circuit is

$$-\frac{1}{A_V} = \frac{V_1}{V_2}$$

Therefore

$$V_1 = \frac{V_2}{A_V}$$

Substituting and factoring gives us

$$i_2 = \frac{V_2 - \left(-\dfrac{V_2}{A_V}\right)}{X_{C_C}} = \frac{V_2\left(1 + \dfrac{1}{A_V}\right)}{X_{C_C}}$$

Now, rearranging and expanding gives us

$$i_2 = \frac{V_2}{\left(\dfrac{X_{C_C}}{\left(1 + \dfrac{1}{A_V}\right)}\right)} = \frac{V_2}{\left(\dfrac{X_{C_C}}{\left(\dfrac{(A_V) + 1}{A_V}\right)}\right)}$$

The effects of any reactance, as seen by the output circuit, would be

$$X_{C_{out(Miller)}} = \frac{X_{C_C}}{\left(\dfrac{(A_V + 1)}{A_V}\right)}$$

Since

$$C_{out(Miller)} = \frac{1}{2\pi f X_{C_{out(Miller)}}}$$

And

$$C_{OB} = \frac{1}{2\pi f X_{C_C}}$$

Substituting and rearranging gives us

$$\frac{1}{2\pi f C_{out(Miller)}} = \frac{1}{2\pi f C_{OB}\left(\dfrac{(A_V + 1)}{A_V}\right)}$$

Now, inverting and multiplying by 2πf gives us

$$C_{out(Miller)} = \left(\frac{(A_V + 1)}{A_V}\right)C_{OB}$$

EXAMPLE 12.5

Figure A.4 will aid us in developing the proof:

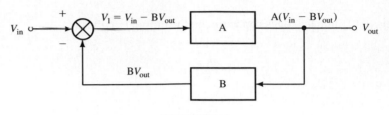

FIGURE A.4

We will start this proof with the basic equation for closed-looped gain.

$$A_{v(CL)} = \frac{V_{out}}{V_{in}}$$

In Figure A.4, V_{out} equals $A(V_{in} - BV_{out})$; isolating V_{out} on one side of the equation gives

$$V_{out} = \frac{AV_{in}}{1 + AB}$$

Also in Figure A.4, V_1 equals $V_{in} - BV_{out}$; rearranging and isolating V_{in} gives

$$V_{in} = V_1 + BV_{out}$$

Substituting these results into the original equation gives

$$A_{v(CL)} = \frac{\left(\dfrac{AV_{in}}{1 + AB}\right)}{V_1 + BV_{out}} = \frac{\left(\dfrac{A(V_1 + BV_{out})}{1 + AB}\right)}{V_1 + BV_{out}}$$

Next, inverting and multiplying

$$A_{v(CL)} = \left(\frac{A(V_1 + BV_{out})}{1 + AB}\right) = \left(\frac{1}{V_1 + BV_{out}}\right)$$

Now, we have the closed-loop gain equation for the noninverting amplifier

$$A_{v(CL)} \frac{A}{1 + AB}$$

Figure A.5 will aid us in developing the proofs for the Wien-bridge oscillator:

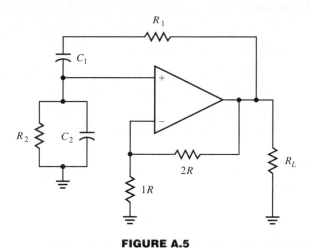

FIGURE A.5

We will start this proof with the reduced feedback voltage at the critical frequency.

$$A_v = \frac{V_{out}}{V_{in}} = \frac{\left(\dfrac{R_1\left(-jX_{C_1}\right)}{\left(R_2 - jX_{C_2}\right)}\right)}{\left(R_2 - jX_{C_2}\right) + \left(\dfrac{R_1\left(-jX_{C_1}\right)}{\left(R_2 - jX_{C_2}\right)}\right)}$$

Since $R_1 = R_2$ and $C_1 = C_2$ for a Wien-bridge oscillator, the subscripts can be dropped. Therefore the gain equation becomes

$$A_v = \frac{\left(\dfrac{R\left(-jX_C\right)}{\left(R - jX_C\right)}\right)}{\left(R - jX_C\right) + \left(\dfrac{R\left(-jX_C\right)}{\left(R - jX_C\right)}\right)}$$

Now, if we multiply both the top and bottom of our equation by j, the base equation reduces to

$$A_v = \frac{\left(\dfrac{RX_C}{\left(R - jX_C\right)}\right)}{j\left(R - jX_C\right) + \left(\dfrac{RX_C}{\left(R - jX_C\right)}\right)}$$

Multiplying our equation again by $(R - jX_C)$ and expanding gives us

$$A_v = \frac{RX_C}{j\left(R - jX_C\right)^2 + RX_C} = \frac{RX_C}{jR^2 + 2RX_C - jX_C^2 + RX_C}$$

Now, grouping like terms in the denominator of our equation gives us

$$A_v = \frac{RX_C}{j\left(R^2 - X_C^2\right) + 3RX_C}$$

Since at resonance the phase shift between resistance and reactance must be 0°, there cannot be any j terms. Therefore at resonance the gain equation becomes:

$$A_v = \frac{V_{out}}{V_{in}} = \frac{RX_C}{3\,RX_C} = \frac{1}{3}$$

From basic AC theory, resonance occurs at the point where resistance and reactance are equal. Therefore, looking at the above derivation resonance will occur where

$$R^2 - X_C^2 = 0$$

Now, rearranging and reducing each term gives us

$$R^2 = X_C^2$$

Therefore

$$R = X_c$$

Starting with the basic equation for determining X_C, we have

$$X_C = \frac{1}{2\pi f C}$$

Rearranging gives us

$$f = \frac{1}{2\pi X_C C}$$

As shown above, at resonance $R = X_C$ therefore the equation becomes:

$$f_r = \frac{1}{2\pi RC}$$

EXAMPLE 15.6

Figure A.6 will aid us in developing the proof for the frequency response of a low-pass filter.

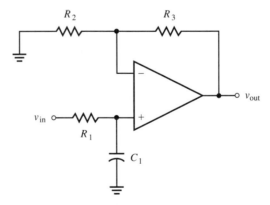

FIGURE A.6

Removing the low-pass filter from the non-inverting input and examining its output voltage, which is the ratio of capacitive reactance to total filter impedance, gives

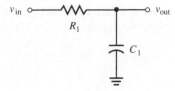

FIGURE A.7

And

$$A_{(filter)} = \frac{V_{out(filter)}}{V_{in(filter)}} = \frac{X_c}{Z_{(filter)}} = \frac{\left(\dfrac{1}{j\omega C}\right)}{\left(R + \dfrac{1}{j\omega C}\right)}$$

Now, multiplying denominator and numerator by jωC and rearranging gives

$$A_{(filter)} = \frac{\left(\dfrac{1}{j\omega C}\right)}{\left(R + \dfrac{1}{j\omega C}\right)} \left(\frac{j\omega C}{j\omega C}\right) = \frac{1}{1 + j\omega RC}$$

Now, since the critical frequency of the filter is

$$f_C = \frac{1}{2\pi RC}$$

And since ω = 2πf, the equation becomes

$$A_{(filter)} = \frac{1}{1 + j\dfrac{1}{f_C}}$$

Since $V_{out(filter)}$ varies with frequency, as does the ratio of X_C and $Z_{(filter)}$, the equation becomes

$$A_{(filter)} = \frac{1}{1 + j\left(\dfrac{f}{f_C}\right)}$$

Now, rearrange the denominator and cancel the j operator

$$1 + j\left(\frac{f}{f_C}\right) = \left[1 + \left(\frac{f}{f_C}\right)^2\right]^{\frac{1}{2}}$$

Therefore, the equation for a single pole filter becomes

$$A_{(filter)} = \frac{1}{\left[1 + \left(\dfrac{f}{f_C}\right)^2\right]^{\frac{1}{2}}}$$

Now, to complete the proof, if two identical filter sections are cascaded, where f and f_C of both filters are the same then

$$A'_{(filter)} = (A_{(filter_1)})(A_{(filter_2)})\ldots\ldots(A_{(filter_n)})$$

Or

$$A'_{(filter)} = (A_{(filter)})^n$$

Finally, substituting these results into our previous equation, the overall gain for identical cascaded low pass filters becomes

$$A_{(filter)} = \frac{1}{\left[\left(1 + \left(\frac{f}{f_C}\right)^2\right)^{\frac{1}{2}}\right]^n}$$

EXAMPLE 17.1

This proof concentrates on the oscillator frequency of an astable multivibrator. In an attempt to minimize confusion, the assumption that each transistor is on for one half of the output pulse will be made; therefore, $R_1 = R_2$ and $C_1 = C_2$. Additionally, to avoid having the transistors being driven into hard cutoff or saturation, the charge on the capacitors must fluctuate between 1/3 VCC and 2/3 VCC.

With the above stated, we will start this proof with the general equation for capacitor exponential curves.

$$V_c = V_s + \left(V_i - V_s\right) e^{-\frac{t}{RC}}$$

Expanding and rearranging the above equation gives us

$$\frac{V_c - V_s}{V_i - V_s} = e^{-\frac{t}{RC}}$$

Now, taking the natural log of the equation and rearranging gives

$$-\frac{t}{RC} = \ln\left(\frac{V_c - V_s}{V_i - V_s}\right)$$

Multiplying both sides of the equation by $-RC$ gives

$$t = RC\left[\ln\left(\frac{V_i - V_s}{V_c - V_s}\right)\right]$$

Finally, multiplying inside the parenthesis by -1 yields

$$t = RC\left[\ln\left(\frac{V_s - V_i}{V_s - V_c}\right)\right]$$

Next, referring back to our original assumptions, the time required to complete one cycle for an astable multivibrator is given by

$$t = \left(R_1C_1 + R_2C_2\right)\left[\ln\left(\frac{V_s - \frac{1}{3}V_s}{V_s - \frac{2}{3}V_s}\right)\right]$$

Where

    $R_1C_1 + R_2C_2$ = total RC time constant

    V_s = the charging voltage

    $1/3\ V_s$ = initial charge on capacitors

    $2/3\ V_s$ = final charge on capacitors

Now rearranging the equation gives

$$t = \left(R_1C_1 + R_2C_2\right)\left[\ln\left(\frac{\frac{3}{3}V_s - \frac{1}{3}V_s}{\frac{3}{3}V_s - \frac{2}{3}V_s}\right)\right] = \left(R_1C_1 + R_2C_2\right)\left[\ln\left(\frac{\frac{2}{3}V_s}{\frac{1}{3}V_s}\right)\right]$$

Inverting within the parentheses and then simplifying the equation gives us

$$t = (R_1C_1 + R_2C_2)\ln(2)$$

Since the natural log of 2 is approximately 0.7 our equation becomes

$$t = 0.7(R_1C_1 + R_2C_2)$$

Finally, inverting both sides of the equation to get frequency gives

$$f = \frac{1}{0.7\left(R_1C_1 + R_2C_2\right)}$$

 National
Semiconductor

LM741 Operational Amplifier

General Description

The LM741 series are general purpose operational amplifiers which feature improved performance over industry standards like the LM709. They are direct, plug-in replacements for the 709C, LM201, MC1439 and 748 in most applications.

The amplifiers offer many features which make their application nearly foolproof: overload protection on the input and output, no latch-up when the common mode range is exceeded, as well as freedom from oscillations.

The LM741C/LM741E are identical to the LM741/LM741A except that the LM741C/LM741E have their performance guaranteed over a 0°C to +70°C temperature range, instead of −55°C to +125°C.

Schematic Diagram

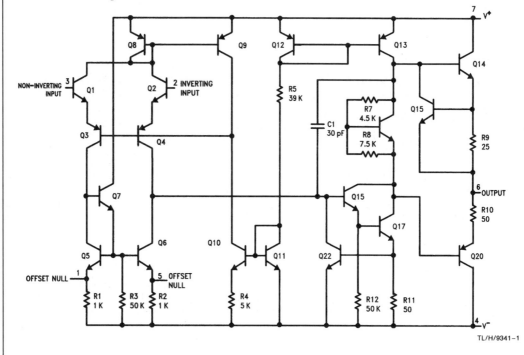

TL/H/9341–1

Offset Nulling Circuit

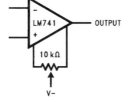

TL/H/9341–7

Absolute Maximum Ratings

If Military/Aerospace specified devices are required, please contact the National Semiconductor Sales Office/Distributors for availability and specifications. (Note 5)

	LM741A	LM741E	LM741	LM741C
Supply Voltage	±22V	±22V	±22V	±18V
Power Dissipation (Note 1)	500 mW	500 mW	500 mW	500 mW
Differential Input Voltage	±30V	±30V	±30V	±30V
Input Voltage (Note 2)	±15V	±15V	±15V	±15V
Output Short Circuit Duration	Continuous	Continuous	Continuous	Continuous
Operating Temperature Range	−55°C to +125°C	0°C to +70°C	−55°C to +125°C	0°C to +70°C
Storage Temperature Range	−65°C to +150°C	−65°C to +150°C	−65°C to +150°C	−65°C to +150°C
Junction Temperature	150°C	100°C	150°C	100°C
Soldering Information				
N-Package (10 seconds)	260°C	260°C	260°C	260°C
J- or H-Package (10 seconds)	300°C	300°C	300°C	300°C
M-Package				
Vapor Phase (60 seconds)	215°C	215°C	215°C	215°C
Infrared (15 seconds)	215°C	215°C	215°C	215°C

See AN-450 "Surface Mounting Methods and Their Effect on Product Reliability" for other methods of soldering surface mount devices.

| ESD Tolerance (Note 6) | 400V | 400V | 400V | 400V |

Electrical Characteristics (Note 3)

Parameter	Conditions	LM741A/LM741E			LM741			LM741C			Units
		Min	Typ	Max	Min	Typ	Max	Min	Typ	Max	
Input Offset Voltage	$T_A = 25°C$ $R_S \leq 10\ k\Omega$ $R_S \leq 50\Omega$		0.8	3.0	1.0	5.0		2.0	6.0		mV mV
	$T_{AMIN} \leq T_A \leq T_{AMAX}$ $R_S \leq 50\Omega$ $R_S \leq 10\ k\Omega$			4.0			6.0			7.5	mV mV
Average Input Offset Voltage Drift				15							$\mu V/°C$
Input Offset Voltage Adjustment Range	$T_A = 25°C,\ V_S = \pm20V$	±10				±15			±15		mV
Input Offset Current	$T_A = 25°C$		3.0	30		20	200		20	200	nA
	$T_{AMIN} \leq T_A \leq T_{AMAX}$			70		85	500			300	nA
Average Input Offset Current Drift				0.5							nA/°C
Input Bias Current	$T_A = 25°C$		30	80		80	500		80	500	nA
	$T_{AMIN} \leq T_A \leq T_{AMAX}$			0.210			1.5			0.8	μA
Input Resistance	$T_A = 25°C,\ V_S = \pm20V$	1.0	6.0		0.3	2.0		0.3	2.0		$M\Omega$
	$T_{AMIN} \leq T_A \leq T_{AMAX},$ $V_S = \pm20V$	0.5									$M\Omega$
Input Voltage Range	$T_A = 25°C$							±12	±13		V
	$T_{AMIN} \leq T_A \leq T_{AMAX}$				±12	±13					V
Large Signal Voltage Gain	$T_A = 25°C,\ R_L \geq 2\ k\Omega$ $V_S = \pm20V,\ V_O = \pm15V$ $V_S = \pm15V,\ V_O = \pm10V$	50			50	200		20	200		V/mV V/mV
	$T_{AMIN} \leq T_A \leq T_{AMAX},$ $R_L \geq 2\ k\Omega,$ $V_S = \pm20V,\ V_O = \pm15V$ $V_S = \pm15V,\ V_O = \pm10V$ $V_S = \pm5V,\ V_O = \pm2V$	32 10			25			15			V/mV V/mV V/mV

Electrical Characteristics (Note 3) (Continued)

Parameter	Conditions	LM741A/LM741E			LM741			LM741C			Units
		Min	Typ	Max	Min	Typ	Max	Min	Typ	Max	
Output Voltage Swing	$V_S = \pm 20V$ $R_L \geq 10\ k\Omega$ $R_L \geq 2\ k\Omega$	± 16 ± 15									V V
	$V_S = \pm 15V$ $R_L \geq 10\ k\Omega$ $R_L \geq 2\ k\Omega$				± 12 ± 10	± 14 ± 13		± 12 ± 10	± 14 ± 13		V V
Output Short Circuit Current	$T_A = 25°C$ $T_{AMIN} \leq T_A \leq T_{AMAX}$	10 10	25	35 40		25			25		mA mA
Common-Mode Rejection Ratio	$T_{AMIN} \leq T_A \leq T_{AMAX}$ $R_S \leq 10\ k\Omega, V_{CM} = \pm 12V$ $R_S \leq 50\Omega, V_{CM} = \pm 12V$	80	95		70	90		70	90		dB dB
Supply Voltage Rejection Ratio	$T_{AMIN} \leq T_A \leq T_{AMAX}$, $V_S = \pm 20V$ to $V_S = \pm 5V$ $R_S \leq 50\Omega$ $R_S \leq 10\ k\Omega$	86	96		77	96		77	96		dB dB

Transient Response
  Rise Time
  Overshoot
	$T_A = 25°C$, Unity Gain		0.25
			6.0
				0.8
				20
						0.3
						5
									0.3
									5
											μs
											%

Bandwidth (Note 4)	$T_A = 25°C$	0.437	1.5								MHz
Slew Rate	$T_A = 25°C$, Unity Gain	0.3	0.7			0.5			0.5		$V/\mu s$
Supply Current	$T_A = 25°C$					1.7	2.8		1.7	2.8	mA
Power Consumption	$T_A = 25°C$ $V_S = \pm 20V$ $V_S = \pm 15V$		80	150		50	85		50	85	mW mW
LM741A	$V_S = \pm 20V$ $T_A = T_{AMIN}$ $T_A = T_{AMAX}$			165 135							mW mW
LM741E	$V_S = \pm 20V$ $T_A = T_{AMIN}$ $T_A = T_{AMAX}$			150 150							mW mW
LM741	$V_S = \pm 15V$ $T_A = T_{AMIN}$ $T_A = T_{AMAX}$					60 45	100 75				mW mW

Note 1: For operation at elevated temperatures, these devices must be derated based on thermal resistance, and I_j max. (listed under "Absolute Maximum Ratings"). $T_j = T_A + (\theta_{jA}\ P_D)$.

Thermal Resistance	Cerdip (J)	DIP (N)	HO8 (H)	SO-8 (M)
θ_{jA} (Junction to Ambient)	100°C/W	100°C/W	170°C/W	195°C/W
θ_{jC} (Junction to Case)	N/A	N/A	25°C/W	N/A

Note 2: For supply voltages less than $\pm 15V$, the absolute maximum input voltage is equal to the supply voltage.

Note 3: Unless otherwise specified, these specifications apply for $V_S = \pm 15V$, $-55°C \leq T_A \leq +125°C$ (LM741/LM741A). For the LM741C/LM741E, these specifications are limited to $0°C \leq T_A \leq +70°C$.

Note 4: Calculated value from: BW (MHz) = 0.35/Rise Time(μs).

Note 5: For military specifications see RETS741X for LM741 and RETS741AX for LM741A.

Note 6: Human body model, 1.5 kΩ in series with 100 pF.

Connection Diagrams

Metal Can Package

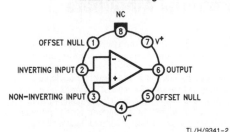

TL/H/9341–2

**Order Number LM741H, LM741H/883*, LM741AH/883
LM741CH or LM741EH
See NS Package Number H08C**

Ceramic Dual-In-Line Package

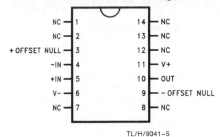

TL/H/9341–5

Order Number LM741J-14/883*, LM741AJ-14/883
See NS Package Number J14A**

*also available per JM38510/10101
**also available per JM38510/10102

Dual-In-Line or S.O. Package

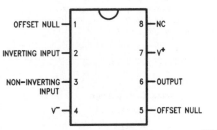

TL/H/9341–3

**Order Number LM741J, LM741J/883, LM741CJ,
LM741CM, LM741CN or LM741EN
See NS Package Number J08A, M08A or N08E**

Ceramic Flatpak

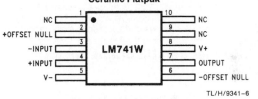

TL/H/9341–6

**Order Number LM741W/883
See NS Package Number W10A**

*LM741H is available per JM38510/10101

Semiconductor

LM101A/LM201A/LM301A Operational Amplifiers

General Description

The LM101A series are general purpose operational amplifiers which feature improved performance over industry standards like the LM709. Advanced processing techniques make possible an order of magnitude reduction in input currents, and a redesign of the biasing circuitry reduces the temperature drift of input current. Improved specifications include:

- Offset voltage 3 mV maximum over temperature (LM101A/LM201A)
- Input current 100 nA maximum over temperature (LM101A/LM201A)
- Offset current 20 nA maximum over temperature (LM101A/LM201A)
- Guaranteed drift characteristics
- Offsets guaranteed over entire common mode and supply voltage ranges
- Slew rate of 10V/μs as a summing amplifier

This amplifier offers many features which make its application nearly foolproof: overload protection on the input and output, no latch-up when the common mode range is exceeded, and freedom from oscillations and compensation with a single 30 pF capacitor. It has advantages over internally compensated amplifiers in that the frequency compensation can be tailored to the particular application. For example, in low frequency circuits it can be overcompensated for increased stability margin. Or the compensation can be optimized to give more than a factor of ten improvement in high frequency performance for most applications.

In addition, the device provides better accuracy and lower noise in high impedance circuitry. The low input currents also make it particularly well suited for long interval integrators or timers, sample and hold circuits and low frequency waveform generators. Further, replacing circuits where matched transistor pairs buffer the inputs of conventional IC op amps, it can give lower offset voltage and a drift at a lower cost.

The LM101A is guaranteed over a temperature range of $-55°C$ to $+125°C$, the LM201A from $-25°C$ to $+85°C$, and the LM301A from 0°C to $+70°C$.

Connection Diagrams (Top View)

Dual-In-Line Package

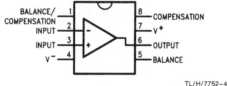

TL/H/7752–4

**Order Number LM101AJ, LM101J/883*,
LM301AJ, LM201AN or LM301AN
See NS Package Number J08A or N08A**

Metal Can Package

Note: Pin 4 connected to case.

TL/H/7752–2

**Order Number LM101AH,
LM101AH/883*, LM201AH or LM301AH
See NS Package Number H08C**

Ceramic Flatpack Package

LM101W

TL/H/7752–4

**Order Number LM101AW/883 or LM101W/883
See NS Package Number W10A**

Dual-In-Line Package

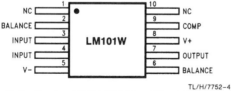

TL/H/7752–3

**Order Number LM101AJ-14/883*
See NS Package Number J14A**

*Available per JM38510/10103.

Reprinted with permission of National Semiconductor

Absolute Maximum Ratings

If Military/Aerospace specified devices are required, please contact the National Semiconductor Sales Office/ Distributors for availability and specifications.

	LM101A/LM201A	LM301A
Supply Voltage	±22V	±18V
Differential Input Voltage	±30V	±30V
Input Voltage (Note 1)	±15V	±15V
Output Short Circuit Duration (Note 2)	Continuous	Continuous
Operating Ambient Temp. Range	−55°C to +125°C (LM101A)	0°C to +70°C
	−25°C to +85°C (LM201A)	
T_J Max		
H-Package	150°C	100°C
N-Package	150°C	100°C
J-Package	150°C	100°C
Power Dissipation at T_A = 25°C		
H-Package (Still Air)	500 mW	300 mW
(400 LF/Min Air Flow)	1200 mW	700 mW
N-Package	900 mW	500 mW
J-Package	1000 mW	650 mW
Thermal Resistance (Typical) θ_{jA}		
H-Package (Still Air)	165°C/W	165°C/W
(400 LF/Min Air Flow)	67°C/W	67°C/W
N Package	135°C/W	135°C/W
J-Package	110°C/W	110°CmW
(Typical) θ_{jC}		
H-Package	25°C/W	25°C/W
Storage Temperature Range	−65°C to +150°C	−65°C to +150°C
Lead Temperature (Soldering, 10 sec.)		
Metal Can or Ceramic	300°C	300°C
Plastic	260°C	260°C
ESD Tolerance (Note 5)	2000V	2000V

Electrical Characteristics (Note 3) $T_A = T_J$

Parameter	Conditions		LM101A/LM201A			LM301A			Units
			Min	Typ	Max	Min	Typ	Max	
Input Offset Voltage	T_A = 25°C, $R_S \leq 50\,k\Omega$			0.7	2.0		2.0	7.5	mV
Input Offset Current	T_A = 25°C			1.5	10		3.0	50	nA
Input Bias Current	T_A = 25°C			30	75		70	250	nA
Input Resistance	T_A = 25°C		1.5	4.0		0.5	2.0		MΩ
Supply Current	T_A = 25°C	$V_S = \pm20V$		1.8	3.0				mA
		$V_S = \pm15V$					1.8	3.0	mA
Large Signal Voltage Gain	T_A = 25°C, $V_S = \pm15V$ $V_{OUT} = \pm10V$, $R_L \geq 2\,k\Omega$		50	160		25	160		V/mV
Input Offset Voltage	$R_S \leq 50\,k\Omega$				3.0			10	mV
Average Temperature Coefficient of Input Offset Voltage	$R_S \leq 50\,k\Omega$			3.0	15		6.0	30	μV/°C
Input Offset Current					20			70	nA
Average Temperature Coefficient of Input Offset Current	$25°C \leq T_A \leq T_{MAX}$			0.01	0.1		0.01	0.3	nA/°C
	$T_{MIN} \leq T_A \leq 25°C$			0.02	0.2		0.02	0.6	nA/°C

Reprinted with permission of National Semiconductor

Electrical Characteristics (Note 3) $T_A = T_J$ (Continued)

Parameter	Conditions		LM101A/LM201A			LM301A			Units
			Min	Typ	Max	Min	Typ	Max	
Input Bias Current					0.1			0.3	μA
Supply Current	$T_A = T_{MAX}$, $V_S = \pm 20V$			1.2	2.5				mA
Large Signal Voltage Gain	$V_S = \pm 15V$, $V_{OUT} = \pm 10V$ $R_L \geq 2k$		25			15			V/mV
Output Voltage Swing	$V_S = \pm 15V$	$R_L = 10\ k\Omega$	± 12	± 14		± 12	± 14		V
		$R_L = 2\ k\Omega$	± 10	± 13		± 10	± 13		V
Input Voltage Range	$V_S = \pm 20V$		± 15						V
	$V_S = \pm 15V$			$+15, -13$		± 12	$+15, -13$		V
Common-Mode Rejection Ratio	$R_S \leq 50\ k\Omega$		80	96		70	90		dB
Supply Voltage Rejection Ratio	$R_S \leq 50\ k\Omega$		80	96		70	96		dB

Note 1: For supply voltages less than $\pm 15V$, the absolute maximum input voltage is equal to the supply voltage.

Note 2: Continuous short circuit is allowed for case temperatures to 125°C and ambient temperatures to 75°C for LM101A/LM201A, and 70°C and 55°C respectively for LM301A.

Note 3: Unless otherwise specified, these specifications apply for C1 = 30 pF, $\pm 5V \leq V_S \leq \pm 20V$ and $-55°C \leq T_A \leq +125°C$ (LM101A), $\pm 5V \leq V_S \leq \pm 20V$ and $-25°C \leq T_A \leq +85°C$ (LM201A), $\pm 5V \leq V_S \leq \pm 15V$ and $0°C \leq T_A \leq +70°C$ (LM301A).

Note 4: Refer to RETS101AX for LM101A military specifications and RETS101X for LM101 military specifications.

Note 5: Human body model, 100 pF discharged through 1.5 kΩ.

Guaranteed Performance Characteristics LM101A/LM201A

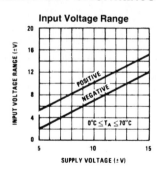

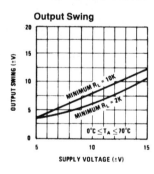

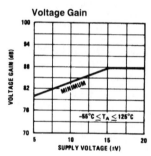

TL/H/7752–5

Guaranteed Performance Characteristics LM301A

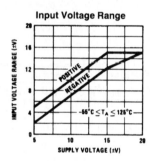

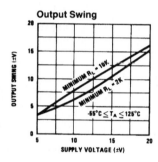

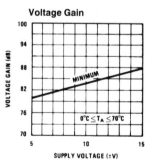

TL/H/7752–6

LM709
Operational Amplifier

General Description

The LM709 series is a monolithic operational amplifier intended for general-purpose applications. Operation is completely specified over the range of voltages commonly used for these devices. The design, in addition to providing high gain, minimizes both offset voltage and bias currents. Further, the class-B output stage gives a large output capability with minimum power drain.

External components are used to frequency compensate the amplifier. Although the unity-gain compensation network specified will make the amplifier unconditionally stable in all feedback configurations, compensation can be tailored to optimize high-frequency performance for any gain setting.

The LM709C is the commercial-industrial version of the LM709. It is identical to the LM709 except that it is specified for operation from 0°C to +70°C.

Connection Diagrams

Metal Can Package

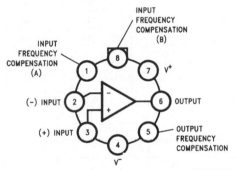

TL/H/11477–4

Order Number LM709AH, LM709H or LM709CH
See NS Package Number H08C

Dual-In-Line Package

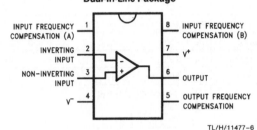

TL/H/11477–6

Order Number LM709CN-8
See NS Package Number N08E

Dual-In-Line Package

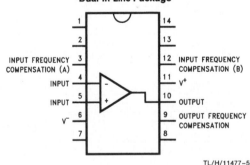

TL/H/11477–5

Order Number LM709CN
See NS Package Number N14A

Absolute Maximum Ratings (Note 3)

If Military/Aerospace specified devices are required, please contact the National Semiconductor Sales Office/Distributors for availability and specifications.

Supply Voltage	
LM709/LM709A/LM709C	±18V
Power Dissipation (Note 1)	
LM709/LM709A	300 mW
LM709C	250 mW
Differential Input Voltage	
LM709/LM709A/LM709C	±5V
Input Voltage	
LM709/LM709A/LM709C	±10V
Output Short-Circuit Duration (T$_A$ = +25°C)	
LM709/LM709A/LM709C	5 seconds
Storage Temperature Range	
LM709/LM709A/LM709C	−65°C to +150°C
Lead Temperature (Soldering, 10 sec.)	
LM709/LM709A/LM709C	300°C

Operating Ratings (Note 3)

Junction Temperature Range (Note 1)	
LM709/LM709A	−55°C to +150°C
LM709C	0°C to +100°C
Thermal Resistance (θ_{JA})	
H Package	150°C/W, (θ_{JC}) 45°C/W
8-Pin N Package	134°C/W
14-Pin N Package	109°C/W

Electrical Characteristics (Note 2)

Parameter	Conditions	LM709A			LM709			LM709C			Units
		Min	Typ	Max	Min	Typ	Max	Min	Typ	Max	
Input Offset Voltage	T$_A$ = 25°C, R$_S$ ≤ 10 kΩ		0.6	2.0		1.0	5.0		2.0	7.5	mV
Input Bias Current	T$_A$ = 25°C		100	200		200	500		300	1500	nA
Input Offset Current	T$_A$ = 25°C		10	50		50	200		100	500	nA
Input Resistance	T$_A$ = 25°C	350	700		150	400		50	250		kΩ
Output Resistance	T$_A$ = 25°C		150			150			150		Ω
Supply Current	T$_A$ = 25°C, V$_S$ = ±15V		2.5	3.6		2.6	5.5		2.6	6.6	mA
Transient Response Risetime Overshoot	V$_{IN}$ = 20 mV, C$_L$ ≤ 100 pF T$_A$ = 25°C		1.5 30			0.3 10	1.0 30		0.3 10	1.0 30	μs %
Slew Rate	T$_A$ = 25°C		0.25			0.25			0.25		V/μs
Input Offset Voltage	R$_S$ ≤ 10 kΩ			3.0			6.0			10	mV
Average Temperature Coefficient of Input Offset Voltage	R$_S$ = 50Ω T$_A$ = 25°C to T$_{MAX}$ T$_A$ = 25°C to T$_{MIN}$ R$_S$ = 10 kΩ T$_A$ = 25°C to T$_{MAX}$ T$_A$ = 25°C to T$_{MIN}$		1.8 1.8 2.0 4.8	10 10 15 25		3.0 6.0			6.0 12		μV/°C
Large Signal Voltage Gain	V$_S$ = ±15V, R$_L$ ≥ 2 kΩ V$_{OUT}$ = ±10V	25		70	25	45	70	15	45		V/mV
Output Voltage Swing	V$_S$ = ±15V, R$_L$ = 10 kΩ V$_S$ = ±15V, R$_L$ = 2 kΩ	±12 ±10	±14 ±13		±12 ±10	±14 ±13		±12 ±10	±14 ±13		V
Input Voltage Range	V$_S$ = ±15V	±8			±8	±10		±8	±10		V
Common-Mode Rejection Ratio	R$_S$ ≤ 10 kΩ	80	110		70	90		65	90		dB
Supply Voltage Rejection Ratio	R$_S$ ≤ 10 kΩ		40	100		25	150		25	200	μV/V
Input Offset Current	T$_A$ = T$_{MAX}$ T$_A$ = T$_{MIN}$		3.5 40	50 250		20 100	200 500		75 125	400 750	nA
Input Bias Current	T$_A$ = T$_{MIN}$		0.3	0.6		0.5	1.5		0.36	2.0	μA
Input Resistance	T$_A$ = T$_{MIN}$	85	170		40	100		50	250		kΩ

Note 1: For operating at elevated temperatures, the device must be derated based on a 150°C maximum junction temperature for LM709/LM709A and 100°C maximum for L709C. For operating at elevated temperatures, the device must be derated based on thermal resistance θ_{JA}, T$_{J(MAX)}$ and T$_A$.

Note 2: These specifications apply for −55°C ≤ T$_A$ ≤ +125°C for the LM709/LM709A and 0°C ≤ T$_A$ ≤ +70°C for the LM709C with the following conditions: ±9V ≤ V$_S$ ≤ ±15V, C1 = 5000 pF, R1 = 1.5 kΩ, C2 = 200 pF and R2 = 51Ω.

Note 3: Absolute Maximum Ratings indicate limits which if exceeded may result in damage. Operating Ratings are conditions where the device is expected to be functional but not necessarily within the guaranteed performance limits. For guaranteed specifications and test conditions, see the Electrical Characteristics.

Schematic Diagram**

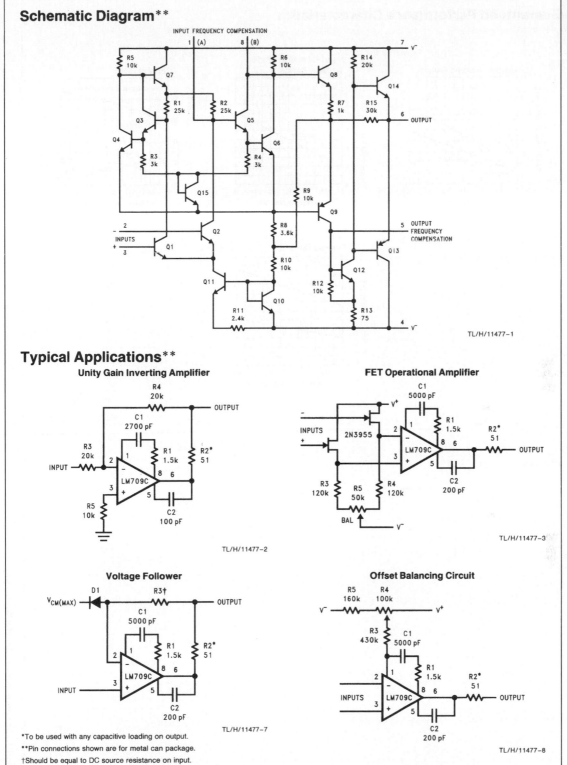

Typical Applications**

Unity Gain Inverting Amplifier

TL/H/11477–2

FET Operational Amplifier

TL/H/11477–3

Voltage Follower

TL/H/11477–7

Offset Balancing Circuit

TL/H/11477–8

*To be used with any capacitive loading on output.

**Pin connections shown are for metal can package.

†Should be equal to DC source resistance on input.

Guaranteed Performance Characteristics

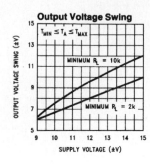

Output Voltage Swing

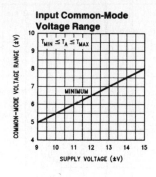

Input Common-Mode Voltage Range

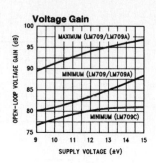

Voltage Gain

Supply Current

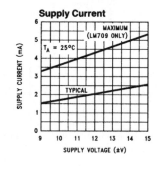

TL/H/11477–9

National Semiconductor

LM139/LM239/LM339/LM2901/LM3302
Low Power Low Offset Voltage Quad Comparators

General Description

The LM139 series consists of four independent precision voltage comparators with an offset voltage specification as low as 2 mV max for all four comparators. These were designed specifically to operate from a single power supply over a wide range of voltages. Operation from split power supplies is also possible and the low power supply current drain is independent of the magnitude of the power supply voltage. These comparators also have a unique characteristic in that the input common-mode voltage range includes ground, even though operated from a single power supply voltage.

Application areas include limit comparators, simple analog to digital converters; pulse, squarewave and time delay generators; wide range VCO; MOS clock timers; multivibrators and high voltage digital logic gates. The LM139 series was designed to directly interface with TTL and CMOS. When operated from both plus and minus power supplies, they will directly interface with MOS logic— where the low power drain of the LM339 is a distinct advantage over standard comparators.

Advantages

- High precision comparators
- Reduced V_{OS} drift over temperature

- Eliminates need for dual supplies
- Allows sensing near GND
- Compatible with all forms of logic
- Power drain suitable for battery operation

Features

- Wide supply voltage range
  LM139 series,   2 V_{DC} to 36 V_{DC} or
                             ± 1 V_{DC} to ± 18 V_{DC}
  LM139A series, LM2901   2 V_{DC} to 28 V_{DC}
  LM3302   or ± 1 V_{DC} to ± 14 V_{DC}
- Very low supply current drain (0.8 mA) — independent of supply voltage
- Low input biasing current   25 nA
- Low input offset current ± 5 nA
  and offset voltage ± 3 mV
- Input common-mode voltage range includes GND
- Differential input voltage range equal to the power supply voltage
- Low output saturation voltage   250 mV at 4 mA
- Output voltage compatible with TTL, DTL, ECL, MOS and CMOS logic systems

Connection Diagrams

Dual-In-Line Package

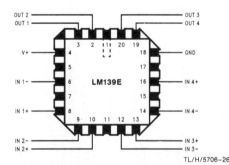

TOP VIEW   TL/H/5706–2

Order Number LM139J, LM139J/883*, LM139AJ, LM139AJ/883, LM239J, LM239AJ, LM339J, LM339AJ or LM2901J**
See NS Package Number J14A
Order Number LM339AM, LM339M or LM2901M
See NS Package Number M14A
Order Number LM339N, LM339AN, LM2901N or LM3302N
See NS Package Number N14A

*Available per JM38510/11201
**Available per SMD# 5962-8873901

TL/H/5706-26

Order Number LM139AE/883 or LM139E/883
See NS Package Number E20A

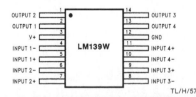

TL/H/5706–27

Order Number LM139AW/883 or LM139W/883*
See NS Package Number W14B

Absolute Maximum Ratings

If Military/Aerospace specified devices are required, please contact the National Semiconductor Sales Office/Distributors for availability and specifications. (Note 10)

	LM139/LM239/LM339 LM139A/LM239A/LM339A LM2901	LM3302
Supply Voltage, V+	36 V_{DC} or ±18 V_{DC}	28 V_{DC} or ±14 V_{DC}
Differential Input Voltage (Note 8)	36 V_{DC}	28 V_{DC}
Input Voltage	−0.3 V_{DC} to +36 V_{DC}	−0.3 V_{DC} to +28 V_{DC}
Input Current ($V_{IN} < −0.3\ V_{DC}$), (Note 3)	50 mA	50 mA
Power Dissipation (Note 1)		
Molded DIP	1050 mW	1050 mW
Cavity DIP	1190 mW	
Small Outline Package	760 mW	
Output Short-Circuit to GND, (Note 2)	Continuous	Continuous
Storage Temperature Range	−65°C to +150°C	−65°C to +150°C
Lead Temperature (Soldering, 10 seconds)	260°C	260°C

Operating Temperature Range
LM339/LM339A — 0°C to +70°C
LM239/LM239A — −25°C to +85°C
LM2901 — −40°C to +85°C
LM139/LM139A — −55°C to +125°C
LM3302 — −40°C to +85°C

Soldering Information
Dual-In-Line Package
Soldering (10 seconds) — 260°C
Small Outline Package
Vapor Phase (60 seconds) — 215°C
Infrared (15 seconds) — 220°C
See AN-450 "Surface Mounting Methods and Their Effect on Product Reliability" for other methods of soldering surface mount devices.
ESD rating (1.5 kΩ in series with 100 pF) — 600V

Electrical Characteristics ($V+ = 5\ V_{DC}$, $T_A = 25°C$, unless otherwise stated)

Parameter	Conditions	LM139A Min	Typ	Max	LM239A, LM339A Min	Typ	Max	LM139 Min	Typ	Max	LM239, LM339 Min	Typ	Max	LM2901 Min	Typ	Max	LM3302 Min	Typ	Max	Units
Input Offset Voltage	(Note 9)		1.0	2.0		1.0	2.0		2.0	5.0		2.0	5.0		2.0	7.0		3	20	mV_{DC}
Input Bias Current	$I_{IN(+)}$ or $I_{IN(−)}$ with Output in Linear Range, (Note 5), $V_{CM}=0V$		25	100		25	250		25	100		25	250		25	250		25	500	nA_{DC}
Input Offset Current	$I_{IN(+)}−I_{IN(−)}$, $V_{CM}=0V$		3.0	25		5.0	50		3.0	25		5.0	50		5	50		3	100	nA_{DC}
Input Common-Mode Voltage Range	$V+=30\ V_{DC}$ (LM3302, $V+=28\ V_{DC}$) (Note 6)	0		$V+−1.5$	0		$V+−1.5$	0		$V+−1.5$	0		$V+−1.5$	0		$V+−1.5$	0		$V+−1.5$	V_{DC}
Supply Current	$R_L=∞$ on all Comparators, $R_L=∞$, $V+=36V$, (LM3302, $V+=28\ V_{DC}$)		0.8	2.0		0.8	2.0		0.8	2.0		0.8	2.0		0.8	2.0		0.8	2.0	mA_{DC}
			1.0	2.5		1.0	2.5		1.0	2.5		1.0	2.5		1.0	2.5		1.0	2.5	mA_{DC}
Voltage Gain	$R_L≥15\ kΩ$, $V+=15\ V_{DC}$ $V_O=1\ V_{DC}$ to 11 V_{DC}	50	200		50	200		50	200		50	200		25	100		2	30		V/mV
Large Signal Response Time	$V_{IN}=$TTL Logic Swing, $V_{REF}=1.4\ V_{DC}$, $V_{RL}=5\ V_{DC}$, $R_L=5.1\ kΩ$		300			300			300			300			300			300		ns
Response Time	$V_{RL}=5\ V_{DC}$, $R_L=5.1\ kΩ$, (Note 7)		1.3			1.3			1.3			1.3			1.3			1.3		μs
Output Sink Current	$V_{IN(−)}=1\ V_{DC}$, $V_{IN(+)}=0$, $V_O≤1.5\ V_{DC}$	6.0	16		6.0	16		6.0	16		6.0	16		6.0	16		6.0	16		mA_{DC}

Reprinted with permission of National Semiconductor

Electrical Characteristics ($V^+ = 5$ V_{DC}, $T_A = 25°C$, unless otherwise stated) (Continued)

Parameter	Conditions	LM139A			LM239A, LM339A			LM139			LM239, LM339			LM2901			LM3302			Units
		Min	Typ	Max	Min	Typ	Max	Min	Typ	Max	Min	Typ	Max	Min	Typ	Max	Min	Typ	Max	
Saturation Voltage	$V_{IN(-)} = 1$ V_{DC}, $V_{IN(+)} = 0$, $I_{SINK} \leq 4$ mA		250	400		250	400		250	400		250	400		250	400		250	500	mV_{DC}
Output Leakage Current	$V_{IN(+)} = 1$ V_{DC}, $V_{IN(-)} = 0$, $V_O = 5$ V_{DC}		0.1			0.1			0.1			0.1			0.1			0.1		nA_{DC}

Electrical Characteristics ($V^+ = 5.0$ V_{DC}, Note 4)

Parameter	Conditions	LM139A			LM239A, LM339A			LM139			LM239, LM339			LM2901			LM3302			Units
		Min	Typ	Max	Min	Typ	Max	Min	Typ	Max	Min	Typ	Max	Min	Typ	Max	Min	Typ	Max	
Input Offset Voltage	(Note 9)			4.0			4.0			9.0			9.0		9	15			40	mV_{DC}
Input Offset Current	$I_{IN(+)} - I_{IN(-)}$, $V_{CM} = 0V$			100			150			100			150		50	200			300	nA_{DC}
Input Bias Current	$I_{IN(+)}$ or $I_{IN(-)}$ with Output in Linear Range, $V_{CM} = 0V$ (Note 5)			300			400			300			400		200	500			1000	nA_{DC}
Input Common-Mode Voltage Range	$V^+ = 30$ V_{DC} (LM3302, $V^+ = 28$ V_{DC}) (Note 6)	0		$V^+ - 2.0$	0		$V^+ - 2.0$	0		$V^+ - 2.0$	0		$V^+ - 2.0$	0		$V^+ - 2.0$	0		$V^+ - 2.0$	V_{DC}
Saturation Voltage	$V_{IN(-)} = 1$ V_{DC}, $V_{IN(+)} = 0$, $I_{SINK} \leq 4$ mA			700			700			700			700		400	700			700	mV_{DC}
Output Leakage Current	$V_{IN(+)} = 1$ V_{DC}, $V_{IN(-)} = 0$, $V_O = 30$ V_{DC}, (LM3302, $V_O = 28$ V_{DC})			1.0			1.0			1.0			1.0			1.0			1.0	μA_{DC}
Differential Input Voltage	Keep all V_{IN}'s ≥ 0 V_{DC} (or V^-, if used), (Note 8)			36			36			36			36			36			28	V_{DC}

Note 1: For operating at high temperatures, the LM339/LM339A, LM2901, LM139, LM3302 must be derated based on a 125°C maximum junction temperature and a thermal resistance of 95°C/W which applies for the device soldered in a printed circuit board, operating in a still air ambient. The LM239 and LM139 must be derated based on a 150°C maximum junction temperature. The low bias dissipation and the "ON-OFF" characteristic of the outputs keeps the chip dissipation very small ($P_D \leq 100$ mW), provided the output transistors are allowed to saturate.

Note 2: Short circuits from the output to V^+ can cause excessive heating and eventual destruction. When considering short circuits to ground, the maximum output current is approximately 20 mA independent of the magnitude of V^+.

Note 3: This input current will only exist when the voltage at any of the input leads is driven negative. It is due to the collector-base junction of the input PNP transistors becoming forward biased and thereby acting as input diode clamps. In addition to this diode action, there is also lateral NPN parasitic transistor action on the IC chip. This transistor action can cause the output voltages of the comparators to go to the V^+ voltage level (or to ground for a large overdrive) for the time duration that an input is driven negative. This is not destructive and normal output states will re-establish when the input voltage, which was negative, again returns to a value greater than -0.3 V_{DC} (at 25°C).

Note 4: These specifications are limited to $-55°C \leq T_A \leq +125°C$, for the LM139/LM139A. With the LM239/LM239A, all temperature specifications are limited to $-25°C \leq T_A \leq +85°C$, the LM339/LM339A temperature specifications are limited to $0°C \leq T_A \leq +70°C$, and the LM2901, LM3302 temperature range is $-40°C \leq T_A \leq +85°C$.

Note 5: The direction of the input current is out of the IC due to the PNP input stage. This current is essentially constant, independent of the state of the output so no loading change exists on the reference or input lines.

Note 6: The input common-mode voltage or either input signal voltage should not be allowed to go negative by more than 0.3V. The upper end of the common-mode voltage range is $V^+ -1.5V$ at 25°C, but either or both inputs can go to $+30$ V_{DC} without damage (25V for LM3302), independent of the magnitude of V^+.

Note 7: The response time specified is a 100 mV input step with 5 mV overdrive. For larger overdrive signals 300 ns can be obtained, see typical performance characteristics section.

Note 8: Positive excursions of input voltage may exceed the power supply level. As long as the other voltage remains within the common-mode range, the comparator will provide a proper output state. The low input voltage state must not be less than -0.3 V_{DC} (or 0.3 V_{DC} below the magnitude of the negative power supply, if used) (at 25°C).

Note 9: At output switch point, $V_O \cong 1.4$ V_{DC}, $R_S = 0\Omega$ with V^+ from 5 V_{DC} to 30 V_{DC}; and over the full input common-mode range (0 V_{DC} to $V^+ -1.5$ V_{DC}), at 25°C. For LM3302, V^+ from 5 V_{DC} to 28 V_{DC}.

Note 10: Refer to RETS139AX for LM139A military specifications and to RETS139X for LM139 military specifications.

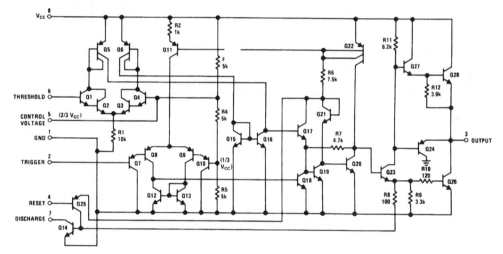

LM555/LM555C Timer

General Description

The LM555 is a highly stable device for generating accurate time delays or oscillation. Additional terminals are provided for triggering or resetting if desired. In the time delay mode of operation, the time is precisely controlled by one external resistor and capacitor. For astable operation as an oscillator, the free running frequency and duty cycle are accurately controlled with two external resistors and one capacitor. The circuit may be triggered and reset on falling waveforms, and the output circuit can source or sink up to 200 mA or drive TTL circuits.

Features

- Direct replacement for SE555/NE555
- Timing from microseconds through hours
- Operates in both astable and monostable modes

- Adjustable duty cycle
- Output can source or sink 200 mA
- Output and supply TTL compatible
- Temperature stability better than 0.005% per °C
- Normally on and normally off output

Applications

- Precision timing
- Pulse generation
- Sequential timing
- Time delay generation
- Pulse width modulation
- Pulse position modulation
- Linear ramp generator

Schematic Diagram

TL/H/7851–1

Reprinted with permission of National Semiconductor

Absolute Maximum Ratings

If Military/Aerospace specified devices are required, please contact the National Semiconductor Sales Office/Distributors for availability and specifications.

Supply Voltage	+18V
Power Dissipation (Note 1)	
LM555H, LM555CH	760 mW
LM555, LM555CN	1180 mW
Operating Temperature Ranges	
LM555C	0°C to +70°C
LM555	−55°C to +125°C

Storage Temperature Range	−65°C to +150°C
Soldering Information	
Dual-In-Line Package	
Soldering (10 Seconds)	260°C
Small Outline Package	
Vapor Phase (60 Seconds)	215°C
Infrared (15 Seconds)	220°C

See AN-450 "Surface Mounting Methods and Their Effect on Product Reliability" for other methods of soldering surface mount devices.

Electrical Characteristics (T_A = 25°C, V_{CC} = +5V to +15V, unless othewise specified)

Parameter	Conditions	Limits						Units
		LM555			LM555C			
		Min	Typ	Max	Min	Typ	Max	
Supply Voltage		4.5		18	4.5		16	V
Supply Current	V_{CC} = 5V, R_L = ∞		3	5		3	6	mA
	V_{CC} = 15V, R_L = ∞		10	12		10	15	mA
	(Low State) (Note 2)							
Timing Error, Monostable								
Initial Accuracy			0.5			1		%
Drift with Temperature	R_A = 1k to 100 kΩ,		30			50		ppm/°C
	C = 0.1 μF, (Note 3)							
Accuracy over Temperature			1.5			1.5		%
Drift with Supply			0.05			0.1		%/V
Timing Error, Astable								
Initial Accuracy			1.5			2.25		%
Drift with Temperature	R_A, R_B = 1k to 100 kΩ,		90			150		ppm/°C
	C = 0.1 μF, (Note 3)							
Accuracy over Temperature			2.5			3.0		%
Drift with Supply			0.15			0.30		%/V
Threshold Voltage			0.667			0.667		x V_{CC}
Trigger Voltage	V_{CC} = 15V	4.8	5	5.2		5		V
	V_{CC} = 5V	1.45	1.67	1.9		1.67		V
Trigger Current			0.01	0.5		0.5	0.9	μA
Reset Voltage		0.4	0.5	1	0.4	0.5	1	V
Reset Current			0.1	0.4		0.1	0.4	mA
Threshold Current	(Note 4)		0.1	0.25		0.1	0.25	μA
Control Voltage Level	V_{CC} = 15V	9.6	10	10.4	9	10	11	V
	V_{CC} = 5V	2.9	3.33	3.8	2.6	3.33	4	V
Pin 7 Leakage Output High			1	100		1	100	nA
Pin 7 Sat (Note 5)								
Output Low	V_{CC} = 15V, I_7 = 15 mA		150			180		mV
Output Low	V_{CC} = 4.5V, I_7 = 4.5 mA		70	100		80	200	mV

Electrical Characteristics T_A = 25°C, V_{CC} = +5V to +15V, (unless othewise specified) (Continued)

Parameter	Conditions	Limits						Units
		LM555			LM555C			
		Min	Typ	Max	Min	Typ	Max	
Output Voltage Drop (Low)	V_{CC} = 15V							
	I_{SINK} = 10 mA		0.1	0.15		0.1	0.25	V
	I_{SINK} = 50 mA		0.4	0.5		0.4	0.75	V
	I_{SINK} = 100 mA		2	2.2		2	2.5	V
	I_{SINK} = 200 mA		2.5			2.5		V
	V_{CC} = 5V							
	I_{SINK} = 8 mA		0.1	0.25				V
	I_{SINK} = 5 mA					0.25	0.35	V
Output Voltage Drop (High)	I_{SOURCE} = 200 mA, V_{CC} = 15V		12.5			12.5		V
	I_{SOURCE} = 100 mA, V_{CC} = 15V	13	13.3		12.75	13.3		V
	V_{CC} = 5V	3	3.3		2.75	3.3		V
Rise Time of Output			100			100		ns
Fall Time of Output			100			100		ns

Note 1: For operating at elevated temperatures the device must be derated above 25°C based on a +150°C maximum junction temperature and a thermal resistance of 164°c/w (T0-5), 106°c/w (DIP) and 170°c/w (S0-8) junction to ambient.

Note 2: Supply current when output high typically 1 mA less at V_{CC} = 5V.

Note 3: Tested at V_{CC} = 5V and V_{CC} = 15V.

Note 4: This will determine the maximum value of R_A + R_B for 15V operation. The maximum total (R_A + R_B) is 20 MΩ.

Note 5: No protection against excessive pin 7 current is necessary providing the package dissipation rating will not be exceeded.

Note 6: Refer to RETS555X drawing of military LM555H and LM555J versions for specifications.

Connection Diagrams

Metal Can Package

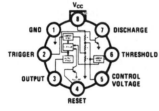

TL/H/7851–2

Top View

Order Number LM555H or LM555CH
See NS Package Number H08C

Dual-In-Line and Small Outline Packages

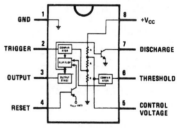

TL/H/7851–3

Top View

Order Number LM555J, LM555CJ,
LM555CM or LM555CN
See NS Package Number J08A, M08A or N08E

n-channel JFETs
designed for . . .

Siliconix

- ■ **Analog Switches**
- ■ **Commutators**
- ■ **Choppers**
- ■ **Integrator Reset Switch**

**Performance Curves NCA
See Section 4**

BENEFITS

- ● Low Insertion Loss and High Accuracy in Test Systems
  $r_{DS(on)} < 25\ \Omega$ (2N4856, 59)
- ● High Off-Isolation
  $I_{D(off)} < 250$ pA
- ● High Speed
  $t_{ON} < 9$ ns

*ABSOLUTE MAXIMUM RATINGS (25°C)

Reverse Gate-Drain or Gate-Source Voltage,
  2N4856-58 . . . . –40 V
Reverse Gate-Drain or Gate-Source Voltage,
  2N4859–61 . . . . –30 V
Gate Current . . . . 50 mA
Total Device Dissipation at 25°C Case Temperature
  (Derate 10 mW/°C) . . . . 1.8 W
Storage Temperature Range . . . . –65 to +200°C
Lead Temperature
  (1/16″ from case for 10 seconds) . . . . 300°C

TO-18
See Section 6

*ELECTRICAL CHARACTERISTICS (25°C unless otherwise noted)

		Characteristic		2N4856 2N4859 Min	Max	2N4857 2N4860 Min	Max	2N4858 2N4861 Min	Max	Unit	Test Conditions	
1		BV$_{GSS}$ Gate-Source Breakdown Voltage	2N4856-58	–40		–40		–40		V	$I_G = -1\ \mu A$, $V_{DS} = 0$	
2			2N4859-61	–30		–30		–30				
3		I$_{GSS}$ Gate Reverse Current	2N4856-58		–250		–250		–250	pA	$V_{GS} = -20$ V, $V_{DS} = 0$	
4					–500		–500		–500	nA		150°C
5	S T A T I C		2N4859-61		–250		–250		–250	pA	$V_{GS} = -15$ V, $V_{DS} = 0$	
6					–500		–500		–500	nA		150°C
7		I$_{D(off)}$ Drain Cutoff Current			250		250		250	pA	$V_{DS} = 15$ V, $V_{GS} = -10$ V	
8					500		500		500	nA		150°C
9		V$_{GS(off)}$ Gate-Source Cutoff Voltage		–4	–10	–2	–6	–0.8	–4	V	$V_{DS} = 15$ V, $I_D = 0.5$ nA	
10		I$_{DSS}$ Saturation Drain Current (Note 1)		50		20	100	8	80	mA	$V_{DS} = 15$ V, $V_{GS} = 0$	
11		V$_{DS(on)}$ Drain-Source ON Voltage			0.75 (20)		0.50 (10)		0.50 (5)	V (mA)	$V_{GS} = 0$, $I_D = (\ \ )$	
12	D Y N	r$_{ds(on)}$ Drain-Source ON Resistance			25		40		60	Ω	$V_{GS} = 0$, $I_D = 0$	f = 1 kHz
13		C$_{iss}$ Common-Source Input Capacitance			18		18		18	pF	$V_{DS} = 0$, $V_{GS} = -10$ V	f = 1 MHz
14		C$_{rss}$ Common-Source Reverse Transfer Capacitance			8		8		8			
15	S W I T C H I N G	t$_{d(on)}$ Turn-ON Delay Time			6 (20) [–10]		6 (10) [–6]		10 (5) [–4]	ns (mA) [V]	$V_{DD} = 10$ V, $V_{GS(on)} = 0$, $I_{D(on)} = (\ \ )$, $V_{GS(off)} = [\ ]$	R$_L$ = { 464 Ω, 2N4856, 59; 953 Ω, 2N4857, 60; 1910 Ω, 2N4858, 61 }
16		t$_r$ Rise Time			3 (20) [–10]		4 (10) [–6]		10 (5) [–4]	ns (mA) [V]		
17		t$_{off}$ Turn-OFF Time			25 (20) [–10]		50 (10) [–6]		100 (5) [–4]	ns (mA) [V]		

*JEDEC registered data.

NOTE:
1. Pulse test required, pulsewidth = 100 µs, duty cycle ≤ 10%.

$$R_L = \frac{V_{DD} - V_{DS(ON)}}{I_{D(ON)}}$$

INPUT PULSE
RISE TIME 0.25 ns
FALL TIME 0.75 ns
PULSE WIDTH 100 ns
PULSE DUTY CYCLE < 10%

SAMPLING SCOPE
RISE TIME 0.75 ns
INPUT RESISTANCE 1 M
INPUT CAPACITANCE 2.5 pF

NCA

Reprinted with permission of Siliconix

n-channel JFETs
designed for . . .

⬛ Siliconix

- ■ **Analog Switches**
- ■ **Commutators**
- ■ **Choppers**
- ■ **Integrator Reset Switch**

BENEFITS

- ● Low Insertion Loss and High Accuracy in Test Systems
  $r_{DS(on)} < 25\ \Omega$ (2N4856A, 59A)
- ● High Off-Isolation
  $I_{D(off)} < 250$ pA
- ● Short Sample and Hold Aperture Time
  $C_{rss} < 4$ pF
- ● High Speed
  $t_{ON} < 8$ ns

TO-18
See Section 6

*ABSOLUTE MAXIMUM RATINGS (25°C)

Reverse Gate-Drain or Gate-Source Voltage,
  2N4856A-58A . . . . −40 V
Reverse Gate-Drain or Gate-Source Voltage,
  2N4859A-61A . . . . −30 V
Gate Current . . . . 50 mA
Total Device Dissipation at 25°C Case Temperature
  (Derate 10 mW/°C) . . . . 1.8 W
Storage Temperature Range . . . . −65 to +200°C
Lead Temperature
  (1/16" from case for 10 seconds) . . . . 300°C

*ELECTRICAL CHARACTERISTICS (25°C unless otherwise noted)

		Characteristic		2N4856A 2N4859A Min	Max	2N4857A 2N4860A Min	Max	2N4858A 2N4861A Min	Max	Unit	Test Conditions
1	BV_GSS	Gate-Source Breakdown Voltage	2N4856A-58A	−40		−40		−40		V	$I_G = -1\ \mu A$, $V_{DS} = 0$
2			2N4859A-61A	−30		−30		−30			
3	I_GSS	Gate Reverse Current	2N4856A-58A		−250		−250		−250	pA	$V_{GS} = -20$ V, $V_{DS} = 0$
4					−500		−500		−500	nA	150°C
5			2N4859A-61A		−250		−250		−250	pA	$V_{GS} = -15$ V, $V_{DS} = 0$
6					−500		−500		−500	nA	150°C
7	I_D(off)	Drain Cutoff Current			250		250		250	pA	$V_{DS} = 15$ V, $V_{GS} = -10$ V
8					500		500		500	nA	150°C
9	V_GS(off)	Gate Source Cutoff Voltage		−4	−10	−2	−6	−0.8	−4	V	$V_{DS} = 15$ V, $I_D = 0.5$ nA
10	I_DSS	Saturation Drain Current (Note 1)		50		20	100	8	80	mA	$V_{DS} = 15$ V, $V_{GS} = 0$
11	V_DS(on)	Drain-Source ON Voltage			0.75 (20)		0.50 (10)		0.50 (5)	V (mA)	$V_{GS} = 0$, $I_D = (\ )$
12	r_ds(on)	Drain-Source ON Resistance			25		40		60	Ω	$V_{GS} = 0$, $I_D = 0$ f = 1 kHz
13	C_iss	Common-Source Input Capacitance			10		10		10	pF	$V_{DS} = 0$, $V_{GS} = -10$ V f = 1 MHz
14	C_rss	Common-Source Reverse Transfer Capacitance			4		3.5		3.5		
15	t_d(on)	Turn-ON Delay Time			5 (20) [−10]		6 (10) [−6]		8 (5) [−4]	ns (mA) [V]	$V_{DD} = 10$ V, $V_{GS(on)} = 0$, $I_{D(on)} = (\ )$, $V_{GS(off)} = [\ ]$
16	t_r	Rise Time			3 (20) [−10]		4 (10) [−6]		8 (5) [−4]	ns (mA) [V]	
17	t_off	Turn-OFF Time			20 (20) [−10]		40 (10) [−6]		80 (5) [−4]	ns (mA) [V]	

Column groups marked STATIC (rows 1–14) and DYN SWITCHING (rows 12–17).

Test Conditions for switching: $R_L = \begin{cases} 464\ \Omega, \text{2N4856A, 59A} \\ 953\ \Omega, \text{2N4857A, 60A} \\ 1910\ \Omega, \text{2N4858A, 61A} \end{cases}$

NCA

*JEDEC registered data.

NOTE:
1. Pulse test required, pulsewidth = 100 μs, duty cycle ≤ 10%.

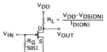

$$R_L = \frac{V_{DD} - V_{DS(ON)}}{I_{D(ON)}}$$

INPUT PULSE
RISE TIME 0.25 ns
FALL TIME 0.75 ns
PULSE WIDTH 100 ns
PULSE DUTY CYCLE < 10%

SAMPLING SCOPE
RISE TIME 0.75 ns
INPUT RESISTANCE 1 M
INPUT CAPACITANCE 2.5 pF

SEMICONDUCTOR
TECHNICAL DATA

1N1204 is a
Motorola Preferred Device

MEDIUM-CURRENT SILICON RECTIFIERS

Silicon rectifiers for medium-current applications requiring:

● High Current Surge —
   240 Amperes @ T_J = 190°C

● Peak Performance at Elevated Temperature —
   12 Amperes @ T_C = 150°C

MEDIUM-CURRENT SILICON RECTIFIERS

50-600 VOLTS
12 AMPERES

DIFFUSED JUNCTION

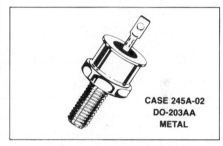

**CASE 245A-02
DO-203AA
METAL**

*MAXIMUM RATINGS

Characteristic	Symbol	1N 1199	1N 1200	1N 1202	1N 1204	1N 1206	Unit
Peak Repetitive Reverse Voltage Working Peak Reverse Voltage DC Blocking Voltage	V_{RRM} V_{RWM} V_R	50	100	200	400	600	Volts
Average Rectified Forward Current (Single phase, resistive load, 60 Hz, T_C = 150°C)	I_O			12			Amp
Non-Repetitive Peak Surge Current (Surge applied at rated load conditions, half wave, single phase, 60 Hz)	I_{FSM}			240 (for 1 cycle)			Amp
Operating Junction Temperature Range	T_J			-65 to +190			°C

*THERMAL CHARACTERISTICS

Characteristic	Symbol	Max	Unit
Thermal Resistance, Junction to Case	$R_{\theta JC}$	2.0	°C/W

*ELECTRICAL CHARACTERISTICS

Characteristic and Conditions	Symbol	Max	Unit
Maximum Instantaneous Forward Voltage (i_F = 40 A, T_C = 25°C)	v_F	1.8	Volts
Maximum Instantaneous Reverse Current (Rated voltage, T_C = 150°C)	i_R	10	mA

*Indicates JEDEC registered data.

MECHANICAL CHARACTERISTICS

CASE: Welded, hermetically sealed construction

FINISH: All external surfaces are corrosion-resistant and the terminal lead is readily solderable

POLARITY: Cathode to case (reverse polarity units are available and denoted by an "R" suffix, i.e., 1N1202R)

MOUNTING POSITION: Any

MOUNTING TORQUE: 15 in-lb max

MAXIMUM TERMINAL TEMPERATURE FOR SOLDERING PURPOSES: 275°C for 10 seconds at 3 kg tension.

WEIGHT: 6 grams (approx.)

**1N1199A
thru
1N1206A**

1N1204A is a
Motorola Preferred Device

MEDIUM-CURRENT SILICON RECTIFIERS

Silicon rectifiers for medium-current applications requiring:

- High Current Surge —
  240 Amperes @ T_J = 200°C

- Peak Performance at Elevated Temperature —
  12 Amperes @ T_C = 150°C

MEDIUM-CURRENT SILICON RECTIFIERS

50-600 VOLTS
12 AMPERES

DIFFUSED JUNCTION

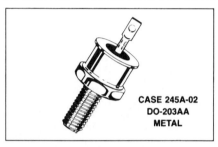

**CASE 245A-02
DO-203AA
METAL**

*MAXIMUM RATINGS

Characteristic	Symbol	1N 1199A	1N 1200A	1N 1202A	1N 1204A	1N 1206A	Unit
Peak Repetitive Reverse Voltage Working Peak Reverse Voltage DC Blocking Voltage	V_{RRM} V_{RWM} V_R	50	100	200	400	600	Volts
Non-Repetitive Peak Reverse Voltage (Halfwave, single phase, 60 Hz peak)	V_{RSM}	100	200	350	600	800	Volts
Average Rectified Forward Current (Single phase, resistive load, 60 Hz, T_C = 150°C)	I_O	← 12 →					Amp
Non-Repetitive Peak Surge Current (Surge applied at rated load conditions, half wave, single phase, 60 Hz)	I_{FSM}	← 240 (for 1 cycle) →					Amp
Operating and Storage Junction Temperature Range	T_J, T_{stg}	← -65 to +200 →					°C

*THERMAL CHARACTERISTICS

Characteristic	Symbol	Max	Unit
Thermal Resistance, Junction to Case	$R_{\theta JC}$	2.0	°C/W

*ELECTRICAL CHARACTERISTICS

Characteristic and Conditions	Symbol	Max	Unit
Maximum Instantaneous Forward Voltage (i_F = 40 A, T_C = 25°C)	v_F	1.35	Volts
Maximum Average Reverse Current at Rated Conditions 1N1199A 1N1200A 1N1202A 1N1204A 1N1206A	I_{RO}	3.0 2.5 2.0 1.5 1.0	mA

*Indicates JEDEC registered data.

MECHANICAL CHARACTERISTICS

CASE: Welded, hermetically sealed construction
FINISH: All external surfaces are corrosion-resistant and the terminal lead is readily solderable
POLARITY: Cathode to case (reverse polarity units are available and denoted by an "R" suffix, i.e., 1N1202RA)
MOUNTING POSITION: Any
MOUNTING TORQUE: 15 in-lb max
MAXIMUM TERMINAL TEMPERATURE FOR SOLDERING PURPOSES: 275°C for 10 seconds at 3 kg tension.
WEIGHT: 6 grams (approx.)

Reprinted with permission of Motorola

1N746A thru 1N759A, 1N957B thru 1N992B, 1N4370A thru 1N4372A

ELECTRICAL CHARACTERISTICS ($T_A = 25°C$, $V_F = 1.5$ V Max at 200 mA for all types)

Type Number (Note 1)	Nominal Zener Voltage V_Z @ I_{ZT} (Note 2) Volts	Test Current I_{ZT} mA	Maximum Zener Impedance Z_{ZT} @ I_{ZT} (Note 3) Ohms	Maximum DC Zener Current I_{ZM} (Note 4) mA	Maximum Reverse Leakage Current	
					$T_A = 25°C$ I_R @ $V_R = 1$ V μA	$T_A = 150°C$ I_R @ $V_R = 1$ V μA
1N4370A	2.4	20	30	150	100	200
1N4371A	2.7	20	30	135	75	150
1N4372A	3	20	29	120	50	100
1N746A	3.3	20	28	110	10	30
1N747A	3.6	20	24	100	10	30
1N748A	3.9	20	23	95	10	30
1N749A	4.3	20	22	85	2	30
1N750A	4.7	20	19	75	2	30
1N751A	5.1	20	17	70	1	20
1N752A	5.6	20	11	65	1	20
1N753A	6.2	20	7	60	0.1	20
1N754A	6.8	20	5	55	0.1	20
1N755A	7.5	20	6	50	0.1	20
1N756A	8.2	20	8	45	0.1	20
1N757A	9.1	20	10	40	0.1	20
1N758A	10	20	17	35	0.1	20
1N759A	12	20	30	30	0.1	20

Type Number (Note 1)	Nominal Zener Voltage V_Z (Note 2) Volts	Test Current I_{ZT} mA	Maximum Zener Impedance (Note 3)			Maximum DC Zener Current I_{ZM} (Note 4) mA	Maximum Reverse Current	
			Z_{ZT} @ I_{ZT} Ohms	Z_{ZK} @ I_{ZK} Ohms	I_{ZK} mA		I_R Maximum μA	Test Voltage Vdc V_R
1N957B	6.8	18.5	4.5	700	1	47	150	5.2
1N958B	7.5	16.5	5.5	700	0.5	42	75	5.7
1N959B	8.2	15	6.5	700	0.5	38	50	6.2
1N960B	9.1	14	7.5	700	0.5	35	25	6.9
1N961B	10	12.5	8.5	700	0.25	32	10	7.6
1N962B	11	11.5	9.5	700	0.25	28	5	8.4
1N963B	12	10.5	11.5	700	0.25	26	5	9.1
1N964B	13	9.5	13	700	0.25	24	5	9.9
1N965B	15	8.5	16	700	0.25	21	5	11.4
1N966B	16	7.8	17	700	0.25	19	5	12.2
1N967B	18	7	21	750	0.25	17	5	13.7
1N968B	20	6.2	25	750	0.25	15	5	15.2
1N969B	22	5.6	29	750	0.25	14	5	16.7
1N970B	24	5.2	33	750	0.25	13	5	18.2
1N971B	27	4.6	41	750	0.25	11	5	20.6
1N972B	30	4.2	49	1000	0.25	10	5	22.8
1N973B	33	3.8	58	1000	0.25	9.2	5	25.1
1N974B	36	3.4	70	1000	0.25	8.5	5	27.4
1N975B	39	3.2	80	1000	0.25	7.8	5	29.7
1N976B	43	3	93	1500	0.25	7	5	32.7
1N977B	47	2.7	105	1500	0.25	6.4	5	35.8
1N978B	51	2.5	125	1500	0.25	5.9	5	38.8
1N979B	56	2.2	150	2000	0.25	5.4	5	42.6
1N980B	62	2	185	2000	0.25	4.9	5	47.1

MAXIMUM RATINGS

Rating	Symbol	Value	Unit
Collector-Emitter Voltage	V_{CEO}	40	Vdc
Collector-Base Voltge	V_{CBO}	60	Vdc
Emitter-Base Voltage	V_{EBO}	6.0	Vdc
Collector Current — Continuous	I_C	200	mAdc
Total Device Dissipation @ T_A = 25°C Derate above 25°C	P_D	625 5.0	mW mW/°C
*Total Device Dissipation @ T_C = 25°C Derate above 25°C	P_D	1.5 12	Watts mW/°C
Operating and Storage Junction Temperature Range	T_J, T_{stg}	− 55 to + 150	°C

*THERMAL CHARACTERISTICS

Characteristic	Symbol	Max	Unit
Thermal Resistance, Junction to Ambient	$R_{\theta JA}$	200	°C/W
Thermal Resistance, Junction to Case	$R_{\theta JC}$	83.3	°C/W

*Indicates Data in addition to JEDEC Requirements.

2N3903
2N3904★

CASE 29-04, STYLE 1
TO-92 (TO-226AA)

3 Collector

2 Base

1 Emitter

GENERAL PURPOSE TRANSISTORS

NPN SILICON

★This is a Motorola
designated preferred device.

ELECTRICAL CHARACTERISTICS (T_A = 25°C unless otherwise noted.)

Characteristic		Symbol	Min	Max	Unit
OFF CHARACTERISTICS					
Collector-Emitter Breakdown Voltage(1) (I_C = 1.0 mAdc, I_B = 0)		$V_{(BR)CEO}$	40	—	Vdc
Collector-Base Breakdown Voltage (I_C = 10 μAdc, I_E = 0)		$V_{(BR)CBO}$	60	—	Vdc
Emitter-Base Breakdown Voltage (I_E = 10 μAdc, I_C = 0)		$V_{(BR)EBO}$	6.0	—	Vdc
Base Cutoff Current (V_{CE} = 30 Vdc, V_{EB} = 3.0 Vdc)		I_{BL}	—	50	nAdc
Collector Cutoff Current (V_{CE} = 30 Vdc, V_{EB} = 3.0 Vdc)		I_{CEX}	—	50	nAdc
ON CHARACTERISTICS					
DC Current Gain(1)		h_{FE}			—
(I_C = 0.1 mAdc, V_{CE} = 1.0 Vdc)	2N3903 2N3904		20 40	— —	
(I_C = 1.0 mAdc, V_{CE} = 1.0 Vdc)	2N3903 2N3904		35 70	— —	
(I_C = 10 mAdc, V_{CE} = 1.0 Vdc)	2N3903 2N3904		50 100	150 300	
(I_C = 50 mAdc, V_{CE} = 1.0 Vdc)	2N3903 2N3904		30 60	— —	
(I_C = 100 mAdc, V_{CE} = 1.0 Vdc)	2N3903 2N3904		15 30	— —	
Collector-Emitter Saturation Voltage(1) (I_C = 10 mAdc, I_B = 1.0 mAdc) (I_C = 50 mAdc, I_B = 5.0 mAdc)		$V_{CE(sat)}$	— —	0.2 0.3	Vdc
Base-Emitter Saturation Voltage(1) (I_C = 10 mAdc, I_B = 1.0 mAdc) (I_C = 50 mAdc, I_B = 5.0 mAdc)		$V_{BE(sat)}$	0.65 —	0.85 0.95	Vdc
SMALL-SIGNAL CHARACTERISTICS					
Current-Gain — Bandwidth Product (I_C = 10 mAdc, V_{CE} = 20 Vdc, f = 100 MHz)	2N3903 2N3904	f_T	250 300	— —	MHz

ELECTRICAL CHARACTERISTICS (continued) (T_A = 25°C unless otherwise noted.)

Characteristic		Symbol	Min	Max	Unit
Output Capacitance (V_{CB} = 5.0 Vdc, I_E = 0, f = 1.0 MHz)		C_{obo}	—	4.0	pF
Input Capacitance (V_{EB} = 0.5 Vdc, I_C = 0, f = 1.0 MHz)		C_{ibo}	—	8.0	pF
Input Impedance (I_C = 1.0 mAdc, V_{CE} = 10 Vdc, f = 1.0 kHz)	2N3903	h_{ie}	1.0	8.0	k ohms
	2N3904		1.0	10	
Voltage Feedback Ratio (I_C = 1.0 mAdc, V_{CE} = 10 Vdc, f = 1.0 kHz)	2N3903	h_{re}	0.1	5.0	X 10^{-4}
	2N3904		0.5	8.0	
Small-Signal Current Gain (I_C = 1.0 mAdc, V_{CE} = 10 Vdc, f = 1.0 kHz)	2N3903	h_{fe}	50	200	—
	2N3904		100	400	
Output Admittance (I_C = 1.0 mAdc, V_{CE} = 10 Vdc, f = 1.0 kHz)		h_{oe}	1.0	40	μmhos
Noise Figure (I_C = 100 μAdc, V_{CE} = 5.0 Vdc, R_S = 1.0 k ohms, f = 1.0 kHz)	2N3903	NF	—	6.0	dB
	2N3904		—	5.0	

SWITCHING CHARACTERISTICS

Delay Time	(V_{CC} = 3.0 Vdc, V_{BE} = 0.5 Vdc,		t_d	—	35	ns
Rise Time	I_C = 10 mAdc, I_{B1} = 1.0 mAdc)		t_r	—	35	ns
Storage Time	(V_{CC} = 3.0 Vdc, I_C = 10 mAdc,	2N3903	t_s	—	175	ns
	I_{B1} = I_{B2} = 1.0 mAdc)	2N3904		—	200	ns
Fall Time			t_f	—	50	ns

(1) Pulse Test: Pulse Width ≤ 300 μs, Duty Cycle ≤ 2.0%.

FIGURE 1 – DELAY AND RISE TIME EQUIVALENT TEST CIRCUIT

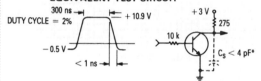

FIGURE 2 – STORAGE AND FALL TIME EQUIVALENT TEST CIRCUIT

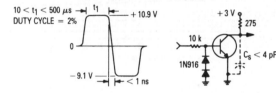

*Total shunt capacitance of test jig and connectors

TYPICAL TRANSIENT CHARACTERISTICS

—— T_J = 25°C - - - T_J = 125°C

FIGURE 3 – CAPACITANCE

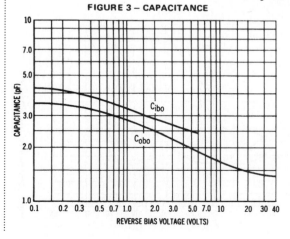

FIGURE 4 – CHARGE DATA

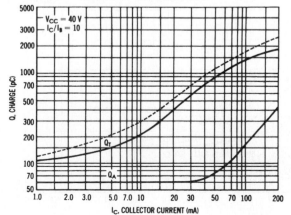

Reprinted with permission of Motorola

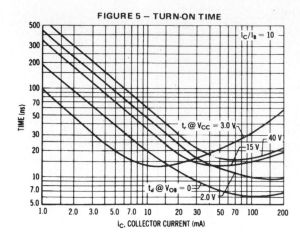

FIGURE 5 – TURN-ON TIME

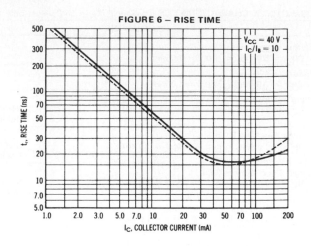

FIGURE 6 – RISE TIME

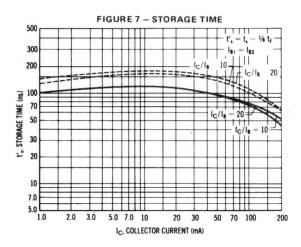

FIGURE 7 – STORAGE TIME

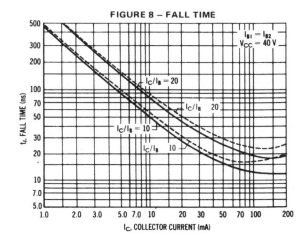

FIGURE 8 – FALL TIME

TYPICAL AUDIO SMALL-SIGNAL CHARACTERISTICS
NOISE FIGURE VARIATIONS
$V_{CE} = 5.0$ Vdc, $T_A = 25^{\circ}C$,
Bandwidth = 1.0 Hz

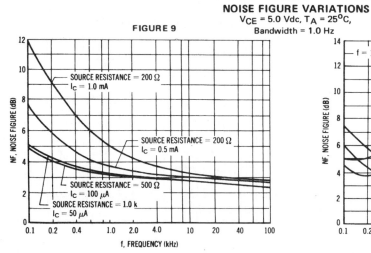

FIGURE 9

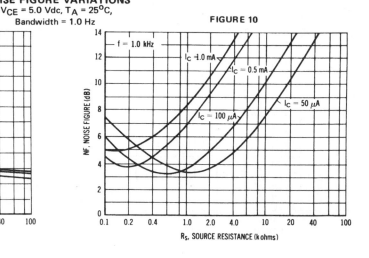

FIGURE 10

Reprinted with permission of Motorola

MAXIMUM RATINGS

Rating	Symbol	Value	Unit
Drain-Source Voltage	V_{DS}	25	Vdc
Drain-Gate Voltage	V_{DG}	25	Vdc
Gate-Source Voltage	V_{GS}	−25	Vdc
Gate Current	I_G	10	mAdc
Total Device Dissipation @ T_A = 25°C Derate above 25°C	P_D	350 2.8	mW mW/°C
Junction Temperature Range	T_J	125	°C
Storage Temperature Range	T_{stg}	−65 to +150	°C

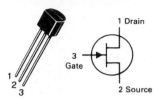

MPF102

CASE 29-04, STYLE 5
TO-92 (TO-226AA)

1 Drain
3 Gate
1
2
3
2 Source

JFET
VHF AMPLIFIER

N-CHANNEL — DEPLETION

Refer to 2N5484 for graphs.

ELECTRICAL CHARACTERISTICS (T_A = 25°C unless otherwise noted.)

Characteristic	Symbol	Min	Max	Unit
OFF CHARACTERISTICS				
Gate-Source Breakdown Voltage (I_G = −10 μAdc, V_{DS} = 0)	$V_{(BR)GSS}$	−25	—	Vdc
Gate Reverse Current (V_{GS} = −15 Vdc, V_{DS} = 0) (V_{GS} = −15 Vdc, V_{DS} = 0, T_A = 100°C)	I_{GSS}	— —	−2.0 −2.0	nAdc μAdc
Gate Source Cutoff Voltage (V_{DS} = 15 Vdc, I_D = 2.0 nAdc)	$V_{GS(off)}$	—	−8.0	Vdc
Gate Source Voltage (V_{DS} = 15 Vdc, I_D = 0.2 mAdc)	V_{GS}	−0.5	−7.5	Vdc
ON CHARACTERISTICS				
Zero-Gate-Voltage Drain Current* (V_{DS} = 15 Vdc, V_{GS} = 0 Vdc)	I_{DSS}	2.0	20	mAdc
SMALL-SIGNAL CHARACTERISTICS				
Forward Transfer Admittance* (V_{DS} = 15 Vdc, V_{GS} = 0, f = 1.0 kHz) (V_{DS} = 15 Vdc, V_{GS} = 0, f = 100 MHz)	$\lvert y_{fs} \rvert$	2000 1600	7500 —	μmhos
Input Admittance (V_{DS} = 15 Vdc, V_{GS} = 0, f = 100 MHz)	$Re(y_{is})$	—	800	μmhos
Output Conductance (V_{DS} = 15 Vdc, V_{GS} = 0, f = 100 MHz)	$Re(y_{os})$	—	200	μmhos
Input Capacitance (V_{DS} = 15 Vdc, V_{GS} = 0, f = 1.0 MHz)	C_{iss}	—	7.0	pF
Reverse Transfer Capacitance (V_{DS} = 15 Vdc, V_{GS} = 0, f = 1.0 MHz)	C_{rss}	—	3.0	pF

*Pulse Test: Pulse Width ≤ 630 ms; Duty Cycle ≤ 10%.

Practice Problem 1.1

2.9268mA
20 KΩ
25 V

Practice Problem 1.2

$15V$
$11v$

Practice Problem 1.5

$6.667V$ With Load Attached
$5V$ With no load
$25\ K\Omega$

Practice Problem 1.7

$3.6mA$

Practice Problem 1.9

.33%
$500mA$
$430mA$
$291.6mA$

Practice Problem 1.13

$235mS$

Practice Problem 1.15

$7.124V$

Review Section 1.1

1. Ohms law
3. Directly
5. The movement of electron through a circuit from – to +
7. False – series circuits have the same I throughout.
9. $R = \dfrac{V}{I} \quad \dfrac{24\ V}{2.4\ MA} = 10\ K\Omega \quad (C)$

Review Section 1.2

1. Equal
3. Node
5. (b) smaller than
7. In a firm voltage divider the load must be a minimum of 10 times the resistance of the voltage divider resistance it is connected across.
9. False – A voltage divider is a series circuit.

Review Section 1.3

1. Parallel
3. The total current entering any node equals the total current leaving the node.
5. A constant current source is an energy source that provides a constant value of current to a load.
7. False
9. (b) 12.0809mA

Review Section 1.4

1. Thevenin's Theorem states any circuit may be reduced to a single voltage source (V_{TH}) in series with a single resistance (R_{TH}) to provide the equivalent circuit for a given load.
3. True (R_{TH} & R_L)
5. A decrease in R_L = a decrease in R_T = I increase = increase V_{RL}.
7. False
9. 0. V
11. False

Review Section 1.5

1. Kirchhoff's voltage law states that the sum of all voltage drops around any closed loop must equal zero.
3. False – Loops with no power source will have a parallel set of branches with equal but opposite polarity voltage drops.
5. (d)

Review Section 1.6

1. More than one voltage source.
3. False – The AC and DC voltages and currents may be algebraically combined.
5. Add aiding currents and subtract opposing currents.
7. The center of the peak to peak AC voltage is not at zero volts but at a specific DC (+ or –) level.

Troubleshooting Question

1. To repair malfunctioning electronics equipment by checking the operation of the actual circuit against the calculated operation.
3. Test instruments such as analog VOMS, DMM, and oscilloscopes.
5. Voltage is measured with a voltmeter connected parallel to a component with circuit power turned on.
7. False – All test instruments have limitation to their use.
9. The 500kΩ load is in parallel with the meter impedance of 1mΩ giving an equivalent of 333.3mΩ

Problems

1. $V_x = I \times R$, $\quad I = \dfrac{V}{R}$, $\quad R + \dfrac{V}{I}$

3. Current flow is the movement of electrons through a circuit from negative to positive.

5. $22.6\,V$
   $32.19\,V$
   $38.36\,V$
   $6.8\,V$

6. $6.8\,mA$

7. $45.6\,K\Omega$
   $91.2\,V$

8. $960\,\Omega$

9. $7.66\,mA$

11. $1\,K\Omega$
    $2.4\,K\Omega$
    $20\,mA$
    $26.4\,V$
    $21.6\,V$

13. $5.785\,V$

15. a). A firm voltage divider is one with a load resistance at least 10 times the value of the resistance of the divider the load is in parallel with.
    b). A stiff voltage divider is one with a load resistance at least 100 times the value of the resistance of the divider the load is in parallel with.

17. a) $206.136\,V$

17. b) $206.136\,V$

19. $20.882\,mA$

25. 5 Loops

27. $+ R_7 + R_6 + R_3 - R_4 - R_5 = 0$

29. $3.983\,mA$

31. 1_{VP}

33. $6.34\,V$ $\quad$ $17.3\ ma\ A$

35. $15\ m\,A$

Review Section 2.1

1. Gas, Liquid, and Solid.
3. Electrons
5. When an electron moves from one atom, to another, a vacancy is created in the outer orbital shell.
7. Valence shell
9. Free, Valence
11. Copper, Silver, Gold

Review Section 2.2

1. Semiconductor
3. Light, Temperature, adding impurities
5. Eight
7. Elemental

Review Section 2.3

1. Increase
3. Extrinsic semiconductor
5. Five
7. Increase

9. Trivalent
11. Holes
13. Free electrons
15. Holes, free electrons

Review Section 2.4

1. The process of charge carriers moving from regions of high carrier concentrations to regions of low carrier concentration.
3. Positive, negative
5. Depletion
7. Cathode, anode

Review Section 2.5

1. Narrower
3. When the diode is reverse biased, the action of the free electrons and holes cause the depletion region (barrier region) to become larger, which decreases the amount of current that can flow through the diode.
5. As a voltage is applied to the diode, causing it to become forward biased, the resistance will decrease until the point that it reaches the break-over point, or knee voltage (.7V for silicon, .3V for germanium). This can be seen by looking at figure 2-20d. At this time, the bulk resistance of the diode, is the only resistance left which allows the current to increase.

Review Section 2.6

3. Zener diode

Review Section 2.7

1. The resistance will remain at the "bulk" resistance value.
3. Reverse

Practice Problem 3.1

$49.47\,V$.
$.329\,A$

Practice Problem 3.3

$13.489\ VDC$
$8.993\,mA$

Review Section 3.1

1. Reciprocal
3. $V_{avg} = \text{Voltage Peak} \times 0.636$
5. $V_{AC} = \text{Voltage Peak} \times 0.707$
7. $V_S = \eta \times V_{pk} = 120\ V_{ac} \times .5 = 60\ V_p$
9. Block one – Power Transformers
11. The function of the filter is to Elimate the ripple from the D.C. voltage.
    The function of the regulator is to provide a constant voltage to the load.

Review Section 3.2

1. Anode, Cathode
3. 1
5. 180 degrees
7. When the top of T1 secondary is postive, D1 will conduct produceing load voltage, and load current. When the top of T1 secondary goes negative, D1 will reverse-biased and will not conduct, no current will flow to the load.
9. Negative
11. 360 degrees
13. When the voltage is positive at Tp A, with respect to the center tap, diode D1 is foward biased – allowing current to flow from center tap through the load, and D1. At this time D2 is reversed biased, no conduction. When the voltage is negative at Tp A, diode D2 is forward biased allowing current to flow through the load.
15. False
17. C two times
19. Half wave $= 180$ degrees of V_S, $V_L = .318 \times V_S$
    Full wave $= 360$ degrees of V_S, $V_L = .636 \times V_S$
    Bridge rectifier $= 360$ degrees of V_S, $V_L = .318 \times V_S$

Review Section 3.3

1. Filter section
3. Ripple
5. 2 times secondary peak voltage
7. True
9. False
11. Most of the ripple voltage is dropped across R1
13. π Filter
15. Capacitor input filter

Review Section 3.4

1. Reversed
5. 1.785 A
9. Output is less

Review Section 3.5

1. Twice
3. False
5. $680V$
7. 170 VDC or 1/2 of V_{out}
9. Current goes down
11. 509
13. High voltage for cathode ray tubes

Review Section 3.6

1. Signal clipper
5. Shunt clipper
7. True

Review Section 3.7

1. 0.7
3. +10V, −10V

Review Section 3.8

1. DC filters, baseline stabilizers, DC restoration, DC inserters
5. Series with source voltage, and parallel to load
7. Negative
9. Negative

Review Section 3.9

1. An unbalanced load is a condition in which the resistance of the load is greater for one half-cycle, than the resistance during the other half-cycle.

What's Wrong With This Circuit?

1. Bridge rectifier is open (one half of the bridge rectifier)
3. Short D1, or D2 in bridge rectifier

Troubleshooting Questions

1. 1. Remove all jewelry
   2. Do not touch any exposed circuit.
   3. Insert test probes with one hand only
   4. Take your time.
   5. Be certain about what you are doing, if you are not sure about a procedure, ask for help
3. A. Short transformer
   B. Shorted cap
   C. Shorted diode
   D. Shorted load
5. Increased ripple, lower output voltage

Problems

1. 75Ω
3. 1.9mA
5. A
7. A. Half wave
   B. Fullwave
   C. Fullwave bridge
9. $10.3V_{pk}$
11. $18\,V_{pk}$
    $3.335mA$
15. The function of R1 is to limit current.
17. $3.279V$
19. 398.909Ω
23. Current
25. $2.105\,\mu A$
27. $666.66k\Omega$
29. A. The DC return circuit has a resistor added to allow a path for the capacitor to discharge while the diode is reverse biased.
    B. The type of diode circuit the DC return would be used in would be a circuit with an unbalanced load.

Review Section 4.1

1. Signal transmission, video and audio signal reproduction and voltage regulation.
3. N   P
5. Emitter
7. P

Review Section 4.2

1. Forward
3. Collector current

Review Section 4.3

1.

$$\frac{1c}{1b} = B$$

Less Than 1 ∞

$$\frac{1e}{1c} = \infty$$

95 to 99 % = ∞

Range 5 to 800 = B

Review Section 4.4

1. Voltage × Current
3. True
5. False
7. False
9. $1B \cdot = 1C$
    $1E \cdot \infty = 1C$
    $1E = 1C + 1B$

Practice Problem 4.1

1. $28.11 \, mW$

Practice Problem 4.1

1. Vb $\rightarrow$ Voltage base to ground
    Vc $\rightarrow$ Voltage collector to ground
    Ve $\rightarrow$ Voltage emitter to ground
3. Beta

Practice Problem 4.3

$1.28 \, mA$
$V_E = 1.9 \, V$
$V_B = 2.6 \, V$
$V_{RC} = 8.69 \, V$
$V_C = 11.3 \, V$

Practice Problem 4.5

$V_{CE} = 6.46 \, v$
$P = 59.5 \, mW$

Practice Problem 4.7

$R_{B(EQ)} = 2 \, K\Omega$
$I_E = 4.77 \, mA$

$I_B = 95.3 \, \mu A$
$V_E = OV$
$V_C = 10.23 \, V$
$V_B = +.7 \, v$
$V_{CE} = 10.23 \, V$
$P = 48.797 \, mW$

Review Section 4.6

1. Collector and emitter
5. $V_B = V_E + .7 \, v$
7. Characteristics of emitter bias are increase stability over the base bias but with the cost of a dual power supply.
9. False
11. True
13. True

Practice Problem 4.8

$V_B = 3.03 \, V$
$R_{TH} = 8.48 \, K\Omega$
$R_{BASE} = 84.85 \, \Omega$
$V_E = 2.33 \, V$
$I_C = 1.94 \, mA$
$V_C = 6.8 \, V$

Practice Problem 4.9

$V_B = 2.56 \, V$
$R_{TH} = 19.19 \, K\Omega$
$R_{BASE} = 127.9 \, \Omega$
$V_E = 1.86 \, V$
$I_C = 845 \mu A$
$V_C = 9.85 \, v$

Practice Problem 4.11

15.15Ω
Voltage divider is stiff – nearly firm.

Review Section 4.7

1. Voltage divider
3. Parallel
5. $V_C = V_{CC} - I_C \times R_C$

Review Section 4.8

1. To keep the same ground (lower) and positive (upper) reference points and prevents the scematic from becoming cluttered.
3. Electrons flow out of the PNP emitter terminal.

Review Section 4.9

1. $I_{C(SAT)} = 5.9 \, mA$
    $V_{CE(OFF)} = 12 \, v$
3. 2.95mA

What's Wrong With This Circuit?

1. $R_{BI} \downarrow = I_C \uparrow$

   Move Toward Saturation.
3. If R_{B1} opens $V_B \quad \downarrow = I_B \downarrow = I_C \downarrow = V_{RC} \downarrow = V_C \uparrow$
7. Open base resistor $= 1_B = 0$ and transistor is cut off.
9. Transistor is at fault
11. Emitter shorted to ground
13. Base resistor is open
15. Bad transistor

Practice Problem 5.1

$100 V_{pp}$
200
20,000

Practice Problem 5.7

$2.26\ V = (AC\ v_{out})$
221.224Ω
$2.26 mA$
100

Practice Problem 5.9

1. $156 = 43.862$dB $\quad 34 = 30.629$dB $\quad 89 = 38.987$dB $\quad 908$
   $= 59.161$dB

Practice Problem 5.10

1. 34dB $= 50.118 \quad 16.5$dB $= 66.834 \quad 22$dB $= 12.589 \quad 9$dB
   $= 2.818$

Review Section 5.1

1. Gain is defined as "The measure of how much larger an output signal is, than an input signal" it can be based on voltage, current, or power.
3. B
5. Output is taken from collector and ground. It provides voltage, current, and power gain. It has medium input impedance, and high output impedance. The common-emitter can provide high voltage, current, or power gain.
7. Low, high
9. Coupling capacitors, direct, transformer

Review Section 5.2

1. DC equivalent circuit, AC equivalent, and AC emitter.
3. Open
5. Parallel
7. The AC emitter resistance value changes as circuit conditions change.

Review Section 5.3

1. Shorts
3. False
5. $IE = \dfrac{VE}{RE}$
7. $V_{RC} = 9.40V$
9. False
11. $V_B = 8.25V \quad V_E = 7.55V \quad IF = 2.287mA \quad r'_e = 10.931\Omega$
    $V_{RE} = 10.748 \quad V_C = 1.251V$
13. Voltage gain will decrease because the AC sees RC in parallel with the load.
15. There swamping resistor is used to stabilize the over all gain and reduce the changes caused by temperature variations this also helps to reduce distortion.
17. There is a 180 degree phase shift from input to output.

Review Section 5.4

1. A stiff voltage divider is when R_{th} is 100 times less than $(B_{DC} \times RE)$
5. $I_B = 1.3113mA \quad I_E = 2.185V \quad V_B = 2.885V$
7. The AC input signal only supplies current to the base of the transistor. The collector, supplies (adds) current to the emitter r'_e. This current is equal to IB $\times$ Beta this means that $Z_{in\ Base} = \dfrac{B \times i_b \times r'e}{i_b} = B \times r'_e$

$Z_{in(base)} + B_{(1e + i'e)}$

Review Section 5.5

1. $I_E = 2.608mA \quad r'_e = 9.585\Omega \quad A_v = 125.189$
3. $A_v = \dfrac{rc}{r'e}$
5. If temperature goes up the DC Beta will increase, which produces more collector current. As the collector current increases the collector to emitter voltage decreases which in turn causes a decrease in base careened. This will offset the increase in collector current.

Review Section 5.6

1. Emitter follower amplifier.
3. Input $=$ Base $+$ Ground $\quad$ output $=$ Remitter $+$ Ground
5. IE$= 45.454$ mA
7. High Low
9. In phase, no phase shift

Review Section 5.7

1. (a) No phase shift from input to output. (b) Current gain approximately 1. (c) Low input impedance (d) good voltage gain.
3. $V_B = 0 \quad V_C = 3.917V \quad V_E = 9.494mV \quad I_C = 2.765$ m A
   $r'_e = 9.038\Omega$
5. $9.498mV$
7. High Frequency applications.

Review Section 5.8

1. False
3. True
5. False

Review Section 5.9

1. 1500.000   20 ×   5 × 15
3. 20 Log $A_v = A_v = 47.958$
5. 210 = 46.444dB   5860 = 95.369dB   18 = 25.105dB
   14 =22.922dB
7. The capacitor C2 isolates (blocks) the DC voltages from each stage.

What's Wrong With This Circuit?

1. Coupling capacitor C1 open
3. Coupling capacitor C3 open

Troubleshooting Questions

1. To check a bypass capacitor, you should place the os-cilloscope probe at the "top" of the capacitor. You should not see the AC signal, if the capacitor is good. If you see the AC signal, the capacitor is not shorting the AC signal to ground, which tells you the capacitor is probably open.
3. To troubleshoot a multistage circuit. Place a signal into the first stage, and check the output of the last stage with an oscilloscope, if the output is not correct, then check the input of the last stage If the input is not correct, move to the input of the previous stage. When you find the stage that has a "good" input sig-nal, and a bad output signal, you can then check that stage, with your oscilloscope to fing where the signal becomes distorted, or stops. The next step is to re-move the signal and the check the DC votages in that stage.

Practice Problem 6.1

5.3V

Practice Problem 6.1

$V_{(cutoff)} = 7.49$ V, $I_{C\,sat} = 69.$ mA

Practice Problem 6.3

IE = 14.3 mA

Practice Problem 6.7

IE = 11.28 mA
There is no change

Practice Problem 6.11

$A_v = 183.4$

Practice Problem 6.13

$V_{CE(cutoff)} = 15$ V
$I_{C(sat)} = 7.75$ mA

Practice Problem 6.15

$f_r = 503.3$ kHz

Practice Problem 6.17

$P_L = 1.27$ W

Review Section 6.1

1. Quiet
3. 0, 180, 360
5. Saturation, cutoff
7. True
9. False

Review Section 6.2

1. Current, Voltage or power
3. distortion
5. False

Review Section 6.3

1. True
3. Distortion   efficiency
5. True
7. $P_L = .5 \times V_{CEO} \times I_{CEQ}$
   .5 = 50% (with load line at center)
   $V_{CEQ} =$ Voltage, collector to emmiter
   $I_{CEQ} =$ collector current at Q point
9. False

Review Section 6.4

1. False max of 180 degrees
3. NPN, PNP
5. True one PNP and one NPN
7. Push-pull
9. True

Review Section 6.5

1. Crossover distortion
3. Gain, distortion
5. True
7. False
11. Common collector

Review Section 6.6

1. False
3. Efficiency, power
5. False
7. Output
9. When the transistor is conducting
11. 100

Review Section 6.7

1. Cascaded
3. RC coupling, transformer coupling, direct coupling and tuned coupling
5. Capacitive reactance
7. Capacitive or inductive
9. Tuned coupling

What's Wrong With These Circuits?

1. Shorted Q1 (Base to emitter)
3. Short Q2 (base to collector)

Troubleshooting Questions

1. Burnt, broken discolored
3. Distortion problems
5. Midpoint, midpoint

Problems

1. $V_E = 7.5$ V $V_B = 6.8$ V $V_C = 17.2$ V
7. $P_Q = 70.72$ mW
9. 25%
11. $V_E = 10.$V $V_B = 9.3$ V $V_C = 20.$ V
17. $P_Q = 99.51$ mW
19. 25%
21. 10 mA
23. $V_E = 9$ V
25. $V_{BQ2} = 8.3$ V
27. 25.15 mA
29. 99.9%
31. 18 V
33. $f_r = 11.2539$ MHz
35. $R_{(acparallel)} = 49.96$ kΩ
37. Using a value of 10Ω $P_L = 1.44$ mW

Practice Problem 7.1

3.00V

Practice Problem 7.3

1.150 mA

Practice Problem 7.5

10.68 V

Practice Problem 7.7

4.221 V
6.221 V
8.779 V

Practice Problem 7.9

800μ mho

Practice Problem 7.11

75.mV

Review Section 7.1

1. False
3. Electrons, Holes
5. True
7. b.
9. Output

Review Section 7.2

1. A = Manufacturing
3. Base Biasing
5. A = resistor
7. True

Review Section 7.3

1. Current source
3. The current through both Q1 and Q2 will be the same which produces a stable ID allowing only VGS to change.

Review Section 7.4

1. Gate, Source
3. When drain current reaches saturation
5. B

Review Section 7.5

1. False
3. Multiple stage
5. False
7. Gate

Review Section 7.6

1. Source follower
3. C diode
5. False
7. Logic switch

What's Wrong With These Circuits?

1. Open C1
3. Open 1MΩ resistor

Troubleshooting Questions

1. True
3. False
5. False

Problems

1. 120kΩ
3. 4000μ mho
5. PD = 7.810 mW
7. 10.369 mW
9. PD = 153mW
11. PD = 373.612mW
13. PD ≈ 30 mW,   ID ≈ 1.9mA,   VDS ≈ 14.V
15. VS = 11.99V

17. 11.949mW
19. 3.8mA
21. Av = 1.65
23. 8.499kΩ
25. 18V
27. 200Ω

Practice Problem 8.1

K = 1.778 mA/V^2

Practice Problem 8.3

ID = 2.667 mA

Practice Problem 8.5

V_{GS} = 833.641 mV
ID = 5.841 mA
VRD = 5.841V
VD = 5.158V
g_{mo} = 2000µmho
g_m = 2416µumho
Av = 2.4168

Practice Problem 8.7

VG = 5.406V
ID = 1.812mA
VRD = 4.894V
VD = 7.105V
Av = 2.43

Practice Problem 8.9

$$K = 3 \frac{mA}{V^2}$$

Review Section 8.1

1. The MOSFET has a gate that is insulated from the remainder of the device.
3. Depletion, enhancement
5. Forward
7. The MOSFET's insulated gate provides higher input impedances, very high gate rersistance, reduce's leakage current, prevents temperature from affecting leakage currents, and is not affected by the polarity of bias on the gate.

Review Section 8.2 and 8.3

1. False
3. False
5. The D MOSFET has an insulated gate which allows it to operate in either the depletion mode or the enhancement mode.
   The D MOSFET can control the width of the channel with a negative gate voltage, the more negative the gate voltage, the smaller the drain current. When the gate voltage is negative enough, the drain current will stop flowing.
   When a positive gate voltage is applied, the number of free electrons flowing through the channel increases. The more positive the gate voltage, the greater the conduction from source to drain.
7. Inversion layer
9. Positive
11. Gate source

Review Section 8.4

1. JFET
3. source resistor

Review Section 8.5

1. Voltage divider, gate

Review Section 8.6

1. Short
3. Dual gate
5. When the gate of the MOSFET is grounded the drain to source resistance can be as high as 10,000,000,000,000 Ω = 10,000,000 MΩ. With this high resistance, no signal, or voltage can pass. When a voltage is appled to the gate, the drain to source resistance will drop to around 25 to 100 Ω, and will allow signals to pass in either direction.
7. The MOSFET is used in digital applications because of its low power consumption, and small size. It can act like a switch, inverter, and can drive an active load.

Troubleshooting Questions

1. They are shipped in antistatic, non-conductive foam.
3. Ground
5. Remove it from the circuit

Problems

1. Using a VGS of 0, ID = 10mA
3. Using a VGS of 0, gm = 24000µmho
5. 16.666MΩ
7. VGS = 3.333V
9. ID = 14.517mA
11. VD = 5.483V
13. ID = 22.500mA
15. ID = 10mA

17. $K = 80 \frac{\mu A}{V^2}$

Practice Problem 9.1

$V_{leakage}$ = 800µV

Practice Problem 9.3

Minimum pulse width time = 245 ηs

Practice Problem 9.5

LTP = 2.185V

Practice Problem 9.7

$V_p = 14.2\,V_p$
Yes. RE is within the acceptable range frequency is 1.549 kHz

Practice Problem 9.9

Yes. Q2 saturates.

Review Section 9.1

1. Conduct
3. **Internal resistance,** prevents the transistor from being a complete short when it is turned on.
   Leakage resistance, allows current to leak through the transistor when it is turned off.
   Capacitance, does not allow the transistor to switch instantaneously.
5. V_{CE}(sat) is when both junctions of the transistor are forward biased. The voltage is approximately .2v.
7. **Delay time** is the time required for the collector current to reach 10% of its final value after the base current begins to flow.
   Rise time is the time required for the transistor's current to change from 10% to 90% of the final output current.
   Fall time is the time required to drop from 90% of the output level to 10% of the output level, after the base current is removed.
   Storage time is the time required for the transistor's output to drop from 100% to 90%, after the base current is removed.
   Decay time is the time required for the transistor to switch from on to off.

Review Section 9.2

1. The switching time of the transistor is decreased, this action will cause the transistor to increase the V_{CE} drop, this also increases the power consumed by the transistor.
3. The base emitter diode is forward biased and reversed biased during normal switching. During the process of forward biasing, the charge carriers stored in the previously reversed biased state must be removed, causing a delay in the forward bias state, when the diode is reversed biased the charge carriers become stored within the diode junction.
5. After the base voltage is removed from the transistor the charged capacitor discharges through the base resistor producing a "negative" voltage. This negative voltage removes the charge carriers from the collector - base junction which allows the device to turn off faster.
7. If the speed up capacitor is too large in value it will not be able to completely discharge during the set-

tling time. The next input pulse will not see the capacitor as a short, therefore the transistor will not see the high inrush current, causing a delay in the transistor's switching time.
If the speedup capacitor is too small, it will be fully discharged before the transistor is turned off. This action will increase the delay time required for the transistor to switch states.

Review Section 9.3

1. Input, output
3. Squaring, amplitude
5. Q1 and Q2 of the Schmitt trigger will switch states, causing a rapid change in the output of the trigger.
7. The trip points must be crossed, upper trip point, and the lower trip point.

Review Section 9.4

1. R_{DS} (on), and I_D (cutoff)
3. MOSFETs have a slower switching time.

Review Section 9.5

1. I_P = Peak Current V_P = Peak voltage V_V = Valley voltage I_V = Valley current
3. The intrinsic standoff is the ratio of the lower resistance to the total resistance, before the UJT is turned on. The equation is:

$$\frac{R_{B1}}{\left(R_{B1} + R_{B2}\right)}$$

5. Inter-base resistance is the resistance for the N-channel, when the emitter is open
7. Any attempts to increase current after the valley has been reached cause a increase in voltage, and saturate the junction
9. Base e logs

Review Section 9.6

1. $V = A_{sin} \times 2\,ft$
   A = amplitude, f = frequency in hertz, t = some value of time.
3. True
5. To detect the leading and trailing edges of pulse trains.
7. False
9. A ramp generator used with a constant current source provides an improved linearity at the output, over a normal ramp generator, even though it requires more circuitry.

Review Section 9.7

1. The action of the "totem pole" configuration is to insure that the output of this device will be TTL compatible. The output of the "totem pole" is determined by the input to each transistor

3. A logic "one", on, or a true $= 2.4V - 5V$
5. The use of a transistor to control a AC motor can be accomplished by using the output of the transistor to energize a relay, which closes the AC contacts which are connected to the AC motor.
7. High frequency applications.

What's Wrong With This Circuit

1. Leaky coupling capacitor
3. Shorted coupling capacitor

Problems

1. 11.99992V
3. 80nA
5. 236μA
7. 420ns
9. 4.761pf
11. 10V
13. 10V
15. 8.322V
19. Q2 is biased on
21. The voltage across C1 will increase
23. 219.696mV
25. 10.9V
27. 6.500kΩ
29. 13V

Practice Problem 10.1

r′e $=12.909\Omega$

Practice Problem 10.3

10kHz

Practice Problem 10.5

$C_{in(MILLER)} = 640pF$
$C_{out(MILLER)} = 42.666pF$

Practice Problem 10.7

15.455Hz

Practice Problem 10.9

$Av = 43.710$

Review Section 10.1

1. Feedback can correct for distortion when the quality of the feedback signal is compared with the quality of the input signal, then using the comparison, make adjustments to correct any errors's of distortion.
3. The amplifier may generate an output signal.

Review Section 10.2

1. V_{TH}, R_{TH}
3. h_{fe} is used to represent Beta
   h_i is a measure of resistance
   h_o is a measure of conductance
   h_f is a ratio

Review Section 10.3

1. Bandwidth can be described as the point where the output power decreases by 3dB on both sides of the output frequency.
   $BW = f_H - f_L$, f_H is the high dominant frequency, f_L is the lower dominant frequency
3. The X_c goes up.
5. $C'e = \dfrac{1}{2\pi(f_T \times r'e)}$

7. At low frequency the gate diode will have a large X_c.

Review Section 10.4

1. The differential amplifier has good temperature stability, it can direct couple and amplify both DC and AC signals. It also rejects unwanted noise on both inputs.
3. "Tail resistor" refers to the resistor that is common to both transistors in the differential amplifier, it is normally referred to as RE.
   "Tail current" is the current that flows through the tail resistor.
5. The input offset current will decrease to 0.
7. The common-mode signal gain $= A_{cm} = \dfrac{R_c}{2R_E}$

   The common-mode rejection ratio $= \dfrac{A_d}{A_{cm}}$

What's Wrong With this Circuit

1. Yes, the circuit will function.

Troubleshooting Questions

1. A signal generator, an oscilloscope
3. Time is converted to frequency by $f = \dfrac{1}{t}$

Problems

1. 889.166 Hz
3. 1.010 kHz
5. $Av = 138.932$
7. 642.137 MHz
9. 1.010 kHz
11. 12.166 V
13. $AV_1 = 233.333$ (unloaded)
15. $AV = 32.428$ k
17. $r'e_1 = r'e_2 = 14.601$ Ω

Practice Problem 11.1

According to the equation $A(v_1 - v_2) = V_0$, V_0 should = 200V. However, this voltage is above the input voltage, which means that the op-amp will saturate which gives us approximately V_0 = to $(-VEE + 1V) = -21V$.

Practice Problem 11.3

For typical value, $I_{B1} = I_{B2} = 80nA$
For maximum value, $I_{B1} = I_{B2} = 500nA$

Practice Problem 11.5

$V_{out} = -11V$

Practice Problem 11.7

$f_{unity} = 600,000$

Practice Problem 11.8

$A = 5$

Practice Problem 11.11

$f_{max} = 79.577kHz$

Review Section 11.1

3. False
5. Open-loop gain
7. 0.0 V
9. To allow the output to swing both positive and negative in equal amounts.
11. False

Review Section 11.3

1. f_{unity}
3. False
5. $f_{max} = \dfrac{\text{slew rate}}{2\pi \times V_p}$
7. decibels (dB)

What's Wrong With this Circuit

1. The op-amp is defective

Problems

1. 1V
3. 100 μV
5. Negative
7. Nulling
9. $I_{B1} = 650nA$ $I_{B2} = 350nA$

11. A = 2
13. 71.428ns
15. $f_{max} = 53.051kHz$
17. The output offset voltage can be eliminated by using Negative feedback.
19. $V_{out} = 62.5mV$
21. 80 dB

Practice Problem 12.1

$V_{out} = 276$ mV

Practice Problem 12.3

$V_{out} = 1.84$ V

Practice Problem 12.5

$V_{V(cl)} = 49.666$

Practice Problem 12.7

The error voltage is 1.469 V

Practice Problem 12.9

a. $V_{out} = 10$ mV $A = 100$
b. $V_{out} = 200$ mV $A = 100$
c. $V_{out} = 500$ mV $A = 100$

Practice Problem 12.11

$Z_{in(eff)} = 4$ GΩ

Practice Problem 12.13

$A = 4.545$

Practice Problem 12.15

$V_{out} = 1.5V$

Practice Problem 12.17

$A_I = 21$

Practice Problem 12.19

$I_{out} = 1$ mA

Practice Problem 12.21

$f_{unity} = 750$ kHz

Practice Problem 12.23

$V_{out} = -10.678$ V

Practice Problem 12.25

$V_{out} = 6.25$ V

Review Section 12.1

1. Negative
3. 1 series-parallel
   2 series-series
   3 parallel-series
   4 parallel-parallel
5. A_{cl} = closed-loop gain
7. Non-inverting voltage amplifier
9. a. inverting voltage amplifier
   b. current-to-voltage transducer

Review Section 12.2

1. Differential amplifier
3. Bootstrapping is when the op-amp and the feedback circuitry try to make both inputs equal.
5. 0 V
7. The open-loop gain times the feedback ratio must be larger than 10.

Review Section 12.3

1. R_f
3. Virtual ground is a point in a circuit that is at ground potential, but is not connected to ground directly.
5. $A_{v(cl)} = 20$
7. Look at R_f and R_{in}, if R_f is larger than R_{in}. If it is then the circuit is supposed to be acting as a current-to-voltage source.
9. The characteristics of an ideal current-to-voltage transducer are, $Z_{in(ef)} = 0$ $Z_{out} = 0$

Review Section 12.4

1. If the load is across the feedback network and has a ground connection, then the feedback is parallel; if not the feedback is series.
3. R_L is in series with both R_f and R_b; while R_f is in parallel with R_b.
5. Infinite
7. The input impedance is zero and the output impedance is infinite.

Review Section 12.5

1. Voltage   current
3. 50 mV
5. Infinity

Review Section 12.5

1. Unity
3. Extremely high input impedance   impedance buffer
5. The gain-bandwidth product is constant. If gain goes up, then bandwidth must go down.
7. A summing amplifier

Troubleshooting Questions

1. False
3. False

What's Wrong With This Circuit

1. The 1 kΩ resistor is open.

Problems

1. $V_{out} = 11.4$ mV
3. B = 0.0909
5. 1 GΩ
7. 18.407, B = 0.0543   A = 18.407   yes.
9. $V_{out} = 1.322$ V error voltage = 14.132 μV
11. $V_{out} = 10$.V
13. Virtual ground is located at the junction of R_f, R_{in}.
15. R_f and R_L are essentially in parallel.
17. $R_{out(miller)} = 0\Omega$
21. 0Ω
23. parallel-series
25. 0Ω
27. $I_{out} = 5.113$ mA
29. $V_{out} = 18.168$ V
31. $I_{out} = 5.113$ mA
33. $A_I = 18.407$
35. Voltage-to-current transducer
37. $V_{RL} = 2.136$ V
39. Infinite
41. $f_{unity} = 750$ kHz
43. $V_{out} = 4.00$ V

Practice Problem 13.1

a. 1.591 Hz   b. 28.420 kHz
c. 10.261Hz   d. 15.392 Khz

Practice Problem 13.3

Phase angle = 20.6954
Z = 19.241 kΩ

Practice Problem 13.5

Z = 54.203 kΩ
Phase angle = 29.875

Practice Problem 13.7

$f_r = 503.292$ Hz

Practice Problem 13.9

$f_r = 871.727$ kHz

Practice Problem 13.11

$Q_{cir} = 11.458$
f_r will decrease to 1.6015 MHz. This is a change of approximately 68 kHz.

Practice Problem 13.13

PW = 1.500 ms
f = 333.333 Hz

Review Section 13.1

1. In phase with
3. The feedback circuit
5. equal 1

Review Section 13.2

1. $f_c = \dfrac{1}{2\pi RC}$
3. Frequency gain
5. "Lag network" is when the output voltage is developed across a capacitor, as shown in **FIGURE 13.8**. This voltage will lag the input voltage.
7. Parallel
9. Bandstop

Review Section 13.3

1. 1 MHz
3. True

Review Section 13.4

1. Bandstop, or notch filter
3. True

Review Section 13.5

1. Capacitor
3. $f_r = \dfrac{1}{\left(2\pi\sqrt{(LC)}\right)}$
5. False
7. Inductor
9. Transformer

Review Section 13.6

1. Capacitor
3. $T = 2RC\ln\left[\dfrac{(1+B)}{(1-B)}\right]$

Review Section 13.7

1. Distort
3. Proportional
5. Capacitor
7. Relaxation oscillator integrator
9. 1. Q_1 must stay off during the entire time the input pulse is low.
   2. The collector of Q_2 must remain a constant current source.

3. Q_1 must be given enough time to fully discharge C_2 before the next negative input pulse is applied to the base.

What's Wrong With This Circuit

The capacitor is probably leaking

Problems

1. Lag network
3. The phase angle = 20.556
5. f_c = 338.627 Hz
9. 5.0329 kHz
11. Bandstop
13. One
15. f_c = 28.420 kHz
17. 1/3
19. Capacitor
21. B = .2
23. Inductor
25. f_r = 19.392 kHz
27. d. square wave
29. V = *(I X t)* = 3 V
31. V = 4.255 V

Practice Problem 14.1

$VD_{(cir)} = 6\,\mu V$

Practice Problem 14.3

Assuming that RlC time constant is 100 times the value of the discharge time, f = 189.725MHz

Practice Problem 14.9

UTP = 998mV   LTP = 1.1mV

Practice Problem 14.11

UTP = 650.mV   LTP = −650mV

Practice Problem 14.13

UTP = 4.794V   LTP = 435.897mV

Practice Problem 14.17

UTP = 4.7V   LTP = 2.7V

Practice Problem 14.19

The feedback will be removed causing the circuit to act like a comparator with no hystersis with a reference voltage of 0V.

Review Section 14.1

1. Buffer
3. Sacrifice

5. True
7. 100, DC
9. Output bounding is when the output is clipped at a specific voltage level
11. False

Review Section 14.2

1. False
3. Slew rate
5. Op-amps comparator
7. False
9. False
11. False
13. False

Review Section 14.3

1. "Window" when used in context with the window comparator, is the voltage level between the lower trip point and the upper trip point.
3. True

What's Wrong With This Circuit

There is not enough information given to identify the exact problem. The first thing that should be checked is the supply voltages, then the input reference voltage. If these are correct then the comparator should be replaced.

Troubleshooting Questions

1.   1. provide gain
     2. provide buffering
     3. reduce the effective diode barrier voltage
3. Shorted
5. True

Problems

3. Assuming a gain of 100,00 . −14V
5. 0V
7. UTP = 395.267 mV   LTP = −395.295mV
9. UTP = 718.667mV   LTP = −718.667mV
11. UTP = 1.569V   LTP = −1.569V
13. UTP = 778.991mV   LTP = −605.172mV
15. UTP = 1.227V   LTP = −1.227V

Practice Problem 15.1

BW = f$_2$ − f$_1$ = 2KHz − 150Hz = 1.85KHz

Practice Problem 15.3

$$S = \frac{f_4 - f_3}{f_2 - f_1} = \frac{1100Hz - 1Hz}{1000Hz - 10Hz} = 1.1$$

Practice Problem 15.5

442.1Hz

Practice Problem 15.7

700Hz
19.2$K\Omega$
0.5547
832.1mV
6.157V
2.382
4.15$K\Omega$
0.3077
1.894V
4.512V

Practice Problem 15.11

5.53

Practice Problem 15.13

1.54KHz
4.82KHz
3.28KHz

Practice Problem 15.15

795.8$m\Omega - F$

Practice Problem 15.17

9.9$u\Omega - F$
9.41$u\Omega - F$
8.34$u\Omega - F$

Review Section 15.1

1:   The two basic types of configurations see with filters are the lattice and ladder arrangements.
3:   The area between the cutoff frequency and the pass-band frequency is called the transition region.
5:   If the phase shift varies linearly with frequency, the filter produces a linear-phase.
7:   A pole is a single RC combination.
9:   The larger the Q factor the narrower the bandwidth, the smaller the Q factor the wider the bandwidth.
11: The transition region is the portion of the frequency response curve that starts at the critical frequency and extends to the point where the minimum amount of attenuation has been reached.

Review Section 15.2

1:   We describe filters by their shape of characteristic curves and layout of filter components.
3:   Low-pass, high-pass, band-pass, band-reject
5:   Between DC and the upper cutoff frequency.
7:   Above the lower cutoff frequency and below the upper cutoff frequency.
11: The differences between the Butterworth, Chebyshev, and Bessel filters are: amplitude response, phase response, frequency response, attenuation outside the passband, time delay within the passband.

Review Section 15.3

1: Filters are called active or passive because of the type of devices they employ. Active filters use devices such as transistors, FETs, and operational amplifiers; passive filters use only RC and RL devices.
3: True
5: Stability, considerable gain, positive and negative feedback inputs.
7: $f_u - f_L$
9: Summing amplifier

Review Section 15.4

1: The miller effect is a reduction in the amplitude of the output voltage due to the high frequency response of the internal feedback capacitor (C_{OB})
3: Simplicity of design, fewer precision parts.
5: The cutoff frequency (f_c), for the Chebyshev filter equals the frequency at which the amplitude response falls out of the ripple band on its way into the stopband.
7: For their theoretical infinite attenuation outside the passband.
9: Low Q, passband losses
11: False – State variable filters can provide low-pass, high-pass, and band pass outputs from the same circuit.

Troubleshooting Questions

1: Verify the type of filter; check the input signal and supply voltages; isolate the filter.
3: True
5: Separate
7: After troubleshooting the passive components of any active filter and not finding any defects, replace the operational amplifier.

Problems

1: 6dB/octave
3: Emitter-Follower Circuit therefore $A_v \cong 1$
$V_{out} = (1)V_{in} = (1)$ (3V) = 3V @ midband gain
@ −3dB:
$V_{out} = 0.707(V_{in}) = 0.707(3V) = 2.12V$

5: $f_c = \dfrac{1}{2\pi RC} = \dfrac{1}{2\pi(10\,K\Omega)(0.001uF)} = 15.92KHz$

7: Decrease
9: $1.26KHz$

11: $A_v = \dfrac{22\,K\Omega}{22\,K\Omega} + 1 = 2$

13: $11.96KHz$
$393.75KHz$
$673.24KHz$
$2.61MHz$
15: 6dB/Octave

Practice Problem 16.1

$\dfrac{810\Omega @ 50°C}{2\,K\Omega @ 25°C}$

$3.04K\Omega @ 50°C$

Practice Problem 16.3

0.004
15

Practice Problem 16.5

22.28
67.52

Practice Problem 16.7

$687.5uA$
−6.88V

Review Section 16.1

1: A transducer connects the physical world to the world of electronics.
3: When using optics couplers to sense radiant energy sources such as incandescent lamps, neon bulbs, light-emitting diodes, and lasers serves as sources of light energy.
5: A photoresistor changes its resistance with a change in light intensity
7: Phototransistors amplify the current generated by light.
9: The Wheatstone gauge gives the advantage of minimizing the effects of resistance changes versus temperature.

Review Section 16.2

1: The elimination of undesired signals by differential amplifiers is called (common-mode) rejection.
3: The ration of the differential gain (A_{vd}) to common-mode gain (A_{vc}).
5: Instrumentation amplifiers amplify small differential signals produced by transducers.
7: To reduce the effects of any loading on the source.
9: (b) 120 dB

Review Section 16.3

1: The output of a transducer is an analog signal which manufactures must convert to digital so that their computers can process the data being received.
3: In a weighted resistor network each input affects the output differently; this is accomplished through doubling the size of each resistor thus halving the effect of each input current at the summing note.
5: A/D converters incorporate amplifiers and are susceptible to impedance, drift, nonlinearity, and noise characteristics.

Review Section 16.4

1: Reverse biasing a photodiode causes a high device (resistance) that (prevents) current flow.
3: The optoisolator shown in Figure 16.34 provides line isolation and high-speed coupling.
5: The differential amplifier used in a balanced thermocouple amplifier circuit has large input impedance, and a high voltage gain.

Troubleshooting Questions

1: True
3: False – A leaky capacitor provides a high resistance path.
5: False

Problems

1: $V_{out} = V_d A_v = (1.67 V) (10) = 16.7 V$
3: $V_{in} = (200 uA) (12 K\Omega) = 2.4 V$
$V_{out} = A_{CL} V_{in} = (1) (2.4 V) = 2.4 V$

5: $A_{vc} = \dfrac{V_{oc}}{V_{ic}} = \dfrac{0.45 V}{2.0 V} = 0.225$

7: $V_{oc} = A_{vc} V_{ic} = (0.225) (1.0 V) = 225 mV$

9: $A_v = \dfrac{2R_2}{R_1} + 1 = \dfrac{2(22 K\Omega)}{10 K\Omega} + 1 = 5.4$

11: Using three op-amps provides the advantage of high input impedance that greatly reduces source loading and unity common-mode gain without close resistor matching.
13: As seen in chapters examples R1 + R2 equal 2R which is in parallel with R3 (2R) which is in series with R which is in parallel with 2R. As such the output voltage from any node is equal to 1/2n times the reference voltage where n equal the node weight with reference to the op-amp input.
Therefore:

$$V_{out} = \left[\frac{1}{2^4}(10V) + \frac{1}{2^3}(10V) \right] = \left[\frac{10V}{16} + \frac{10V}{8} \right] = 1.875V$$

Practice Problem 17.1

$\overline{Q} = Q_1$ & $Q = Q_2$
$Q = Q_1$ & $Q = Q_2$
$Q = Q_2$ & $Q = Q_1$

Practice Problem 17.3

A negative input pulse applied to the base of Q_1 will cause the transistors to switch states again.

Practice Problem 17.5

$t_d = RC = (12 K\Omega) (0.1 uF) = 1.2 mS$

Practice Problem 17.7

$T_P = 1.1 RC = 1.1 (560 \Omega) (0.15 uF) = 92.4 uS$

Review Section 17.1

1: False – Astable multivibrators have no stable states
3: False – Adding resistors and diodes improves rise time.
5: Coupling
7: No

Review Section 17.2

1: (b) Two
3: A triggering pulse is a positive or negative input signal used to push a multivibrator into a desired state.
5: True
7: Clock pulses trigger flip-flop action.

Review Section 17.3

1: (b) One
3: Monostable multivibrators generate one output pulse for every triggering pulse.
5: Adjustable-width pulse generators; Pulse shapers; delay elements
7: Monostable multivibrators typically use (negative) going trigger pulses.

Review Section 17.4

1: The 555 timer is popular because of the pulse widths it can generate and because the circuit can operate with a minimum of external components.
3: 555 timers can operate from supply voltages that range from 4.5 to 18 V. The height of the output pulse corresponds to the supply voltage used.
5: The 555's internal structure is made up of a voltage-divider circuit which supplies the reference voltages for the two high-gain comparators that control a set-reset flip-flop. Also, there is an internal transistor switch which functions as part of the 555's timing circuitry.
7: In Figure 17.22 a 555 timer is configured as an monostable multivibrator. The upper comparator and the discharge connects to the junction between the resistor and capacitor.

Review Section 17.5

1: Bistable multivibrators are stable weather the output is high or low, Monostables generate pulses and shape waveforms. Astable multivibrators oscillate and provide square wave outputs for timing circuits.
3: In the automobile alarm circuit shown in Figure 17.25, timer A provides a delay that allows the driver to enter the automobile.
5: (b) Monostable

Troubleshooting Questions

1: Troubleshooting a 555 timer circuit is a matter of deductions.
3: A capacitor and resistor in parallel.
5: A leaky capacitor causes a DECREASE in the output frequency.

Problems

1: None
3: $105uS$
5: $3.25KHz$
9: False – Bistable multivibrators require a clocking pulse to change from one state to another state.
11: Resistive coupling connects the two transistors to each other.
13: The supply voltage level and the level of the input triggering pulse.
15: Another name for the circuit is one-shot.
19: The input trigger is (negative) going.
21: $460.8Hz$
23: The output frequency would decrease if R_1 was increased in value.
25: $2.42mS$

Practice Problem 18.3

$15.8uS$

Practice Problem 18.5

If R_2 is increased it will cause the triac to fire sooner, which will cause the average current to increase.

Practice Problem 18.7

$14.04mA$
$960mV$

Review Section 18.1

1: True
3: Thyristors are four-layer semiconductors.
5: Close

Review Section 18.2

1: SCRs have three terminals
3: One
5: False – The dv/dt effect can cause the SCR to trigger falsely.
7: LASCRs are light-activated.
9: If the SCR is operating under DC conditions the Anode-To-Cathode supply voltage must be removed. However, for AC conditions the Anode-To-Cathode voltage must be sufficiently decreased.

Review Section 18.3

1: True
3: Triacs often work as AC power controllers.

5: The terminals of Triac are labeled as MT1, MT2, and gate.
7: Triggering of a Triac may take the form of DC, rectified AC, AC, OR pulses.

Review Section 18.4

1: Gate turn-off switches have characteristics that allow a small reverse gate current to turn of the device.
3: A LASCR is a light activated device that operates as a switch.
5: Light triggers a LASCR on.
7: SBSs work as full-wave triggering devices for triacs.
9: True

Review Section 18.5

1: Unijunction transistors work in triggering circuits, oscillator circuits, and timing circuits.
3: (d) one PN junction
5: a) The interbase resistance is the sum on the internal base resistances of the UJT device.
   b) It is the ratio of the emitter-basel resistance to the interbase resistance.
   c) The peak-point voltage is the sum of the voltage across R_{B1} and the forward voltage drop across the junction.
7: The unijunction transistor has the disadvantage of being unilateral to current flow. Thus a UJT requires a DC voltage to operater.

Review Section 18.6

1: Capacitor C_1
3: SCR_1 will be turned off when SCR_2 is triggered on. This will connecting capacitor C_1 across SCR_1, which will momentarily reverse-bias the anode and cathode voltage turning off SCR_1.
5: To protect DC circuits from overvoltage or overcurrent conditions.
7: Overcurrent protection
9: In Figure 18.52, time delay is provided by the UJT circuit.

Problems

3: First, by applying a forward bias voltage in excess of the breakover voltage across the anode-to-cathode. Second, applying the proper trigger level to the gate.
5: The minimum current required to cause a thyristor to conduct is called the latching current.
7: Current-controlled devices, rectifiers, and latching switches.
9: The purpose of the snubber circuit is to prevent false triggering of an SCR.
11: Resistance-Capacitance triggering.

15: False – triacs were developed to improve control over AC power.

19: A diac is triggered by forward voltage.

Practice Problem 19.1

$$\text{Load} - \text{Regulation} = \left(\frac{V_{nl} - V_{fl}}{V_{nl}}\right) X100$$

$$= \left(\frac{15V - 14.5V}{15}V\right) X100 = 3.33\%$$

Practice Problem 19.3

$$V_{out} = I_F R_{ADJ} = (0.03A)(100K\Omega) = 3KV$$

Practice Problem 19.5

$$\frac{I_x}{I_R} = \frac{V_D + V_{BE} + V_{R4}}{V_D + V_{R4}}$$

$$= \frac{0.7V + 0.7V + 1.3V}{0.7V + 1.3V} = \frac{2.7V}{2.0V} = 1.35$$

$$\frac{I_{SC}}{I_R} = \frac{V_D + V_{BE}}{V_D + V_{R4}} = \frac{0.7V + 0.7V}{0.7V + 1.3V} = \frac{1.4V}{2.0V} = 0.70$$

Practice Problem 19.7

$$R_c = \frac{V}{I} = \frac{50V}{0.3A} = 166.7\Omega$$

If load resistance is greater than crossover resistance the supply will regulate voltage; however, if load resistance is less than crossover resistance the supply will regulate current.

Review Section 19.1

1: In a simple power supply its output voltage will vary with load current and line voltage, therefore it cannot supply the exact DC voltages necessary to many applications.

3: A common design scheme to protect voltage regulators against overvoltage is a crowbar protection circuit.

7: Regulation depends upon a feedback circuit, which senses a change in DC output voltage. Without a DC reference voltage regulation could change dramatically.

9: False – Recovery time is the amount of time required for either the regulated voltage or current to return to a specified limit.

Review Section 19.2

1: False – in a series pass regulator the pass transistor is connected in series with the load.

3: Current overloads and short circuits can cause the pass-transistor to break down due to excessive power dissipation.

5: True

7: For a shunt regulator to operate properly the pass transistors must be selected to operate within the maximum voltage, current, and dissipated power limits set by the reference resistor and within the output voltage and load current specifications of the circuit.

Review Section 19.3

1: In a conventional series regulator the pass transistor conducts continuously over the linear operating range. Therefore the pass transistor must dissipate power continuously, because of this power dissipating requirement conventional series pass regulators cannot handle high-current power supply requirements.

3: If the output voltage decreases the difference voltage will cause conduction time to increase. However, if the output voltage rises the conduction time will decrease.

5: Switching regulators can deliver more current to the load than is seen at the unregulated input, this characteristic makes them valuable for computer applications.

9: False – The fall and rise time determine how long it will take the switching transistor to compensate for any changes detected in the output voltage.

11: $f_{osc} = \dfrac{1.1}{R_T C_T}$

Review Section 19.4

1: If the output of a regulator circuit shorts to ground it will cause excessive current to flow through the pass transistor, which can destroy the device.

3: Foldback current limiting is a form of protection against excessive current that shuts the regulator off if an overload or short circuit occurs. As such, foldback current limiting helps maintain load current requirements while protecting the pass transistor from drawing more than its designed amount of maximum current.

5: The differential amplifier improves the current-sensing ability of the foldback current limiting circuit because of its greater sensitivity.

Review Section 19.5

1: See figure 19-24

3: True

5: False – The reference voltage determines when the regulator circuit should switch to compensate for any changes in the load requirements.

7: Some applications, such as personal computers, require that the load *current* remain constant as the load *resistance* changes.

9: The driver amplifier controls the pass-transistor.

Review Section 19.6

1: The RCA hybrid IC's ability to regulate voltage well is due to its temperature compensation for the voltage reference, a differential amplifier, a series pass transistor and an overcurrent protective circuit all within a single IC.

3: Three-terminal regulator general characteristics are: load regulation, ripple rejection, load current requirements, input current requirements, and power dissipation.

7: $V_{out} = 1.25V\left(\dfrac{R_2}{R_1} + 1\right)$

9: Figure 19.35 shows an LM317 three-terminal regulator configured as a constant current source.

11: The popularity of the three-terminal regulator stems form its simplicity, low-cost, high reliability, and performance.

Problems

1:
$$\text{Load Regulation} = \left(\frac{V_{NL} - V_{FL}}{V_{NL}}\right)X100$$
$$= \left(\frac{20V - 19V}{20}\right)X100 = 5\%$$

3: $V_{out(max)}$ regulated $= V_Z = 6.1\,V$

5:
$$\frac{V_{REF}}{V_0} = \frac{R_{CAL}}{R_{ADJ}} \therefore V_0 = \left(\frac{R_{ADJ}}{R_{CAL}}\right)V_{REF}$$
$$= \left(\frac{1.96\,K\Omega}{1\,K\Omega}\right)6.1V = 11.96V$$

7: $V_{OUT(MIN)} = 12V - (12V)(0.01) = 11.88\,V$
$V_{OUT(MAX)} = 12V + (12V)(0.01) = 12.12\,V$

9:
$$\text{Ripple Rejection Ratio} = \frac{\text{Ripple Outside Voltage}}{\text{Ripple Input Voltage}}$$
$$= \frac{200\mu V}{2V} = 6.458\,V$$

Practice Problem 20.1

Carrier Frequency = 790KHz
Upper Sideband Frequency = 790KHz + 5KHz = 795KHz
Lower Sideband Frequency = 790KHz – 5KHz = 785KHz

Practice Problem 20.3

$$f = \frac{1}{2\pi\sqrt{LC}}$$
$$= \frac{1}{2\pi\sqrt{(10uH)(0.0022uF)}} = 1.073\,MH$$

Practice Problem 20.5

$$f_c = \frac{1}{2\pi RC}$$
$$= \frac{1}{2\pi(1.5\,K\Omega)(0.033uF)} = 3.215\,KHz$$

Practice Problem 20.7

$$V_f = \frac{f_2 - f_1}{V_2 - V_1} = \frac{12KHz - 8KHz}{3V - 1.5V} = 2.667\,KHz/V$$

Review Section 20.1

1: See Table 20.1

3: In amplitude modulation the carrier signals instantaneous amplitude is changed by a modulating signal, but not its frequency.

5: In frequency modulation the frequency of the carrier signal is changed by the modulating signal while the amplitude remains constant.

7: FM sideband frequencies are spaced at multiples of the modulating frequency from the carrier.

Review Section 20.2

1: An oscillator is a circuit that generates AC.

3: The distinguishing characteristics of the Hartley oscillator are its tapped inductors or center-taped transformer with a single parallel capacitor.

5: The phase-shift oscillators distinguishing characteristics are three RC networks connected in series.

7: The two types of AM modulation are high-level modulation and low-level modulation.

Review Section 20.3

1: The detector rectifies the RF waveform and recovers the smooth envelope with low-pass filtering. Thus, RF signals become attenuated while audio frequencies are allowed to pass.

3: Foster-Seely detectors; ratio detectors

5: C_2 is used to couple the primary voltage at the center frequency produced by L_1 and C_1 to the radio-frequency choke L_4.

Review Section 20.4

1: Phase comparator; low-pass filter; an amplifier; and a voltage-controlled oscillator.
3: Free-running; capture; and phase-lock
5: The output is taken from the voltage-controlled oscillator.

Review Section 20.5

1: Decrease
3: A varactor diode is used as a voltage variable capacitance.
5: The gunn diode.

Troubleshooting Questions

1: Determine the symptom; make initial checks; locate the faulty stage; locate the faulty circuit; find the defect.
3: Whenever there is a chance that the injected signal will change the DC voltages of the circuit under test.

Problems

1: The VHF frequency range is 30MHz to 300MHz
3: Carrier Frequency = 99.9MHz

First Sideband Frequencies = 99.9MHz ± 15KHz = 99.915MHz & 99.885MHz

5: The feedback circuit must provide a signal of proper phase to the input.
The feedback circuit must provide a signal of proper amplitude to the input.
7: $f = 3 \times 10\text{MHz} = 30\text{MHz}$
9: Since the input frequency is above resonance (10.8MHz > 10.7MHz), the phase angle between inductor L_3 and L_4 voltage is less than the phase angle between the L_2 and L_4 voltage. Therefore, V_{R2} is greater than V_{R1} and the output voltage will be negative.

11: $\Theta_{rad} = \left(\dfrac{\pi}{180°} \right)\Theta = \left(\dfrac{\pi}{180°} \right)30° = 0.5236_{rad}$

13: The three steps the PLL goes through are free-running, capture, and phase lock.
15: The minimum negative resistance current is called valley current (IV).

WHAT DO ALL THE ABBREVIATIONS MEAN?
As we have studied electronic circuits, we have encountered abbreviations for each part of the circuit, signal reference points, and resistances. Here is a list of the abbreviations we have used and their meanings.

I_E	Emitter current for DC signal conditions
I_B	Base current for DC signal conditions
I_C	Collector current for DC signal conditions
V_{EE}	Emitter bias or supply voltage
V_{cc}	Collector bias or supply voltage
V_{BB}	Base bias or supply voltage
V_E	DC voltage at the emitter terminal to ground
V_B	DC voltage at the base terminal to ground
V_C	DC voltage at the collector terminal to ground
r_s	Internal resistance of source voltage
r'_e	Emitter resistance for AC signal conditions
r_b	Base resistance for AC signal conditions
r_c	Collector resistance for AC conditions
A_v	Voltage gain
A_1	Current gain
A_p	Power gain
v_{in}	Applied input voltage
v_{out}	Output voltage
R_E	DC emitter resistance in a transistor circuit
V_{TH}	Thevenin voltage
R_{TH}	Thevenin resistance
i_e	AC emitter current
Z	Impedance, measured in ohms
dB	Decibel
V	DC voltage
v	AC voltage

AC analysis, of the differential amplifier, 491–93
Acceptor impurities, 68
AC emitter resistance, 223
AC output voltage, 280
    clipping, 280–81
AC theory review, 88–89
Active diode circuits, 652–63
    active clampers, 661–63
    active half-wave rectifiers, 653–54
    active limiters, 658–61
    active peak detector, 656–58
    half-wave rectifiers with gain, 654–56
    output bounding, 660
    troubleshooting, 693–95
Active filters, 707–58
    band-pass, 735–37
    band-reject, 739–40
        twin-T, 749–50
    cascading filters, 737–39, 740
    classification of
        band-pass, 718–20
        band-reject, 720–21
        Bessel, 725–27, 727 tab, 728
        Butterworth, 721–24, 727 tab, 728
        Chebyshev, 724–25, 727 tab, 728
        choosing the right filter, 721
        high-pass, 717–18
        low-pass, 715–17
        notch, 720
        passive filter network, 715
    defining filter performance, 708–14
        balance, 708
        bandwidth, 709–12
        center frequency point, 710–11
        critical frequency, 712
        frequency, 709
        frequency loss, 713
        ladder configuration, 708
        lattice configuration, 708
        phase shift, 713
        pi (π) configuration, 708
        poles, 712
        rise time t_r, 714
        selectivity, 713
        settling time t_s, 714
        time and filters, 714
    high-pass, 734–35
        VCVS, 748–49
    low-pass, 729–34
    state-variable, 750–51
    troubleshooting, 752–54
    using Miller-effect feedback, 741–52
    voltage-controlled voltage source (VCVS), 744–46
    vs passive, 728–29
Active limiter, 658–61
Alpha (α), 146–48
Ammeter, 49

Amplifiers, small signal
    capacitor coupling, 217
    circuit configurations, 215–16
    common-base, 254–58
        using voltage divider bias with, 257–58
    common-collector, 248–54
        input impedance of, 251–53
    common-emitter, 226–48
        adding a swamping resistor, 235–37
        calculating gain, 226–29
        design factors, 237
        how load effects gain, 232–34
        phase inversion, 237
        swamping out AC emitter resistance, 234–35
        that uses collector feedback bias, 244, 244–47
        using DC emitter bias, 230–32
        using voltage divider bias input impedance, 238–44
    defined, 213
    equivalent circuits, 218–26
        AC, 222–23
        DC, 219–21
    gain
        concept of, 214–15
        stage v. transistor, 259–60
    multistage, 217–18, 260–67
        troubleshooting, 267–70
    single-stage, 217–18
        See also Differential amplifiers; Op amps; Power amplifiers
Amplitude modulation (AM), 931–35, 943
    detectors, 946–47
Anode, 73
Antimony, 67
Armstrong oscillator, 625–26
Arsenic, 67
Astable multivibrators, 802–7
Atomic theory, 60–61
Avalanche breakdown, 80
Average current (I_{avg}), 88
Average voltage (V_{avg}), 88

Balance, 708
Band-pass filters, 107, 611
    See also Active filters
Band-reject filters. See Active filters
Bandstop filter, 612
Bandwidth
    active filters, 709–12
    BJT and JFET amplifiers, 473–88
        bandwidth defined, 474
        BJT high-frequency analysis, 480–81
        BJT low-frequency signal analysis, 476–80

finding dominant critical frequency of a circuit, 476
        gain-bandwidth product, 480
        half-power points, 474
        highpass circuit, 474
        JFET high-frequency analysis, 486–87
        JFET low-frequency signal analysis, 484–85
        lagging RC networks, 475–76
        leading RC networks, 474–75
        lowpass circuit, 475
        Miller's theorem, 481–84
Barrier potential, 73
Base-bias
    improving, 163–64
    prototype transistor, 203–4
    using, 156–63
Bessel filter, 721, 725–27, 727 tab, 728
    low-pass VCVS values, 746 tab, 747–48
Beta (β), 146–47, 147–48
    box, 152–55
        bias model, 153
        flowchart, 153
    minimizing the effects of, 164–89
    thermal runway, 148–49
Biased clippers, 116–19
Biasing
    base bias
        improving, 163–64
        using, 156–63
    Class AB amplifiers, 303–4
    Class B amplifier, 299
    DC, 142
    defined, 75
    differential amplifiers, 490–91
    forward, 74, 75–76
    minimizing effects of Beta
        adding RE, 164–69
        collector feedback bias, 174–79
        using emitter bias, 170–74
        using voltage divider bias, 179–84
    of the PNP transistor, 189–95
    reverse, 76
    transistors, 144–45
    the UJT, 867–71
        See also Junction field effect transistors (JFETs):biasing
Bilateral trigger diodes, 866
Bipolar junction transistors (BJTs), 141–211
    alpha, 146–48
    base bias, 156–63
        improving, 163–64
    Beta box, 152–55
    beta thermal runway, 148–49
    biasing, 144–45
        of the PNP transistor, 189–95
            collector feedback, 192–93

drawing it upside down, 193–95
breakdown, 151–52
collector-base junction, voltage to reverse bias, 150
collector curves, 195–200
   DC load lines, 197–99
controlling current with, 146
DC cutoff, 151
feedback, effects of, 177–79
minimizing the effects of Beta, 164–89
   adding a feedback resistor (R_E), 164–69
   collector feedback bias, 174–77
   setting up the voltage divider, 185–89
   using emitter bias, 170–74
   using voltage divider bias, 179–84
power dissipation in transistor circuits, 149–50
saturation, 151
structure of, 142–43
as two back-to-back diodes, 143–44
voltage from base to emitter, 150
   See also Bandwidth: BJT and JFET amplifiers; Troubleshooting: transistors
Bistable multivibrators, 802–7
Bode plot, 524
Bootstrapping, 543
Breakdown
   diodes, 80
   region, drain curves, 343
   voltage, 79–80, 151–52
      limiting, 120–22
Bridge rectifier, 97–101
Butterworth filters, 721–24, 727*tab*, 728
   high-pass, 734–35
   low-pass, 729–34
      VCVS values, 746*tab*, 747

Capacitance, 81
Capacitor input filter, 106
Capture and lock (PLL), 953
Carrier deviation, 935–36
Cascade, 324
Cascading, 260–61
   differential amplifiers, 777–78
   filters, 737–39, 740
Cascode amplifier, 406–8
Cathode, 73
Center frequency point, 710
Center-tapped full-wave rectifier circuits, 93–97
Channel, JFETs, defined, 338
Chebyshev filter, 721, 724–25, 727*tab*, 728
   low-pass VCVS values, 746*tab*, 747–48
Circuit analysis
   current dividers and Kirchhoff's current law, 15–21
   redrawing complex circuits, 27–29
   using Kirchhoff's voltage law, 35–40

using Ohm's law, 3–6
using superposition theorem, 40–47
using Thevenin's theorem, 22–35
   RC time constant, 30
   Thevenin equivalent circuit, 22–23
   Thevenizing circuits, 24–31
      unbalanced bridge, 32–34
voltage dividers, 6–15
Clamper action, Class C amplifiers, 312–14
Clampers, 661–63
Clapp oscillator, 624
Class A amplifiers, 282–94
   AC analysis, 284–85
   DC analysis, 283–84
   delivering power to the load, 292–93
   distortion, 282–83
   power dissipation, 290–92
   stage efficiency, 293–94
Class AB amplifiers, 302–11
   biasing, 303–4
   crossover distortion, 302
   Darlington pairs, 306–11
   quasicomplementary-symmetry, 304–6
Class B amplifiers, 295–302
   biasing, 299
   complementary operation, 295–96
   current mirror concept, 299–300
   dissipating and delivering power, 297–98
   operating with AC and DC conditions, 300–301
   push-pull circuits, 296–97
   temperature stability, 300
Class C amplifiers, 312–24
   circuit operation, 312–14
   clamper action, 312–14
   coupling stages together, 324–29
   dissipating and delivering power with, 320–23
   efficiency and, 323
   emitter breakdown, 316–17
   operating conditions, 314–16
   sine wave, maintaining, 319–20
   tuned, 317–19
Clippers, 59
Clipping, 280–81
Closed-loop circuits and Kirchhoff's voltage law, 35–37
CMOS analog applications, MOSFETs, 408–9
Collector-base junction, 142
Collector curves, 195–200
Collector feedback bias
   analyzing a common-emitter circuit, 244, 244–47
   minimizing the effects of Beta, 174–77
   for the PNP transistor, 192–93
Collector feedback prototype transistor, 206
Collector to base, emitter open (V_{CBO}), 152

Colpitts oscillator, 618–24, 630, 631, 938, 940–41
Common-base circuit, 217–16, 254–58
   using voltage divider bias with, 257–58
Common-collector circuit, 216, 248–54
   input impedance of, 251–53
Common-drain JFET amplifier, 370–72
Common-emitter circuits, 215–16
Common-gate JFET amplifier, 372–74
Common-mode rejection, 775–77
   ratio (CMRR), 494–95, 529–30
Common-source JFET amplifier, 366–70
Commutating capacitor. *See* Speed-up capacitor
Commutation. *See* Triacs:commutation
Comparators, 664–66
   crossing detectors, 665–66
   troubleshooting, 695–97
   See also Schmitt triggers; Window comparators
Conductance (G), 362
Conductors, 63
Constant current source, 18–19
   for a differential amplifier, 495–96
Converters, 784–92
Copper atom, 62
Coupling amplifier stages, 324–29
Covalent bonding, 65
Critical frequency, 598, 601–3, 712
Crossing detectors, 665–66
   with hysteresis:Schmitt triggers, 666–68
Crossover distortion, 302
Crystal-controlled Oscillators, 629–31
Crystals
   semiconductor, 64–65
      extrinsic, 66–67
      growth of, 65–66
      intrinsic, 66. *See also* Oscillators:crystal controlled
Current dividers and Kirchhoff's current law, 15–21
Current feedback. *See* negative feedback:series-series
Current gain(A_i), 214
   parallel-series Op-amp circuits, 568–71
Current mirror concept, 299–300
Current source bias, 359–61
Current-to-voltage transducers, 562–64

Darlington pairs, 306–11
   with AC signal referenced to ground, 327–28
Data sheets, 133–35
DC (Direct current)
   amplification with differential amplifiers, 489–90
   cutoff, 151
   drift, 489
   load lines, 197–99
   effect of adding bias, 303–4
   emitter bias and, 199

static switch, 877
Decay time(t(dec)), 424
Decibel (dB), 261
Delay time, 423
Depletion bias, 395–97
Depletion mode, JFETs, 340
Depletion MOSFETs. *See* MOSFETs: depletion (D) MOSFETs
Depletion region, 72–73
Detectors, 946–50
    AM, 946–47
    FM, 947
    Foster-Seely FM, 947–49
    ratio, 949–50
Diac, 866
Dielectric properties of silicon dioxide, 390
Differential amplifiers, 485–96, 773–84
    AC analysis of, 491–93
    biasing, 490–91
    cascading, 777–78
    common-mode rejection, 775
    DC amplification with, 489–90
        DC drift, 489
    difference-mode signal, 775
    instrumentation, 778
        troubleshooting, 795
        using precision resistors, 778, 779
        using three op-amps, 778–81
        with bootstrap power supply, 781–82
    inverting input, 492
    noninverting input, 491
    output configurations, 773–77
    parameters, 493–96
        common mode rejection ratio, 494–95
        constant current source, 495–96
        input bias current, 493
        input offset current, 493
        input offset voltage, 494
        output offset voltage, 494
    "tail current", 490
    "tail resistor", 490
    troubleshooting, 496–98
Differentiator circuit, 451–52
Differentiators, 631–34
Diffusion, 70–71
Digital applications, MOSFETs, 409–10
Digital multimeter, 48
Digital-to-analog converter (DAC), 583, 784–92
    R-2R ladder network, 786–92
    weighted-resistor network, 784–86
Diodes
    applications
        AC theory review, 88–89
        breakdown voltage limiting, 120–22
        diode drop, 119
        filtering
            adding a resistor in series with filter capacitor, 105–6

a full-wave rectifier circuit with a capacitor, 104–5
            a half-wave rectifier circuit with a capacitor, 102–4
            types of filters, defining, 106–7
        power supply, 88
            input filters, 102–7
            regulating, 107–11
                using a zener diode, 108–11
        as rectifiers
            center-tapped full-wave rectifier circuits, 93–97
            full-wave bridge rectifier circuits, 97–101
            full-wave rectifier circuits, 93
            half-wave rectifier circuits, 91–93
        as signal clampers, 123–28
        as signal clippers, 115–22
            adding another shunt arm, 119–20
            biased, 116–19
        transformer review, 89–90
        unbalanced loads and the DC return, 129
        voltage multipliers, 111–15
            doubler, 111–13
            triplers and quadruplers, 113–14
    characteristics of, 77–81
        avalanche breakdown, 80
        breakdown voltage, 80
        curves, 77–79
        peak inverse voltage, 79–80
        peak reverse voltage rating, 77
    conduction
        within forward biasing, 74, 75–76
        within reverse biasing, 76
    controlled avalanche rectifiers, 80
    defined, 59
    forward current, 76
    minority current, 76
        carriers, 76
    operation of
        capacitance, 81–83
        resistance, 80–81
        voltage drops, 81
        zener diodes, 81–83
    pictorial representation, 73
    PN junction, 71–72
    reverse current, 76
    schematic diagram of, 73
    See also High-frequency diodes
Direct coupling, 326–27
Distortion, 281–82
    and negative feedback, 470
Dominant critical frequency, 476
Donors, 67
Double L filter, 106
Drain, 387
    JFETs, 338, 342–44
Drain feedback bias in an E MOSFET circuit, 402–5

Drift, 894
Dual-gate MOSFETs, 407–8
Dual in-line packages(DIPS), 505
Dual polarity power supply, 508
Dual-tracking regulator, 923
Duty cycle, 422
DV/DT effect, 847–50
Dynamic microphone, 771–73
Dynamic testing, 960

Efficiency
    and Class C amplifiers, 323
    defined, 282
Electrical conduction, law of, 61
Electricity defined, 63
Electron current flow, 4
Electronic crowbar, 878–79
Electrons, 60
Elements, 60
Emitter-base junction, 142
Emitter bias
    minimizing the effect of Beta, 170–74
    prototype transistor, 204–5
Emitter breakdown, Class C amplifiers, 316–17
Emitter-follower circuit. *See* Common-collector circuit
Enhancement MOSFETs. *See* MOSFETs: enhancement (E MOSFETs)
Epitaxial growth, 66
Equipotential points, 28
Equivalent circuits, 218–26
    AC, 222–23
        emitter resistance, 223
    DC, 219–21
Equivalent resistance (R_{eq}), 11–13

Feedback
    control loop, 547–51
    node, 175
    resistor (R_E), minimizing the effect of Beta, 164–69
FET. *See* Field effect transistor (FET)
Field effect transistor (FET), 337
    switching
        JFET, 439–41
        MOSFET, 441
Filters and filtering
    adding a resistor in series with filter capacitor, 105–6
    a full-wave rectifier circuit with a capacitor, 104–5
    a half-wave rectifier circuit with a capacitor, 102–4
    filtering defined, 707
    ripple voltage, 103
    types of filters, defining, 106–7
    See also Active filters
Firm voltage dividers, 185–86
555 timer, 820–25
    used in an automobile alarm circuit, 826, 827

Foldback current-limiting, 906–9
Forward current, 76
Forward gain, LC oscillators, 620–21
Forward transfer admittance (yfs), 362
Foster-Seely FM detectors, 947–49
Frequency (hertz), 88
    and filters, 709
    spectrum, 930tab
Frequency modulation (FM), 935–37, 944
    detectors, 947
    PLL demodulator, 953–55
Frequency multiplier, 319–20, 943
Frequency response, 450
    curves, 709–21
Frequency stability of oscillation, 942–43
Full-wave phase control circuit, 878
Full-wave rectifier circuits, 93

Gain
    concept of, 214–15
    gain stage v. transistor, 259–60
    of multistage amplifiers, 260–67
Gain-bandwidth product, 480
    Op-amps, 524–27
Galvonometer, 32–33
Gate, JFETs, 338
    bias, 347–49
Germanium, 73
Gunn diodes, 958, 959

Half-power points, 474
Half-wave rectifier circuits, 91–93,
    653–54
Half-wave signal, 92
Harmonic, 319
Hartley oscillators, 624–25, 938, 939–40
High-frequency analysis
    BJTs, 480–81
    JFETs, 486–87
High-frequency diodes, 955–60
    Gunn, 958, 959
    IMPATT, 958, 959
    microwave power, 958–60
    tunnel diodes, 955–56
    tunneling, 955
    valley voltage, 956
    varactor, 956–58
Highpass circuit. See Leading RC networks
High-pass filter, 107. See also Active filters
Holding current (I_{HX}), 841
Holes, 61
Hybrid (h) parameters, 471–73
Hysteresis
    loops, 679–89
    voltage, 435

IMPATT diodes, 958, 959
Induced channel, 392
Input bias current, 493
    of Op-amps, 513, 517–20
Input impedance
    of a common-collector circuit, 251–53

of a common-emitter that uses voltage
        divider bias, 238–44
Op-amp circuits, 512–13, 551–52,
        564–67, 571–77
Input offset current, 493
Input offset voltage, 494
    Op-amps, 520–21
Instrumentation amplifiers. See Differential
        amplifiers: instrumentation
Insulated gate field effect transistor
        (IGFET). See MOSFETs
Insulators, 63
Integrated circuits, 488
    voltage regulators, 913–20
    See also Differential amplifiers; Op-
        amps
Integrator circuit, 452–53
Integrators, 634–37
Interbase resistance (R_{BB}), 442
Intrinsic standoff ratio (ISR), 442
Inversion layer, 392
Inverted double L filter, 106
Inverter, 410
Inverting
    amplifier, 559–62
        current amplifier, 567–73
    input, 492
Ions, 61

Junction field effect transistors (JFETs),
        337–78
    basic operation, 328–46
        connecting supply voltages to,
            339–40
        gate-to-source supply voltage,
            341–42
        JFET drain, 342–43
            curves, 343–44
        source-to-drain supply voltage,
            340–41
        square law device, 345
    biasing, 346–59
        current source bias, 359–61
        gate bias and self-bias, comparison
            of, 353–54
        matching JFET characteristics, 346
        using gate bias, 347–49
        using self-bias, 349–54
        with a voltage divider, 354–59
    depletion mode, 340
    difference between JFETs and BJTs,
        338
    switching, 439–41
    transconductance, 361–64
        determining
            at the Q point, 363
            at the saturated drain current,
                362–63
    troubleshooting, 377–78
    used as amplifiers, 364–74
        common-drain, 370–72
        common-gate, 372–74

common-source, 366–70
    voltage gains finding the, 366
    voltage gains of BJTs and JFETs
        compared, 365–66
used as linear and logic switches, 376
used as source followers, 374–76
used as variable resistors, 376
    See also Bandwidth: BJT and JFET
        amplifiers

Kirchhoff's current law, 15–21
Kirchhoff's voltage law, 35–40
    for checking answers, 37
    developing simple equations, 38–39

Lagging RC networks, 475–76
Latching current (I_L), 841
Lattice, 65
LC filter circuits, 610–13
Lead and lag circuits, 598–610
Leading RC networks, 474–75
Leakage current, 419–20
Light-activated SCRs (LASCRs), 864, 865
    triggering circuit, 879–80
Linear and logic switches, JFETs, 376
Linear regulators. See Regulators: linear
Line regulation, 890–91
Load, unbalanced, and the DC return, 129
Loading effects, LC oscillators, 621–24
Load lines, 276–81
Load regulation, 888–90
Lower trip point (LTP), 432
Low-frequency analysis
    BJTs, 476–80
    JFETs, 484–85
Lowpass circuit. See Lagging RC networks
Low-pass filters, 107. See also Active filters

Magnetic phonograph cartridge, 771–73
Majority carriers, 69–70
Matter, 60
Metal oxide semiconductor field effect
        transistor. See MOSFETs
Meters. See Test instruments
Microwave power diodes, 958–60
Miller-effect feedback, 741–52
Miller's theorem, 481–84
Milliammeter, 49
Minority carriers, 69–70
Minority current, 76
Modulation, 930–37
    amplitude modulation (AM), 931–35
        sidebands, 934–35
        generation, 935
    concept of, 930–31
    defined, 930
    emitter and collector, 944–45
    frequency modulation (FM), 935–37
        center frequency, 935
        frequency (carrier) deviation,
            935–36
        sidebands, 937

frequency spectrum, 930
modulators, 943–44
Molecules, 60
Monostable multivibrators, 811–20
MOSFETs, 385–414
  advantages of, 386–87
  block diagram of, 386
  cascode amplifier, 406–8
  CMOS analog applications, 408–9
  depletion-type MOSFETs (D
       MOSFETs), 387–91
    biasing and circuit analysis, 395–97
    drain curves, 389
    N-channel, 388, 389
    operation, 390–91
    P-channel, 388, 389
    transconductance curve for, 390
    using voltage divider bias, 397–99
    using zero bias, 399–400
  digital applications, 409–10
  drain, 387
  dual-gate, 407–8
  enhancement-type MOSFETs (E
       MOSFETs), 387, 391–94
    calculate constant K, 393–94
    drain feedback bias, 402–5
    N-channel, 391, 392
    P-channel, 391, 392
    transconductance curve, 393
    using voltage divider bias, 400–402
  inversion layer, 392
  inverter, 410
  source, 387
  switching, 441
  threshold voltage (Vth), 392
  troubleshooting, 411
  values and parameters for, 386 *tab*
Multivibrators, 801–36
  the 555 timer, 820–25
    internal structure of, 821
    operating as an astable
       multivibrator, 822–24
    operating as an monostable
       multivibrator, 822–24
    troubleshooting, 831–32
    using in an automobile alarm
       circuit, 826, 827
  astable, 802–7
    analyzing a simple circuit, 803–6
    improving rise time, 806–7
  bistable, 807–11
  generating pulses and square waves
      with, 826
  monostable, 811–20
    analysis of a simple circuit, 817–20
    maximum duty cycle, 813–15
    noise immunity, 816–17
    pulse width, 815–16
    timing, 815
    triggering, 812–13
    using to design an oscilloscope-
       triggered time base, 826–30

triggering pulse, 808

N-channel D MOSFETs, 388, 389
N-channel JFETs, 338
Negative alternation, 88
Negative feedback, 470
  parallel-parallel, 540–41, 558–67
  series-parallel, 538–39
    voltage amplifier, 542–58
  series-series, 540
Negative peak clippers, 115
Negative resistance region, 444
Neutrons, 60
Node, 10
Noise generator, 960–62
Noninverting AC voltage amplifier,
    579–81
Noninverting input, 491
Nonlinear distortion, 345
Notch filter, 612
NPN transistors. *See* Bipolar junction
    transistors (BJTs)
N-type semiconductor materials, 67
  in a Shockley diode, 840–41
Nulling, 522

Ohmic region, 343
Ohmmeter, 47–49
Ohm's law, 3–6
One-shot multivibrators. *See* Monostable
    multivibrators
Op-amp circuits, 537–93
  applications
    instrumentation, 584
    noninverting AC voltage amplifier,
      579–81
    summing amplifiers, 581–84
    voltage follower, 577–79
  bootstrapping, 543
  differentiator, 633
  integrator, 635
  loop gain, 538
  parallel-parallel feedback, 540–41,
    558–67
    current-to-voltage transducers,
      562–64
    input and output impedance,
      564–67
    inverting voltage amplifier, 559–62
  parallel-series feedback, 541
    inverting current amplifier, 567–73
    current gain, 568–71
    input and output impedance,
      571–73
  series-parallel feedback, 538–39
    noninverting voltage amplifier,
      542–58
    feedback control loop, 547–51
    ideal analysis, 542–47
    input impedance, 551–52
    output impedance, 552–53
    sacrifice factor, 555–58

voltage gain, 553–55
  series-series feedback, 540
    voltage-to-current transducer,
      573–77
  troubleshooting, 585–88
  virtual ground, 558
    See also Active diode circuits; Op
      amps
Op-amps, 504–34
  basic operation, 509–12
  electrical parameters and impedances
    of input bias current, 513, 517–20
    of input impedance, 512–13
    of input offset voltage, 520–21
    of nulling, 522
    of output offset voltage, 522–23
  frequency-dependent characteristics
    Bode plot, 524
    common-mode rejection ratio,
      529–30
    gain-bandwidth product, 524–27
    output impedance, 523–24
    power bandwidth, 528–29
    roll off, 525
    slew rate, 527–28
      distortion, 528
  linear applications, 506
  output terminal, 509
  positively saturated op-amp, open loop
      configuration, 507
  power supply requirements and
      limitations, 508–9
  schematic symbol for, 507
  specification sheet, 514–16
  troubleshooting, 530–31. *See also* Op
      amp circuits
Operational amplifiers. *See* Op-amps
Optoisolators, 769–70
  in a high-speed line receiver, 793
Oscillators, 595–631
  crystal-controlled, 629–31
    Colpitts, 630
    Pierce, 630
  emitter and collector modulation,
    944–45
  frequency multipliers, 943
  frequency selection circuits
    bandpass filter, 611
    bandstop filter, 612
    LC filter circuits:parallel resonance,
      610–13
    lead and lag networks, 598–10
      determining critical frequency,
        601–3
      lead-lag combinations, 610
      phase shift, 603–9
    notch filter, 612
  general theory, 596–97
  LC Armstrong, 625–26
  LC Clapp, 624
  LC Colpitts, 618–24
    feedback ratio, 620

forward gain, 620–21
frequency of oscillations, 619
loading effects, 621–24
positive feedback, 619
LC Hartley, 624–25
modulators, 943–45
relaxation, 626–29
sinusoidal, 596, 938–43
crystal, 938, 943
frequency stability, 942–43
Hartley, 938, 939–40
phase-shift, 938, 941–42
troubleshooting, 641–42
twin-T filter, 616–18
Wien-bridge, 613–16
with JFET control, 615–16
Oscilloscope, 49
digitizing, 48
using a triggered time base, 826–31
Output current, series-series feedback Op-amp circuits, 574
Output impedance, Op-amp circuits, 523–24, 552–53, 564–67, 571–77
Output offset voltage, 522–23
Overvoltage protection, voltage regulators, 892–93

Parallel branches, 9
Parallel resonance, 610–13
Passband, 709
P-channel D MOSFETs, 388, 389
depletion-type MOSFETs (D MOSFETs) drain curves for, 389
drain, 387
source, 387
P-channel JFETs, 338
Peak detectors, 656–58
Peak inverse voltage, 79–80
Peak negative voltage ($-V_{peak}$), 88
Peak positive voltage ($+V_{peak}$), 88
Peak reverse voltage rating, 77
Peak-to-peak voltage (V_{p-p}), 43
Peak voltage(V_p), 43
Pentavalent doping, 67
Period, 88
Personal computer switching power supply, 921–23
Phase-locked loops (PLL), 950–55
capture and lock, 953
components
amplifier and filter, 952
phase comparator, 951
voltage-controlled oscillator (VCO), 951–52
FM demodulator, 953–55
Phase shift, 603–10
filters and, 713
Phase-shift oscillators, 938, 941–42
Phosphorus, 67
Photodiodes, 768

automobile lamp dimmer, 792–93
Photoresistors, 766–68
Phototransistors, 768–69
Pierce Oscillator, 630, 631
Pi (π) filter, 106
Pinch-off voltage, JFETs, 341
PNP transistors. *See* Bipolar junction transistors
Positive alternation, 88
Positive clipper. *See* Active diode circuits:half-wave rectifier with gain
Positive feedback
defined, 595
LC oscillators, 619
Positive peak clippers, 115
Power amplifiers, 275–335
classifying, 281–82
coupling stages together, 324–29
direct coupling, 326–27
efficiency, 282
load lines, 276–81
AC cutoff and saturation voltages, 276–80
AC output voltage, maximum amplitude, 280
clipping, 280–81
RC coupling, 324–25
referencing direct-coupled AC signal to ground, 327–28
transformer coupling, 325, 326
troubleshooting, 329–30
tuned coupling, 328–29
See also Class A amplifiers; Class B amplifiers; Class C amplifiers; Class AB amplifiers
Power bandwidth, Op-amps, 528–29
Power dissipation, voltage regulators, 893
Power gain (A_p), 214
Power supplies, 88
dual polarity, 508
input filters, 102–7
regulating, 107–11
troubleshooting, 130–32
voltage regulating, current regulating, 911–13
Power supply rejection ratio, 781
Precision time-delay circuit, 880–81
Programmable unijunction transistor (PUT), 874–77
Protons, 60
P-type semiconductor materials, 68–69
in a Shockley diode, 840–41
Pulse repetition frequency, 438
Pulses, 826
Pulse-width modulated step-down converter, 903–5
Push-pull circuits, Class B amplifiers, 296–97

Q point, 276

determining transconductance at, 363
Quasicomplementary-symmetry amplifier, 304–6
Quiescent, defined, 276

R-2R ladder network, 786–92
Ramp signals, 451
generators, 453–58
Ratio detectors, 949–50
RC coupling of amplifiers, 324–25
RC integrator circuit, 635
RC time constant, 30
Recombination, 69–70
Recovery time, 430, 893
Rectangular waveform, 802
Regulation, 107
See also Regulators; Voltage regulation
Regulators
applications
automatic shutdown, 921
switching power supply for a PC, 921–23
current regulators, 909–13
power supplies, 909–11
voltage regulating, current regulating power supplies, 911–13
foldback current limiting, 906–9
IC voltage regulators, 913–20
for precision power supplies, 913, 914
RCA hybrid IC circuit, 913–15
series, 913–15
three-terminal, 915–17
adjustable, 917–18
used as constant current sources, 918–19
used as high current sources, 919–20
linear, 894–900
series pass, 894–98
current limiting for transistors, 241–62
transistor characteristics, 896
shunt, 898–900
switching, 900–905
configurations, 902–3
pulse-width modulated step-down converter, 903–5
troubleshooting, 923–24
See also Voltage regulation and regulators
Relaxation oscillators, 626–29
Resistance, diode, 80–81
Reverse current, 76
Ripple, 709
Ripple rejection ratio, 893–94
Rise time (t_r), 423, 714
in astable multivibrators, 806–7
Roll off, 525
Root-mean-square (rms), 89
of an AC voltage, 292

Sacrifice factor, 555–57
Safety procedures, 130
Saturated drain current, 362–63
Saturation, 151
   JFETs, 341
   voltages, 420, 422
Sawtooth generators, 637–41
Schmitt triggers, 432–39, 666–89
   crossing detectors with hysteresis,
       666–68
   inverting, 668–73
     adjusting the trip points, 670–73
   noninverting, 673–79
     adjusting the trip points, 677–79
   with nonzero centered hysteresis loops,
       679–89
   trip points, 667–89
   troubleshooting, 695–97
Scientific calculator, using, 604–10
Self-bias, JFETs, 349–54
Semiconductors
   acceptor impurities, 68
   adding impurities to, 64
   barrier potential, 71, 73
   basic atomic theory, 60–61
   covalent bonding, 65
   crystals, 64–65
     extrinsic, 66–67
     growth of, 65–66
     intrinsic, 66
   defined, 64
   depletion region, 71, 72–73
   diffusion, 70–71
     current, 72
   diodes
     characteristics of, 77–81
       avalanche breakdown, 80
       breakdown, 80
       curve, 77–79
       peak in verse voltage, 79–80
     conduction within, 74–77
       forward biasing, 74, 75–76
       reverse biasing, 76
     controlled avalanche rectifiers, 80
     operation of
       capacitance, 81
       resistance, 80–81
       voltage drops, 81
       zener diodes, 81–83
     pictorial representation, 73
     PN junction, 71–72
     schematic diagram, 73
   doping, 66–67
   electron flow, 61–63
   elemental, 64
   intermetallic or compound, 64
   law of electrical conduction, 61
   majority carriers, 68
   materials, 64
     N-type, 67
     P-type, 68–69
   minority carriers, 68

pentavalent doping, 67
PN junction, 71–72
   recombination, 68–69
   trivalent doping, 68
Series aiding, 103
Series circuits
   current flow and Ohm's law, 4
   resistance and Ohm's law, 5
   voltage and Ohm's law, 5
Series opposing, 103
Series pass regulators, 894–98
Series strings, 9
Settling time (t_s), 430, 714
Shunt
   clippers, 115–16
   regulators, 898–900
Sidebands
   AM, 934–35
   FM, 937
Signal
   clampers, 123–28
   clippers, 115–22
   injector, 960–62
   tracing, 962
Silicon bilateral switch (SBS), 864–65
Silicon-controlled rectifiers (SCRs), 634,
       843–57
   basic operation, 843–44
   defining the specifications, 844–46
   DV/DT effect, 847–50
   heat sinks, 852–53
   light-activated, 864, 865
   phase control, 853–57
   reverse-bias conditions for, 851–53
   triggering, 847
   turning off, 851
   turning on, 850
Sine waves, 449–50
Single in-line packages (SIPS), 505
Sinusoidal oscillators. *See* Oscillators:
       sinusoidal
Slew rate, 527–28
   distortion, 528
Slope, 634
Small signal transistor parameters, 470–73
   hybrid or h, 471–73
Solid state devices, def., 59
Source, 387
   JFETs, 338
Source followers, JFETs, 370, 374–76
Source-to-drain supply voltage, JFETs,
       340–41
Spectrum analyzer, 473
Speed-up capacitor, 426–32
Square law device, JFETs, 345
Square waves, 450–51, 826
State-variable filter, 750–51
Static testing, 960
Step-down transformer, 90
Step-up transformer, 90
Step voltage change, 423
Stiff voltage dividers, 185–86

Stopband, 709
Storage time, 424
Strain gauge, 770–71
   circuit, 782–83
Summing amplifiers, 581–84
Superposition theorem, 40–47
Surface mount devices, 505
Switches
   DC static, 877
   gate turn-off, 864
   high-speed or electronic crowbar,
       878–79
   LASCRs, 864
   linear and logic, JFETs, 376
   silicon bilateral (SBS), 864–65
Switching regulators. *See* Regulators:
       switching
Symmetrical limiter, 121–22

Tail
   current, 490
   resistor, 490
Tank circuits, 319–22
Test instruments
   ammeter, 49
   digital multimeters, 48
   galvanometer, 32–33
   milliammeter, 49
   ohmmeter, 47–49
   oscilloscope, 48, 49
   voltmeter, 49
   volt-ohm-milliammeter, 48
   zero adjust knob, 49
Thermal runway and beta, 148–49
Thermistors, 763–65
   circuits, 782, 783
Thermocouples, 765–66
   balanced amplifier, 793–94
Thevenin's theorem, 22–35
   RC time constant, 30
   series parallel circuit, 23–24
   Thevenin equivalent circuit, 22, 22–23
   thevenizing a circuit, 24–31
     unbalanced bridge circuit, 32–34
Threshold voltage (V(th)),392
   defined, 376
Thyristors, 839–84
   applications
     basic triac control circuit, 877
     DC static switch, 877–78
     full-wave phase control circuit,
       878
     high-speed switch or electronic
       crowbar, 878–79
     LASCR triggering circuit, 879–80
     precision time-delay circuit, 880–81
   bilateral trigger diodes (diacs), 866
   breakover voltage, 841
   gate turn-off switch, 864
   light-activated SCRs, 864, 865
   Shockley diode, 840–43
   silicon bilateral switch (SBS), 864–65

silicon-controlled rectifiers (SCRs), 843–57
  basic operation, 843–44
  defining the specifications, 844–46
  DV/DT effect, 847–50
  heat sinks, 852–53
  phase control, 853–57
  reverse-bias conditions for, 851–53
  triggering, 847
  turning the SCR off, 851
  turning the SCR on, 850
  triacs, 857–64
    commutation, 859–62
    properties of, 859
    schematic symbol for, 858
    triggering methods, 859
  troubleshooting, 881–82
  See also Unijunction transistors (UJTs)
Time delay circuit, 453
"Totem pole" configuration, 458–60
Transconductance, JFETs, 361–64
Transducers, 562–64, 762–73
  dynamic microphones, 771–73
  introduction, 761
  magnetic phonograph cartridge, 771–73
  negative temperature coefficient, 763
  optoisolators, 769–70
  optoisolators in high-speed line receiver, 793
  photodiodes, 768
    automobile lamp dimmer, 792–93
  photoresistors, 766–68
  phototransistors, 768–69
  positive temperature coefficient, 763
  strain gauge, 770–71
    gauge circuit, 782–83
  thermistors, 763–65
    circuits, 782, 783
  thermocouples, 765–66
    balanced amplifier, 793–94
Transformer
  coupling, 325, 326
  review, 89–90
Transients, 376
Transistors. See Bipolar junction transistors; Junction field effect transistors (JFETs)
Transistor switching, 417–66
  BJT switch, 430–32
  common-emitter configuration, 418–19
  decay time, 424
  delay time, 423
  duty cycle, 422
  frequency response, 450
  generating signals
    differentiator circuit, 451–52
    integrator circuit, 452–53
    ramp and triangle, 451

ramp generators, 453–58
sine waves, 449–50
square waves, 450–51
time delay circuit, 453
hysteresis voltage, 435
interbase resistance(RBB), 442
interfacing between voltage levels, 460–61
intrinsic standoff ratio (ISR), 442
JFET, 439–41
lower trip point(LTP), 432
MOSFET, 441
negative resistance region, 444
operations, 418–25
  leakage current, 419–20
  saturation voltages, 420, 422
  speeds, 422–25
    adding speed-up capacitor, 426–32
  pulse train, 438
  recovery time, 430
  rise time (t_r), 423
Schmitt triggers, 432–39
  analyzing for the lower threshold point, 435–38
  analyzing for the upper threshold point, 434–35
  input-output characteristics, 438–39
specification sheet, 421
step voltage change, 423
storage time, 424
"totem pole" configuration, 458–60
transistor-transistor logic (TTL), 458–60
troubleshooting, 461–62
UJT operation, 441–49
  negative resistance, 443–44
  relaxation oscillator, 444–49
upper trip point(UTP), 432
Transistor-transistor logic (TTL), 458–60
Triacs, 634, 857–64
  basic control circuit, 877
  commutation, 859–62
  properties, 859
  schematic symbol for, 858
  triggering methods, 859
Triangle signals, 451
Triangular waveform generator, 637
Triggering, 808, 812–13
  circuit, LASCR, 879–80
  circuit, UJTs, 871–73
  thyristors, 847, 859
Trip points, 627, 667–89, 692–93
Trivalent doping, 68
Troubleshooting
  555 timers, 831–32
  active diode circuits, 693–95
  active filters, 752–54
  comparators and Schmitt triggers, 695–97

differential amplifiers, 496–98
dynamic, 960
faults in power supply circuits, 130–32
instrumentation amplifiers, 795
JFETs, 377–78
method of, five steps, 960
MOSFETs, 411
multistage amplifiers, 267–70
Op-amp circuits, 585–88
  checking the feedback network, 587
  if op-amp is defective, 587
  if R_f opens, 586
  if R_{in} opens, 586–87
Op-amps, 530–31
oscillators, 641–42
power amplifiers, 329–30
regulators, 923–24
safety procedures, 130
signal injection, 960–62
signal tracing, 962
static, 960
thyristors, 881–82
transistors, 200–208
  assume nothing, 201
  base-bias prototype, 203–4
  checking the transistors, 201–3
  check power supply, 200
  collector feedback prototype, 206
  emitter-bias prototype, 204–5
  use eyes and nose, 200–201
  voltage divider prototype, 205–6
transistor switching, 461–62
waveshaping circuits, 641–42
Tuned Class C amplifiers, 317–19
Tuned coupling, 328–29
Tunneling, 955
Turns ratio(γ), 89–90
Twin-T filter, 616–18, 749–50

Unbalanced bridge circuit
  thevenizing, 32–34
Unbalanced loads, 129
Unijunction transistors (UJTs), 441–49, 867–77
  biasing the UJT, 867–71
  characteristic curves, 869, 870
  negative resistance, 443–44
  parameters for, 870 tab
  programmable (PUT), 874–77
    maximum ratings for, 875 tab
  relaxation oscillator, 444–49
  synchronizing, 873–74
  triggering circuit, 871–73
Unity-gain amplifier. See Voltage follower
Upper trip point (UTP), 432

Valence crystals, 65
Valley voltage, 956
Varactor diodes, 956–58
Variable resistors, JFETs, 376
Virtual ground, 558

Voltage
  clampers, 59
  divider, 6–15
    biasing JFETs with, 354–59
    bias with a D MOSFET, 397–99
    bias with E MOSFETs, 400–402
    general equation, 8
    loaded, 11
    to minimize the effects of Beta,
        179–89
    parallel branches, 9
    preventing loading effects, 13–15
      firm voltage dividers, 13–14
      stiff voltage dividers, 13,
          14–15
    prototype transistor, 205–6
    series string, 9
    stiff and firm, 185–86
    unknown load, 10–11
  drop, 81
  follower, 577–79
  gain (A_v), 214
    of a JFET amplifier, 366
    of BJTs and JFETs compared,
        365–66
    Op-amp circuits, 553–55
  gate-to-source supply, 341–42
  multipliers, 111–15

pinch-off (V_p), 341
source-to-drain supply, 340–41
threshold, 376
Voltage-controlled oscillator (VCO),
    951–52
Voltage-controlled voltage source filters
    (VCVS), 744–46
  high-pass, 748–49
Voltage-dependent resistor, 904
Voltage regulation and regulators,
    888–94
  line regulation, 890–91
  load regulation, 888–90
  recovery time, 893
  voltage regulating, current regulating
      power supply, 911–13
  voltage regulators, 891–92
    drift, 894
    overvoltage protection for, 892–93
    power dissipation, 893
    ripple rejection ratio, 893–94
  zener diodes, 82. *See also* Regulators
Voltmeter, 49
Volt-ohm-milliammeter, 48

Waveshaping circuits, 596, 631–41
  differentiators, 631–34
    Op-amp, 633

integrators, 634–37
sawtooth generators, 637–41
  bootstrap, 640–41
  constant current, 638–40
  RC ramp, 638
triangular waveform generator, 637
troubleshooting, 641–42
Weighted-resistor network, 784–86
Wheatstone bridge, 771, 773, 774, 783
Wien-bridge oscillator, 613–16
Window comparators, 689–93
  determining trip points for, 692–93

Zener diode
  as a regulator, 108–11
  breakdown voltage limiting circuit of,
      120
  limiter circuit, 660–61
  operation, 81–83
  symmetrical limiter, 121–22
  voltage-current characteristic curve of,
      121
Zener impedance, 108
Zener knee, 81–82
Zener voltage, 79–80
Zero adjust knob, 49
Zero bias, with a D MOSFET, 399–400
Zero crossing points, 276